Intermediate
Algebra

An Applied Approach

Intermediate Algebra

An Applied Approach

THIRD EDITION

Richard N. Aufmann

Palomar College, California

Vernon C. Barker

Palomar College, California

HOUGHTON MIFFLIN COMPANY   Boston

Dallas   Geneva, Illinois
Palo Alto   Princeton, New Jersey

Chapter Opener Art designed and rendered by Daniel P. Derdula.

Most Interest Grabber Art designed and illustrated by Daniel P. Derdula, with airbrushing by Linda Phinney on Chapters 2 and 9, and calligraphy by Susan Fong on Chapter 4. Linda Phinney is credited for illustrating Chapter 6.

All interior Math Figures rendered by Network Graphics (135 Fell Court, Hauppauge, New York 11788).

Cover design by Slide Graphics of New England, Inc.

Interior design by George McLean.

Printed in the U.S.A.

ISBN Numbers:
Text: 0-395-43193-X
Instructor's Annotated Edition: 0-395-57073-5
Solutions Manual: 0-395-57074-3
Instructor's Resource Manual/Testing Program: 0-395-57078-6
Test Bank for the Computerized Test Generator: 0-395-57075-1
Transparencies: 0-395-57076-X
Videos: 0-395-57077-8

ABCDEFGHIJ-VH-9876543210

Contents

3   Polynomials   *101*

4   Rational Expressions   *165*

Content and Format © 1991 HMCo.

5   Exponents and Radicals   *217*

Content and Format © 1991 HMCo.

8   Functions and Relations   *361*

9   Conic Sections   *411*

Content and Format © 1991 HMCo.

Preface

The third edition of *Intermediate Algebra: An Applied Approach* provides mathematically sound and comprehensive coverage of the topics considered essential in an intermediate algebra course. Our strategy in preparing this revision has been to build on the successful features of the second edition, features designed to enhance the student's mastery of math skills. In addition, we have expanded the ancillary package for both the instructor and the student by adding transparencies and videos.

Features

The Interactive Approach

Instructors have long recognized the need for a text that requires the student to use a skill as it is being taught. *Intermediate Algebra: An Applied Approach* uses an interactive technique that meets this need. Each section is divided into objectives, and every objective contains one or more sets of matched-pair examples. The first example in each set is worked out; the second example is not. By solving this second problem, the student interacts with the text. The complete worked-out solutions to these examples are provided in an appendix at the end of the book, so the student can obtain immediate feedback on and reinforcement of the skill being learned.

Emphasis on Problem-Solving Strategies

Intermediate Algebra: An Applied Approach features a carefully developed approach to problem solving that emphasizes developing strategies to solve problems. For each type of word problem contained in the text, the student is prompted to use a "strategy step" before performing the actual manipulation of numbers and variables. By developing problem-solving strategies, the student will know better how to analyze and solve those word problems encountered in an intermediate algebra course.

Applications

The traditional approach to teaching or reviewing algebra covers only the straightforward manipulation of numbers and variables and thereby fails to teach students the practical value of algebra. By contrast, *Intermediate Algebra: An Applied Approach* emphasizes applications. Wherever appropriate, the last objective in each section presents applications that require the student to use the skills covered in that section to solve practical problems. Also, most of Chapter 2, *"First-Degree Equations and Inequalities,"* and portions of several other chapters are devoted entirely to applications. This carefully integrated applied approach generates awareness on the student's part of the value of algebra as a real-life tool.

Complete Integrated Learning System Organized by Objectives

Each chapter begins with a list of the learning objectives included within that chapter. Each of the objectives is then restated in the chapter to remind the student of the current topic of discussion. The same objectives that organize the text organize each ancillary. The Solutions Manual, Computerized Test Generator, Computer Tutor™, Videos, Transparencies, Test Bank, and the Printed Testing Program have all been prepared so that both the student and instructor can easily connect all of the different aids.

Exercises

There are more than 6000 exercises in the text, grouped in the following categories:

- **End-of-section exercise sets,** which are keyed to the corresponding learning objectives, provide ample practice and review of each skill.
- **Chapter review exercises,** which appear at the end of each chapter, help the student integrate all of the skills presented in the chapter.
- **Chapter tests,** which appear at the end of each chapter, are typical one-hour exams that the student can use to prepare for an in-class test.
- **Cumulative review exercises,** which appear at the end of each chapter (beginning with Chapter 2), help the student retain math skills learned in earlier chapters.
- **Final exam,** which follows the last chapter, can be used as a review item or practice final.
- **Calculator exercises,** which appear in the end-of-section exercises, are included throughout the text. These exercises provide the student with the opportunity to practice using a hand-held calculator and are identified by a special color box printed over the exercise number for each calculator exercise.

Calculator and Computer Enrichment Topics

Each chapter also contains optional calculator or computer enrichment topics. Calculator topics provide the student with valuable key-stroking instructions and practice in using a hand-held calculator. Computer topics correspond directly to the programs found on the Math ACE (Additional Computer Exercises) Disk. These topics range from solving first-degree equations to graphing linear equations in two variables.

New To This Edition

Topical Coverage

Chapter 1, a review of the real numbers, has been condensed.

The chapter on *Linear Equations in Two Variables* has been moved to precede Chapter 7, *Quadratic Equations,* to better connect the graphing skills developed in each chapter.

In Chapter 4, negative exponents are introduced before division of monomials. With this change, a single rule for division of monomials can be stated.

The coverage of factoring has been expanded to include the ac method of factoring as an alternative to factoring by using trial factors.

Coverage of conic sections has been increased. Chapter 8, *Graphing Functions and Relations,* has been broken up into two chapters, *Functions and Relations* and *Conic Sections.* This material has been greatly expanded in the Third Edition.

Keeping with our commitment to applications, over two hundred word problems have been rewritten to reflect contemporary situations.

Chapter Review and Chapter Test

There is now a Chapter Review and a Chapter Test at the end of each chapter. Problems in the Chapter Review are grouped according to the section heads for that chapter. Problems in the Chapter Test are not divided into sections or objectives. Instead, the objective references are given in the Answer Section at the back of the book so the student will know which objective to restudy if necessary.

New Testing Program

Both the Computerized Testing Program and the Printed Testing Program have been completely rewritten to provide instructors with the option of creating countless new tests.

Supplements For The Student

Two computerized study aids, the Computer Tutor™ and the Math ACE (Additional Computer Exercises) Disk have been carefully designed for the student. The Computer Tutor™ has been expanded to include nine "you-try-it" examples with specific help screens for each lesson.

The COMPUTER TUTOR™

The Computer Tutor™ is an interactive instructional microcomputer program for student use. Each learning objective in the text is supported by a lesson on the Computer Tutor™. As a reminder of this, a small computer icon appears to the right of each objective title in the text. Lessons on the tutor provide additional instruction and practice and can be used in several ways: (1) to cover material the student missed because of absence from class; (2) to repeat instruction on the skill or concept that the student has not yet mastered; or (3) to review material in preparation for examinations. This tutorial program is available for the IBM PC and compatible computers.

Math ACE (Additional Computer Exercises) Disk

The Math ACE Disk contains a number of computational and drill-and-practice programs that correspond to selected Calculator and Computer Enrichment Topics in the text. These programs are available for the IBM PC and compatible computers.

Supplements For The Instructor

Intermediate Algebra: An Applied Approach has an unusually complete set of teaching aids for the instructor.

Instructor's Annotated Edition

The Instructor's Annotated Edition is an exact replica of the student text except that the answers to all of the exercises are printed in color next to the problems.

Solutions Manual

The Solutions Manual contains worked-out solutions for all end-of-section exercise sets, chapter reviews, chapter tests, cumulative reviews, and the final exam.

Instructor's Resource Manual/Testing Program

The Instructor's Resource Manual/Testing Program contains the printed testing program, which is the first of three sources of testing material available to users of *Intermediate Algebra: An Applied Approach*. Eight printed tests (in two formats—free response and multiple choice) are provided for each chapter, as are cumulative and final exams. In addition, the Instructor's Manual includes the documentation for all the software ancillaries (ACE, the Computer Tutor, and the Instructor's Computerized Test Generator) as well as suggested course sequences and class assignments.

Instructor's Computerized Test Generator

The Instructor's Computerized Test Generator is the second source of testing material for use with *Intermediate Algebra: An Applied Approach*. The database contains over 1900 new test items. These questions are unique to the test generator and do not repeat items provided in the Instructor's Resource Manual/Testing Program. Organized according to the keyed objectives in the text, the Test Generator is designed to produce an unlimited number of tests for each chapter of the text, including cumulative tests and final exams. It is available for the IBM PC or compatible computers with editing capabilities for all non-graphic questions.

Printed Test Bank

The Printed Test Bank, the third component of the testing materials, is a print-out of all items in the Instructor's Computerized Test Generator. Instructors using the Test Generator can use the test bank to select specific items from the database. Instructors who do not have access to a computer can use the test bank to select items to be included on a test being prepared by hand.

Videotapes

Approximately 50 half-hour videotape lessons accompany *Intermediate Algebra: An Applied Approach*. These lessons follow the format and style of the text and are closely tied to specific sections of the text.

Transparencies

Approximately 250 transparencies accompany *Intermediate Algebra: An Applied Approach*. These transparencies contain the complete solution to every "you-try-it" example in the text.

Acknowledgments

The authors would like to thank the people who have reviewed this manuscript and provided many valuable suggestions:

Geoffrey Akst
Borough of Manhattan Community College, NY

Betty Jo Baker
Lansing Community College, MI

Judith Brower
North Idaho College, ID

Mary Cabral
Middlesex Community College, MA

Carmy Carranga
Indiana University of Pennsylvania, PA

Patricia Confort
Roger Williams College, RI

Michael Contino
California State University, CA

Bob C. Denton
Orange Coast College, CA

Sharon Edgmon
Bakersfield College, CA

Joel Greenstein
New York Technical College, NY

Bonnie-Lou Guertin
Mt. Wachusett Community College, MA

Lynn Hartsell
Dona Ana Branch Community College, NM

Ida E. Hendricks
Shepherd College, WV

Diana L. Hestwood
Minneapolis Community College, MN

Pamela Hunt
Paris Junior College, TX

Arlene Jesky
Rose State College, OK

Sue Korsak
New Mexico State University, NM

Randy Leifson
Pierce College, WA

Virginia M. Licata
Camden County College, NJ

Judy Liles
North Harris County College-South Campus, TX

David Longshore
Victor Valley College, CA

Margaret Luciano
Broome Community College, NY

Carl C. Maneri
Wright State University, OH

Rudy Maglio
DePaul University, IL

Patricia McCann
Franklin University, OH

Dale Miller
Highland Park Community College, MI

Judy Miller
Delta College, MI

Ellen Milosheff
Triton College, IL

Scott L. Mortensen
Dixie College, UT

Michelle Mosman
Des Moines Area Community College, IA

Linda Murphy
Northern Essex Community College, MA

Wendell Neal
Houston Community College, TX

Doris Nice
University of Wisconsin-Parkside, WI

Kent Pearce
Texas Tech University, TX

Sue Porter
Davenport College, MI

Jack Rotman
Lansing Community College, MI

Karen Schwitters
Seminole Community College, FL

Dorothy Smith
Del Mar College, TX

James T. Sullivan
Massachusetts Bay Community College, MA

Lana Taylor
Siena Heights College, MI

William N. Thomas, Jr.
University of Toledo, OH

Dana Mignogna Thompson
Mount Aloysius Junior College, PA

Paul Treuer
University of Minnesota-Duluth, MN

Thomas Wentland
Columbus College, GA

William T. Wheeler
Abraham Baldwin College, GA

Harvey Wilensky
San Diego Miramar College, CA

Professor Warren Wise
Blue Ridge Community College, VA

Wayne Wolfe
Orange Coast College, CA

Kenneth Word
Central Texas College, TX

Justane Valudez-Ortiz
San Jose College, CA

To the Student

Many students feel that they will never understand math while others appear to do very well with little effort. Oftentimes what makes the difference is that successful students take an active role in the learning process.

Learning mathematics requires your *active* participation. Although doing homework is one way you can actively participate, it is not the only way. First, you must attend class regularly and become an active participant. Secondly, you must become actively involved with the textbook.

Intermediate Algebra: An Applied Approach was written and designed with you in mind as a participant. Here are some suggestions on how to use the features of this textbook.

There are 12 chapters in this text. Each chapter is divided into sections and each section is subdivided into learning objectives. Each learning objective is labeled with a letter from A–E.

First, read each objective statement carefully so you will understand the learning goal that is being presented. Next, read the objective material carefully, being sure to note each bold word. These words indicate important concepts that you should familiarize yourself with. Study each in-text example carefully, noting the techniques and strategies used to solve the example.

You will then come to the key learning feature of this text, the *boxed examples*. These examples have been designed to aid you in a very specific way. Notice that in each example box, the example on the left is completely worked out and the example on the right is not. The reason for this is that *you* are expected to work the right-hand example (in the space provided) in order to immediately test your understanding of the material you have just studied.

You should study the worked-out example carefully by working through each step presented. This allows you to focus on each step and reinforces the technique for solving that type of problem. You can then use the worked-out example as a model for solving similar problems.

Then you should solve the right-hand example using the problem-solving techniques that you have just studied. When you have completed your solution, check your work by turning to the page in the appendix where the complete solution can be found. The page number on which the solution appears is printed at the bottom of the example box in the right-hand corner. By checking your solution, you will know immediately whether or not you fully understand the skill you just studied.

When you have completed studying an objective, do all of the exercises in the exercise set that correspond with that objective. The exercises will be labeled with the same letter as the objective. Algebra is a subject that needs to be learned in small sections and practiced continually in order to be mastered. Doing all of the exercises in each exercise set will help you master the problem-solving techniques necessary for success.

Once you have completed the exercises to an objective, you should check your answers to the odd-numbered exercises with those found in the back of the book.

After completing a chapter, read the Chapter Summary. This summary highlights the important topics covered in the chapter. Following the Chapter Summary are Chapter Review Exercises, a Chapter Test, and a Cumulative Review (beginning with Chapter 2). Doing the review exercises is an important way of testing your understanding of the chapter. The answer to each review exercise is in an appendix at back of the book. Each answer is followed by a reference that tells which objective that exercise was taken from. For example, (4.2B) means Section 4.2, Objective B. After checking your answers, restudy any objective that you missed. It may be very helpful to retry some of the exercises for that objective to reinforce your problem-solving techniques.

The Chapter Test should be used to prepare for an exam. We suggest that you try the Chapter Test a few days before your actual exam. Take the test in a quiet place and try to complete the test in the same amount of time you will be allowed for your exam. When taking the Chapter Test, practice the strategies of successful test takers: 1) scan the entire test to get a feel for the questions; 2) read the directions carefully; 3) work the problems that are easiest for you first; and perhaps most importantly, 4) try to stay calm.

When you have completed the Chapter Test, check your answers. If you missed a question, review the material in that objective and rework some of the exercises from that objective. This will strengthen your ability to perform the skills in that objective.

The Cumulative Review allows you to refresh the skills you have learned in previous chapters. This is very important in mathematics. By consistently reviewing previous materials, you will retain the previous skills as you build new ones.

Remember, to be successful, attend class regularly; read the textbook carefully; actively participate in class; work with your textbook using the boxed examples for immediate feedback and reinforcement of each skill; do all the homework assignments; review constantly; and work carefully.

Intermediate Algebra

An Applied Approach

1

Review of Real Numbers

OBJECTIVES

- ▶ To find the absolute value and the additive inverse of a number
- ▶ To add, subtract, multiply, and divide integers
- ▶ To evaluate exponential expressions
- ▶ To add, subtract, multiply, and divide rational numbers
- ▶ To use the Order of Operations Agreement to simplify expressions
- ▶ To evaluate a variable expression
- ▶ To use and identify the Properties of the Real Numbers
- ▶ To simplify a variable expression
- ▶ To translate a verbal expression into a variable expression and then simplify the resulting expression
- ▶ To find the union and intersection of sets
- ▶ To graph the solution set of an inequality in one variable

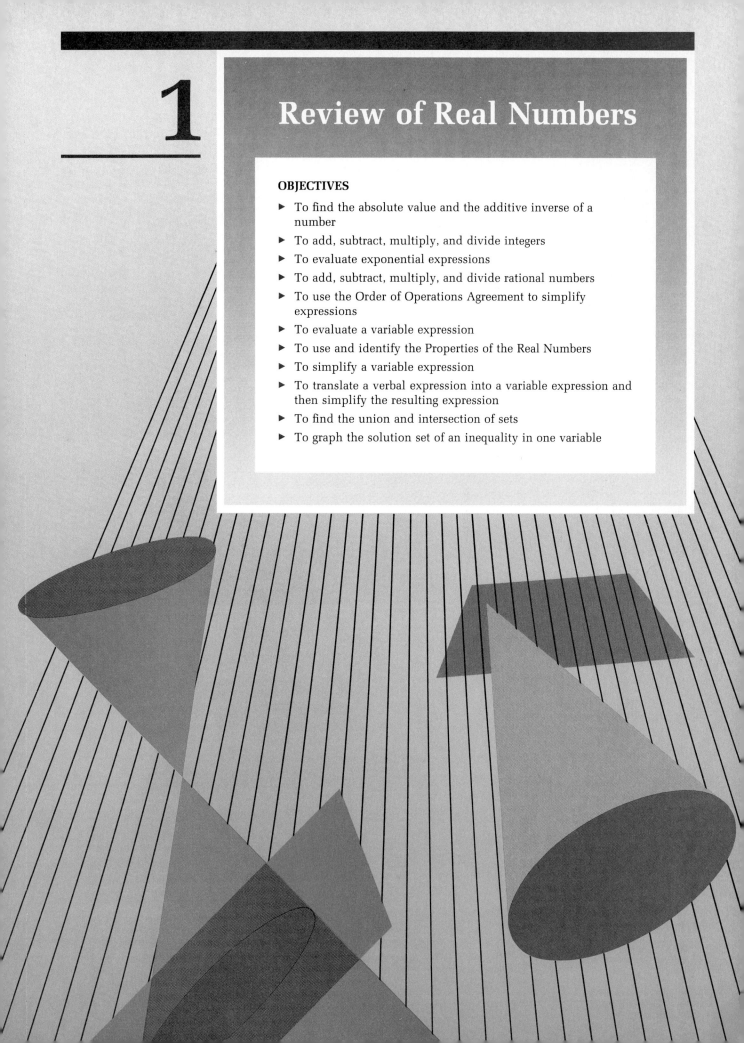

Early Egyptian Number System

The early Egyptian type of picture writing shown at the right is known as hieroglyphics.

The Egyptian hieroglyphic method of representing numbers differs from our modern version in an important way. For the hieroglyphic number, the symbol indicated the value. For example, the symbol ∩ meant 10. The symbol was repeated to get larger values.

The markings at the right represent the number 743. Each vertical stroke represents 1, each ∩ represents 10, and each 𝟗 represents 100.

There are 3 ones, 4 tens, and 7 hundreds representing the number 743.

3 + 40 + 700

743

In our system, the position of a number is important. The number 5 in 356 means 5 tens but the number 5 in 3517 means 5 hundreds. Our system is called a positional number system.

Consider the hieroglyphic number at the right. Notice that the ∩ is missing. For the early Egyptians, when a certain group of ten was not needed, it was just omitted.

There are 4 ones and 5 hundreds. Thus, the number 504 is represented by this group of markings.

4 + 500

504

In a positional system of notation like ours, a zero is used to show that a certain group of ten is not needed. This may seem like a fairly simple idea, but it was not until near the end of the 7th century that a zero was introduced to the number system.

SECTION 1.1   Operations on the Integers

Objective A

To find the absolute value and the additive inverse of a number

The **integers** are . . . $-4, -3, -2, -1, 0, 1, 2, 3, 4, \ldots$

The three dots before and after the list of integers mean that the list continues without end, and that there is no smallest integer and there is no largest integer.

The integers can be shown on the number line.

The integers to the left of zero on the number line are called **negative integers.** The integers to the right of zero are called **positive integers.** Zero is neither a positive nor a negative integer.

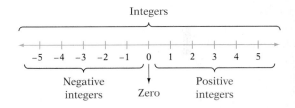

The positive integers are also called the **natural numbers.**

$1, 2, 3, 4, \ldots$

The positive integers and zero are called the **whole numbers.**

$0, 1, 2, 3, \ldots$

The **graph** of an integer is shown by placing a heavy dot on the number line directly above the integer.

The graph of -3 on the number line

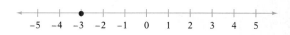

Just as the word "it" is used in language to stand for an object, a letter of the alphabet can be used in mathematics to stand for a number. Such a letter is called a **variable.**

A number line can be used to show the relative order of two integers a and b. If a is to the left of b on the number line, then a is less than b ($a < b$). If a is to the right of b on the number line, then a is greater than b ($a > b$).

Negative 2 is greater than negative 4.

$-2 > -4$

Negative 1 is less than 2.

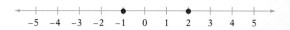

$-1 < 2$

Two numbers that are the same distance from zero on the number line but on opposite sides of zero are **opposite numbers,** or **opposites.** The opposite of a number is called its **additive inverse.**

−3 is the additive inverse of 3.

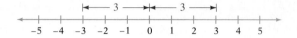

The additive inverse of −4 is 4.

The **absolute value** of a number is a measure of its distance from zero on the number line. Therefore, the absolute value of a number is a positive number or zero. The symbol for absolute value is | |.

The absolute value of a positive number is the number itself.

$$|7| = 7$$

The absolute value of a negative number is its additive inverse.

$$|-7| = 7$$

The absolute value of zero is zero.

$$|0| = 0$$

Evaluate $-|-15|$.

$$-|-15| = -15$$

The absolute value sign does not affect the negative sign in front of the absolute value sign.

Example 1   Find the additive inverse.
  a. −16   b. 46

Solution   a. 16   b. −46

Example 2   Find the additive inverse.
  a. 18   b. −32

Your solution   a. −18   b. 32

Example 3   Evaluate.
  a. $|-8|$ b. $-|-29|$

Solution   a. 8   b. −29

Example 4   Evaluate.
  a. $|-11|$ b. $-|51|$

Your solution   a. 11   b. −51

Solutions on p. A9

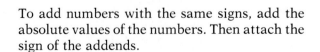

Objective B   **To add, subtract, multiply, and divide integers**

To add numbers with the same signs, add the absolute values of the numbers. Then attach the sign of the addends.

$$6 + 8 = 14$$
$$-6 + (-8) = -14$$

To add numbers with different signs, find the difference between the absolute values of the numbers. Then attach the sign of the number with the greater absolute value.

$$-2 + 8 = 6$$
$$2 + (-8) = -6$$

Subtraction is defined as the addition of the additive inverse.

If a and b are integers, then $\boldsymbol{a - b = a + (-b)}$.

$-3 - 7 = -3 + (-7) = -10$
$-3 - (-7) = -3 + 7 = 4$

Simplify: $-2 + (-3) - 6 + (-8)$

This problem contains both addition and subtraction.	$-2 + (-3) - 6 + (-8)$
Rewrite subtraction as the addition of the additive inverse.	$-2 + (-3) + (-6) + (-8)$
Add the first two numbers.	$-5 + (-6) + (-8)$
Add the sum to the third number.	$-11 + (-8)$
Continue until all the numbers have been added.	-19
To multiply numbers with the same signs, multiply the absolute values of the factors. The product is positive.	$8 \cdot 4 = 32$ $(-8)(-4) = 32$
To multiply numbers with different signs, multiply the absolute values of the factors. The product is negative.	$(-8) \cdot 4 = -32$ $8 \cdot (-4) = -32$
The quotient of two numbers with the same signs is positive.	$10 \div 2 = 5$ $-10 \div (-2) = 5$
The quotient of two numbers with different signs is negative.	$-10 \div 2 = -5$ $10 \div (-2) = -5$

Example 5   Simplify: $5 - 9 + (-34)$

Solution
$5 - 9 + (-34)$
$5 + (-9) + (-34)$
$-4 + (-34)$
-38

Example 6   Simplify: $-7 + (-13) - (-8)$

Your solution -12

Example 7   Simplify: $(-2)(-6)(-12)$

Solution
$(-2)(-6)(-12)$
$12(-12)$
-144

Example 8   Simplify: $(-2)(3)(-4)(8)$

Your solution 192

Example 9   Simplify: $|-120| \div |4|$

Solution
$|-120| \div |4|$
30

Example 10   Simplify: $-64 \div |-32|$

Your solution -2

Solutions on p. A9

| Objective C | **To evaluate exponential expressions** |

Repeated multiplication of the same factor can be written using an exponent.

$2 \cdot 2 \cdot 2 \cdot 2 \cdot 2 \cdot 2 = 2^6$ ←—**Exponent** $b \cdot b \cdot b \cdot b \cdot b = b^5$ ←—**Exponent**
                    ↑—**Base**   ↑—**Base**

The **exponent** indicates how many times the factor, called the **base**, occurs in the multiplication. The multiplication $2 \cdot 2 \cdot 2 \cdot 2 \cdot 2 \cdot 2$ is in **factored form.** The exponential expression 2^6 is in **exponential form.**

2^1 is read "the first power of two" or just "two."

2^2 is read "the second power of two" or "two squared."

2^3 is read "the third power of two" or "two cubed."

2^4 is read "the fourth power of two."

2^5 is read "the fifth power of two."

b^5 is read "the fifth power of b."

Usually the exponent 1 is not written.

nth Power of b

If b is an integer and n is a positive integer, the nth power of b is defined as the product of n factors of b.
$$b^n = \underbrace{b \cdot b \cdot b \cdots b}_{n \text{ factors}}$$

To evaluate an exponential expression, write each factor as many times as indicated by the exponent. Then multiply.

$5^4 = 5 \cdot 5 \cdot 5 \cdot 5 = 625$

$3^2 \cdot 5^3 = (3 \cdot 3)(5 \cdot 5 \cdot 5) = 9 \cdot 125 = 1125$

Evaluate $(-3)^4$ and -3^4.

Note that the negative of a number is taken to a power only when the negative sign is *inside* the parentheses.

$(-3)^4 = (-3)(-3)(-3)(-3) = 81$

$-3^4 = -(3 \cdot 3 \cdot 3 \cdot 3) = -81$

Example 11   Evaluate: $(-2)^3 \cdot 3^2$

Solution $(-2)^3 \cdot 3^2$
$(-2)(-2)(-2) \cdot (3)(3)$
$-8 \cdot 9$
-72

Example 12   Evaluate: $-3^3 \cdot 2^2$

Your solution -108

Solution on p. A9

1.1 EXERCISES

▶ **Objective A**

Find the additive inverse.

1. -2
2

2. -8
8

3. 4
-4

4. 6
-6

5. -75
75

6. 83
-83

7. 51
-51

8. -126
126

9. -83
83

10. -42
42

Evaluate.

11. $|3|$
3

12. $|8|$
8

13. $|-4|$
4

14. $|-6|$
6

15. $-|-32|$
-32

16. $-|-16|$
-16

17. $-|-22|$
-22

18. $-|126|$
-126

19. $|-436|$
436

20. $|-16|$
16

▶ **Objective B**

Simplify.

21. $-18 + (-12)$
-30

22. $-18 - 7$
-25

23. $5 - 22$
-17

24. $16 \cdot (-60)$
-960

25. $3 \cdot 4 \cdot (-8)$
-96

26. $18 \cdot 0 \cdot (-7)$
0

27. $18 \div (-3)$
-6

28. $25 \div (-5)$
-5

29. $-60 \div (-12)$
5

30. $(-9)(-2)(-3)(10)$
-540

31. $-20(35)(-16)$
11,200

32. $54(19)(-82)$
$-84{,}132$

33. $-271(-365)$
98,915

34. $|(-16)(10)|$
160

35. $|12(-8)|$
96

36. $|7 - 18|$
11

37. $|15 - (-8)|$
23

38. $|-16 - (-20)|$
4

39. $|-56 \div 8|$
7

40. $|81 \div (-9)|$
9

Simplify.

41. $|-153 \div (-9)|$
17

42. $|-4| - |-2|$
2

43. $-|-8| + |-4|$
-4

44. $-|-16| - |24|$
-40

45. $-30 + (-16) - 14 - 2$
-62

46. $3 - (-2) + (-8) - 11$
-14

47. $-2 + (-19) - 16 + 12$
-25

48. $-6 + (-9) - 18 + 32$
-1

49. $13 - |6 - 12|$
7

50. $-9 - |-7 - (-15)|$
-17

51. $738 - 46 + (-105) - 219$
368

52. $-871 - (-387) - 132 - 460$
-1076

53. $-442 \div (-17)$
26

54. $621 \div (-23)$
-27

55. $-4897 \div 59$
-83

▶ **Objective C**

Simplify.

56. 5^3
125

57. 3^4
81

58. -2^3
-8

59. -4^3
-64

60. $(-5)^3$
-125

61. $(-8)^2$
64

62. $2^2 \cdot 3^4$
324

63. $4^2 \cdot 3^3$
432

64. $-2^2 \cdot 3^2$
-36

65. $-3^2 \cdot 5^3$
-1125

66. $(-2)^3(-3)^2$
-72

67. $(-4)^3(-2)^3$
512

68. $4 \cdot 2^3 \cdot 3^3$
864

69. $-4(-3)^2(4^2)$
-576

70. $2^2 \cdot (-10)(-2)^2$
-160

71. $-3(-2)^2(-5)$
60

72. $(-3)^3 \cdot 15 \cdot (-2)^4$
-6480

73. $-4 \cdot 2^2 \cdot 5^5$
$-50,000$

74. $2^5 \cdot (-3)^4 \cdot 4^5$
$2,654,208$

75. $4^4(-3)^5(6)^2$
$-2,239,488$

SECTION 1.2   Operations on the Real Numbers

Objective A   **To add, subtract, multiply, and divide rational numbers**

A **rational number** is the quotient of two integers. Therefore, a rational number is a number which can be written in the form $\frac{a}{b}$, where a and b are integers and b is not zero. A rational number written in this way is commonly called a fraction. Because an integer can be written as the quotient of the integer and 1, every integer is a rational number. For example, $6 = \frac{6}{1}$. A number written in decimal notation is also a rational number. For example, $0.7 = \frac{7}{10}$.

A rational number written as a fraction can be written in decimal notation.

Write $\frac{3}{16}$ as a decimal.

Write $\frac{6}{11}$ as a decimal.

The fraction bar can be read "÷".

$$\frac{3}{16} = 3 \div 16$$

$$\begin{array}{r} 0.1875 \\ 16\overline{)3.0000} \\ -1\,6 \\ \hline 1\,40 \\ -1\,28 \\ \hline 120 \\ -112 \\ \hline 80 \\ -80 \\ \hline 0 \end{array}$$ ← This is called a **terminating decimal.**

0 ← The remainder is zero.

$$\frac{3}{16} = 0.1875$$

$$\begin{array}{r} 0.5454\ldots \\ 11\overline{)6.0000} \\ -5\,5 \\ \hline 50 \\ -44 \\ \hline 60 \\ -55 \\ \hline 50 \\ -44 \\ \hline 6 \end{array}$$ ← This is called a **repeating decimal.**

6 ← The remainder is never zero.

$$\frac{6}{11} = 0.\overline{54}$$ The bar over the digits 54 is used to show that these digits repeat.

Every rational number can be written as a terminating or repeating decimal.

Some numbers, for example $\sqrt{7}$ and π, have decimal representations which never terminate or repeat. These numbers are called **irrational numbers.**

$\sqrt{7} = 2.6457513\ldots$

$\pi = 3.141592654\ldots$

The rational numbers and the irrational numbers taken together are called the **real numbers.**

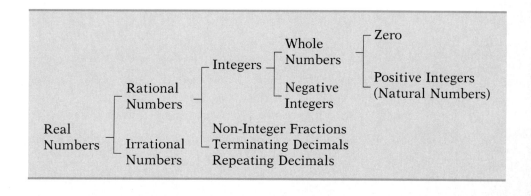

Sum of Two Fractions

> The sum of two fractions with a common denominator is the sum of the numerators over the common denominator.
>
> $$\frac{a}{c} + \frac{b}{c} = \frac{a+b}{c}$$

To add fractions, first rewrite the fractions as equivalent fractions, using the least common multiple (LCM) of the denominators as the common denominator. Then add the numerators and place the sum over the common denominator. Write the answer in simplest form.

Simplify: $-\frac{5}{8} + \frac{7}{12}$

The LCM = 24. Write equivalent fractions using the LCM as the common denominator. Then add the numerators.

$$-\frac{5}{8} + \frac{7}{12} = -\frac{15}{24} + \frac{14}{24} = \frac{-15+14}{24} = -\frac{1}{24}$$

The Product of Two Fractions

> The product of two fractions is the product of the numerators over the product of the denominators.
>
> $$\frac{a}{b} \cdot \frac{c}{d} = \frac{ac}{bd}$$

Simplify: $\frac{5}{12} \cdot \frac{8}{15}$

Multiply the numerators and multiply the denominators.

$$\frac{5}{12} \cdot \frac{8}{15} = \frac{5 \cdot 8}{12 \cdot 15} = \frac{40}{180}$$

Write the answer in simplest form.

$$= \frac{20 \cdot 2}{20 \cdot 9} = \frac{2}{9}$$

If you have difficulty writing the fraction in simplest form, try finding the prime factorization of the numerator and denominator, then divide by the common factors. For the last example, you could do the following.

$$\frac{5 \cdot 8}{12 \cdot 15} = \frac{\overset{1}{\cancel{5}} \cdot (\overset{1}{\cancel{2}} \cdot \overset{1}{\cancel{2}} \cdot 2)}{(\underset{1}{\cancel{2}} \cdot \underset{1}{\cancel{2}} \cdot 3) \cdot (3 \cdot \underset{1}{\cancel{5}})} = \frac{2}{3 \cdot 3} = \frac{2}{9}$$

The **reciprocal** of a fraction is a fraction with the numerator and denominator interchanged.

Quotient of Two Fractions

> To divide two fractions, multiply by the reciprocal of the divisor.
>
> $$\frac{a}{b} \div \frac{c}{d} = \frac{a}{b} \cdot \frac{d}{c}$$

Simplify: $\frac{3}{8} \div \frac{9}{16}$

Multiply by the reciprocal of the divisor.

$$\frac{3}{8} \div \frac{9}{16} = \frac{3}{8} \cdot \frac{16}{9} = \frac{48}{72} = \frac{2}{3}$$

To add or subtract decimals, write the numbers so that the decimal points are in a vertical line. Then proceed as in the addition or subtraction of integers. Write the decimal point in the answer directly below the decimal points in the problem.

Simplify: $-18.429 + 6.3927$

Since the signs are different subtract.

$$\begin{array}{r} 18.4290 \\ -\;\;6.3927 \\ \hline 12.0363 \end{array}$$

Attach the sign of the number with the greater absolute value.

$-18.429 + 6.3927 = -12.0363$

To multiply decimals, multiply as in integers. Write the decimal point in the product so that the number of decimal places in the product equals the sum of the decimal places in the factors.

$$\begin{array}{r} 31.7 \quad \text{1 decimal place} \\ \times\;0.059 \quad \text{3 decimal places} \\ \hline 2853 \\ 1585\;\;\; \\ \hline 1.8703 \quad \text{4 decimal places} \end{array}$$

To divide decimals, move the decimal point in the divisor to make it a whole number. Move the decimal point in the dividend the same number of places to the right. Place the decimal point in the quotient directly over the decimal point in the dividend. Then divide as in whole numbers.

The symbol $\approx$ is used to indicate that the quotient is an approximate value after being rounded off.

$$\begin{array}{r} 2.371 \approx 2.37 \\ 1.72\,\overline{)4.07\,900} \\ -3\,44 \\ \hline 63\,9 \\ -51\,6 \\ \hline 12\,30 \\ -12\,04 \\ \hline 260 \\ -172 \\ \hline 88 \end{array}$$

Example 1 Simplify: $\dfrac{3}{8} + \dfrac{5}{12} - \dfrac{9}{16}$

Solution $\dfrac{3}{8} + \dfrac{5}{12} - \dfrac{9}{16} = \dfrac{18}{48} + \dfrac{20}{48} - \dfrac{27}{48}$

$\qquad\qquad = \dfrac{18 + 20 - 27}{48} = \dfrac{11}{48}$

Example 2 Simplify: $\dfrac{5}{6} - \dfrac{3}{8} + \dfrac{7}{9}$

Your solution $\dfrac{89}{72}$

Example 3 Simplify: $6.329 - 12.49$

Solution
$$\begin{array}{r} 12.490 \\ -\;\;6.329 \\ \hline 6.161 \end{array}$$

$6.329 - 12.49 = -6.161$

Example 4 Simplify: $-8.729 + 12.094$

Your solution 3.365

Example 5 Simplify: $-\dfrac{5}{12} \cdot \dfrac{4}{25}$

Solution $-\dfrac{5}{12} \cdot \dfrac{4}{25} = -\dfrac{5 \cdot 4}{12 \cdot 25}$

$\qquad\qquad = -\dfrac{20}{300} = -\dfrac{1}{15}$

Example 6 Simplify $\dfrac{5}{8} \div \left(-\dfrac{15}{40}\right)$

Your solution $-\dfrac{5}{3}$

Solutions on p. A9

Example 7   Simplify: $0.0527 \div (-0.27)$
Round to the nearest
hundredth.

Solution

$$0.195 \approx 0.20$$

$$0.27.\overline{)0.05.270}$$
$$\underline{-2\ 7}$$
$$2\ 57$$
$$\underline{-2\ 43}$$
$$140$$
$$\underline{-135}$$
$$5$$

$0.0527 \div (-0.27) \approx -0.20$

Example 8   Simplify: $-4.027 \cdot 0.49$
Round to the nearest
hundredth.

Your solution -1.97

Solution on p. A9

Objective B

To use the Order of Operations Agreement to simplify expressions

Evaluate: $16 + 4 \cdot 2$

In this problem, there are two arithmetic operations, addition and multiplication. The operations could be performed in different orders.

Add first.	$\underline{16 + 4}\ \cdot 2$	Multiply first.	$16 + \underset{\frown}{4 \cdot 2}$
Then multiply.	$\underline{20}\ \ \cdot 2$	Then add.	$\underline{16 +\ \ 8}$
	40		24

In order to prevent more than one answer to the same problem, an Order of Operations Agreement is followed.

The Order of Operations Agreement

Step 1   Perform operations inside grouping symbols. Grouping symbols include parentheses ( ), brackets [ ], and the fraction bar.
Step 2   Simplify exponential expressions.
Step 3   Do multiplication and division as they occur from left to right.
Step 4   Do addition and subtraction as they occur from left to right.

Simplify: $8 - \frac{22 - 2}{4 + 1} \div 2^2$

$$8 - \frac{22 - 2}{4 + 1} \div 2^2$$

Perform operations above and below the fraction bar.

$$8 - \frac{20}{5} \div 2^2$$

Simplify exponential expressions.

$$8 - \frac{20}{5} \div 4$$

Do multiplication and division as they occur from left to right.

$$8 - 4 \div 4$$

$$8 - 1$$

Do addition and subtraction as they occur from left to right.

7

Content and Format © 1991 HMCo.

One or more of the above steps may not be needed to simplify an expression. In that case, proceed to the next step in the Order of Operations Agreement.

When an expression has grouping symbols inside grouping symbols, perform the operations inside the inner grouping symbols first.

Simplify: $4.3 - [(25 - 9) \div 2]^2$

Perform operations inside grouping symbols.	$4.3 - [(25 - 9) \div 2]^2$ $4.3 - [16 \div 2]^2$
Simplify exponential expressions.	$4.3 - [8]^2$
Perform addition and subtraction as they occur from left to right.	$4.3 - 64$
	-59.7

A **complex fraction** is a fraction whose numerator or denominator contains one or more fractions. Examples of complex fractions are shown at the right.

$$\frac{\frac{2}{3}}{\frac{1}{3}}, \quad \frac{\frac{2}{5} + 1}{\frac{7}{8}} \leftarrow \text{Main Fraction Bar}$$

One way to simplify a complex fraction is to perform operations above and below the main fraction bar using the Order of Operations Agreement.

Simplify: $\dfrac{\frac{3}{4} - 1}{\frac{5}{8}}$

$$\frac{\frac{3}{4} - 1}{\frac{5}{8}}$$

Perform operations above and below the main fraction bar.	$\dfrac{\frac{3}{4} - \frac{4}{4}}{\frac{5}{8}}$ $\dfrac{-\frac{1}{4}}{\frac{5}{8}}$
Multiply the numerator of the complex fraction by the reciprocal of the denominator of the complex fraction.	$-\dfrac{1}{4} \cdot \dfrac{8}{5}$ $-\dfrac{1 \cdot 8}{4 \cdot 5}$ $-\dfrac{8}{20}$ $-\dfrac{2}{5}$

Example 9   Simplify:
$(-1.2)^3 - 8.4 \div 2.1$

Solution $(-1.2)^3 - 8.4 \div 2.1$
$-1.728 - 8.4 \div 2.1$
$-1.728 - 4$
-5.728

Example 10   Simplify:
$(3.81 - 1.41)^2 \div 0.036 - 1.89$

Your solution   158.11

Example 11   Simplify:
$\left(\frac{1}{2}\right)^3 - \left[\left(\frac{2}{3} + \frac{1}{4}\right) \div \frac{5}{6}\right]$

Solution $\left(\frac{1}{2}\right)^3 - \left[\left(\frac{2}{3} + \frac{1}{4}\right) \div \frac{5}{6}\right]$

$\left(\frac{1}{2}\right)^3 - \left[\frac{11}{12} \div \frac{5}{6}\right]$

$\left(\frac{1}{2}\right)^3 - \left[\frac{11}{12} \cdot \frac{6}{5}\right]$

$\left(\frac{1}{2}\right)^3 - \frac{11}{10}$

$\frac{1}{8} - \frac{11}{10}$

$-\frac{39}{40}$

Example 12   Simplify:
$\frac{1}{3} + \frac{5}{8} \div \frac{15}{16} - \frac{7}{12}$

Your solution $\frac{5}{12}$

Example 13   Simplify:

$9 \cdot \dfrac{\frac{5}{6} - 2}{\frac{3}{8}} \div \frac{7}{6}$

Solution $9 \cdot \dfrac{\frac{5}{6} - 2}{\frac{3}{8}} \div \frac{7}{6}$

$9 \cdot \dfrac{-\frac{7}{6}}{\frac{3}{8}} \div \frac{7}{6}$

$9 \cdot \left(-\frac{7}{6} \cdot \frac{8}{3}\right) \div \frac{7}{6}$

$9 \cdot \left(-\frac{28}{9}\right) \div \frac{7}{6}$

$-28 \div \frac{7}{6}$

$-28 \cdot \frac{6}{7}$

-24

Example 14   Simplify: $\dfrac{11}{12} - \dfrac{\frac{5}{4}}{2 - \frac{7}{2}} \cdot \frac{3}{4}$

Your solution $\frac{37}{24}$

1.2 EXERCISES

▶ **Objective A**

Simplify.

1. $\frac{7}{12} + \frac{5}{16}$

$\frac{43}{48}$

2. $\frac{3}{8} - \frac{5}{12}$

$-\frac{1}{24}$

3. $-\frac{5}{9} - \frac{14}{15}$

$-\frac{67}{45}$

4. $\frac{1}{2} + \frac{1}{7} - \frac{5}{8}$

$\frac{1}{56}$

5. $-\frac{1}{3} + \frac{5}{9} - \frac{7}{12}$

$-\frac{13}{36}$

6. $\frac{1}{3} + \frac{19}{24} - \frac{7}{8}$

$\frac{1}{4}$

7. $\frac{2}{3} - \frac{5}{12} + \frac{5}{24}$

$\frac{11}{24}$

8. $-\frac{7}{10} + \frac{4}{5} + \frac{5}{6}$

$\frac{14}{15}$

9. $\frac{5}{8} - \frac{7}{12} + \frac{1}{2}$

$\frac{13}{24}$

10. $-\frac{1}{3} \cdot \frac{5}{8}$

$-\frac{5}{24}$

11. $\left(\frac{6}{35}\right)\left(-\frac{5}{16}\right)$

$-\frac{3}{56}$

12. $\frac{2}{3}\left(-\frac{9}{20}\right) \cdot \frac{5}{12}$

$-\frac{1}{8}$

13. $-\frac{8}{15} \div \frac{4}{5}$

$-\frac{2}{3}$

14. $-\frac{2}{3} \div \left(-\frac{6}{7}\right)$

$\frac{7}{9}$

15. $-\frac{11}{24} \div \frac{7}{12}$

$-\frac{11}{14}$

16. $\frac{7}{9} \div \left(-\frac{14}{27}\right)$

$-\frac{3}{2}$

17. $\left(-\frac{5}{12}\right)\left(\frac{4}{35}\right)\left(\frac{7}{8}\right)$

$-\frac{1}{24}$

18. $\frac{6}{35}\left(-\frac{7}{40}\right)\left(-\frac{8}{21}\right)$

$\frac{2}{175}$

19. $-14.27 + 1.296$

-12.974

20. $-0.4355 + 172.5$

172.0645

21. $1.832 - 7.84$

-6.008

22. $(3.52)(4.7)$

16.544

23. $(0.03)(10.5)(6.1)$

1.9215

24. $(1.2)(3.1)(-6.4)$

-23.808

25. $5.418 \div (-0.9)$

-6.02

26. $-0.2645 \div (-0.023)$

11.5

27. $-0.4355 \div 0.065$

-6.7

28. $-6.58 - 3.97 + 0.875$

-9.675

29. $8.8 + 3.072 - 56.457$

-44.585

Simplify. Round to the nearest hundredth.

30. $38.241 \div [-(-6.027)] - 7.453$

-1.11

31. $-9.0508 - (-3.177) + 24.77$

18.90

Simplify. Round to the nearest hundredth.

32. $-287.3069 \div 0.1415$
-2030.44

33. $6472.3018 \div (-3.59)$
-1802.87

▶ **Objective B**

Simplify.

34. $5 - 3(8 \div 4)^2$
-7

35. $4^2 - (5 - 2)^2 \cdot 3$
-11

36. $16 - \dfrac{2^2 - 5}{3^2 + 2}$
$\dfrac{177}{11}$

37. $\dfrac{4(5 - 2)}{4^2 - 2^2} \div 4$
$\dfrac{1}{4}$

38. $\dfrac{3 + \frac{2}{3}}{\frac{11}{16}}$
$\dfrac{16}{3}$

39. $\dfrac{\frac{11}{14}}{4 - \frac{6}{7}}$
$\dfrac{1}{4}$

40. $5[(2 - 4) \cdot 3 - 2]$
-40

41. $2[(16 \div 8) - (-2)] + 4$
12

42. $16 - 4\left(\dfrac{8 - 2}{3 - 6}\right) \div \dfrac{1}{2}$
32

43. $25 \div 5\left(\dfrac{16 + 8}{-2^2 + 8}\right) - 5$
25

44. $6[3 - (-4 + 2) \div 2]$
24

45. $12 - 4[2 - (-3 + 5) - 8]$
44

46. $\dfrac{1}{2} - \left(\dfrac{2}{3} \div \dfrac{5}{9}\right) + \dfrac{5}{6}$
$\dfrac{2}{15}$

47. $\left(-\dfrac{3}{5}\right)^2 - \dfrac{3}{5} \cdot \dfrac{5}{9} + \dfrac{7}{10}$
$\dfrac{109}{150}$

48. $\dfrac{1}{2} - \dfrac{\frac{17}{25}}{4 - \frac{3}{5}} \div \dfrac{1}{5}$
$-\dfrac{1}{2}$

49. $\dfrac{3}{4} + \dfrac{3 - \frac{7}{9}}{\frac{5}{6}} \cdot \dfrac{2}{3}$
$\dfrac{91}{36}$

50. $\dfrac{2}{3} - \left[\dfrac{3}{8} + \dfrac{5}{6}\right] \div \dfrac{3}{5}$
$-\dfrac{97}{72}$

51. $\dfrac{3}{4} \div \left[\dfrac{5}{8} - \dfrac{5}{12}\right] + 2$
$\dfrac{28}{5}$

52. $0.4(1.2 - 2.3)^2 + 5.8$
6.284

53. $5.4 - (0.3)^2 \div 0.09$
4.4

54. $1.75 \div 0.25 - (1.25)^2$
5.4375

55. $(3.5 - 4.2)^2 - 3.50 \div 2.5$
-0.91

56. $25.76 \div (6.96 - 3.27)^2$
1.891878

57. $(3.09 - 4.77)^3 - 4.07 \cdot 3.66$
-19.637832

SECTION 1.3   Variable Expressions

Objective A   To evaluate a variable expression

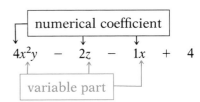

An expression which contains one or more variables is a **variable expression.**

A variable expression is shown at the right. The expression has 4 addends, which are called the **terms** of the expression. The variable expression has 3 variable terms and 1 constant term.

$$\overbrace{}^{\text{4 terms}}$$
$$4x^2y \; \overbrace{- \; 2z \; - \; x}^{} \; \overbrace{+ \; 4}^{}$$
$$\underbrace{}_{\text{variable terms}} \qquad \underbrace{}_{\substack{\text{constant} \\ \text{term}}}$$

Each variable term is composed of a **numerical coefficient** and a **variable part.** When the numerical coefficient is 1 or −1, the 1 is usually not written.

$$\boxed{\text{numerical coefficient}}$$
$$4x^2y \; - \; 2z \; - \; 1x \; + \; 4$$
$$\boxed{\text{variable part}}$$

Replacing the variable in a variable expression by its value and then simplifying the resulting expression is called **evaluating the variable expression.**

Evaluate $a^2 - (ab - c)$ when $a = -2$, $b = 3$, and $c = -4$.

Replace each variable with its value.

Use the Order of Operations Agreement to simplify the resulting numerical expression.

$$a^2 - (ab - c)$$
$$(-2)^2 - [(-2)(3) - (-4)]$$
$$(-2)^2 - [-6 - (-4)]$$
$$(-2)^2 - [-2]$$
$$4 - [-2]$$
$$6$$

Example 1   Evaluate $ab - (b - c)$ when $a = 2$, $b = -3$, and $c = -4$.

Solution $ab - (b - c)$
$2(-3) - [(-3) - (-4)]$
$2(-3) - 1$
-7

Example 2   Evaluate $(b - c)^2 \div ab$ when $a = -3$, $b = 2$, and $c = -4$.

Your solution -24

Example 3   Evaluate $|2x - 7|$ when $x = 2$.

Solution $|2x - 7|$
$|2 \cdot 2 - 7|$
$|4 - 7|$
$|-3|$
3

Example 4   Evaluate $|3 - 4x|$ when $x = -2$.

Your solution 11

Solutions on p. A10

| Objective B | **To use and identify the Properties of the Real Numbers** |

The Properties of the Real Numbers describe the way operations on numbers can be performed. Following is a list of some of the real number properties and an example of each property.

Properties of the Real Numbers

The Commutative Property of Addition

$$a + b = b + a$$

$3 + 2 = 2 + 3$
$5 = 5$

The Commutative Property of Multiplication

$$a \cdot b = b \cdot a$$

$(3)(-2) = (-2)(3)$
$-6 = -6$

The Associative Property of Addition

$$(a + b) + c = a + (b + c)$$

$(3 + 4) + 5 = 3 + (4 + 5)$
$7 + 5 = 3 + 9$
$12 = 12$

The Associative Property of Multiplication

$$(a \cdot b) \cdot c = a \cdot (b \cdot c)$$

$(3 \cdot 4) \cdot 5 = 3 \cdot (4 \cdot 5)$
$12 \cdot 5 = 3 \cdot 20$
$60 = 60$

The Addition Property of Zero

$$a + 0 = 0 + a = a$$

$3 + 0 = 0 + 3 = 3$

The Multiplication Property of Zero

$$a \cdot 0 = 0 \cdot a = 0$$

$3 \cdot 0 = 0 \cdot 3 = 0$

The Multiplication Property of One

$$a \cdot 1 = 1 \cdot a = a$$

$5 \cdot 1 = 1 \cdot 5 = 5$

The Inverse Property of Addition

$$a + (-a) = (-a) + a = 0$$

$4 + (-4) = (-4) + 4 = 0$

$-a$ is called the **additive inverse** of a. a is the additive inverse of $-a$. The sum of a number and its additive inverse is 0.

Content and Format © 1991 HMCo.

The Inverse Property of Multiplication

$$a \cdot \frac{1}{a} = \frac{1}{a} \cdot a = 1, a \neq 0$$

$$(4)\left(\tfrac{1}{4}\right) = \left(\tfrac{1}{4}\right)(4) = 1$$

$\frac{1}{a}$ is called the reciprocal of a. $\frac{1}{a}$ is also called the **multiplicative inverse** of a. The product of a number and its multiplicative inverse is 1.

The Distributive Property

$$a(b + c) = ab + ac$$

$$3(4 + 5) = 3 \cdot 4 + 3 \cdot 5$$
$$3 \cdot 9 = 12 + 15$$
$$27 = 27$$

Division Properties of Zero and One

Zero divided by any number other than zero is zero.	$\frac{0}{a} = 0, a \neq 0$
Division by zero is not defined.	$\frac{a}{0}$ is undefined.
Any number other than zero divided by itself is one.	$\frac{a}{a} = 1, a \neq 0$

Example 5 Complete the statement by using the Commutative Property of Multiplication.

$$(x)\left(\tfrac{1}{4}\right) = (?)(x)$$

Solution $(x)\left(\tfrac{1}{4}\right) = \left(\tfrac{1}{4}\right)(x)$

Example 6 Complete the statement by using the Inverse Property of Addition.

$$3x + ? = 0$$

Your solution $-3x$

Example 7 Identify the property that justifies the statement.

$$3(x + 4) = 3x + 12$$

Solution The Distributive Property

Example 8 Identify the property that justifies the statement.

$$(a + 3b) + c = a + (3b + c)$$

Your solution The Associative Property of Addition

Solutions on p. A10

| Objective C | **To simplify a variable expression** |

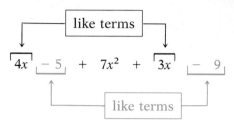

Like terms of a variable expression are the terms with the same variable part.

Constant terms are like terms.

$$\overset{\fbox{like terms}}{\fbox{$4x$}\,\boxed{-5}\,+\,7x^2\,+\,\boxed{3x}\,\boxed{-9}}$$

like terms

To simplify a variable expression **combine** like terms, by using the Distributive Property to add the coefficients.

$3x + 2x$
$(3 + 2)x$
$5x$

The Distributive Property is also used to remove parentheses from a variable expression so that like terms can be combined.

Simplify: $2(x + y) + 3(y - 3x)$

Use the Distributive Property to remove parentheses.

$2(x + y) + 3(y - 3x)$
$2x + 2y + 3y - 9x$

Use the Commutative and Associative Properties of Addition to rearrange and group like terms.

$(2x - 9x) + (2y + 3y)$ Do this step mentally.

Combine like terms.

$-7x + 5y$

Simplify: $\frac{1}{4}[10 - (4x - 2)] - x$

$\frac{1}{4}[10 - (4x - 2)] - x$

Use the Distributive Property to remove parentheses.

$\frac{1}{4}[10 - 4x + 2] - x$

Use the Commutative and Associative Properties of Addition to rearrange and group like terms.

$\frac{1}{4}[(10 + 2) - 4x] - x$ Do this step mentally.

Combine like terms.

$\frac{1}{4}[12 - 4x] - x$

Use the Distributive Property to remove brackets.

$3 - x - x$

Combine like terms.

$3 - 2x$

Example 9   Simplify:
$(3x - 2y + 1) - (2x - 5y + 4)$

Solution $(3x - 2y + 1) - (2x - 5y + 4)$
$3x - 2y + 1 - 2x + 5y - 4$
$x + 3y - 3$

Example 10   Simplify:
$(2x + xy - y) - (5x - 7xy + y)$

Your solution $-3x + 8xy - 2y$

Solution on p. A10

Example 11 Simplify:
$4y - 2[x - 3(x + y) - 5y]$

Solution $4y - 2[x - 3(x + y) - 5y]$
$4y - 2[x - 3x - 3y - 5y]$
$4y - 2[-2x - 8y]$
$4y + 4x + 16y$
$4x + 20y$

Example 12 Simplify:
$2x - 3[y - 3(x - 2y + 4)]$

Your solution $11x - 21y + 36$

Solution on p. A10

Objective D

To translate a verbal expression into a variable expression and then simplify the resulting expression

One of the major skills required in applied mathematics is to translate a verbal expression into a variable expression. This requires recognizing the phrases that translate into mathematical operations. A partial list of the verbal phrases used to indicate the different mathematical operations is given below.

Addition	added to	5 added to x	$5 + x$
	more than	2 more than t	$t + 2$
	the sum of	the sum of s and r	$s + r$
	increased by	z increased by 7	$z + 7$
	the total of	the total of 6 and b	$6 + b$
Subtraction	minus	y minus 8	$y - 8$
	less than	5 less than p	$p - 5$
	decreased by	n decreased by 1	$n - 1$
	the difference between	the difference between t and 4	$t - 4$
Multiplication	times	9 times y	$9y$
	of	one third of p	$\frac{1}{3}p$
	the product of	the product of x and y	xy
	multiplied by	t multiplied by 43	$43t$
Division	divided by	a divided by 6	$\frac{a}{6}$
	the quotient of	the quotient of s and t	$\frac{s}{t}$
	the ratio of	the ratio of r to 5	$\frac{r}{5}$
Power	the square of	the square of b	b^2
	the cube of	the cube of x	x^3

In most applications which involve translating phrases into variable expressions, the variable to be used is not given. To translate these phrases, a variable must be assigned to an unknown quantity before the variable expression can be written. After translating a verbal expression into a variable expression, simplify the variable expression by using the Addition, Multiplication, and Distributive Properties.

Translate and simplify "the total of five times a number and twice the difference between the number and three."

Assign a variable to one of the unknown quantities.	the unknown number: n
Use the assigned variable to write an expression for any other unknown quantity.	five times the number: $5n$ the difference between the number and three: $n - 3$ twice the difference between the number and three: $2(n - 3)$
Use the assigned variable to write the variable expression.	$5n + 2(n - 3)$
Simplify the variable expression.	$5n + 2n - 6$ $7n - 6$

Example 13

Translate and simplify "the sum of three consecutive integers."

Solution

the first integer: n
the next consecutive integer: $n + 1$
the next consecutive integer: $n + 2$

$n + (n + 1) + (n + 2)$
$3n + 3$

Example 14

Translate and simplify "one half of the sum of two consecutive integers."

Your solution

$\frac{1}{2}[n + (n + 1)]$

$n + \frac{1}{2}$

Example 15

Translate and simplify "fifteen minus one half of the sum of a number and ten."

Solution

the unknown number: n
the sum of the number and ten: $n + 10$
one half of the sum of the number and ten: $\frac{1}{2}(n + 10)$

$15 - \frac{1}{2}(n + 10)$

$15 - \frac{1}{2}n - 5$

$10 - \frac{1}{2}n$

Example 16

Translate and simplify "the sum of three eighths of a number and five twelfths of the number."

Your solution

$\frac{3}{8}n + \frac{5}{12}n$

$\frac{19}{24}n$

Solutions on p. A10

1.3   EXERCISES

▶ **Objective A**

Evaluate the variable expression when $a = 2$, $b = 3$, $c = -1$, and $d = -4$.

1. $ab + dc$
10

2. $2ab - 3dc$
0

3. $4cd \div a^2$
4

4. $b^2 - (d - c)^2$
0

5. $(b - 2a)^2 + c$
0

6. $(b - d)^2 \div (b - d)$
7

7. $(bc + a)^2 \div (d - b)$
$-\frac{1}{7}$

8. $\frac{1}{3}b^3 - \frac{1}{4}d^3$
25

9. $\frac{1}{4}a^4 - \frac{1}{6}bc$
$\frac{9}{2}$

10. $2b^2 \div \frac{ad}{2}$
$-\frac{9}{2}$

11. $\frac{3ac}{-4} - c^2$
$\frac{1}{2}$

12. $\frac{2d - 2a}{2bc}$
2

13. $\frac{3b - 5c}{3a - c}$
2

14. $\frac{2d - a}{b - 2c}$
-2

15. $\frac{a - d}{b + c}$
3

16. $|a^2 + d|$
0

17. $-a|a + 2d|$
-12

18. $d|b - 2d|$
-44

19. $\frac{2a - 4d}{3b - c}$
2

20. $\frac{3d - b}{b - 2c}$
-3

21. $-3d \div \left|\frac{ab - 4c}{2b + c}\right|$
6

22. $-2bc + \left|\frac{bc + d}{ab - c}\right|$
7

23. $2(d - b) \div (3a - c)$
-2

24. $(d - 4a)^2 \div c^3$
-144

Evaluate the variable expression when $a = 1.5$, $b = -2.4$, and $c = -0.5$.

25. $b^2 - ac$
6.51

26. $(a - b)^2 \div c^2$
60.84

27. $\frac{a^2}{c} - (b - a)^2$
-19.71

▶ **Objective B**

Use the given Property of the Real Numbers to complete the statement.

28. The Commutative Property of Multiplication

$3 \cdot 4 = 4 \cdot ?$
3

29. The Commutative Property of Addition

$7 + 15 = ? + 7$
15

30. The Associative Property of Addition

$(3 + 4) + 5 = ? + (4 + 5)$
3

31. The Associative Property of Multiplication

$(3 \cdot 4) \cdot 5 = 3 \cdot (? \cdot 5)$
4

Use the given Property of the Real Numbers to complete the statement.

32. The Division Properties of Zero

$\frac{5}{?}$ is undefined.

0

33. The Multiplication Property of Zero

$5 \cdot ? = 0$

0

34. The Distributive Property

$3(x + 2) = 3x + ?$

6

35. The Distributive Property

$5(y + 4) = ? \cdot y + 20$

5

36. The Division Properties of Zero

$\frac{?}{-6} = 0$

0

37. The Inverse Property of Addition

$(x + y) + ? = 0$

$[-(x + y)]$

38. The Inverse Property of Multiplication

$\frac{1}{mn}(mn) = ?$

1

39. The Multiplication Property of One

$? \cdot 1 = x$

x

40. The Associative Property of Multiplication

$2(3x) = ? \cdot x$

$(2 \cdot 3)$

41. The Commutative Property of Addition

$ab + bc = bc + ?$

ab

Identify the property that justifies the statement.

42. $\frac{0}{-5} = 0$

The Division Property of Zero

43. $-8 + 8 = 0$

The Inverse Property of Addition

44. $(-12)\left(-\frac{1}{12}\right) = 1$

The Inverse Property of Multiplication

45. $(3 \cdot 4) \cdot 2 = 2 \cdot (3 \cdot 4)$

The Commutative Property of Multiplication

46. $y + 0 = y$

The Addition Property of Zero

47. $2x + (5y + 8) = (2x + 5y) + 8$

The Associative Property of Addition

48. $\frac{-9}{0}$ is undefined.

The Division Property of Zero

49. $(x + y)z = xz + yz$

The Distributive Property

50. $6(x + y) = 6x + 6y$

The Distributive Property

51. $(-12y)(0) = 0$

The Multiplication Property of Zero

52. $(ab)c = a(bc)$

The Associative Property of Multiplication

53. $(x + y) + z = (y + x) + z$

The Commutative Property of Addition

Content and Format © 1991 HMCo.

▶ **Objective C**

Simplify.

54. $5x + 7x$
$12x$

55. $3x + 10x$
$13x$

56. $-8ab - 5ab$
$-13ab$

57. $-2x + 5x - 7x$
$-4x$

58. $3x - 5x + 9x$
$7x$

59. $-2a + 7b + 9a$
$7a + 7b$

60. $5b - 8a - 12b$
$-8a - 7b$

61. $12\left(\frac{1}{12}x\right)$
x

62. $\frac{1}{3}(3y)$
y

63. $-3(x - 2)$
$-3x + 6$

64. $-5(x - 9)$
$-5x + 45$

65. $(x + 2)5$
$5x + 10$

66. $-(x + y)$
$-x - y$

67. $-(-x - y)$
$x + y$

68. $3(-2a + 3a - 5)$
$3a - 15$

69. $3(x - 2y) - 5$
$3x - 6y - 5$

70. $4x - 3(2y - 5)$
$4x - 6y + 15$

71. $-2a - 3(3a - 7)$
$-11a + 21$

72. $3x - 2(5x - 7)$
$-7x + 14$

73. $2x - 3(x - 2y)$
$-x + 6y$

74. $3[a - 5(5 - 3a)]$
$48a - 75$

75. $5[-2 - 6(a - 5)]$
$140 - 30a$

76. $3[x - 2(x + 2y)]$
$-3x - 12y$

77. $5[y - 3(y - 2x)]$
$-10y + 30x$

78. $-2(x - 3y) + 2(3y - 5x)$
$-12x + 12y$

79. $4(-a - 2b) - 2(3a - 5b)$
$-10a + 2b$

80. $5(3a - 2b) - 3(-6a + 5b)$
$33a - 25b$

81. $-7(2a - b) + 2(-3b + a)$
$-12a + b$

82. $3x - 2[y - 2(x + 3[2x + 3y])]$
$31x + 34y$

83. $2x - 4[x - 4(y - 2[5y + 3])]$
$-2x - 144y - 96$

84. $4 - 2(7x - 2y) - 3(-2x + 3y)$
$4 - 8x - 5y$

85. $3x + 8(x - 4) - 3(2x - y)$
$5x - 32 + 3y$

86. $\frac{1}{3}[8x - 2(x - 12) + 3]$
$2x + 9$

87. $\frac{1}{4}[14x - 3(x - 8) - 7x]$
$x + 6$

▶ **Objective D**

Translate into a variable expression. Then simplify.

88. a number minus the sum of the number and two
$n - (n + 2)$; -2

89. a number decreased by the difference between five and the number
$n - (5 - n)$; $2n - 5$

90. the sum of one third of a number and four fifths of the number
$\frac{1}{3}n + \frac{4}{5}n$; $\frac{17}{15}n$

91. the difference between three eighths of a number and one sixth of the number
$\frac{3}{8}n - \frac{1}{6}n$; $\frac{5}{24}n$

92. five times the product of eight and a number
$5(8n)$; $40n$

93. a number increased by two thirds of the number
$n + \frac{2}{3}n$; $\frac{5}{3}n$

94. the difference between the product of seventeen and a number and twice the number
$17n - 2n$; $15n$

95. one half of the total of six times a number and twenty-two
$\frac{1}{2}(6n + 22)$; $3n + 11$

96. three times the total of two consecutive integers
$3[n + (n + 1)]$; $6n + 3$

97. the sum of the second and third of three consecutive integers
$(n + 1) + (n + 2)$; $2n + 3$

98. thirty less than the product of six and the sum of the first and third of three consecutive integers
$6[n + (n + 2)] - 30$; $12n - 18$

99. twelve more than the product of four and the second of three consecutive integers
$4(n + 1) + 12$; $4n + 16$

100. three more than the sum of the first and third of three consecutive integers
$[n + (n + 2)] + 3$; $2n + 5$

101. three times the quotient of twice a number and six
$3\left(\frac{2n}{6}\right)$; n

102. the sum of five times a number and twelve added to the product of fifteen and the number
$(5n + 12) + 15n$; $20n + 12$

103. four less than twice the sum of a number and eleven
$2(n + 11) - 4$; $2n + 18$

104. twice a number plus the product of two more than the number and eight
$2n + (n + 2)(8)$; $10n + 16$

105. twenty minus the product of four more than a number and twelve
$20 - (n + 4)(12)$; $-12n - 28$

106. a number added to the product of five plus the number and four
$n + (5 + n)(4)$; $5n + 20$

107. a number plus the product of the number minus twelve and three
$n + (n - 12)(3)$; $4n - 36$

SECTION 1.4   Sets

Objective A   **To find the union and intersection of sets**

A **set** is a collection of objects. The objects in a set are the **elements** of the set.

A set can be written in various ways. The **roster method** of writing a set encloses a list of the elements of the set in braces.

The set of the three planets nearest the sun is shown at the right.   {Mercury, Venus, Earth}

Use the roster method to write the set of even natural numbers less than 10.

A set is frequently designated by a capital letter. $A = \{2, 4, 6, 8\}$

This is an example of a **finite set.** All the elements of the set can be listed.

Use the roster method to write the set of natural numbers greater than 5.

The three dots mean that the pattern of numbers continues without end. $B = \{6, 7, 8, 9, 10, 11, \ldots\}$

This is an example of an **infinite set.** It is impossible to list all the elements of the set.

The symbol "$\in$" means "is an element of."

$9 \in B$ is read "9 is an element of set B."

Given $A = \{-1, 5, 9\}$, then $-1 \in A$, $5 \in A$, and $9 \in A$.

$12 \notin A$ is read "12 is not an element of A."

The **empty set,** or **null set,** is the set which contains no elements. The symbol $\varnothing$ or { } is used to represent the empty set. (Note: It is incorrect to write $\{\varnothing\}$ as the empty set.)

The set of trees over 1000 ft tall is the empty set.

A second method of representing a set is **set builder notation.** The set of all integers greater than -3 would be written: $\{x | x > -3, x \text{ is an integer}\}$, read "the set of all x such that x is greater than -3, x is an integer." Set builder notation can be used to describe almost any set, but it is especially useful when writing infinite sets.

Use set builder notation to write the set of real numbers less than 5.

"$x \in$ real numbers" is read "x is an element of the real numbers." $\{x | x < 5, x \in \text{real numbers}\}$

This is another example of an infinite set. It is impossible to list all the elements in this set. For example, 3.7, $-\frac{3}{8}$, and -101.0102 are elements of this set.

Just as operations such as addition and multiplication are performed on real numbers, there are various set operations.

Union of Two Sets

The **union** of two sets, written $A \cup B$, is the set of all elements which belong to either A **or** to B.

$$A \cup B = \{x \mid x \in A \text{ or } x \in B\}$$

Find $A \cup B$, given $A = \{2, 3, 5, 7\}$ and $B = \{0, 1, 2, 3, 4\}$.

The union of two sets contains all the elements that belong to either A or to B. The elements that belong to both sets are listed only once.

$A \cup B = \{0, 1, 2, 3, 4, 5, 7\}$

Intersection of Two Sets

The **intersection** of two sets, written $A \cap B$, is the set of all elements that are common to both A **and** to B.

$$A \cap B = \{x \mid x \in A \text{ and } x \in B\}$$

Find $A \cap B$, given $A = \{2, 3, 5, 7\}$ and $B = \{0, 1, 2, 3, 4\}$.

The intersection of two sets contains all the elements that are common to both A and B.

$A \cap B = \{2, 3\}$

Example 1

Use the roster method to write the set of prime numbers less than 15.

Solution

$A = \{2, 3, 5, 7, 11, 13\}$

Example 2

Use the roster method to write the set of even natural numbers less than 12.

Your solution

$A = \{2, 4, 6, 8, 10\}$

Example 3

Write $\{x \mid x < 10, x \text{ is a natural number}\}$ by using the roster method.

Solution

$A = \{1, 2, 3, 4, 5, 6, 7, 8, 9\}$

Example 4

Write $\{x \mid x < 10, x \text{ is a positive odd integer}\}$ by using the roster method.

Your solution

$A = \{1, 3, 5, 7, 9\}$

Example 5

Is $-7 \in \{x \mid x > -5, x \in \text{real numbers}\}$?

Solution

-7 is not greater than -5.

No, -7 is not an element of the set.

Example 6

Is $0.35 \in \{x \mid x < 2, x \in \text{real numbers}\}$?

Your solution

Yes

Example 7

Find $C \cup D$, given $C = \{1, 5, 9, 13, 17\}$ and $D = \{3, 5, 7, 9, 11\}$.

Solution

$C \cup D = \{1, 3, 5, 7, 9, 11, 13, 17\}$

Example 8

Find $A \cup C$ given $A = \{-2, -1, 0, 1, 2\}$ and $C = \{-5, -1, 0, 1, 5\}$.

Your solution

$A \cup C = \{-5, -2, -1, 0, 1, 2, 5\}$

Solutions on p. A10

Example 9

Find $A \cap B$, given $A = \{8, 10, 12, 14\}$ and $B = \{7, 8, 9, 10\}$.

Solution

$A \cap B = \{8, 10\}$

Example 10

Find $D \cap C$, given $D = \{1, 3, 5, 7, 9\}$ and $C = \{3, 4, 5, 6, 7\}$.

Your solution

$D \cap C = \{3, 5, 7\}$

Example 11

Find $A \cap B$, given
$A = \{x | x$ is a natural number$\}$ and
$B = \{x | x$ is a negative integer$\}$.

Solution

There are no natural numbers that are also negative numbers.

$A \cap B = \varnothing$

Example 12

Find $E \cap F$, given
$E = \{x | x$ is an odd integer$\}$ and
$F = \{x | x$ is an even integer$\}$.

Your solution

$E \cap F = \varnothing$

Solutions on p. A10

| **Objective B** | **To graph the solution set of an inequality in one variable** |

An **inequality** expresses the relative order of two mathematical expressions. The symbols $>$, $<$, $\le$ (less than or equal to), and $\ge$ (greater than or equal to) are used to write inequalities.

$$7 > -2$$
$$2x < 4$$
$$2x - 3y \ge 6$$
$$x^2 - 2x - 3 \le 0$$

The **solution set of an inequality** is the set of real numbers which make the inequality true. The solution set can be graphed on a number line.

Graph the solution set of $x > -2$.

The solution set is the real numbers greater than -2. The circle on the graph indicates that -2 is not included in the solution set.

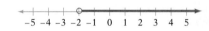

The solution set is written $\{x | x > -2, x \in$ real numbers$\}$.

Graph the solution set of $x \ge -2$.

The dot at -2 indicates that -2 is included in the solution set.

The solution set is written $\{x | x \ge -2, x \in$ real numbers$\}$.

For the remainder of this unit, all variables will represent real numbers. Using this convention, the solution set in the last example would be written $\{x | x \ge -2\}$.

The union of two sets is the set of all elements belonging to either one or the other of the two sets.

Graph the solution set of $\{x|x \le -1\} \cup \{x|x > 3\}$.

The solution set is the set of all numbers that are either less than or equal to -1 or greater than 3.

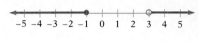

The solution set is written $\{x|x \le -1 \text{ or } x > 3\}$.

Graph the solution set of $\{x|x > 2\} \cup \{x|x > 4\}$.

The solution set is the set of all numbers that are either greater than 2 or greater than 4.

The solution set is written $\{x|x > 2\}$.

The intersection of two sets is the set that contains the elements common to both sets.

Graph the solution set of $\{x|x > -2\} \cap \{x|x < 5\}$.

The solution set is the set of numbers that are greater than -2 and less than 5.

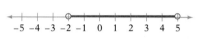

The solution set can be written $\{x|x > -2 \text{ and } x < 5\}$. However, it is more commonly written $\{x|-2 < x < 5\}$.

Graph the solution set of $\{x|x < 4\} \cap \{x|x < 5\}$.

The solution set is the set of numbers that are less than 4 and less than 5.

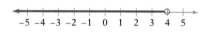

The solution set is written $\{x|x < 4\}$.

Example 13   Graph the solution set of $x \le 3$.

Solution   The solution set is $\{x|x \le 3\}$.

Example 14   Graph the solution set of $x > -3$.

Your solution

Example 15   Graph the solution set of $\{x|x < 0\} \cap \{x|x > -3\}$.

Solution   The solution set is $\{x|-3 < x < 0\}$.

Example 16   Graph the solution set of $\{x|x > -3\} \cap \{x|x < 4\}$.

Your solution

Example 17   Graph the solution set of $\{x|x < 0\} \cup \{x|x > 3\}$.

Solution   The solution set is $\{x|x < 0 \text{ or } x > 3\}$

Example 18   Graph the solution set of $\{x|x \ge 1\} \cup \{x|x \le -3\}$.

Your solution

Solutions on p. A10

█████ **1.4** █ **EXERCISES**

▶ **Objective A**

Use the roster method to write the set.

1. the integers between -3 and 5
$A = \{-2, -1, 0, 1, 2, 3, 4\}$

2. the integers between -4 and 0
$A = \{-3, -2, -1\}$

3. the even natural numbers less than 13
$A = \{2, 4, 6, 8, 10, 12\}$

4. the odd natural numbers less than 13
$A = \{1, 3, 5, 7, 9, 11\}$

5. the prime numbers between 2 and 5
$A = \{3\}$

6. the prime numbers between 30 and 40
$A = \{31, 37\}$

7. the positive integers less than 20 which are divisible by 3
$A = \{3, 6, 9, 12, 15, 18\}$

8. the perfect square integers less than 100
$A = \{1, 4, 9, 16, 25, 36, 49, 64, 81\}$

Use set builder notation to write the set.

9. the integers greater than 4
$\{x \mid x > 4, x \text{ is an integer}\}$

10. the integers less than -2
$\{x \mid x < -2, x \text{ is an integer}\}$

11. the real numbers greater than or equal to 1
$\{x \mid x \geq 1, x \in \text{real numbers}\}$

12. the real numbers less than or equal to -3
$\{x \mid x \leq -3, x \in \text{real numbers}\}$

13. the integers between -2 and 5
$\{x \mid -2 < x < 5, x \text{ is an integer}\}$

14. the integers between -5 and 2
$\{x \mid -5 < x < 2, x \text{ is an integer}\}$

15. the real numbers between 0 and 1
$\{x \mid 0 < x < 1, x \in \text{real numbers}\}$

16. the real numbers between -2 and 4
$\{x \mid -2 < x < 4, x \in \text{real numbers}\}$

Find $A \cup B$.

17. $A = \{1, 4, 9\}, B = \{2, 4, 6\}$
$A \cup B = \{1, 2, 4, 6, 9\}$

18. $A = \{-1, 0, 1\}, B = \{0, 1, 2\}$
$A \cup B = \{-1, 0, 1, 2\}$

19. $A = \{2, 3, 5, 8\}, B = \{9, 10\}$
$A \cup B = \{2, 3, 5, 8, 9, 10\}$

20. $A = \{1, 3, 5, 7\}, B = \{2, 4, 6, 8\}$
$A \cup B = \{1, 2, 3, 4, 5, 6, 7, 8\}$

21. $A = \{-4, -2, 0, 2, 4\}, B = \{0, 4, 8\}$
$A \cup B = \{-4, -2, 0, 2, 4, 8\}$

22. $A = \{-3, -2, -1\}, B = \{-2, -1, 0, 1\}$
$A \cup B = \{-3, -2, -1, 0, 1\}$

23. $A = \{1, 2, 3, 4, 5\}, B = \{3, 4, 5\}$
$A \cup B = \{1, 2, 3, 4, 5\}$

24. $A = \{2, 4\}, B = \{0, 1, 2, 3, 4, 5\}$
$A \cup B = \{0, 1, 2, 3, 4, 5\}$

Find $A \cap B$.

25. $A = \{6, 12, 18\}$, $B = \{3, 6, 9\}$
$A \cap B = \{6\}$

26. $A = \{-4, 0, 4\}$, $B = \{-2, 0, 2\}$
$A \cap B = \{0\}$

27. $A = \{1, 5, 10, 20\}$, $B = \{5, 10, 15, 20\}$
$A \cap B = \{5, 10, 20\}$

28. $A = \{-9, -5, 0, 7\}$, $B = \{-7, -5, 0, 5, 7\}$
$A \cap B = \{-5, 0, 7\}$

29. $A = \{1, 2, 4, 8\}$, $B = \{3, 5, 6, 7\}$
$A \cap B = \varnothing$

30. $A = \{-3, -2, -1, 0\}$, $B = \{1, 2, 3, 4\}$
$A \cap B = \varnothing$

31. $A = \{2, 4, 6, 8, 10\}$, $B = \{4, 6\}$
$A \cap B = \{4, 6\}$

32. $A = \{1, 3, 5, 7, 9\}$, $B = \{1, 9\}$
$A \cap B = \{1, 9\}$

▶ Objective B

Graph the solution set.

33. $\{x \mid x \geq 3\}$

34. $\{x \mid x \leq -2\}$

35. $\{x \mid x > 0\}$

36. $\{x \mid x < 4\}$

37. $\{x \mid x > 1\} \cup \{x \mid x < -1\}$

38. $\{x \mid x \leq 2\} \cup \{x \mid x > 4\}$

39. $\{x \mid x \leq 2\} \cap \{x \mid x \geq 0\}$

40. $\{x \mid x > -1\} \cap \{x \mid x \leq 4\}$
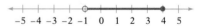

41. $\{x \mid x > 1\} \cap \{x \mid x \geq -2\}$

42. $\{x \mid x < 4\} \cap \{x \mid x \leq 0\}$

43. $\{x \mid x > 2\} \cup \{x \mid x > 1\}$

44. $\{x \mid x < -2\} \cup \{x \mid x < -4\}$

Calculators and Computers

Euclidean Algorithm   An algorithm is a method or procedure which repeats the same sequence of steps over and over. For example, to divide two whole numbers, the division algorithm is used.

$$
\begin{array}{r}
2 \\
3\overline{)747} \\
-6 \\
\hline
14
\end{array}
$$

1. Estimate the quotient. $3\overline{)7}^{\,2}$
2. Multiply the quotient times the divisor. $(2 \times 3 = 6)$
3. Subtract. $(7 - 6)$
4. "Bring down" the next digit. (4)

$$
\begin{array}{r}
249 \\
3\overline{)747} \\
-6 \\
\hline
14 \\
-12 \\
\hline
27 \\
-27 \\
\hline
0
\end{array}
$$

The division problem is completed by repeating these four steps until there are no numbers to "bring down." The algorithm is the four steps: estimate, multiply, subtract, and bring down.

The Euclidean Algorithm is a procedure to find the greatest common divisor (GCD) of two numbers.

Find the GCD of 15 and 25 using the Euclidean Algorithm.

$$
\begin{array}{r}
1 \\
15\overline{)25} \\
-15 \\
\hline
10
\end{array}
$$

1. Divide the smaller number into the larger number.

$$
\begin{array}{r}
1 \\
10\overline{)15} \\
-10 \\
\hline
5
\end{array}
$$

2. Divide the remainder (10) into the divisor (15).

$$
\begin{array}{r}
2 \\
5\overline{)10} \\
-10 \\
\hline
0
\end{array}
$$

3. Repeat Step 2 until the remainder is zero.

4. The divisor in the last division is the GCD.

The GCD is 5.

The Euclidean Algorithm is the four listed steps. A computer program can be written to carry out this procedure for any two positive integers. One such program is on the Math ACE Disk.

Find the GCD of 5 and 8 using the Euclidean Algorithm. You may use the program on the Math ACE Disk to check your answer.

Chapter Summary

Key Words   The *integers* are. . . , $-4, -3, -2, -1, 0, 1, 2, 3, 4, . . .$

The *negative integers* are the integers . . . , $-4, -3, -2, -1$.

The *positive integers* are the integers $1, 2, 3, 4, . . .$ The positive integers are also called the *natural numbers*.

The positive integers and zero are called the *whole numbers*.

A *rational number* is a number of the form $\frac{a}{b}$, where a and b are integers and b is not equal to zero.

An *irrational number* is a number whose decimal representation never terminates nor repeats.

The rational numbers and the irrational numbers taken together are called the *real numbers*.

The *absolute value* of a number is a measure of its distance from zero on a number line.

The expression a^n is in *exponential form*, where a is the base and n is the exponent.

A *complex fraction* is a fraction whose numerator or denominator contains one or more fractions.

The *Order of Operations Agreement* is used to simplify numerical expressions.

A *variable expression* is an expression which contains one or more variables.

The *terms* of a variable expression are the addends of the expression.

A *variable term* is composed of a numerical coefficient and a variable part.

The *additive inverse* of a number is the opposite of the number.

Content and Format © 1991 HMCo.

The *multiplicative inverse* of a number is the reciprocal of the number.

A *set* is a collection of objects. The objects of the set are the *elements* of the set.

The *roster method* of writing a set encloses a list of the elements of the set in braces.

The *empty set*, or *null set* written $\varnothing$ or { }, is the set that contains no elements.

A *finite set* is a set in which the elements can be counted.

An *infinite set* is a set in which it is impossible to list all of the elements.

The *union* of two sets, written $A \cup B$, is the set that contains all of the elements of A and all of the elements of B. The elements that are in both set A and set B are listed only once.

The *intersection* of two sets, written $A \cap B$, is the set that contains the elements that are common to both A and B.

An *inequality* is an expression that contains the symbol $<$, $>$, $\leq$, or $\geq$.

The *solution set of an inequality* is the set of real numbers that make the inequality true.

Essential Rules	**The Commutative Property of Addition**	If a and b are real numbers, then $a + b = b + a$.
	The Associative Property of Addition	If a, b, and c are real numbers, then $a + (b + c) = (a + b) + c$.
	The Commutative Property of Multiplication	If a and b are real numbers, then $ab = ba$.
	The Associative Property of Multiplication	If a, b, and c are real numbers, then $a(bc) = (ab)c$.
	The Addition Property of Zero	If a is a real number, then $a + 0 = 0 + a = a$.

The Multiplication Property of One

If a is a real number, then
$a \cdot 1 = 1 \cdot a = a$.

The Inverse Property of Addition

If a is a real number, then
$a + (-a) = (-a) + a = 0$.

The Inverse Property of Multiplication

If a is a non-zero number, then
$a \cdot \frac{1}{a} = \frac{1}{a} \cdot a = 1$.

The Distributive Property

If a, b, and c are real numbers, then
$a(b + c) = ab + ac$.

Chapter Review

SECTION 1

1. Find the additive inverse of 23.
-23

2. Evaluate: $-|-5|$
-5

3. Simplify: $-10 - (-3) - 8$
-15

4. Simplify: $-204 \div (-17)$
12

5. Simplify: $18 - |-12 + 8|$
14

6. Simplify: $-2 \cdot (4^2) \cdot (-3)^2$
-288

SECTION 2

7. Simplify: $-\dfrac{3}{8} + \dfrac{3}{5} - \dfrac{1}{6}$
$\dfrac{7}{120}$

8. Simplify: $\dfrac{3}{5}\left[-\dfrac{10}{21}\right]\left[-\dfrac{7}{15}\right]$
$\dfrac{2}{15}$

9. Simplify: $-\dfrac{3}{8} \div \dfrac{3}{5}$
$-\dfrac{5}{8}$

10. Simplify: $-4.07 + 2.3 - 1.07$
-2.84

11. Simplify: $-3.286 \div (-1.06)$
3.1

12. Simplify: $20 \div \dfrac{3^2 - 2^2}{3^2 + 2^2}$
52

SECTION 3

13. Evaluate $2a^2 - \dfrac{3b}{a}$ when
$a = -3$ and $b = 2$.
20

14. Evaluate $(a - 2b^2) \div ab$
when $a = 4$ and $b = -3$.
$\dfrac{7}{6}$

15. Use the Distributive Property to complete the
statement.
$6x - 21y = ?\ (2x - 7y)$
3

16. Use the Commutative Property of Addition to
complete the statement.
$3(x + y) = 3(? + x)$
y

17. Use the Commutative Property of Multiplication to complete the statement.
$(ab)14 = 14?$
ab

18. Use the Associative Property of Addition to complete the statement.
$3 + (4 + y) = (3 + ?) + y$
4

19. Identify the property that justifies the statement.
$(-4) + 4 = 0$
Inverse Property of Addition

20. Identify the property that justifies the statement.
$2(3x) = (2 \cdot 3)x$
The Associative Property of Multiplication

21. Simplify:
$-2(x - 3) + 4(2 - x)$
$-6x + 14$

22. Simplify:
$4y - 3[x - 2(3 - 2x) - 4y]$
$16y - 15x + 18$

23. Translate and simplify "The sum of three consecutive odd integers."
$x + (x + 2) + (x + 4); 3x + 6$

24. Translate and simplify "Twelve minus the quotient of three more than a number and the number."
$12 - \left(\dfrac{x + 3}{x}\right); \dfrac{11x - 3}{x}$

SECTION 4

25. Find $A \cup B$ given
$A = \{1, 3, 5, 7\}$ and $B = \{2, 4, 6, 8\}$.
$A \cup B = \{1, 2, 3, 4, 5, 6, 7, 8\}$

26. Find $A \cap B$ given
$A = \{0, 1, 2, 3\}$ and $B = \{2, 3, 4, 5\}$.
$A \cap B = \{2, 3\}$

27. Graph the solution set of $\{x|x \geq -3\}$.

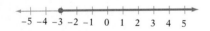

28. Graph the solution set of $\{x|x < 1\}$.

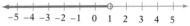

29. Graph the solution set of
$\{x|x \leq -3\} \cup \{x|x > 0\}$.

30. Graph the solution set of
$\{x|x \leq 4\} \cap \{x|x > -2\}$.

Chapter Test

1. Find the additive inverse of -12.
    12   [1.1A]

2. Evaluate: $-|-2|$
    -2 [1.1A]

3. Simplify: $2 - (-12) + 3 - 5$
    12   [1.1B]

4. Simplify: $(-2)(-3)(-5)$
    -30 [1.1B]

5. Simplify: $-180 \div 12$
    -15 [1.1B]

6. Simplify: $|-3 - (-5)|$
    2   [1.1B]

7. Simplify: $-5^2 \cdot 4$
    -100 [1.1C]

8. Simplify: $(-2)^3(-3)^2$
    -72 [1.1C]

9. Simplify: $\frac{2}{3} - \frac{5}{12} + \frac{4}{9}$
    $\frac{25}{36}$ [1.2A]

10. Simplify: $\left(-\frac{2}{3}\right)\left(\frac{9}{15}\right)\left(\frac{10}{27}\right)$
    $-\frac{4}{27}$ [1.2A]

11. Simplify: $4.27 - 6.98 + 1.3$
    -1.41 [1.2A]

12. Simplify: $-15.092 \div 3.08$
    -4.9 [1.2A]

13. Simplify: $12 - 4\left(\frac{5^2 - 1}{3}\right) \div 16$
    10   [1.2B]

14. Simplify: $8 - 4(2 - 3)^2 \div 2$
    6   [1.2B]

15. Evaluate $(a - b)^2 \div (2b + 1)$ when $a = 2$ and $b = -3$.
    -5 [1.3A]

16. Evaluate $\frac{b^2 - c^2}{a - 2c}$ when $a = 2$, $b = 3$, and $c = -1$.
    2   [1.3A]

17. Use the Commutative Property of Addition to complete the statement.
    $(3 + 4) + 2 = (? + 3) + 2$
    4   [1.3B]

18. Identify the property that justifies the statement.
    $-2(x + y) = -2x - 2y$
    The Distributive Property   [1.3B]

19. Simplify:
$3x - 2(x - y) - 3(y - 4x)$
$13x - y$ [1.3C]

20. Simplify:
$2x - 4[2 - 3(x + 4y) - 2]$
$14x + 48y$ [1.3C]

21. Translate and simplify "thirteen decreased by the product of three less than a number and nine."
$13 - (n - 3)(9)$; $40 - 9n$ [1.3D]

22. Translate and simplify "five times the sum of two consecutive integers."
$5[n + (n + 1)]$; $10n + 5$ [1.3D]

23. Find $A \cup B$, given $A = \{1, 3, 5, 7\}$ and $B = \{2, 3, 4, 5\}$.
$A \cup B = \{1, 2, 3, 4, 5, 7\}$ [1.4A]

24. Find $A \cup B$, given $A = \{-2, -1, 0, 1, 2, 3\}$ and $B = \{-1, 0, 1\}$.
$A \cup B = \{-2, -1, 0, 1, 2, 3\}$ [1.4A]

25. Find $A \cap B$, given $A = \{1, 3, 5, 7\}$ and $B = \{5, 7, 9, 11\}$.
$A \cap B = \{5, 7\}$ [1.4A]

26. Find $A \cap B$, given $A = \{-3, -2, -1, 0, 1, 2, 3\}$ and $B = \{-1, 0, 1\}$.
$A \cap B = \{-1, 0, 1\}$ [1.4A]

27. Graph the solution set of $\{x \mid x \leq -1\}$.
[1.4B]

28. Graph the solution set of $\{x \mid x > 3\}$.
[1.4B]

29. Graph the solution set of $\{x \mid x \leq 3\} \cup \{x \mid x < -2\}$.
[1.4B]

30. Graph the solution set of $\{x \mid x < 3\} \cap \{x \mid x > -2\}$.
[1.4B]

2

First-Degree Equations and Inequalities

OBJECTIVES

▶ To solve an equation using the Addition or the Multiplication Property of Equations

▶ To solve an equation using both the Addition and Multiplication Properties of Equations

▶ To solve an equation containing parentheses

▶ To solve an inequality in one variable

▶ To solve a compound inequality

▶ To solve application problems

▶ To solve an absolute value equation

▶ To solve an absolute value inequality

▶ To solve application problems

▶ To solve integer problems

▶ To solve coin and stamp problems

▶ To solve value mixture problems

▶ To solve uniform motion problems

▶ To solve investment problems

▶ To solve percent mixture problems

Moscow Papyrus

Most of the early history of mathematics can be traced to the Egyptians and Babylonians. This early work began around 3000 B.C. The Babylonians used clay tablets and a type of writing called *cuneiform* (wedge-shape) to record their thoughts and discoveries. Here are the symbols for 10, 1, and subtraction.

Simple groupings of these symbols were used to represent numbers less than 60. The numbers 32 and 28 are shown below.

The Egyptians used papyrus, a plant reed dried and pounded thin, and hieratic writing that was derived from hieroglyphics to make records. A number of papyrus documents have been discovered over the years. One particularly famous one is called the Moscow Papyrus, which dates from 1850 B.C. It is approximately 18 feet long and three inches wide and contains 25 problems.

The Moscow Papyrus dealt with solving practical problems which were related to geometry, food preparation, and grain allotments. Here is one of the problems:

The width of a rectangle is $\frac{3}{4}$ the length and the area is 12.

Find the dimensions of the rectangle.

You might recognize this as a type of problem you have solved before. Word problems are very old indeed. This one is around 3900 years old.

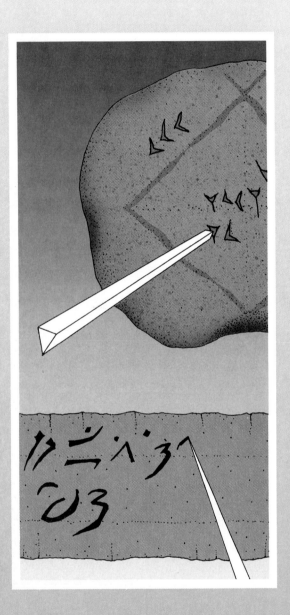

SECTION 2.1   # Solving First-Degree Equations

Objective A   **To solve an equation using the Addition or the Multiplication Property of Equations**

An **equation** expresses the equality of two mathematical expressions. The expressions can be either numerical or variable expressions.

$$\left. \begin{array}{l} 2 + 8 = 10 \\ x + 8 = 11 \\ x^2 + 2y = 7 \end{array} \right\} \text{Equations}$$

The equation at the right is a **conditional equation.** The equation is true if the variable is replaced by 3. The equation is false if the variable is replaced by 4.

$x + 2 = 5$	Conditional Equation
$3 + 2 = 5$	A true equation
$4 + 2 = 5$	A false equation

The replacement values of the variable that will make an equation true are called the **roots** or the **solutions** of the equation.

The solution of the equation $x + 2 = 5$ is 3.

The equation at the right is an **identity.** Any replacement for x will result in a true equation.

$x + 2 = x + 2$ Identity

The equation at the right has **no solution,** since there is no number that equals itself plus one. Any replacement value for x will result in a false equation.

$x = x + 1$ No solution

Each of the equations at the right is a **first-degree equation in one variable.** All variables have an exponent of 1.

$$\left. \begin{array}{l} x + 2 = 12 \\ 3y - 2 = 5y \\ 3(a + 2) = 14a \end{array} \right. \quad \begin{array}{l} \text{First-Degree} \\ \text{Equations} \end{array}$$

To **solve** an equation means to find a root or solution of the equation. The simplest equation to solve is an equation of the form **variable = constant,** since the constant is the solution.

If $x = 3$, then 3 is the solution of the equation, since $3 = 3$ is a true equation.

In solving an equation, the goal is to rewrite the given equation in the form *variable = constant*. The Addition Property of Equations can be used to rewrite an equation in this form.

The Addition Property of Equations

If a, b, and c are algebraic expressions, then the equation $a = b$ has the same solutions as the equation $a + c = b + c$.

The Addition Property of Equations states that the same quantity can be added to each side of an equation without changing the solution of the equation. This property is used to remove a term from one side of the equation by adding the opposite of that term to each side of the equation.

Solve: $x - 3 = 7$

Add the opposite of the constant term -3 to each side of the equation. After simplifying, the equation will be in the form variable = constant.

$$x - 3 = 7$$
$$x - 3 + 3 = 7 + 3$$
$$x + 0 = 10$$
$$x = 10$$

variable = constant

Check: $x - 3 = 7$
$$10 - 3 = 7$$
$$7 = 7$$

The solution is 10.

Because subtraction is defined in terms of addition, the Addition Property of Equations allows the same number to be subtracted from each side of an equation.

Solve: $x + \frac{7}{12} = \frac{1}{2}$

Add the opposite of the constant $\frac{7}{12}$ to each side of the equation.

This is equivalent to subtracting $\frac{7}{12}$ from each side of the equation.

$$x + \frac{7}{12} = \frac{1}{2}$$
$$x + \frac{7}{12} - \frac{7}{12} = \frac{1}{2} - \frac{7}{12}$$
$$x + 0 = \frac{6}{12} - \frac{7}{12}$$
$$x = -\frac{1}{12}$$

You should check this solution.   The solution is $-\frac{1}{12}$.

The Multiplication Property of Equations can also be used to rewrite an equation in the form variable = constant.

The Multiplication Property of Equations

If a, b, and c are algebraic expressions and $c \neq 0$, then the equation $a = b$ has the same solution as the equation $ac = bc$.

This property states that each side of an equation can be multiplied by the same nonzero number without changing the solutions of the equation.

Recall that the goal of solving an equation is to rewrite the equation in the form variable = constant. The Multiplication Property of Equations is used to rewrite an equation in this form by multiplying each side by the reciprocal of the coefficient.

Solve: $-\frac{3}{4}x = 12$

Multiply each side of the equation by $-\frac{4}{3}$, the reciprocal of $-\frac{3}{4}$. After simplifying, the equation will be in the form variable = constant.

$$-\frac{3}{4}x = 12$$
$$\left(-\frac{4}{3}\right)\left(-\frac{3}{4}\right)x = \left(-\frac{4}{3}\right)12$$
$$1x = -16$$
$$x = -16$$

Check: $-\frac{3}{4}(-16) = 12$
$$12 = 12$$

The solution is -16.

Content and Format © 1991 HMCo.

Because division is defined in terms of multiplication, the Multiplication Property of Equations allows each side of an equation to be divided by the same nonzero quantity.

Solve: $-5x = 9$

Multiply each side of the equation by the reciprocal of -5. This is equivalent to dividing each side of the equation by -5.

$$-5x = 9$$
$$\frac{-5x}{-5} = \frac{9}{-5}$$
$$1x = -\frac{9}{5}$$
$$x = -\frac{9}{5}$$

The solution is $-\frac{9}{5}$.

You should check the solution.

When using the Multiplication Property of Equations it is usually easier to multiply each side of the equation by the reciprocal of the coefficient when the coefficient is a fraction. Divide each side of the equation by the coefficient when the coefficient is an integer or decimal.

Example 1 Solve: $x - 7 = -12$

Solution
$$x - 7 = -12$$
$$x - 7 + 7 = -12 + 7$$
$$x = -5$$

The solution is -5.

Example 2 Solve: $x + 4 = -3$

Your solution -7

Example 3 Solve: $-\frac{2x}{7} = \frac{4}{21}$

Solution
$$-\frac{2x}{7} = \frac{4}{21}$$
$$\left(-\frac{7}{2}\right)\left(-\frac{2}{7}x\right) = \left(-\frac{7}{2}\right)\left(\frac{4}{21}\right)$$
$$x = -\frac{2}{3}$$

The solution is $-\frac{2}{3}$.

Example 4 Solve: $-3x = 18$

Your solution -6

Solutions on p. A11

Objective B	**To solve an equation using both the Addition and the Multiplication Properties of Equations**	

In solving an equation, frequently the application of both the Addition and the Multiplication Properties of Equations is required.

Solve: $4x - 3 + 2x = 8 + 9x - 12$

Simplify each side of the equation by combining like terms.

$$4x - 3 + 2x = 8 + 9x - 12$$
$$6x - 3 = -4 + 9x$$

Subtract $9x$ from each side of the equation. Then simplify.

$$6x - 3 - 9x = -4 + 9x - 9x$$
$$-3x - 3 = -4$$

Add 3 to each side of the equation. Then simplify.

$$-3x - 3 + 3 = -4 + 3$$
$$-3x = -1$$

Divide each side of the equation by the coefficient -3. Then simplify.

$$\frac{-3x}{-3} = \frac{-1}{-3}$$

$$x = \frac{1}{3}$$

Write the solution.

The solution is $\frac{1}{3}$.

Check:

$$4x - 3 + 2x = 8 + 9x - 12$$

$4\left(\frac{1}{3}\right) - 3 + 2\left(\frac{1}{3}\right)$	$8 + 9\left(\frac{1}{3}\right) - 12$
$\frac{4}{3} - 3 + \frac{2}{3}$	$8 + 3 - 12$
$-\frac{5}{3} + \frac{2}{3}$	$11 - 12$

$$-1 = -1$$

Example 5   Solve: $3x - 5 = x + 2 - 7x$

Solution
$$3x - 5 = x + 2 - 7x$$
$$3x - 5 = -6x + 2$$
$$3x - 5 + 6x = -6x + 2 + 6x$$
$$9x - 5 = 2$$
$$9x - 5 + 5 = 2 + 5$$
$$9x = 7$$
$$\frac{9x}{9} = \frac{7}{9}$$
$$x = \frac{7}{9}$$
The solution is $\frac{7}{9}$.

Example 6   Solve: $6x - 5 - 3x = 14 - 5x$

Your solution $\frac{19}{8}$

Solution on p. A11

| **Objective C** | **To solve an equation containing parentheses** |

When an equation contains parentheses, one of the steps required in solving the equation involves the use of the Distributive Property.

Solve: $3(x - 2) + 3 = 2(6 - x)$

Use the Distributive Property to remove parentheses. Then simplify.

$$3(x - 2) + 3 = 2(6 - x)$$
$$3x - 6 + 3 = 12 - 2x$$
$$3x - 3 = 12 - 2x$$

Add $2x$ to each side of the equation. Then simplify.

$$3x - 3 + 2x = 12 - 2x + 2x$$ Do this step mentally.

$$5x - 3 = 12$$

Add 3 to each side of the equation. Then simplify.

$$5x - 3 + 3 = 12 + 3$$ Do this step mentally.

$$5x = 15$$

Divide each side of the equation by the coefficient 5. Then simplify.

$$\frac{5x}{5} = \frac{15}{5}$$
$$x = 3$$

Write the solution.

The solution is 3.

Check: $3(x - 2) + 3 = 2(6 - x)$

$3(3 - 2) + 3$	$2(6 - 3)$
$3(1) + 3$	$2(3)$
$3 + 3$	6

$$6 = 6$$

To solve an equation containing fractions, first clear the denominators by multiplying each side of the equation by the least common multiple (LCM) of the denominators.

Solve: $\frac{x}{2} - \frac{7}{9} = \frac{x}{6} + \frac{2}{3}$

Find the LCM of the denominators.

The LCM of 2, 9, 6, and 3 is 18.

$$\frac{x}{2} - \frac{7}{9} = \frac{x}{6} + \frac{2}{3}$$

Multiply each side of the equation by the LCM.

$$18\left(\frac{x}{2} - \frac{7}{9}\right) = 18\left(\frac{x}{6} + \frac{2}{3}\right)$$

Use the Distributive Property to remove parentheses. Then simplify.

$$\frac{18x}{2} - \frac{18 \cdot 7}{9} = \frac{18x}{6} + \frac{18 \cdot 2}{3}$$
$$9x - 14 = 3x + 12$$

Subtract $3x$ from each side of the equation. Then simplify.

$$9x - 14 - 3x = 3x + 12 - 3x$$ Do this step mentally.

$$6x - 14 = 12$$

Add 14 to each side of the equation. Then simplify.

$$6x - 14 + 14 = 12 + 14$$ Do this step mentally.

$$6x = 26$$

Divide each side of the equation by the coefficient of x. Then simplify.

$$\frac{6x}{6} = \frac{26}{6}$$

$$x = \frac{13}{3}$$

Write the solution.

The solution is $\frac{13}{3}$.

Check:

$$\frac{x}{2} - \frac{7}{9} = \frac{x}{6} + \frac{2}{3}$$

$$\frac{\frac{13}{3}}{2} - \frac{7}{9} \quad \Bigg| \quad \frac{\frac{13}{3}}{6} + \frac{2}{3}$$

$$\frac{13}{6} - \frac{7}{9} \quad \Bigg| \quad \frac{13}{18} + \frac{2}{3}$$

$$\frac{39}{18} - \frac{14}{18} \quad \Bigg| \quad \frac{13}{18} + \frac{12}{18}$$

$$\frac{25}{18} = \frac{25}{18}$$

Example 7 Solve:
$$5(2x - 7) + 2 = 3(4 - x) - 12$$

Solution
$$5(2x - 7) + 2 = 3(4 - x) - 12$$
$$10x - 35 + 2 = 12 - 3x - 12$$
$$10x - 33 = -3x$$
$$-33 = -13x$$
$$\frac{-33}{-13} = \frac{-13x}{-13}$$
$$\frac{33}{13} = x$$

The solution is $\frac{33}{13}$.

Example 8 Solve:
$$6(5 - x) - 12 = 2x - 3(4 + x)$$

Your solution   6

Example 9 Solve:
$$\frac{3x - 2}{12} - \frac{x}{9} = \frac{2x}{4}$$

Solution The LCM of 12, 9, and 4 is 36.

$$\frac{3x - 2}{12} - \frac{x}{9} = \frac{2x}{4}$$

$$36\left(\frac{3x - 2}{12} - \frac{x}{9}\right) = 36\left(\frac{2x}{4}\right)$$

$$\frac{36(3x - 2)}{12} - \frac{36x}{9} = \frac{36(2x)}{4}$$

$$3(3x - 2) - 4x = 9(2x)$$
$$9x - 6 - 4x = 18x$$
$$5x - 6 = 18x$$
$$-6 = 13x$$
$$\frac{-6}{13} = \frac{13x}{13}$$
$$-\frac{6}{13} = x$$

The solution is $-\frac{6}{13}$.

Example 10 Solve:
$$\frac{2x - 7}{3} - \frac{5x + 4}{5} = \frac{-x - 4}{30}$$

Your solution -10

Solutions on p. A11

2.1   EXERCISES

▶ **Objective A**

Solve and check:

1. $x - 2 = 7$
9

2. $x - 8 = 4$
12

3. $a + 3 = -7$
-10

4. $a + 5 = -2$
-7

5. $b - 3 = -5$
-2

6. $10 + m = 2$
-8

7. $-7 = x + 8$
-15

8. $-12 = x - 3$
-9

9. $3x = 12$
4

10. $8x = 4$
$\frac{1}{2}$

11. $-3x = 2$
$-\frac{2}{3}$

12. $-5a = 7$
$-\frac{7}{5}$

13. $-\frac{3}{2} + x = \frac{4}{3}$
$\frac{17}{6}$

14. $\frac{2}{7} + x = \frac{17}{21}$
$\frac{11}{21}$

15. $x + \frac{2}{3} = \frac{5}{6}$
$\frac{1}{6}$

16. $\frac{5}{8} - y = \frac{3}{4}$
$-\frac{1}{8}$

17. $\frac{2}{3}y = 5$
$\frac{15}{2}$

18. $\frac{3}{5}y = 12$
20

19. $-\frac{5}{8}x = \frac{4}{5}$
$-\frac{32}{25}$

20. $-\frac{5}{12}y = \frac{7}{16}$
$-\frac{21}{20}$

21. $-\frac{3b}{5} = -\frac{3}{5}$
1

22. $-\frac{7b}{12} = \frac{7}{8}$
$-\frac{3}{2}$

23. $-\frac{2}{3}x = -\frac{5}{8}$
$\frac{15}{16}$

24. $-\frac{3}{4}x = -\frac{4}{7}$
$\frac{16}{21}$

25. $-\frac{5}{8}x = 40$
-64

26. $\frac{2}{7}y = -8$
-28

27. $-\frac{5}{6}y = -\frac{25}{36}$
$\frac{5}{6}$

28. $-\frac{15}{24}x = -\frac{10}{27}$
$\frac{16}{27}$

29. $b - 14.72 = -18.45$
-3.73

30. $y + 4.29 = -3.17$
-7.46

31. $b + 2.009 = 1.8276$
-0.1814

32. $-0.29y = 4.495$
-15.5

33. $3.5t = 4.69$
1.34

34. $5.8z = 3.886$
0.67

▶ **Objective B**

Solve and check:

35. $3x + 5x = 12$
$\frac{3}{2}$

36. $2x - 7x = 15$
-3

37. $2x - 4 = 12$
8

38. $5x + 9 = 6$
$-\frac{3}{5}$

39. $3y - 5y = 0$
0

40. $2y - 9 = -9$
0

Solve and check:

41. $4x - 6 = 3x$
6

42. $2a - 7 = 5a$
$-\frac{7}{3}$

43. $7x + 12 = 9x$
6

44. $3x - 12 = 5x$
-6

45. $4x + 2 = 4x$
No solution

46. $3m - 7 = 3m$
No solution

47. $2x + 2 = 3x + 5$
-3

48. $7x - 9 = 3 - 4x$
$\frac{12}{11}$

49. $2 - 3t = 3t - 4$
1

50. $7 - 5t = 2t - 9$
$\frac{16}{7}$

51. $3b - 2b = 4 - 2b$
$\frac{4}{3}$

52. $3x - 5 = 6 - 9x$
$\frac{11}{12}$

53. $3x + 7 = 3 + 7x$
1

54. $\frac{5}{8}b - 3 = 12$
24

55. $\frac{1}{3} - 2b = 3$
$-\frac{4}{3}$

56. $b + \frac{1}{5}b = 2$
$\frac{5}{3}$

57. $\frac{2}{3}y + y = \frac{2}{3}$
$\frac{2}{5}$

58. $\frac{2}{3}y - 5 = 7$
18

59. $\frac{7}{12}x - 3 = 4$
12

60. $3.24a + 7.14 = 5.34a$
3.4

61. $5.3y + 0.35 = 5.02y$
-1.25

62. $22.946 - 4.13x = 6.3x$
2.2

63. $3x - 2x + 7 = 12 - 4x$
1

64. $2x - 9x + 3 = 6 - 5x$
$-\frac{3}{2}$

65. $2t - 7 + 5t = 16 - 8t$
$\frac{23}{15}$

66. $-2t + 8 - 4t = -8t + 2$
-3

67. $7 + 8y - 12 = 3y - 8 + 5y$
No solution

68. $2y - 4 + 8y = 7y - 8 + 3y$
No solution

69. $2x - 5 + 7x = 11 - 3x + 4x$
2

70. $9 + 4x - 12 = -3x + 5x + 8$
$\frac{11}{2}$

Solve and check:

71. $-30 + 2x + 12 = 8x - 3 - 5x$
-15

72. $-3x + 8 + 9x = -12 + 3x - 4$
-8

73. $1.2b - 3.8b = 1.6b + 4.494$
-1.07

74. $6.7a - 9.2a = 6.55a - 3.91865$
0.433

▶ **Objective C**

Solve and check:

75. $2x + 2(x + 1) = 10$
2

76. $2x + 3(x - 5) = 15$
6

77. $2(a - 3) = 2(4 - 2a)$
$\frac{7}{3}$

78. $5(2 - b) = -3(b - 3)$
$\frac{1}{2}$

79. $3 - 2(y - 3) = 4y - 7$
$\frac{8}{3}$

80. $3(y - 5) - 5y = 2y + 9$
-6

81. $2(3x - 2) - 5x = 3 - 2x$
$\frac{7}{3}$

82. $4 - 3x = 7x - 2(3 - x)$
$\frac{5}{6}$

83. $4(x - 2) + 2 = 4x - 2(2 - x)$
-1

84. $2x - 3(x - 4) = 2(3 - 2x) + 2$
$-\frac{4}{3}$

85. $8 - 5(4 - 3x) = 2(4 - x) - 8x$
$\frac{4}{5}$

86. $-3x - 2(4 + 5x) = 14 - 3(2x - 3)$
$-\frac{31}{7}$

87. $3[2 - 3(y - 2)] = 12$
$\frac{4}{3}$

88. $3y = 2[5 - 3(2 - y)]$
$\frac{2}{3}$

89. $4[3 + 5(3 - x) + 2x] = 6 - 2x$
$\frac{33}{5}$

90. $2[4 + 2(5 - x) - 2x] = 4x - 7$
$\frac{35}{12}$

91. $3[4 - 2(a - 2)] = 3(2 - 4a)$
-3

92. $2[3 - 2(z + 4)] = 3(4 - z)$
-22

Solve and check:

93. $-3(x - 2) = 2[x - 4(x - 2) + x]$
10

94. $3[x - (2 - x) - 2x] = 3(4 - x)$
6

95. $\frac{2}{9}t - \frac{5}{6} = \frac{1}{12}t$
6

96. $\frac{3}{4}t - \frac{7}{12}t = 1$
6

97. $\frac{2}{3}x - \frac{5}{6}x - 3 = \frac{1}{2}x - 5$
3

98. $\frac{1}{2}x - \frac{3}{4}x + \frac{5}{8} = \frac{3}{2}x - \frac{5}{2}$
$\frac{25}{14}$

99. $\frac{3x - 2}{4} - 3x = 12$
$-\frac{50}{9}$

100. $\frac{2a - 9}{5} + 3 = 2a$
$\frac{3}{4}$

101. $\frac{5 - 2x}{5} + \frac{x - 4}{10} = \frac{3}{10}$
1

102. $\frac{2x - 5}{12} - \frac{3 - x}{6} = \frac{11}{12}$
$\frac{11}{2}$

103. $\frac{x - 2}{4} - \frac{x + 5}{6} = \frac{5x - 2}{9}$
$-\frac{40}{17}$

104. $\frac{2x - 1}{4} + \frac{3x + 4}{8} = \frac{1 - 4x}{12}$
$-\frac{4}{29}$

105. $\frac{5x - 2}{6} = \frac{4 - x}{9} - \frac{3x - 4}{12}$
$\frac{40}{43}$

106. $\frac{3x - 7}{4} + \frac{3 - 5x}{6} = \frac{3 - 6x}{10}$
3

107. $0.492 - 0.34(y - 2.5) = 3.2(4.2y - 5.8)$
1.444267

108. $2.9(x - 3.4) = -1.432 - 6.7(x - 4.4)$
3.94875

109. If $5x - 7 = 2x + 2$, evaluate $7x + 4$.
25

110. If $3x - 5 = 9x + 4$, evaluate $6x - 3$.
-12

111. If $2x - 5(x + 1) = 7$, evaluate $x^2 - 1$.
15

112. If $3(2x + 1) = 5 - 2(x - 2)$
evaluate $2x^2 + 1$.
$\frac{17}{8}$

113. If $4 - 3(2x + 3) = 5 - 4x$ evaluate
$x^2 - 2x$.
35

114. If $5 - 2(4x - 1) = 3x + 7$
evaluate $x^4 - x^2$.
0

115. If $7x - 4(3x - 5) = 2 - 3(2x + 1)$
evaluate $\frac{x}{x + 7}$.
$\frac{3}{2}$

116. If $4x - 2(5x - 3) = 3 - 4(2x - 1)$
evaluate $-4x^2 + 2x - 1$.
-1

SECTION 2.2   First-Degree Inequalities

Objective A   To solve an inequality in one variable

The **solution set of an inequality** is a set of numbers, each element of which, when substituted for the variable, results in a true inequality.

The inequality at the right is true if the variable is replaced by 3, -1.98, or $\frac{2}{3}$.

$$x - 1 < 4$$
$$3 - 1 < 4$$
$$-1.98 - 1 < 4$$
$$\tfrac{2}{3} - 1 < 4$$

There are many values of the variable x which will make the inequality $x - 1 < 4$ true. The solution set of the inequality is any number less than 5. The solution set can be written in set builder notation as $\{x \mid x < 5\}$.

The graph of the solution set of $x - 1 < 4$ is shown to the right.

$$\begin{array}{ccccccccccc} -5 & -4 & -3 & -2 & -1 & 0 & 1 & 2 & 3 & 4 & 5 \end{array}$$

In solving an inequality, the Addition and Multiplication Properties of Inequalities are used to rewrite the inequality in the form *variable < constant* or *variable > constant*.

The Addition Property of Inequalities

> If $a > b$, then $a + c > b + c$.
> If $a < b$, then $a + c < b + c$.

The Addition Property of Inequalities states that the same number can be added to each side of an inequality without changing the solution set of the inequality. This property is also true for an inequality which contains the symbol $\leq$ or $\geq$.

The Addition Property of Inequalities is used to remove a term from one side of an inequality by adding the additive inverse of that term to each side of the inequality. Because subtraction is defined in terms of addition, the same number can be subtracted from each side of an inequality without changing the solution set of the inequality.

Solve: $x + 2 \geq 4$

Subtract 2 from each side of the inequality. Simplify.

$$x + 2 \geq 4$$
$$x + 2 - 2 \geq 4 - 2$$
$$x \geq 2$$

Write the solution set.

$$\{x \mid x \geq 2\}$$

Solve: $3x - 4 < 2x - 1$

Subtract $2x$ from each side of the inequality. Simplify.

$$3x - 4 < 2x - 1$$

$$\boxed{3x - 4 - 2x < 2x - 1 - 2x}$$ Do this step mentally.

$$x - 4 < -1$$

Add 4 to each side of the inequality. Simplify.

$$\boxed{x - 4 + 4 < -1 + 4}$$ Do this step mentally.

$$x < 3$$

Write the solution set. $\{x \mid x < 3\}$

The Multiplication Property of Inequalities is used to remove a coefficient from one side of an inequality by multiplying each side of the inequality by the reciprocal of the coefficient.

The Multiplication Property of Inequalities

Rule 1   If $a > b$ and $c > 0$, then $ac > bc$.
         If $a < b$ and $c > 0$, then $ac < bc$.

Rule 2   If $a > b$ and $c < 0$, then $ac < bc$.
         If $a < b$ and $c < 0$, then $ac > bc$.

Here are some examples of this property.

Rule 1		**Rule 2**	
$3 > 2$	$2 < 5$	$3 > 2$	$2 < 5$
$3(4) > 2(4)$	$2(4) < 5(4)$	$3(-4) < 2(-4)$	$2(-4) > 5(-4)$
$12 > 8$	$8 < 20$	$-12 < -8$	$-8 > -20$

Rule 1 states that when each side of an inequality is multiplied by a positive number, the inequality symbol remains the same. However, Rule 2 states that when each side of an inequality is multiplied by a negative number, the inequality symbol must be reversed. Because division is defined in terms of multiplication, when each side of an inequality is divided by a *positive* number, the inequality symbol remains the same. When each side of an inequality is divided by a *negative* number, the inequality symbol must be reversed.

The Multiplication Property of Inequalities is true for the symbols $\leq$ and $\geq$.

Solve: $-3x > 9$

Divide each side of the inequality by the coefficient -3. Because -3 is a negative number, the inequality symbol must be reversed.

$$-3x > 9$$

$$\frac{-3x}{-3} < \frac{9}{-3}$$

Simplify.

$$x < -3$$

Write the solution set.

$$\{x \mid x < -3\}$$

Solve: $3x + 2 < -4$

Subtract 2 from each side of the inequality.

$$3x + 2 < -4$$
$$3x < -6$$

Divide each side of the inequality by the coefficient 3. Simplify.

$$\frac{3x}{3} < \frac{-6}{3}$$
$$x < -2$$

Write the solution set.

$\{x | x < -2\}$

Solve: $2x - 9 > 4x + 5$

Subtract $4x$ from each side of the inequality.

$$2x - 9 > 4x + 5$$
$$-2x - 9 > 5$$

Add 9 to each side of the inequality.

$$-2x > 14$$

Divide each side of the inequality by the coefficient -2. Reverse the inequality symbol. Simplify.

$$\frac{-2x}{-2} < \frac{14}{-2}$$
$$x < -7$$

Write the solution set.

$\{x | x < -7\}$

Solve: $5(x - 2) \geq 9x - 3(2x - 4)$

Use the Distributive Property to remove parentheses. Simplify.

$$5(x - 2) \geq 9x - 3(2x - 4)$$
$$5x - 10 \geq 9x - 6x + 12$$
$$5x - 10 \geq 3x + 12$$

Subtract $3x$ from each side of the inequality.

$$2x - 10 \geq 12$$

Add 10 to each side of the inequality.

$$2x \geq 22$$

Divide each side of the inequality by the coefficient 2. Simplify.

$$\frac{2x}{2} \geq \frac{22}{2}$$
$$x \geq 11$$

Write the solution set.

$\{x | x \geq 11\}$

Example 1   Solve: $x + 3 > 4x + 6$

Solution
$$x + 3 > 4x + 6$$
$$-3x + 3 > 6$$
$$-3x > 3$$
$$\frac{-3x}{-3} < \frac{3}{-3}$$
$$x < -1$$
$$\{x | x < -1\}$$

Example 2   Solve: $2x - 1 < 6x + 7$

Your solution $\{x | x > -2\}$

Solution on p. A11

Example 3 Solve:
$$3x - 5 \le 3 - 2(3x + 1)$$

Solution
$$3x - 5 \le 3 - 2(3x + 1)$$
$$3x - 5 \le 3 - 6x - 2$$
$$3x - 5 \le 1 - 6x$$
$$9x - 5 \le 1$$
$$9x \le 6$$
$$\frac{9x}{9} \le \frac{6}{9}$$
$$x \le \frac{2}{3}$$
$$\left\{x \mid x \le \frac{2}{3}\right\}$$

Example 4 Solve:
$$5x - 2 \le 4 - 3(x - 2)$$

Your solution
$$\left\{x \mid x \le \frac{3}{2}\right\}$$

Solution on p. A11

| Objective B | **To solve a compound inequality** |

A **compound inequality** is formed by joining two inequalities with a connective word such as "and" or "or". The inequalities at the right are compound inequalities.

$$2x < 4 \text{ and } 3x - 2 > -8$$

$$2x + 3 > 5 \text{ or } x + 2 < 5$$

The solution set of a compound inequality with the connective word **and** is the set of all elements common to the solution sets of each inequality. Therefore, it is the intersection of the solution sets of the two inequalities.

Solve: $2x < 6$ and $3x + 2 > -4$

Solve each inequality.

$$\begin{array}{cc} 2x < 6 & \text{and} \quad 3x + 2 > -4 \\ x < 3 & \quad 3x > -6 \\ & \quad x > -2 \end{array}$$

Find the intersection of the solution sets.

$$\{x \mid x < 3\} \qquad \{x \mid x > -2\}$$
$$\{x \mid x < 3\} \cap \{x \mid x > -2\} = \{x \mid -2 < x < 3\}$$

Solve: $-3 < 2x + 1 < 5$

This inequality is equivalent to the compound inequality shown at the right.

$$-3 < 2x + 1 < 5$$

$$\begin{array}{cc} -3 < 2x + 1 & \text{and} \quad 2x + 1 < 5 \\ -4 < 2x & \quad 2x < 4 \end{array}$$

Solve each inequality.

$$\begin{array}{cc} -2 < x & \quad x < 2 \end{array}$$

Find the intersection of the solution sets.

$$\{x \mid x > -2\} \qquad \{x \mid x < 2\}$$
$$\{x \mid x > -2\} \cap \{x \mid x < 2\} = \{x \mid -2 < x < 2\}$$

There is an alternate method for solving the inequality in the last example.

Solve: $-3 < 2x + 1 < 5$

Subtract 1 from each of the three parts of the inequality.

$$-3 < 2x + 1 < 5$$
$$-3 - 1 < 2x + 1 - 1 < 5 - 1$$
$$-4 < 2x < 4$$

Divide each of the three parts of the inequality by the coefficient 2.

$$\frac{-4}{2} < \frac{2x}{2} < \frac{4}{2}$$
$$-2 < x < 2$$

Write the solution set.

$$\{x \mid -2 < x < 2\}$$

The solution set of a compound inequality with the connective word **or** is the union of the solution sets of the two inequalities.

Solve: $2x + 3 > 7$ or $4x - 1 < 3$

Solve each inequality.

$$2x + 3 > 7 \quad \text{or} \quad 4x - 1 < 3$$
$$2x > 4 \qquad\qquad 4x < 4$$
$$x > 2 \qquad\qquad\quad x < 1$$

Find the union of the solution sets.

$$\{x \mid x > 2\} \qquad \{x \mid x < 1\}$$
$$\{x \mid x > 2\} \cup \{x \mid x < 1\} =$$
$$\{x \mid x > 2 \text{ or } x < 1\}$$

Example 5

Solve: $1 < 3x - 5 < 4$

Solution

$$1 < 3x - 5 < 4$$
$$1 + 5 < 3x - 5 + 5 < 4 + 5$$
$$6 < 3x < 9$$
$$\frac{6}{3} < \frac{3x}{3} < \frac{9}{3}$$
$$2 < x < 3$$

$$\{x \mid 2 < x < 3\}$$

Example 6

Solve: $-2 \le 5x + 3 \le 13$

Your solution

$$\{x \mid -1 \le x \le 2\}$$

Example 7

Solve: $11 - 2x > -3$ and $7 - 3x < 4$

Solution

$$11 - 2x > -3 \quad \text{and} \quad 7 - 3x < 4$$
$$-2x > -14 \qquad\qquad -3x < -3$$
$$x < 7 \qquad\qquad\qquad x > 1$$
$$\{x \mid x < 7\} \qquad\qquad \{x \mid x > 1\}$$
$$\{x \mid x < 7\} \cap \{x \mid x > 1\} = \{x \mid 1 < x < 7\}$$

Example 8

Solve: $2 - 3x > 11$ or $5 + 2x > 7$

Your solution

$$\{x \mid x < -3 \quad \text{or} \quad x > 1\}$$

Solutions on p. A12

Objective C   **To solve Application Problems**

Example 9

Company A rents cars for $6 a day and 14¢ for every mile driven. Company B rents cars for $12 a day and 8¢ for every mile driven. You want to rent a car for 5 days. How many miles can you drive a Company A car during the 5 days if it is to cost less than a Company B car?

Strategy

To find the number of miles, write and solve an inequality using N to represent the number of miles.

Solution

Cost of car A < Cost of car B

$$6(5) + 0.14N < 12(5) + 0.08N$$
$$30 + 0.14N < 60 + 0.08N$$
$$30 + 0.06N < 60$$
$$0.06N < 30$$
$$N < 500$$

It is less expensive to rent from Company A if the car is driven less than 500 mi.

Example 10

The base of a triangle is 12 in. and the height is $x + 2$ in. Express as an integer the maximum height of the triangle when the area is less than 50 in².

Your strategy

Your solution

8 in.

Example 11

Find three consecutive positive odd integers whose sum is between 27 and 51.

Strategy

To find the three consecutive odd integers, write and solve a compound inequality using x to represent the first odd integer.

Solution

$$\begin{matrix} \text{Lower limit} \\ \text{of the sum} \end{matrix} < \text{sum} < \begin{matrix} \text{upper limit} \\ \text{of the sum} \end{matrix}$$

$$27 < x + (x + 2) + (x + 4) < 51$$
$$27 < 3x + 6 < 51$$
$$27 - 6 < 3x + 6 - 6 < 51 - 6$$
$$21 < 3x < 45$$
$$\frac{21}{3} < \frac{3x}{3} < \frac{45}{3}$$
$$7 < x < 15$$

The three odd integers are 9, 11, and 13; or 11, 13, and 15; or 13, 15, and 17

Example 12

An average score of 80 to 89 in a history course receives a B. A student has grades of 72, 94, 83, and 70 on four exams. Find the range of scores on the fifth exam that will give the student a B for the course.

Your strategy

Your solution

$81 \le N \le 100$

Solutions on p. A12

Content and Format © 1991 HMCo.

| **2.2** | **EXERCISES** |

▶ **Objective A**

Solve:

1. $x - 3 < 2$
$\{x|x < 5\}$

2. $x + 4 \geq 2$
$\{x|x \geq -2\}$

3. $4x \leq 8$
$\{x|x \leq 2\}$

4. $6x > 12$
$\{x|x > 2\}$

5. $-2x > 8$
$\{x|x < -4\}$

6. $-3x \leq -9$
$\{x|x \geq 3\}$

7. $3x - 1 > 2x + 2$
$\{x|x > 3\}$

8. $5x + 2 \geq 4x - 1$
$\{x|x \geq -3\}$

9. $2x - 1 > 7$
$\{x|x > 4\}$

10. $3x + 2 < 8$
$\{x|x < 2\}$

11. $5x - 2 \leq 8$
$\{x|x \leq 2\}$

12. $4x + 3 \leq -1$
$\{x|x \leq -1\}$

13. $6x + 3 > 4x - 1$
$\{x|x > -2\}$

14. $7x + 4 < 2x - 6$
$\{x|x < -2\}$

15. $8x + 1 \geq 2x + 13$
$\{x|x \geq 2\}$

16. $5x - 4 < 2x + 5$
$\{x|x < 3\}$

17. $4 - 3x < 10$
$\{x|x > -2\}$

18. $2 - 5x > 7$
$\{x|x < -1\}$

19. $7 - 2x \geq 1$
$\{x|x \leq 3\}$

20. $3 - 5x \leq 18$
$\{x|x \geq -3\}$

21. $-3 - 4x > -11$
$\{x|x < 2\}$

22. $-2 - x < 7$
$\{x|x > -9\}$

23. $4x - 2 < x - 11$
$\{x|x < -3\}$

24. $6x + 5 \leq x - 10$
$\{x|x \leq -3\}$

25. $x + 7 \geq 4x - 8$
$\{x|x \leq 5\}$

26. $3x + 1 \leq 7x - 15$
$\{x|x \geq 4\}$

27. $3x + 2 \leq 7x + 4$
$\left\{x|x \geq -\frac{1}{2}\right\}$

28. $3x - 5 \geq -2x + 5$
$\{x|x \geq 2\}$

29. $\frac{3}{5}x - 2 < \frac{3}{10} - x$
$\left\{x|x < \frac{23}{16}\right\}$

30. $\frac{5}{6}x - \frac{1}{6} < x - 4$
$\{x|x > 23\}$

Solve:

31. $\frac{2}{3}x - \frac{3}{2} < \frac{7}{6} - \frac{1}{3}x$

$\left\{x \mid x < \frac{8}{3}\right\}$

32. $\frac{7}{12}x - \frac{3}{2} < \frac{2}{3}x + \frac{5}{6}$

$\{x \mid x > -28\}$

33. $\frac{1}{2}x - \frac{3}{4} < \frac{7}{4}x - 2$

$\{x \mid x > 1\}$

34. $6 - 2(x - 4) \le 2x + 10$

$\{x \mid x \ge 1\}$

35. $4(2x - 1) > 3x - 2(3x - 5)$

$\left\{x \mid x > \frac{14}{11}\right\}$

36. $2(1 - 3x) - 4 > 10 + 3(1 - x)$

$\{x \mid x < -5\}$

37. $2 - 5(x + 1) \ge 3(x - 1) - 8$

$\{x \mid x \le 1\}$

38. $7 + 2(4 - x) < 9 - 3(6 + x)$

$\{x \mid x < -24\}$

39. $3(4x + 3) \le 7 - 4(x - 2)$

$\left\{x \mid x \le \frac{3}{8}\right\}$

40. $2 - 2(7 - 2x) < 3(3 - x)$

$\{x \mid x < 3\}$

41. $3 + 2(x + 5) \ge x + 5(x + 1) + 1$

$\left\{x \mid x \le \frac{7}{4}\right\}$

42. $10 - 13(2 - x) < 5(3x - 2)$

$\{x \mid x > -3\}$

43. $3 - 4(x + 2) \le 6 + 4(2x + 1)$

$\left\{x \mid x \ge -\frac{5}{4}\right\}$

44. $3x - 2(3x - 5) \le 2 - 5(x - 4)$

$\{x \mid x \le 6\}$

45. $12 - 2(3x - 2) \ge 5x - 2(5 - x)$

$\{x \mid x \le 2\}$

▶ **Objective B**

Solve:

46. $3x < 6$ and $x + 2 > 1$

$\{x \mid -1 < x < 2\}$

47. $x - 3 \le 1$ and $2x \ge -4$

$\{x \mid -2 \le x \le 4\}$

48. $x + 2 \ge 5$ or $3x \le 3$

$\{x \mid x \ge 3 \text{ or } x \le 1\}$

49. $2x < 6$ or $x - 4 > 1$

$\{x \mid x < 3 \text{ or } x > 5\}$

50. $-2x > -8$ and $-3x < 6$

$\{x \mid -2 < x < 4\}$

51. $\frac{1}{2}x > -2$ and $5x < 10$

$\{x \mid -4 < x < 2\}$

Solve:

52. $\frac{1}{3}x < -1$ or $2x > 0$
$\{x \mid x < -3 \text{ or } x > 0\}$

53. $\frac{2}{3}x > 4$ or $2x < -8$
$\{x \mid x > 6 \text{ or } x < -4\}$

54. $x + 4 \geq 5$ and $2x \geq 6$
$\{x \mid x \geq 3\}$

55. $3x < -9$ and $x - 2 < 2$
$\{x \mid x < -3\}$

56. $-5x > 10$ and $x + 1 > 6$
$\varnothing$

57. $7x < 14$ and $1 - x < 4$
$\{x \mid -3 < x < 2\}$

Solve:

58. $2x - 3 > 1$ and $3x - 1 < 2$
$\varnothing$

59. $4x + 1 < 5$ and $4x + 7 > -1$
$\{x \mid -2 < x < 1\}$

60. $3x + 7 < 10$ or $2x - 1 > 5$
$\{x \mid x < 1 \text{ or } x > 3\}$

61. $6x - 2 < -14$ or $5x + 1 > 11$
$\{x \mid x < -2 \text{ or } x > 2\}$

62. $-5 < 3x + 4 < 16$
$\{x \mid -3 < x < 4\}$

63. $5 < 4x - 3 < 21$
$\{x \mid 2 < x < 6\}$

64. $0 < 2x - 6 < 4$
$\{x \mid 3 < x < 5\}$

65. $-2 < 3x + 7 < 1$
$\{x \mid -3 < x < -2\}$

66. $4x - 1 > 11$ or $4x - 1 \leq -11$
$\left\{x \mid x > 3 \text{ or } x \leq -\frac{5}{2}\right\}$

67. $3x - 5 > 10$ or $3x - 5 < -10$
$\left\{x \mid x > 5 \text{ or } x < -\frac{5}{3}\right\}$

68. $2x - 3 \geq 5$ and $3x - 1 > 11$
$\{x \mid x > 4\}$

69. $6x - 2 < 5$ or $7x - 5 < 16$
$\{x \mid x < 3\}$

70. $9x - 2 < 7$ and $3x - 5 > 10$
$\varnothing$

71. $8x + 2 \leq -14$ and $4x - 2 > 10$
$\varnothing$

Solve:

72. $3x - 11 < 4$ or $4x + 9 \geq 1$
The solution set is the
set of real numbers.

73. $5x + 12 \geq 2$ or $7x - 1 \leq 13$
The solution set is the
set of real numbers.

74. $-6 \leq 5x + 14 \leq 24$
$\{x|-4 \leq x \leq 2\}$

75. $3 \leq 7x - 14 \leq 31$
$\left\{x|\frac{17}{7} \leq x \leq \frac{45}{7}\right\}$

76. $3 - 2x > 7$ and $5x + 2 > -18$
$\{x|-4 < x < -2\}$

77. $1 - 3x < 16$ and $1 - 3x > -16$
$\left\{x|-5 < x < \frac{17}{3}\right\}$

78. $5 - 4x > 21$ or $7x - 2 > 19$
$\{x|x < -4 \text{ or } x > 3\}$

79. $6x + 5 < -1$ or $1 - 2x < 7$
The solution set is the
set of real numbers.

80. $3 - 7x \leq 31$ and $5 - 4x > 1$
$\{x|-4 \leq x < 1\}$

81. $9 - x \geq 7$ and $9 - 2x < 3$
$\varnothing$

▶ Objective C *Application Problems*

Solve:

82. Five times the difference between a number and two is greater than the quotient of two times the number and three. Find the smallest integer that will satisfy the inequality.
3

83. Two times the difference between a number and eight is less than or equal to five times the sum of the number and four. Find the smallest number that will satisfy the inequality.
-12

Solve:

84. The length of a rectangle is 2 ft more than four times the width. Express as an integer the maximum width of the rectangle when the perimeter is less than 34 ft.
2 ft

85. The length of a rectangle is five cm less than twice the width. Express as an integer the maximum width of the rectangle when the perimeter is less than 60 cm.
11 cm

86. A cellular phone company offers its customers a rate of $99.00 for up to 200 minutes per month of cellular phone time, or a rate of $35.00 per month plus $0.40 for each minute of cellular phone time. For how many minutes per month can a customer who chooses the second option use a cellular phone before the charges exceed the first option?
Less than 160 minutes

87. A cellular phone company offers its customers a rate of $36.20 per month plus $0.40 for each minute of cellular phone time, or $20 per month plus $0.76 for each minute of cellular phone time. For how many minutes can a customer who chooses the second option use a cellular phone before the charges exceed the first option?
Less than 45 minutes

88. You can rent a car from Company A for $15 a day and 10¢ a mile or from Company B for $10 a day and 24¢ per mile. You want to rent a car for one week. How many miles can you drive a Company B car during the week if it is to cost you less than a Company A car?
Less than 250 mi

89. Agency A rents cars for $11 a day and 12¢ for every mile driven. Agency B rents cars for $8 per day and 18¢ for every mile driven. You want to rent a car for one week. How many miles can you drive an Agency B car during the week if it is to cost you less than an Agency A car?
Less than 350 mi

90. The temperature range for a week is between 14° F and 77° F. Find the temperature range in Celsius degrees. F = 9C/5 + 32
$-10° < C < 25°$

91. The temperature range for a week in a mountain town was between 0° C and 30° C. Find the temperature range in Fahrenheit degrees. C = 5(F − 32)/9
$32° < F < 86°$

Solve.

92. You are a sales account executive earning $1200 per month plus 6% commission on the amount of sales. Your goal is to earn a minimum of $6000 per month. What amount of sales will enable you to earn $6000 or more per month?
$80,000 or more

93. An account executive earns $1000 per month, plus 5% commission on the amount of sales. The executive's goal is to earn a minimum of $3200 per month. What amount of sales will enable the executive to earn $3200 or more per month?
$44,000 or more

94. You have a choice of two types of checking accounts. One account has a charge of $6 per month plus 2¢ per check. The second account has a charge of $2 per month and 7¢ per check. If you choose the second account, how many checks can you write if it is to cost less than the first type of account?
Less than 80 checks

95. A bank offers two types of checking accounts. One account has a charge of $4 per month plus 4¢ per check. The second account has a charge of $1 per month and 10¢ per check. How many checks can a customer who has the second type of account write if it is to cost the customer less than the first type of account?
Less than 50 checks

96. An average score of 90 or above in a history class receives an A grade. You have grades of 95, 89, and 81 on three exams. Find the range of scores on the fourth exam that will give you an A grade for the course.
$95 \le x \le 100$

97. An average of 70 to 79 in a mathematics class receives a C grade. A student has grades of 56, 91, 83, and 62 on four tests. Find the range of scores on the fifth test that will give the student a C for the course.
$58 \le x \le 100$

98. Find four consecutive integers whose sum is between 62 and 78.
15, 16, 17, 18 or 16, 17, 18, 19 or 17, 18, 19, 20

99. Find three consecutive even integers whose sum is between 30 and 51.
10, 12, 14 or 12, 14, 16 or 14, 16, 18

| SECTION 2.3 | **Absolute Value Equations and Inequalities** |

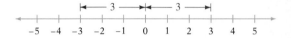

| Objective A | **To solve an absolute value equation** |

The **absolute value** of a number is its distance from zero on the number line. Distance is always a positive number or zero. Therefore, the absolute value of a number is always a positive number or zero.

The distance from 0 to 3 or from 0 to −3 is 3 units.

$|3| = 3$ $\qquad$ $|-3| = 3$

Absolute value can also be used to represent the distance between any two points on the number line. The **distance between two points** on the number line is the absolute value of the difference between the coordinates of the two points.

The distance between point a and point b is given by $|b - a|$.

The distance between 4 and −3 on the number line is 7 units. Note that the order in which the coordinates are subtracted does not affect the distance.

$$\begin{aligned} \text{Distance} &= |-3 - 4| \\ &= |-7| \\ &= 7 \end{aligned} \qquad \begin{aligned} \text{Distance} &= |4 - (-3)| \\ &= |7| \\ &= 7 \end{aligned}$$

For any two numbers a and b, $|\mathbf{b - a}| = |\mathbf{a - b}|$.

An equation containing an absolute value symbol is called an **absolute value equation.**

$$|x| = 3$$
$$|x + 2| = 8$$
$$|3x - 4| = 5x - 9$$

Absolute Value Equations

If $a \geq 0$ and $|x| = a$, then $x = a$ or $x = -a$, since $|a| = a$ and $|-a| = a$.

Given $|x| = 3$, then $x = 3$ or $x = -3$, since $|3| = 3$ and $|-3| = 3$.

Solve: $|x + 2| = 8$

Remove the absolute value sign and rewrite as two equations.

$$|x + 2| = 8$$
$$x + 2 = 8 \qquad x + 2 = -8$$

Solve each equation.

$$x = 6 \qquad x = -10$$

Write the solutions.

The solutions are 6 and −10.

Check:
$$\begin{array}{c|c} |x + 2| = 8 & |x + 2| = 8 \\ |6 + 2| \mid 8 & |-10 + 2| \mid 8 \\ |8| \mid 8 & |-8| \mid 8 \\ 8 = 8 & 8 = 8 \end{array}$$

Solve: $|5 - 3x| - 8 = -4$

$$|5 - 3x| - 8 = -4$$

Solve for the absolute value.

$$|5 - 3x| = 4$$

Remove the absolute value sign and rewrite as two equations.

$$5 - 3x = 4 \qquad 5 - 3x = -4$$

Solve each equation.

$$-3x = -1 \qquad -3x = -9$$
$$x = \frac{1}{3} \qquad x = 3$$

Write the solution.

The solutions are $\frac{1}{3}$ and 3.

$\frac{1}{3}$ and 3 check as solutions.

Example 1

Solve: $|2 - x| = 12$

Solution

$$|2 - x| = 12$$
$$2 - x = 12 \qquad 2 - x = -12$$
$$-x = 10 \qquad -x = -14$$
$$x = -10 \qquad x = 14$$

The solutions are -10 and 14.

Example 2

Solve: $|2x - 3| = 5$

Your solution

4 and -1

Example 3

Solve: $|2x| = -4$

Solution

$$|2x| = -4$$

There is no solution to this equation because the absolute value of a number must be non-negative.

Example 4

Solve: $|x - 3| = -2$

Your solution

No solution

Solutions on p. A12

Content and Format © 1991 HMCo.

Example 5

Solve: $3 - |2x - 4| = -5$

Solution

$$3 - |2x - 4| = -5$$
$$-|2x - 4| = -8$$
$$|2x - 4| = 8$$

$$2x - 4 = 8 \qquad 2x - 4 = -8$$
$$2x = 12 \qquad 2x = -4$$
$$x = 6 \qquad x = -2$$

The solutions are 6 and -2.

Example 6

Solve: $5 - |3x + 5| = 3$

Your solution

-1 and $-\frac{7}{3}$

Solution on p. A12

| Objective B | **To solve an absolute value inequality** |

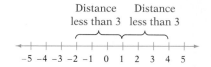

Recall that absolute value represents the distance between two points. For example, the solutions of the absolute value equation $|x - 1| = 3$, are the numbers whose distance from 1 is 3. Therefore, the solutions are -2 and 4.

The solutions of the absolute value inequality $|x - 1| < 3$ are the numbers whose distance from 1 is less than 3. Therefore, the solutions are the numbers greater than -2 and less than 4. The solution set is $\{x \mid -2 < x < 4\}$.

Distance less than 3   Distance less than 3

$-5\ -4\ -3\ -2\ -1\ \ 0\ \ 1\ \ 2\ \ 3\ \ 4\ \ 5$

To solve an absolute value inequality of the form $|ax + b| < c$, solve the equivalent compound inequality $-c < ax + b < c$.

Solve: $|3x - 1| < 5$

Solve the equivalent compound inequality.

$$|3x - 1| < 5$$
$$-5 < 3x - 1 < 5$$
$$-5 + 1 < 3x - 1 + 1 < 5 + 1$$
$$-4 < 3x < 6$$
$$\frac{-4}{3} < \frac{3x}{3} < \frac{6}{3}$$
$$-\frac{4}{3} < x < 2$$

Write the solution set.

$$\left\{ x \mid -\frac{4}{3} < x < 2 \right\}$$

The solutions of the absolute value inequality $|x + 1| > 2$ are the numbers whose distance from -1 is greater than 2. Therefore, the solutions are the numbers that are less than -3 or greater than 1. The solution set is $\{x \mid x < -3 \text{ or } x > 1\}$.

Distance greater than 2   Distance greater than 2

$-5\ -4\ -3\ -2\ -1\ \ 0\ \ 1\ \ 2\ \ 3\ \ 4\ \ 5$

To solve an absolute value inequality of the form $|ax + b| > c$, solve the equivalent compound inequality $ax + b < -c$ or $ax + b > c$.

Solve: $|3 - 2x| > 1$

Solve the equivalent compound inequality.

$$|3 - 2x| > 1$$
$$3 - 2x < -1 \quad \text{or} \quad 3 - 2x > 1$$
$$-2x < -4 \qquad\qquad -2x > -2$$
$$x > 2 \qquad\qquad\quad x < 1$$
$$\{x \,|\, x > 2\} \qquad\qquad \{x \,|\, x < 1\}$$

Find the union of the solution sets.

$$\{x \,|\, x > 2\} \cup \{x \,|\, x < 1\} = \{x \,|\, x > 2 \quad \text{or} \quad x < 1\}$$

Example 7   Solve: $|4x - 3| < 5$

Solution   Solve the equivalent compound inequality.

$$-5 < 4x - 3 < 5$$
$$-5 + 3 < 4x - 3 + 3 < 5 + 3$$
$$-2 < 4x < 8$$
$$\frac{-2}{4} < \frac{4x}{4} < \frac{8}{4}$$
$$-\frac{1}{2} < x < 2$$
$$\left\{ x \,\middle|\, -\frac{1}{2} < x < 2 \right\}$$

Example 8   Solve: $|3x + 2| < 8$

Your solution $\left\{ x \,\middle|\, -\frac{10}{3} < x < 2 \right\}$

Example 9   Solve: $|x - 3| < 0$

Solution   The absolute value of a number must be non-negative. The solution set is the empty set.

Example 10   Solve: $|3x - 7| < 0$

Your solution $\varnothing$

Solutions on pp. A12–A13

Content and Format © 1991 HMCo.

Example 11   Solve: $|2x - 1| > 7$

Example 12   Solve: $|5x + 3| > 8$

Solution   Solve the equivalent compound inequality.

$2x - 1 < -7$ or $2x - 1 > 7$
$\quad\ 2x < -6 \qquad\qquad 2x > 8$
$\quad\ \ x < -3 \qquad\qquad\ \ x > 4$
$\{x \mid x < -3\} \qquad\quad \{x \mid x > 4\}$
$\{x \mid x < -3\} \cup \{x \mid x > 4\} =$
$\{x \mid x < -3 \text{ or } x > 4\}$

Your solution $\left\{x \mid x < -\frac{11}{5} \text{ or } x > 1\right\}$

Solution on p. A13

Objective C   **To solve application problems**

The **tolerance** of a component, or part, is the acceptable amount by which the component may vary from a given measurement. For example, the diameter of a piston may vary from the given measurement of 9 cm by 0.001 cm. This is written as 9 cm ± 0.001 cm, read "9 centimeters plus or minus 0.001 centimeters." The maximum diameter, or **upper limit**, of the piston is 9 cm + 0.001 cm = 9.001 cm. The minimum diameter, or **lower limit**, is 9 cm − 0.001 cm = 8.999 cm.

The lower and upper limits of the diameter of the piston could also be found by solving the absolute value inequality $|d - 9| \le 0.001$, where d is the diameter of the piston.

$$|d - 9| \le 0.001$$
$$-0.001 \le d - 9 \le 0.001$$
$$-0.001 + 9 \le d - 9 + 9 \le 0.001 + 9$$
$$8.999 \le d \le 9.001$$

The lower and upper limits of the diameter of the piston are 8.999 cm and 9.001 cm.

Example 13

A doctor has prescribed 2 cc of medication for a patient. The tolerance is 0.03 cc. Find the lower and upper limits of the amount of medication to be given.

Strategy

Let p represent the prescribed amount of medication, T the tolerance, and m the given amount of medication. Solve the absolute value inequality $|m - p| \leq T$ for m.

Solution

$|m - p| \leq T$
$|m - 2| \leq 0.03$

$-0.03 \leq m - 2 \leq 0.03$
$-0.03 + 2 \leq m - 2 + 2 \leq 0.03 + 2$
$1.97 \leq m \leq 2.03$

The lower and upper limits of the amount of medication to be given to the patient are 1.97 cc and 2.03 cc.

Example 14

A machinist must make a bushing that has a tolerance of 0.003 in. The diameter of the bushing is 2.55 in. Find the lower and upper limits of the diameter of the bushing.

Your strategy

Your solution

Lower limit: 2.547 in.
Upper limit: 2.553 in.

Solution on p. A13

Content and Format © 1991 HMCo.

2.3   **EXERCISES**

▶ **Objective A**

Solve:

1. $|x| = 7$
7 and -7

2. $|a| = 2$
2 and -2

3. $|b| = 4$
4 and -4

4. $|c| = 12$
12 and -12

5. $|-y| = 6$
6 and -6

6. $|-t| = 3$
3 and -3

7. $|-a| = 7$
7 and -7

8. $|-x| = 3$
3 and -3

9. $|x| = -4$
No solution

10. $|y| = -3$
No solution

11. $|-t| = -3$
No solution

12. $|-y| = -2$
No solution

13. $|x + 2| = 3$
1 and -5

14. $|x + 5| = 2$
-3 and -7

15. $|y - 5| = 3$
8 and 2

16. $|y - 8| = 4$
12 and 4

17. $|a - 2| = 0$
2

18. $|a + 7| = 0$
-7

19. $|x - 2| = -4$
No solution

20. $|x + 8| = -2$
No solution

21. $|2x - 5| = 4$
$\frac{9}{2}$ and $\frac{1}{2}$

22. $|3x - 9| = 3$
4 and 2

23. $|3x - 2| = 7$
3 and $-\frac{5}{3}$

24. $|5x + 8| = 2$
$-\frac{6}{5}$ and -2

25. $|2x + 5| = 8$
$\frac{3}{2}$ and $-\frac{13}{2}$

26. $|2x - 3| = 5$
4 and -1

27. $|4 - 2x| = 6$
-1 and 5

28. $|7 - 3x| = 5$
$\frac{2}{3}$ and 4

29. $|2 - 3y| = 4$
$-\frac{2}{3}$ and 2

30. $|4 - 9x| = 14$
$-\frac{10}{9}$ and 2

31. $|4 - 3x| = 4$
0 and $\frac{8}{3}$

32. $|2 - 5x| = 2$
0 and $\frac{4}{5}$

33. $|3 - 4x| = 9$
$-\frac{3}{2}$ and 3

34. $|2 - 5x| = 3$
$-\frac{1}{5}$ and 1

35. $|2x - 3| = 0$
$\frac{3}{2}$

36. $|5x + 5| = 0$
-1

37. $|3x - 2| = -4$
No solution

38. $|2x + 5| = -2$
No solution

39. $|x - 2| - 2 = 3$
7 and -3

Solve:

40. $|x - 9| - 3 = 2$
14 and 4

41. $|3a + 2| - 4 = 4$
2 and $-\frac{10}{3}$

42. $|2a + 9| + 4 = 5$
-4 and -5

43. $|2 - y| + 3 = 4$
1 and 3

44. $|8 - y| - 3 = 1$
4 and 12

45. $|2x - 3| + 3 = 3$
$\frac{3}{2}$

46. $|4x - 7| - 5 = -5$
$\frac{7}{4}$

47. $|2x - 3| + 4 = -4$
No solution

48. $|3x - 2| + 1 = -1$
No solution

49. $|6x - 5| - 2 = 4$
$\frac{11}{6}$ and $-\frac{1}{6}$

50. $|4b + 3| - 2 = 7$
$\frac{3}{2}$ and -3

51. $|3t + 2| + 3 = 4$
$-\frac{1}{3}$ and -1

52. $|5x - 2| + 5 = 7$
$\frac{4}{5}$ and 0

53. $3 - |x - 4| = 5$
No solution

54. $2 - |x - 5| = 4$
No solution

55. $8 - |2x - 3| = 5$
3 and 0

56. $8 - |3x + 2| = 3$
1 and $-\frac{7}{3}$

57. $|2 - 3x| + 7 = 2$
No solution

58. $|1 - 5a| + 2 = 3$
0 and $\frac{2}{5}$

59. $|8 - 3x| - 3 = 2$
1 and $\frac{13}{3}$

60. $|6 - 5b| - 4 = 3$
$-\frac{1}{5}$ and $\frac{13}{5}$

61. $|2x - 8| + 12 = 2$
No solution

62. $|3x - 4| + 8 = 3$
No solution

63. $2 + |3x - 4| = 5$
$\frac{7}{3}$ and $\frac{1}{3}$

64. $5 + |2x + 1| = 8$
1 and -2

65. $5 - |2x + 1| = 5$
$-\frac{1}{2}$

66. $3 - |5x + 3| = 3$
$-\frac{3}{5}$

67. $6 - |2x + 4| = 3$
$-\frac{1}{2}$ and $-\frac{7}{2}$

68. $8 - |3x - 2| = 5$
$\frac{5}{3}$ and $-\frac{1}{3}$

69. $8 - |1 - 3x| = -1$
$-\frac{8}{3}$ and $\frac{10}{3}$

70. $3 - |3 - 5x| = -2$
$-\frac{2}{5}$ and $\frac{8}{5}$

71. $5 + |2 - x| = 3$
No solution

72. $6 + |3 - 2x| = 2$
No solution

▶ **Objective B**

Solve.

73. $|x| > 3$
$\{x | x > 3 \text{ or } x < -3\}$

74. $|x| < 5$
$\{x | -5 < x < 5\}$

75. $|x + 1| > 2$
$\{x | x > 1 \text{ or } x < -3\}$

76. $|x - 2| > 1$
$\{x | x > 3 \text{ or } x < 1\}$

77. $|x - 5| \le 1$
$\{x | 4 \le x \le 6\}$

78. $|x - 4| \le 3$
$\{x | 1 \le x \le 7\}$

79. $|2 - x| \ge 3$
$\{x | x \ge 5 \text{ or } x \le -1\}$

80. $|3 - x| \ge 2$
$\{x | x \le 1 \text{ or } x \ge 5\}$

81. $|2x + 1| < 5$
$\{x | -3 < x < 2\}$

82. $|3x - 2| < 4$
$\{x | -\frac{2}{3} < x < 2\}$

83. $|5x + 2| > 12$
$\{x | x > 2 \text{ or } x < -\frac{14}{5}\}$

84. $|7x - 1| > 13$
$\{x | x > 2 \text{ or } x < -\frac{12}{7}\}$

85. $|4x - 3| \le -2$
$\varnothing$

86. $|5x + 1| \le -4$
$\varnothing$

87. $|2x + 7| > -5$
The solution set is the set of real numbers.

88. $|3x - 1| > -4$
The solution set is the set of real numbers.

89. $|4 - 3x| \ge 5$
$\{x | x \le -\frac{1}{3} \text{ or } x \ge 3\}$

90. $|7 - 2x| > 9$
$\{x | x < -1 \text{ or } x > 8\}$

91. $|5 - 4x| \le 13$
$\{x | -2 \le x \le \frac{9}{2}\}$

92. $|3 - 7x| < 17$
$\{x | -2 < x < \frac{20}{7}\}$

93. $|6 - 3x| \le 0$
$\{x | x = 2\}$

94. $|10 - 5x| \ge 0$
The solution set is the set of real numbers.

95. $|2 - 9x| > 20$
$\{x | x < -2 \text{ or } x > \frac{22}{9}\}$

96. $|5x - 1| < 16$
$\{x | -3 < x < \frac{17}{5}\}$

97. $|2x - 3| + 2 < 8$
$\{x | -\frac{3}{2} < x < \frac{9}{2}\}$

98. $|3x - 5| + 1 < 7$
$\{x | -\frac{1}{3} < x < \frac{11}{3}\}$

99. $|2 - 5x| - 4 > -2$
$\{x | x < 0 \text{ or } x > \frac{4}{5}\}$

100. $|4 - 2x| - 9 > -3$
$\{x | x < -1 \text{ or } x > 5\}$

101. $8 - |2x - 5| < 3$
$\{x | x > 5 \text{ or } x < 0\}$

102. $12 - |3x - 4| > 7$
$\{x | -\frac{1}{3} < x < 3\}$

▶ Objective C   *Application Problems*

Solve.

103. A doctor has prescribed 4 cc of medication for a patient. The tolerance is 0.05 cc. Find the lower and upper limits of the amount of medication to be given.
3.95 cc, 4.05 cc

104. A doctor has prescribed 3 cc of medication for a patient. The tolerance is 0.04 cc. Find the lower and upper limits of the amount of medication to be given.
2.96 cc, 3.04 cc

105. A machinist must make a bushing that has a tolerance of 0.002 in. The diameter of the bushing is 2.65 in. Find the lower and upper limits of the diameter of the bushing.
2.648 in., 2.652 in.

106. The diameter of a bushing is 2.45 in. The bushing has a tolerance of 0.001 in. Find the lower and upper limits of the diameter of the bushing.
2.449 in., 2.451 in.

107. A piston rod for an automobile is $9\frac{5}{8}$ in. long with a tolerance of $\frac{1}{32}$ in. Find the lower and upper limits of the length of the piston rod.
$9\frac{19}{32}$ in., $9\frac{21}{32}$ in.

108. A piston rod for an automobile is $9\frac{3}{8}$ in. long with a tolerance of $\frac{1}{64}$ in. Find the lower and upper limits of the length of the piston rod.
$9\frac{23}{64}$ in., $9\frac{25}{64}$ in.

The tolerance of the resistors used in electronics is given as a percent. Use your calculator for the following exercises.

109. Find the lower and upper limits of a 29,000-ohm resistor with a 2% tolerance.
28,420 ohms, 29,580 ohms

110. Find the lower and upper limits of a 15,000-ohm resistor with a 10% tolerance.
13,500 ohms, 16,500 ohms

111. Find the lower and upper limits of a 25,000-ohm resistor with a 5% tolerance.
23,750 ohms, 26,250 ohms

112. Find the lower and upper limits of a 56-ohm resistor with a 5% tolerance.
53.2 ohms, 58.8 ohms

Content and Format © 1991 HMCo.

SECTION 2.4

Applications: Puzzle Problems

Objective A

To solve integer problems

An equation states that two mathematical expressions are equal. Therefore, to translate a sentence into an equation requires recognition of the words or phrases which mean "equals." A partial list of these phrases includes "is," "is equal to," "amounts to," and "represents."

Once the sentence is translated into an equation, the equation can be solved by rewriting the equation in the form *variable = constant*.

Recall that an **even integer** is an integer that is divisible by 2. An **odd integer** is an integer that is not divisible by 2.

Consecutive integers are integers which follow one another in order. Examples of consecutive integers are shown at the right. (Assume that the variable *n* represents an integer.)

$$8, 9, 10$$
$$-3, -2, -1$$
$$n, n + 1, n + 2$$

Examples of **consecutive even integers** are shown at the right. (Assume that the variable *n* represents an even integer.)

$$16, 18, 20$$
$$-6, -4, -2$$
$$n, n + 2, n + 4$$

Examples of **consecutive odd integers** are shown at the right. (Assume that the variable *n* represents an odd integer.)

$$11, 13, 15$$
$$-23, -21, -19$$
$$n, n + 2, n + 4$$

The sum of three consecutive even integers is seventy-eight. Find the integers.

Strategy for Solving an Integer Problem

> Let a variable represent one of the integers. Express each of the other integers in terms of that variable. Remember that for consecutive integer problems, consecutive integers will differ by 1. Consecutive even or consecutive odd integers will differ by 2.

Represent three consecutive even integers.

First even integer: n
Second even integer: $n + 2$
Third even integer: $n + 4$

Determine the relationship among the integers.

The sum of the three even integers is 78.

$$n + (n + 2) + (n + 4) = 78$$
$$3n + 6 = 78$$
$$3n = 72$$
$$n = 24$$

$$n + 2 = 24 + 2 = 26$$
$$n + 4 = 24 + 4 = 28$$

The three consecutive even integers are 24, 26, and 28.

Example 1

One number is four more than another number. The sum of the two numbers is sixty-six. Find the two numbers.

Strategy

- The smaller number: n
  The larger number: $n + 4$
- The sum of the numbers is 66.
  $n + (n + 4) = 66$

Solution

$$n + (n + 4) = 66$$
$$2n + 4 = 66$$
$$2n = 62$$
$$n = 31$$

$$n + 4 = 31 + 4 = 35$$

The numbers are 31 and 35.

Example 2

The sum of three numbers is eighty-one. The second number is twice the first number, and the third number is three less than four times the first number. Find the numbers.

Your strategy

Your solution

12, 24, and 45

Example 3

Five times the first of three consecutive even integers is five more than the product of four and the third integer. Find the integers.

Strategy

- First even integer: n
  Second even integer: $n + 2$
  Third even integer: $n + 4$
- Five times the first integer equals five more than the product of four and the third integer.

  $$5n = 4(n + 4) + 5$$

Solution

$$5n = 4(n + 4) + 5$$
$$5n = 4n + 16 + 5$$
$$5n = 4n + 21$$
$$n = 21$$

Since 21 is not an even integer, there is no solution.

Example 4

Find three consecutive odd integers such that three times the sum of the first two integers is ten more than the product of the third integer and four.

Your strategy

Your solution

No solution

Solutions on p. A13

| Objective B | **To solve coin and stamp problems** |

In solving problems dealing with coins or stamps of different values, it is necessary to represent the value of the coins or stamps in the same unit of money. The unit of money is frequently cents. For example:

The value of five 8¢ stamps is $5 \cdot 8$, or 40 cents.
The value of four 20¢ stamps is $4 \cdot 20$, or 80 cents.
The value of n 10¢ stamps is $n \cdot 10$, or $10n$ cents.

A collection of stamps consists of 5¢, 13¢, and 18¢ stamps. The number of 13¢ stamps is two more than three times the number of 5¢ stamps. The number of 18¢ stamps is five less than the number of 13¢ stamps. The total value of all the stamps is $1.68. Find the number of 18¢ stamps.

Strategy for Solving a Stamp Problem

> For each denomination of stamp, write a numerical or variable expression for the number of stamps, the value of the stamp, and the total value of the stamps in cents. The results can be recorded in a table.

The number of 5¢ stamps: x
The number of 13¢ stamps: $3x + 2$
The number of 18¢ stamps: $(3x + 2) - 5 = 3x - 3$

Stamp	Number of stamps	·	Value of stamp in cents	=	Total value in cents
5¢	x	·	5	=	$5x$
13¢	$3x + 2$	·	13	=	$13(3x + 2)$
18¢	$3x - 3$	·	18	=	$18(3x - 3)$

> Determine the relationship between the total values of the stamps. Use the fact that the sum of the total values of each denomination of stamp is equal to the total value of all the stamps.

The sum of the total values of each denomination of stamp is equal to the total value of all the stamps (168 cents).

$$5x + 13(3x + 2) + 18(3x - 3) = 168$$
$$5x + 39x + 26 + 54x - 54 = 168$$
$$98x - 28 = 168$$
$$98x = 196$$
$$x = 2$$

The number of 18¢ stamps is $3x - 3$.
Replace x by 2 and evaluate. $3x - 3 = 3(2) - 3 = 3$

There are three 18¢ stamps in the collection.

Example 5

A coin bank contains $1.80 in nickels and dimes; in all, there are twenty-two coins in the bank. Find the number of nickels and the number of dimes in the bank.

Example 6

A collection of stamps contains 3¢, 10¢, and 15¢ stamps. The number of 10¢ stamps is two more than twice the number of 3¢ stamps. There are three times as many 15¢ stamps as there are 3¢ stamps. The total value of the stamps is $1.56. Find the number of 15¢ stamps.

Strategy

- Number of nickels: x
  Number of dimes: $22 - x$

Coin	Number	Value	Total value
Nickel	x	5	$5x$
Dime	$22 - x$	10	$10(22 - x)$

- The sum of the total values of each denomination of coin equals the total value of all the coins (180 cents).

  $5x + 10(22 - x) = 180$

Your strategy

Solution

$$5x + 10(22 - x) = 180$$
$$5x + 220 - 10x = 180$$
$$-5x + 220 = 180$$
$$-5x = -40$$
$$x = 8$$

$22 - x = 22 - 8 = 14$

The bank contains 8 nickels and 14 dimes.

Your solution

6 stamps

Solution on p. A14

Content and Format © 1991 HMCo.

2.4   EXERCISES

▶ **Objective A   *Application Problems***

Solve:

1. What number must be added to the numerator and denominator of $\frac{5}{7}$ to produce the fraction $\frac{4}{5}$?
    3

2. What number must be added to the numerator and denominator of $\frac{6}{11}$ to produce the fraction $\frac{2}{3}$?
    4

3. The sum of two integers is 10. Three times the larger integer is three less than eight times the smaller integer. Find the integers.
    3 and 7

4. The sum of two integers is thirty. Eight times the smaller integer is six more than five times the larger integer. Find the integers.
    12 and 18

5. One integer is eight less than another integer. The sum of the two integers is fifty. Find the integers.
    21 and 29

6. One integer is four more than another integer. The sum of the integers is twenty-six. Find the integers.
    11 and 15

7. The sum of three numbers is one hundred twenty-three. The second number is two more than twice the first number. The third number is five less than the product of three and the first number. Find the three numbers.
    21, 44, and 58

8. The sum of three numbers is forty-two. The second number is twice the first number and the third number is three less than the second number. Find the three numbers.
    9, 18, and 15

9. The sum of three consecutive integers is negative fifty-seven. Find the integers.
    −20, −19, and −18

10. The sum of three consecutive integers is one hundred twenty-nine. Find the integers.
    42, 43, and 44

11. Five times the smallest of three consecutive odd integers is ten more than twice the largest. Find the integers.
    No Solution

12. Find three consecutive even integers such that twice the sum of the first and third integers is twenty-one more than the second integer.
    No Solution

13. Find three consecutive odd integers such that three times the middle integer is seven more than the sum of the first and third integers.
    5, 7, 9

14. Find three consecutive even integers such that four times the sum of the first and third integers is twenty less than six times the middle integer.
    −12, −10, and −8

▶ Objective B   *Application Problems*

15. A collection of 53 coins has a value of $3.70. The collection contains only nickels and dimes. Find the number of dimes in the collection.
21 dimes

16. A collection of 22 coins has a value of $4.75. The collection contains dimes and quarters. Find the number of quarters in the collection.
17 quarters

17. A coin bank contains 22 coins in nickels, dimes, and quarters. There are four times as many dimes as quarters. The value of the coins is $2.30. How many dimes are in the bank?
12 dimes

18. A coin collection contains nickels, dimes, and quarters. There are twice as many dimes as quarters and seven more nickels than dimes. The total value of all the coins is $2.00. How many quarters are in the collection?
3 quarters

19. A cashier has $730 in twenty-dollar bills and five-dollar bills. In all, the cashier has 68 bills. How many twenty-dollar bills does the cashier have?
26 twenty-dollar bills

20. A department store uses twice as many five-dollar bills in conducting its daily business as ten-dollar bills. $2500 was obtained in five- and ten-dollar bills for the day's business. How many five-dollar bills were obtained?
250 five-dollar bills

21. A stamp collector has some 15¢ stamps and some 20¢ stamps. The number of 15¢ stamps is eight less than three times the number of 20¢ stamps. The total value is $4. Find the number of each type of stamp in the collection.
20¢ stamps: 8; 15¢ stamps: 16

22. An office has some 20¢ stamps and some 28¢ stamps. Altogether the office has 140 stamps for a total value of $31.20. How many of each type stamp does the office have?
20¢ stamps: 100; 28¢ stamps: 40

23. A stamp collection consists of 3¢, 8¢, and 13¢ stamps. The number of 8¢ stamps is three less than twice the number of 3¢ stamps. The number of 13¢ stamps is twice the number of 8¢ stamps. The total value of all the stamps is $2.53. Find the number of 3¢ stamps in the collection.
5, 3¢ stamps

24. An account executive bought 330 stamps for $79.50. The purchase included 15¢ stamps, 20¢ stamps, and 40¢ stamps. The number of 20¢ stamps is four times the number of 15¢ stamps. How many 40¢ stamps were purchased?
80, 40¢ stamps

25. A stamp collector has 8¢, 13¢, and 18¢ stamps. The collector has twice as many 8¢ stamps as 18¢ stamps. There are three more 13¢ than 18¢ stamps. The total value of the stamps in the collection is $3.68. Find the number of 18¢ stamps in the collection.
7, 18¢ stamps

26. A stamp collection consists of 3¢, 12¢, and 15¢ stamps. The number of 3¢ stamps is five times the number of 12¢ stamps. The number of 15¢ stamps is four less than the number of 12¢ stamps. The total value of the stamps in the collection is $3.18. Find the number of 15¢ stamps in the collection.
5, 15¢ stamps

SECTION 2.5

Applications: Value Mixture and Uniform Motion Problems

Objective A   **To solve value mixture problems**

A **value mixture problem** involves combining two ingredients which have different prices into a single blend. For example, a coffee manufacturer may blend two types of coffee into a single blend.

The solution of a value mixture problem is based upon the equation $AC = V$, where A is the amount of the ingredient, C is the cost per unit of the ingredient, and V is the value of the ingredient.

How many pounds of peanuts which cost $2.25 per pound must be mixed with 40 lb of cashews which cost $6.00 per pound to make a mixture which cost $3.50 per pound?

Strategy for Solving a Value Mixture Problem

For each ingredient in the mixture, write a numerical or variable expression for the amount of the ingredient used, the unit cost of the ingredient, and the value of the amount used. For the mixture, write a numerical or variable expression for the amount, the unit cost of the mixture, and the value of the amount. The results can be recorded in a table.

Amount of peanuts: x
Amount of cashews: 40

	Amount, A	·	Unit cost, C	=	Value, V
Peanuts	x	·	$2.25	=	2.25x
Cashews	40	·	$6.00	=	6.00(40)
Mixture	40 + x	·	$3.50	=	3.50(40 + x)

Determine how the values of each ingredient are related. Use the fact that the sum of the values of each ingredient is equal to the value of the mixture.

The sum of the values of the peanuts and the cashews is equal to the value of the mixture.

$$2.25x + 6.00(40) = 3.50(40 + x)$$
$$2.25x + 240 = 140 + 3.50x$$
$$-1.25x + 240 = 140$$
$$-1.25x = -100$$
$$x = 80$$

The mixture must contain 80 lb of peanuts.

Example 1

How many ounces of a gold alloy which cost $320 an ounce must be mixed with 100 oz of an alloy which cost $100 an ounce to make a mixture which cost $160 an ounce?

Example 2

A butcher combined hamburger which cost $3.00 per pound with hamburger which cost $1.80 per pound. How many pounds of each were used to make a 75-pound mixture to sell for $2.20 per pound?

Strategy

- Ounces of $320 gold alloy: x
  Ounces of $100 gold alloy: 100

	Amount	Cost	Value
$320 alloy	x	320	320x
$100 alloy	100	100	100(100)
Mixture	$x + 100$	160	160($x + 100$)

- The sum of the values before mixing equals the value after mixing.

  $$320x + 100(100) = 160(x + 100)$$

Your strategy

Solution

$$320x + 100(100) = 160(x + 100)$$
$$320x + 10{,}000 = 160x + 16{,}000$$
$$160x + 10{,}000 = 16{,}000$$
$$160x = 6000$$
$$x = 37.5$$

The mixture must contain 37.5 oz of the $320 gold alloy.

Your solution

25 lb of $3.00 hamburger
50 lb of $1.80 hamburger

Solution on p. A14

| **Objective B** | **To solve uniform motion problems** |

A car that travels constantly in a straight line at 55 mph is in uniform motion. **Uniform motion** means that the speed of an object does not change.

The solution of a uniform motion problem is based upon the equation $rt = d$, where r is the rate of travel, t is the time spent traveling, and d is the distance traveled.

An executive has an appointment 785 mi from the office. The executive takes a helicopter from the office to the airport and a plane from the airport to the business appointment. The helicopter averages 70 mph and the plane averages 500 mph. The total time spent traveling was 2 h. Find the distance from the executive's office to the airport.

Strategy for Solving a Uniform Motion Problem

> For each object, write a numerical or variable expression for the distance, rate, and time. The results can be recorded in a table.

The total time of travel was 2 h.

Unknown time in the helicopter: t
Time in the plane: $2 - t$

	Rate, r	·	*Time, t*	=	*Distance, d*
Helicopter	70	·	t	=	$70t$
Plane	500	·	$2 - t$	=	$500(2 - t)$

> Determine how the distances traveled by each object are related. For example, the total distance traveled by both objects may be known, or it may be known that the two objects traveled the same distance.

The total distance traveled is 785 mi.

$$70t + 500(2 - t) = 785$$
$$70t + 1000 - 500t = 785$$
$$-430t + 1000 = 785$$
$$-430t = -215$$
$$t = 0.5$$

The time spent traveling from the office to the airport in the helicopter is 0.5 h. To find the distance between these two points, substitute the values r and t into the equation $rt = d$.

$$rt = d$$
$$70 \cdot 0.5 = d$$
$$35 = d$$

The distance from the office to the airport is 35 mi.

Example 3

A long distance runner started a course running at an average speed of 6 mph. One and one half hours later, a cyclist traveled the same course at an average speed of 12 mph. How long after the runner started did the cyclist overtake the runner?

Strategy

- Unknown time for the cyclist: t
  Time for the runner: $t + 1.5$

	Rate	Time	Distance
Runner	6	$t + 1.5$	$6(t + 1.5)$
Cyclist	12	t	$12t$

- The runner and the cyclist travel the same distance.

  $6(t + 1.5) = 12t$

Solution

$$6(t + 1.5) = 12t$$
$$6t + 9 = 12t$$
$$9 = 6t$$
$$\frac{3}{2} = t$$

The cyclist traveled for 1.5 h.

$t + 1.5 = 1.5 + 1.5 = 3$

The cyclist overtook the runner 3 h after the runner started.

Example 4

Two small planes start from the same point and fly in opposite directions. The first plane is flying 30 mph faster than the second plane. In 4 h the planes are 1160 mi apart. Find the rate of each plane.

Your strategy

Your solution

1st plane: 160 mph
2nd plane: 130 mph

Solution on p. A14

2.5   EXERCISES

▶ Objective A   *Application Problems*

1. Forty pounds of cashews costing $5.60 per pound were mixed with 100 pounds of peanuts costing $1.89 per pound. Find the selling price of the resulting mixture.
   $2.95/lb

2. A coffee merchant combines coffee costing $6 per pound with coffee costing $3.50 per pound. How many pounds of each should be used to make 25 pounds of a blend costing $4.25 per pound?
   7.5 lb of $6.00 blend; 17.5 lb of $3.50 blend

3. Adult tickets for a play cost $4.00 and children's tickets cost $1.00. For one performance, 420 tickets were sold. Receipts for the performance were $1410. Find the number of adult tickets sold.
   330 adult tickets

4. Tickets for a school play sold for $2.50 for each adult and $1.00 for each child. The total receipts for 113 tickets sold were $221. Find the number of adult tickets sold.
   72 adult tickets

5. A restaurant mixes fifty liters of pure maple syrup which cost $9.50 per liter with imitation maple syrup which costs $4.00 per liter. How much imitation maple syrup is needed to make a mixture that costs $5.00 per liter?
   225 L

6. To make a flour mix, a miller combined soybeans which cost $8.50 per bushel with wheat which cost $4.50 per bushel. How many bushels of each were used to make a mixture of 1000 bushels to sell for $5.50 per bushel?
   250 bushels of soybeans; 750 bushels of wheat

7. A goldsmith combined pure gold which cost $400 per ounce with an alloy of gold costing $150 per ounce. How many ounces of each were used to make 50 oz of gold alloy to sell for $250 per ounce?
   20 oz of pure gold; 30 oz of the alloy

8. A silversmith combined pure silver which cost $5.20 an ounce with 50 oz of a silver alloy which cost $2.80 an ounce. How many ounces of the pure silver were used to make an alloy of silver selling for $4.40 an ounce?
   100 oz

9. A tea mixture was made from 40 lb of tea costing $5.40 per pound and 60 lb of tea costing $3.25 per pound. Find the selling price of the tea mixture.
   $4.11/lb

10. Find the selling price per ounce of a face cream mixture made from 100 oz of face cream which cost $3.46 per ounce and 60 oz of face cream which cost $12.50 per ounce.
    $6.85/oz

11. A fruitstand owner combined cranberry juice which cost $4.60 per gallon with 50 gallon of apple juice which cost $2.24 per gallon. How much cranberry juice was used to make cranapple juice to sell for $3.00 per gallon? Round to the nearest tenth.
    23.8 gal

12. Walnuts which cost $4.05 per kilogram were mixed with cashews which cost $7.25 per kilogram. How many kilograms of each were used to make a 50-kilogram mixture to sell for $6.25 per kilogram? Round to the nearest tenth.
    15.6 kg of walnuts; 34.4 kg of cashews

▶ Objective B   *Application Problems*

13. A car traveling at 56 mph overtakes a cyclist who, traveling at 14 mph had a 1.5 h-head start. How far from the starting point does the car overtake the cyclist?
28 mi

14. A helicopter traveling 130 mph overtakes a speeding car traveling 80 mph. The car had a one-half hour head start. How far from the starting point did the helicopter overtake the car?
104 mi

15. Two planes are 1620 mi apart and traveling toward each other. One plane is traveling 120 mph faster than the other plane. The planes meet in 1.5 h. Find the speed of each plane.
1st plane: 480 mph; 2nd plane: 600 mph

16. Two cars are 310 mi apart and traveling toward each other. One car travels 8 mph faster than the other car. The cars meet in 2.5 h. Find the speed of each car.
1st car: 58 mph; 2nd car: 66 mph

17. A ferry leaves a harbor and travels to a resort island at an average speed of 20 mph. On the return trip, the ferry travels at an average speed of 12 mph due to fog. The total time for the trip is 5 h. How far is the island from the harbor?
37.5 mi

18. A commuter plane provides transportation from an international airport to the surrounding cities. One commuter plane averaged 250 mph flying to a city and 150 mph returning to the international airport. The total flying time was 4 h. Find the distance between the two airports.
375 mi

19. Two planes start from the same point and fly in opposite directions. The first plane is flying 50 mph slower than the second plane. In 2.5 h the planes are 1400 mi apart. Find the rate of each plane.
1st plane: 255 mph; 2nd plane: 305 mph

20. Two hikers start from the same point and hike in opposite directions around a lake whose shoreline is 13 mi. One hiker walks 0.5 mph faster than the other hiker. How fast did each hiker walk if they meet in 2 h?
3 mph, 3.5 mph

21. A student rode a bicycle to the repair shop and then walked home. The student averaged 12 mph riding to the shop and 3 mph walking home. The round trip took one hour. How far is it between the student's home and the bicycle shop?
2.4 mi

22. A passenger train leaves a depot 1.5 h after a freight train leaves the same depot. The passenger train is traveling 18 mph faster than the freight train. Find the rate of each train if the passenger train overtakes the freight train in 2.5 h.
Freight train: 30 mph; Passenger train: 48 mph

23. A plane leaves an airport at 3 P.M. At 4 P.M. another plane leaves the same airport traveling in the same direction at a speed 150 mph faster than the first plane. Four hours after the first plane took off the second plane is 250 miles ahead of the first plane. How far did the second plane travel?
1050 mi

24. A jogger and a cyclist set out at 9 A.M. from the same point headed in the same direction. The average speed of the cyclist is four times the speed of the jogger. In 2 h, the cyclist is 33 mi ahead of the jogger. How far did the cyclist ride?
44 mi

| | SECTION 2.6 | | # Applications: Problems Involving Percent |

Objective A   ### To solve investment problems

The annual simple interest that an investment earns is given by the equation $Pr = I$, where P is the principal, or the amount invested, r is the simple interest rate, and I is the simple interest. The solution of an investment problem is based upon this equation.

You have a total of $8000 invested in two simple interest accounts. On one account, the money market fund, the annual simple interest rate is 11.5%. On the second account, the bond fund, the annual simple interest rate is 9.75%. The total annual interest earned by the two accounts is $823.75. How much do you have invested in each account?

Strategy for Solving a Problem Involving Money
Deposited in Two Simple Interest Accounts

> For each amount invested, use the equation $Pr = I$. Write a numerical or variable expression for the principal, the interest rate, and the interest earned. The results can be recorded in a table.

The total amount invested is $8000.   Amount invested at 11.5%: x
                                           Amount invested at 9.75%: $8000 - x$

	Principal, P	·	Interest rate, r	=	Interest earned, I
Amount at 11.5%	x	·	0.115	=	$0.115x$
Amount at 9.75%	$8000 - x$	·	0.0975	=	$0.0975(8000 - x)$

> Determine how the amounts of interest earned on each amount are related. For example, the total interest earned by both accounts may be known or it may be known that the interest earned on one account is equal to the interest earned on the other account.

The total annual interest earned is
$823.75.

$$0.115x + 0.0975(8000 - x) = 823.75$$
$$0.115x + 780 - 0.0975x = 823.75$$
$$0.0175x + 780 = 823.75$$
$$0.0175x = 43.75$$
$$x = 2500$$

The amount invested at 9.75% is $8000 - x$.
Replace x by 2500 and evaluate. $8000 - x = 8000 - 2500 = 5500$

The amount invested at 11.5% is $2500.
The amount invested at 9.75% is $5500.

Example 1

An investment of $4000 is made at an annual simple interest rate of 10.9%. How much additional money must be invested at an annual simple interest rate of 14.5% so that the total interest earned is 12% of the total investment?

Strategy

- Additional amount to be invested at 14.5%: x

	Principal	*Rate*	*Interest*
Amount at 10.9%	4000	0.109	0.109(4000)
Amount at 14.5%	x	0.145	0.145x
Amount at 12%	4000 + x	0.12	0.12(4000 + x)

- The sum of the interest earned by the two investments equals the interest earned by the total investment.

$$0.109(4000) + 0.145x = 0.12(4000 + x)$$

Solution

$$
\begin{aligned}
0.109(4000) + 0.145x &= 0.12(4000 + x) \\
436 + 0.145x &= 480 + 0.12x \\
436 + 0.025x &= 480 \\
0.025x &= 44 \\
x &= 1760
\end{aligned}
$$

$1760 must be invested at an annual simple interest rate of 14.5%.

Example 2

An investment of $3500 is made at an annual simple interest rate of 13.2%. How much additional money must be invested at an annual simple interest rate of 11.5% so that the total interest earned is $1037?

Your strategy

Your solution

$5000

Solution on p. A15

Content and Format © 1991 HMCo.

| Objective B | **To solve percent mixture problems** |

The amount of a substance in a solution or alloy can be given as a percent of the total solution or alloy. For example, a 10% hydrogen peroxide solution means that 10% of the total solution is hydrogen peroxide. The remaining 90% is water.

The solution of a percent mixture problem is based upon the equation $Ar = Q$, where A is the amount of solution or alloy, r is the percent of concentration, and Q is the quantity of a substance in the solution or alloy.

A chemist mixes an 11% acid solution with a 4% acid solution. How many milliliters of each solution should the chemist use to make a 700-milliliter solution which is 6% acid?

Strategy for Solving a Percent Mixture Problem

> For each solution, use the equation $Ar = Q$. Write a numerical or variable expression for the amount of solution, percent of concentration, and the quantity of the substance in the solution. The results can be recorded in a table.

The total amount of solution is 700 ml.   Amount of 11% solution: x
Amount of 4% solution: $700 - x$

	Amount of solution, A	·	*Percent of concentration, r*	=	*Quantity of substance, Q*
11% solution	x	·	0.11	=	$0.11x$
4% solution	$700 - x$	·	0.04	=	$0.04(700 - x)$
6% solution	700	·	0.06	=	$0.06(700)$

> Determine how the quantities of the substance in each solution are related. Use the fact that the sum of the quantities of the substances being mixed is equal to the quantity of the substance after mixing.

The sum of the quantities of the substances in the 11% solution and the 4% solution is equal to the quantity of the substance in the 6% solution.

$$0.11x + 0.04(700 - x) = 0.06(700)$$
$$0.11x + 28 - 0.04x = 42$$
$$0.07x + 28 = 42$$
$$0.07x = 14$$
$$x = 200$$

The amount of 4% solution is $700 - x$.
Replace x by 200 and evaluate.

$$700 - x = 700 - 200 = 500$$

The chemist should use 200 ml of the 11% solution and 500 ml of the 4% solution.

Example 3

How many grams of pure acid must be added to 60 g of an 8% acid solution to make a 20% acid solution?

Example 4

A butcher has some hamburger which is 22% fat and some which is 12% fat. How many pounds of each should be mixed to make 80 lb of hamburger which is 18% fat?

Strategy

- Grams of pure acid: x

	Amount	Percent	Quantity
Pure Acid (100%)	x	1.00	x
8%	60	0.08	0.08(60)
20%	$60 + x$	0.20	$0.20(60 + x)$

- The sum of the quantities before mixing equals the quantity after mixing.

$$x + 0.08(60) = 0.20(60 + x)$$

Your strategy

Solution

$$
\begin{aligned}
x + 0.08(60) &= 0.20(60 + x) \\
x + 4.8 &= 12 + 0.20x \\
0.8x + 4.8 &= 12 \\
0.8x &= 7.2 \\
x &= 9
\end{aligned}
$$

To make the 20% acid solution, 9 g of pure acid must be used.

Your solution

48 lb of the hamburger that is 22% fat

32 lb of the hamburger that is 12% fat

Solution on p. A15

Content and Format © 1991 HMCo.

2.6   EXERCISES

▶ **Objective A   *Application Problems***

1. Two investments earn an annual income of $1069. One investment is in a 7.2% tax-free annual simple interest account and the other investment is in a 9.8% simple interest CD. The total amount invested is $12,500. How much is invested in each account?
$6000 @ 7.2%, $6500 @ 9.8%

2. Two investments earn an annual income of $765. One investment earns an annual simple interest rate of 8.5% while the other investment earns an annual simple rate of 10.2%. The total amount invested is $8000. How much is invested in each account?
$3000 @ 8.5%, $5000 @ 10.2%

3. An investment club invested $5000 at an annual simple interest rate of 8.4%. How much additional money must be invested at an annual simple interest rate of 10.5% so that the total interest earned will be 9% of the total investment?
$2000

4. An investment of $4500 is made at an annual simple interest rate of 7.8%. How much additional money must be invested at an annual simple interest rate of 11% so that the total interest earned is 9% of the total investment?
$2700

5. An account executive deposited $42,000 into two simple interest accounts. On the tax-free account the annual simple interest rate is 7%, while on the money market fund the annual simple interest rate is 9.8%. How much should be invested in each account so that the interest earned by each account is the same?
$24,500 @ 7%, $17,500 @ 9.8%

6. An investment club invested $10,800 into two simple interest accounts. On one account the annual simple interest rate is 8.2%. On the other, the annual simple interest rate is 10.25%. How much should be invested in each account so that the interest earned by each account is the same?
$6000 @ 8.2%, $4800 @ 10.25%

7. A financial manager recommended an investment plan in which 30% of a client's investment be placed in a 7.5% annual simple interest tax-free account, 45% be placed in 9% high-grade bonds, and the remainder in an 11.5% high-risk investment. The total interest earned from the investments would be $2293.75. Find the total amount to be invested.
$25,000

8. The manager of a trust account invests 25% of a client's account in a money market fund which earns 8% annual simple interest, 40% in bonds which earn 10.5% annual simple interest, and the remainder in trust deeds which earn 13% annual simple interest. How much should be invested in each type of investment so that the total interest earned is $4300?
$10,000 @ 8%, $16,000 @ 10.5%, $14,000 @ 13%

9. An investment club invested $12,000 in two accounts. One investment earned 12.6% annual simple interest, while the other investment lost 5%. The total earnings from both investments were $104. Find the amount invested at 12.6%.
$4000

10. A total of $18,000 is invested in two accounts. One investment earned 11.2% annual simple interest while the other investment lost 4.7%. The total earnings from both investments were $1062. Find the amount invested at 11.2%.
$12,000

▶ **Objective B**   *Application Problems*

11. How many quarts of water must be added to 5 quarts of an 80% antifreeze solution to make a 50% antifreeze solution?
3 qt

12. How many milliliters of alcohol must be added to 200 ml of a 25% iodine solution to make a 10% iodine solution?
300 ml

13. A goldsmith mixed 10 g of a 50% gold alloy with 40 g of a 15% gold alloy. What is the percent concentration of the resulting alloy?
22%

14. A silversmith mixed 30 g of a 60% silver alloy with 70 g of a 20% silver alloy. What is the percent concentration of the resulting alloy?
32%

15. How many ounces of pure water must be added to 75 oz of an 8% salt solution to make a 5% salt solution?
45 oz

16. How many pounds of a 15% aluminum alloy must be mixed with 500 lb of a 22% aluminum alloy to make a 20% aluminum alloy?
200 lb

17. A hospital staff mixed 75% disinfectant solution with a 25% disinfectant solution. How many liters of each were used to make 20 L of a 40% disinfectant solution?
6 L of 75% solution, 14 L of 25% solution

18. A butcher has some hamburger which is 21% fat and some hamburger which is 15% fat. How many pounds of each should be mixed to make 84 lb of hamburger which is 17% fat?
28 lb of 21% fat, 56 lb of 15% fat

19. How much water must be evaporated from 8 gallons of an 8% salt solution in order to obtain a 12% solution?
$2\frac{2}{3}$ gal

20. How much water must be evaporated from 6 quarts of a 50% antifreeze solution to produce a 75% solution?
2 qt

21. A car radiator contains 12 qt of a 40% antifreeze solution. How many quarts will have to be replaced with pure antifreeze if the resulting solution is to be 60% antifreeze?
4 qt

22. A student mixed 40 ml of a 4% hydrogen peroxide solution with 25 ml of a 12% solution. Find the percent concentration of the resulting mixture. Round to the nearest tenth of a percent.
7.1%

Calculators and Computers

Solving a First-Degree Equation

The program SOLVE A FIRST-DEGREE EQUATION on the Math ACE Disk will allow you to practice solving the following three types of equations:

1. $ax + b = c$
2. $ax + b = cx + d$
3. Equations with parentheses

After you select the type of equation you want to practice, a problem will be displayed on the screen. Using paper and pencil, solve the problem. When you are ready, press the RETURN key and the complete solution will be displayed. Compare your solution with the displayed solution. All answers are rounded to the nearest hundredth.

When you finish a problem, you may continue practicing the type of problem you have selected, return to the main menu and select a different type, or quit the program.

Chapter Summary

Key Words

An *equation* expresses the equality of two mathematical expressions.

The simplest equation is an equation of the form *variable = constant*.

A *solution* or *root* of an equation is a replacement value for the variable which will make the equation true.

To *solve* an equation means to find its solutions.

An equation of the form $ax + b = c$, $a \neq 0$, is called a *first-degree equation*.

Consecutive integers are integers that follow one another in order.

A *compound inequality* is formed by joining two inequalities with a connective word such as "and" or "or".

The *absolute value* of a number is its distance from zero on the number line.

An *absolute value equation* contains an absolute value symbol.

A *value mixture problem* involves combining two ingredients that have different prices into a single blend.

Uniform motion means that the speed of an object does not change.

Essential Rules

The Addition Property of Equations	If $a = b$, then $a + c = b + c$.
The Multiplication Property of Equations	If $a = b$ and $c \neq 0$, then $ac = bc$.
Addition Property of Inequalities	If $a > b$, then $a + c > b + c$. If $a < b$, then $a + c < b + c$.

Multiplication Property of Inequalities	**Rule 1** If $a > b$ and $c > 0$, then $ac > bc$.
	         If $a < b$ and $c > 0$, then $ac < bc$.
	
	Rule 2 If $a > b$ and $c < 0$, then $ac < bc$.
	         If $a < b$ and $c < 0$, then $ac > bc$.

Value Mixture Equation	Value $=$ amount $\times$ unit cost V $=$ A $\times$ C
Uniform Motion Equation	Distance $=$ rate $\times$ time d $=$ r $\times$ t
Annual Simple Interest Equation	Simple Interest $=$ Principal $\times$ simple interest rate I $=$ P $\times$ r
Percent Mixture Equation	Quantity $=$ amount $\times$ percent concentration Q $=$ A $\times$ r

Chapter Review

SECTION 1

1. Solve: $x + 4 = -5$
-9

2. Solve: $-\frac{2}{3}x = \frac{4}{9}$
$-\frac{2}{3}$

3. Solve: $3t - 3 + 2t = 7t - 15$
6

4. Solve: $\frac{3}{5}x - 3 = 2x + 5$
$-\frac{40}{7}$

5. Solve: $2(a - 3) = 5(4 - 3a)$
$\frac{26}{17}$

6. Solve: $2x - (3 - 2x) = 4 - 3(4 - 2x)$
$\frac{5}{2}$

7. Solve: $\frac{1}{2}x - \frac{5}{8} = \frac{3}{4}x + \frac{3}{2}$
$-\frac{17}{2}$

8. Solve $\frac{2x - 3}{3} + 2 = \frac{2 - 3x}{5}$
$-\frac{9}{19}$

SECTION 2

9. Solve: $3x - 7 > -2$
$\left\{x \mid x > \frac{5}{3}\right\}$

10. Solve: $3x < 4$ and $x + 2 > -1$
$\left\{x \mid -3 < x < \frac{4}{3}\right\}$

11. Solve: $5x - 2 > 8$ or $3x + 2 < -4$
$\{x \mid x > 2 \text{ or } x < -2\}$

12. Solve: $3x - 2 > x - 4$ or $7x - 5 < 3x + 3$
The solution set is the set of real numbers.

13. A sales executive earns $800 per month plus 4% commission on the amount of sales. The executive's goal is to earn $3000 per month. What amount of sales will enable the executive to earn $3000 or more per month?
$55,000 or more

14. An average score of 80 to 90 in a psychology class receives a B grade. A student has grades of 92, 66, 72, and 88 on four tests. Find the range of scores on the fifth test that will give the student a B for the course.
$82 \le x \le 100$

SECTION 3

15. Solve: $|x - 4| - 8 = -3$
-1 and 9

16. Solve: $6 + |3x - 3| = 2$
No Solution

17. Solve: $|2x - 5| < 3$
$\{x|1 < x < 4\}$

18. Solve: $|4x - 5| \geq 3$
$\left\{x|x \geq 2 \text{ or } x \leq \frac{1}{2}\right\}$

19. The diameter of a bushing is 2.75 in. The bushing has a tolerance of 0.003 in. Find the lower and upper limits of the diameter of the bushing.
2.747 in., 2.753 in.

20. A doctor has prescribed 2 cc of medication for a patient. The tolerance is 0.25 cc. Find the lower and upper limits of the amount of medication to be given.
1.75 cc, 2.25 cc

SECTION 4

21. The sum of two integers is twenty. Five times the smaller integer is two more than twice the larger integer. Find the integers.
6 and 14

22. A coin collection contains 30 coins in nickels, dimes, and quarters. There are three more dimes than nickels. The value of the coins is $3.55. Find the number of quarters in the collection.
10 quarters

SECTION 5

23. A grocer mixed apple juice which cost $3.20 per gallon with 40 gallons of cranberry juice which cost $5.50 per gallon. How much apple juice was used to make cranapple juice to sell for $4.20 per gallon?
52 gal

24. Two planes are 1680 mi apart and traveling toward each other. One plane is traveling 80 mph faster than the other plane. The planes meet in 1.75 h. Find the speed of each plane.
520 mph, 440 mph

SECTION 6

25. Two investments earn an annual income of $635. One investment is earning an annual simple interest rate of 10.5% and the other investment is earning an annual simple interest rate of 6.4%. The total investment is $8000. Find the amount invested in each account.
$3000 at 10.5%, $5000 at 6.4%

26. An alloy containing 30% tin is mixed with an alloy containing 70% tin. How many pounds of each were used to make 500 lb of an alloy containing 40% tin?
375 lb of 30% tin, 125 lb of 70% tin

Chapter Test

1. Solve: $x - 2 = -4$

   -2 [2.1A]

2. Solve: $b + \frac{3}{4} = \frac{5}{8}$

   $-\frac{1}{8}$ [2.1A]

3. Solve: $-\frac{3}{4}y = -\frac{5}{8}$

   $\frac{5}{6}$ [2.1A]

4. Solve: $3x - 5 = 7$

   4 [2.1A]

5. Solve: $\frac{3}{4}y - 2 = 6$

   $\frac{32}{3}$ [2.1B]

6. Solve: $2x - 3 - 5x = 8 + 2x - 10$

   $-\frac{1}{5}$ [2.1B]

7. Solve: $2[a - (2 - 3a) - 4] = a - 5$

   1 [2.1C]

8. Solve: $\frac{2}{3}t - \frac{5}{6}t = 4$

   -24 [2.1B]

9. Solve: $\frac{2x + 1}{3} - \frac{3x + 4}{6} = \frac{5x - 9}{9}$

   $\frac{12}{7}$ [2.1C]

10. Solve: $3x - 2 \geq 6x + 7$

    $\{x | x \leq -3\}$ [2.2A]

11. Solve: $4 - 3(x + 2) < 2(2x + 3) - 1$

    $\{x | x > -1\}$ [2.2A]

12. Solve: $4x - 1 > 5$ or $2 - 3x < 8$

    $\{x | x > -2\}$ [2.2B]

13. Solve: $4 - 3x \geq 7$ and $2x + 3 \geq 7$

    $\varnothing$ [2.2B]

14. Solve: $|3 - 5x| = 12$

    3 and $-\frac{9}{5}$ [2.3A]

15. Solve: $2 - |2x - 5| = -7$

    7 and -2 [2.3A]

16. Solve: $|3x - 5| \leq 4$

    $\{x | \frac{1}{3} \leq x \leq 3\}$ [2.3B]

17. Solve: $|4x - 3| > 5$

    $\{x | x > 2$ or $x < -\frac{1}{2}\}$ [2.3B]

18. Agency A rents cars for $12 a day and 10¢ for every mile driven. Agency B rents cars for $24 per day with unlimited mileage. How many miles per day can you drive an agency A car if it is to cost you less than an agency B car?
Less than 120 mi   [2.2C]

19. A doctor prescribed 3 cc of medication for a patient. The tolerance is 0.10 cc. Find the lower and upper limits of the amounts of medication to be given.
2.90 cc, 3.10 cc   [2.3C]

20. The sum of two integers is fifteen. Eight times the smaller integer is one less than three times the larger integer. Find the integers.
4, 11 [2.4A]

21. A stamp collection contains 11¢, 15¢, and 24¢ stamps. There are twice as many 11¢ stamps as 15¢ stamps. There are 30 stamps in all with a value of $4.40. How many 24¢ stamps are in the collection?
6, 24¢ stamps   [2.4B]

22. A butcher combines 100 lb of hamburger which costs $1.60 per pound with 60 lb of hamburger which costs $3.20 per pound. Find the selling price of the hamburger mixture.
$2.20   [2.5A]

23. A jogger runs a distance at a speed of 8 mph and returns the same distance running at a speed of 6 mph. Find the total distance that the jogger ran if the total time running was one hour and forty-five minutes.
12 mi [2.5B]

24. An investment of $12,000 is deposited into two simple interest accounts. On one account the annual simple interest rate is 7.8%. On the other, the annual simple interest rate is 9%. The total interest earned for one year is $1020. How much was invested in each account?
$5000 @ 7.8%, $7000 @ 9%   [2.6A]

25. How many ounces of pure water must be added to 60 oz of an 8% salt solution to make a 3% salt solution?
100 oz   [2.6B]

Cumulative Review

1. Simplify: $-4 - (-3) - 8 + (-2)$
   -11 [1.1B]

2. Simplify: $-2^2 \cdot 3^3$
   -108 [1.1C]

3. Simplify: $4 - (2 - 5)^2 \div 3 + 2$
   3 [1.1C]

4. Simplify: $4 \div \dfrac{\frac{3}{8} - 1}{5} \cdot 2$
   -64 [1.2A]

5. Evaluate $2a^2 - (b - c)^2$ when $a = 2$, $b = 3$, and $c = -1$.
   -8 [1.3A]

6. Evaluate $\frac{a - b^2}{b - c}$ when $a = 2$, $b = -3$, and $c = 4$.
   1 [1.3A]

7. Identify the property that justifies the statement.
   $(2x + 3y) + 2 = (3y + 2x) + 2$
   Commutative Property of Addition [1.3B]

8. Translate and simplify "the sum of three times a number and six added to the product of three and a number."
   $(3n + 6) + 3n$; $6n + 6$ [1.3D]

9. Simplify: $3x - 2[x - 3(2 - 3x) + 5]$
   $-17x + 2$ [1.3C]

10. Simplify: $5[y - 2(3 - 2y) + 6]$
    $25y$ [1.3C]

11. Find $A \cap B$, given $A = \{-4, -2, 0, 2\}$ and $B = \{-4, 0, 4, 8\}$.
    $\{-4, 0\}$ [1.4A]

12. Graph the solution set of $\{x | x \le 3\} \cap \{x | x > -1\}$.

    [number line from -5 to 5] [1.4B]

13. Solve: $4 - 3x = -2$
    $x = 2$ [2.1A]

14. Solve: $-\frac{5}{6}b = -\frac{5}{12}$
    $b = \frac{1}{2}$ [2.1A]

15. Solve: $2x + 5 = 5x + 2$
    $x = 1$ [2.1B]

16. Solve: $\frac{5}{12}x - 3 = 7$
    $x = 24$ [2.1B]

17. Solve: $2[3 - 2(3 - 2x)] = 2(3 + x)$
    $x = 2$ [2.1C]

18. Solve: $3[2x - 3(4 - x)] = 2(1 - 2x)$
    $x = 2$ [2.1C]

19. Solve: $\frac{1}{2}y - \frac{2}{3}y + \frac{5}{12} = \frac{3}{4}y - \frac{1}{2}$
    $y = 1$ [2.1C]

20. Solve: $\frac{3x - 1}{4} - \frac{4x - 1}{12} = \frac{3 + 5x}{8}$
    $x = -\frac{13}{5}$ [2.1C]

21. Solve: $3 - 2(2x - 1) \ge 3(2x - 2) + 1$
$\{x | x \le 1\}$ [2.2A]

22. Solve: $3x + 2 \le 5$ and $x + 5 \ge 1$
$\{x | -4 \le x \le 1\}$ [2.2B]

23. Solve: $|3 - 2x| = 5$
-1 and 4   [2.3A]

24. Solve: $3 - |2x - 3| = -8$
7 and -4 [2.3A]

25. Solve: $|3x - 1| > 5$
$\{x | x < -\frac{4}{3}$ or $x > 2\}$ [2.3B]

26. Solve: $|2x - 4| < 8$
$\{x | -2 < x < 6\}$ [2.3B]

27. A bank offers two types of checking accounts. One account has a charge of $5 per month plus 4¢ for each check. The second account has a charge of $2 per month plus 10¢ for each check. How many checks can a customer who has the second type of account write if it is to cost the customer less than the first type of account?
Less than 50 checks   [2.2C]

28. Four times the sum of the first and third of three consecutive odd integers is one less than seven times the middle integer. Find the first integer.
-3 [2.4A]

29. A coin purse contains dimes and quarters. The number of dimes is five less than twice the number of quarters. The total value of the coins is $4.00. Find the number of dimes in the coin purse.
15 dimes   [2.4B]

30. A silversmith combined pure silver which cost $8.50 an ounce with 100 oz of a silver alloy which cost $4.00 an ounce. How many ounces of pure silver were used to make an alloy of silver selling for $6.00 an ounce?
80 oz   [2.5A]

31. Two planes are 1400 mi apart and traveling toward each other. One plane is traveling 120 mph faster than the other plane. The planes meet in 2.5 h. Find the speed of the slower plane.
220 mph   [2.5B]

32. An investment club has $15,000 invested in two accounts. One investment earned 8.5% annual simple interest while the other earned 13.5% annual simple interest. The amount of interest earned in one year was $1575. How much was invested in the 13.5% account?
$6000   [2.6A]

33. How many liters of a 12% acid solution must be mixed with 4 L of a 5% acid solution to make an 8% acid solution?
3 L   [2.6B]

3

Polynomials

OBJECTIVES

▶ To add or subtract polynomials
▶ To multiply monomials
▶ To divide monomials and simplify expressions containing integer exponents
▶ To write a number using scientific notation
▶ To solve application problems
▶ To multiply a polynomial by a monomial
▶ To multiply two polynomials
▶ To multiply polynomials that have special products
▶ To solve application problems
▶ To factor a monomial from a polynomial
▶ To factor by grouping
▶ To factor a trinomial of the form $x^2 + bx + c$
▶ To factor $ax^2 + bx + c$
▶ To factor the difference of two perfect squares or a perfect square trinomial
▶ To factor the sum or the difference of two cubes
▶ To factor a trinomial that is quadratic in form
▶ To factor completely
▶ To solve an equation by factoring

Origins of the Word Algebra

The word *algebra* has its origins in an Arabic book written around 825 A.D. titled *Hisab al-jabr w' almuqa-balah*, by al-Khowarizmi. The word *al-jabr*, which literally translated means reunion, was written as the word *algebra* in Latin translations of al-Khowarizmi's work and became synonymous with equations and the solutions of equations. It is interesting to note that an early meaning of the Spanish word *algebrista* was "bonesetter" or "reuniter of broken bones."

There is actually a second contribution to our language of mathematics by al-Khowarizmi. One of the translations of his work into Latin shortened his name to *Algoritmi*. A further modification of this word gives us our present word "algorithm." An algorithm is a procedure or set of instructions which are used to solve different types of problems. Computer scientists use algorithms when writing computer programs.

A further historical note is not about the word algebra, but about Omar Khayyam, a Persian who probably read al-Khowarizmi's work. Omar Khayyam is especially noted as a poet and the author of the *Rubiat*. However, he was also an excellent mathematician and astronomer and made many contributions to mathematics.

Operations on Polynomials

Objective A

To add or subtract polynomials

A **monomial** is a number, a variable, or a product of a number and variables.

The examples at the right are monomials. The **degree of a monomial** is the sum of the exponents of the variables.

x	degree 1 $(x = x^1)$
$3x^2$	degree 2
$4x^2y$	degree 3
$6x^3y^4z^2$	degree 9
x^n	degree n

In this chapter, the variable n is considered a positive integer when used as an exponent. The degree of a nonzero constant term is zero.

6　　degree 0

$\frac{1}{x}$ is not a monomial because a variable appears in the denominator.

A **polynomial** is a variable expression in which the terms are monomials.

A **binomial** is a polynomial of two terms.　　$5x^2y + 6x$

A **trinomial** is a polynomial of three terms.　　$3x^2 + 9xy - 5y$

Polynomials with more than three terms do not have special names.

The **degree of a polynomial** is the greatest of the degrees of any of its terms.

$$3x + 2 \quad\quad 3x^2 + 2x - 4 \quad\quad 4x^3y^2 + 6x^4 - y \quad\quad 3x^{2n} - 5x^n + 2$$
$$\text{degree 1} \quad\quad \text{degree 2} \quad\quad\quad \text{degree 5} \quad\quad\quad\quad \text{degree } 2n$$

The terms of a polynomial in one variable are usually arranged so that the exponents of the variable decrease from left to right. This is called **descending order.**

$$2x^2 - x + 8 \quad\quad 3y^5 - 3y^3 + y^2 - 12$$

For a polynomial in more than one variable, descending order may refer to any one of the variables.

The polynomial at the right is shown first in descending order of the x variable and then in descending order of the y variable.

$2x^2 + 3xy + 5y^2$

$5y^2 + 3xy + 2x^2$

The **additive inverse** of the polynomial $x^2 + 5x - 4$ is $-(x^2 + 5x - 4)$.

To simplify the additive inverse of a polynomial, change the sign of every term inside the parentheses.

$-(x^2 + 5x - 4) = -x^2 - 5x + 4$

Polynomials can be added by combining like terms. Either a vertical or horizontal format can be used.

Simplify $(3x^2 + 2x - 7) + (7x^3 - 3 + 4x^2)$. Use a horizontal format.

Use the Commutative and Associative Properties of Addition to rearrange and group like terms.
$$(3x^2 + 2x - 7) + (7x^3 - 3 + 4x^2)$$
$$7x^3 + (3x^2 + 4x^2) + 2x + (-7 - 3)$$

Combine like terms.
Write the polynomial in descending order.
$$7x^3 + 7x^2 + 2x - 10$$

Simplify $(4x^2 + 5x - 3) + (7x^3 - 7x + 1) + (2x - 3x^2 + 4x^3 + 1)$.
Use a vertical format.

Arrange the terms of each polynomial in descending order with like terms in the same column.

Add the terms in each column.

$$\begin{array}{r} 4x^2 + 5x - 3 \\ 7x^3 \qquad - 7x + 1 \\ \underline{4x^3 - 3x^2 + 2x + 1} \\ 11x^3 + \; x^2 \qquad - 1 \end{array}$$

To subtract two polynomials, add the additive inverse of the second polynomial to the first.

Simplify $(3x^2 - 7xy + y^2) - (-4x^2 + 7xy - 3y^2)$. Use a horizontal format.

Rewrite subtraction as the addition of the additive inverse.
$$(3x^2 - 7xy + y^2) - (-4x^2 + 7xy - 3y^2)$$
$$(3x^2 - 7xy + y^2) + (4x^2 - 7xy + 3y^2)$$

Combine like terms.
$$7x^2 - 14xy + 4y^2$$

Simplify $(6x^2 - 3x + 7) - (3x^2 - 5x + 12)$. Use a vertical format.

Rewrite subtraction as the addition of the additive inverse.
$$(6x^2 - 3x + 7) + (-3x^2 + 5x - 12)$$

Arrange the terms of each polynomial in descending order with like terms in the same column.
$$\begin{array}{r} 6x^2 - 3x + \; 7 \\ \underline{-3x^2 + 5x - 12} \\ 3x^2 + 2x - \; 5 \end{array}$$

Combine the terms in each column.

Example 1

Simplify:
$(4x^2 - 3xy + 7y^2) + (-3x^2 + 7xy + y^2)$
Use a vertical format.

Solution

$$\begin{array}{r} 4x^2 - 3xy + 7y^2 \\ \underline{-3x^2 + 7xy + \; y^2} \\ x^2 + 4xy + 8y^2 \end{array}$$

Example 2

Simplify:
$(-3x^2 - 4x + 9) + (-5x^2 - 7x + 1)$
Use a vertical format.

Your solution

$$-8x^2 - 11x + 10$$

Solution on p. A15

Example 3

Simplify: $(3x^2 - 2x + 4) - (7x^2 + 3x - 12)$
Use a vertical format.

Add the additive inverse of
$(7x^2 + 3x - 12)$ to $(3x^2 - 2x + 4)$.

$$\begin{array}{r} 3x^2 - 2x + 4 \\ -7x^2 - 3x + 12 \\ \hline -4x^2 - 5x + 16 \end{array}$$

Example 4

Simplify: $(-5x^2 + 2x - 3) - (6x^2 + 3x - 7)$
Use a vertical format.

Your solution

$-11x^2 - x + 4$

Example 5

Simplify:
$(2x^{2n} - 3x^n + 7) - (3x^{2n} + 3x^n + 5)$
Use a horizontal format.

Solution

$(2x^{2n} - 3x^n + 7) - (3x^{2n} + 3x^n + 5)$
$(2x^{2n} - 3x^n + 7) + (-3x^{2n} - 3x^n - 5)$
$-x^{2n} - 6x^n + 2$

Example 6

Simplify:
$(5x^{2n} - 3x^n - 7) - (-2x^{2n} - 5x^n + 8)$
Use a horizontal format.

Your solution

$7x^{2n} + 2x^n - 15$

Solutions on p. A15

Objective B   **To multiply monomials**

3

In an exponential expression, the exponent indicates the number of times the base occurs as a factor.

The product of exponential expressions with the *same* base can be simplified by adding the exponents.

$$x^3 \cdot x^4 = (x \cdot x \cdot x) \cdot (x \cdot x \cdot x \cdot x) = x^7$$
$$x^3 \cdot x^4 = x^{3+4} = x^7$$

Rule for Multiplying Exponential Expressions

If m and n are integers, then $x^m \cdot x^n = x^{m+n}$.

Simplify: $(5a^2b^4)(2ab^5)$

Use the Commutative and Associative Properties to re-arrange and group factors.

$$(5a^2b^4)(2ab^5) = (5 \cdot 2)(a^2 \cdot a)(b^4 \cdot b^5)$$

$$= 10a^{2+1}b^{4+5}$$ Do this step mentally.

Multiply variables with like bases by adding the exponents.

$$= 10a^3b^9$$

A power of an exponential expression can be simplied by multiplying the exponents.

$$(x^4)^3 = x^4 \cdot x^4 \cdot x^4 = x^{4+4+4} = x^{12}$$
$$(x^4)^3 = x^{4 \cdot 3} = x^{12}$$

Rule for Simplifying Powers of Exponential Expressions

If m and n are integers, then $(x^m)^n = x^{m \cdot n}$.

Simplify: $(x^4)^5$

Multiply the exponents.

$$(x^4)^5 = x^{4 \cdot 5}$$ Do this step mentally.

$$= x^{20}$$

A power of the product of exponential expressions can be simplied by multiplying each exponent inside the parentheses by the exponent outside the parentheses.

$$(x^3 \cdot y^4)^2 = (x^3 \cdot y^4)(x^3 \cdot y^4) = x^3 \cdot x^3 \cdot y^4 \cdot y^4 = x^6y^8$$
$$(x^3 \cdot y^4)^2 = x^{3 \cdot 2} \cdot y^{4 \cdot 2} = x^6y^8$$

Rule for Simplifying Powers of Products

If m, n, and p are integers, then $(x^m \cdot y^n)^p = x^{m \cdot p}y^{n \cdot p}$.

Simplify: $(2a^3b^4)^3$

Multiply each exponent inside the parentheses by the exponent outside the parentheses.

$$(2a^3b^4)^3 = 2^{1 \cdot 3}a^{3 \cdot 3}b^{4 \cdot 3}$$ Do this step mentally.

$$= 2^3a^9b^{12}$$
$$= 8a^9b^{12}$$

Example 7

Simplify: $(-3ab^2)(-10a^2b^4)(a^2b)$

Solution

$(-3ab^2)(-10a^2b^4)(a^2b) = 30a^5b^7$

Example 8

Simplify: $(7xy^3)(-5x^2y^2)(-xy^2)$

Your solution

$35x^4y^7$

Example 9

Simplify: $(2xy^2)(-3xy^4)^3$

Solution

$$\begin{aligned}(2xy^2)(-3xy^4)^3 &= (2xy^2)[(-3)^3x^3y^{12}]\\ &= (2xy^2)(-27x^3y^{12})\\ &= -54x^4y^{14}\end{aligned}$$

Example 10

Simplify: $(-3a^2b^4)(-2ab^3)^4$

Your solution

$-48a^6b^{16}$

Example 11

Simplify: $(x^{n+2})^5$

Solution

$(x^{n+2})^5 = x^{5n+10}$

Example 12

Simplify: $(y^{n-3})^2$

Your solution

y^{2n-6}

Example 13

Simplify: $[(2xy^2)^2]^3$

Solution

$$\begin{aligned}[(2xy^2)^2]^3 &= (2xy^2)^{2\cdot3} = (2xy^2)^6\\ &= 2^6x^6y^{12} = 64x^6y^{12}\end{aligned}$$

Example 14

Simplify: $[(ab^3)^3]^4$

Your solution

$a^{12}b^{36}$

Example 15

Simplify: $(2ab)(3a)^2 + 5a(2a^2b)$

Solution

$$\begin{aligned}(2ab)(3a)^2 + 5a(2a^2b) &= (2ab)(3^2a^2) + 10a^3b\\ &= (2ab)(9a^2) + 10a^3b\\ &= 18a^3b + 10a^3b\\ &= 28a^3b\end{aligned}$$

Example 16

Simplify: $6a^2(2a) + 3a(2a^2)$

Your solution

$18a^3$

Content and Format © 1991 HMCo.

Solutions on pp. A15–A16

| **Objective C** | **To divide monomials and simplify expressions containing integer exponents** |

The quotient of two exponential expressions with the same base can be simplified by writing each expression in factored form, dividing by the common factors, and then writing the result with an exponent.

$$\frac{a^6}{a^2} = \frac{\overset{1}{\cancel{a}} \cdot \overset{1}{\cancel{a}} \cdot a \cdot a \cdot a \cdot a}{\underset{1}{\cancel{a}} \cdot \underset{1}{\cancel{a}}} = a^4$$

Note that subtracting the exponents gives the same result.

$$\frac{a^6}{a^2} = a^{6-2} = a^4$$

To divide two monomials with the same base, subtract the exponents of the like bases.

Simplify: $\frac{a^6 b^2}{a^3 b}$

Subtract the exponents of the like bases.

$$\frac{a^6 b^2}{a^3 b} = a^{6-3} b^{2-1} = a^3 b$$

Simplify: $\frac{r^8 s^6}{r^7 s}$

Subtract the exponents of the like bases.

$$\frac{r^8 s^6}{r^7 s} = r^{8-7} s^{6-1} = r s^5$$

Recall that for any number $a \neq 0$, $\frac{a}{a} = 1$. This property is true for exponential expressions as well. For example, for $a \neq 0$, $\frac{a^4}{a^4} = 1$.

This expression also can be simplified by subtracting the exponents of the like bases.

$$\frac{a^4}{a^4} = a^{4-4} = a^0$$

Because $\frac{a^4}{a^4} = 1$ and $\frac{a^4}{a^4} = a^{4-4} = a^0$, the definition of zero as an exponent is given as follows.

Definition of Zero as an Exponent

If $a \neq 0$, then $a^0 = 1$.

Note in this definition that $a \neq 0$. The expression 0^0 is not defined.

Simplify: $(2a - b)^0$, $2a - b \neq 0$.

Any nonzero expression to the zero power is 1.

$$(2a - b)^0 = 1$$

Simplify: $-(4a^3 b^4)^0$, $a \neq 0$, $b \neq 0$

Any nonzero expression to the zero power is 1. Because the negative sign is outside the parenthesis, the answer is -1.

$$-(4a^3 b^4)^0 = -(1) = -1$$

Examine the quotient $\frac{a^4}{a^7}$.

Write the numerator and denominator in factored form. Divide by the common factors.

$$\frac{a^4}{a^7} = \frac{\overset{1}{\cancel{a}} \cdot \overset{1}{\cancel{a}} \cdot \overset{1}{\cancel{a}} \cdot \overset{1}{\cancel{a}}}{\underset{1}{\cancel{a}} \cdot \underset{1}{\cancel{a}} \cdot \underset{1}{\cancel{a}} \cdot \underset{1}{\cancel{a}} \cdot a \cdot a \cdot a} = \frac{1}{a^3}$$

Now simplify the same expression by subtracting the exponents on the like bases. $\dfrac{a^4}{a^7} = a^{4-7} = a^{-3}$

Since $\dfrac{a^4}{a^7} = \dfrac{1}{a^3}$ and $\dfrac{a^4}{a^7} = a^{-3}$, the definition of a negative exponent is given as follows.

Definition of Negative Exponent

If n is a positive integer and $a \neq 0$, then
$$a^{-n} = \frac{1}{a^n} \quad \text{and} \quad a^n = \frac{1}{a^{-n}}.$$

Write 4^{-3} with a positive exponent then evaluate.

Write the expression with a positive exponent. $4^{-3} = \dfrac{1}{4^3}$

Evaluate. $= \dfrac{1}{64}$

Now that negative exponents have been defined, the rule for dividing exponential expressions can be stated.

Rule for Dividing Exponential Expressions

If m and n are integers and $a \neq 0$, then $\dfrac{a^m}{a^n} = a^{m-n}$.

Write $\dfrac{2^{-3}}{2^2}$ with a positive exponent and then evaluate.

Write the expression with a positive exponent. $\dfrac{2^{-3}}{2^2} = 2^{-3-2} = 2^{-5} = \dfrac{1}{2^5}$

Evaluate. $= \dfrac{1}{32}$

An exponential expression is in simplest form when it only contains positive exponents.

Simplify: $\dfrac{3x^6 y^{-6}}{6xy^2}$

Divide variables with like bases by subtracting the exponents. Write the result with positive exponents.
$$\frac{3x^6 y^{-6}}{6xy^2} = \frac{x^{6-1}y^{-6-2}}{2}$$
$$= \frac{x^5 y^{-8}}{2} = \frac{x^5}{2y^8}$$

Simplify: $\dfrac{ab^5}{a^{-4}b^6}$

Divide variables with like bases by subtracting the exponents.
$$\frac{ab^5}{a^{-4}b^6} = a^{1-(-4)}b^{5-6} = a^5 b^{-1}$$
$$= \frac{a^5}{b}$$

The rules for multiplying exponential expressions and powers of exponential expressions are true for all integers. These rules are stated here.

Rules of Exponents

If m, n, and p are integers, then

$$x^m \cdot x^n = x^{m+n} \qquad (x^m)^n = x^{mn} \qquad (x^m y^n)^p = x^{mp} y^{np}$$

$$\frac{x^m}{x^n} = x^{m-n}, \, x \neq 0 \qquad x^0 = 1, \, x \neq 0 \qquad \left(\frac{x^m}{y^n}\right)^p = \frac{x^{mp}}{y^{np}}, \, y \neq 0$$

Simplify: $(x^{-2}y)^{-1}(2x^2y^{-3})^2$

Use the Rules of Exponents to simplify powers.

$$(x^{-2}y)^{-1}(2x^2y^{-3})^2 = (x^2y^{-1})(2^2x^4y^{-6})$$

Use the Rules of Exponents to add exponents of like bases.

$$= 4x^6y^{-7}$$

Write the result without a negative exponent.

$$= \frac{4x^6}{y^7}$$

Simplify: $\frac{(3x^{-3}y^2)^{-2}}{(x^3y^{-2})^{-3}}$

Use the Rules of Exponents.

$$\frac{(3x^{-3}y^2)^{-2}}{(x^3y^{-2})^{-3}} = \frac{3^{-2}x^6y^{-4}}{x^{-9}y^6}$$

$$= 3^{-2}x^{15}y^{-10}$$

$$= \frac{x^{15}}{3^2y^{10}} = \frac{x^{15}}{9y^{10}}$$

Simplify: $x^{-1}y + xy^{-1}$

Use the Rule of Negative Exponents to rewrite the expression without negative exponents.

$$x^{-1}y + xy^{-1} = \frac{y}{x} + \frac{x}{y}$$

Write each fraction in terms of the LCM of the denominators.

$$= \frac{y^2}{xy} + \frac{x^2}{xy}$$

Add the two fractions.

$$= \frac{y^2 + x^2}{xy}$$

Example 17

Write $\frac{4^{-3}}{4^{-2}}$ with a positive exponent. Then evaluate.

Solution

$\frac{4^{-3}}{4^{-2}} = 4^{-3-(-2)} = 4^{-1} = \frac{1}{4}$

Example 19

Simplify: $(a^3b^{-4})^2(3a^{-2}b^{-3})^{-1}$

Solution

$(a^3b^{-4})^2(3a^{-2}b^{-3})^{-1} = (a^6b^{-8})(3^{-1}a^2b^3)$

$= 3^{-1}a^8b^{-5}$

$= \frac{a^8}{3b^5}$

Example 18

Write $\frac{3^2}{3^{-3}}$ with a positive exponent. Then evaluate.

Your solution

243

Example 20

Simplify: $(4r^{-2}t^{-1})^{-2}(r^4t^{-1})^3$

Solution

$\frac{r^{16}}{16t}$

Solutions on p. A16

Example 21

Simplify: $\dfrac{-28x^6z^{-3}}{42x^{-1}z^4}$

Solution

$\dfrac{-28x^6z^{-3}}{42x^{-1}z^4} = -\dfrac{14 \cdot 2x^{6-(-1)}z^{-3-4}}{14 \cdot 3} = -\dfrac{2x^7}{3z^7}$

Example 22

Simplify: $\dfrac{20r^{-2}t^{-5}}{-16r^{-3}s^{-2}}$

Your solution

$-\dfrac{5rs^2}{4t^5}$

Example 23

Simplify: $\dfrac{(3a^{-1}b^4)^{-3}}{(6^{-1}a^{-3}b^{-4})^3}$

Solution

$\dfrac{(3a^{-1}b^4)^{-3}}{(6^{-1}a^{-3}b^{-4})^3} = \dfrac{3^{-3}a^3b^{-12}}{6^{-3}a^{-9}b^{-12}} = 3^{-3} \cdot 6^3a^{12}b^0$

$= \dfrac{6^3a^{12}}{3^3} = \dfrac{216a^{12}}{27} = 8a^{12}$

Example 24

Simplify: $\dfrac{(9u^{-6}v^4)^{-1}}{(6u^{-3}v^{-2})^{-2}}$

Your solution

$\dfrac{4}{v^8}$

Example 25

Simplify: $\left[-\dfrac{8r^{-1}t^2}{6r^4t^{-2}} \right]^{-3}$

Solution

$\left[-\dfrac{8r^{-1}t^2}{6r^4t^{-2}} \right]^{-3} = \left[-\dfrac{4r^{-5}t^4}{3} \right]^{-3}$

$= -\dfrac{4^{-3}r^{15}t^{-12}}{3^{-3}} = -\dfrac{27r^{15}}{64t^{12}}$

Example 26

Simplify: $\left[\dfrac{4a^4b^{-2}}{8a^5b^{-1}} \right]^{-2}$

Your solution

$4a^2b^2$

Example 27

Simplify: $\left(\dfrac{-3y^4}{4x^6} \right)^3 \left(\dfrac{2x}{y^2} \right)^4$

Solution

$\left(\dfrac{-3y^4}{4x^6} \right)^3 \left(\dfrac{2x}{y^2} \right)^4 = \left(\dfrac{(-3)^3y^{12}}{4^3x^{18}} \right)\left(\dfrac{2^4x^4}{y^8} \right)$

$= \left(\dfrac{-27y^{12}}{64x^{18}} \right)\left(\dfrac{16x^4}{y^8} \right) = -\dfrac{27 \cdot 16x^4y^{12}}{64x^{18}y^8}$

$= -\dfrac{27y^4}{4x^{14}}$

Example 28

Simplify: $\left(\dfrac{4a^2}{5b^3} \right)^3 \left(\dfrac{-10a^3}{2b^8} \right)^2$

Your solution

$\dfrac{64a^{12}}{5b^{25}}$

Example 29

Simplify: $\dfrac{x^{4n}}{x^{2n}}$

Solution

$\dfrac{x^{4n}}{x^{2n}} = x^{4n-2n} = x^{2n}$

Example 30

Simplify: $\dfrac{y^{4n}}{y^{3n}}$

Your solution

y^{n}

Example 31

Simplify: $\dfrac{x^{4n-2}}{x^{2n-5}}$

Solution

$\dfrac{x^{4n-2}}{x^{2n-5}} = x^{4n-2-(2n-5)}$
$= x^{4n-2-2n+5} = x^{2n+3}$

Example 32

Simplify: $\dfrac{a^{2n+1}}{a^{n+3}}$

Your solution

a^{n-2}

Solutions on p. A16

| Objective D | **To write a number using scientific notation** |

Integer exponents are used to represent the very large and very small numbers encountered in the fields of science and engineering. For example, the mass of the electron is 0.00000000000000000000000000009 g. Numbers such as this one are difficult to read and write, so a more convenient system for writing such numbers has been developed. It is called **scientific notation.**

To express a number in scientific notation, write the number as the product of a number between 1 and 10 and a power of 10. The form for scientific notation is **a × 10ⁿ**, where $1 \le a < 10$.

For numbers greater than 10, move the decimal point to the right of the first digit. The exponent n is positive and equal to the number of places the decimal point has been moved.

$$965{,}000 \quad\quad = 9.65 \times 10^{5}$$
$$3{,}600{,}000 \quad\quad = 3.6 \times 10^{6}$$
$$92{,}000{,}000{,}000 = 9.2 \times 10^{10}$$

For numbers less than 1, move the decimal point to the right of the first non-zero digit. The exponent n is negative. The absolute value of the exponent is equal to the number of places the decimal point has been moved.

$$0.0002 \quad\quad = 2 \times 10^{-4}$$
$$0.0000000974 \quad = 9.74 \times 10^{-8}$$
$$0.000000000086 = 8.6 \times 10^{-11}$$

Converting a number written in scientific notation to decimal notation requires moving the decimal point.

When the exponent is positive, move the decimal point to the right the same number of places as the exponent.

$1.32 \times 10^4 = 13{,}200$

$1.4 \times 10^8\ = 140{,}000{,}000$

When the exponent is negative, move the decimal point to the left the same number of places as the absolute value of the exponent.

$1.32 \times 10^{-2} = 0.0132$

$1.4 \times 10^{-4}\ = 0.00014$

Numerical calculations involving numbers which have more digits than the hand-held calculator is able to handle can be performed using scientific notation.

Simplify: $\dfrac{220{,}000 \times 0.000000092}{0.0000011}$

Write the numbers in scientific notation.

$$\frac{220{,}000 \times 0.000000092}{0.0000011} = \frac{2.2 \times 10^5 \times 9.2 \times 10^{-8}}{1.1 \times 10^{-6}}$$

Simplify.

$$= \frac{(2.2)(9.2) \times 10^{5+(-8)-(-6)}}{1.1}$$

$$= 18.4 \times 10^3 = 18{,}400$$

Example 33   Write 0.000041 in scientific notation.

Solution $0.000041 = 4.1 \times 10^{-5}$

Example 34   Write 942,000,000 in scientific notation.

Your solution 9.42×10^8

Example 35   Write 3.3×10^7 in decimal notation.

Solution $3.3 \times 10^7 = 33{,}000{,}000$

Example 36   Write 2.7×10^{-5} in decimal notation.

Your solution 0.000027

Example 37   Simplify:
$\dfrac{2{,}400{,}000{,}000 \times 0.0000063}{0.00009 \times 480}$

Solution $\dfrac{2{,}400{,}000{,}000 \times 0.0000063}{0.00009 \times 480}$

$$= \frac{2.4 \times 10^9 \times 6.3 \times 10^{-6}}{9 \times 10^{-5} \times 4.8 \times 10^2}$$

$$= \frac{(2.4)(6.3) \times 10^{9+(-6)-(-5)-2}}{(9)(4.8)}$$

$$= 0.35 \times 10^6 = 350{,}000$$

Example 38   Simplify:
$\dfrac{5{,}600{,}000 \times 0.000000081}{900 \times 0.000000028}$

Your solution $18{,}000$

Solutions on p. A16

Objective E   To solve application problems

3

Example 39

How many miles does light travel in one day? The speed of light is 186,000 mi/s. Write the answer in scientific notation.

Strategy

To find the distance traveled:

- Write the speed of light in scientific notation.
- Write the number of seconds in one day in scientific notation.
- Use the equation $d = rt$, where r is the speed of light and t is the number of seconds in one day.

Solution

$186,000 = 1.86 \times 10^5$
$24 \cdot 60 \cdot 60 = 86,400 = 8.64 \times 10^4$
$d = rt$
$d = (1.86 \times 10^5)(8.64 \times 10^4)$
$d = 1.86 \times 8.64 \times 10^9$
$d = 16.0704 \times 10^9$
$d = 1.60704 \times 10^{10}$

Light travels 1.60704×10^{10} mi in one day.

Example 40

A computer can do an arithmetic operation in 1×10^{-7} s. How many arithmetic operations can the computer perform in one minute? Write the answer in scientific notation.

Your strategy

Your solution

6×10^8 operations

Solution on p. A16

Content and Format © 1991 HMCo.

| 3.1 | **EXERCISES** |

▶ **Objective A**

Simplify: Use a vertical format.

1. $(5x^2 + 2x - 7) + (x^2 - 8x + 12)$
$6x^2 - 6x + 5$

2. $(3x^2 - 2x + 7) + (-3x^2 + 2x - 12)$
-5

3. $(x^2 - 3xy + y^2) + (2x^2 - 3y^2)$
$3x^2 - 3xy - 2y^2$

4. $(3x^2 + 2y^2) + (-5x^2 + 2xy - 3y^2)$
$-2x^2 + 2xy - y^2$

5. $(x^2 - 3x + 8) - (2x^2 - 3x + 7)$
$-x^2 + 1$

6. $(2x^2 + 3x - 7) - (5x^2 - 8x - 2)$
$-3x^2 + 11x - 5$

7. $(x^{2n} + 7x^n - 3) + (-x^{2n} + 2x^n + 8)$
$9x^n + 5$

8. $(2x^{2n} - x^n - 1) + (5x^{2n} + 7x^n + 1)$
$7x^{2n} + 6x^n$

Simplify: Use a horizontal format.

9. $(3y^3 - 7y) + (2y^2 - 8y + 2)$
$3y^3 + 2y^2 - 15y + 2$

10. $(-2y^2 - 4y - 12) + (5y^2 - 5y)$
$3y^2 - 9y - 12$

11. $(2a^2 - 3a - 7) - (-5a^2 - 2a - 9)$
$7a^2 - a + 2$

12. $(3a^2 - 9a) - (-5a^2 + 7a - 6)$
$8a^2 - 16a + 6$

13. $(3x^4 - 3x^3 - x^2) + (3x^3 - 7x^2 + 2x)$
$3x^4 - 8x^2 + 2x$

14. $(3x^4 - 2x + 1) + (3x^3 - 5x - 8)$
$3x^4 + 3x^3 - 7x - 7$

15. $(3a^3 - 5a^2 - 6) + (-3a^3 - 6a + 2)$
$-5a^2 - 6a - 4$

16. $(6a^3 - 2a^2 - 12) + (6a^2 + 3a + 9)$
$6a^3 + 4a^2 + 3a - 3$

17. $(b^{2n} - b^n - 3) - (2b^{2n} - 3b^n + 4)$
$-b^{2n} + 2b^n - 7$

18. $(x^{2n} - x^n + 2) - (3x^{2n} - x^n + 5)$
$-2x^{2n} - 3$

Simplify: Use a horizontal format.

19. $(4x^2 - 3x - 9) - (-2x^3 - 3x^2 + 6x) - (5x^3 - 3x^2 + 9x + 2)$
$-3x^3 + 10x^2 - 18x - 11$

20. $(2a^3 - 2a^2b + 2ab^2 + 3b^3) - (4a^2b - 3ab^2 - 5b^3) + (3a^3 - 2a^2b + 5ab^2 - b^3)$
$5a^3 - 8a^2b + 10ab^2 + 7b^3$

▶ **Objective B**

Simplify:

21. $(ab^3)(a^3b)$
a^4b^4

22. $(-2ab^4)(-3a^2b^4)$
$6a^3b^8$

23. $(9xy^2)(-2x^2y^2)$
$-18x^3y^4$

24. $(x^2y)^2$
x^4y^2

25. $(x^2y^4)^4$
x^8y^{16}

26. $(-2ab^2)^3$
$-8a^3b^6$

27. $(-3x^2y^3)^4$
$81x^8y^{12}$

28. $(2^2a^2b^3)^3$
$64a^6b^9$

29. $(3^3a^5b^3)^2$
$729a^{10}b^6$

30. $(xy)(x^2y)^4$
x^9y^5

31. $(x^2y^2)(xy^3)^3$
x^5y^{11}

32. $[(2x)^4]^2$
$256x^8$

33. $[(3x)^3]^2$
$729x^6$

34. $[(x^2y)^4]^5$
$x^{40}y^{20}$

35. $[(ab)^3]^6$
$a^{18}b^{18}$

36. $[(2ab)^3]^2$
$64a^6b^6$

37. $[(2xy)^3]^4$
$4096x^{12}y^{12}$

38. $[(3x^2y^3)^2]^2$
$81x^8y^{12}$

39. $[(2a^4b^3)^3]^2$
$64a^{24}b^{18}$

40. $y^n \cdot y^{2n}$
y^{3n}

41. $x^n \cdot x^{n+1}$
x^{2n+1}

42. $y^{2n} \cdot y^{4n+1}$
y^{6n+1}

43. $y^{3n} \cdot y^{3n-2}$
y^{6n-2}

44. $(a^n)^{2n}$
a^{2n^2}

45. $(a^{n-3})^{2n}$
a^{2n^2-6n}

46. $(y^{2n-1})^3$
y^{6n-3}

47. $(x^{3n+2})^5$
x^{15n+10}

48. $(b^{2n-1})^n$
b^{2n^2-n}

49. $(2xy)(-3x^2yz)(x^2y^3z^3)$
$-6x^5y^5z^4$

50. $(x^2z^4)(2xyz^4)(-3x^3y^2)$
$-6x^6y^3z^8$

51. $(3b^5)(2ab^2)(-2ab^2c^2)$
$-12a^2b^9c^2$

Simplify:

52. $(-c^3)(-2a^2bc)(3a^2b)$
$6a^4b^2c^4$

53. $(-2x^2y^3z)(3x^2yz^4)$
$-6x^4y^4z^5$

54. $(2a^2b)^3(-3ab^4)^2$
$72a^8b^{11}$

55. $(-3ab^3)^3(-2^2a^2b)^2$
$-432a^7b^{11}$

56. $(4ab)^2(-2ab^2c^3)^3$
$-128a^5b^8c^9$

57. $(-2ab^2)(-3a^4b^5)^3$
$54a^{13}b^{17}$

▶ **Objective C**

Simplify:

58. 2^{-3}
$\frac{1}{8}$

59. $\frac{1}{3^{-5}}$
243

60. $\frac{1}{x^{-4}}$
x^4

61. $\frac{1}{y^{-3}}$
y^3

62. $\frac{2x^{-2}}{y^4}$
$\frac{2}{x^2y^4}$

63. $\frac{a^3}{4b^{-2}}$
$\frac{a^3b^2}{4}$

64. $x^{-3}y$
$\frac{y}{x^3}$

65. xy^{-4}
$\frac{x}{y^4}$

66. $-5x^0$
-5

67. $\frac{1}{2x^0}$
$\frac{1}{2}$

68. $\frac{(2x)^0}{-2^3}$
$-\frac{1}{8}$

69. $\frac{-3^{-2}}{(2y)^0}$
$-\frac{1}{9}$

70. $(3x^{-2})^2$
$\frac{9}{x^4}$

71. $(5x^2)^{-3}$
$\frac{1}{125x^6}$

72. $x^{-4}x^4$
1

73. $x^{-3}x^{-5}$
$\frac{1}{x^8}$

74. $a^{-2}\cdot a^4$
a^2

75. $a^{-5}\cdot a^7$
a^2

76. $\frac{x^{-3}}{x^2}$
$\frac{1}{x^5}$

77. $\frac{x^4}{x^{-5}}$
x^9

78. $\frac{y^{-7}}{y^{-8}}$
y

79. $\frac{y^{-2}}{y^6}$
$\frac{1}{y^8}$

80. $(x^2y^{-4})^2$
$\frac{x^4}{y^8}$

81. $(x^3y^5)^{-2}$
$\frac{1}{x^6y^{10}}$

82. $\frac{x^{-2}y^{-11}}{xy^{-2}}$
$\frac{1}{x^3y^9}$

83. $\frac{x^4y^3}{x^{-1}y^{-2}}$
x^5y^5

84. $\frac{a^{-1}b^{-3}}{a^4b^{-5}}$
$\frac{b^2}{a^5}$

85. $\frac{a^6b^{-4}}{a^{-2}b^5}$
$\frac{a^8}{b^9}$

86. $(2a^{-1})^{-2}(2a^{-1})^4$
$\frac{4}{a^2}$

87. $(3a)^{-3}(9a^{-1})^{-2}$
$\frac{1}{2187a}$

88. $(x^{-2}y)^2(xy)^{-2}$
$\frac{1}{x^6}$

89. $(x^{-1}y^2)^{-3}(x^2y^{-4})^{-3}$
$\frac{y^6}{x^3}$

Simplify:

90. $\dfrac{50b^{10}}{70b^5}$

$\dfrac{5b^5}{7}$

91. $\dfrac{x^3y^6}{x^6y^2}$

$\dfrac{y^4}{x^3}$

92. $\dfrac{x^{17}y^5}{-x^7y^{10}}$

$-\dfrac{x^{10}}{y^5}$

93. $\dfrac{-6x^2y}{12x^4y}$

$-\dfrac{1}{2x^2}$

94. $\dfrac{-2x^4y^2}{-6x^4y^4}$

$\dfrac{1}{3y^2}$

95. $\dfrac{25a^2b^{12}}{10a^5b^7}$

$\dfrac{5b^5}{2a^3}$

96. $\dfrac{-48ab^{10}}{32a^4b^3}$

$-\dfrac{3b^7}{2a^3}$

97. $\dfrac{a^2b^3c^7}{a^6bc^5}$

$\dfrac{b^2c^2}{a^4}$

98. $\dfrac{a^5b^6c}{a^9b^2c^4}$

$\dfrac{b^4}{a^4c^3}$

99. $\dfrac{(2x^2y)^2}{8x^2y^3}$

$\dfrac{x^2}{2y}$

100. $\dfrac{(3xy^2)^2}{9x^3y^4}$

$\dfrac{1}{x}$

101. $\dfrac{25x^6y^7}{(5x^2y^3)^2}$

x^2y

102. $\dfrac{2x^2y^4}{(3xy^2)^3}$

$\dfrac{2}{27xy^2}$

103. $\dfrac{-3ab^2}{(9a^2b^4)^3}$

$-\dfrac{1}{243a^5b^{10}}$

104. $\left(\dfrac{-12a^2b^3}{9a^5b^9}\right)^3$

$-\dfrac{64}{27a^9b^{18}}$

105. $\left(\dfrac{12x^3y^2z}{18xy^3z^4}\right)^4$

$\dfrac{16x^8}{81y^4z^{12}}$

106. $\dfrac{(4x^2y)^2}{(2xy^3)^3}$

$\dfrac{2x}{y^7}$

107. $\dfrac{(3a^2b)^3}{(-6ab^3)^2}$

$\dfrac{3a^4}{4b^3}$

108. $\dfrac{(-4x^2y^3)^2}{(2xy^2)^3}$

$2x$

109. $\dfrac{(-3a^2b^3)^2}{(-2ab^4)^3}$

$-\dfrac{9a}{8b^6}$

110. $\dfrac{(-4xy^3)^3}{(-2x^7y)^4}$

$-\dfrac{4y^5}{x^{25}}$

111. $\dfrac{(-8x^2y^2)^4}{(16x^3y^7)^2}$

$\dfrac{16x^2}{y^6}$

112. $\dfrac{a^{5n}}{a^{3n}}$

a^{2n}

113. $\dfrac{b^{6n}}{b^{10n}}$

$\dfrac{1}{b^{4n}}$

114. $\dfrac{-x^{5n}}{x^{2n}}$

$-x^{3n}$

115. $\dfrac{y^{2n}}{-y^{8n}}$

$-\dfrac{1}{y^{6n}}$

116. $\dfrac{x^{2n-1}}{x^{n-3}}$

x^{n+2}

117. $\dfrac{y^{3n+2}}{y^{2n+4}}$

y^{n-2}

118. $\dfrac{a^{3n}b^n}{a^nb^{2n}}$

$\dfrac{a^{2n}}{b^n}$

119. $\dfrac{x^ny^{3n}}{x^ny^{5n}}$

$\dfrac{1}{y^{2n}}$

120. $\dfrac{a^{3n-2}b^{n+1}}{a^{2n+1}b^{2n+2}}$

$\dfrac{a^{n-3}}{b^{n+1}}$

121. $\dfrac{x^{2n-1}y^{n-3}}{x^{n+4}y^{n+3}}$

$\dfrac{x^{n-5}}{y^6}$

122. $\left(\dfrac{4^{-2}xy^{-3}}{x^{-3}y}\right)^3\left(\dfrac{8^{-1}x^{-2}y}{x^4y^{-1}}\right)^{-2}$

$\dfrac{x^{24}}{64y^{16}}$

123. $\left(\dfrac{9ab^{-2}}{8a^{-2}b}\right)^{-2}\left(\dfrac{3a^{-2}b}{2a^2b^{-2}}\right)^3$

$\dfrac{8b^{15}}{3a^{18}}$

124. $\left(\dfrac{2ab^{-1}}{ab}\right)^{-1}\left(\dfrac{3a^{-2}b}{a^2b^2}\right)^{-2}$

$\dfrac{a^8b^4}{18}$

125. $1 + (1 + (1 + 2^{-1})^{-1})^{-1}$

$\dfrac{8}{5}$

126. $2 - (2 - (2 - 2^{-1})^{-1})^{-1}$

$\dfrac{5}{4}$

127. $\dfrac{x^{-1} + y^{-1}}{x^{-1} - y^{-1}}$

$\dfrac{y + x}{y - x}$

128. $\dfrac{x^{-1} + y}{x^{-1} - y}$

$\dfrac{1 + xy}{1 - xy}$

129. $(a + b)^{-1}$

$\dfrac{1}{a + b}$

130. $\dfrac{1}{(a - b)^{-1}}$

$a - b$

131. $a + a^{-1}b$

$\dfrac{a^2 + b}{a}$

132. $x^{-1} + y^{-1}$

$\dfrac{y + x}{xy}$

Content and Format © 1991 HMCo.

Simplify:

133. $\dfrac{\frac{x^{-1}}{y^{-1}} + \frac{y}{x}}{\frac{2y}{x}}$

134. $\dfrac{x^{-1}y^{-1} + xy}{\frac{1 + x^2y^2}{xy}}$

135. $\dfrac{\frac{x^{-1}}{y} + \frac{y^{-1}}{x}}{\frac{2}{xy}}$

▶ **Objective D**

Write in scientific notation.

136. 0.00000467
4.67×10^{-6}

137. 0.00000005
5×10^{-8}

138. 0.00000000017
1.7×10^{-10}

139. 4,300,000
4.3×10^{6}

140. 200,000,000,000
2×10^{11}

141. 9,800,000,000
9.8×10^{9}

Write in decimal notation.

142. 1.23×10^{-7}
0.000000123

143. 6.2×10^{-12}
0.0000000000062

144. 8.2×10^{15}
8,200,000,000,000,000

145. 6.34×10^{5}
634,000

146. 3.9×10^{-2}
0.039

147. 4.35×10^{9}
4,350,000,000

Simplify:

148. $(3 \times 10^{-12})(5 \times 10^{16})$
150,000

149. $(8.9 \times 10^{-5})(3.2 \times 10^{-6})$
0.0000000002848

150. $(0.0000065)(3,200,000,000,000)$
20,800,000

151. $(480,000)(0.0000000096)$
0.004608

152. $\dfrac{9 \times 10^{-3}}{6 \times 10^{5}}$
0.000000015

153. $\dfrac{2.7 \times 10^{4}}{3 \times 10^{-6}}$
9,000,000,000

154. $\dfrac{0.0089}{500,000,000}$
0.0000000000178

155. $\dfrac{4,800}{0.00000024}$
20,000,000,000

156. $\dfrac{0.00056}{0.000000000004}$
140,000,000

157. $\dfrac{0.000000346}{0.0000005}$
0.692

158. $\dfrac{(3.2 \times 10^{-11})(2.9 \times 10^{15})}{8.1 \times 10^{-3}}$
11,456,790

159. $\dfrac{(6.9 \times 10^{27})(8.2 \times 10^{-13})}{4.1 \times 10^{15}}$
1.38

160. $\dfrac{(0.00000004)(84,000)}{(0.0003)(1,400,000)}$
0.000008

161. $\dfrac{(720)(0.0000000039)}{(26,000,000,000)(0.018)}$
0.000000000000006

▶ Objective E   *Application Problems*

Solve: Write the answer in scientific notation.

162. How many kilometers does light travel in one day? The speed of light is 300,000 km/s.
2.592×10^{10} km

163. How many meters does light travel in 8 h? The speed of light is 300,000,000 m/s.
8.64×10^{12} m

164. A computer can do an arithmetic operation in 3×10^{-6} s. How many arithmetic operations can the computer perform in one minute?
2×10^{7} operations

165. A computer can do an arithmetic operation in 9×10^{-7} s. How many arithmetic operations can the computer perform in one hour?
4×10^{9} operations

166. The national debt is 3×10^{12} dollars. How much would each American citizen have to pay in order to pay off the national debt? Use a figure of two hundred fifty million for the number of citizens.
1.2×10^{4} dollars

167. A high-speed centrifuge makes 9×10^{7} revolutions each minute. Find the time in seconds for the centrifuge to make one revolution.
$6.\overline{6} \times 10^{-7}$ s

168. How long does it take light to travel to the earth from the sun? The sun is 9.3×10^{7} mi from the earth and light travels 1.86×10^{5} mi/s.
5×10^{2} s

169. The mass of the earth is 5.9×10^{27} g. The mass of the sun is 2×10^{33} g. How many times heavier is the sun than the earth?
3.38983×10^{5} times heavier

170. The distance to the sun is 9.3×10^{7} mi. A satellite leaves the earth traveling at a constant speed of 1×10^{5} mi/h. How long does it take for the satellite to reach the sun?
9.3×10^{2} h

171. The weight of 31 million orchid seeds is one ounce. Find the weight of one orchid seed.
3.225806×10^{-8} oz

172. One light year, an astronomical unit of distance, is the distance that light will travel in one year. Light travels 1.86×10^{5} mi/s. Find the measure of one light year in miles. Use a 360-day year.
5.785344×10^{12} mi

173. The light from a nearby star, Alpha Centauri, takes 4.3 years to reach the earth. Light travels at 1.86×10^{5} mi/s. How far is Alpha Centauri from the earth? Use a 360-day year.
2.4876979×10^{13} mi

SECTION 3.2	**Multiplication of Polynomials**

Objective A	**To multiply a polynomial by a monomial**

To multiply a polynomial by a monomial, use the Distributive Property and the Rule for Multiplying Exponential Expressions.

Simplify: $-5x(x^2 - 2x + 3)$

$$-5x(x^2 - 2x + 3)$$

Use the Distributive Property.
$$\boxed{-5x(x^2) - (-5x)(2x) + (-5x)(3)}$$ Do this step mentally.

Use the Rule for Multiplying Exponential Expressions.
$$-5x^3 + 10x^2 - 15x$$

Simplify: $x^2 - x[3 - x(x - 2) + 3]$

$$x^2 - x[3 - x(x - 2) + 3]$$

Use the Distributive Property to remove the inner grouping symbols.
$$x^2 - x[3 - x^2 + 2x + 3]$$

Combine like terms.
$$x^2 - x[6 - x^2 + 2x]$$

Use the Distributive Property to remove the brackets.
$$x^2 - 6x + x^3 - 2x^2$$

Combine like terms and write the polynomial in descending order.
$$x^3 - x^2 - 6x$$

Simplify: $x^n(x^n - x^2 + 1)$

$$x^n(x^n - x^2 + 1)$$

Use the Distributive Property.
$$x^n(x^n) - (x^n)(x^2) + (x^n)(1)$$

Use the Rule for Multiplying Exponential Expressions.
$$x^{2n} - x^{n+2} + x^n$$

Example 1

Simplify: $(3a^2 - 2a + 4)(-3a)$

Solution

$(3a^2 - 2a + 4)(-3a) = -9a^3 + 6a^2 - 12a$

Example 2

Simplify: $(2b^2 - 7b - 8)(-5b)$

Your solution

$-10b^3 + 35b^2 + 40b$

Solution on p. A17

Example 3

Simplify: $y - 3y[y - 2(3y - 6) + 2]$

Solution

$y - 3y[y - 2(3y - 6) + 2]$
$= y - 3y[y - 6y + 12 + 2]$
$= y - 3y[-5y + 14]$
$= y + 15y^2 - 42y$
$= 15y^2 - 41y$

Example 4

Simplify: $x^2 - 2x[x - x(4x - 5) + x^2]$

Your solution

$6x^3 - 11x^2$

Example 5

Simplify: $x^{n+2}(x^{n-1} + 2x - 1)$

Solution

$x^{n+2}(x^{n-1} + 2x - 1)$
$= x^{n+2}(x^{n-1}) + (x^{n+2})(2x) - (x^{n+2})(1)$
$= x^{n+2+(n-1)} + 2x^{n+2+1} - x^{n+2}$
$= x^{2n+1} + 2x^{n+3} - x^{n+2}$

Example 6

Simplify: $y^{n+3}(y^{n-2} - 3y^2 + 2)$

Your solution

$y^{2n+1} - 3y^{n+5} + 2y^{n+3}$

Solutions on p. A17

| Objective B | **To multiply two polynomials** |

The product of two polynomials is the polynomial obtained by multiplying each term of one polynomial by each term of the other polynomial and then combining like terms.

Multiply: $(2x^2 - 2x + 1)(3x + 2)$

Use the Distributive Property to multiply the trinomial by each term of the binomial.

$(2x^2 - 2x + 1)(3x + 2)$
$(2x^2 - 2x + 1)(3x) + (2x^2 - 2x + 1)(2)$

Use the Distributive Property to remove parentheses.

$(6x^3 - 6x^2 + 3x) + (4x^2 - 4x + 2)$

Combine like terms.

$6x^3 - 2x^2 - x + 2$

A convenient method of multiplying two polynomials is to use a vertical format, similar to that used for multiplication of whole numbers.

Multiply: $(3x^2 - 4x + 8)(2x - 7)$

$$
\begin{array}{r}
3x^2 - 4x + 8 \\
2x - 7 \\
\hline
\end{array}
$$

Like terms are in the same column.
Combine like terms.

$-21x^2 + 28x - 56 = -7(3x^2 - 4x + 8)$
$6x^3 - 8x^2 + 16x \quad = 2x(3x^2 - 4x + 8)$
$6x^3 - 29x^2 + 44x - 56$

It is frequently necessary to find the product of two binomials. The product can be found by using a method called **FOIL,** which is based upon the Distributive Property. The letters of FOIL stand for **F**irst, **O**uter, **I**nner, and **L**ast.

Simplify: $(3x - 2)(2x + 5)$

Multiply the **F**irst terms.	$(3x - 2)(2x + 5)$	$3x \cdot 2x = 6x^2$
Multiply the **O**uter terms.	$(3x - 2)(2x + 5)$	$3x \cdot 5 = 15x$
Multiply the **I**nner terms.	$(3x - 2)(2x + 5)$	$-2 \cdot 2x = -4x$
Multiply the **L**ast terms.	$(3x - 2)(2x + 5)$	$-2 \cdot 5 = -10$

$$\qquad\qquad\qquad\qquad\qquad\qquad\qquad F \quad\;\; O \quad\;\; I \quad\;\; L$$

Add the products. $\qquad\qquad (3x - 2)(2x + 5) \qquad = \qquad 6x^2 + 15x - 4x - 10$

Combine like terms. $\qquad\qquad\qquad\qquad\qquad\qquad\qquad = \qquad 6x^2 + 11x - 10$

Simplify: $(6x - 5)(3x - 4)$

$(6x - 5)(3x - 4)$ $\boxed{= 6x(3x) + 6x(-4) + (-5)(3x) + (-5)(-4)}$ Do this step mentally.

$\qquad\qquad\qquad\qquad = 18x^2 - 24x - 15x + 20$

$\qquad\qquad\qquad\qquad = 18x^2 - 39x + 20$

Example 7

Simplify: $(4a^3 - 3a + 7)(a - 5)$

Solution

$$
\begin{array}{r}
4a^3 - 3a + 7 \\
a - 5 \\
\hline
-20a^3 \qquad\quad + 15a - 35 \\
4a^4 \qquad\quad - 3a^2 + 7a \\
\hline
4a^4 - 20a^3 - 3a^2 + 22a - 35
\end{array}
$$

Example 8

Simplify: $(-2b^2 + 5b - 4)(-3b + 2)$

Your solution

$6b^3 - 19b^2 + 22b - 8$

Example 9

Simplify: $(5a - 3b)(2a + 7b)$

Solution

$(5a - 3b)(2a + 7b)$
$= 10a^2 + 35ab - 6ab - 21b^2$
$= 10a^2 + 29ab - 21b^2$

Example 10

Simplify: $(3x - 4)(2x - 3)$

Your solution

$6x^2 - 17x + 12$

Example 11

Simplify: $(a^n - 2b^n)(3a^n - b^n)$

Solution

$(a^n - 2b^n)(3a^n - b^n)$
$= 3a^{2n} - a^n b^n - 6a^n b^n + 2b^{2n}$
$= 3a^{2n} - 7a^n b^n + 2b^{2n}$

Example 12

Simplify: $(2x^n + y^n)(x^n - 4y^n)$

Your solution

$2x^{2n} - 7x^n y^n - 4y^{2n}$

Solutions on p. A17

| Objective C | **To multiply polynomials that have special products** | |

Using FOIL, a pattern for the product of the sum and difference of two terms and for the square of a binomial can be found.

The Sum and Difference of Two Terms $(a + b)(a - b) = a^2 - ab + ab - b^2$
$$= a^2 - b^2$$

Square of the first term ⟶

Square of the second term ⟶

The Square of a Binomial $(a + b)^2 = (a + b)(a + b) = a^2 + ab + ab + b^2$
$$= a^2 + 2ab + b^2$$

Square of the first term ⟶

Twice the product of the two terms ⟶

Square of the last term ⟶

Simplify: $(4x + 3)(4x - 3)$

$(4x + 3)(4x - 3)$ is the sum and difference of two terms.

$(4x + 3)(4x - 3)$ $\boxed{= (4x)^2 - 3^2}$ Do this step mentally.

$$= 16x^2 - 9$$

Simplify: $(2x - 3y)^2$

$(2x - 3y)^2$ is the square of a binomial.

$(2x - 3y)^2$ $\boxed{= (2x)^2 + 2(2x)(-3y) + (-3y)^2}$ Do this step mentally.

$$= 4x^2 - 12xy + 9y^2$$

Example 13

Simplify: $(2a - 3)(2a + 3)$

Solution

$(2a - 3)(2a + 3) = 4a^2 - 9$

Example 14

Simplify: $(3x - 7)(3x + 7)$

Your solution

$9x^2 - 49$

Example 15

Simplify: $(x^n + 5)(x^n - 5)$

Solution

$(x^n + 5)(x^n - 5) = x^{2n} - 25$

Example 16

Simplify: $(2x^n + 3)(2x^n - 3)$

Your solution

$4x^{2n} - 9$

Solutions on p. A17

Example 17

Simplify: $(2x + 7y)^2$

Solution

$(2x + 7y)^2 = 4x^2 + 28xy + 49y^2$

Example 18

Simplify: $(3x - 4y)^2$

Your solution

$9x^2 - 24xy + 16y^2$

Example 19

Simplify: $(x^{2n} - 2)^2$

Solution

$(x^{2n} - 2)^2 = x^{4n} - 4x^{2n} + 4$

Example 20

Simplify: $(2x^n - 8)^2$

Your solution

$4x^{2n} - 32x^n + 64$

Solutions on p. A17

Objective D   **To solve application problems**

Example 21

The length of a rectangle is $(2x + 3)$ ft. The width is $(x - 5)$ ft. Find the area of the rectangle in terms of the variable x.

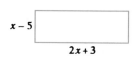

Strategy

To find the area, replace the variables L and w in the equation $A = L \cdot w$ by the given values and solve for A.

Solution

$A = L \cdot w$
$A = (2x + 3)(x - 5)$
$A = 2x^2 - 10x + 3x - 15$
$A = 2x^2 - 7x - 15$

The area is $(2x^2 - 7x - 15)$ ft^2.

Example 22

The base of a triangle is $(2x + 6)$ ft. The height is $(x - 4)$ ft. Find the area of the triangle in terms of the variable x.

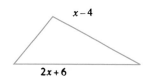

Your strategy

Your solution

$(x^2 - x - 12)$ ft^2

Solution on p. A17

Example 23

The corners are cut from a rectangular piece of cardboard measuring 8 in. by 12 in. The sides are folded up to make a box. Find the volume of the box in terms of the variable x, where x is the length of the side of the square cut from each corner of the rectangle.

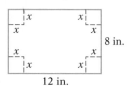

8 in.

12 in.

Strategy

Length of the box: $12 - 2x$
Width of the box: $8 - 2x$
Height of the box: x
To find the volume, replace the variables L, w, and h in the equation $V = L \cdot w \cdot h$ and solve for V.

Solution

$V = L \cdot w \cdot h$
$V = (12 - 2x)(8 - 2x)x$
$V = (96 - 24x - 16x + 4x^2)x$
$V = (96 - 40x + 4x^2)x$
$V = 96x - 40x^2 + 4x^3$
$V = 4x^3 - 40x^2 + 96x$

The volume is $(4x^3 - 40x^2 + 96x)$ in^3.

Example 24

Find the volume of the rectangular solid shown in the diagram below. All dimensions are in feet.

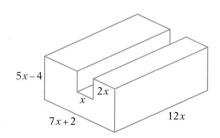

$5x - 4$

$2x$

x

$7x + 2$

$12x$

Your strategy

Your solution

$(396x^3 - 216x^2 - 96x)$ ft^3

Example 25

The radius of a circle is $(3x - 2)$ cm. Find the area of the circle in terms of the variable x. Use $\pi \approx 3.14$.

Strategy

To find the area, replace the variable r in the equation $A = \pi r^2$ by the given value and solve for A.

Solution

$A = \pi r^2$
$A = 3.14(3x - 2)^2$
$A = 3.14(9x^2 - 12x + 4)$
$A = 28.26x^2 - 37.68x + 12.56$

The area is $(28.26x^2 - 37.68x + 12.56)$ cm^2.

Example 26

The radius of a circle is $(2x + 3)$ cm. Find the area of the circle in terms of the variable x. Use $\pi \approx 3.14$.

Your strategy

Your solution

$(12.56x^2 + 37.68x + 28.26)$ cm^2

Solutions on pp. A17–A18

3.2   EXERCISES

▶ **Objective A**

Simplify:

1. $2x(x - 3)$
$2x^2 - 6x$

2. $2a(2a + 4)$
$4a^2 + 8a$

3. $3x^2(2x^2 - x)$
$6x^4 - 3x^3$

4. $-4y^2(4y - 6y^2)$
$-16y^3 + 24y^4$

5. $3xy(2x - 3y)$
$6x^2y - 9xy^2$

6. $-4ab(5a - 3b)$
$-20a^2b + 12ab^2$

7. $x^n(x + 1)$
$x^{n+1} + x^n$

8. $y^n(y^{2n} - 3)$
$y^{3n} - 3y^n$

9. $x^n(x^n + y^n)$
$x^{2n} + x^ny^n$

10. $x - 2x(x - 2)$
$-2x^2 + 5x$

11. $2b + 4b(2 - b)$
$-4b^2 + 10b$

12. $-2y(3 - y) + 2y^2$
$4y^2 - 6y$

13. $-2a^2(3a^2 - 2a + 3)$
$-6a^4 + 4a^3 - 6a^2$

14. $4b(3b^3 - 12b^2 - 6)$
$12b^4 - 48b^3 - 24b$

15. $3b(3b^4 - 3b^2 + 8)$
$9b^5 - 9b^3 + 24b$

16. $(2x^2 - 3x - 7)(-2x^2)$
$-4x^4 + 6x^3 + 14x^2$

17. $(-3y^2 - 4y + 2)(y^2)$
$-3y^4 - 4y^3 + 2y^2$

18. $(6b^4 - 5b^2 - 3)(-2b^3)$
$-12b^7 + 10b^5 + 6b^3$

19. $-5x^2(4 - 3x + 3x^2 + 4x^3)$
$-20x^2 + 15x^3 - 15x^4 - 20x^5$

20. $-2y^2(3 - 2y - 3y^2 + 2y^3)$
$-6y^2 + 4y^3 + 6y^4 - 4y^5$

21. $-2x^2y(x^2 - 3xy + 2y^2)$
$-2x^4y + 6x^3y^2 - 4x^2y^3$

22. $3ab^2(3a^2 - 2ab + 4b^2)$
$9a^3b^2 - 6a^2b^3 + 12ab^4$

23. $x^n(x^{2n} + x^n + x)$
$x^{3n} + x^{2n} + x^{n+1}$

24. $x^{2n}(x^{2n-2} + x^{2n} + x)$
$x^{4n-2} + x^{4n} + x^{2n+1}$

25. $a^{n+1}(a^n - 3a + 2)$
$a^{2n+1} - 3a^{n+2} + 2a^{n+1}$

26. $a^{n+4}(a^{n-2} + 5a^2 - 3)$
$a^{2n+2} + 5a^{n+6} - 3a^{n+4}$

27. $2y^2 - y[3 - 2(y - 4) - y]$
$5y^2 - 11y$

28. $3x^2 - x[x - 2(3x - 4)]$
$8x^2 - 8x$

29. $2y - 3[y - 2y(y - 3) + 4y]$
$6y^2 - 31y$

30. $4a^2 - 2a[3 - a(2 - a + a^2)]$
$2a^4 - 2a^3 + 8a^2 - 6a$

▶ **Objective B**

Simplify:

31. $(x - 2)(x + 7)$
$x^2 + 5x - 14$

32. $(y + 8)(y + 3)$
$y^2 + 11y + 24$

33. $(2y - 3)(4y + 7)$
$8y^2 + 2y - 21$

34. $(5x - 7)(3x - 8)$
$15x^2 - 61x + 56$

35. $2(2x - 3y)(2x + 5y)$
$8x^2 + 8xy - 30y^2$

36. $-3(7x - 3y)(2x - 9y)$
$-42x^2 + 207xy - 81y^2$

37. $-3(2a - 3b)(5a + 4b)$
$-30a^2 + 21ab + 36b^2$

38. $-2(3a - 5b)(a + 7b)$
$-6a^2 - 32ab + 70b^2$

39. $(5a + 2b)(3a + 7b)$
$15a^2 + 41ab + 14b^2$

40. $(5x + 9y)(3x + 2y)$
$15x^2 + 37xy + 18y^2$

41. $(3x - 7y)(7x + 2y)$
$21x^2 - 43xy - 14y^2$

42. $(5x - 9y)(6x - 5y)$
$30x^2 - 79xy + 45y^2$

43. $(xy + 4)(xy - 3)$
$x^2y^2 + xy - 12$

44. $(xy - 5)(2xy + 7)$
$2x^2y^2 - 3xy - 35$

45. $(2x^2 - 5)(x^2 - 5)$
$2x^4 - 15x^2 + 25$

46. $(x^2 - 4)(x^2 - 6)$
$x^4 - 10x^2 + 24$

47. $(5x^2 - 5y)(2x^2 - y)$
$10x^4 - 15x^2y + 5y^2$

48. $(x^2 - 2y^2)(x^2 + 4y^2)$
$x^4 + 2x^2y^2 - 8y^4$

49. $(x^n + 2)(x^n - 3)$
$x^{2n} - x^n - 6$

50. $(x^n - 4)(x^n - 5)$
$x^{2n} - 9x^n + 20$

51. $(2a^n - 3)(3a^n + 5)$
$6a^{2n} + a^n - 15$

52. $(5b^n - 1)(2b^n + 4)$
$10b^{2n} + 18b^n - 4$

53. $(2a^n - b^n)(3a^n + 2b^n)$
$6a^{2n} + a^nb^n - 2b^{2n}$

54. $(3x^n + b^n)(x^n + 2b^n)$
$3x^{2n} + 7b^nx^n + 2b^{2n}$

55. $(x - 2)(x^2 - 3x + 7)$
$x^3 - 5x^2 + 13x - 14$

56. $(x + 3)(x^2 + 5x - 8)$
$x^3 + 8x^2 + 7x - 24$

57. $(x + 5)(x^3 - 3x + 4)$
$x^4 + 5x^3 - 3x^2 - 11x + 20$

58. $(a + 2)(a^3 - 3a^2 + 7)$
$a^4 - a^3 - 6a^2 + 7a + 14$

59. $(2a - 3b)(5a^2 - 6ab + 4b^2)$
$10a^3 - 27a^2b + 26ab^2 - 12b^3$

60. $(3a + b)(2a^2 - 5ab - 3b^2)$
$6a^3 - 13a^2b - 14ab^2 - 3b^3$

61. $(2y^2 - 1)(y^3 - 5y^2 - 3)$
$2y^5 - 10y^4 - y^3 - y^2 + 3$

62. $(2b^2 - 3)(3b^2 - 3b + 6)$
$6b^4 - 6b^3 + 3b^2 + 9b - 18$

63. $(2x - 5)(2x^4 - 3x^3 - 2x + 9)$
$4x^5 - 16x^4 + 15x^3 - 4x^2 + 28x - 45$

64. $(2a - 5)(3a^4 - 3a^2 + 2a - 5)$
$6a^5 - 15a^4 - 6a^3 + 19a^2 - 20a + 25$

Simplify:

65. $(x^2 + 2x - 3)(x^2 - 5x + 7)$
$x^4 - 3x^3 - 6x^2 + 29x - 21$

66. $(x^2 - 3x + 1)(x^2 - 2x + 7)$
$x^4 - 5x^3 + 14x^2 - 23x + 7$

67. $(a - 2)(2a - 3)(a + 7)$
$2a^3 + 7a^2 - 43a + 42$

68. $(b - 3)(3b - 2)(b - 1)$
$3b^3 - 14b^2 + 17b - 6$

69. $(x^n + 1)(x^{2n} + x^n + 1)$
$x^{3n} + 2x^{2n} + 2x^n + 1$

70. $(a^{2n} - 3)(a^{5n} - a^{2n} + a^n)$
$a^{7n} - 3a^{5n} - a^{4n} + a^{3n} + 3a^{2n} - 3a^n$

71. $(x^n + y^n)(x^n - 2x^n y^n + 3y^n)$
$x^{2n} - 2x^{2n}y^n + 4x^n y^n - 2x^n y^{2n} + 3y^{2n}$

72. $(x^n - y^n)(x^{2n} - 3x^n y^n - y^{2n})$
$x^{3n} - 4x^{2n}y^n + 2x^n y^{2n} + y^{3n}$

▶ **Objective C**

Simplify:

73. $(3x - 2)(3x + 2)$
$9x^2 - 4$

74. $(4y + 1)(4y - 1)$
$16y^2 - 1$

75. $(6 - x)(6 + x)$
$36 - x^2$

76. $(10 + b)(10 - b)$
$100 - b^2$

77. $(2a - 3b)(2a + 3b)$
$4a^2 - 9b^2$

78. $(5x - 7y)(5x + 7y)$
$25x^2 - 49y^2$

79. $(x^2 + 1)(x^2 - 1)$
$x^4 - 1$

80. $(x^2 + y^2)(x^2 - y^2)$
$x^4 - y^4$

81. $(x^n + 3)(x^n - 3)$
$x^{2n} - 9$

82. $(x^n + y^n)(x^n - y^n)$
$x^{2n} - y^{2n}$

83. $(x - 5)^2$
$x^2 - 10x + 25$

84. $(y + 2)^2$
$y^2 + 4y + 4$

85. $(3a + 5b)^2$
$9a^2 + 30ab + 25b^2$

86. $(5x - 4y)^2$
$25x^2 - 40xy + 16y^2$

87. $(x^2 - 3)^2$
$x^4 - 6x^2 + 9$

88. $(x^2 + y^2)^2$
$x^4 + 2x^2 y^2 + y^4$

89. $(2x^2 - 3y^2)^2$
$4x^4 - 12x^2 y^2 + 9y^4$

90. $(x^n - 1)^2$
$x^{2n} - 2x^n + 1$

91. $(a^n - b^n)^2$
$a^{2n} - 2a^n b^n + b^{2n}$

92. $(2x^n + 5y^n)^2$
$4x^{2n} + 20x^n y^n + 25y^{2n}$

93. $y^2 - (x - y)^2$
$-x^2 + 2xy$

94. $a^2 + (a + b)^2$
$2a^2 + 2ab + b^2$

95. $(x - y)^2 - (x + y)^2$
$-4xy$

96. $(a + b)^2 + (a - b)^2$
$2a^2 + 2b^2$

▶ **Objective D** *Application Problems*

Solve:

97. The length of a rectangle is $(3x - 2)$ ft. The width is $(x + 4)$ ft. Find the area of the rectangle in terms of the variable x.
$(3x^2 + 10x - 8)$ ft^2

98. The base of a triangle is $(x - 4)$ ft. The height is $(3x + 2)$ ft. Find the area of the triangle in terms of the variable x.
$(\frac{3}{2}x^2 - 5x - 4)$ ft^2

99. Find the area of the figure shown below. All dimensions given are in meters.
$(x^2 + 3x)$ m^2

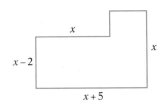

100. Find the area of the figure shown below. All dimensions given are in feet.
$(x^2 + 12x + 16)$ ft^2

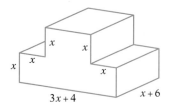

101. The length of the side of a cube is $(x + 3)$ cm. Find the volume of the cube in terms of the variable x.
$(x^3 + 9x^2 + 27x + 27)$ cm^3

102. The length of a box is $(3x + 2)$ cm, the width is $(x - 4)$ cm, and the height is x cm. Find the volume of the box in terms of the variable x.
$(3x^3 - 10x^2 - 8x)$ cm^3

103. Find the volume of the figure shown below. All dimensions given are in inches.
$(2x^3)$ in^3

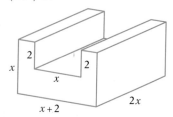

104. Find the volume of the figure shown below. All dimensions given are in centimeters.
$(4x^3 + 32x^2 + 48x)$ cm^3

105. The radius of a circle is $(5x + 4)$ in. Find the area of the circle in terms of the variable x. Use $\pi \approx 3.14$.
$(78.5x^2 + 125.6x + 50.24)$ in^2

106. The radius of a circle is $(x - 2)$ in. Find the area of the circle in terms of the variable x. Use $\pi \approx 3.14$.
$(3.14x^2 - 12.56x + 12.56)$ in^2

| SECTION **3.3** | **Factoring Polynomials** |

Objective A   **To factor a monomial from a polynomial**

The greatest common factor GCF of two or more monomials is the product of each common factor with the smallest exponent.

$$16a^4b = 2^4a^4b$$
$$40a^2b^5 = 2^3 \cdot 5a^2b^5$$
$$\text{GCF} = 2^3a^2b = 8a^2b$$

Note that the exponent of each variable in the GCF is the same as the smallest exponent of the variable in either of the monomials.

To **factor a polynomial** means to write the polynomial as a product of other polynomials.

In the example at the right, $3x$ is the GCF of the terms $3x^2$ and $6x$. $3x$ is a **common monomial factor** of the terms of the binomial. $x - 2$ is a **binomial factor** of $3x^2 - 6x$.

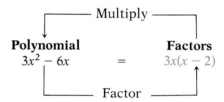

Factor: $4x^3y^2 + 12x^3y + 20xy^2$

Find the GCF of the terms of the polynomial.

The GCF of $4x^3y^2$, $12x^3y$, and $20xy^2$ is $4xy$.

Rewrite each term of the polynomial as a product with the GCF as one of the factors.

$4x^3y^2 + 12x^3y + 20xy^2 =$

$\overline{| \; 4xy(x^2y) + 4xy(3x^2) + 4xy(5y) = |}$ Do this step mentally.

Use the Distributive Property to write the polynomial as a product of factors.

$4xy(x^2y + 3x^2 + 5y)$

Example 1

Factor: $4x^2y^2 - 6xy^2 + 12xy^3$

Solution

The GCF of $4x^2y^2$, $6xy^2$, and $12xy^3$ is $2xy^2$.
$4x^2y^2 - 6xy^2 + 12xy^3 = 2xy^2(2x - 3 + 6y)$

Example 2

Factor: $3x^3y - 6x^2y^2 - 3xy^3$

Your solution

$3xy(x^2 - 2xy - y^2)$

Solution on p. A18

Example 3

Factor: $x^{2n} + x^{n+1} + x^n$

Solution

The GCF of x^{2n}, x^{n+1}, and x^n is x^n.

$x^{2n} + x^{n+1} + x^n = x^n(x^n + x + 1)$

Example 4

Factor: $6t^{2n} - 9t^n$

Your solution

$3t^n(2t^n - 3)$

Solution on p. A18

Objective B   To factor by grouping

3

In the examples at the right, the binomials in parentheses are called binomial factors.

$4x^4(2x - 3)$
$-2r^2s(5r + 2s)$

The Distributive Property is used to factor a common binomial factor from an expression.

Factor: $4a(2b + 3) - 5(2b + 3)$

The common binomial factor is $(2b + 3)$. Use the Distributive property to write the expression as a product of factors.

$4a(2b + 3) - 5(2b + 3)$

$= (2b + 3)(4a - 5)$

Consider the binomial $x - y$. Factoring -1 from this binomial gives

$$x - y = -(y - x)$$

This equation is sometimes used to factor a common binomial from an expression.

Factor: $6r(r - s) - 7(s - r)$

Rewrite the expression as a sum of terms which have a common binomial factor. Use $s - r = -(r - s)$.

$6r(r - s) - 7(s - r)$
$6r(r - s) + 7(r - s)$
$(r - s)(6r + 7)$

Some polynomials can be factored by grouping terms so that a common binomial factor is found.

Factor: $3xz - 4yz - 3xa + 4ya$

Group the first two terms and the last two terms. Note that $-3xa + 4ya = -(3xa - 4ya)$.

$3xz - 4yz - 3xa + 4ya$
$(3xz - 4yz) - (3xa - 4ya)$

Factor the GCF from each group.

$z(3x - 4y) - a(3x - 4y)$

Write the expression as the product of factors.

$(3x - 4y)(z - a)$

Factor: $8y^2 + 4y - 6ay - 3a$

Group the first two terms and the last two terms. Note that $-6ay - 3a = -(6ay + 3a)$.

$$8y^2 + 4y - 6ay - 3a$$
$$(8y^2 + 4y) - (6ay + 3a)$$

Factor the GCF from each group.

$$4y(2y + 1) - 3a(2y + 1)$$

Write the expression as the product of factors.

$$(2y + 1)(4y - 3a)$$

Example 5

Factor: $x^2(5y - 2) - 7(5y - 2)$

Solution

$x^2(5y - 2) - 7(5y - 2) = (5y - 2)(x^2 - 7)$

Example 6

Factor: $3(6x - 7y) - 2x^2(6x - 7y)$

Your solution

$(6x - 7y)(3 - 2x^2)$

Example 7

Factor: $15x^2 + 6x - 5xz - 2z$

Solution

$$15x^2 + 6x - 5xz - 2z$$
$$= (15x^2 + 6x) - (5xz + 2z)$$
$$= 3x(5x + 2) - z(5x + 2)$$
$$= (5x + 2)(3x - z)$$

Example 8

Factor: $4a^2 - 6a - 6ax + 9x$

Your solution

$(2a - 3)(2a - 3x)$

Solutions on p. A18

| Objective C | To factor a trinomial of the form $x^2 + bx + c$ |

A **quadratic trinomial** is a trinomial of the form $ax^2 + bx + c$, where a, b, and c are nonzero constants. The degree of a quadratic trinomial is 2. Here are examples of quadratic trinomials:

$$4x^2 - 3x - 7 \qquad z^2 + z + 10 \qquad 2y^2 + 4y - 9$$
$$(a = 4, b = -3, c = -7) \qquad (a = 1, b = 1, c = 10) \qquad (a = 2, b = 4, c = -9)$$

To factor a **quadratic trinomial** means to express the trinomial as the product of two binomials. For example,

Trinomial		Factored Form
$2x^2 - x - 1$	$=$	$(2x + 1)(x - 1)$
$y^2 - 3y + 2$	$=$	$(y - 1)(y - 2)$

In this objective, trinomials of the form $x^2 + bx + c$ ($a = 1$) will be factored. The next objective deals with trinomials where $a \neq 1$.

The method by which factors of a trinomial are found is based upon FOIL. Consider the following binomial products, noting the relationship between the constant terms of the binomials and the terms of the trinomial.

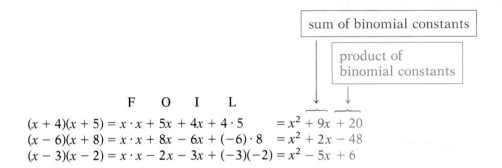

$$
\begin{array}{ll}
 \text{F} \quad\ \text{O} \quad\ \text{I} \quad\ \text{L} & \\
(x + 4)(x + 5) = x \cdot x + 5x + 4x + 4 \cdot 5 & = x^2 + 9x + 20 \\
(x - 6)(x + 8) = x \cdot x + 8x - 6x + (-6) \cdot 8 & = x^2 + 2x - 48 \\
(x - 3)(x - 2) = x \cdot x - 2x - 3x + (-3)(-2) & = x^2 - 5x + 6
\end{array}
$$

Observe two important points from these examples.

1. The constant term of the trinomial is the *product* of the constant terms of the binomial. The coefficient of x in the trinomial is the *sum* of the constant terms of the binomial.

2. When the constant term of the trinomial is positive, the constant terms of the binomial have the *same* sign. When the constant term of the trinomial is negative, the constant terms of the binomial have *opposite* signs.

Factor: $x^2 - 7x + 12$

The constant term is positive. The signs of the binomial constants will be the same.

Find two negative factors of 12 whose sum is -7.

Factors	Sum
$-1, -12$	-13
$-2, -6$	-8
$-3, -4$	-7

Write the trinomial in factored form. $x^2 - 7x + 12 = (x - 3)(x - 4)$

Check: $(x - 3)(x - 4) = x^2 - 4x - 3x + 12 = x^2 - 7x + 12$

Factor: $y^2 + 10y - 24$

The constant term is negative. The signs of the binomial constants will be opposite.

Find two factors, with opposite signs, of -24 whose sum is 10. All of the possible factors are shown at the right. In practice, when the correct pair is found, the remaining choices need not be checked.

Factors	Sum
$-1,\quad 24$	23
$1,\ -24$	-23
$-2,\quad 12$	10
$2,\ -12$	-10
$-3,\quad 8$	5
$3,\ -8$	-5

Write the trinomial in factored form. $y^2 + 10y - 24 = (y - 2)(y + 12)$

Check: $(y - 2)(y + 12) = y^2 + 12y - 2y - 24 = y^2 + 10y - 24$

When only integers are used, some trinomials do not factor. For example, to factor $x^2 + 11x + 5$, it would be necessary to find two positive integers whose product is 5 and whose sum is 11. This is not possible, since the only positive factors of 5 are 1 and 5, and the sum of 1 and 5 is 6. The polynomial $x^2 + 11x + 5$ is a prime polynomial. The polynomial is said to be nonfactorable over the integers. Binomials of the form $x + a$ or $x - a$ are also prime polynomials.

Example 9

Factor: $x^2 + 2x - 4$

Solution

There are no factors of -4 whose signs are opposite and whose sum is 2.
The trinomial is nonfactorable over the integers.

Example 11

Factor: $10 - 3x - x^2$

Solution

$10 - 3x - x^2 = (5 + x)(2 - x)$

Example 10

Factor: $x^2 - x - 20$

Your solution

$(x + 4)(x - 5)$

Example 12

Factor: $x^2 + 5xy + 6y^2$

Your solution

$(x + 2y)(x + 3y)$

Solutions on p. A18

| Objective D | **To factor $ax^2 + bx + c$** | |

There are various methods of factoring trinomials of the form $ax^2 + bx + c$, where $a \neq 1$. Factoring by using trial factors and factoring by grouping will be discussed in this objective. Factoring by using trial factors is illustrated first.

To use the trial factor method, use the factors of a and the factors of c to write all of the possible binomial factors of the trinomial. Then use FOIL to determine the correct factorization. To reduce the number of trial factors that must be considered, remember the following.

1. Use the signs of the constant term and the coefficient of x in the trinomial to determine the signs of the binomial factors. If the constant term is positive, the signs of the binomial factors will be the same as the sign of the coefficient of x in the trinomial. If the sign of the constant term is negative, the constant terms in the binomial will have opposite signs.

2. If the terms of the trinomial do not have a common factor, then the terms in either one of the binomial factors will not have a common factor.

Factor: $3x^2 - 8x + 4$

The terms have no common factor. The constant term is positive. The coefficient of x is negative. The binomial constants will be negative.	*Factors of 3* (coefficient of x^2) 1, 3	*Factors of 4* (constant term) $-1, -4$ $-2, -2$

Write trial factors. Use the *Outer* and *Inner* products of FOIL to determine the middle term of the trinomial.	*Trial Factors* $(x - 1)(3x - 4)$ $(x - 4)(3x - 1)$ $(x - 2)(3x - 2)$	*Middle Term* $-4x - 3x = -7x$ $-x - 12x = -13x$ $-2x - 6x = -8x$

Write the trinomial in factored form. $3x^2 - 8x + 4 = (x - 2)(3x - 2)$

Factor: $10 - x - 2x^2$

The terms have no common factor. The coefficient of x^2 is -2. The signs of the binomials will be opposites.	*Factors of 10* (constant term) 1, 10 2, 5	*Factors of -2* (coefficient of x^2) 1, -2 $-1,\ \ 2$

Write trial factors. Use the *Outer* and *Inner* products of FOIL to determine the middle term of the trinomial.	*Trial Factors* $(1 + x)(10 - 2x)$ $(1 - x)(10 + 2x)$ $(2 + x)(5 - 2x)$ $(2 - x)(5 + 2x)$	*Middle Term* common factor common factor $-4x + 5x = x$ $4x - 5x = -x$

Write the trinomial in factored form. $10 - x - 2x^2 = (2 - x)(5 + 2x)$

Factor: $10y^3 + 44y^2 - 30y$

The GCF is $2y$. Factor the GCF from the terms.	$10y^3 + 44y^2 - 30y = 2y(5y^2 + 22y - 15)$

Factor the trinomial. The constant term is negative. The binomial constants will have opposite signs.	*Factors of 5* (coefficient of y^2) 1, 5	*Factors of -15* (constant term) $-1,\ \ \ 15$ $1, -15$ $-3,\ \ \ 5$ $3, -5$

Write trial factors. Use the *Outer* and *Inner* products of FOIL to determine the middle term of the trinomial. It is not necessary to test trial factors which have a common factor.	*Trial Factors* $(y - 1)(5y + 15)$ $(y + 15)(5y - 1)$ $(y + 1)(5y - 15)$ $(y - 15)(5y + 1)$ $(y - 3)(5y + 5)$ $(y + 5)(5y - 3)$ $(y + 3)(5y - 5)$ $(y - 5)(5y + 3)$	*Middle Term* Common factor $-y + 75y = 74y$ Common factor $y - 75y = -74y$ Common factor $-3y + 25y = 22y$ Common factor $3y - 25y = -22y$

Write the trinomial in factored form. $10y^3 + 44y^2 - 30y = 2y(y + 5)(5y - 3)$

For this example, all the trial factors were listed. Once the correct factors have been found, the remaining trial factors can be omitted.

Trinomials of the form $ax^2 + bx + c$ can also be factored by grouping. This method is an extension of the method discussed in the previous objective.

To factor $ax^2 + bx + c$, first find two factors of $a \cdot c$ whose sum is b. Then use factoring by grouping to write the factorization of the trinomial.

Factor: $3x^2 + 11x + 8$

Find two positive factors of 24 ($ac = 3 \cdot 8$) whose sum is 11, the coefficient of x.

Positive factors of 24	Sum
1, 24	25
2, 12	14
3, 8	11

The required sum has been found. The remaining factors need not be checked.

Use the factors of 24 whose sum is 11 to write $11x$ as $3x + 8x$. Factor by grouping.

$$3x^2 + 11x + 8 = 3x^2 + 3x + 8x + 8$$
$$= (3x^2 + 3x) + (8x + 8)$$
$$= 3x(x + 1) + 8(x + 1)$$
$$= (x + 1)(3x + 8)$$

Check: $(x + 1)(3x + 8) = 3x^2 + 8x + 3x + 8 = 3x^2 + 11x + 8$

Factor: $4z^2 - 17z - 21$

Find two factors of -84 [$ac = 4 \cdot (-21)$] whose sum is -17, the coefficient of z.

When the required sum is found, the remaining factors need not be checked.

Factors of -84	Sum
1, -84	-83
-1, 84	83
2, -42	-40
-2, 42	40
3, -28	-25
-3, 28	25
4, -21	-17

Use the factors of -84 whose sum is -17 to write $-17z$ as $4z - 21z$. Factor by grouping. Recall that $-21z - 21 = -(21z + 21)$.

$$4z^2 - 17z - 21 = 4z^2 + 4z - 21z - 21$$
$$= (4z^2 + 4z) - (21z + 21)$$
$$= 4z(z + 1) - 21(z + 1)$$
$$= (z + 1)(4z - 21)$$

Factor: $3x^2 - 11x + 4$

Find two negative factors of 12 [3 · 4] whose sum is −11.

Factors of 12	Sum
−1, −12	−13
−2, −6	−8
−3, −4	−7

Because no integer factors of 12 have a sum of −11, $3x^2 - 11x + 4$ is nonfactorable over the integers. $3x^2 - 11x + 4$ is a prime polynomial over the integers.

Either method of factoring discussed in this objective will always lead to a correct factorization of trinomials of the form $ax^2 + bx + c$ that are not prime polynomials.

Example 13

Factor: $6x^2 + 11x - 10$

Solution

$6x^2 + 11x - 10 = (2x + 5)(3x - 2)$

Example 14

Factor: $4x^2 + 15x - 4$

Your solution

$(x + 4)(4x - 1)$

Example 15

Factor: $12x^2 - 32x + 5$

Solution

$12x^2 - 32x + 5 = (6x - 1)(2x - 5)$

Example 16

Factor: $10x^2 + 39x + 14$

Your solution

$(2x + 7)(5x + 2)$

Example 17

Factor: $30y + 2xy - 4x^2y$

Solution

The GCF of $30y$, $2xy$, and $4x^2y$ is $2y$.
$30y + 2xy - 4x^2y = 2y(15 + x - 2x^2)$
$= 2y(5 + 2x)(3 - x)$

Example 18

Factor: $3a^3b^3 + 3a^2b^2 - 60ab$

Your solution

$3ab(ab + 5)(ab - 4)$

Solutions on p. A18

3.3 EXERCISES

▶ **Objective A**

Factor:

1. $6a^2 - 15a$
$3a(2a - 5)$

2. $32b^2 + 12b$
$4b(8b + 3)$

3. $4x^3 - 3x^2$
$x^2(4x - 3)$

4. $12a^5b^2 + 16a^4b$
$4a^4b(3ab + 4)$

5. $3a^2 - 10b^3$
Nonfactorable

6. $9x^2 + 14y^4$
Nonfactorable

7. $x^5 - x^3 - x$
$x(x^4 - x^2 - 1)$

8. $y^4 - 3y^2 - 2y$
$y(y^3 - 3y - 2)$

9. $16x^2 - 12x + 24$
$4(4x^2 - 3x + 6)$

10. $2x^5 + 3x^4 - 4x^2$
$x^2(2x^3 + 3x^2 - 4)$

11. $5b^2 - 10b^3 + 25b^4$
$5b^2(1 - 2b + 5b^2)$

12. $x^2y^4 - x^2y - 4x^2$
$x^2(y^4 - y - 4)$

13. $x^{2n} - x^n$
$x^n(x^n - 1)$

14. $a^{5n} + a^{2n}$
$a^{2n}(a^{3n} + 1)$

15. $x^{3n} - x^{2n}$
$x^{2n}(x^n - 1)$

16. $y^{4n} + y^{2n}$
$y^{2n}(y^{2n} + 1)$

17. $a^{2n+2} + a^2$
$a^2(a^{2n} + 1)$

18. $b^{n+5} - b^5$
$b^5(b^n - 1)$

19. $12x^2y^2 - 18x^3y + 24x^2y$
$6x^2y(2y - 3x + 4)$

20. $14a^4b^4 - 42a^3b^3 + 28a^3b^2$
$14a^3b^2(ab^2 - 3b + 2)$

21. $24a^3b^2 - 4a^2b^2 - 16a^2b^4$
$4a^2b^2(6a - 1 - 4b^2)$

22. $10x^2y + 20x^2y^2 + 30x^2y^3$
$10x^2y(1 + 2y + 3y^2)$

23. $y^{2n+2} + y^{n+2} - y^2$
$y^2(y^{2n} + y^n - 1)$

24. $a^{2n+2} + a^{2n+1} + a^n$
$a^n(a^{n+2} + a^{n+1} + 1)$

▶ **Objective B**

Factor:

25. $x(a + 2) - 2(a + 2)$
$(a + 2)(x - 2)$

26. $3(x + y) + a(x + y)$
$(x + y)(3 + a)$

27. $a(x - 2) - b(2 - x)$
$(x - 2)(a + b)$

Factor:

28. $3(a - 7) - b(7 - a)$
$(a - 7)(3 + b)$

29. $x(a - 2b) + y(2b - a)$
$(a - 2b)(x - y)$

30. $b(3 - 2c) - 5(2c - 3)$
$(3 - 2c)(b + 5)$

31. $xy + 4y - 2x - 8$
$(x + 4)(y - 2)$

32. $ab + 7b - 3a - 21$
$(a + 7)(b - 3)$

33. $ax + bx - ay - by$
$(a + b)(x - y)$

34. $2ax - 3ay - 2bx + 3by$
$(2x - 3y)(a - b)$

35. $x^2y - 3x^2 - 2y + 6$
$(y - 3)(x^2 - 2)$

36. $a^2b + 3a^2 + 2b + 6$
$(b + 3)(a^2 + 2)$

37. $6 + 2y + 3x^2 + x^2y$
$(3 + y)(2 + x^2)$

38. $15 + 3b - 5a^2 - a^2b$
$(5 + b)(3 - a^2)$

39. $2ax^2 + bx^2 - 4ay - 2by$
$(2a + b)(x^2 - 2y)$

40. $4a^2x + 2a^2y - 6bx - 3by$
$(2x + y)(2a^2 - 3b)$

41. $6xb + 3ax - 4by - 2ay$
$(2b + a)(3x - 2y)$

42. $a^2x - 3a^2y + 2x - 6y$
$(x - 3y)(a^2 + 2)$

43. $x^ny - 5x^n + y - 5$
$(y - 5)(x^n + 1)$

44. $a^nx^n + 2a^n + x^n + 2$
$(x^n + 2)(a^n + 1)$

▶ **Objective C**

Factor:

45. $x^2 - 8x + 15$
$(x - 3)(x - 5)$

46. $x^2 + 12x + 20$
$(x + 10)(x + 2)$

47. $a^2 + 12a + 11$
$(a + 1)(a + 11)$

48. $a^2 + a - 72$
$(a + 9)(a - 8)$

49. $b^2 + 2b - 35$
$(b + 7)(b - 5)$

50. $a^2 + 7a + 6$
$(a + 6)(a + 1)$

51. $y^2 - 16y + 39$
$(y - 3)(y - 13)$

52. $y^2 - 18y + 72$
$(y - 6)(y - 12)$

53. $b^2 + 4b - 32$
$(b + 8)(b - 4)$

54. $x^2 + x - 132$
$(x + 12)(x - 11)$

55. $a^2 - 15a + 56$
$(a - 7)(a - 8)$

56. $x^2 + 15x + 50$
$(x + 10)(x + 5)$

57. $y^2 + 13y + 12$
$(y + 1)(y + 12)$

58. $b^2 - 6b - 16$
$(b + 2)(b - 8)$

59. $x^2 + 4x - 5$
$(x + 5)(x - 1)$

60. $a^2 - 3ab + 2b^2$
$(a - b)(a - 2b)$

61. $a^2 + 11ab + 30b^2$
$(a + 5b)(a + 6b)$

62. $a^2 + 8ab - 33b^2$
$(a + 11b)(a - 3b)$

Factor:

63. $x^2 - 14xy + 24y^2$
$(x - 2y)(x - 12y)$

64. $x^2 + 5xy + 6y^2$
$(x + 3y)(x + 2y)$

65. $y^2 + 2xy - 63x^2$
$(y + 9x)(y - 7x)$

66. $2 + x - x^2$
$(1 + x)(2 - x)$

67. $21 - 4x - x^2$
$(7 + x)(3 - x)$

68. $5 + 4x - x^2$
$(1 + x)(5 - x)$

69. $50 + 5a - a^2$
$(5 + a)(10 - a)$

70. $x^2 - 5x + 6$
$(x - 3)(x - 2)$

71. $x^2 - 7x - 12$
Nonfactorable

▶ **Objective D**

Factor:

72. $2x^2 + 7x + 3$
$(2x + 1)(x + 3)$

73. $2x^2 - 11x - 40$
$(2x + 5)(x - 8)$

74. $6y^2 + 5y - 6$
$(2y + 3)(3y - 2)$

75. $4y^2 - 15y + 9$
$(4y - 3)(y - 3)$

76. $6b^2 - b - 35$
$(3b + 7)(2b - 5)$

77. $2a^2 + 13a + 6$
$(2a + 1)(a + 6)$

78. $3y^2 - 22y + 39$
$(3y - 13)(y - 3)$

79. $12y^2 - 13y - 72$
Nonfactorable

80. $6a^2 - 26a + 15$
Nonfactorable

81. $5x^2 + 26x + 5$
$(5x + 1)(x + 5)$

82. $4a^2 - a - 5$
$(a + 1)(4a - 5)$

83. $11x^2 - 122x + 11$
$(11x - 1)(x - 11)$

84. $10x^2 - 29x + 10$
$(5x - 2)(2x - 5)$

85. $2x^2 + 5x + 12$
Nonfactorable

86. $4x^2 - 6x + 1$
Nonfactorable

87. $6x^2 + 5xy - 21y^2$
$(2x - 3y)(3x + 7y)$

88. $6x^2 + 41xy - 7y^2$
$(6x - y)(x + 7y)$

89. $4a^2 + 43ab + 63b^2$
$(4a + 7b)(a + 9b)$

90. $7a^2 + 46ab - 21b^2$
$(7a - 3b)(a + 7b)$

91. $10x^2 - 23xy + 12y^2$
$(2x - 3y)(5x - 4y)$

92. $18x^2 + 27xy + 10y^2$
$(6x + 5y)(3x + 2y)$

93. $24 + 13x - 2x^2$
$(3 + 2x)(8 - x)$

94. $6 - 7x - 5x^2$
$(2 + x)(3 - 5x)$

95. $8 - 13x + 6x^2$
Nonfactorable

96. $30 + 17a - 20a^2$
Nonfactorable

97. $15 - 14a - 8a^2$
$(3 - 4a)(5 + 2a)$

98. $35 - 6b - 8b^2$
$(5 + 2b)(7 - 4b)$

Factor:

99. $6x^2 - 29x + 35$
$(3x - 7)(2x - 5)$

100. $4x^3 - 10x^2 + 6x$
$2x(2x - 3)(x - 1)$

101. $9a^3 + 30a^2 - 24a$
$3a(3a - 2)(a + 4)$

102. $12y^3 + 22y^2 - 70y$
$2y(3y - 5)(2y + 7)$

103. $5y^4 - 29y^3 + 20y^2$
$y^2(5y - 4)(y - 5)$

104. $30a^2 + 85ab + 60b^2$
$5(3a + 4b)(2a + 3b)$

105. $4x^3 + 10x^2y - 24xy^2$
$2x(2x - 3y)(x + 4y)$

106. $8a^4 + 37a^3b - 15a^2b^2$
$a^2(a + 5b)(8a - 3b)$

107. $100 - 5x - 5x^2$
$5(5 + x)(4 - x)$

108. $50x^2 + 25x^3 - 12x^4$
$x^2(10 - 3x)(5 + 4x)$

109. $320x - 8x^2 - 4x^3$
$4x(10 + x)(8 - x)$

110. $96y - 16xy - 2x^2y$
$2y(12 + x)(4 - x)$

111. $20x^2 - 38x^3 - 30x^4$
$2x^2(2 - 5x)(5 + 3x)$

112. $4x^2y^2 - 32xy + 60$
$4(xy - 3)(xy - 5)$

113. $a^4b^4 - 3a^3b^3 - 10a^2b^2$
$a^2b^2(ab + 2)(ab - 5)$

114. $2a^2b^4 + 9ab^3 - 18b^2$
$b^2(ab + 6)(2ab - 3)$

115. $90a^2b^2 + 45ab + 10$
$5(18a^2b^2 + 9ab + 2)$

116. $3x^3y^2 + 12x^2y - 96x$
$3x(xy + 8)(xy - 4)$

117. $4x^4 - 45x^2 + 80$
Nonfactorable

118. $x^4 + 2x^2 + 15$
Nonfactorable

119. $2a^5 + 14a^3 + 20a$
$2a(a^2 + 5)(a^2 + 2)$

120. $3b^6 - 9b^4 - 30b^2$
$3b^2(b^2 + 2)(b^2 - 5)$

121. $3x^4y^2 - 39x^2y^2 + 120y^2$
$3y^2(x^2 - 5)(x^2 - 8)$

122. $2a^3b^3 - 10a^2b^2 + 12ab$
$2ab(ab - 2)(ab - 3)$

123. $3x^3y^3 + 6x^2y^2 - 24xy$
$3xy(xy - 2)(xy + 4)$

124. $2x^4 + 14x^2 + 24$
$2(x^2 + 3)(x^2 + 4)$

125. $y^5 - 8y^3 + 15y$
$y(y^2 - 3)(y^2 - 5)$

126. $12x + x^2 - 6x^3$
$x(3 - 2x)(4 + 3x)$

127. $3y - 16y^2 + 16y^3$
$y(3 - 4y)(1 - 4y)$

128. $12x^{2n} - 30x^n + 12$
$6(2x^n - 1)(x^n - 2)$

129. $12y^{2n} - 51y^n + 45$
$3(y^n - 3)(4y^n - 5)$

130. $2x^{3n} + 4x^{2n} - 30x^n$
$2x^n(x^n - 3)(x^n + 5)$

131. $x^{3n} + 10x^{2n} + 16x^n$
$x^n(x^n + 2)(x^n + 8)$

132. $10x^{2n} + 25x^n - 60$
$5(2x^n - 3)(x^n + 4)$

SECTION 3.4 | **Special Factoring**

Objective A | **To factor the difference of two perfect squares or a perfect square trinomial**

The product of a term and itself is called a **perfect square.** The exponents on variables of perfect squares are always even numbers.

Term		Perfect Square
5	$5 \cdot 5 =$	25
x	$x \cdot x =$	x^2
$3y^4$	$3y^4 \cdot 3y^4 =$	$9y^8$
x^n	$x^n \cdot x^n =$	x^{2n}

The **square root** of a perfect square is one of the two equal factors of the perfect square. "$\sqrt{}$" is the symbol for square root. To find the exponent of the square root of a variable term, multiply the exponent by $\frac{1}{2}$.

$\sqrt{25} = 5$
$\sqrt{x^2} = x$
$\sqrt{9y^8} = 3y^4$
$\sqrt{x^{2n}} = x^n$

The difference of two perfect squares is the product of the sum and difference of two terms. The factors of the difference of two squares are the sum and difference of the square roots of the perfect squares.

Factors of the Difference of Two Squares

$$a^2 - b^2 = (a + b)(a - b)$$

The sum of two perfect squares, $a^2 + b^2$, is nonfactorable over the integers.

Factor: $4x^2 - 81y^2$

Write the binomial as the difference of two perfect squares.

$$4x^2 - 81y^2 = (2x)^2 - (9y)^2$$

The factors are the sum and difference of the square roots of the perfect squares.

$$= (2x + 9y)(2x - 9y)$$

A perfect square trinomial is the square of a binomial.

Factors of the Square of a Trinomial

$$a^2 + 2ab + b^2 = (a + b)^2$$
$$a^2 - 2ab + b^2 = (a - b)^2$$

In factoring a perfect square trinomial, remember that the terms of the binomial are the square roots of the perfect squares of the trinomial. The sign in the binomial is the sign of the middle term of the trinomial.

Factor: $x^2 - 14x + 49$

The trinomial is a perfect square.
Write the factors as the square of a binomial. $x^2 - 14x + 49 = (x - 7)^2$

Example 1

Factor: $25x^2 - 1$

Solution

$25x^2 - 1 = (5x)^2 - (1)^2$
$\qquad = (5x + 1)(5x - 1)$

Example 2

Factor: $x^2 - 36y^4$

Your solution

$(x + 6y^2)(x - 6y^2)$

Example 3

Factor: $4x^2 - 20x + 25$

Solution

$4x^2 - 20x + 25 = (2x - 5)^2$

Example 4

Factor: $9x^2 + 12x + 4$

Your solution

$(3x + 2)^2$

Example 5

Factor: $(x + y)^2 - 4$

Solution

$(x + y)^2 - 4 = (x + y)^2 - (2)^2$
$\qquad = (x + y + 2)(x + y - 2)$

Example 6

Factor: $(a + b)^2 - (a - b)^2$

Your Solution

$4ab$

Solutions on p. A18

Objective B	**To factor the sum or the difference of two cubes**

The product of the same three factors is called a **perfect cube.** The exponents on variables of perfect cubes are always divisible by 3.

Term		Perfect Cube
2	$2 \cdot 2 \cdot 2 =$	8
$3y$	$3y \cdot 3y \cdot 3y =$	$27y^3$
y^2	$y^2 \cdot y^2 \cdot y^2 =$	y^6

The **cube root** of a perfect cube is one of the three equal factors of the perfect cube. "$\sqrt[3]{}$" is the symbol for cube root. To find the exponent of the cube root of a variable term, multiply the exponent by $\frac{1}{3}$.

$$\sqrt[3]{8} = 2$$
$$\sqrt[3]{27y^3} = 3y$$
$$\sqrt[3]{y^6} = y^2$$

Factoring the Sum or Difference of Two Cubes

$$a^3 + b^3 = (a + b)(a^2 - ab + b^2)$$

$$a^3 - b^3 = (a - b)(a^2 + ab + b^2)$$

Factor: $8x^3 - 27$

Write the binomial as the difference of two perfect cubes.

$$8x^3 - 27 = (2x)^3 - 3^3$$

$$= (2x - 3)(4x^2 + 6x + 9)$$

The terms of the binomial factor are the cube roots of the perfect cubes. The sign of the binomial factor is the same sign as in the given binomial. The trinomial factor is obtained from the binomial factor.

Square of the first↗ term

Opposite of the product⌐ of the two terms

Square of the last term

Factor: $a^3 + 64y^3$

Write the binomial as the sum of two perfect cubes.
Write the factors.

$$a^3 + 64y^3 = a^3 + (4y)^3$$

$$= (a + 4y)(a^2 - 4ay + 16y^2)$$

Factor: $64y^4 - 125y$

Factor out y, the GCF.

$$64y^4 - 125y = y(64y^3 - 125)$$

Write the binomial as the difference of two cubes.

$$= y[(4y)^3 - 5^3]$$

Write the factors.

$$= y(4y - 5)(16y^2 + 20y + 25)$$

Example 7

Factor: $x^3y^3 - 1$

Solution

$x^3y^3 - 1 = (xy)^3 - 1^3$
$\qquad = (xy - 1)(x^2y^2 + xy + 1)$

Example 8

Factor: $8x^3 + y^3z^3$

Your solution

$(2x + yz)(4x^2 - 2xyz + y^2z^2)$

Example 9

Factor: $(x + y)^3 - x^3$

Solution

$(x + y)^3 - x^3$
$\qquad = [(x + y) - x][(x + y)^2 + x(x + y) + x^2]$
$\qquad = y(x^2 + 2xy + y^2 + x^2 + xy + x^2)$
$\qquad = y(3x^2 + 3xy + y^2)$

Example 10

Factor: $(x - y)^3 + (x + y)^3$

Your solution

$2x(x^2 + 3y^2)$

Solutions on p. A18

Objective C   **To factor a trinomial that is quadratic in form**

Certain trinomials which are not quadratic can be expressed as quadratic trinomials by making suitable variable substitutions. A trinomial is quadratic in form if it can be written as $au^2 + bu + c$.

Each of the trinomials shown below is quadratic in form.

$$x^4 + 5x^2 + 6$$
$$(x^2)^2 + 5(x^2) + 6$$

Let $u = x^2$. $u^2 + 5u + 6$

$$2x^2y^2 + 3xy - 9$$
$$2(xy)^2 + 3(xy) - 9$$

Let $u = xy$. $2u^2 + 3u - 9$

To factor a trinomial that is quadratic in form, the first term in each binomial will be u.

For example, $x^4 + 5x^2 + 6$ is quadratic form when $u = x^2$. To factor the trinomial, think of it as $u^2 + 5u + 6$.

Let $u = x^2$.
Factor.
Replace u by x^2

$$x^4 + 5x^2 + 6 = u^2 + 5u + 6$$
$$= (u + 3)(u + 2)$$
$$= (x^2 + 3)(x^2 + 2)$$

Here is another example.

Factor $x - 2\sqrt{x} - 15$.

Let $u = \sqrt{x}$. Then $u^2 = x$
Factor.
Replace u by $\sqrt{x}$.

$$x - 2\sqrt{x} - 15 = u^2 - 2u - 15$$
$$= (u - 5)(u + 3)$$
$$= (\sqrt{x} - 5)(\sqrt{x} + 3)$$

Example 11

Factor: $6x^2y^2 - xy - 12$

Solution

Let $u = xy$.
$$6x^2y^2 - xy - 12 = 6u^2 - u - 12$$
$$= (3u + 4)(2u - 3)$$
$$= (3xy + 4)(2xy - 3)$$

Example 12

Factor: $3x^4 + 4x^2 - 4$

Your solution

$(x^2 + 2)(3x^2 - 2)$

Solution on p. A18

| Objective D | **To factor completely** |

When factoring a polynomial completely, ask the following questions about the polynomial.

1. Is there a common factor? If so, factor out the GCF.
2. If the polynomial is a binomial, is it the difference of two perfect squares, the sum of two cubes, or the difference of two cubes? If so, factor.
3. If the polynomial is a trinomial, is it a perfect square trinomial or the product of two binomials? If so, factor.
4. Can the polynomial be factored by grouping? If so, factor.
5. Is each factor nonfactorable over the integers? If not, factor.

Example 13

Factor: $x^2y + 2x^2 - y - 2$

Solution

$$
\begin{aligned}
x^2y + 2x^2 - y - 2 &= (x^2y + 2x^2) + (-y - 2) \\
&= x^2(y + 2) - (y + 2) \\
&= (y + 2)(x^2 - 1) \\
&= (y + 2)(x + 1)(x - 1)
\end{aligned}
$$

Example 14

Factor: $4x - 4y - x^3 + x^2y$

Your solution

$(x - y)(2 + x)(2 - x)$

Example 15

Factor: $x^{4n} - y^{4n}$

Solution

$$
\begin{aligned}
x^{4n} - y^{4n} &= (x^{2n})^2 - (y^{2n})^2 \\
&= (x^{2n} + y^{2n})(x^{2n} - y^{2n}) \\
&= (x^{2n} + y^{2n})[(x^n)^2 - (y^n)^2] \\
&= (x^{2n} + y^{2n})(x^n + y^n)(x^n - y^n)
\end{aligned}
$$

Example 16

Factor: $x^{4n} - x^{2n}y^{2n}$

Your solution

$x^{2n}(x^n + y^n)(x^n - y^n)$

Content and Format © 1991 HMCo.

3.4 EXERCISES

▶ **Objective A**

Factor:

1. $x^2 - 16$
$(x + 4)(x - 4)$

2. $y^2 - 49$
$(y + 7)(y - 7)$

3. $4x^2 - 1$
$(2x + 1)(2x - 1)$

4. $81x^2 - 4$
$(9x + 2)(9x - 2)$

5. $16x^2 - 121$
$(4x + 11)(4x - 11)$

6. $49y^2 - 36$
$(7y + 6)(7y - 6)$

7. $1 - 9a^2$
$(1 + 3a)(1 - 3a)$

8. $16 - 81y^2$
$(4 + 9y)(4 - 9y)$

9. $x^2y^2 - 100$
$(xy + 10)(xy - 10)$

10. $a^2b^2 - 25$
$(ab + 5)(ab - 5)$

11. $x^2 + 4$
Nonfactorable

12. $a^2 + 16$
Nonfactorable

13. $25 - a^2b^2$
$(5 + ab)(5 - ab)$

14. $64 - x^2y^2$
$(8 + xy)(8 - xy)$

15. $a^{2n} - 1$
$(a^n + 1)(a^n - 1)$

16. $b^{2n} - 16$
$(b^n + 4)(b^n - 4)$

17. $x^2 - 12x + 36$
$(x - 6)^2$

18. $y^2 - 6y + 9$
$(y - 3)^2$

19. $b^2 - 2b + 1$
$(b - 1)^2$

20. $a^2 + 14a + 49$
$(a + 7)^2$

21. $16x^2 - 40x + 25$
$(4x - 5)^2$

22. $49x^2 + 28x + 4$
$(7x + 2)^2$

23. $4a^2 + 4a - 1$
Nonfactorable

24. $9x^2 + 12x - 4$
Nonfactorable

25. $b^2 + 7b + 14$
Nonfactorable

26. $y^2 - 5y + 25$
Nonfactorable

27. $x^2 + 6xy + 9y^2$
$(x + 3y)^2$

28. $4x^2y^2 + 12xy + 9$
$(2xy + 3)^2$

29. $25a^2 - 40ab + 16b^2$
$(5a - 4b)^2$

30. $4a^2 - 36ab + 81b^2$
$(2a - 9b)^2$

31. $x^{2n} + 6x^n + 9$
$(x^n + 3)^2$

32. $y^{2n} - 16y^n + 64$
$(y^n - 8)^2$

33. $(x - 4)^2 - 9$
$(x - 7)(x - 1)$

34. $16 - (a - 3)^2$
$(7 - a)(1 + a)$

35. $(x - y)^2 - (a + b)^2$
$(x - y + a + b)(x - y - a - b)$

36. $(x - 2y)^2 - (x + y)^2$
$(-3y)(2x - y)$

▶ **Objective B**

Factor:

37. $x^3 - 27$
$(x - 3)(x^2 + 3x + 9)$

38. $y^3 + 125$
$(y + 5)(y^2 - 5y + 25)$

39. $8x^3 - 1$
$(2x - 1)(4x^2 + 2x + 1)$

40. $64a^3 + 27$
$(4a + 3)(16a^2 - 12a + 9)$

41. $x^3 - y^3$
$(x - y)(x^2 + xy + y^2)$

42. $x^3 - 8y^3$
$(x - 2y)(x^2 + 2xy + 4y^2)$

43. $m^3 + n^3$
$(m + n)(m^2 - mn + n^2)$

44. $27a^3 + b^3$
$(3a + b)(9a^2 - 3ab + b^2)$

45. $64x^3 + 1$
$(4x + 1)(16x^2 - 4x + 1)$

46. $1 - 125b^3$
$(1 - 5b)(1 + 5b + 25b^2)$

47. $27x^3 - 8y^3$
$(3x - 2y)(9x^2 + 6xy + 4y^2)$

48. $64x^3 + 27y^3$
$(4x + 3y)(16x^2 - 12xy + 9y^2)$

49. $x^3y^3 + 64$
$(xy + 4)(x^2y^2 - 4xy + 16)$

50. $8x^3y^3 + 27$
$(2xy + 3)(4x^2y^2 - 6xy + 9)$

51. $16x^3 - y^3$
Nonfactorable

52. $27x^3 - 8y^2$
Nonfactorable

53. $8x^3 - 9y^3$
Nonfactorable

54. $27a^3 - 16$
Nonfactorable

55. $(a - b)^3 - b^3$
$(a - 2b)(a^2 - ab + b^2)$

56. $a^3 + (a + b)^3$
$(2a + b)(a^2 + ab + b^2)$

57. $x^{6n} + y^{3n}$
$(x^{2n} + y^n)(x^{4n} - x^{2n}y^n + y^{2n})$

58. $x^{3n} + y^{3n}$
$(x^n + y^n)(x^{2n} - x^ny^n + y^{2n})$

59. $x^{3n} + 8$
$(x^n + 2)(x^{2n} - 2x^n + 4)$

60. $a^{3n} + 64$
$(a^n + 4)(a^{2n} - 4a^n + 16)$

▶ **Objective C**

Factor:

61. $x^2y^2 - 8xy + 15$
$(xy - 3)(xy - 5)$

62. $x^2y^2 - 8xy - 33$
$(xy + 3)(xy - 11)$

63. $x^2y^2 - 17xy + 60$
$(xy - 5)(xy - 12)$

64. $a^2b^2 + 10ab + 24$
$(ab + 6)(ab + 4)$

65. $x^4 - 9x^2 + 18$
$(x^2 - 3)(x^2 - 6)$

66. $y^4 - 6y^2 - 16$
$(y^2 + 2)(y^2 - 8)$

67. $b^4 - 13b^2 - 90$
$(b^2 + 5)(b^2 - 18)$

68. $a^4 + 14a^2 + 45$
$(a^2 + 9)(a^2 + 5)$

69. $x^4y^4 - 8x^2y^2 + 12$
$(x^2y^2 - 2)(x^2y^2 - 6)$

Factor:

70. $a^4b^4 + 11a^2b^2 - 26$
$(a^2b^2 + 13)(a^2b^2 - 2)$

71. $x^{2n} + 3x^n + 2$
$(x^n + 1)(x^n + 2)$

72. $a^{2n} - a^n - 12$
$(a^n + 3)(a^n - 4)$

73. $3x^2y^2 - 14xy + 15$
$(3xy - 5)(xy - 3)$

74. $5x^2y^2 - 59xy + 44$
$(5xy - 4)(xy - 11)$

75. $6a^2b^2 - 23ab + 21$
$(2ab - 3)(3ab - 7)$

76. $10a^2b^2 + 3ab - 7$
$(ab + 1)(10ab - 7)$

77. $2x^4 - 13x^2 - 15$
$(2x^2 - 15)(x^2 + 1)$

78. $3x^4 + 20x^2 + 32$
$(3x^2 + 8)(x^2 + 4)$

79. $2x^{2n} - 7x^n + 3$
$(2x^n - 1)(x^n - 3)$

80. $4x^{2n} + 8x^n - 5$
$(2x^n + 5)(2x^n - 1)$

81. $6a^{2n} + 19a^n + 10$
$(2a^n + 5)(3a^n + 2)$

82. $3y^{2n} - 16y^n + 16$
$(3y^n - 4)(y^n - 4)$

▶ **Objective D**

Factor:

83. $5x^2 + 10x + 5$
$5(x + 1)^2$

84. $12x^2 - 36x + 27$
$3(2x - 3)^2$

85. $3x^4 - 81x$
$3x(x - 3)(x^2 + 3x + 9)$

86. $27a^4 - a$
$a(3a - 1)(9a^2 + 3a + 1)$

87. $7x^2 - 28$
$7(x + 2)(x - 2)$

88. $20x^2 - 5$
$5(2x + 1)(2x - 1)$

89. $y^4 - 10y^3 + 21y^2$
$y^2(y - 3)(y - 7)$

90. $y^5 + 6y^4 - 55y^3$
$y^3(y + 11)(y - 5)$

91. $x^4 - 16$
$(x^2 + 4)(x + 2)(x - 2)$

92. $16x^4 - 81$
$(4x^2 + 9)(2x + 3)(2x - 3)$

93. $8x^5 - 98x^3$
$2x^3(2x + 7)(2x - 7)$

94. $16a - 2a^4$
$2a(2 - a)(4 + 2a + a^2)$

95. $x^3y^3 - x^3$
$x^3(y - 1)(y^2 + y + 1)$

96. $a^3b^6 - b^3$
$b^3(ab - 1)(a^2b^2 + ab + 1)$

97. $x^6y^6 - x^3y^3$
$x^3y^3(xy - 1)(x^2y^2 + xy + 1)$

98. $x^4 - 2x^3 - 35x^2$
$x^2(x + 5)(x - 7)$

99. $x^4 + 15x^3 - 56x^2$
$x^2(x^2 + 15x - 56)$

100. $4x^2 + 4x - 1$
Nonfactorable

Factor:

101. $8x^4 - 40x^3 + 50x^2$
$2x^2(2x - 5)^2$

102. $6x^5 + 74x^4 + 24x^3$
$2x^3(3x + 1)(x + 12)$

103. $x^4 - y^4$
$(x^2 + y^2)(x + y)(x - y)$

104. $16a^4 - b^4$
$(4a^2 + b^2)(2a + b)(2a - b)$

105. $x^6 + y^6$
$(x^2 + y^2)(x^4 - x^2y^2 + y^4)$

106. $x^4 - 5x^2 - 4$
Nonfactorable

107. $a^4 - 25a^2 - 144$
Nonfactorable

108. $3b^5 - 24b^2$
$3b^2(b - 2)(b^2 + 2b + 4)$

109. $16a^4 - 2a$
$2a(2a - 1)(4a^2 + 2a + 1)$

110. $x^4y^2 - 5x^3y^3 + 6x^2y^4$
$x^2y^2(x - 3y)(x - 2y)$

111. $a^4b^2 - 8a^3b^3 - 48a^2b^4$
$a^2b^2(a + 4b)(a - 12b)$

112. $16x^3y + 4x^2y^2 - 42xy^3$
$2xy(2x - 3y)(4x + 7y)$

113. $24a^2b^2 - 14ab^3 - 90b^4$
$2b^2(3a + 5b)(4a - 9b)$

114. $x^3 - 2x^2 - x + 2$
$(x + 1)(x - 1)(x - 2)$

115. $x^3 - 2x^2 - 4x + 8$
$(x - 2)^2(x + 2)$

116. $8xb - 8x - 4b + 4$
$4(2x - 1)(b - 1)$

117. $4xy + 8x + 4y + 8$
$4(y + 2)(x + 1)$

118. $4x^2y^2 - 4x^2 - 9y^2 + 9$
$(2x + 3)(2x - 3)(y + 1)(y - 1)$

119. $4x^4 - x^2 - 4x^2y^2 + y^2$
$(x + y)(x - y)(2x + 1)(2x - 1)$

120. $x^5 - 4x^3 - 8x^2 + 32$
$(x + 2)(x - 2)^2(x^2 + 2x + 4)$

121. $x^6y^3 + x^3 - x^3y^3 - 1$
$(x - 1)(x^2 + x + 1)(xy + 1)(x^2y^2 - xy + 1)$

122. $a^{2n+2} - 6a^{n+2} + 9a^2$
$a^2(a^n - 3)^2$

123. $x^{2n+1} + 2x^{n+1} + x$
$x(x^n + 1)^2$

124. $2x^{n+2} - 7x^{n+1} + 3x^n$
$x^n(2x - 1)(x - 3)$

125. $3b^{n+2} + 4b^{n+1} - 4b^n$
$b^n(3b - 2)(b + 2)$

SECTION 3.5   Solving Equations by Factoring

| Objective A | To solve an equation by factoring |

Consider the equation $ab = 0$. If a is not zero, then b must be zero. Conversely, if b is not zero then a must be zero. This is summarized in the Principle of Zero Products.

Principle of Zero Products

If the product of two factors is zero, then at least one of the factors must be zero.

$$\text{If } ab = 0, \text{ then } a = 0 \text{ or } b = 0.$$

The Principle of Zero Products is used to solve equations.

Solve: $(x - 4)(x + 2) = 0$

By the Principle of Zero Products, if $(x - 4)(x + 2) = 0$, then $x - 4 = 0$ or $x + 2 = 0$.

$$x - 4 = 0 \qquad\qquad x + 2 = 0$$
$$x = 4 \qquad\qquad\quad x = -2$$

Check:

$$\begin{array}{c|c} (x - 4)(x + 2) = 0 & (x - 4)(x + 2) = 0 \\ \hline (4 - 4)(4 + 2)\ \big|\ 0 & (-2 - 4)(-2 + 2)\ \big|\ 0 \\ 0 \cdot 6\ \big|\ 0 & -6 \cdot 0\ \big|\ 0 \\ 0 = 0 & 0 = 0 \end{array}$$

-2 and 4 check as solutions. The solutions are -2 and 4.

An equation of the form $ax^2 + bx + c = 0$, $a \neq 0$, is a **quadratic equation.** A quadratic equation is in standard form when the polynomial is written in descending order and equal to zero.

Some quadratic equations can be solved by factoring and then using the Principle of Zero Products.

Solve: $2x^2 - x = 1$

Write the equation in standard form.
Factor.
Use the Principle of Zero Products.
Solve each equation.

$$2x^2 - x = 1$$
$$2x^2 - x - 1 = 0$$
$$(2x + 1)(x - 1) = 0$$
$$2x + 1 = 0 \qquad x - 1 = 0$$
$$2x = -1 \qquad\qquad x = 1$$
$$x = -\frac{1}{2}$$

The solutions are $-\frac{1}{2}$ and 1. You should check these solutions.

The Principle of Zero Products can be extended to more than two factors. For example, if $abc = 0$, then $a = 0$, $b = 0$, or $c = 0$.

Solve: $x^3 - x^2 - 4x + 4 = 0$

Factor by grouping.

$$x^3 - x^2 - 4x + 4 = 0$$
$$(x^3 - x^2) - (4x - 4) = 0$$
$$x^2(x - 1) - 4(x - 1) = 0$$
$$(x - 1)(x^2 - 4) = 0$$
$$(x - 1)(x - 2)(x + 2) = 0$$

Use the Principle of Zero Products. Solve each equation.

$$x - 1 = 0 \qquad x - 2 = 0 \qquad x + 2 = 0$$
$$x = 1 \qquad x = 2 \qquad x = -2$$

The solutions are -2, 1, and 2.

Example 1

Solve: $x^2 + (x + 2)^2 = 100$

Solution

$$x^2 + (x + 2)^2 = 100$$
$$x^2 + x^2 + 4x + 4 = 100$$
$$2x^2 + 4x + 4 = 100$$
$$2x^2 + 4x - 96 = 0$$
$$2(x - 6)(x + 8) = 0$$
$$x - 6 = 0 \qquad x + 8 = 0$$
$$x = 6 \qquad x = -8$$

The solutions are -8 and 6.

Example 2

Solve: $(x + 4)(x - 1) = 14$

Your solution

-6 and 3

Example 3

Solve: $t^4 - 2t^2 + 1 = 0$

Solution

$$t^4 - 2t^2 + 1 = 0$$
$$(t^2 - 1)(t^2 - 1) = 0$$
$$(t - 1)(t + 1)(t - 1)(t + 1) = 0$$
$$t - 1 = 0, t + 1 = 0, t - 1 = 0, t + 1 = 0$$
$$t = 1, t = -1, t = 1, t = -1$$

The solutions are -1 and 1. These solutions are called **repeated** solutions.

Example 4

Solve $a^4 - 5a^2 + 4 = 0$

Your solution

-2, -1, 1, and 2

Solutions on p. A19

3.5   EXERCISES

▶ **Objective A**

Solve:

1. $(x - 5)(x + 3) = 0$
5 and -3

2. $(x - 2)(x + 6) = 0$
2 and -6

3. $(x + 7)(x - 8) = 0$
-7 and 8

4. $(2x - 3)(x + 7) = 0$
$\frac{3}{2}$ and -7

5. $(3x + 5)(x - 4) = 0$
$-\frac{5}{3}$ and 4

6. $(4x - 1)(3x + 5) = 0$
$\frac{1}{4}$ and $-\frac{5}{3}$

7. $(x + 1)(x - 3)(x + 4) = 0$
-4, -1, and 3

8. $(x - 5)(x - 3)(x + 1) = 0$
-1, 3, and 5

9. $(2x - 1)(x - 3)(3x + 4) = 0$
$-\frac{4}{3}, \frac{1}{2},$ and 3

10. $x(2x + 5)(x - 6) = 0$
$-\frac{5}{2}$, 0, and 6

11. $2x(3x - 2)(x + 4) = 0$
-4, 0, and $\frac{2}{3}$

12. $6x(3x - 7)(x + 7) = 0$
$-7, \frac{7}{3},$ and 0

13. $x^2 + 2x - 15 = 0$
-5 and 3

14. $t^2 + 3t - 10 = 0$
-5 and 2

15. $z^2 - 4z + 3 = 0$
1 and 3

16. $s^2 - 5s + 4 = 0$
1 and 4

17. $p^2 + 3p + 2 = 0$
-1 and -2

18. $v^2 + 6v + 5 = 0$
-1 and -5

19. $x^2 - 6x + 9 = 0$
3

20. $y^2 - 8y + 16 = 0$
4

21. $12y^2 + 8y = 0$
0 and $-\frac{2}{3}$

22. $6x^2 - 9x = 0$
0 and $\frac{3}{2}$

23. $r^2 - 10 = 3r$
-2 and 5

24. $t^2 - 12 = 4t$
6 and -2

25. $3v^2 - 5v + 2 = 0$
$\frac{2}{3}$ and 1

26. $2p^2 - 3p - 2 = 0$
2 and $-\frac{1}{2}$

27. $3s^2 + 8s = 3$
$\frac{1}{3}$ and -3

28. $3x^2 + 5x = 12$
$\frac{4}{3}$ and -3

29. $9z^2 = 12z - 4$
$\frac{2}{3}$

30. $6r^2 = 12 - r$
$-\frac{3}{2}$ and $\frac{4}{3}$

31. $4t^2 = 4t + 3$
$-\frac{1}{2}$ and $\frac{3}{2}$

32. $5y^2 + 11y = 12$
$\frac{4}{5}$ and -3

33. $4v^2 - 4v + 1 = 0$
$\frac{1}{2}$

34. $9s^2 - 6s + 1 = 0$
$\frac{1}{3}$

35. $x^2 - 9 = 0$
-3 and 3

36. $t^2 - 16 = 0$
-4 and 4

Solve:

37. $4y^2 - 1 = 0$
$-\frac{1}{2}$ and $\frac{1}{2}$

38. $9z^2 - 4 = 0$
$-\frac{2}{3}$ and $\frac{2}{3}$

39. $x + 15 = x(x - 1)$
-3 and 5

40. $7y^2 + 11y = y^2 + 72$
$-\frac{9}{2}$ and $\frac{8}{3}$

41. $6z^2 + 17z = -12$
$-\frac{3}{2}$ and $-\frac{4}{3}$

42. $t + 15 = t(t - 1)$
-3 and 5

43. $r + 18 = r(r - 2)$
-3 and 6

44. $x^2 - x - 2 = (2x - 1)(x - 3)$
1 and 5

45. $v^2 + v + 5 = (3v + 2)(v - 4)$
$-1, \frac{13}{2}$

46. $z^2 + 5z - 4 = (2z + 1)(z - 4)$
0 and 12

47. $4x^2 + x - 10 = (x - 2)(x + 1)$
-2 and $\frac{4}{3}$

48. $x^3 + 2x^2 - 15x = 0$
$-5, 0,$ and 3

49. $c^3 + 3c^2 - 10c = 0$
$-5, 0,$ and 2

50. $t^3 - 10t = 3t^2$
$-2, 0,$ and 5

51. $m^3 - 12m = 4m^2$
$-2, 0,$ and 6

52. $3y^3 - 5y^2 + 2y = 0$
$0, \frac{2}{3},$ and 1

53. $2p^3 - 3p^2 - 2p = 0$
$-\frac{1}{2}, 0,$ and 2

54. $y^4 - 13y^2 + 36 = 0$
$-3, -2, 2,$ and 3

55. $k^4 - 10k^2 + 9 = 0$
$-3, -1, 1,$ and 3

56. $z^4 - 18z^2 + 81 = 0$
-3 and 3

57. $y^4 - 8y^2 + 16 = 0$
-2 and 2

58. $x^3 + x^2 - 4x - 4 = 0$
$-2, -1,$ and 2

59. $a^3 + a^2 - 9a - 9 = 0$
$-3, -1,$ and 3

60. $2x^3 - x^2 - 2x + 1 = 0$
$-1, \frac{1}{2},$ and 1

61. $3x^3 + 2x^2 - 12x - 8 = 0$
$-2, -\frac{2}{3},$ and 2

62. $2x^3 + 3x^2 - 18x - 27 = 0$
$-3, -\frac{3}{2},$ and 3

63. $5x^3 + 2x^2 - 20x - 8 = 0$
$-2, -\frac{2}{5},$ and 2

Calculators and Computers

General Factoring   The program GENERAL FACTORING on the Math ACE Disk can be used to practice factoring. You may choose to practice polynomials of the form

$$x^2 + bx + c$$
$$ax^2 + bx + c \quad \text{or}$$
$$x^3 + a^3$$

These choices, along with the option of quitting the program, are given on a menu screen. You may practice for as long as you like with any type of problem. At the end of each problem, you may select to return to the menu screen or to continue practicing.

The program will present you with a polynomial to factor. When you have tried to factor the polynomial using paper and pencil, press the RETURN key on the keyboard. The correct factorization will be displayed.

Chapter Summary

Key Words   A *monomial* is a number, a variable, or a product of numbers and variables.

A *polynomial* is a variable expression in which the terms are monomials.

A *binomial* is a polynomial of *two* terms.

A *trinomial* is a polynomial of *three* terms.

The *degree of a polynomial* in one variable is the greatest of the degrees of any of its terms.

To *factor a polynomial* means to write the polynomial as the product of other polynomials.

A *quadratic trinomial* is a polynomial of the form $ax^2 + bx + c$, where a, b, and c are non-zero constants.

A polynomial is *nonfactorable over the integers* if it does not factor using only integers.

To *factor a quadratic trinomial* of the form $ax^2 + bx + c$ means to express the trinomial as the product of two binomials.

The product of a term and itself is a *perfect square*.

The *square root* of a perfect square is one of the two equal factors of the perfect square.

The product of the same three factors is called a *perfect cube*.

The *cube root* of a perfect cube is one of the three equal factors of the perfect cube.

An equation of the form $ax^2 + bx + c = 0$, $a \neq 0$, is a quadratic equation.

Essential Rules

Rule of Negative Exponents	If n is a positive integer and $x \neq 0$, then $x^{-n} = \frac{1}{x^n}$ and $x^n = \frac{1}{x^{-n}}$.
Rule for Multiplying Exponential Expressions	If m and n are integers, then $x^m \cdot x^n = x^{m+n}$.
Rule for Simplifying Powers of Exponential Expressions	If m and n are integers, then $(x^m)^n = x^{m \cdot n}$.
Rule for Simplifying Powers of Products	If m, n, and p are integers, then $(x^m \cdot y^n)^p = x^{mp}y^{np}$.
Rule for Dividing Exponential Expressions	If m and n are integers and $x \neq 0$, then $\frac{x^m}{x^n} = x^{m-n}$
Rule for Simplifying Powers of Quotients	If m, n, and p are positive integers and $y \neq 0$, then $\left(\frac{x^m}{y^n}\right)^p = \frac{x^{mp}}{y^{np}}$.
The Sum and Difference of Two Terms	$(a + b)(a - b) = a^2 - b^2$
The Square of a Binomial	$(a + b)^2 = a^2 + 2ab + b^2$ $(a - b)^2 = a^2 - 2ab + b^2$
The Sum or Difference of Two Cubes	$a^3 + b^3 = (a + b)(a^2 - ab + b^2)$ $a^3 - b^3 = (a - b)(a^2 + ab + b^2)$
Principle of Zero Products	If $ab = 0$, then $a = 0$ or $b = 0$.

Chapter Review

SECTION 1

1. Simplify $(5x^2 - 8xy + 2y^2) - (x^2 - 3y^2)$.
   $4x^2 - 8xy + 5y^2$

2. Simplify $(-2a^2b^4)^3 (3ab^2)$.
   $-24a^7b^{14}$

3. Simplify $(2x^{-1}y^2z^5)^4 (-3x^3yz^{-3})^2$.
   $144x^2y^{10}z^{14}$

4. Simplify $x^{-1}y + y^{-1}x$.
   $\dfrac{x^2 + y^2}{xy}$

5. Simplify $\dfrac{(2a^4b^{-3}c^2)^3}{(2a^3b^2c^{-1})^4}$.
   $\dfrac{c^{10}}{2b^{17}}$

6. Write 2.54×10^{-3} in decimal notation.
   0.00254

7. The mass of the moon is 3.7×10^{-8} times the mass of the sun. The mass of the sun is 2.19×10^{27} tons. Find the mass of the moon. Write the answer in scientific notation.
   8.103×10^{19} tons

8. The most distant object visible from earth without the aid of a telescope is the Great Galaxy of Andromeda. It takes light from the Great Galaxy of Andromeda 2.2×10^6 years to travel to earth. Light travels about 6.7×10^8 miles per hour. How far from earth is the Great Galaxy of Andromeda? Use a 360-day year.
   1.27×10^{19} miles

SECTION 2

9. Simplify $4x^2y(3x^3y^2 + 2x^{-1}y - 7y^3)$.
   $12x^5y^3 + 8xy^2 - 28x^2y^4$

10. Simplify $5x^2 - 4x[x - (3x + 2) + x]$.
    $9x^2 + 8x$

11. Simplify $(x - 4)(3x + 2)(2x - 3)$.
    $6x^3 - 29x^2 + 14x + 24$

12. Simplify $(5a + 2b)(5a - 2b)$.
    $25a^2 - 4b^2$

13. Simplify $(4x - 3y)^2$.
    $16x^2 - 24xy + 9y^2$

14. The length of a rectangle is $(5x + 3)$ cm. The width is $(2x - 7)$ cm. Find the area of the rectangle in terms of the variable x.
    $(10x^2 - 29x - 21)$ cm^2

SECTION 3

15. Factor $18a^5b^2 - 12a^3b^3 + 30a^2b$.
$6a^2b(3a^3b - 2ab^2 + 5)$

16. Factor $2ax + 4bx - 3ay - 6by$.
$(a + 2b)(2x - 3y)$

17. Factor $12 + x - x^2$.
$(4 - x)(3 + x)$

18. Factor $x^2 - 3x - 40$.
$(x - 8)(x + 5)$

19. Factor $6x^2 - 31x + 18$.
$(3x - 2)(2x - 9)$

20. Factor $24x^2 + 38x + 15$.
$(6x + 5)(4x + 3)$

SECTION 4

21. Factor $x^2y^2 - 9$.
$(xy + 3)(xy - 3)$

22. Factor $4x^2 + 12xy + 9y^2$.
$(2x + 3y)^2$

23. Factor $x^{2n} - 12x^n + 36$.
$(x^n - 6)^2$

24. Factor $64a^3 - 27b^3$.
$(4a - 3b)(16a^2 + 12ab + 9b^2)$

25. Factor $(a + b)^3 + 1$.
$(a + b + 1)(a^2 + 2ab + b^2 - a - b + 1)$

26. Factor $15x^4 + x^2 - 6$.
$(3x^2 + 2)(5x^2 - 3)$

27. Factor $21x^4y^4 + 23x^2y^2 + 6$.
$(7x^2y^2 + 3)(3x^2y^2 + 2)$

28. Factor $3a^6 - 15a^4 - 18a^2$.
$3a^2(a^2 + 1)(a^2 - 6)$

SECTION 5

29. Solve $6x^2 + 60 = 39x$.
$\frac{5}{2}$ and 4

30. Solve $x^3 + 16 = x(x + 16)$.
1, 4, and -4

Content and Format © 1991 HMCo.

Chapter Test

1. Simplify:
   $(6x^3 - 7x^2 + 6x - 7) - (4x^3 - 3x^2 + 7)$
   $2x^3 - 4x^2 + 6x - 14$ [3.1A]

2. Simplify: $(-4a^2b)^3(-ab^4)$
   $64a^7b^7$ [3.1B]

3. Simplify: $\frac{(2a^{-4}b^2)^3}{4a^{-2}b^{-1}}$
   $\frac{2b^7}{a^{10}}$ [3.1C]

4. Write the number 0.000000501 in scientific notation.
   5.01×10^{-7} [3.1D]

5. Write the number of seconds in one week in scientific notation.
   6.048×10^5 [3.1E]

6. Simplify: $-6rs^2(3r - 2s - 3)$
   $-18r^2s^2 + 12rs^3 + 18rs^2$ [3.2A]

7. Simplify: $-5x[3 - 2(2x - 4) - 3x]$
   $35x^2 - 55x$ [3.2A]

8. Simplify: $(3a + 4b)(2a - 7b)$
   $6a^2 - 13ab - 28b^2$ [3.2B]

9. Simplify: $(3t^3 - 4t^2 + 1)(2t^2 - 5)$
   $6t^5 - 8t^4 - 15t^3 + 22t^2 - 5$ [3.2B]

10. Simplify: $(7 - 5x)(7 + 5x)$
    $49 - 25x^2$ [3.2C]

11. Simplify: $(3z - 5)^2$
    $9z^2 - 30z + 25$ [3.2C]

12. The length of a rectangle is $(5x + 1)$ ft. The width is $(2x - 1)$ ft. Find the area of the rectangle in terms of the variable x.
    $(10x^2 - 3x - 1)$ ft^2 [3.2D]

13. Factor: $6a^3b^2 - 4a^2b^2 + 4ab^4$
$2ab^2(3a^2 - 2a + 2b^2)$ [3.3A]

14. Factor: $y^2 + 6y - 72$
$(y + 12)(y - 6)$ [3.3C]

15. Factor: $12x^2 - 11x + 2$
$(3x - 2)(4x - 1)$ [3.3D]

16. Factor: $12 - 17x + 6x^2$
$(3 - 2x)(4 - 3x)$ [3.3D]

17. Factor: $6a^4 - 13a^2 - 5$
$(2a^2 - 5)(3a^2 + 1)$ [3.4C]

18. Factor: $12x^3 + 12x^2 - 45x$
$3x(2x - 3)(2x + 5)$ [3.4D]

19. Factor: $16x^2 - 25$
$(4x \quad 5)(4x + 5)$ [3.4A]

20. Factor: $16t^2 + 24t + 9$
$(4t + 3)^2$ [3.4A]

21. Factor: $27x^3 - 8$
$(3x - 2)(9x^2 + 6x + 4)$ [3.4B]

22. Factor: $6x^2 - 4x - 3xa + 2a$
$(3x - 2)(2x - a)$ [3.3B]

23. Factor: $3x^4 - 23x^2 - 36$
$(3x^2 + 4)(x - 3)(x + 3)$ [3.4C]

24. Solve: $6x^2 = x + 1$
$-\frac{1}{3}$ and $\frac{1}{2}$ [3.5A]

25. Solve: $6x^3 + x^2 - 6x - 1 = 0$
$-1, -\frac{1}{6},$ and 1 [3.5A]

Cumulative Review

1. Simplify: $8 - 2[-3 - (-1)]^2 \div 4$
   6   [1.2B]

2. Evaluate $\frac{2a - b}{b - c}$ when $a = 4$, $b = -2$, and $c = 6$.
   $-\frac{5}{4}$ [1.3A]

3. Identify the property that justifies the statement. $2x + (-2x) = 0$
   Inverse Property of Addition   [1.3B]

4. Simplify: $2x - 4[x - 2(3 - 2x) + 4]$
   $-18x + 8$ [1.3C]

5. Solve: $\frac{2}{3} - y = \frac{5}{6}$
   $y = -\frac{1}{6}$ [2.1A]

6. Solve: $8x - 3 - x = -6 + 3x - 8$
   $x = -\frac{11}{4}$ [2.1B]

7. Solve: $\frac{3x - 5}{3} - 8 = 2$
   $x = \frac{35}{3}$ [2.1C]

8. Solve: $3 - |2 - 3x| = -2$
   -1 and $\frac{7}{3}$ [2.3A]

9. Simplify:
   $(3x^2 - 7xy + 2y^2) - (-5x^2 - 2xy + 3y^2)$
   $8x^2 - 5xy - y^2$ [3.1A]

10. Simplify: $(2ab^2)^2(-3a^2b)^4$
    $324a^{10}b^8$ [3.1B]

11. Simplify: $\frac{(-3x^2y^3)^3}{(-2xy^4)^4}$
    $-\frac{27x^2}{16y^7}$ [3.1C]

12. Simplify: $(4a^{-2}b^3)(2a^3b^{-1})^{-2}$
    $\frac{b^5}{a^8}$ [3.1C]

13. Simplify: $\frac{(5x^3y^{-3}z)^{-2}}{(y^4z^{-2})}$
    $\frac{y^2}{25x^6}$ [3.1C]

14. Simplify: $3 - (3 - 3^{-1})^{-1}$
    $\frac{21}{8}$ [3.1C]

15. Simplify: $3x[8 - 2(3x - 6) + 5x] + 8$
    $-3x^2 + 60x + 8$ [3.2A]

16. Simplify: $(2x + 3)(2x^2 - 3x + 1)$
    $4x^3 - 7x + 3$ [3.2B]

17. Simplify: $(x^n + 1)^2$
$x^{2n} + 2x^n + 1$ [3.2C]

18. Factor: $8x^2 - 26x + 15$
$(4x - 3)(2x - 5)$ [3.3D]

19. Factor: $-4x^3 + 14x^2 - 12x$
$-2x(2x - 3)(x - 2)$ [3.3D]

20. Factor: $4x^2 - 20xy + 25y^2$
$(2x - 5y)^2$ [3.4A]

21. Factor: $a(x - y) - b(y - x)$
$(x - y)(a + b)$ [3.3B]

22. Factor: $x^4 - 16$
$(x - 2)(x + 2)(x^2 + 4)$ [3.4D]

23. Factor: $2x^3 - 16$
$2(x - 2)(x^2 + 2x + 4)$ [3.4D]

24. Factor: $a^3 + 64$
$(a + 4)(a^2 - 4a + 16)$ [3.4B]

25. The sum of two integers is twenty-four. The difference between four times the smaller integer and nine is three less than twice the larger integer. Find the integers.
9 and 15   [2.4A]

26. How many ounces of pure gold which cost $360 per ounce must be mixed with 80 oz of an alloy which cost $120 per ounce to make a mixture which cost $200 per ounce?
40 oz   [2.5A]

27. Two bicycles are 25 miles apart and traveling toward each other. One cyclist is traveling at $\frac{2}{3}$ the rate of the other cyclist. They pass in two hours. Find the rate of each cyclist.
slower cyclist: 5 mph
faster cyclist: 7.5 mph   [2.5B]

28. If $3000 is invested at an annual simple interest rate of 7.5%, how much additional money must be invested at an annual simple interest rate of 10% so that the total interest earned is 9% of the total investment?
$4500   [2.6A]

4

Rational Expressions

OBJECTIVES

▶ To simplify a rational expression
▶ To multiply rational expressions
▶ To divide rational expressions
▶ To divide polynomials
▶ To divide polynomials by using synthetic division
▶ To rewrite rational expressions in terms of a common denominator
▶ To add or subtract rational expressions
▶ To simplify a complex fraction
▶ To solve a proportion
▶ To solve application problems
▶ To solve a fractional equation
▶ To solve a literal equation for one of the variables
▶ To solve work problems
▶ To solve uniform motion problems

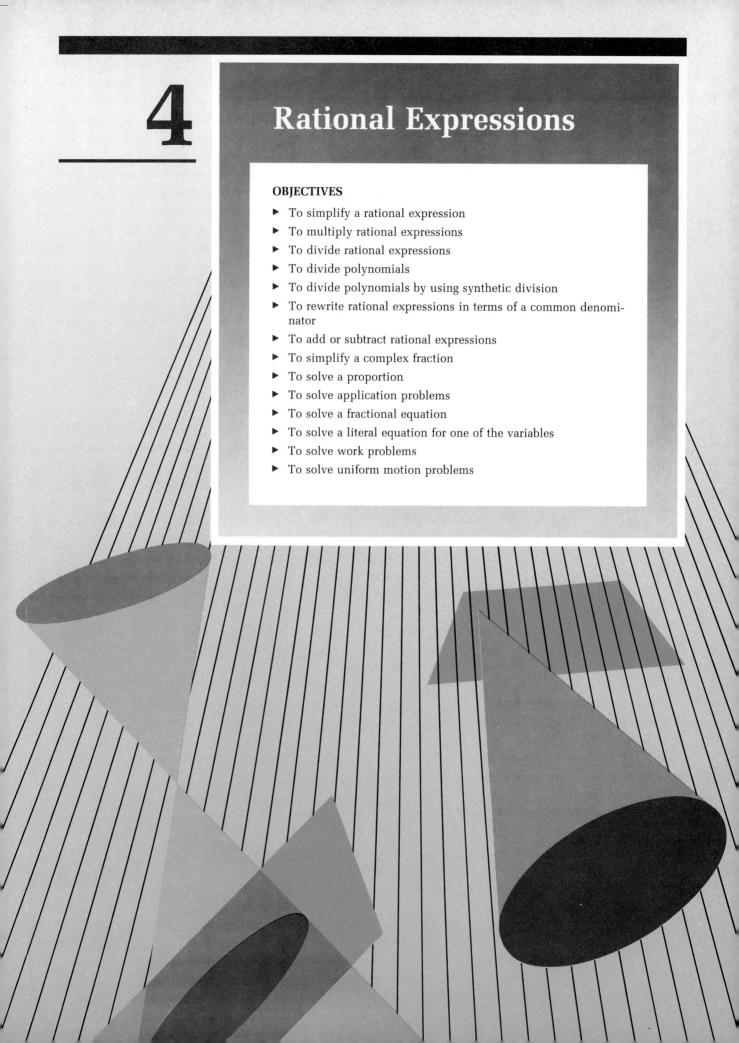

The Abacus

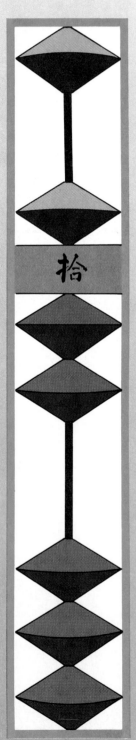

Numerations of the ancient civilizations were unsuited for the purpose of calculation. The abacus is an ancient device used to add, subtract, multiply, and divide, and to calculate square and cube roots. There are many variations of the abacus, but the one shown here has been used in China for many hundreds of years.

The abacus shown here consists of 13 rows of beads. A crossbar separates the beads. Each upper bead represents five units and each lower bead represents one unit. The place holder of each row is shown above the abacus. The first row of beads represents numbers from one to nine, the second row represents numbers from ten to ninety, etc.

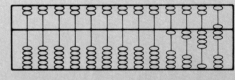

The number shown at the right is 7036.

From the thousands' row—
  one upper bead—5000
  two lower beads—2000

From the hundreds' row—no beads
From the tens' row—three lower
            beads—30
From the ones' row—one upper bead—5
            one lower bead—1

Add: 1247 + 2516

1247 is shown at the right.

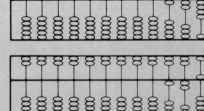

Add 6 to the ones' column

Move the 10 (the upper two beads in the ones' row) to the tens' row (one lower bead). This makes 5 beads in the lower tens' row. Remove 5 beads from the lower row and place one bead in the upper row.

Continue adding in each row until the problem is completed. Carry if necessary.

1247 + 2516 = 3763

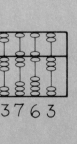

Content and Format © 1991 HMCo.

Multiplication and Division of Rational Expressions

Objective A   **To simplify a rational expression**

A fraction in which the numerator or denominator is a polynomial is called a **rational expression.** Examples of rational expressions are shown at the right.

$$\frac{2}{a}, \quad \frac{x^2 + y}{2x + 3}, \quad \frac{x^2 + 2x - 4}{x^4 - 2x}$$

The algebraic expression shown at the right is not a rational expression, since $\sqrt{x} + 3$ is not a polynomial.

$$\frac{\sqrt{x} + 3}{2}$$

Rational expressions name a real number for each real number replacement of the variable or variables. Thus all properties of real numbers apply to rational expressions. However, when the variables in a rational expression are replaced with real numbers, the resulting denominator must not equal zero, since division by zero is not defined.

For example, the value of x cannot be 3 in the rational expression at the right.

$$\frac{2x - 5}{3x - 9}$$

$$\frac{2 \cdot 3 - 5}{3 \cdot 3 - 9} = \frac{6 - 5}{9 - 9} = \frac{1}{0} \quad \text{Not a real number}$$

A rational expression is in simplest form when the numerator and denominator have no common factors.

Simplify: $\dfrac{x^2 - 25}{x^2 + 13x + 40}$

Factor the numerator and the denominator.

$$\frac{x^2 - 25}{x^2 + 13x + 40} = \frac{(x - 5)(x + 5)}{(x + 5)(x + 8)}$$

Divide by the common factor.

$$= \frac{(x - 5)\overset{1}{\cancel{(x + 5)}}}{\underset{1}{\cancel{(x + 5)}}(x + 8)}$$

Write the answer in simplest form.

$$= \frac{x - 5}{x + 8}$$

Simplify: $\dfrac{12 + 5x - 2x^2}{2x^2 - 3x - 20}$

Factor the numerator and the denominator.

$$\frac{12 + 5x - 2x^2}{2x^2 - 3x - 20} = \frac{(4 - x)(3 + 2x)}{(x - 4)(2x + 5)}$$

Divide by the common factor.
Recall that $4 - x = -(x - 4)$.

$$= \frac{\overset{-1}{\cancel{(4 - x)}}(3 + 2x)}{\underset{1}{\cancel{(x - 4)}}(2x + 5)}$$

Therefore, $\dfrac{4 - x}{x - 4} = \dfrac{-(x - 4)}{x - 4} = \dfrac{-1}{1} = -1$.

Write the answer in simplest form.

$$= -\frac{2x + 3}{2x + 5}$$

Example 1

Simplify: $\dfrac{6x^3 - 9x^2}{12x^2 - 18x}$

Solution

$$\dfrac{6x^3 - 9x^2}{12x^2 - 18x} = \dfrac{3x^2(2x - 3)}{6x(2x - 3)}$$
$$= \dfrac{3x^2(2x - 3)}{6x(2x - 3)} = \dfrac{x}{2}$$

Example 2

Simplify: $\dfrac{6x^4 - 24x^3}{12x^3 - 48x^2}$

Your solution

$\dfrac{x}{2}$

Example 3

Simplify: $\dfrac{12x^3y^2 + 6x^3y^3}{6x^2y^2}$

Solution

$$\dfrac{12x^3y^2 + 6x^3y^3}{6x^2y^2} = \dfrac{6x^3y^2(2 + y)}{6x^2y^2}$$
$$= x(2 + y)$$

Example 4

Simplify: $\dfrac{21a^3b - 14a^3b^2}{7a^2b}$

Your solution

$a(3 - 2b)$

Example 5

Simplify: $\dfrac{2x - 8x^2}{16x^3 - 28x^2 + 6x}$

Solution

$$\dfrac{2x - 8x^2}{16x^3 - 28x^2 + 6x} = \dfrac{2x(1 - 4x)}{2x(8x^2 - 14x + 3)}$$
$$= \dfrac{2x(1 - 4x)}{2x(4x - 1)(2x - 3)}$$
$$= \dfrac{2x(1 - 4x)^{-1}}{2x(4x - 1)_{1}(2x - 3)}$$
$$= -\dfrac{1}{2x - 3}$$

Example 6

Simplify: $\dfrac{20x - 15x^2}{15x^3 - 5x^2 - 20x}$

Your solution

$-\dfrac{1}{x + 1}$

Example 7

Simplify: $\dfrac{x^{2n} + x^n - 2}{x^{2n} - 1}$

Solution

$$\dfrac{x^{2n} + x^n - 2}{x^{2n} - 1} = \dfrac{(x^n - 1)(x^n + 2)}{(x^n - 1)(x^n + 1)}$$
$$= \dfrac{(x^n - 1)^{1}(x^n + 2)}{(x^n - 1)_{1}(x^n + 1)}$$
$$= \dfrac{x^n + 2}{x^n + 1}$$

Example 8

Simplify: $\dfrac{x^{2n} + x^n - 12}{x^{2n} - 3x^n}$

Your solution

$\dfrac{x^n + 4}{x^n}$

| **Objective B** | **To multiply rational expressions** | |

The product of two fractions is a fraction whose numerator is the product of the numerators of the two fractions and whose denominator is the product of the denominators of the two fractions.

$$\frac{a}{b} \cdot \frac{c}{d} = \frac{ac}{bd} \qquad \frac{5}{a+2} \cdot \frac{b-3}{3} = \frac{5(b-3)}{(a+2)3} = \frac{5b-15}{3a+6}$$

The product of two rational expressions can often be simplified by factoring the numerator and the denominator.

Simplify: $\dfrac{x^2 - 2x}{2x^2 + x - 15} \cdot \dfrac{2x^2 - x - 10}{x^2 - 4}$

$$\frac{x^2 - 2x}{2x^2 + x - 15} \cdot \frac{2x^2 - x - 10}{x^2 - 4}$$

Factor the numerator and the denominator of each fraction.

$$= \frac{x(x-2)}{(x+3)(2x-5)} \cdot \frac{(x+2)(2x-5)}{(x+2)(x-2)}$$

Multiply.

$$= \frac{x(x-2)(x+2)(2x-5)}{(x+3)(2x-5)(x+2)(x-2)}$$

Simplify.

$$= \frac{x\overset{1}{\cancel{(x-2)}}\overset{1}{\cancel{(x+2)}}\overset{1}{\cancel{(2x-5)}}}{(x+3)\underset{1}{\cancel{(2x-5)}}\underset{1}{\cancel{(x+2)}}\underset{1}{\cancel{(x-2)}}}$$

Write the answer in simplest form.

$$= \frac{x}{x+3}$$

Example 9

Simplify: $\dfrac{2x^2 - 6x}{3x - 6} \cdot \dfrac{6x - 12}{8x^3 - 12x^2}$

Solution

$$\frac{2x^2 - 6x}{3x - 6} \cdot \frac{6x - 12}{8x^3 - 12x^2} = \frac{2x(x-3)}{3(x-2)} \cdot \frac{6(x-2)}{4x^2(2x-3)}$$

$$= \frac{2x(x-3) \cdot 6(x-2)}{3(x-2) \cdot 4x^2(2x-3)}$$

$$= \frac{2x(x-3) \cdot 6\overset{1}{\cancel{(x-2)}}}{3\underset{1}{\cancel{(x-2)}} \cdot 4x^2(2x-3)}$$

$$= \frac{x-3}{x(2x-3)}$$

Example 10

Simplify:

$$\frac{12 + 5x - 3x^2}{x^2 + 2x - 15} \cdot \frac{2x^2 + x - 45}{3x^2 + 4x}$$

Your solution

$-\dfrac{2x - 9}{x}$

| Objective C | **To divide rational expressions** |

The **reciprocal** of a rational expression is the rational expression with the numerator and denominator interchanged.

Rational Expression $\left\{ \begin{array}{cc} \dfrac{a}{b} & \dfrac{b}{a} \\[2ex] \dfrac{a^2 - 2y}{4} & \dfrac{4}{a^2 - 2y} \end{array} \right\}$ Reciprocal

To divide two rational expressions, multiply by the reciprocal of the divisor.

$$\frac{a}{b} \div \frac{c}{d} = \frac{a}{b} \cdot \frac{d}{c}$$

$$\frac{2}{a} \div \frac{5}{b} = \frac{2}{a} \cdot \frac{b}{5} = \frac{2b}{5a}$$

$$\frac{x + y}{2} \div \frac{x - y}{5} = \frac{x + y}{2} \cdot \frac{5}{x - y} = \frac{(x + y)5}{2(x - y)} = \frac{5x + 5y}{2x - 2y}$$

Example 11

Simplify: $\dfrac{12x^2y^2 - 24xy^2}{5z^2} \div \dfrac{4x^3y - 8x^2y}{3z^4}$

Solution

$\dfrac{12x^2y^2 - 24xy^2}{5z^2} \div \dfrac{4x^3y - 8x^2y}{3z^4}$

$= \dfrac{12x^2y^2 - 24xy^2}{5z^2} \cdot \dfrac{3z^4}{4x^3y - 8x^2y}$

$= \dfrac{12xy^2(x - 2)}{5z^2} \cdot \dfrac{3z^4}{4x^2y(x - 2)} = \dfrac{12xy^2(x - 2)3z^4}{5z^2 \cdot 4x^2y(x - 2)}$

$= \dfrac{12xy^2(\overset{1}{\cancel{x - 2}})3z^4}{5z^2 \cdot 4x^2y(\cancel{x - 2})} = \dfrac{9yz^2}{5x}$

Example 12

Simplify: $\dfrac{6x^2 - 3xy}{10ab^4} \div \dfrac{16x^2y^2 - 8xy^3}{15a^2b^2}$

Your solution

$\dfrac{9a}{16b^2y^2}$

Example 13

Simplify: $\dfrac{3y^2 - 10y + 8}{3y^2 + 8y - 16} \div \dfrac{2y^2 - 7y + 6}{2y^2 + 5y - 12}$

Solution

$\dfrac{3y^2 - 10y + 8}{3y^2 + 8y - 16} \div \dfrac{2y^2 - 7y + 6}{2y^2 + 5y - 12}$

$= \dfrac{3y^2 - 10y + 8}{3y^2 + 8y - 16} \cdot \dfrac{2y^2 + 5y - 12}{2y^2 - 7y + 6}$

$= \dfrac{(y - 2)(3y - 4)}{(3y - 4)(y + 4)} \cdot \dfrac{(y + 4)(2y - 3)}{(y - 2)(2y - 3)}$

$= \dfrac{(y - 2)(3y - 4)(y + 4)(2y - 3)}{(3y - 4)(y + 4)(y - 2)(2y - 3)}$

$= \dfrac{(\overset{1}{\cancel{y - 2}})(\overset{1}{\cancel{3y - 4}})(\overset{1}{\cancel{y + 4}})(\overset{1}{\cancel{2y - 3}})}{(\underset{1}{\cancel{3y - 4}})(\underset{1}{\cancel{y + 4}})(\underset{1}{\cancel{y - 2}})(\underset{1}{\cancel{2y - 3}})} = 1$

Example 14

Simplify: $\dfrac{6x^2 - 7x + 2}{3x^2 + x - 2} \div \dfrac{4x^2 - 8x + 3}{5x^2 + x - 4}$

Your solution

$\dfrac{5x - 4}{2x - 3}$

Solutions on pp. A19–A20

4.1   EXERCISES

▶ **Objective A**

Simplify:

1. $\dfrac{4 - 8x}{4}$

$1 - 2x$

2. $\dfrac{8y + 2}{2}$

$4y + 1$

3. $\dfrac{6x^2 - 2x}{2x}$

$3x - 1$

4. $\dfrac{3y - 12y^2}{3y}$

$1 - 4y$

5. $\dfrac{8x^2(x - 3)}{4x(x - 3)}$

$2x$

6. $\dfrac{16y^4(y + 8)}{12y^3(y + 8)}$

$\dfrac{4y}{3}$

7. $\dfrac{2x - 6}{3x - x^2}$

$-\dfrac{2}{x}$

8. $\dfrac{3a^2 - 6a}{12 - 6a}$

$-\dfrac{a}{2}$

9. $\dfrac{6x^3 - 15x^2}{12x^2 - 30x}$

$\dfrac{x}{2}$

10. $\dfrac{-36a^2 - 48a}{18a^3 + 24a^2}$

$-\dfrac{2}{a}$

11. $\dfrac{a^2 + 4a}{4a - 16}$

In simplest form

12. $\dfrac{3x - 6}{x^2 + 2x}$

In simplest form

13. $\dfrac{16x^3 - 8x^2 + 12x}{4x}$

$4x^2 - 2x + 3$

14. $\dfrac{3x^3y^3 - 12x^2y^2 + 15xy}{3xy}$

$x^2y^2 - 4xy + 5$

15. $\dfrac{-10a^4 - 20a^3 + 30a^2}{-10a^2}$

$a^2 + 2a - 3$

16. $\dfrac{-7a^5 - 14a^4 + 21a^3}{-7a^3}$

$a^2 + 2a - 3$

17. $\dfrac{3x^{3n} - 9x^{2n}}{12x^{2n}}$

$\dfrac{x^n - 3}{4}$

18. $\dfrac{8a^n}{4a^{2n} - 8a^n}$

$\dfrac{2}{a^n - 2}$

19. $\dfrac{x^{2n} + x^n y^n}{x^{2n} - y^{2n}}$

$\dfrac{x^n}{x^n - y^n}$

20. $\dfrac{a^{2n} - b^{2n}}{5a^{3n} + 5a^{2n}b^n}$

$\dfrac{a^n - b^n}{5a^{2n}}$

21. $\dfrac{x^2 - 7x + 12}{x^2 - 9x + 20}$

$\dfrac{x - 3}{x - 5}$

22. $\dfrac{x^2 - x - 20}{x^2 - 2x - 15}$

$\dfrac{x + 4}{x + 3}$

23. $\dfrac{x^2 - xy - 2y^2}{x^2 - 3xy + 2y^2}$

$\dfrac{x + y}{x - y}$

24. $\dfrac{2x^2 + 7xy - 4y^2}{4x^2 - 4xy + y^2}$

$\dfrac{x + 4y}{2x - y}$

25. $\dfrac{6 - x - x^2}{3x^2 - 10x + 8}$

$-\dfrac{x + 3}{3x - 4}$

26. $\dfrac{3x^2 + 10x - 8}{8 - 14x + 3x^2}$

$\dfrac{x + 4}{x - 4}$

27. $\dfrac{14 - 19x - 3x^2}{3x^2 - 23x + 14}$

$-\dfrac{x + 7}{x - 7}$

Simplify:

28. $\dfrac{x^2 + x - 12}{x^2 - x - 12}$

In simplest form

29. $\dfrac{a^2 - 7a + 10}{a^2 + 9a + 14}$

In simplest form

30. $\dfrac{x^2 - 2x}{x^2 + 2x}$

$\dfrac{x - 2}{x + 2}$

31. $\dfrac{a^2 - b^2}{a^3 + b^3}$

$\dfrac{a - b}{a^2 - ab + b^2}$

32. $\dfrac{x^4 - y^4}{x^2 + y^2}$

$x^2 - y^2$

33. $\dfrac{8x^3 - y^3}{4x^2 - y^2}$

$\dfrac{4x^2 + 2xy + y^2}{2x + y}$

34. $\dfrac{a^2 - b^2}{a^3 - b^3}$

$\dfrac{a + b}{a^2 + ab + b^2}$

35. $\dfrac{x^3 + y^3}{3x^3 - 3x^2y + 3xy^2}$

$\dfrac{x + y}{3x}$

36. $\dfrac{3x^3 + 3x^2 + 3x}{9x^3 - 9}$

$\dfrac{x}{3(x - 1)}$

37. $\dfrac{3x^3 - 21x^2 + 30x}{6x^4 - 24x^3 - 30x^2}$

$\dfrac{x - 2}{2x(x + 1)}$

38. $\dfrac{3x^2y - 15xy + 18y}{6x^2y + 6xy - 36y}$

$\dfrac{x - 3}{2(x + 3)}$

39. $\dfrac{x^3 - 4xy^2}{3x^3 - 2x^2y - 8xy^2}$

$\dfrac{x + 2y}{3x + 4y}$

40. $\dfrac{4a^2 - 8ab + 4b^2}{4a^2 - 4b^2}$

$\dfrac{a - b}{a + b}$

41. $\dfrac{4x^3 - 14x^2 + 12x}{24x + 4x^2 - 8x^3}$

$-\dfrac{2x - 3}{2(2x + 3)}$

42. $\dfrac{6x^3 - 15x^2 - 75x}{150x + 30x^2 - 12x^3}$

$-\dfrac{1}{2}$

43. $\dfrac{x^2 - 4}{a(x + 2) - b(x + 2)}$

$\dfrac{x - 2}{a - b}$

44. $\dfrac{x^2(a - 2) - a + 2}{ax^2 - ax}$

$\dfrac{(a - 2)(x + 1)}{ax}$

45. $\dfrac{x^4 + 3x^2 + 2}{x^4 - 1}$

$\dfrac{x^2 + 2}{(x - 1)(x + 1)}$

46. $\dfrac{x^4 - 2x^2 - 3}{x^4 + 2x^2 + 1}$

$\dfrac{x^2 - 3}{x^2 + 1}$

47. $\dfrac{x^2y^2 + 4xy - 21}{x^2y^2 - 10xy + 21}$

$\dfrac{xy + 7}{xy - 7}$

48. $\dfrac{6x^2y^2 + 11xy + 4}{9x^2y^2 + 9xy - 4}$

$\dfrac{2xy + 1}{3xy - 1}$

49. $\dfrac{a^{2n} - a^n - 2}{a^{2n} + 3a^n + 2}$

$\dfrac{a^n - 2}{a^n + 2}$

50. $\dfrac{a^{2n} + a^n - 12}{a^{2n} - 2a^n - 3}$

$\dfrac{a^n + 4}{a^n + 1}$

51. $\dfrac{a^{2n} - 1}{a^{2n} - 2a^n + 1}$

$\dfrac{a^n + 1}{a^n - 1}$

52. $\dfrac{a^{2n} + 2a^nb^n + b^{2n}}{a^{2n} - b^{2n}}$

$\dfrac{a^n + b^n}{a^n - b^n}$

53. $\dfrac{(x - 3) - b(x - 3)}{b(x + 3) - x - 3}$

$-\dfrac{x - 3}{x + 3}$

54. $\dfrac{x^2(a + b) + a + b}{x^4 - 1}$

$\dfrac{a + b}{(x + 1)(x - 1)}$

▶ **Objective B**

Simplify:

55. $\dfrac{27a^2b^5}{16xy^2} \cdot \dfrac{20x^2y^3}{9a^2b}$

$\dfrac{15b^4xy}{4}$

56. $\dfrac{15x^2y^4}{24ab^3} \cdot \dfrac{28a^2b^4}{35xy^4}$

$\dfrac{abx}{2}$

57. $\dfrac{3x - 15}{4x^2 - 2x} \cdot \dfrac{20x^2 - 10x}{15x - 75}$

1

58. $\dfrac{2x^2 + 4x}{8x^2 - 40x} \cdot \dfrac{6x^3 - 30x^2}{3x^2 + 6x}$

$\dfrac{x}{2}$

59. $\dfrac{x^2y^3}{x^2 - 4x - 5} \cdot \dfrac{2x^2 - 13x + 15}{x^4y^3}$

$\dfrac{2x - 3}{x^2(x + 1)}$

60. $\dfrac{2x^2 - 5x + 3}{x^6y^3} \cdot \dfrac{x^4y^4}{2x^2 - x - 3}$

$\dfrac{y(x - 1)}{x^2(x + 1)}$

61. $\dfrac{x^2 - 3x + 2}{x^2 - 8x + 15} \cdot \dfrac{x^2 + x - 12}{8 - 2x - x^2}$

$-\dfrac{x - 1}{x - 5}$

62. $\dfrac{x^2 + x - 6}{12 + x - x^2} \cdot \dfrac{x^2 + x - 20}{x^2 - 4x + 4}$

$-\dfrac{x + 5}{x - 2}$

63. $\dfrac{x^{n+1} + 2x^n}{4x^2 - 6x} \cdot \dfrac{8x^2 - 12x}{x^{n+1} - x^n}$

$\dfrac{2(x + 2)}{x - 1}$

64. $\dfrac{x^{2n} + 2x^n}{x^{n+1} + 2x} \cdot \dfrac{x^2 - 3x}{x^{n+1} - 3x^n}$

1

65. $\dfrac{12 + x - 6x^2}{6x^2 + 29x + 28} \cdot \dfrac{2x^2 + x - 21}{4x^2 - 9}$

$-\dfrac{x - 3}{2x + 3}$

66. $\dfrac{x^2 + 5x + 4}{4 + x - 3x^2} \cdot \dfrac{3x^2 + 2x - 8}{x^2 + 4x}$

$-\dfrac{x + 2}{x}$

67. $\dfrac{x^{2n} - x^n - 6}{x^{2n} + x^n - 2} \cdot \dfrac{x^{2n} - 5x^n - 6}{x^{2n} - 2x^n - 3}$

$\dfrac{x^n - 6}{x^n - 1}$

68. $\dfrac{x^{2n} + 3x^n + 2}{x^{2n} - x^n - 6} \cdot \dfrac{x^{2n} + x^n - 12}{x^{2n} - 1}$

$\dfrac{x^n + 4}{x^n - 1}$

69. $\dfrac{x^3 - y^3}{2x^2 + xy - 3y^2} \cdot \dfrac{2x^2 + 5xy + 3y^2}{x^2 + xy + y^2}$

$x + y$

70. $\dfrac{x^4 - 5x^2 + 4}{3x^2 - 4x - 4} \cdot \dfrac{3x^2 - 10x - 8}{x^2 - 4}$

$\dfrac{(x + 1)(x - 1)(x - 4)}{x - 2}$

71. $\dfrac{x^2 + x - 6}{3x^2 + 5x - 12} \cdot \dfrac{2x^2 - 14x}{16x^2 - 4x} \cdot \dfrac{12x^2 - 19x + 4}{x^2 + 5x - 14}$

$\dfrac{x - 7}{2(x + 7)}$

72. $\dfrac{x^2 - y^2}{x^2 + xy + y^2} \cdot \dfrac{x^2 - xy}{3x^2 - 3xy} \cdot \dfrac{x^3 - y^3}{x^2 - 2xy + y^2}$

$\dfrac{x + y}{3}$

▶ **Objective C**

Simplify:

73. $\dfrac{6x^2y^4}{35a^2b^5} \div \dfrac{12x^3y^3}{7a^4b^5}$

$\dfrac{a^2y}{10x}$

74. $\dfrac{12a^4b^7}{13x^2y^2} \div \dfrac{18a^5b^6}{26xy^3}$

$\dfrac{4by}{3ax}$

75. $\dfrac{2x - 6}{6x^2 - 15x} \div \dfrac{4x^2 - 12x}{18x^3 - 45x^2}$

$\dfrac{3}{2}$

76. $\dfrac{4x^2 - 4y^2}{6x^2y^2} \div \dfrac{3x^2 + 3xy}{2x^2y - 2xy^2}$

$\dfrac{4(x - y)^2}{9x^2y}$

77. $\dfrac{2x^2 - 2y^2}{14x^2y^4} \div \dfrac{x^2 + 2xy + y^2}{35xy^3}$

$\dfrac{5(x - y)}{xy(x + y)}$

78. $\dfrac{8x^3 + 12x^2y}{4x^2 - 9y^2} \div \dfrac{16x^2y^2}{4x^2 - 12xy + 9y^2}$

$\dfrac{2x - 3y}{4y^2}$

79. $\dfrac{x^2 - 8x + 15}{x^2 + 2x - 35} \div \dfrac{15 - 2x - x^2}{x^2 + 9x + 14}$

$-\dfrac{x + 2}{x + 5}$

80. $\dfrac{2x^2 + 13x + 20}{8 - 10x - 3x^2} \div \dfrac{6x^2 - 13x - 5}{9x^2 - 3x - 2}$

$-\dfrac{2x + 5}{2x - 5}$

81. $\dfrac{x^{2n} + x^n}{2x - 2} \div \dfrac{4x^n + 4}{x^{n+1} - x^n}$

$\dfrac{x^{2n}}{8}$

82. $\dfrac{x^{2n} - 4}{4x^n + 8} \div \dfrac{x^{n+1} - 2x}{4x^3 - 12x^2}$

$x(x - 3)$

83. $\dfrac{14 + 17x - 6x^2}{3x^2 + 14x + 8} \div \dfrac{4x^2 - 49}{2x^2 + 15x + 28}$

-1

84. $\dfrac{16x^2 - 9}{6 - 5x - 4x^2} \div \dfrac{16x^2 + 24x + 9}{4x^2 + 11x + 6}$

-1

85. $\dfrac{2x^{2n} - x^n - 6}{x^{2n} - x^n - 2} \div \dfrac{2x^{2n} + x^n - 3}{x^{2n} - 1}$

1

86. $\dfrac{x^{4n} - 1}{x^{2n} + x^n - 2} \div \dfrac{x^{2n} + 1}{x^{2n} + 3x^n + 2}$

$(x^n + 1)^2$

87. $\dfrac{6x^2 + 6x}{3x + 6x^2 + 3x^3} \div \dfrac{x^2 - 1}{1 - x^3}$

$-\dfrac{2(x^2 + x + 1)}{(x + 1)^2}$

88. $\dfrac{x^3 + y^3}{2x^3 + 2x^2y} \div \dfrac{3x^3 - 3x^2y + 3xy^2}{6x^2 - 6y^2}$

$\dfrac{(x + y)(x - y)}{x^3}$

89. $\dfrac{x^2 - 9}{x^2 - x - 6} \div \dfrac{2x^2 + x - 15}{x^2 + 7x + 10} \cdot \dfrac{2x^2 - x - 10}{x^2 - 25}$

$\dfrac{x + 2}{x - 5}$

90. $\dfrac{3x^2 + 10x - 8}{x^2 + 4x + 3} \cdot \dfrac{x^2 + 6x + 9}{3x^2 - 5x + 2} \div \dfrac{x^2 + x - 12}{x^2 - 1}$

$\dfrac{x + 3}{x - 3}$

Division of Polynomials

Objective A

To divide polynomials

Some rational expressions can not be simplified by factoring and dividing by the common factors. In these cases, long division of polynomials is used.

To divide two polynomials, use a method similar to that used for division of whole numbers. To check division of polynomials use

Dividend = (quotient × divisor) + remainder

Simplify: $(x^2 + 5x - 7) \div (x + 3)$

Step 1

$$\begin{array}{r} x \\ x + 3 \overline{)x^2 + 5x - 7} \\ \underline{x^2 + 3x} \downarrow \\ 2x - 7 \end{array}$$

Think: $x \overline{)x^2} = \dfrac{x^2}{x} = x$

Multiply: $x(x + 3) = x^2 + 3x$

Subtract: $(x^2 + 5x) - (x^2 + 3x) = 2x$

Step 2

$$\begin{array}{r} x + 2 \\ x + 3 \overline{)x^2 + 5x - 7} \\ \underline{x^2 + 3x} \\ 2x - 7 \\ \underline{2x + 6} \\ -13 \end{array}$$

Think: $x \overline{)2x} = \dfrac{2x}{x} = 2$

Multiply: $2(x + 3) = 2x + 6$

Subtract: $(2x - 7) - (2x + 6) = -13$

The remainder is -13.

Check: $(x + 2)(x + 3) + (-13) = x^2 + 3x + 2x + 6 - 13 = x^2 + 5x - 7$

$(x^2 + 5x - 7) \div (x + 3) = x + 2 - \dfrac{13}{x + 3}$

Simplify: $\dfrac{6 - 6x^2 + 4x^3}{2x + 3}$

Arrange the terms in descending order. There is no term of x in $4x^3 - 6x^2 + 6$. Insert a zero for the missing term so that like terms will be in columns.

$$\begin{array}{r} 2x^2 - 6x + 9 \\ 2x + 3 \overline{)4x^3 - 6x^2 + 0 + 6} \\ \underline{4x^3 + 6x^2} \\ -12x^2 + 0 \\ \underline{-12x^2 - 18x} \\ 18x + 6 \\ \underline{18x + 27} \\ -21 \end{array}$$

$$\dfrac{4x^3 - 6x^2 + 6}{2x + 3} = 2x^2 - 6x + 9 - \dfrac{21}{2x + 3}$$

Example 1

Simplify: $\dfrac{12x^2 - 11x + 10}{4x - 5}$

Solution

$$
\begin{array}{r}
3x \;+\; 1 \\
4x - 5\overline{)12x^2 - 11x + 10} \\
\underline{12x^2 - 15x} \\
4x + 10 \\
\underline{4x - 5} \\
15
\end{array}
$$

$\dfrac{12x^2 - 11x + 10}{4x - 5} = 3x + 1 + \dfrac{15}{4x - 5}$

Example 2

Simplify: $\dfrac{15x^2 + 17x - 20}{3x + 4}$

Your solution

$5x - 1 - \dfrac{16}{3x + 4}$

Example 3

Simplify: $\dfrac{x^3 + 1}{x + 1}$

Solution

Insert zeros for the missing terms.

$$
\begin{array}{r}
x^2 - x + 1 \\
x + 1\overline{)x^3 + 0 + 0 + 1} \\
\underline{x^3 + x^2} \\
-x^2 + 0 \\
\underline{-x^2 - x} \\
x + 1 \\
\underline{x + 1} \\
0
\end{array}
$$

$\dfrac{x^3 + 1}{x + 1} = x^2 - x + 1$

Example 4

Simplify: $\dfrac{3x^3 + 8x^2 - 6x + 2}{3x - 1}$

Your solution

$x^2 + 3x - 1 + \dfrac{1}{3x - 1}$

Example 5

Simplify:
$(2x^4 - 7x^3 + 3x^2 + 4x - 5) \div (x^2 - 2x - 2)$

Solution

$$
\begin{array}{r}
2x^2 - 3x \;+\; 1 \\
x^2 - 2x - 2\overline{)2x^4 - 7x^3 + 3x^2 + 4x - 5} \\
\underline{2x^4 - 4x^3 - 4x^2} \\
-3x^3 + 7x^2 + 4x \\
\underline{-3x^3 + 6x^2 + 6x} \\
x^2 - 2x - 5 \\
\underline{x^2 - 2x - 2} \\
-3
\end{array}
$$

$(2x^4 - 7x^3 + 3x^2 + 4x - 5) \div (x^2 - 2x - 2)$

$\quad = 2x^2 - 3x + 1 - \dfrac{3}{x^2 - 2x - 2}$

Example 6

Simplify:
$(3x^4 - 11x^3 + 16x^2 - 16x + 8) \div (x^2 - 3x + 2)$

Your solution

$3x^2 - 2x + 4$

Solutions on p. A20

| **Objective B** | **To divide polynomials by using synthetic division** |

Synthetic division is a shorter method of dividing a polynomial by a binomial of the form $x - a$.

Simplify $(3x^2 - 4x + 6) \div (x - 2)$ by using long division.

$$
\begin{array}{r}
3x + 2 \\
x - 2 \overline{)\,3x^2 - 4x + 6} \\
\underline{3x^2 - 6x} \\
2x + 6 \\
\underline{2x - 4} \\
10
\end{array}
$$

$$(3x^2 - 4x + 6) \div (x - 2) = 3x + 2 + \frac{10}{x - 2}$$

The variables may be omitted because the position of a term indicates the power of the term.

$$
\begin{array}{r}
3\quad 2 \\
-2\overline{)\,3\;-4\quad 6} \\
\underline{3\;-6} \\
2\quad 6 \\
\underline{2\;-4} \\
10
\end{array}
$$

The numbers shown in color are exactly the same as the ones above. Removing the colored numbers condenses the vertical spacing.

$$
\begin{array}{r}
3\quad 2 \\
-2\overline{)\,3\;-4\quad 6} \\
\underline{-6\;-4} \\
2\quad 10
\end{array}
$$

The number in color on the top row is the same as the one in the bottom row. Writing the 3 from the top row in the bottom row allows the spacing to be condensed even further.

$$
\begin{array}{c|ccc}
-2 & 3 & -4 & 6 \\
   &   & -6 & -4 \\
\hline
   & 3 & 2 & 10
\end{array}
$$

The quotient is $3x + 2$. The remainder is 10.

$$
\underbrace{3 \qquad 2}_{\substack{\text{Terms of the} \\ \text{quotient}}} \qquad \underbrace{10}_{\text{Remainder}}
$$

By replacing the constant term in the divisor by its additive inverse, terms may be added rather than subtracted. This is illustrated in the following example.

Simplify $(3x^3 + 6x^2 - x - 2) \div (x + 3)$

$x + 3 = x - (-3); a = -3$

Note: The constant term of $x + 3$ is 3. The additive inverse of 3 is -3.

$$
\begin{array}{c|cccc}
-3 & 3 & 6 & -1 & -2 \\
\hline
   & 3
\end{array}
$$

Bring down the 3.

Multiply $(-3) \cdot 3$ and add the product, -9, to 6.

$$
\begin{array}{c|cccc}
-3 & 3 & 6 & -1 & -2 \\
   &   & -9 & & \\
\hline
   & 3 & -3
\end{array}
$$

Multiply $(-3)(-3)$ and add the product, 9, to -1.

$$
\begin{array}{c|cccc}
-3 & 3 & 6 & -1 & -2 \\
   &   & -9 & 9 & \\
\hline
   & 3 & -3 & 8
\end{array}
$$

Multiply $(-3) \cdot 8$ and add the product, -24, to -2.

$$
\begin{array}{c|cccc}
-3 & 3 & 6 & -1 & -2 \\
   &   & -9 & 9 & -24 \\
\hline
   & 3 & -3 & 8 & -26
\end{array}
$$

$$
\underbrace{3 \qquad -3 \qquad 8}_{\substack{\text{Terms of} \\ \text{the quotient}}} \qquad \underbrace{-26}_{\text{Remainder}}
$$

The degree of the quotient is one less than the degree of the dividend.

$$(3x^3 + 6x^2 - x - 2) \div (x + 3) = 3x^2 - 3x + 8 - \frac{26}{x+3}$$

Simplify: $(2x^3 + 3x^2 - 4x + 8) \div (x - 2)$

$x - a = x - 2; a = 2$

Write down the value of a and the coefficients of the dividend. Bring down the 2. Multiply $2 \cdot 2$. Add the product, 4, to 3. Continue until all the coefficients have been used.

$$\begin{array}{c|rrrr} 2 & 2 & 3 & -4 & 8 \\ & & 4 & 14 & 20 \\ \hline & 2 & 7 & 10 & 28 \end{array}$$

Terms of the quotient   Remainder

Write the quotient.

$$(2x^3 + 3x^2 - 4x + 8) \div (x - 2) = 2x^2 + 7x + 10 + \frac{28}{x-2}$$

Example 7

Simplify: $(5x^2 - 3x + 7) \div (x - 1)$

Solution

$$\begin{array}{c|rrr} 1 & 5 & -3 & 7 \\ & & 5 & 2 \\ \hline & 5 & 2 & 9 \end{array}$$

$(5x^2 - 3x + 7) \div (x - 1) = 5x + 2 + \frac{9}{x-1}$

Example 8

Simplify: $(6x^2 + 8x - 5) \div (x + 2)$

Your solution

$6x - 4 + \frac{3}{x+2}$

Example 9

Simplify: $(2x^3 + 4x^2 - 3x + 12) \div (x + 4)$

Solution

$$\begin{array}{c|rrrr} -4 & 2 & 4 & -3 & 12 \\ & & -8 & 16 & -52 \\ \hline & 2 & -4 & 13 & -40 \end{array}$$

$(2x^3 + 4x^2 - 3x + 12) \div (x + 4)$
$= 2x^2 - 4x + 13 - \frac{40}{x+4}$

Example 10

Simplify: $(5x^3 - 12x^2 - 8x + 16) \div (x - 2)$

Your solution

$5x^2 - 2x - 12 - \frac{8}{x-2}$

Example 11

Simplify: $(3x^4 - 8x^2 + 2x + 1) \div (x + 2)$

Solution

Insert a zero for the missing term.

$$\begin{array}{c|rrrrr} -2 & 3 & 0 & -8 & 2 & 1 \\ & & -6 & 12 & -8 & 12 \\ \hline & 3 & -6 & 4 & -6 & 13 \end{array}$$

$(3x^4 - 8x^2 + 2x + 1) \div (x + 2)$
$= 3x^3 - 6x^2 + 4x - 6 + \frac{13}{x+2}$

Example 12

Simplify: $(2x^4 - 3x^3 - 8x^2 - 2) \div (x - 3)$

Your solution

$2x^3 + 3x^2 + x + 3 + \frac{7}{x-3}$

Solutions on p. A20

4.2 **EXERCISES**

▶ **Objective A**

Divide by using long division.

1. $(x^2 + 3x - 40) \div (x - 5)$
    $x + 8$

2. $(x^2 - 14x + 24) \div (x - 2)$
    $x - 12$

3. $(x^3 - 3x^2 + 2) \div (x - 3)$
    $x^2 + \dfrac{2}{x - 3}$

4. $(x^3 + 4x^2 - 8) \div (x + 4)$
    $x^2 - \dfrac{8}{x + 4}$

5. $(6x^2 + 13x + 8) \div (2x + 1)$
    $3x + 5 + \dfrac{3}{2x + 1}$

6. $(12x^2 + 13x - 14) \div (3x - 2)$
    $4x + 7$

7. $(10x^2 + 9x - 5) \div (2x - 1)$
    $5x + 7 + \dfrac{2}{2x - 1}$

8. $(18x^2 - 3x + 2) \div (3x + 2)$
    $6x - 5 + \dfrac{12}{3x + 2}$

9. $(8x^3 - 9) \div (2x - 3)$
    $4x^2 + 6x + 9 + \dfrac{18}{2x - 3}$

10. $(64x^3 + 4) \div (4x + 2)$
    $16x^2 - 8x + 4 - \dfrac{4}{4x + 2}$

11. $(6x^4 - 13x^2 - 4) \div (2x^2 - 5)$
    $3x^2 + 1 + \dfrac{1}{2x^2 - 5}$

12. $(12x^4 - 11x^2 + 10) \div (3x^2 + 1)$
    $4x^2 - 5 + \dfrac{15}{3x^2 + 1}$

13. $\dfrac{-10 - 33x + 3x^3 - 8x^2}{3x + 1}$
    $x^2 - 3x - 10$

14. $\dfrac{10 - 49x + 38x^2 - 8x^3}{1 - 4x}$
    $2x^2 - 9x + 10$

15. $\dfrac{x^3 - 5x^2 + 7x - 4}{x - 3}$
    $x^2 - 2x + 1 - \dfrac{1}{x - 3}$

16. $\dfrac{2x^3 - 3x^2 + 6x + 4}{2x + 1}$
    $x^2 - 2x + 4$

17. $\dfrac{16x^2 - 13x^3 + 2x^4 + 20 - 9x}{x - 5}$
    $2x^3 - 3x^2 + x - 4$

18. $\dfrac{x - x^2 + 5x^3 + 3x^4 - 2}{x + 2}$
    $3x^3 - x^2 + x - 1$

▶ **Objective B**

Divide by using synthetic division.

19. $(2x^2 - 6x - 8) \div (x + 1)$
$2x - 8$

20. $(3x^2 + 19x + 20) \div (x + 5)$
$3x + 4$

21. $(3x^2 - 14x + 16) \div (x - 2)$
$3x - 8$

22. $(4x^2 - 23x + 28) \div (x - 4)$
$4x - 7$

23. $(3x^2 - 4) \div (x - 1)$
$3x + 3 - \frac{1}{x - 1}$

24. $(4x^2 - 8) \div (x - 2)$
$4x + 8 + \frac{8}{x - 2}$

25. $(2x^3 - x^2 + 6x + 9) \div (x + 1)$
$2x^2 - 3x + 9$

26. $(3x^3 + 10x^2 + 6x - 4) \div (x + 2)$
$3x^2 + 4x - 2$

27. $(18 + x - 4x^3) \div (2 - x)$
$4x^2 + 8x + 15 + \frac{12}{x - 2}$

28. $(12 - 3x^2 + x^3) \div (x + 3)$
$x^2 - 6x + 18 - \frac{42}{x + 3}$

29. $(2x^3 + 5x^2 - 5x + 20) \div (x + 4)$
$2x^2 - 3x + 7 - \frac{8}{x + 4}$

30. $(5x^3 + 3x^2 - 17x + 6) \div (x + 2)$
$5x^2 - 7x - 3 + \frac{12}{x + 2}$

31. $\frac{5 + 5x - 8x^2 + 4x^3 - 3x^4}{2 - x}$
$3x^3 + 2x^2 + 12x + 19 + \frac{33}{x - 2}$

32. $\frac{3 - 13x - 5x^2 + 9x^3 - 2x^4}{3 - x}$
$2x^3 - 3x^2 - 4x + 1$

33. $\frac{3x^4 + 3x^3 - x^2 + 3x + 2}{x + 1}$
$3x^3 - x + 4 - \frac{2}{x + 1}$

34. $\frac{4x^4 + 12x^3 - x^2 - x + 2}{x + 3}$
$4x^3 - x + 2 - \frac{4}{x + 3}$

35. $\frac{2x^4 - x^2 + 2}{x - 3}$
$2x^3 + 6x^2 + 17x + 51 + \frac{155}{x - 3}$

36. $\frac{x^4 - 3x^3 - 30}{x + 2}$
$x^3 - 5x^2 + 10x - 20 + \frac{10}{x + 2}$

SECTION 4.3 Addition and Subtraction of Rational Expressions

| Objective A | **To rewrite rational expressions in terms of a common denominator** |

When adding or subtracting rational expressions, it is frequently necessary to express the rational expressions in terms of a common denominator. This common denominator is the LCM of the denominators.

The LCM of two or more polynomials is the simplest polynomial that contains the factors of each polynomial. To find the LCM, first factor each polynomial completely. The LCM is the product of each factor the greatest number of times it occurs in any one factorization.

Find the LCM of $3x^2 + 15x$ and $6x^4 + 24x^3 - 30x^2$.

Factor each polynomial.

$$3x^2 + 15x = 3x(x + 5)$$
$$6x^4 + 24x^3 - 30x^2 = 6x^2(x^2 + 4x - 5)$$
$$= 6x^2(x - 1)(x + 5)$$

The LCM is the product of the LCM of the numerical coefficients and each variable factor the greatest number of times it occurs in any one factorization.

$$\text{LCM} = 6x^2(x - 1)(x + 5)$$

Write the fractions $\frac{x + 2}{x^2 - 2x}$ and $\frac{5x}{3x - 6}$ in terms of the LCM of the denominators.

Find the LCM of the denominators.

The LCM is $3x(x - 2)$.

For each fraction, multiply the numerator and denominator by the factor whose product with the denominator is the LCM.

$$\frac{x + 2}{x^2 - 2x} = \frac{x + 2}{x(x - 2)} \cdot \frac{3}{3} = \frac{3x + 6}{3x(x - 2)}$$

$$\frac{5x}{3x - 6} = \frac{5x}{3(x - 2)} \cdot \frac{x}{x} = \frac{5x^2}{3x(x - 2)}$$

Example 1

Write the fractions $\frac{3x}{x - 1}$ and $\frac{4}{2x + 5}$ in terms of the LCM of the denominators.

Solution

The LCM is $(x - 1)(2x + 5)$.

$$\frac{3x}{x - 1} = \frac{3x}{x - 1} \cdot \frac{2x + 5}{2x + 5} = \frac{6x^2 + 15x}{(x - 1)(2x + 5)}$$

$$\frac{4}{2x + 5} = \frac{4}{2x + 5} \cdot \frac{x - 1}{x - 1} = \frac{4x - 4}{(x - 1)(2x + 5)}$$

Example 2

Write the fractions $\frac{2x}{2x - 5}$ and $\frac{3}{x + 4}$ in terms of the LCM of the denominators.

Your solution

$$\frac{2x^2 + 8x}{(2x - 5)(x + 4)}$$

$$\frac{6x - 15}{(2x - 5)(x + 4)}$$

Solution on p. A21

Example 3

Write the fractions $\frac{2a-3}{a^2-2a}$ and $\frac{a+1}{2a^2-a-6}$ in terms of the LCM of the denominators.

Solution

$a^2 - 2a = a(a-2);$
$2a^2 - a - 6 = (2a+3)(a-2)$

The LCM is $a(a-2)(2a+3)$.

$\frac{2a-3}{a^2-2a} = \frac{2a-3}{a(a-2)} \cdot \frac{2a+3}{2a+3}$

$\qquad = \frac{4a^2-9}{a(a-2)(2a+3)}$

$\frac{a+1}{2a^2-a-6} = \frac{a+1}{(a-2)(2a+3)} \cdot \frac{a}{a}$

$\qquad = \frac{a^2+a}{a(a-2)(2a+3)}$

Example 4

Write the fractions $\frac{3x}{2x^2-11x+15}$ and $\frac{x-2}{x^2-3x}$ in terms of the LCM of the denominators.

Your solution

$\frac{3x^2}{x(x-3)(2x-5)}$

$\frac{2x^2-9x+10}{x(x-3)(2x-5)}$

Example 5

Write the fractions $\frac{2x-3}{3x-x^2}$ and $\frac{3x}{x^2-4x+3}$ in terms of the LCM of the denominators.

Solution

$3x - x^2 = x(3-x) = -x(x-3);$
$x^2 - 4x + 3 = (x-3)(x-1)$

The LCM is $x(x-3)(x-1)$.

$\frac{2x-3}{3x-x^2} = -\frac{2x-3}{x(x-3)} \cdot \frac{x-1}{x-1}$

$\qquad = -\frac{2x^2-5x+3}{x(x-3)(x-1)}$

$\frac{3x}{x^2-4x+3} = \frac{3x}{(x-3)(x-1)} \cdot \frac{x}{x}$

$\qquad = \frac{3x^2}{x(x-3)(x-1)}$

Example 6

Write the fractions $\frac{2x-7}{2x-x^2}$ and $\frac{3x-2}{3x^2-5x-2}$ in terms of the LCM of the denominators.

Your solution

$-\frac{6x^2-19x-7}{x(x-2)(3x+1)}$

$\frac{3x^2-2x}{x(x-2)(3x+1)}$

Solutions on p. A21

Objective B	**To add or subtract rational expressions**

When adding rational expressions in which the denominators are the same, add the numerators. The denominator of the sum is the common denominator.

$$\frac{a}{c} + \frac{b}{c} = \frac{a + b}{c}$$

$$\frac{4x}{15} + \frac{8x}{15} = \frac{4x + 8x}{15} = \frac{12x}{15} = \frac{4x}{5} \qquad \text{Note that the sum is written in simplest form.}$$

$$\frac{a}{a^2 - b^2} + \frac{b}{a^2 - b^2} = \frac{a + b}{a^2 - b^2} = \frac{a + b}{(a - b)(a + b)} = \frac{\overset{1}{(a + b)}}{(a - b)\underset{1}{(a + b)}} = \frac{1}{a - b}$$

When subtracting rational expression with *like* denominators, subtract the numerators. The denominator of the difference is the common denominator. Write the answer in simplest form.

$$\frac{7x - 12}{2x^2 + 5x - 12} - \frac{3x - 6}{2x^2 + 5x - 12} = \frac{(7x - 12) - (3x - 6)}{2x^2 + 5x - 12} = \frac{4x - 6}{2x^2 + 5x - 12}$$

$$= \frac{2\overset{1}{(2x - 3)}}{\underset{1}{(2x - 3)}(x + 4)} = \frac{2}{x + 4}$$

Before two rational expressions with *different* denominators can be added or subtracted, each rational expression must be expressed in terms of a common denominator. This common denominator is the LCM of the denominators of the rational expressions.

Simplify: $\dfrac{x}{x - 3} - \dfrac{x + 1}{x - 2}$

The common denominator is the LCM of the denominators.

The LCM is $(x - 3)(x - 2)$.

Express each fraction in terms of the LCM.

$$\frac{x}{x - 3} - \frac{x + 1}{x - 2} = \frac{x}{x - 3} \cdot \frac{x - 2}{x - 2} - \frac{x + 1}{x - 2} \cdot \frac{x - 3}{x - 3}$$

$$= \frac{x(x - 2) - (x + 1)(x - 3)}{(x - 3)(x - 2)}$$

$$= \frac{(x^2 - 2x) - (x^2 - 2x - 3)}{(x - 3)(x - 2)}$$

$$= \frac{3}{(x - 3)(x - 2)}$$

Simplify: $\dfrac{3x}{2x - 3} + \dfrac{3x + 6}{2x^2 + x - 6}$

Find the LCM of the denominators.

The LCM is $(2x - 3)(x + 2)$.

Express each fraction in terms of the LCM.

$$\frac{3x}{2x - 3} + \frac{3x + 6}{2x^2 + x - 6} = \frac{3x}{2x - 3} \cdot \frac{x + 2}{x + 2} + \frac{3x + 6}{(2x - 3)(x + 2)}$$

$$= \frac{3x(x + 2) + (3x + 6)}{(2x - 3)(x + 2)}$$

$$= \frac{(3x^2 + 6x) + (3x + 6)}{(2x - 3)(x + 2)}$$

$$= \frac{3x^2 + 9x + 6}{(2x - 3)(x + 2)} = \frac{3(x + 2)(x + 1)}{(2x - 3)(x + 2)}$$

$$= \frac{3(x + 1)}{2x - 3}$$

Example 7

Simplify: $\dfrac{2}{x} - \dfrac{3}{x^2} + \dfrac{1}{xy}$

Solution

The LCM is $x^2 y$.

$\dfrac{2}{x} - \dfrac{3}{x^2} + \dfrac{1}{xy} = \dfrac{2}{x} \cdot \dfrac{xy}{xy} - \dfrac{3}{x^2} \cdot \dfrac{y}{y} + \dfrac{1}{xy} \cdot \dfrac{x}{x}$

$\qquad = \dfrac{2xy}{x^2 y} - \dfrac{3y}{x^2 y} + \dfrac{x}{x^2 y} = \dfrac{2xy - 3y + x}{x^2 y}$

Example 8

Simplify: $\dfrac{2}{b} - \dfrac{1}{a} + \dfrac{4}{ab}$

Your solution

$\dfrac{2a - b + 4}{ab}$

Example 9

Simplify: $\dfrac{x}{2x - 4} - \dfrac{4 - x}{x^2 - 2x}$

Solution

$2x - 4 = 2(x - 2); \quad x^2 - 2x = x(x - 2)$

The LCM is $2x(x - 2)$.

$\dfrac{x}{2x - 4} - \dfrac{4 - x}{x^2 - 2x} = \dfrac{x}{2(x - 2)} \cdot \dfrac{x}{x} - \dfrac{4 - x}{x(x - 2)} \cdot \dfrac{2}{2}$

$\qquad = \dfrac{x^2 - (4 - x)2}{2x(x - 2)}$

$\qquad = \dfrac{x^2 - (8 - 2x)}{2x(x - 2)} = \dfrac{x^2 + 2x - 8}{2x(x - 2)}$

$\qquad = \dfrac{(x + 4)(x - 2)}{2x(x - 2)} = \dfrac{(x + 4)\overset{1}{\cancel{(x - 2)}}}{2x\underset{1}{\cancel{(x - 2)}}}$

$\qquad = \dfrac{x + 4}{2x}$

Example 10

Simplify: $\dfrac{a - 3}{a^2 - 5a} + \dfrac{a - 9}{a^2 - 25}$

Your solution

$\dfrac{2a + 3}{a(a + 5)}$

Example 11

Simplify: $\dfrac{x}{x + 1} - \dfrac{2}{x - 2} - \dfrac{3}{x^2 - x - 2}$

Solution

The LCM is $(x + 1)(x - 2)$.

$\dfrac{x}{x + 1} - \dfrac{2}{x - 2} - \dfrac{3}{x^2 - x - 2}$

$= \dfrac{x}{x + 1} \cdot \dfrac{x - 2}{x - 2} - \dfrac{2}{x - 2} \cdot \dfrac{x + 1}{x + 1} - \dfrac{3}{(x + 1)(x - 2)}$

$= \dfrac{x(x - 2) - 2(x + 1) - 3}{(x + 1)(x - 2)}$

$= \dfrac{x^2 - 2x - 2x - 2 - 3}{(x + 1)(x - 2)}$

$= \dfrac{x^2 - 4x - 5}{(x + 1)(x - 2)} = \dfrac{\overset{1}{\cancel{(x + 1)}}(x - 5)}{\underset{1}{\cancel{(x + 1)}}(x - 2)} = \dfrac{x - 5}{x - 2}$

Example 12

Simplify: $\dfrac{2x}{x - 4} - \dfrac{x - 1}{x + 1} + \dfrac{2}{x^2 - 3x - 4}$

Your solution

$\dfrac{x^2 + 7x - 2}{(x - 4)(x + 1)}$

Solutions on p. A21

4.3   EXERCISES

▶ **Objective A**

Write each fraction in terms of the LCM of the denominators.

1. $\dfrac{3}{4x^2y}$, $\dfrac{17}{12xy^4}$

$\dfrac{9y^3}{12x^2y^4}$, $\dfrac{17x}{12x^2y^4}$

2. $\dfrac{5}{16a^3b^3}$, $\dfrac{7}{30a^5b}$

$\dfrac{75a^2}{240a^5b^3}$, $\dfrac{56b^2}{240a^5b^3}$

3. $\dfrac{x-2}{3x(x-2)}$, $\dfrac{3}{6x^2}$

$\dfrac{2x^2-4x}{6x^2(x-2)}$, $\dfrac{3x-6}{6x^2(x-2)}$

4. $\dfrac{5x-1}{4x(2x+1)}$, $\dfrac{2}{5x^3}$

$\dfrac{25x^3-5x^2}{20x^3(2x+1)}$, $\dfrac{16x+8}{20x^3(2x+1)}$

5. $\dfrac{3x-1}{2x^2-10x}$, $-3x$

$\dfrac{3x-1}{2x(x-5)}$, $-\dfrac{6x^3-30x^2}{2x(x-5)}$

6. $\dfrac{4x-3}{3x(x-2)}$, $2x$

$\dfrac{4x-3}{3x(x-2)}$, $\dfrac{6x^3-12x^2}{3x(x-2)}$

7. $\dfrac{3x}{2x-3}$, $\dfrac{5x}{2x+3}$

$\dfrac{6x^2+9x}{(2x-3)(2x+3)}$, $\dfrac{10x^2-15x}{(2x-3)(2x+3)}$

8. $\dfrac{2}{7y-3}$, $\dfrac{-3}{7y+3}$

$\dfrac{14y+6}{(7y-3)(7y+3)}$, $\dfrac{-21y+9}{(7y-3)(7y+3)}$

9. $\dfrac{2x}{x^2-9}$, $\dfrac{x+1}{x-3}$

$\dfrac{2x}{(x+3)(x-3)}$, $\dfrac{x^2+4x+3}{(x+3)(x-3)}$

10. $\dfrac{3x}{16-x^2}$, $\dfrac{2x}{16-4x}$

$\dfrac{12x}{4(4+x)(4-x)}$, $\dfrac{8x+2x^2}{4(4+x)(4-x)}$

11. $\dfrac{3}{3x^2-12y^2}$, $\dfrac{5}{6x-12y}$

$\dfrac{6}{6(x+2y)(x-2y)}$, $\dfrac{5x+10y}{6(x+2y)(x-2y)}$

12. $\dfrac{2x}{x^2-36}$, $\dfrac{x-1}{6x-36}$

$\dfrac{12x}{6(x+6)(x-6)}$, $\dfrac{x^2+5x-6}{6(x+6)(x-6)}$

13. $\dfrac{3x}{x^2-1}$, $\dfrac{5x}{x^2-2x+1}$

$\dfrac{3x^2-3x}{(x+1)(x-1)^2}$, $\dfrac{5x^2+5x}{(x+1)(x-1)^2}$

14. $\dfrac{x^2+2}{x^3-1}$, $\dfrac{3}{x^2+x+1}$

$\dfrac{x^2+2}{(x-1)(x^2+x+1)}$, $\dfrac{3x-3}{(x-1)(x^2+x+1)}$

15. $\dfrac{x-3}{8-x^3}$, $\dfrac{2}{4+2x+x^2}$

$-\dfrac{x-3}{(x-2)(x^2+2x+4)}$,

$\dfrac{2x-4}{(x-2)(x^2+2x+4)}$

16. $\dfrac{2x}{x^2+x-6}$, $\dfrac{-4x}{x^2+5x+6}$

$\dfrac{2x^2+4x}{(x+3)(x-2)(x+2)}$, $-\dfrac{4x^2-8x}{(x+3)(x-2)(x+2)}$

17. $\dfrac{2x}{x^2+2x-3}$, $\dfrac{-x}{x^2+6x+9}$

$\dfrac{2x^2+6x}{(x-1)(x+3)^2}$, $-\dfrac{x^2-x}{(x-1)(x+3)^2}$

18. $\dfrac{3x}{2x^2-x-3}$, $\dfrac{-2x}{2x^2-11x+12}$

$\dfrac{3x^2-12x}{(x+1)(2x-3)(x-4)}$, $-\dfrac{2x^2+2x}{(x+1)(2x-3)(x-4)}$

19. $\dfrac{-4x}{4x^2-16x+15}$, $\dfrac{3x}{6x^2-19x+10}$

$-\dfrac{12x^2-8x}{(2x-3)(2x-5)(3x-2)}$, $\dfrac{6x^2-9x}{(2x-3)(2x-5)(3x-2)}$

20. $\dfrac{3}{2x^2+5x-12}$, $\dfrac{2x}{3-2x}$, $\dfrac{3x-1}{x+4}$

$\dfrac{3}{(2x-3)(x+4)}$, $-\dfrac{2x^2+8x}{(2x-3)(x+4)}$, $\dfrac{6x^2-11x+3}{(2x-3)(x+4)}$

21. $\dfrac{5}{6x^2-17x+12}$, $\dfrac{2x}{4-3x}$, $\dfrac{x+1}{2x-3}$

$\dfrac{5}{(3x-4)(2x-3)}$, $-\dfrac{4x^2-6x}{(3x-4)(2x-3)}$, $\dfrac{3x^2-x-4}{(3x-4)(2x-3)}$

Write each fraction in terms of the LCM of the denominators.

22. $\dfrac{3x}{x-4}, \dfrac{4}{x+5}, \dfrac{x+2}{20-x-x^2}$

$\dfrac{3x^2+15x}{(x+5)(x-4)}, \dfrac{4x-16}{(x+5)(x-4)}, -\dfrac{x+2}{(x+5)(x-4)}$

23. $\dfrac{2x}{x-3}, \dfrac{-2}{x+5}, \dfrac{x-1}{15-2x-x^2}$

$\dfrac{2x^2+10x}{(x-3)(x+5)}, -\dfrac{2x-6}{(x-3)(x+5)}, -\dfrac{x-1}{(x-3)(x+5)}$

24. $\dfrac{2}{x^{2n}-1}, \dfrac{5}{x^{2n}+2x^n+1}$

$\dfrac{2x^n+2}{(x^n-1)(x^n+1)^2}, \dfrac{5x^n-5}{(x^n-1)(x^n+1)^2}$

25. $\dfrac{x-5}{x^{2n}+3x^n+2}, \dfrac{2x}{x^n+2}$

$\dfrac{x-5}{(x^n+1)(x^n+2)}, \dfrac{2x^{n+1}+2x}{(x^n+1)(x^n+2)}$

▶ **Objective B**

Simplify:

26. $\dfrac{3}{2xy}-\dfrac{7}{2xy}-\dfrac{9}{2xy}$

$-\dfrac{13}{2xy}$

27. $-\dfrac{3}{4x^2}+\dfrac{8}{4x^2}-\dfrac{3}{4x^2}$

$\dfrac{1}{2x^2}$

28. $\dfrac{x}{x^2-3x+2}-\dfrac{2}{x^2-3x+2}$

$\dfrac{1}{x-1}$

29. $\dfrac{3x}{3x^2+x-10}-\dfrac{5}{3x^2+x-10}$

$\dfrac{1}{x+2}$

30. $\dfrac{3}{2x^2y}-\dfrac{8}{5x}-\dfrac{9}{10xy}$

$\dfrac{15-16xy-9x}{10x^2y}$

31. $\dfrac{2}{5ab}-\dfrac{3}{10a^2b}+\dfrac{4}{15ab^2}$

$\dfrac{12ab-9b+8a}{30a^2b^2}$

32. $\dfrac{2}{3x}-\dfrac{3}{2xy}+\dfrac{4}{5xy}-\dfrac{5}{6x}$

$-\dfrac{5y+21}{30xy}$

33. $\dfrac{3}{4ab}-\dfrac{2}{5a}+\dfrac{3}{10b}-\dfrac{5}{8ab}$

$\dfrac{5-16b+12a}{40ab}$

34. $\dfrac{2x-1}{12x}-\dfrac{3x+4}{9x}$

$-\dfrac{6x+19}{36x}$

35. $\dfrac{3x-4}{6x}-\dfrac{2x-5}{4x}$

$\dfrac{7}{12x}$

36. $\dfrac{3x+2}{4x^2y}-\dfrac{y-5}{6xy^2}$

$\dfrac{7xy+6y+10x}{12x^2y^2}$

37. $\dfrac{2y-4}{5xy^2}+\dfrac{3-2x}{10x^2y}$

$\dfrac{2xy-8x+3y}{10x^2y^2}$

38. $\dfrac{2x}{x-3}-\dfrac{3x}{x-5}$

$-\dfrac{x^2+x}{(x-3)(x-5)}$

39. $\dfrac{3a}{a-2}-\dfrac{5a}{a+1}$

$-\dfrac{2a^2-13a}{(a-2)(a+1)}$

40. $\dfrac{3}{2a-3}+\dfrac{2a}{3-2a}$

-1

41. $\dfrac{x}{2x-5}-\dfrac{2}{5x-2}$

$\dfrac{5x^2-6x+10}{(2x-5)(5x-2)}$

42. $\dfrac{1}{x+h}-\dfrac{1}{h}$

$-\dfrac{x}{h(x+h)}$

43. $\dfrac{1}{a-b}+\dfrac{1}{b}$

$\dfrac{a}{b(a-b)}$

44. $\dfrac{2}{x}-3-\dfrac{10}{x-4}$

$-\dfrac{3x^2-4x+8}{x(x-4)}$

45. $\dfrac{6a}{a-3}-5+\dfrac{3}{a}$

$\dfrac{a^2+18a-9}{a(a-3)}$

46. $\dfrac{1}{2x-3}-\dfrac{5}{2x}+1$

$\dfrac{4x^2-14x+15}{2x(2x-3)}$

Simplify:

47. $\dfrac{5}{x} - \dfrac{5x}{5 - 6x} + 2$

$\dfrac{17x^2 + 20x - 25}{x(6x - 5)}$

48. $\dfrac{3}{x^2 - 1} + \dfrac{2x}{x^2 + 2x + 1}$

$\dfrac{2x^2 + x + 3}{(x - 1)(x + 1)^2}$

49. $\dfrac{1}{x^2 - 6x + 9} - \dfrac{1}{x^2 - 9}$

$\dfrac{6}{(x + 3)(x - 3)^2}$

50. $\dfrac{x}{x + 3} - \dfrac{3 - x}{x^2 - 9}$

$\dfrac{x + 1}{x + 3}$

51. $\dfrac{1}{x + 2} - \dfrac{3x}{x^2 + 4x + 4}$

$-\dfrac{2x - 2}{(x + 2)^2}$

52. $\dfrac{2x - 3}{x + 5} - \dfrac{x^2 - 4x - 19}{x^2 + 8x + 15}$

$\dfrac{x + 2}{x + 3}$

53. $\dfrac{-3x^2 + 8x + 2}{x^2 + 2x - 8} - \dfrac{2x - 5}{x + 4}$

$-\dfrac{5x^2 - 17x + 8}{(x + 4)(x - 2)}$

54. $\dfrac{x^n}{x^{2n} - 1} - \dfrac{2}{x^n + 1}$

$-\dfrac{x^n - 2}{(x^n + 1)(x^n - 1)}$

55. $\dfrac{2}{x^n - 1} + \dfrac{x^n}{x^{2n} - 1}$

$\dfrac{3x^n + 2}{(x^n + 1)(x^n - 1)}$

56. $\dfrac{2}{x^n - 1} - \dfrac{6}{x^{2n} + x^n - 2}$

$\dfrac{2}{x^n + 2}$

57. $\dfrac{2x^n - 6}{x^{2n} - x^n - 6} + \dfrac{x^n}{x^n + 2}$

1

58. $\dfrac{2x - 2}{4x^2 - 9} - \dfrac{5}{3 - 2x}$

$\dfrac{12x + 13}{(2x + 3)(2x - 3)}$

59. $\dfrac{x^2 + 4}{4x^2 - 36} - \dfrac{13}{x + 3}$

$\dfrac{x^2 - 52x + 160}{4(x + 3)(x - 3)}$

60. $\dfrac{x - 2}{x + 1} - \dfrac{3 - 12x}{2x^2 - x - 3}$

$\dfrac{2x + 3}{2x - 3}$

61. $\dfrac{3x - 4}{4x + 1} + \dfrac{3x + 6}{4x^2 + 9x + 2}$

$\dfrac{3x - 1}{4x + 1}$

62. $\dfrac{x + 1}{x^2 + x - 6} - \dfrac{x + 2}{x^2 + 4x + 3}$

$\dfrac{2x + 5}{(x + 3)(x - 2)(x + 1)}$

63. $\dfrac{x + 1}{x^2 + x - 12} - \dfrac{x - 3}{x^2 + 7x + 12}$

$\dfrac{10x - 6}{(x + 3)(x + 4)(x - 3)}$

64. $\dfrac{x - 1}{2x^2 + 11x + 12} + \dfrac{2x}{2x^2 - 3x - 9}$

$\dfrac{3x^2 + 4x + 3}{(x + 4)(2x + 3)(x - 3)}$

65. $\dfrac{x - 2}{4x^2 + 4x - 3} + \dfrac{3 - 2x}{6x^2 + x - 2}$

$-\dfrac{x^2 + 4x - 5}{(2x + 3)(2x - 1)(3x + 2)}$

66. $\dfrac{x}{x - 3} - \dfrac{2}{x + 4} - \dfrac{14}{x^2 + x - 12}$

$\dfrac{x - 2}{x - 3}$

67. $\dfrac{x^2}{x^2 + x - 2} + \dfrac{3}{x - 1} - \dfrac{4}{x + 2}$

$\dfrac{x^2 - x + 10}{(x + 2)(x - 1)}$

68. $\dfrac{x^2 + 6x}{x^2 + 3x - 18} - \dfrac{2x - 1}{x + 6} + \dfrac{x - 2}{3 - x}$

$-\dfrac{2x^2 - 9x - 9}{(x + 6)(x - 3)}$

69. $\dfrac{2x^2 - 2x}{x^2 - 2x - 15} - \dfrac{2}{x + 3} + \dfrac{x}{5 - x}$

$\dfrac{x - 2}{x + 3}$

Simplify:

70. $\dfrac{4 - 20x}{6x^2 + 11x - 10} - \dfrac{4}{2 - 3x} + \dfrac{x}{2x + 5}$

$\dfrac{3x^2 - 14x + 24}{(3x - 2)(2x + 5)}$

71. $\dfrac{x}{4x - 1} + \dfrac{2}{2x + 1} + \dfrac{6}{8x^2 + 2x - 1}$

$\dfrac{x + 4}{4x - 1}$

72. $\dfrac{7 - 4x}{2x^2 - 9x + 10} + \dfrac{x - 3}{x - 2} - \dfrac{x + 1}{2x - 5}$

$\dfrac{x - 12}{2x - 5}$

73. $\dfrac{x}{3x + 4} + \dfrac{3x + 2}{x - 5} - \dfrac{7x^2 + 24x + 28}{3x^2 - 11x - 20}$

1

74. $\dfrac{32x - 9}{2x^2 + 7x - 15} + \dfrac{x - 2}{3 - 2x} + \dfrac{3x + 2}{x + 5}$

$\dfrac{5x - 1}{2x - 3}$

75. $\dfrac{x + 1}{1 - 2x} - \dfrac{x + 3}{4x - 3} + \dfrac{10x^2 + 7x - 9}{8x^2 - 10x + 3}$

$\dfrac{x + 1}{2x - 1}$

76. $\dfrac{x^2}{x^3 - 8} - \dfrac{x + 2}{x^2 + 2x + 4}$

$\dfrac{4}{(x - 2)(x^2 + 2x + 4)}$

77. $\dfrac{2x}{4x^2 + 2x + 1} + \dfrac{4x + 1}{8x^3 - 1}$

$\dfrac{1}{2x - 1}$

78. $\dfrac{2x^2}{x^4 - 1} - \dfrac{1}{x^2 - 1} + \dfrac{1}{x^2 + 1}$

$\dfrac{2}{x^2 + 1}$

79. $\dfrac{x^2 - 12}{x^4 - 16} + \dfrac{1}{x^2 - 4} - \dfrac{1}{x^2 + 4}$

$\dfrac{1}{x^2 + 4}$

80. $\left[\dfrac{x + 8}{4} + \dfrac{4}{x}\right] \div \dfrac{x + 4}{16x^2}$

$4x(x + 4)$

81. $\left[\dfrac{a - 3}{a^2} - \dfrac{a - 3}{9}\right] \div \dfrac{a^2 - 9}{3a}$

$\dfrac{3 - a}{3a}$

82. $\dfrac{3}{x - 2} - \dfrac{x^2 + x}{2x^3 + 3x^2} \cdot \dfrac{2x^2 + x - 3}{x^2 + 3x + 2}$

$\dfrac{2x^2 + 9x - 2}{x(x - 2)(x + 2)}$

83. $\dfrac{x^2 - 4x + 4}{2x + 1} \cdot \dfrac{2x^2 + x}{x^3 - 4x} - \dfrac{3x - 2}{x + 1}$

$-\dfrac{2x^2 + 5x - 2}{(x + 2)(x + 1)}$

84. $\left[\dfrac{x - y}{x^2} - \dfrac{x - y}{y^2}\right] \div \dfrac{x^2 - y^2}{xy}$

$\dfrac{y - x}{xy}$

85. $\left[\dfrac{a - 2b}{b} + \dfrac{b}{a}\right] \div \left[\dfrac{b + a}{a} - \dfrac{2a}{b}\right]$

$\dfrac{b - a}{b + 2a}$

86. $\dfrac{2}{x - 3} - \dfrac{x}{x^2 - x - 6} \cdot \dfrac{x^2 - 2x - 3}{x^2 - x}$

$\dfrac{x^2 + 4x - 1}{(x - 3)(x + 2)(x - 1)}$

87. $\dfrac{2x}{x^2 - x - 6} - \dfrac{6x - 6}{2x^2 - 9x + 9} \div \dfrac{x^2 + x - 2}{2x - 3}$

$\dfrac{2}{x + 2}$

SECTION 4.4 | Complex Fractions

Objective A **To simplify a complex fraction**

A **complex fraction** is a fraction whose numerator or denominator contains one or more fractions. Examples of complex fractions are shown below.

$$\frac{5}{2 + \frac{1}{2}}, \qquad \frac{5 + \frac{1}{y}}{5 - \frac{1}{y}}, \qquad \frac{x + 4 + \frac{1}{x + 2}}{x - 2 + \frac{1}{x + 2}}$$

Simplify: $\dfrac{\frac{1}{x} + \frac{1}{y}}{\frac{1}{x} - \frac{1}{y}}$

Find the LCM of the denominators of the fractions in the numerator and denominator.

The LCM of x and y is xy.

Multiply the numerator and denominator of the complex fraction by the LCM.

$$\frac{\frac{1}{x} + \frac{1}{y}}{\frac{1}{x} - \frac{1}{y}} = \frac{\frac{1}{x} + \frac{1}{y}}{\frac{1}{x} - \frac{1}{y}} \cdot \frac{xy}{xy}$$

$$= \frac{\frac{1}{x} \cdot xy + \frac{1}{y} \cdot xy}{\frac{1}{x} \cdot xy - \frac{1}{y} \cdot xy} = \frac{y + x}{y - x}$$

Example 1

Simplify: $\dfrac{1 + \frac{1}{y - 2}}{1 - \frac{2}{y + 1}}$

Solution

The LCM of $y - 2$ and $y + 1$ is $(y - 2)(y + 1)$.

$$\frac{1 + \frac{1}{y - 2}}{1 - \frac{2}{y + 1}} \cdot \frac{(y - 2)(y + 1)}{(y - 2)(y + 1)}$$

$$= \frac{(y - 2)(y + 1) + \frac{1}{(y - 2)}(y - 2)(y + 1)}{(y - 2)(y + 1) - \frac{2}{y + 1}(y - 2)(y + 1)}$$

$$= \frac{y^2 - y - 2 + y + 1}{y^2 - y - 2 - 2y + 4} = \frac{y^2 - 1}{y^2 - 3y + 2} = \frac{(y - 1)(y + 1)}{(y - 1)(y - 2)}$$

$$= \frac{y + 1}{y - 2}$$

Example 2

Simplify: $\dfrac{1 - \frac{2}{x + 3}}{1 - \frac{1}{x + 2}}$

Your solution

$\frac{x + 2}{x + 3}$

Solution on p. A21

Example 3

Simplify: $\dfrac{2x - 1 + \dfrac{7}{x+4}}{3x - 8 + \dfrac{17}{x+4}}$

Solution

The LCM is $x + 4$.

$$\frac{2x - 1 + \dfrac{7}{x+4}}{3x - 8 + \dfrac{17}{x+4}} = \frac{2x - 1 + \dfrac{7}{x+4}}{3x - 8 + \dfrac{17}{x+4}} \cdot \frac{x+4}{x+4}$$

$$= \frac{(2x - 1)(x + 4) + \dfrac{7}{x+4}(x+4)}{(3x - 8)(x + 4) + \dfrac{17}{x+4}(x+4)}$$

$$= \frac{2x^2 + 7x - 4 + 7}{3x^2 + 4x - 32 + 17} = \frac{2x^2 + 7x + 3}{3x^2 + 4x - 15}$$

$$= \frac{(2x + 1)(x + 3)}{(3x - 5)(x + 3)} = \frac{(2x + 1)\overset{1}{\cancel{(x + 3)}}}{(3x - 5)\underset{1}{\cancel{(x + 3)}}}$$

$$= \frac{2x + 1}{3x - 5}$$

Example 4

Simplify: $\dfrac{2x + 5 + \dfrac{14}{x-3}}{4x + 16 + \dfrac{49}{x-3}}$

Your solution

$\dfrac{x - 1}{2x + 1}$

Example 5

Simplify: $1 + \dfrac{a}{2 + \dfrac{1}{a}}$

Solution

The LCM of the denominators

of the complex fraction $\dfrac{a}{2 + \dfrac{1}{a}}$ is a.

$$1 + \frac{a}{2 + \dfrac{1}{a}} = 1 + \frac{a}{2 + \dfrac{1}{a}} \cdot \frac{a}{a}$$

$$= 1 + \frac{a \cdot a}{2 \cdot a + \dfrac{1}{a} \cdot a} = 1 + \frac{a^2}{2a + 1}$$

The LCM is $2a + 1$.

$$1 + \frac{a^2}{2a + 1} = \frac{2a + 1}{2a + 1} + \frac{a^2}{2a + 1}$$

$$= \frac{2a + 1 + a^2}{2a + 1} = \frac{a^2 + 2a + 1}{2a + 1}$$

$$= \frac{(a + 1)^2}{2a + 1}$$

Example 6

Simplify: $2 - \dfrac{1}{2 - \dfrac{1}{x}}$

Your solution

$\dfrac{3x - 2}{2x - 1}$

Solutions on pp. A21–A22

4.4   EXERCISES

▶ **Objective A**

Simplify:

1. $\dfrac{2 - \frac{1}{3}}{4 + \frac{11}{3}}$

$\dfrac{5}{23}$

2. $\dfrac{3 + \frac{5}{2}}{8 - \frac{3}{2}}$

$\dfrac{11}{13}$

3. $\dfrac{3 - \frac{2}{3}}{5 + \frac{5}{6}}$

$\dfrac{2}{5}$

4. $\dfrac{5 - \frac{3}{4}}{2 + \frac{1}{2}}$

$\dfrac{17}{10}$

5. $\dfrac{1 + \frac{1}{x}}{1 - \frac{1}{x^2}}$

$\dfrac{x}{x - 1}$

6. $\dfrac{\frac{1}{y^2} - 1}{1 + \frac{1}{y}}$

$\dfrac{1 - y}{y}$

7. $\dfrac{a - 2}{\frac{4}{a} - a}$

$-\dfrac{a}{a + 2}$

8. $\dfrac{\frac{25}{a} - a}{5 + a}$

$\dfrac{5 - a}{a}$

9. $\dfrac{\frac{1}{a^2} - \frac{1}{a}}{\frac{1}{a^2} + \frac{1}{a}}$

$-\dfrac{a - 1}{a + 1}$

10. $\dfrac{\frac{1}{b} + \frac{1}{2}}{\frac{4}{b^2} - 1}$

$\dfrac{b}{2(2 - b)}$

11. $\dfrac{2 - \frac{4}{x + 2}}{5 - \frac{10}{x + 2}}$

$\dfrac{2}{5}$

12. $\dfrac{4 + \frac{12}{2x - 3}}{5 + \frac{15}{2x - 3}}$

$\dfrac{4}{5}$

13. $\dfrac{\frac{3}{2a - 3} + 2}{\frac{-6}{2a - 3} - 4}$

$-\dfrac{1}{2}$

14. $\dfrac{\frac{-5}{b - 5} - 3}{\frac{10}{b - 5} + 6}$

$-\dfrac{1}{2}$

15. $\dfrac{\frac{x}{x + 1} - \frac{1}{x}}{\frac{x}{x + 1} + \frac{1}{x}}$

$\dfrac{x^2 - x - 1}{x^2 + x + 1}$

16. $\dfrac{\frac{2a}{a - 1} - \frac{3}{a}}{\frac{1}{a - 1} + \frac{2}{a}}$

$\dfrac{2a^2 - 3a + 3}{3a - 2}$

17. $\dfrac{1 - \frac{1}{x} - \frac{6}{x^2}}{1 - \frac{4}{x} + \frac{3}{x^2}}$

$\dfrac{x + 2}{x - 1}$

18. $\dfrac{1 - \frac{3}{x} - \frac{10}{x^2}}{1 + \frac{11}{x} + \frac{18}{x^2}}$

$\dfrac{x - 5}{x + 9}$

19. $\dfrac{1 + \frac{1}{x} - \frac{12}{x^2}}{\frac{9}{x^2} + \frac{3}{x} - 2}$

$-\dfrac{x + 4}{2x + 3}$

20. $\dfrac{\frac{15}{x^2} - \frac{2}{x} - 1}{\frac{4}{x^2} - \frac{5}{x} + 4}$

$-\dfrac{x^2 + 2x - 15}{4x^2 - 5x + 4}$

Simplify:

21. $\dfrac{6 + \dfrac{2}{x} - \dfrac{20}{x^2}}{3 - \dfrac{17}{x} + \dfrac{20}{x^2}}$

$\dfrac{2(x + 2)}{x - 4}$

22. $\dfrac{3 + \dfrac{19}{x} + \dfrac{20}{x^2}}{6 + \dfrac{5}{x} - \dfrac{4}{x^2}}$

$\dfrac{x + 5}{2x - 1}$

23. $\dfrac{1 - \dfrac{1}{x - 4}}{1 - \dfrac{6}{x + 1}}$

$\dfrac{x + 1}{x - 4}$

24. $\dfrac{1 + \dfrac{3}{x + 2}}{1 + \dfrac{6}{x - 1}}$

$\dfrac{x - 1}{x + 2}$

25. $\dfrac{1 - \dfrac{2}{x - 3}}{1 + \dfrac{3}{2 - x}}$

$\dfrac{x - 2}{x - 3}$

26. $\dfrac{1 + \dfrac{x}{x + 1}}{1 + \dfrac{x - 1}{x + 2}}$

$\dfrac{x + 2}{x + 1}$

27. $\dfrac{x - 4 + \dfrac{9}{2x + 3}}{x + 3 - \dfrac{5}{2x + 3}}$

$\dfrac{x - 3}{x + 4}$

28. $\dfrac{2x - 3 - \dfrac{10}{4x - 5}}{3x + 2 + \dfrac{11}{4x - 5}}$

$\dfrac{2x - 5}{3x - 1}$

29. $\dfrac{3x - 2 - \dfrac{5}{2x - 1}}{x - 6 + \dfrac{9}{2x - 1}}$

$\dfrac{3x + 1}{x - 5}$

30. $\dfrac{x + 4 - \dfrac{7}{2x - 5}}{2x + 7 - \dfrac{28}{2x - 5}}$

$\dfrac{x - 3}{2x - 7}$

31. $\dfrac{\dfrac{1}{a} - \dfrac{3}{a - 2}}{\dfrac{2}{a} + \dfrac{5}{a - 2}}$

$-\dfrac{2a + 2}{7a - 4}$

32. $\dfrac{\dfrac{2}{b} - \dfrac{5}{b + 3}}{\dfrac{3}{b} + \dfrac{3}{b + 3}}$

$-\dfrac{b - 2}{2b + 3}$

33. $\dfrac{\dfrac{1}{y^2} - \dfrac{1}{xy} - \dfrac{2}{x^2}}{\dfrac{1}{y^2} - \dfrac{3}{xy} + \dfrac{2}{x^2}}$

$\dfrac{x + y}{x - y}$

34. $\dfrac{\dfrac{2}{b^2} - \dfrac{5}{ab} - \dfrac{3}{a^2}}{\dfrac{2}{b^2} + \dfrac{7}{ab} + \dfrac{3}{a^2}}$

$\dfrac{a - 3b}{a + 3b}$

35. $\dfrac{\dfrac{x - 1}{x + 1} - \dfrac{x + 1}{x - 1}}{\dfrac{x - 1}{x + 1} + \dfrac{x + 1}{x - 1}}$

$-\dfrac{2x}{x^2 + 1}$

36. $\dfrac{\dfrac{y}{y + 2} - \dfrac{y}{y - 2}}{\dfrac{y}{y + 2} + \dfrac{y}{y - 2}}$

$-\dfrac{2}{y}$

37. $a + \dfrac{a}{a + \dfrac{1}{a}}$

$\dfrac{a^3 + a^2 + a}{a^2 + 1}$

38. $4 - \dfrac{2}{2 - \dfrac{3}{x}}$

$\dfrac{6x - 12}{2x - 3}$

39. $a - \dfrac{a}{1 - \dfrac{a}{1 - a}}$

$-\dfrac{a^2}{1 - 2a}$

40. $3 - \dfrac{3}{3 - \dfrac{3}{3 - x}}$

$\dfrac{3 - 2x}{2 - x}$

41. $3 - \dfrac{2}{1 - \dfrac{2}{3 - \dfrac{2}{x}}}$

$-\dfrac{3x + 2}{x - 2}$

42. $a + \dfrac{a}{2 + \dfrac{1}{1 - \dfrac{2}{a}}}$

$\dfrac{2a(2a - 3)}{3a - 4}$

SECTION 4.5   Ratio and Proportion

Objective A   **To solve a proportion**

Quantities such as 3 feet, 5 liters, and 2 miles are number quantities written with units. In these examples the units are feet, liters, and miles.

A **ratio** is the quotient of two quantities that have the same unit.

The weekly wages of a painter are $425. The painter spends $50 a week for food. The ratio of the wages spent for food to the total weekly wages is written:

$$\frac{\$50}{\$425} = \frac{50}{425} = \frac{2}{17}$$ A ratio is in simplest form when the two numbers do not have a common factor. Note that the units are not written.

A **rate** is the quotient of two quantities which have different units.

A car travels 180 mi on 3 gal of gas. The miles-to-gallons rate is:

$$\frac{180 \text{ mi}}{3 \text{ gal}} = \frac{60 \text{ mi}}{1 \text{ gal}}$$ A rate is in simplest form when the two numbers do not have a common factor. The units are written as part of the rate.

A **proportion** is an equation that states the equality of two ratios or rates. For example, $\frac{90 \text{ km}}{4 \text{ L}} = \frac{45 \text{ km}}{2 \text{ L}}$ and $\frac{3}{4} = \frac{x+2}{16}$ are proportions.

Recall that an equation containing fractions can be solved by multiplying each side of the equation by the LCM of the denominators.

Solve the proportion $\frac{2}{7} = \frac{x}{5}$.

$$\frac{2}{7} = \frac{x}{5}$$

Multiply each side of the proportion by the LCM of the denominators.

$$\frac{2}{7} \cdot 35 = \frac{x}{5} \cdot 35$$

Solve the equation.

$$10 = 7x$$

$$\frac{10}{7} = x$$

The solution is $\frac{10}{7}$.

Example 1   Solve the proportion $\frac{3}{12} = \frac{5}{x+5}$.

Solution
$$\frac{3}{12} = \frac{5}{x+5}$$
$$\frac{3}{12} \cdot 12(x+5) = \frac{5}{x+5} \cdot 12(x+5)$$
$$3(x+5) = 5 \cdot 12$$
$$3x + 15 = 60$$
$$3x = 45$$
$$x = 15$$
The solution is 15.

Example 2   Solve the proportion $\frac{5}{x-2} = \frac{3}{4}$.

Your solution $\frac{26}{3}$

Solution on p. A22

Example 3

Solve the proportion $\frac{3}{x-2} = \frac{4}{2x+1}$.

Solution

$$\frac{3}{x-2} = \frac{4}{2x+1}$$

$$\frac{3}{x-2}(2x+1)(x-2) = \frac{4}{2x+1}(2x+1)(x-2)$$

$$3(2x+1) = 4(x-2)$$
$$6x+3 = 4x-8$$
$$2x+3 = -8$$
$$2x = -11$$
$$x = -\frac{11}{2}$$

The solution is $-\frac{11}{2}$.

Example 4

Solve the proportion $\frac{5}{2x-3} = \frac{-2}{x+1}$.

Your solution

$\frac{1}{9}$

Solution on p. A22

Objective B   To solve application problems

Example 5

A stock investment of 50 shares pays a dividend of $106. At this rate, how many additional shares are required to earn a dividend of $424?

Strategy

To find the additional number of shares which are required, write and solve a proportion using x to represent the additional number of shares. Then $50 + x$ is the total number of shares of stock.

Solution

$$\frac{\$106}{50} = \frac{\$424}{50+x}$$

$$\frac{53}{25} = \frac{424}{50+x}$$

$$\frac{53}{25}(25)(50+x) = \frac{424}{50+x}(25)(50+x)$$

$$53(50+x) = 424(25)$$
$$2650 + 53x = 10,600$$
$$53x = 7950$$
$$x = 150$$

An additional 150 shares of stock are required.

Example 6

Two pounds of cashews cost $3.10. At this rate, how much would 15 pounds of cashews cost? Round to the nearest cent.

Your strategy

Your solution

$23.25

Solution on p. A22

4.5   EXERCISES

▶ **Objective A**

Solve the proportion.

1. $\dfrac{x}{30} = \dfrac{3}{10}$

9

2. $\dfrac{5}{15} = \dfrac{x}{75}$

25

3. $\dfrac{2}{x} = \dfrac{8}{30}$

$\dfrac{15}{2}$

4. $\dfrac{80}{16} = \dfrac{15}{x}$

3

5. $\dfrac{x+1}{10} = \dfrac{2}{5}$

3

6. $\dfrac{5-x}{10} = \dfrac{3}{2}$

-10

7. $\dfrac{4}{x+2} = \dfrac{3}{4}$

$\dfrac{10}{3}$

8. $\dfrac{8}{3} = \dfrac{24}{x+3}$

6

9. $\dfrac{x}{4} = \dfrac{x-2}{8}$

-2

10. $\dfrac{8}{x-5} = \dfrac{3}{x}$

-3

11. $\dfrac{16}{2-x} = \dfrac{4}{x}$

$\dfrac{2}{5}$

12. $\dfrac{6}{x-5} = \dfrac{1}{x}$

-1

13. $\dfrac{8}{x-2} = \dfrac{4}{x+1}$

-4

14. $\dfrac{4}{x-4} = \dfrac{2}{x-2}$

0

15. $\dfrac{x}{3} = \dfrac{x+1}{7}$

$\dfrac{3}{4}$

16. $\dfrac{x-3}{2} = \dfrac{3+x}{5}$

7

17. $\dfrac{8}{3x-2} = \dfrac{2}{2x+1}$

$-\dfrac{6}{5}$

18. $\dfrac{3}{2x-4} = \dfrac{-5}{x+2}$

$\dfrac{14}{13}$

19. $\dfrac{3x+1}{3x-4} = \dfrac{x}{x-2}$

-2

20. $\dfrac{x-2}{x-5} = \dfrac{2x}{2x+5}$

$\dfrac{10}{11}$

▶ **Objective B**   *Application Problems*

Solve.

21. The real estate tax for a house which cost $80,000 is $1200. At this rate, what is the value of a house for which the real estate tax is $1725?
$115,000

22. The license fee for a car which cost $6000 was $72. At the same rate, what is the license fee for a car which cost $8200?
$98.40

23. In a wildlife preserve, 50 ducks are captured, tagged, and then released. Later 120 ducks are captured and two of the 120 ducks are found to have tags. Estimate the number of ducks in the preserve.
3000 ducks

24. A pre-election survey showed that 5 out of every 8 voters would vote in an election. At this rate, how many people would be expected to vote in a city of 224,000?
140,000 people

Solve:

25. A quality control inspector found 8 defective transistors in a shipment of 3000 transistors. At this rate, how many transistors would be defective in a shipment of 24,000 transistors?
64 transistors

26. A contractor estimated that 12 ft^2 of window space will be allowed for every 150 ft^2 of floor space. Using this estimate, how much window space will be allowed for 3300 ft^2 of floor space?
264 ft^2

27. The scale on an architectural drawing is $\frac{1}{4}$ in. represents one foot. Find the dimensions of a room that measures $4\frac{1}{2}$ in. by 6 in. on the drawing.
18 ft by 24 ft

28. One hundred forty-four ceramic tiles are required to tile a 25-square-foot area. At this rate, how many tiles are required to tile 275 ft^2?
1584 tiles

29. One and one-half ounces of a medication are required for a 140-pound adult. At the same rate, how many additional ounces of medication are required for a 210-pound adult?
0.75 oz

30. Eight ounces of an insecticide is mixed with 24 gal of water to make a spray for spraying an orange grove. How much additional insecticide is required to be mixed with 60 gal of water?
12 oz

31. A stock investment of 200 shares pays a dividend of $240. At this rate, how many additional shares are required to earn a dividend of $600?
300 shares

32. An investment of $4000 earns $460 each year. At the same rate, how much additional money must be invested to earn $805 each year?
$3000

33. A farmer estimates that 8250 bushels of corn can be harvested from 150 acres of land. Using this estimate, how many additional acres are needed to harvest 11,000 bushels of corn?
50 acres

34. A caterer estimates that 2 gal of fruit punch will serve 25 people. How much additional punch is necessary to serve 60 people?
2.8 gal

35. A contractor estimated that 40 ft^3 of cement is required to make a 120 ft^2 concrete floor. Using this estimate, how many additional cubic feet of cement would be required to make a 150 ft^2 concrete floor?
10 ft^3

36. A computer printer can print a 600-word document in 25 s. At this rate, how many seconds are required to print a document which contains 780 words?
32.5 s

SECTION 4.6   Rational Equations

Objective A   To solve a fractional equation

To solve an equation containing fractions, **clear denominators** by multiplying each side of the equation by the LCM of the denominators. Then solve for the variable.

Occasionally, a value of the variable that appears to be a solution will make one of the denominators zero. In this case, the equation has no solution for that value of the variable.

Solve: $\dfrac{3x}{x-5} = x + \dfrac{12}{x-5}$

$$\frac{3x}{x-5} = x + \frac{12}{x-5}$$

Multiply each side of the equation by the LCM of the denominators.

$$(x-5)\left(\frac{3x}{x-5}\right) = (x-5)\left(x + \frac{12}{x-5}\right)$$

Simplify.

$$3x = (x-5)x + (x-5)\left(\frac{12}{x-5}\right)$$
$$3x = x^2 - 5x + 12$$

Solve the quadratic equation by factoring.

$$0 = x^2 - 8x + 12$$
$$0 = (x-2)(x-6)$$
$$x = 2 \quad \text{or} \quad x = 6$$

2 and 6 check as solutions.

The solutions are 2 and 6.

Example 1

Solve: $\dfrac{3x}{x-3} = 2 + \dfrac{9}{x-3}$

Solution

$$\frac{3x}{x-3} = 2 + \frac{9}{x-3}$$

$$(x-3)\left(\frac{3x}{x-3}\right) = (x-3)\left(2 + \frac{9}{x-3}\right)$$

$$3x = (x-3)2 + (x-3)\left(\frac{9}{x-3}\right)$$

$$3x = 2x - 6 + 9$$
$$3x = 2x + 3$$
$$x = 3$$

3 does not check as a solution.

The equation has no solution.

Example 2

Solve: $\dfrac{x}{x-2} + x = \dfrac{6}{x-2}$

Your solution

−2, 3

Solution on p. **A23**

| Objective B | **To solve a literal equation for one of the variables** | |

A **literal equation** is an equation which contains more than one variable. Examples of literal equations are shown at the right.

$$3x - 2y = 4$$
$$v^2 = v_0^2 + 2as$$

Formulas are used to express a relationship among physical quantities. A **formula** is a literal equation which states rules about measurement. Examples of formulas are shown at the right.

$$s = vt - 16t^2 \quad \text{(Physics)}$$
$$c^2 = a^2 + b^2 \quad \text{(Geometry)}$$
$$I = P(1 + r)^t \quad \text{(Business)}$$

The Addition and Multiplication Properties of Equations can be used to solve a literal equation for one of the variables. The goal is to rewrite the equation so that the variable being solved for is alone on one side of the equation and all the other numbers and variables are on the other side.

Solve $C = \frac{5}{9}(F - 32)$ for F.

$$C = \frac{5}{9}(F - 32)$$

Use the Distributive Property to remove parentheses.

$$C = \frac{5}{9}F - \frac{160}{9}$$

Add the additive inverse of the constant term $-\frac{160}{9}$ to each side of the equation.

$$C + \frac{160}{9} = \frac{5}{9}F$$

Multiply each side of the equation by the reciprocal of the coefficient $\frac{5}{9}$. Then simplify.

$$\frac{9}{5}\left(C + \frac{160}{9}\right) = \frac{9}{5}\left(\frac{5}{9}F\right)$$

$$\frac{9}{5}C + 32 = F$$

Example 3

Solve $A = P + Prt$ for P.

Solution

$$A = P + Prt$$
$$A = (1 + rt)P \quad \text{Factoring}$$
$$\frac{A}{1 + rt} = \frac{(1 + rt)P}{1 + rt}$$

$$\frac{A}{1 + rt} = P$$

Example 4

Solve $\frac{1}{R_1} + \frac{1}{R_2} = \frac{1}{R}$ for R.

Your solution

$$R = \frac{R_1 R_2}{R_2 + R_1}$$

Solution on p. A23

| **Objective C** | **To solve work problems** |

If a mason can build a retaining wall in 12 h, then in 1 h the mason can build $\frac{1}{12}$ of the wall. The mason's rate of work is $\frac{1}{12}$ of the wall each hour. The **rate of work** is that part of a task which is completed in one unit of time. If an apprentice can build the wall in x hours, the rate of work for the apprentice is $\frac{1}{x}$ of the wall each hour.

In solving a work problem, the goal is to determine the time it takes to complete a task. The basic equation that is used to solve work problems is:

Rate of work × time worked = part of task completed

For example, if a pipe can fill a tank in 5 h, then in 2 h the pipe will fill $\frac{1}{5} \times 2 = \frac{2}{5}$ of the tank. In t hours, the pipe will fill $\frac{1}{5} \times t = \frac{t}{5}$ of the tank.

A mason can build a wall in 10 h. An apprentice can build a wall in 15 h. How long will it take to build the wall when they work together?

Strategy for Solving a Work Problem

> For each person or machine, write a numerical or variable expression for the rate of work, the time worked, and the part of the task completed. The results can be recorded in a table.

Unknown time to build the wall working together: t

	Rate of Work	·	*Time Worked*	=	*Part of Task Completed*
Mason	$\frac{1}{10}$	·	t	=	$\frac{t}{10}$
Apprentice	$\frac{1}{15}$	·	t	=	$\frac{t}{15}$

> Determine how the parts of the task completed are related. Use the fact that the sum of the parts of the task completed must equal 1, the complete task.

The sum of the part of the task completed by the mason and the part of the task completed by the apprentice is 1.

$$\frac{t}{10} + \frac{t}{15} = 1$$

$$30\left(\frac{t}{10} + \frac{t}{15}\right) = 30(1)$$

$$3t + 2t = 30$$

$$5t = 30$$

$$t = 6$$

Working together, they will build the wall in 6 h.

Example 5

An electrician requires 12·h to wire a house. The electrician's apprentice can wire a house in 16 h. After working alone on one job for 4 h, the electrician quits and the apprentice completes the task. How long does it take the apprentice to finish wiring the house?

Strategy

■ Time required for the apprentice to finish wiring the house: t

	Rate	Time	Part
Electrician	$\frac{1}{12}$	4	$\frac{4}{12}$
Apprentice	$\frac{1}{16}$	t	$\frac{t}{16}$

■ The sum of the part of the task completed by the electrician and the part of the task completed by the apprentice is 1.

$$\frac{4}{12} + \frac{t}{16} = 1$$

Solution

$$\frac{4}{12} + \frac{t}{16} = 1$$

$$\frac{1}{3} + \frac{t}{16} = 1$$

$$48\left(\frac{1}{3} + \frac{t}{16}\right) = 48(1)$$

$$16 + 3t = 48$$

$$3t = 32$$

$$t = \frac{32}{3}$$

It will take the apprentice $10\frac{2}{3}$ h to finish wiring the house.

Example 6

Two water pipes can fill a tank with water in 6 h. The larger pipe working alone can fill the tank in 9 h. How long will it take the smaller pipe working alone to fill the tank?

Your strategy

Your solution

18 h

Solution on p. A23

Content and Format © 1991 HMCo.

Objective D

To solve uniform motion problems

A car that travels constantly in a straight line at 55 mph is in uniform motion. **Uniform motion** means that the speed of an object does not change.

The basic equation used to solve uniform motion problems is:

Distance = rate × time

An alternate form of this equation can be written by solving the equation for time.

$$\frac{\textbf{Distance}}{\textbf{Rate}} = \textbf{time}$$

This form of the equation is useful when the total time of travel for two objects or the time of travel between two points is known.

A motorist drove 150 mi on country roads before driving 50 mi of mountain roads. The rate of speed on the country roads was three times the rate on the mountain roads. The time spent traveling the 200 mi was 5 h. Find the rate of the motorist on the country roads.

Strategy for Solving a Uniform Motion Problem

> For each object, write a numerical or variable expression for the distance, rate, and time. The results can be recorded in a table.

The unknown rate of speed on the mountain roads: r
Rate of speed on the country roads: $3r$

	Distance	÷	Rate	=	Time
Country roads	150	÷	$3r$	=	$\frac{150}{3r}$
Mountain roads	50	÷	r	=	$\frac{50}{r}$

> Determine how the times traveled by each object are related. For example, it may be known that the times are equal or the total time may be known.

The total time of the trip is 5 h.

$$\frac{150}{3r} + \frac{50}{r} = 5$$

$$3r\left(\frac{150}{3r} + \frac{50}{r}\right) = 3r(5)$$

$$150 + 150 = 15r$$
$$300 = 15r$$
$$20 = r$$

The rate of speed on the country roads was $3r$.
Replace r with 20 and evaluate.

$$3r = 3(20) = 60$$

The rate of speed on the country roads was 60 mph.

Example 7

A marketing executive traveled 810 mi on a corporate jet in the same amount of time as it took to travel an additional 162 mi by helicopter. The rate of the jet was 360 mph faster than the rate of the helicopter. Find the rate of the jet.

Example 8

A plane can fly at a rate of 150 mph in calm air. Traveling with the wind, the plane flew 700 mi in the same amount of time as it flew 500 mi against the wind. Find the rate of the wind.

Strategy

■ Rate of the helicopter: r
Rate of the jet: $r + 360$

	Distance	Rate	Time
Jet	810	$r + 360$	$\frac{810}{r+360}$
Helicopter	162	r	$\frac{162}{r}$

■ The time traveled by jet is equal to the time traveled by helicopter.

$$\frac{810}{r+360} = \frac{162}{r}$$

Your strategy

Solution

$$\frac{810}{r+360} = \frac{162}{r}$$

$$r(r+360)\left(\frac{810}{r+360}\right) = r(r+360)\left(\frac{162}{r}\right)$$

$$810r = (r+360)162$$
$$810r = 162r + 58{,}320$$
$$648r = 58{,}320$$
$$r = 90$$
$$r + 360 = 90 + 360 = 450$$

The rate of the jet was 450 mph.

Your solution

25 mph

Solution on p. A23

4.6   EXERCISES

▶ **Objective A**

Solve:

1. $\frac{x}{2} + \frac{5}{6} = \frac{x}{3}$

-5

2. $\frac{x}{5} - \frac{2}{9} = \frac{x}{15}$

$\frac{5}{3}$

3. $\frac{8}{2x - 1} = 2$

$\frac{5}{2}$

4. $3 = \frac{18}{3x - 4}$

$\frac{10}{3}$

5. $1 - \frac{3}{y} = 4$

-1

6. $7 + \frac{6}{y} = 5$

-3

7. $\frac{3}{x - 2} = \frac{4}{x}$

8

8. $\frac{5}{x} = \frac{2}{x + 3}$

-5

9. $\frac{6}{2y + 3} = \frac{6}{y}$

-3

10. $\frac{3}{x - 4} + 2 = \frac{5}{x - 4}$

5

11. $\frac{5}{y + 3} - 2 = \frac{7}{y + 3}$

-4

12. $5 + \frac{8}{a - 2} = \frac{4a}{a - 2}$

No solution

13. $\frac{-4}{a - 4} = 3 - \frac{a}{a - 4}$

No solution

14. $\frac{x}{x + 1} - x = \frac{-4}{x + 1}$

$-2, 2$

15. $\frac{2x}{x + 2} + 3x = \frac{-5}{x + 2}$

$-1, -\frac{5}{3}$

16. $\frac{x}{x - 3} + x = \frac{3x - 4}{x - 3}$

$1, 4$

17. $\frac{x}{2x - 9} - 3x = \frac{10}{9 - 2x}$

$-\frac{1}{3}, 5$

18. $\frac{2}{4y^2 - 9} + \frac{1}{2y - 3} = \frac{3}{2y + 3}$

$\frac{7}{2}$

19. $\frac{5}{x - 2} - \frac{2}{x + 2} = \frac{3}{x^2 - 4}$

$-\frac{11}{3}$

20. $\frac{5}{x^2 - 7x + 12} = \frac{2}{x - 3} + \frac{5}{x - 4}$

No solution

21. $\frac{9}{x^2 + 7x + 10} = \frac{5}{x + 2} - \frac{3}{x + 5}$

No solution

▶ **Objective B**

Solve the formula for the given variable.

22. $P = 2L + 2w$; w (Geometry)

$w = \frac{P - 2L}{2}$

23. $F = \frac{9}{5}C + 32$; C (Temperature Conversion)

$C = \frac{5F - 160}{9}$

24. $S = C - rC$; C (Business)

$C = \frac{S}{1 - r}$

25. $A = P + Prt$; P (Business)

$P = \frac{A}{1 + rt}$

Solve the formula for the given variable.

26. $PV = nRT$; R (Chemistry)

$R = \dfrac{PV}{nT}$

27. $A = \frac{1}{2}bh$; h (Geometry)

$h = \dfrac{2A}{b}$

28. $F = \dfrac{Gm_1 m_2}{r^2}$; m_2 (Physics)

$m_2 = \dfrac{Fr^2}{Gm_1}$

29. $\dfrac{P_1 V_1}{T_1} = \dfrac{P_2 V_2}{T_2}$; P_2 (Chemistry)

$P_2 = \dfrac{P_1 V_1 T_2}{T_1 V_2}$

30. $I = \dfrac{E}{R + r}$; R (Physics)

$R = \dfrac{E - Ir}{I}$

31. $S = V_0 t - 16t^2$; V_0 (Physics)

$V_0 = \dfrac{S + 16t^2}{t}$

32. $A = \frac{1}{2}h(b_1 + b_2)$; b_2 (Geometry)

$b_2 = \dfrac{2A - hb_1}{h}$

33. $V = \frac{1}{3}\pi r^2 h$; h (Geometry)

$h = \dfrac{3V}{\pi r^2}$

34. $\dfrac{1}{R} = \dfrac{1}{R_1} + \dfrac{1}{R_2}$; R_2 (Physics)

$R_2 = \dfrac{RR_1}{R_1 - R}$

35. $\dfrac{1}{f} = \dfrac{1}{a} + \dfrac{1}{b}$; b (Physics)

$b = \dfrac{af}{a - f}$

36. $a_n = a_1 + (n - 1)d$; d (Mathematics)

$d = \dfrac{a_n - a_1}{n - 1}$

37. $P = \dfrac{R - C}{n}$; R (Business)

$R = Pn + C$

38. $S = 2wh + 2wL + 2Lh$; h (Geometry)

$h = \dfrac{S - 2wL}{2w + 2L}$

39. $S = 2\pi r^2 + 2\pi rh$; h (Geometry)

$h = \dfrac{S - 2\pi r^2}{2\pi r}$

▶ Objective C   *Application Problems*

Solve:

40. One member of a gardening team can landscape a new lawn in 36 h. The other member of the team can do the job in 45 h. How long would it take to landscape the lawn when both gardeners work together?
20 h

41. One solar heating panel can raise the temperature of water 1° in 40 min. A second solar heating panel can raise the temperature 1° in 60 min. How long would it take to raise the temperature of the water 1° when both solar panels are operating?
24 min

Solve:

42. A mason can construct a retaining wall in 16 h. The mason's apprentice can do the job in 24 h. How long would it take to construct a wall when they work together?
9.6 h

43. One printer can print the paychecks for the employees of a company in 80 min. A second printer can print the checks in 120 min. How long would it take to print the checks with both printers operating?
48 min

44. An experienced electrician can wire a room twice as fast as an apprentice electrician. Working together, the electricians can wire a room in 3 h. How long would it take the apprentice working alone to wire a room?
9 h

45. A new machine can package transistors three times faster than an older machine. Working together, the machines can package the transistors in 6 h. How long would it take the new machine working alone to package the transistors?
8 h

46. A new printer can print checks four times faster than an old printer. The old printer can print the checks in 40 min. How long would it take to print the checks when both printers are operating?
8 min

47. One member of a telephone crew can wire new telephone lines in 6 h, while it would take 9 h for the other member of the crew to do the job. How long would it take to wire new telephone lines when both members of the crew are working together?
3.6 h

48. An experienced bricklayer can work twice as fast as an apprentice bricklayer. After working together on a job for 6 h, the experienced bricklayer quit. The apprentice required 10 more hours to finish the job. How long would it take the experienced bricklayer working alone to do the job?
14 h

49. A roofer requires 8 h to shingle a roof. After the roofer and an apprentice work on a roof for 2 h, the roofer moves on to another job. The apprentice requires 10 more hours to finish the job. How long would it take the apprentice working alone to do the job?
16 h

50. A welder requires 20 h to do a job. After the welder and an apprentice work on a job for 8 h, the welder quits. The apprentice finishes the job in 13 h. How long would it take the apprentice working alone to do the job?
35 h

Solve:

51. The larger of the two printers being used to print the payroll for a major corporation requires 30 min to print the payroll. After both printers have been operating for 10 min, the larger printer malfunctions. The smaller printer requires 40 more minutes to complete the payroll. How long would it take the smaller printer working alone to print the payroll?
75 min

52. Three machines are filling soda bottles. The machines can fill the daily quota of soda bottles in 10 h, 12 h, and 15 h, respectively. How long would it take to fill the daily quota of soda bottles when all three machines are working?
4 h

53. With both hot and cold water running, a bathtub can be filled in 8 min. The drain will empty the tub in 10 min. A child turns both faucets on and leaves the drain open. How long will it be before the bathtub starts to overflow?
40 min

54. The inlet pipe can fill a water tank in 45 min. The outlet pipe can empty the tank in 30 min. How long would it take to empty a full tank when both pipes are open?
90 min

55. Three computers can print out a task in 20 min, 30 min, and 60 min, respectively. How long would it take to complete the task when all three computers are working?
10 min

56. Water from a tank is being used for irrigation at the same time as the tank is being filled. The two inlet pipes can fill the tank in 5 h and 10 h, respectively. The outlet pipe can empty the tank in 20 h. How long will it take to fill the tank when all three pipes are open?
4 h

57. An oil tank has two inlet pipes and one outlet pipe. One inlet pipe can fill the tank in 12 h, and the other inlet pipe can fill the tank in 20 h. The outlet pipe can empty the tank in 10 h. How long would it take to fill the tank when all three pipes are open?
30 h

58. Two clerks are addressing advertising envelopes for a company. One clerk can address one envelop every 30 seconds while it takes 40 seconds for the second clerk to address one envelop. How long will it take them, working together, to address 140 envelopes?
2400 s

59. Two painters who work at the same rate can each paint one room of an office building in 4 hours. After working together for one hour on one room of the office building, a third painter joins them and together they finish the room in half an hour. How long would it take the third painter to paint the room alone?
2 h

Content and Format © 1991 HMCo.

▶ **Objective D**   *Application Problems*

Solve:

60. The rate of a bicyclist is 7 mph faster than the rate of a long distance runner. The bicyclist travels 30 mi in the same amount of time as the runner travels 16 mi. Find the rate of the runner.
8 mph

61. A passenger train travels 240 mi in the same amount of time as a freight train travels 168 mi. The rate of the passenger train is 18 mph faster than the rate of the freight train. Find the rate of each train.
passenger train: 60 mph; freight train: 42 mph

62. A commercial jet travels 1800 mi in the same amount of time as a corporate jet travels 1350 mi. The rate of the commercial jet is 150 mph faster than the rate of the corporate jet. Find the rate of each jet.
corporate jet: 450 mph; commercial jet: 600 mph

63. An express bus travels 310 mi in the same amount of time as a car travels 275 mi. The rate of the car is 7 mph less than the rate of the bus. Find the rate of the bus.
62 mph

64. A sales executive traveled 55 mi by car and then an additional 990 mi by plane. The rate of the plane was six times faster than the rate of the car. The total time of the trip was 4 h. Find the rate of the plane.
330 mph

65. A cyclist rode 40 mi before having a flat tire and then walking 5 mi to a service station. The cycling rate was four times faster than the walking rate. The time spent cycling and walking was 5 h. Find the rate at which the cyclist was riding.
12 mph

66. A motorcycle travels 196 mi in the same amount of time as a car travels 161 mi. The rate of the motorcycle is 10 mph faster than the rate of the car. Find the rate of the motorcycle.
56 mph

67. A cabin cruiser travels 20 mi in the same amount of time as a power boat travels 45 mi. The rate of the cabin cruiser is 10 mph less than the rate of the power boat. Find the rate of the cabin cruiser.
8 mph

68. A cyclist and a jogger start from a town at the same time and head for a destination 30 mi away. The rate of the cyclist is twice the rate of the jogger. The cyclist arrives 3 h ahead of the jogger. Find the rate of the cyclist.
10 mph

69. An express train and a car leave a town at 3 P.M., and head for a town 280 mi away. The rate of the express train is twice the rate of the car. The train arrives 4 h ahead of the car. Find the rate of the train.
70 mph

Solve:

70. A canoe can travel 8 mph in still water. Rowing with the current of a river, the canoe can travel 15 mi in the same amount of time as it takes to travel 9 mi against the current. Find the rate of the current.
2 mph

71. A tour boat used for river excursions can travel 7 mph in calm water. The amount of time it takes to travel 20 mi with the current is the same amount of time as it takes to travel 8 mi against the current. Find the rate of the current.
3 mph

72. A jet can fly at a rate of 525 mph in calm air. Traveling with the wind, the plane flew 1815 mi in the same amount of time as it flew 1335 mi against the wind. Find the rate of the wind.
80 mph

73. A plane can fly at a rate of 175 mph in calm air. Traveling with the wind, the plane flew 615 mi in the same amount of time as it flew 435 mi against the wind. Find the rate of the wind.
30 mph

74. A single-engine plane and a car start from a town at 6 A.M. and head for a town 450 mi away. The rate of the plane is three times the rate of the car. The plane arrives 6 h ahead of the car. Find the rate of the plane.
150 mph

75. A single-engine plane and a commercial jet leave an airport at 10 A.M. and head for an airport 720 mi away. The rate of the jet is four times the rate of the single-engine plane. The single-engine plane arrives 4.5 h after the jet. Find the rate of each plane.
single-engine plane: 120 mph; jet: 480 mph

76. A jet can travel 550 mph in calm air. Flying with the wind, the jet can travel 3059 mi in the same amount of time as it takes to fly 2450 mi against the wind. Find the rate of the wind. Round to the nearest hundredth.
60.80 mph

77. A twin-engine plane can travel 180 mph in calm air. Flying with the wind the plane can travel 900 mi in the same amount of time as it takes to fly 500 mi against the wind. Find the rate of the wind. Round to the nearest hundredth.
51.43 mph

78. A river excursion motorboat can travel 6 mph in calm water. A recent trip of 16 miles down a river and then returning took 6 h. Find the rate of the river's current.
2 mph

79. A pilot can fly a plane at 125 mph in calm air. A recent trip of 300 mi flying with the wind and 300 mi returning against the wind took 5 h. Find the rate of the wind.
25 mph

Calculators and Computers

Rational Expressions The program RATIONAL EXPRESSIONS on the Math ACE Disk will give you additional practice in multiplying and dividing rational expressions. There are three levels of difficulty, with the first level the easiest of the problems and the third level the most difficult problems. You may choose the level you wish to practice.

After you choose a level of difficulty, a problem will be displayed on the screen. Using paper and pencil, simplify the expression. When you are ready, press the RETURN key. The correct solution will be displayed.

After each problem you will have the choice of continuing the same level of problems, returning to the menu to change the level, or quitting the program.

Chapter Summary

Key Words A fraction in which the numerator and denominator are polynomials is called a *rational expression.*

A rational expression is in *simplest form* when the numerator and denominator have no common factors.

The *least common multiple* (LCM) of two or more polynomials is the simplest polynomial that contains the factors of each polynomial.

Synthetic division is a shorter method of dividing a polynomial by a binomial of the form $x - a$. This method uses only the coefficients of the variable terms.

The *reciprocal* of a rational expression is the rational expression with the numerator and denominator interchanged.

A *complex fraction* is a fraction whose numerator or denominator contains one or more fractions.

A *ratio* is the quotient of two quantities which have the same unit.

A *rate* is the quotient of two quantities which have different units.

A *proportion* is an equation which states the equality of two ratios or rates.

A *literal equation* is an equation which contains more than one variable.

A *formula* is a literal equation which states rules about measurements.

Essential Rules   *Dividend = (quotient × divisor) + remainder*

To multiply fractions: $\dfrac{a}{b} \cdot \dfrac{c}{d} = \dfrac{ac}{bd}$

To divide fractions: $\dfrac{a}{b} \div \dfrac{c}{d} = \dfrac{a}{b} \cdot \dfrac{d}{c}$

To add fractions: $\dfrac{a}{c} + \dfrac{b}{c} = \dfrac{a+b}{c}$

To subtract fractions: $\dfrac{a}{c} - \dfrac{b}{c} = \dfrac{a-b}{c}$

Equation for Work Problems: $\text{Rate of work} \times \text{time} = \text{part of task completed}$

Uniform Motion Equation: Distance = rate × time

Chapter Review

SECTION 1

1. Simplify: $\dfrac{3x^4 + 11x^2 - 4}{3x^4 + 13x^2 + 4}$

$\dfrac{3x^2 - 1}{3x^2 + 1}$

2. Simplify: $\dfrac{x^3 - 27}{x^2 - 9}$

$\dfrac{x^2 + 3x + 9}{x + 3}$

3. Simplify: $\dfrac{a^6b^4 + a^4b^6}{a^5b^4 - a^4b^4} \cdot \dfrac{a^2 - b^2}{a^4 - b^4}$

$\dfrac{1}{a - 1}$

4. Simplify: $\dfrac{x^3 - 8}{x^3 + 2x^2 + 4x} \cdot \dfrac{x^3 + 2x^2}{x^2 - 4}$

x

5. Simplify: $\dfrac{27x^3 - 8}{9x^3 + 6x^2 + 4x} \div \dfrac{9x^2 - 12x + 4}{9x^2 - 4}$

$\dfrac{3x + 2}{x}$

6. Simplify: $\dfrac{x^{n+1} + x}{x^{2n} - 1} \div \dfrac{x^{n+2} - x^2}{x^{2n} - 2x^n + 1}$

$\dfrac{1}{x}$

SECTION 2

7. Simplify: $\dfrac{15x^2 + 2x - 2}{3x - 2}$

$5x + 4 + \dfrac{6}{3x - 2}$

8. Simplify: $\dfrac{12x^2 - 16x - 7}{6x + 1}$

$2x - 3 - \dfrac{4}{6x + 1}$

9. Simplify: $\dfrac{4x^3 + 27x^2 + 10x + 2}{x + 6}$

$4x^2 + 3x - 8 + \dfrac{50}{x + 6}$

10. Simplify: $\dfrac{x^4 - 4}{x - 4}$

$x^3 + 4x^2 + 16x + 64 + \dfrac{252}{x - 4}$

SECTION 3

11. Write each fraction in terms of the LCM of the denominators.

$\dfrac{4x}{4x - 1}, \dfrac{3x - 1}{4x + 1}$

$\dfrac{16x^2 + 4x}{(4x - 1)(4x + 1)}, \dfrac{12x^2 - 7x + 1}{(4x - 1)(4x + 1)}$

12. Write each fraction in terms of the LCM of the denominators.

$\dfrac{x - 3}{x - 5}, \dfrac{x}{x^2 - 9x + 20}, \dfrac{1}{4 - x}$

$\dfrac{x^2 - 7x + 12}{(x - 5)(x - 4)}, \dfrac{x}{(x - 5)(x - 4)}, -\dfrac{x - 5}{(x - 5)(x - 4)}$

13. Simplify: $\dfrac{5}{3a^2b^3} + \dfrac{7}{8ab^4}$

$\dfrac{21a + 40b}{24a^2b^4}$

14. Simplify: $\dfrac{3x^2 + 2}{x^2 - 4} - \dfrac{9x - x^2}{x^2 - 4}$

$\dfrac{4x - 1}{x + 2}$

15. Simplify: $\dfrac{6x}{3x^2 - 7x + 2} - \dfrac{2}{3x - 1} + \dfrac{3x}{x - 2}$

$\dfrac{9x^2 + x + 4}{(3x - 1)(x - 2)}$

16. Simplify: $\dfrac{x}{x - 3} - 4 - \dfrac{2x - 5}{x + 2}$

$-\dfrac{5x^2 - 17x - 9}{(x - 3)(x + 2)}$

SECTION 4

17. Simplify: $\dfrac{x - 6 + \dfrac{6}{x-1}}{x + 3 - \dfrac{12}{x-1}}$

$\dfrac{x-4}{x+5}$

18. Simplify: $x + \dfrac{\dfrac{4}{x} - 1}{\dfrac{1}{x} - \dfrac{3}{x^2}}$

$\dfrac{x}{x-3}$

19. Simplify: $3 + \dfrac{1}{1 + \dfrac{1}{1 + \dfrac{1}{x}}}$

$\dfrac{7x+4}{2x+1}$

20. Simplify: $\dfrac{\dfrac{3x+4}{3x-4} + \dfrac{3x-4}{3x+4}}{\dfrac{3x-4}{3x+4} - \dfrac{3x+4}{3x-4}}$

$-\dfrac{9x^2 + 16}{24x}$

SECTION 5

21. Solve $\dfrac{10}{5x+3} = \dfrac{2}{10x-3}$.

$\dfrac{2}{5}$

22. Solve $\dfrac{3x-2}{x+6} = \dfrac{3x+1}{x+9}$.

4

23. On a certain map, 2.5 inches represents 10 miles. How many miles would be represented by 12 inches?
48 miles

24. A student reads 2 pages of text in 5 minutes. At the same rate, how long will it take to read 150 pages?
375 min

SECTION 6

25. Solve: $\dfrac{30}{x^2 + 5x + 4} + \dfrac{10}{x+4} = \dfrac{4}{x+1}$

No solution

26. Solve: $\dfrac{6}{2x-3} = \dfrac{5}{x+5} + \dfrac{5}{2x^2 + 7x - 15}$

10

27. Solve $Q = \dfrac{N-S}{N}$ for N.

$N = \dfrac{S}{1-Q}$

28. Solve $S = \dfrac{a}{1-r}$ for r.

$r = \dfrac{S-a}{S}$

29. An electrician requires 65 minutes to install a ceiling fan. The electrician and an apprentice working together take 40 minutes to install the fan. How long would it take the apprentice working alone to install the ceiling fan?
104 min

30. A helicopter travels 9 mi in the same amount of time as an airplane travels 10 mi. The rate of the airplane is 20 mph faster than the rate of the helicopter. Find the rate of the helicopter.
180 mph

Chapter Test

1. Simplify: $\dfrac{v^3 - 4v}{2v^2 - 5v + 2}$

   $\dfrac{v(v + 2)}{2v - 1}$ [4.1A]

2. Simplify: $\dfrac{2a^2 - 8a + 8}{4 + 4a - 3a^2}$

   $-\dfrac{a - 2}{2 - 3a}$ [4.1A]

3. Simplify: $\dfrac{3x^2 - 12}{5x - 15} \cdot \dfrac{2x^2 - 18}{x^2 + 5x + 6}$

   $\dfrac{6(x - 2)}{5}$ [4.1B]

4. Simplify: $\dfrac{x^2 + x - 6}{x^2 + 7x + 12} \div \dfrac{x^2 - 3x + 2}{x^2 + 6x + 8}$

   $\dfrac{x + 2}{x - 1}$ [4.1C]

5. Simplify: $\dfrac{2x^2 - x - 3}{2x^2 - 5x + 3} \div \dfrac{3x^2 - x - 4}{x^2 - 1}$

   $\dfrac{x + 1}{3x - 4}$ [4.1C]

6. Simplify: $(12t^3 + 16t^2 - 9) \div (4t + 4)$

   $3t^2 + t - 1 - \dfrac{5}{4t + 4}$ [4.2A]

7. Simplify: $\dfrac{3x^4 - 2x^2 + x + 2}{x - 1}$

   $3x^3 + 3x^2 + x + 2 + \dfrac{4}{x - 1}$ [4.2B]

8. Write each fraction in terms of the LCM of the denominators.

   $\dfrac{x + 1}{x^2 + x - 6}, \dfrac{2x}{x^2 - 9}$

   $\dfrac{x^2 - 2x - 3}{(x + 3)(x - 3)(x + 2)}, \dfrac{2x^2 - 4x}{(x + 3)(x - 3)(x + 2)}$ [4.3A]

9. Simplify: $\dfrac{2x - 1}{x + 2} - \dfrac{x}{x - 3}$

   $\dfrac{x^2 - 9x + 3}{(x + 2)(x - 3)}$ [4.3B]

10. Simplify: $\dfrac{x + 2}{x^2 + 3x - 4} - \dfrac{2x}{x^2 - 1}$

    $\dfrac{-x^2 - 5x + 2}{(x + 1)(x + 4)(x - 1)}$ [4.3B]

11. Simplify: $\dfrac{1 - \dfrac{1}{x} - \dfrac{12}{x^2}}{1 + \dfrac{6}{x} + \dfrac{9}{x^2}}$

    $\dfrac{x - 4}{x + 3}$ [4.4A]

12. Simplify: $\dfrac{1 - \dfrac{1}{x + 2}}{1 - \dfrac{3}{x + 4}}$

    $\dfrac{x + 4}{x + 2}$ [4.4A]

13. Solve the proportion $\dfrac{3}{x + 1} = \dfrac{2}{x}$.

    2   [4.5A]

14. Solve: $\dfrac{4x}{x + 1} - x = \dfrac{2}{x + 1}$

    1, 2   [4.6A]

15. Solve: $\dfrac{4x}{2x-1} = 2 - \dfrac{1}{2x-1}$

No solution   [4.6A]

16. Solve $ax = bx + c$ for x.

$x = \dfrac{c}{a-b}$ [4.6B]

17. Solve $\dfrac{1}{r} = \dfrac{1}{2} - \dfrac{2}{t}$ for t.

$t = \dfrac{4r}{r-2}$ [4.6B]

18. An interior designer uses 2 rolls of wallpaper for every 45 ft^2 of wall space in an office. At this rate, how many rolls of wallpaper are needed for an office that has 315 ft^2 of wall space?
14 rolls   [4.5B]

19. One landscaper can till the soil for a lawn in 30 min while it takes a second landscaper 15 min to do the same job. How long would it take to till the soil for the lawn with both landscapers working together?
10 min   [4.6C]

20. A cyclist travels 20 mi in the same amount of time as a hiker walks 6 mi. The rate of the cyclist is 7 mph faster than the rate of the hiker. Find the rate of the cyclist.
10 mph   [4.6D]

Cumulative Review

1. Simplify: $8 - 4[-3 - (-2)]^2 \div 5$

    $\frac{36}{5}$ [1.2B]

2. Evaluate $3a^2 - (b^2 - c)^2$ when $a = 2$, $b = -3$, and $c = 1$.

    -52 [1.3A]

3. Solve: $-\frac{2}{3}y = -\frac{4}{9}$

    $y = \frac{2}{3}$ [2.1A]

4. Solve: $\frac{2x - 3}{6} - \frac{x}{9} = \frac{x - 4}{3}$

    $x = 7\frac{1}{2}$ [2.1C]

5. Solve: $5 - |x - 4| = 2$

    7 and 1   [2.3A]

6. Solve: $(x - 2)(x + 1) = 4$

    -2 and 3   [3.5A]

7. Simplify: $\dfrac{(2a^{-2}b^3)^{-2}}{(4a)^{-1}}$

    $\frac{a^5}{b^6}$ [3.1C]

8. Solve: $x - 3(1 - 2x) \geq 1 - 4(2 - 2x)$

    $\{x \mid x \leq 4\}$ [2.2A]

9. Simplify: $(2a^2 - 3a + 1)(-2a^2)$

    $-4a^4 + 6a^3 - 2a^2$ [3.2A]

10. Factor: $2x^{2n} + 3x^n - 2$

    $(2x^n - 1)(x^n + 2)$ [3.4C]

11. Factor: $x^3y^3 - 27$

    $(xy - 3)(x^2y^2 + 3xy + 9)$ [3.4B]

12. Simplify: $\dfrac{x^4 + x^3y - 6x^2y^2}{x^3 - 2x^2y}$

    $x + 3y$ [4.1A]

13. Simplify: $\dfrac{4x^3 + 2x^2 - 10x + 1}{x - 2}$

    $4x^2 + 10x + 10 + \frac{21}{x - 2}$ [4.2B]

14. Simplify: $\dfrac{16x^2 - 9y^2}{16x^2y - 12xy^2} \div \dfrac{4x^2 - xy - 3y^2}{12x^2y^2}$

    $\frac{3xy}{x - y}$ [4.1C]

15. Write each fraction in terms of the LCM of the denominators. $\dfrac{xy}{2x^2 + 2x}$, $\dfrac{2}{2x^4 - 2x^3 - 4x^2}$

$\dfrac{x^3y - 2x^2y}{2x^2(x + 1)(x - 2)}$; $\dfrac{2}{2x^2(x + 1)(x - 2)}$ [4.3A]

16. Simplify: $\dfrac{5x}{3x^2 - x - 2} - \dfrac{2x}{x^2 - 1}$

$-\dfrac{x}{(3x + 2)(x + 1)}$ [4.3B]

17. Simplify: $\dfrac{x - 4 + \dfrac{5}{x + 2}}{x + 2 - \dfrac{1}{x + 2}}$

$\dfrac{x - 3}{x + 3}$ [4.4A]

18. Solve the proportion. $\dfrac{2}{x - 3} = \dfrac{5}{2x - 3}$

$x = 9$ [4.5A]

19. Solve: $\dfrac{3}{x^2 - 36} = \dfrac{2}{x - 6} - \dfrac{5}{x + 6}$

$x = 13$ [4.6A]

20. Solve $I = \dfrac{E}{R + r}$ for r.

$r = \dfrac{E - IR}{I}$ [4.6B]

21. Simplify: $(1 - x^{-1})^{-1}$

$\dfrac{x}{x - 1}$ [3.1C]

22. The sum of two integers is 15. Five times the smaller integer is five more than twice the larger integer. Find the integers.
smaller integer: 5
larger integer: 10   [2.4A]

23. How many pounds of almonds which cost $5.40 per pound must be mixed with 50 lb of peanuts which cost $2.60 per pound to make a mixture which cost $4.00 per pound?
50 lb   [2.5A]

24. A pre-election survey showed that three out of five voters would vote in an election. At this rate, how many people would be expected to vote in a city of 125,000?
75,000   [4.5B]

25. A new computer can work six times faster than an older computer. Working together, the computers can complete a job in 12 min. How long would it take the new computer working alone to do the job?
14 min   [4.6C]

26. A plane can fly at a rate of 300 mph in calm air. Traveling with the wind, the plane flew 900 mi in the same amount of time as it flew 600 mi against the wind. Find the rate of the wind.
60 mph   [4.6D]

5 Exponents and Radicals

OBJECTIVES

▶ To simplify expressions with rational exponents

▶ To write exponential expressions as radical expressions and to write radical expressions as exponential expressions

▶ To simplify expressions of the form $\sqrt[n]{a^n}$

▶ To simplify radical expressions

▶ To add or subtract radical expressions

▶ To multiply radical expressions

▶ To divide radical expressions

▶ To simplify a complex number

▶ To add or subtract complex numbers

▶ To multiply complex numbers

▶ To divide complex numbers

▶ To solve an equation containing one or more radical expressions

▶ To solve application problems

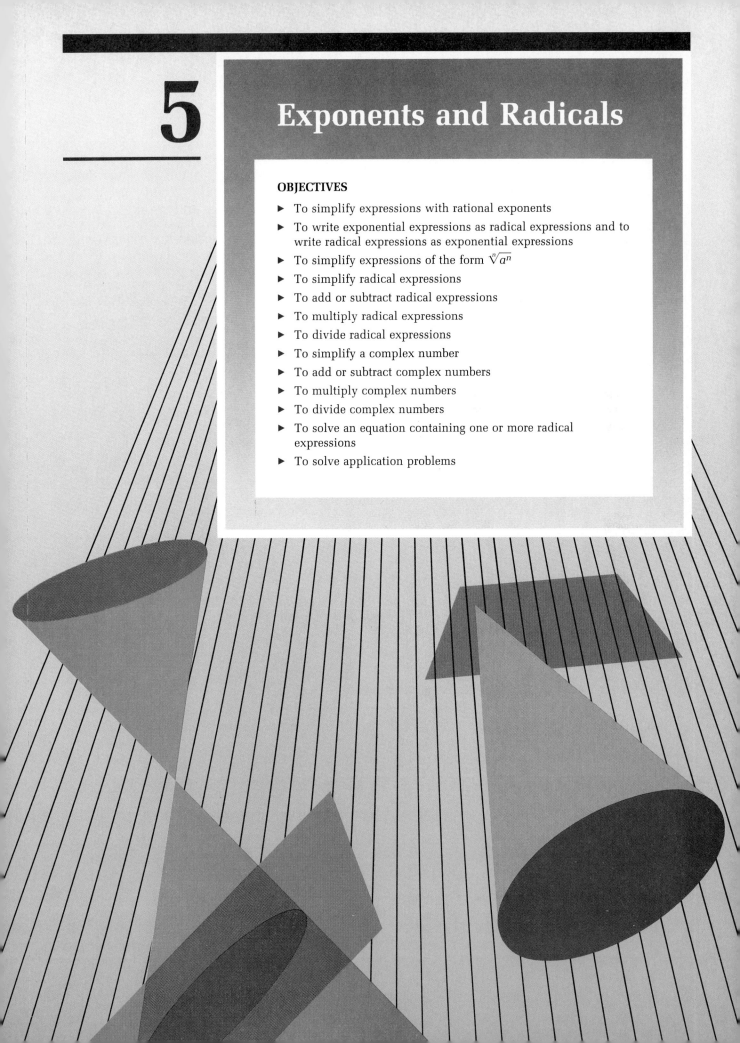

Golden Ratio

The golden rectangle fascinated the early Greeks and appeared in much of their architecture. They considered this particular rectangle to be the most pleasing to the eye, and consequently, when used in the design of a building, would make the structure pleasant to see.

The golden rectangle is constructed from a square by drawing a line from the midpoint of the base of the square to the opposite vertex. Now extend the base of the square, starting from the midpoint, the length of the line. The resulting rectangle is called the golden rectangle.

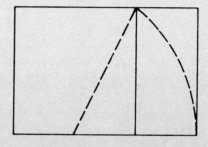

The Parthenon in Athens, Greece, is the classic example of the use of the golden rectangle in Greek architecture. A rendering of the Parthenon is shown here.

SECTION 5.1	Rational Exponents and Radical Expressions

Objective A	**To simplify expressions with rational exponents**

In this section, the definition of an exponent is extended beyond integers so that any rational number can be used as an exponent. The definition is made so that the Rules of Exponents hold true for rational exponents.

Consider the expression $(a^{1/n})^n$ for $a > 0$ and n a positive integer. Now simplify assuming the Power Rule of Exponents is true.

$$(a^{1/n})^n = a^{\frac{1}{n} \cdot n} = a^1 = a$$

Since $(a^{1/n})^n = a$, the number $a^{1/n}$ is the number whose nth power is a.

If $a > 0$ and n is a positive integer, then $a^{1/n}$ is called the nth root of a.

$25^{1/2} = 5$ because $(5)^2 = 25$. $8^{1/3} = 2$ because $(2)^3 = 8$.

In the expression $a^{1/n}$, if a is a negative number and n is a positive even integer, then $a^{1/n}$ is not a real number.

$(-4)^{1/2}$ is not a real number, since there is no real number whose 2nd power is -4.

When n is a positive odd integer, a can be a positive or a negative number.

$(-27)^{1/3} = -3$ because $(-3)^3 = -27$.

Using the definition of $a^{1/n}$ and the Rules of Exponents, it is possible to define any exponential expression that contains a rational exponent.

If m and n are positive integers and $a^{1/n}$ is a real number, then
$$a^{m/n} = (a^{1/n})^m.$$

The expression $a^{m/n}$ can also be written $a^{m/n} = a^{m \cdot \frac{1}{n}} = (a^m)^{1/n}$.

As shown above, expressions which contain rational exponents do not always represent real numbers when the base of the exponential expression is a negative number. For this reason, all variables in this chapter represent positive numbers unless otherwise stated.

Simplify: $27^{2/3}$

Rewrite 27 as 3^3.

$27^{2/3} = (3^3)^{2/3}$

Use the Rule for Powers of Exponential Expressions.

$\boxed{= 3^{3(2/3)}}$ Do this step mentally.

Simplify.

$= 3^2$

$= 9$

Simplify: $32^{-2/5}$

Rewrite 32 as 2^5.

$32^{-2/5} = (2^5)^{-2/5}$

Use the Rule for Powers of Exponential Expressions.

$= 2^{-2}$

Use the Rule of Negative Exponents.

$= \dfrac{1}{2^2}$

Simplify.

$= \dfrac{1}{4}$

Simplify: $a^{1/2} \cdot a^{2/3} \cdot a^{-1/4}$

Use the Rule for Multiplying Exponential Expressions.

$a^{1/2} \cdot a^{2/3} \cdot a^{-1/4} \boxed{= a^{1/2 + 2/3 - 1/4}}$ Do this step mentally.

$= a^{6/12 + 8/12 - 3/12}$

Simplify.

$= a^{11/12}$

Simplify: $(x^6 y^4)^{3/2}$

Use the Rule for Simplifying Powers of Products.

$(x^6 y^4)^{3/2} \boxed{= x^{6(3/2)} y^{4(3/2)}}$ Do this step mentally.

Simplify.

$= x^9 y^6$

Simplify: $\left(\dfrac{8a^3 b^{-4}}{64a^{-9} b^2}\right)^{2/3}$

Rewrite 8 as 2^3 and 64 as 2^6.

$\left(\dfrac{8a^3 b^{-4}}{64a^{-9} b^2}\right)^{2/3} = \left(\dfrac{2^3 a^3 b^{-4}}{2^6 a^{-9} b^2}\right)^{2/3}$

Use the Rule for Dividing Exponential Expressions.

$= (2^{-3} a^{12} b^{-6})^{2/3}$

Use the Rule for Powers of Quotients.

$= 2^{-2} a^8 b^{-4}$

Use the Rule of Negative Exponents and simplify.

$= \dfrac{a^8}{2^2 b^4} = \dfrac{a^8}{4b^4}$

Example 1 Simplify: $64^{-2/3}$

Solution $64^{-2/3} = (2^6)^{-2/3} = 2^{-4}$

$= \dfrac{1}{2^4} = \dfrac{1}{16}$

Example 2 Simplify: $16^{-3/4}$

Your solution $\dfrac{1}{8}$

Example 3 Simplify: $(-49)^{3/2}$

Solution The base of the exponential expression is a negative number, while the denominator of the exponent is a positive even number.

Therefore, $(-49)^{3/2}$ is not a real number.

Example 4 Simplify: $(-81)^{3/4}$

Your solution Not a real number

Example 5 Simplify: $(x^{1/2}y^{-3/2}z^{1/4})^{-3/2}$

Solution $(x^{1/2}y^{-3/2}z^{1/4})^{-3/2}$

$= x^{-3/4}y^{9/4}z^{-3/8}$

$= \dfrac{y^{9/4}}{x^{3/4}z^{3/8}}$

Example 6 Simplify: $(x^{3/4}y^{1/2}z^{-2/3})^{-4/3}$

Your solution $\dfrac{z^{8/9}}{xy^{2/3}}$

Example 7 Simplify: $\dfrac{x^{1/2}y^{-5/4}}{x^{-4/3}y^{1/3}}$

Solution $\dfrac{x^{1/2}y^{-5/4}}{x^{-4/3}y^{1/3}}$

$= x^{3/6 - (-8/6)}y^{-15/12 - 4/12}$

$= x^{11/6}y^{-19/12} = \dfrac{x^{11/6}}{y^{19/12}}$

Example 8 Simplify: $\left(\dfrac{16a^{-2}b^{4/3}}{9a^4b^{-2/3}}\right)^{-1/2}$

Your solution $\dfrac{3a^3}{4b}$

Solutions on p. A24

Objective B

To write exponential expressions as radical expressions and to write radical expressions as exponential expressions

Recall that $a^{1/n}$ is the nth root of a. The expression $\sqrt[n]{a}$ is another symbol for the nth root of a.

If a is a real number, then $a^{1/n} = \sqrt[n]{a}$.

In the expression $\sqrt[n]{a}$, the symbol $\sqrt{}$ is called a **radical,** n is the **index** of the radical, and a is the **radicand.** When $n = 2$, the radical expression represents a square root and the index 2 is usually not written.

An exponential expression with a rational exponent can be written as a radical expression.

If $a^{1/n}$ is a real number, then $a^{m/n} = a^{m \cdot 1/n} = (a^m)^{1/n} = \sqrt[n]{a^m}$.

The expression $a^{m/n}$ can also be written $a^{m/n} = a^{1/n \cdot m} = (\sqrt[n]{a})^m$.

The exponential expression at the right has been written as a radical expression.

$$y^{2/3} = (y^2)^{1/3}$$
$$= \sqrt[3]{y^2}$$

The radical expressions at the right have been written as exponential expressions.

$$\sqrt[5]{x^6} = (x^6)^{1/5} = x^{6/5}$$
$$\sqrt{17} = (17)^{1/2} = 17^{1/2}$$

Write $(5x)^{2/5}$ as a radical expression.

The denominator of the rational exponent is the index of the radical. The numerator is the power of the radicand. Simplify.

$$(5x)^{2/5} = \sqrt[5]{(5x)^2}$$
$$= \sqrt[5]{25x^2}$$

Write $\sqrt[3]{x^4}$ as an exponential expression with a rational exponent.

The index of the radical is the denominator of the rational exponent. The power of the radicand is the numerator of the rational exponent.

$$\sqrt[3]{x^4} = (x^4)^{1/3} = x^{4/3}$$

Write $\sqrt[3]{a^3 + b^3}$ as an exponential expression with a rational exponent.

$$\sqrt[3]{a^3 + b^3} = (a^3 + b^3)^{1/3}$$

Note that $(a^3 + b^3)^{1/3} \ne a + b$.

Example 9   Write $(3x)^{5/4}$ as a radical expression.

Solution $(3x)^{5/4} = \sqrt[4]{(3x)^5} = \sqrt[4]{243x^5}$

Example 10   Write $(2x^3)^{3/4}$ as a radical expression.

Your solution $\sqrt[4]{8x^9}$

Example 11   Write $-2x^{2/3}$ as a radical expression.

Solution $-2x^{2/3} = -2(x^2)^{1/3} = -2\sqrt[3]{x^2}$

Example 12   Write $-5a^{5/6}$ as a radical expression.

Your solution $-5\sqrt[6]{a^5}$

Example 13   Write $\sqrt[4]{3a}$ as an exponential expression.

Solution $\sqrt[4]{3a} = (3a)^{1/4}$

Example 14   Write $\sqrt[3]{3ab}$ as an exponential expression.

Your solution $(3ab)^{1/3}$

Example 15   Write $\sqrt{a^2 - b^2}$ as an exponential expression.

Solution $\sqrt{a^2 - b^2} = (a^2 - b^2)^{1/2}$

Example 16   Write $\sqrt[4]{x^4 + y^4}$ as an exponential expression.

Your solution $(x^4 + y^4)^{1/4}$

Solutions on p. A24

Objective C   To simplify expressions of the form $\sqrt[n]{a^n}$

Every positive number has two square roots, one a positive and one a negative number. For example, because $(5)^2 = 25$ and $(-5)^2 = 25$, there are two square roots of 25, 5 and -5.

The symbol $\sqrt{}$ is used to indicate the positive or **principle square root.** To indicate the negative square root of a number, a negative sign is placed in front of the radical.	$\sqrt{25} = 5$ $-\sqrt{25} = -5$
The square root of zero is zero.	$\sqrt{0} = 0$
The square root of a negative number is not a real number, since the square of a real number must be positive.	$\sqrt{-25}$ is not a real number.
The square root of a squared negative number is a positive number.	$\sqrt{(-5)^2} = \sqrt{25} = 5$

For any real number a, $\sqrt{a^2} = |a|$ and $-\sqrt{a^2} = -|a|$.
If a is a positive real number, then $\sqrt{a^2} = a$ and $(\sqrt{a})^2 = a$.

The cube root of a positive number is positive.	$\sqrt[3]{8} = 2$, since $2^3 = 8$.
The cube root of a negative number is negative.	$\sqrt[3]{-8} = -2$, since $(-2)^3 = -8$.

For any real number a, $\sqrt[3]{a^3} = a$.

The following properties hold true for finding the nth root of a real number.

If n is an even integer, then $\sqrt[n]{a^n} = |a|$ and $-\sqrt[n]{a^n} = -|a|$. If n is an odd integer, then $\sqrt[n]{a^n} = a$.

For example,

$$\sqrt[6]{y^6} = |y| \qquad -\sqrt[12]{x^{12}} = -|x| \qquad \sqrt[5]{b^5} = b$$

Since it has been stated that all variables in this unit represent positive numbers unless otherwise stated, it is not necessary to use the absolute value signs.

Simplify: $\sqrt{121}$

Write the prime factorization of the radicand in exponential form.

$$\sqrt{121} = \sqrt{11^2}$$

Write the radical expression as an exponential expression.

$$= (11^2)^{1/2}$$

Do this step mentally.

Use the Rule for Powers of Exponential Expressions.

$$= 11$$

Simplify: $\sqrt[3]{x^6 y^6}$

The radicand is a perfect cube since the exponents on the variables are divisible by 3. Write the radical expression as an exponential expression.

$$\sqrt[3]{x^6 y^6} \quad = (x^6 y^6)^{1/3}$$

Do this step mentally.

Use the Rule for Simplifying Powers of Products.

$$= x^2 y^2$$

Simplify: $\sqrt[5]{x^{15}}$

The radicand is a perfect fifth power since 15 is divisible by 5. Write the radical expression as an exponential expression.

$$\sqrt[5]{x^{15}} \quad = (x^{15})^{1/5}$$

Do this step mentally.

Use the Rule for Simplifying Powers of Exponential Expressions.

$$= x^3$$

Simplify: $-\sqrt[4]{x^{12}}$

The radicand is a perfect fourth power since 12 is divisible by 4.
Write the radical expression as an exponential expression.

$$-\sqrt[4]{x^{12}} \quad = -(x^{12})^{1/4}$$

Do this step mentally.

Use the Rule for Simplifying Powers of Exponential Expressions.

$$= -x^3$$

Example 17   Simplify: $\sqrt{49x^2 y^{12}}$

Solution $\sqrt{49x^2 y^{12}} = \sqrt{7^2 x^2 y^{12}}$
$$= 7xy^6$$

Example 18   Simplify: $\sqrt{121x^{10} y^4}$

Your solution $11x^5 y^2$

Example 19   Simplify: $-\sqrt[4]{16a^4 b^8}$

Solution $-\sqrt[4]{16a^4 b^8} = -\sqrt[4]{2^4 a^4 b^8}$
$$= -2ab^2$$

Example 20   Simplify: $\sqrt[3]{-125a^6 b^9}$

Your solution $-5a^2 b^3$

Solutions on p. A24

Content and Format © 1991 HMCo.

5.1 EXERCISES

▶ **Objective A**

Simplify:

1. $8^{1/3}$
2

2. $16^{1/2}$
4

3. $9^{3/2}$
27

4. $25^{3/2}$
125

5. $27^{-2/3}$
$\frac{1}{9}$

6. $64^{-1/3}$
$\frac{1}{4}$

7. $32^{2/5}$
4

8. $16^{3/4}$
8

9. $(-25)^{5/2}$
Not a real number

10. $(-36)^{1/4}$
Not a real number

11. $\left(\frac{25}{49}\right)^{-3/2}$
$\frac{343}{125}$

12. $\left(\frac{8}{27}\right)^{-2/3}$
$\frac{9}{4}$

13. $x^{1/2}x^{1/2}$
x

14. $a^{1/3}a^{5/3}$
a^2

15. $y^{-1/4}y^{3/4}$
$y^{1/2}$

16. $x^{2/5} \cdot x^{-4/5}$
$\frac{1}{x^{2/5}}$

17. $x^{-2/3} \cdot x^{3/4}$
$x^{1/12}$

18. $x \cdot x^{-1/2}$
$x^{1/2}$

19. $a^{1/3} \cdot a^{3/4} \cdot a^{-1/2}$
$a^{7/12}$

20. $y^{-1/6} \cdot y^{2/3} \cdot y^{1/2}$
y

21. $\frac{a^{1/2}}{a^{3/2}}$
$\frac{1}{a}$

22. $\frac{b^{1/3}}{b^{4/3}}$
$\frac{1}{b}$

23. $\frac{y^{-3/4}}{y^{1/4}}$
$\frac{1}{y}$

24. $\frac{x^{-3/5}}{x^{1/5}}$
$\frac{1}{x^{4/5}}$

25. $\frac{y^{2/3}}{y^{-5/6}}$
$y^{3/2}$

26. $\frac{b^{3/4}}{b^{-3/2}}$
$b^{9/4}$

27. $(x^2)^{-1/2}$
$\frac{1}{x}$

28. $(a^8)^{-3/4}$
$\frac{1}{a^6}$

29. $(x^{-2/3})^6$
$\frac{1}{x^4}$

30. $(y^{-5/6})^{12}$
$\frac{1}{y^{10}}$

31. $(a^{-1/2})^{-2}$
a

32. $(b^{-2/3})^{-6}$
b^4

33. $(x^{-3/8})^{-4/5}$
$x^{3/10}$

34. $(y^{-3/2})^{-2/9}$
$y^{1/3}$

35. $(a^{1/2} \cdot a)^2$
a^3

36. $(b^{2/3} \cdot b^{1/6})^6$
b^5

37. $(x^{-1/2} \cdot x^{3/4})^{-2}$
$\frac{1}{x^{1/2}}$

38. $(a^{1/2} \cdot a^{-2})^3$
$\frac{1}{a^{9/2}}$

39. $(y^{-1/2} \cdot y^{3/2})^{2/3}$
$y^{2/3}$

40. $(b^{-2/3} \cdot b^{1/4})^{-4/3}$
$b^{5/9}$

41. $(x^8y^2)^{1/2}$
x^4y

42. $(a^3b^9)^{2/3}$
a^2b^6

Simplify:

43. $(x^4y^2z^6)^{3/2}$
$x^6y^3z^9$

44. $(a^8b^4c^4)^{3/4}$
$a^6b^3c^3$

45. $(x^{-3}y^6)^{-1/3}$
$\dfrac{x}{y^2}$

46. $(a^2b^{-6})^{-1/2}$
$\dfrac{b^3}{a}$

47. $(x^{-2}y^{1/3})^{-3/4}$
$\dfrac{x^{3/2}}{y^{1/4}}$

48. $(a^{-2/3}b^{2/3})^{3/2}$
$\dfrac{b}{a}$

49. $\left(\dfrac{x^{1/2}}{y^{-2}}\right)^4$
x^2y^8

50. $\left(\dfrac{b^{-3/4}}{a^{-1/2}}\right)^8$
$\dfrac{a^4}{b^6}$

51. $\dfrac{x^{1/4}\cdot x^{-1/2}}{x^{2/3}}$
$\dfrac{1}{x^{11/12}}$

52. $\dfrac{b^{1/2}\cdot b^{-3/4}}{b^{1/4}}$
$\dfrac{1}{b^{1/2}}$

53. $\left(\dfrac{y^{2/3}\cdot y^{-5/6}}{y^{1/9}}\right)^9$
$\dfrac{1}{y^{5/2}}$

54. $\left(\dfrac{a^{1/3}\cdot a^{-2/3}}{a^{1/2}}\right)^4$
$\dfrac{1}{a^{10/3}}$

55. $\left(\dfrac{b^2\cdot b^{-3/4}}{b^{-1/2}}\right)^{-1/2}$
$\dfrac{1}{b^{7/8}}$

56. $\dfrac{(x^{-5/6}\cdot x^3)^{-2/3}}{x^{4/3}}$
$\dfrac{1}{x^{25/9}}$

57. $(a^{2/3}b^2)^6(a^3b^3)^{1/3}$
a^5b^{13}

58. $(x^3y^{-1/2})^{-2}(x^{-3}y^2)^{1/6}$
$\dfrac{y^{4/3}}{x^{13/2}}$

59. $(16m^{-2}n^4)^{-1/2}(mn^{1/2})$
$\dfrac{m^2}{4n^{3/2}}$

60. $(27m^3n^{-6})^{1/3}(m^{-1/3}n^{5/6})^6$
$\dfrac{3n^3}{m}$

61. $\left(\dfrac{x^{1/2}y^{-3/4}}{y^{2/3}}\right)^{-6}$
$\dfrac{y^{17/2}}{x^3}$

62. $\left(\dfrac{x^{1/2}y^{-5/4}}{y^{-3/4}}\right)^{-4}$
$\dfrac{y^2}{x^2}$

63. $\left(\dfrac{2^{-6}b^{-3}}{a^{-1/2}}\right)^{-2/3}$
$\dfrac{16b^2}{a^{1/3}}$

64. $\left(\dfrac{49c^{5/3}}{a^{-1/4}b^{5/6}}\right)^{-3/2}$
$\dfrac{b^{5/4}}{343a^{3/8}c^{5/2}}$

65. $y^{3/2}(y^{1/2}-y^{-1/2})$
y^2-y

66. $y^{3/5}(y^{2/5}+y^{-3/5})$
$y+1$

67. $a^{-1/4}(a^{5/4}-a^{9/4})$
$a-a^2$

68. $x^{4/3}(x^{2/3}+x^{-1/3})$
x^2+x

69. $x^n\cdot x^{3n}$
x^{4n}

70. $a^{2n}\cdot a^{-5n}$
$\dfrac{1}{a^{3n}}$

71. $x^n\cdot x^{n/2}$
$x^{3n/2}$

72. $a^{n/2}\cdot a^{-n/3}$
$a^{n/6}$

73. $\dfrac{y^{n/2}}{y^{-n}}$
$y^{3n/2}$

74. $\dfrac{b^{m/3}}{b^m}$
$\dfrac{1}{b^{2m/3}}$

75. $(x^{2n})^n$
x^{2n^2}

76. $(x^{5n})^{2n}$
x^{10n^2}

77. $(x^{n/4}y^{n/8})^8$
$x^{2n}y^n$

78. $(x^{n/2}y^{n/3})^6$
$x^{3n}y^{2n}$

▶ Objective B

Rewrite the exponential expression as a radical expression.

79. $3^{1/4}$
$\sqrt[4]{3}$

80. $5^{1/2}$
$\sqrt{5}$

81. $a^{3/2}$
$\sqrt{a^3}$

82. $b^{4/3}$
$\sqrt[3]{b^4}$

83. $(2t)^{5/2}$
$\sqrt{32t^5}$

84. $(3x)^{2/3}$
$\sqrt[3]{9x^2}$

85. $-2x^{2/3}$
$-2\sqrt[3]{x^2}$

86. $-3a^{2/5}$
$-3\sqrt[5]{a^2}$

87. $(a^2b)^{2/3}$
$\sqrt[3]{a^4b^2}$

88. $(x^2y^3)^{3/4}$
$\sqrt[4]{x^6y^9}$

89. $(a^2b^4)^{3/5}$
$\sqrt[5]{a^6b^{12}}$

90. $(a^3b^7)^{3/2}$
$\sqrt{a^9b^{21}}$

91. $(4x-3)^{3/4}$
$\sqrt[4]{(4x-3)^3}$

92. $(3x-2)^{1/3}$
$\sqrt[3]{3x-2}$

93. $x^{-2/3}$
$\dfrac{1}{\sqrt[3]{x^2}}$

94. $b^{-3/4}$
$\dfrac{1}{\sqrt[4]{b^3}}$

Rewrite the radical expression as an exponential expression.

95. $\sqrt{14}$
$14^{1/2}$

96. $\sqrt{7}$
$7^{1/2}$

97. $\sqrt[3]{x}$
$x^{1/3}$

98. $\sqrt[4]{y}$
$y^{1/4}$

99. $\sqrt[3]{x^4}$
$x^{4/3}$

100. $\sqrt[4]{a^3}$
$a^{3/4}$

101. $\sqrt[5]{b^3}$
$b^{3/5}$

102. $\sqrt[4]{b^5}$
$b^{5/4}$

103. $\sqrt[3]{2x^2}$
$(2x^2)^{1/3}$

104. $\sqrt[5]{4y^7}$
$(4y^7)^{1/5}$

105. $-\sqrt{3x^5}$
$-(3x^5)^{1/2}$

106. $-\sqrt[4]{4x^5}$
$-(4x^5)^{1/4}$

107. $3x\sqrt[3]{y^2}$
$3xy^{2/3}$

108. $2y\sqrt{x^3}$
$2yx^{3/2}$

109. $\sqrt{a^2-2}$
$(a^2-2)^{1/2}$

110. $\sqrt{3-y^2}$
$(3-y^2)^{1/2}$

▶ **Objective C**

Simplify:

111. $\sqrt{x^{16}}$
x^8

112. $\sqrt{y^{14}}$
y^7

113. $-\sqrt{x^8}$
$--x^4$

114. $-\sqrt{a^6}$
$-a^3$

115. $\sqrt[3]{x^3y^9}$
xy^3

116. $\sqrt[3]{a^6b^{12}}$
a^2b^4

117. $-\sqrt[3]{x^{15}y^3}$
$-x^5y$

118. $-\sqrt[3]{a^9b^9}$
$-a^3b^3$

119. $\sqrt{16a^4b^{12}}$
$4a^2b^6$

120. $\sqrt{25x^8y^2}$
$5x^4y$

121. $\sqrt{-16x^4y^2}$
Not a real number

122. $\sqrt{-9a^4b^8}$
Not a real number

123. $\sqrt[3]{27x^9}$
$3x^3$

124. $\sqrt[3]{8a^{21}b^6}$
$2a^7b^2$

125. $\sqrt[3]{-64x^9y^{12}}$
$-4x^3y^4$

126. $\sqrt[3]{-27a^3b^{15}}$
$-3ab^5$

127. $-\sqrt[4]{x^8y^{12}}$
$-x^2y^3$

128. $-\sqrt[4]{a^{16}b^4}$
$-a^4b$

129. $\sqrt[5]{x^{20}y^{10}}$
x^4y^2

130. $\sqrt[5]{a^5b^{25}}$
ab^5

131. $\sqrt[4]{81x^4y^{20}}$
$3xy^5$

132. $\sqrt[4]{16a^8b^{20}}$
$2a^2b^5$

133. $\sqrt[5]{32a^5b^{10}}$
$2ab^2$

134. $\sqrt[5]{-32x^{15}y^{20}}$
$-2x^3y^4$

SECTION 5.2 — Operations on Radical Expressions

Objective A — To simplify radical expressions

If a number is not a perfect power, its root can only be approximated, for example, $\sqrt{5}$ and $\sqrt[3]{3}$. These numbers are **irrational numbers.** Their decimal representations never terminate or repeat.

$$\sqrt{5} = 2.2360679\ldots \qquad \sqrt[3]{3} = 1.4422495\ldots$$

The approximate square roots and cube roots of the positive integers up to 100 can be found in the Appendix on page A3. The roots have been rounded to the nearest thousandth.

A radical expression is in simplest form when the radicand contains no factor that is a perfect power. The Product Property of Radicals is used to simplify radical expressions whose radicands are not perfect powers.

The Product Property of Radicals

If $\sqrt[n]{a}$ and $\sqrt[n]{b}$ are positive real numbers, then $\sqrt[n]{ab} = \sqrt[n]{a} \cdot \sqrt[n]{b}$ and $\sqrt[n]{a} \cdot \sqrt[n]{b} = \sqrt[n]{ab}$.

Simplify: $\sqrt{48}$

Write the prime factorization of the radicand in exponential form.

$$\sqrt{48} = \sqrt{2^4 \cdot 3}$$

Use the Product Property of Radicals to write the expression as a product.

$$= \sqrt{2^4}\sqrt{3}$$

Simplify.

$$= 2^2\sqrt{3}$$
$$= 4\sqrt{3}$$

Simplify: $\sqrt[3]{x^7}$

Write the radicand as the product of a perfect cube and a factor which does not contain a perfect cube.

$$\sqrt[3]{x^7} = \sqrt[3]{x^6 \cdot x}$$

Use the Product Property of Radicals to write the expression as a product.

$$= \sqrt[3]{x^6}\sqrt[3]{x}$$

Simplify.

$$= x^2\sqrt[3]{x}$$

Simplify: $\sqrt[4]{32x^7}$

Write the prime factorization of the coefficient of the radicand in exponential form.

$$\sqrt[4]{32x^7} = \sqrt[4]{2^5x^7}$$

Write the radicand as the product of a perfect fourth power and factors which do not contain a perfect fourth power.

$$= \sqrt[4]{2^4x^4(2x^3)}$$

Use the Product Property of Radicals to write the expression as a product.

$$= \sqrt[4]{2^4x^4}\sqrt[4]{2x^3}$$

Simplify.

$$= 2x\sqrt[4]{2x^3}$$

Simplify $\sqrt{175}$. Then find the decimal approximation. Round to the nearest thousandth.

Write the prime factorization of the radicand in exponential form.

$$\sqrt{175} = \sqrt{5^2 \cdot 7}$$

Use the Product Property of Radicals to write the expression as a product.

$$= \sqrt{5^2}\sqrt{7}$$

Simplify.

$$= 5\sqrt{7}$$

Replace the radical expression by the decimal approximation found on page A3.

$$\approx 5(2.646)$$

Simplify.

$$= 13.230$$

Example 1 Simplify: $\sqrt[4]{x^9}$

Solution $\sqrt[4]{x^9} = \sqrt[4]{x^8 \cdot x} = \sqrt[4]{x^8}\sqrt[4]{x}$
$= x^2\sqrt[4]{x}$

Example 2 Simplify: $\sqrt[5]{x^7}$

Your solution $x\sqrt[5]{x^2}$

Example 3 Simplify: $\sqrt[3]{-27a^5b^{12}}$

Solution $\sqrt[3]{-27a^5b^{12}}$
$= \sqrt[3]{(-3)^3a^5b^{12}}$
$= \sqrt[3]{(-3)^3a^3b^{12}(a^2)}$
$= \sqrt[3]{(-3)^3a^3b^{12}}\sqrt[3]{a^2}$
$= -3ab^4\sqrt[3]{a^2}$

Example 4 Simplify: $\sqrt[3]{-64x^8y^{18}}$

Your solution $-4x^2y^6\sqrt[3]{x^2}$

Example 5 Simplify $\sqrt{180}$. Then find the decimal approximation. Use the table on page A3.

Solution $\sqrt{180}$
$= \sqrt{3^2 \cdot 2^2 \cdot 5}$
$= \sqrt{3^2 \cdot 2^2}\sqrt{5} = 3 \cdot 2\sqrt{5}$
$= 6\sqrt{5} \approx 6(2.236) = 13.416$

Example 6 Simplify $\sqrt{216}$. Then find the decimal approximation.

Your solution 14.697

Solutions on p. A24

Content and Format © 1991 HMCo.

Objective B

To add or subtract radical expressions

The Distributive Property is used to simplify the sum or difference of radical expressions that have the same radicand and the same index. For example,

$$3\sqrt{5} + 8\sqrt{5} = (3 + 8)\sqrt{5} = 11\sqrt{5}$$

$$2\sqrt[3]{3x} - 9\sqrt[3]{3x} = (2 - 9)\sqrt[3]{3x} = -7\sqrt[3]{3x}$$

Radical expressions that are in simplest form and have unlike radicands or different indices cannot be simplified by the Distributive Property. The expressions below cannot be simplified by the Distributive Property.

$$3\sqrt[4]{2} - 6\sqrt[4]{3} \qquad 2\sqrt[4]{4x} + 3\sqrt[3]{4x}$$

Simplify: $3\sqrt{32x^2} - 2x\sqrt{2} + \sqrt{128x^2}$

First simplify each term. Then combine like terms by using the Distributive Property.

$$
\begin{aligned}
3\sqrt{32x^2} - 2x\sqrt{2} + \sqrt{128x^2} &= 3\sqrt{2^5x^2} - 2x\sqrt{2} + \sqrt{2^7x^2} \\
&= 3\sqrt{2^4x^2}\sqrt{2} - 2x\sqrt{2} + \sqrt{2^6x^2}\sqrt{2} \\
&= 3 \cdot 2^2x\sqrt{2} - 2x\sqrt{2} + 2^3x\sqrt{2} \\
&= 12x\sqrt{2} - 2x\sqrt{2} + 8x\sqrt{2} \\
&= 18x\sqrt{2}
\end{aligned}
$$

Example 7

Simplify: $5b\sqrt[4]{32a^7b^5} - 2a\sqrt[4]{162a^3b^9}$

Solution

$5b\sqrt[4]{32a^7b^5} - 2a\sqrt[4]{162a^3b^9}$

$\quad = 5b\sqrt[4]{2^5a^7b^5} - 2a\sqrt[4]{3^4 \cdot 2a^3b^9}$

$\quad = 5b\sqrt[4]{2^4a^4b^4}\sqrt[4]{2a^3b} - 2a\sqrt[4]{3^4b^8}\sqrt[4]{2a^3b}$

$\quad = 5b \cdot 2ab\sqrt[4]{2a^3b} - 2a \cdot 3b^2\sqrt[4]{2a^3b}$

$\quad = 10ab^2\sqrt[4]{2a^3b} - 6ab^2\sqrt[4]{2a^3b}$

$\quad = 4ab^2\sqrt[4]{2a^3b}$

Example 8

Simplify: $3xy\sqrt[3]{81x^5y} - \sqrt[3]{192x^8y^4}$

Your solution

$5x^2y\sqrt[3]{3x^2y}$

Solution on p. A24

| Objective C | **To multiply radical expressions** |

The Product Property of Radicals is used to multiply radical expressions with the same index.

$$\sqrt{3x} \cdot \sqrt{5y} = \sqrt{3x \cdot 5y} = \sqrt{15xy}$$

Simplify: $\sqrt[3]{2a^5b}\sqrt[3]{16a^2b^2}$

Use the Product Property of Radicals to multiply the radicands.

$$\sqrt[3]{2a^5b}\sqrt[3]{16a^2b^2} = \sqrt[3]{32a^7b^3}$$

Simplify.

$$= \sqrt[3]{2^5a^7b^3}$$
$$= \sqrt[3]{2^3a^6b^3}\sqrt[3]{2^2a}$$
$$= 2a^2b\sqrt[3]{4a}$$

Simplify: $\sqrt{2x}(\sqrt{8x} - \sqrt{3})$

Use the Distributive Property to remove parentheses.

$$\sqrt{2x}(\sqrt{8x} - \sqrt{3}) = \sqrt{16x^2} - \sqrt{6x}$$

Simplify.

$$= \sqrt{2^4x^2} - \sqrt{6x}$$
$$= 2^2x - \sqrt{6x}$$
$$= 4x - \sqrt{6x}$$

Simplify: $(2\sqrt{5} - 3)(3\sqrt{5} + 4)$

Use the FOIL method to remove parentheses.

$$(2\sqrt{5} - 3)(3\sqrt{5} + 4) = 6(\sqrt{5})^2 + 8\sqrt{5} - 9\sqrt{5} - 12$$
$$= 30 + 8\sqrt{5} - 9\sqrt{5} - 12$$

Combine like terms.

$$= 18 - \sqrt{5}$$

Simplify: $(4\sqrt{a} - \sqrt{b})(2\sqrt{a} + 5\sqrt{b})$

Use FOIL.

$$(4\sqrt{a} - \sqrt{b})(2\sqrt{a} + 5\sqrt{b}) = 8(\sqrt{a})^2 + 20\sqrt{ab} - 2\sqrt{ab} - 5(\sqrt{b})^2$$
$$= 8a + 18\sqrt{ab} - 5b$$

The expressions $a + b$ and $a - b$ are **conjugates** of each other. Recall that $(a + b)(a - b) = a^2 - b^2$. This identity is used to simplify conjugate radical expressions.

Simplify: $(\sqrt{11} - 3)(\sqrt{11} + 3)$

The radical expressions are conjugates.

$$(\sqrt{11} - 3)(\sqrt{11} + 3) = (\sqrt{11})^2 - 3^2 = 11 - 9$$
$$= 2$$

Example 9

Simplify: $\sqrt{3x}(\sqrt{27x^2} - \sqrt{3x})$

Solution

$$\sqrt{3x}(\sqrt{27x^2} - \sqrt{3x}) = \sqrt{81x^3} - \sqrt{9x^2}$$
$$= \sqrt{3^4x^3} - \sqrt{3^2x^2}$$
$$= \sqrt{3^4x^2}\sqrt{x} - \sqrt{3^2x^2}$$
$$= 3^2x\sqrt{x} - 3x$$
$$= 9x\sqrt{x} - 3x$$

Example 10

Simplify: $\sqrt{5b}(\sqrt{3b} - \sqrt{10})$

Your solution

$b\sqrt{15} - 5\sqrt{2b}$

Example 11

Simplify: $(2\sqrt[3]{x} - 3)(3\sqrt[3]{x} - 4)$

Solution

$$(2\sqrt[3]{x} - 3)(3\sqrt[3]{x} - 4)$$
$$= 6\sqrt[3]{x^2} - 8\sqrt[3]{x} - 9\sqrt[3]{x} + 12$$
$$= 6\sqrt[3]{x^2} - 17\sqrt[3]{x} + 12$$

Example 12

Simplify: $(2\sqrt[3]{2x} - 3)(\sqrt[3]{2x} - 5)$

Your solution

$2\sqrt[3]{4x^2} - 13\sqrt[3]{2x} + 15$

Example 13

Simplify: $(2\sqrt{x} - \sqrt{2y})(2\sqrt{x} + \sqrt{2y})$

Solution

$$(2\sqrt{x} - \sqrt{2y})(2\sqrt{x} + \sqrt{2y})$$
$$= (2\sqrt{x})^2 - (\sqrt{2y})^2$$
$$= 4x - 2y$$

Example 14

Simplify: $(\sqrt{a} - 3\sqrt{y})(\sqrt{a} + 3\sqrt{y})$

Your solution

$a - 9y$

Objective D	**To divide radical expressions**

The Quotient Property of Radicals is used to divide radical expressions with the same index.

The Quotient Property of Radicals

If $\sqrt[n]{a}$ and $\sqrt[n]{b}$ are real numbers and $b \neq 0$, then

$$\sqrt[n]{\dfrac{a}{b}} = \dfrac{\sqrt[n]{a}}{\sqrt[n]{b}} \quad \text{and} \quad \dfrac{\sqrt[n]{a}}{\sqrt[n]{b}} = \sqrt[n]{\dfrac{a}{b}}.$$

Simplify: $\sqrt[3]{\dfrac{81x^5}{y^6}}$

Use the Quotient Property of Radicals.
$$\sqrt[3]{\dfrac{81x^5}{y^6}} = \dfrac{\sqrt[3]{81x^5}}{\sqrt[3]{y^6}}$$

Simplify each radical expression.
$$= \dfrac{\sqrt[3]{3^4 x^5}}{\sqrt[3]{y^6}}$$

$$= \dfrac{\sqrt[3]{3^3 x^3}\sqrt[3]{3x^2}}{\sqrt[3]{y^6}}$$

$$= \dfrac{3x\sqrt[3]{3x^2}}{y^2}$$

Simplify: $\dfrac{\sqrt{5a^4 b^7 c^2}}{\sqrt{ab^3 c}}$

Use the Quotient Property of Radicals.
$$\dfrac{\sqrt{5a^4 b^7 c^2}}{\sqrt{ab^3 c}} = \sqrt{\dfrac{5a^4 b^7 c^2}{ab^3 c}}$$

Simplify the radicand.
$$= \sqrt{5a^3 b^4 c}$$

$$= \sqrt{a^2 b^4}\sqrt{5ac}$$

$$= ab^2\sqrt{5ac}$$

A radical expression is in simplest form when no radical remains in the denominator of the radical expression. The procedure used to remove a radical from the denominator is called **rationalizing the denominator.**

Simplify: $\dfrac{2}{\sqrt{x}}$

Multiply the numerator and the denominator by 1 in the form of $\dfrac{\sqrt{x}}{\sqrt{x}}$.
$$\dfrac{2}{\sqrt{x}} = \dfrac{2}{\sqrt{x}} \cdot \dfrac{\sqrt{x}}{\sqrt{x}}$$

Simplify.
$$= \dfrac{2\sqrt{x}}{(\sqrt{x})^2}$$

$$= \dfrac{2\sqrt{x}}{x}$$

Simplify: $\dfrac{3x}{\sqrt[3]{4x}}$

$\sqrt[3]{4x} = \sqrt[3]{2^2x}.$ $\sqrt[3]{2^2x} \cdot \sqrt[3]{2x^2} = \sqrt[3]{2^3x^3}$,
a perfect cube. Multiply the numerator and
denominator by $\dfrac{\sqrt[3]{2x^2}}{\sqrt[3]{2x^2}}$ which equals 1.

Then simplify.

$\dfrac{3x}{\sqrt[3]{4x}} = \dfrac{3x}{\sqrt[3]{2^2x}} \cdot \dfrac{\sqrt[3]{2x^2}}{\sqrt[3]{2x^2}}$

$= \dfrac{3x\sqrt[3]{2x^2}}{\sqrt[3]{2^3x^3}} = \dfrac{3x\sqrt[3]{2x^2}}{2x}$

$= \dfrac{3\sqrt[3]{2x^2}}{2}$

Simplify: $\dfrac{\sqrt{x} - \sqrt{y}}{\sqrt{x} + \sqrt{y}}$

To simplify a fraction that has a binomial
radical expression in the denominator,
multiply the numerator and denominator by the conjugate of the denominator.
Then simplify.

$\dfrac{\sqrt{x} - \sqrt{y}}{\sqrt{x} + \sqrt{y}} = \dfrac{\sqrt{x} - \sqrt{y}}{\sqrt{x} + \sqrt{y}} \cdot \dfrac{\sqrt{x} - \sqrt{y}}{\sqrt{x} - \sqrt{y}}$

$= \dfrac{(\sqrt{x})^2 - \sqrt{xy} - \sqrt{xy} + (\sqrt{y})^2}{(\sqrt{x})^2 - (\sqrt{y})^2}$

$= \dfrac{x - 2\sqrt{xy} + y}{x - y}$

Example 15

Simplify: $\dfrac{5}{\sqrt{5x}}$

Solution

$\dfrac{5}{\sqrt{5x}} = \dfrac{5}{\sqrt{5x}} \cdot \dfrac{\sqrt{5x}}{\sqrt{5x}}$

$= \dfrac{5\sqrt{5x}}{\sqrt{5^2x^2}} = \dfrac{5\sqrt{5x}}{5x}$

$= \dfrac{\sqrt{5x}}{x}$

Example 16

Simplify: $\dfrac{y}{\sqrt{3y}}$

Your solution

$\dfrac{\sqrt{3y}}{3}$

Example 17

Simplify: $\dfrac{3}{\sqrt[4]{2x}}$

Solution

$\dfrac{3}{\sqrt[4]{2x}} = \dfrac{3}{\sqrt[4]{2x}} \cdot \dfrac{\sqrt[4]{2^3x^3}}{\sqrt[4]{2^3x^3}}$

$= \dfrac{3\sqrt[4]{8x^3}}{\sqrt[4]{2^4x^4}}$

$= \dfrac{3\sqrt[4]{8x^3}}{2x}$

Example 18

Simplify: $\dfrac{3}{\sqrt[3]{3x^2}}$

Your solution

$\dfrac{\sqrt[3]{9x}}{x}$

Solutions on p. A24

Example 19

Simplify: $\dfrac{3}{5 - 2\sqrt{3}}$

Solution

$$\dfrac{3}{5 - 2\sqrt{3}} \cdot \dfrac{5 + 2\sqrt{3}}{5 + 2\sqrt{3}} = \dfrac{15 + 6\sqrt{3}}{5^2 - (2\sqrt{3})^2}$$

$$= \dfrac{15 + 6\sqrt{3}}{25 - 12}$$

$$= \dfrac{15 + 6\sqrt{3}}{13}$$

Example 20

Simplify: $\dfrac{3 + \sqrt{6}}{2 - \sqrt{6}}$

Your solution

$-\dfrac{12 + 5\sqrt{6}}{2}$

Example 21

Simplify: $\dfrac{3 + \sqrt{y}}{3 - \sqrt{y}}$

Solution

$$\dfrac{3 + \sqrt{y}}{3 - \sqrt{y}} = \dfrac{3 + \sqrt{y}}{3 - \sqrt{y}} \cdot \dfrac{3 + \sqrt{y}}{3 + \sqrt{y}}$$

$$= \dfrac{3^2 + 3\sqrt{y} + 3\sqrt{y} + (\sqrt{y})^2}{3^2 - (\sqrt{y})^2}$$

$$= \dfrac{9 + 6\sqrt{y} + y}{9 - y}$$

Example 22

Simplify: $\dfrac{\sqrt{2} + \sqrt{x}}{\sqrt{2} - \sqrt{x}}$

Your solution

$\dfrac{2 + 2\sqrt{2x} + x}{2 - x}$

Solutions on pp. A24–A25

5.2   EXERCISES

▶ **Objective A**

Simplify:

1. $\sqrt{x^4y^3z^5}$
$x^2yz^2\sqrt{yz}$

2. $\sqrt{x^3y^6z^9}$
$xy^3z^4\sqrt{xz}$

3. $\sqrt{8a^3b^8}$
$2ab^4\sqrt{2a}$

4. $\sqrt{24a^9b^6}$
$2a^4b^3\sqrt{6a}$

5. $\sqrt{45x^2y^3z^5}$
$3xyz^2\sqrt{5yz}$

6. $\sqrt{60xy^7z^{12}}$
$2y^3z^6\sqrt{15xy}$

7. $\sqrt{-9x^3}$
Not a real
number

8. $\sqrt{-x^2y^5}$
Not a real
number

9. $\sqrt[3]{a^{16}b^8}$
$a^5b^2\sqrt[3]{ab^2}$

10. $\sqrt[3]{a^5b^8}$
$ab^2\sqrt[3]{a^2b^2}$

11. $\sqrt[3]{-125x^2y^4}$
$-5y\sqrt[3]{x^2y}$

12. $\sqrt[3]{-216x^5y^9}$
$-6xy^3\sqrt[3]{x^2}$

13. $\sqrt[3]{a^4b^5c^6}$
$abc^2\sqrt[3]{ab^2}$

14. $\sqrt[3]{a^8b^{11}c^{15}}$
$a^2b^3c^5\sqrt[3]{a^2b^2}$

15. $\sqrt[4]{16x^9y^5}$
$2x^2y\sqrt[4]{xy}$

16. $\sqrt[4]{64x^8y^{10}}$
$2x^2y^2\sqrt[4]{4y^2}$

Simplify. Then find the decimal approximation.

17. $\sqrt{600}$
24.49

18. $\sqrt{432}$
20.784

19. $\sqrt{539}$
23.219

20. $\sqrt{320}$
17.888

21. $\sqrt{845}$
29.068

22. $\sqrt{468}$
21.636

23. $\sqrt{600,000}$
774.6

24. $\sqrt{120,000}$
346.4

▶ **Objective B**

Simplify:

25. $2\sqrt{x} - 8\sqrt{x}$
$-6\sqrt{x}$

26. $3\sqrt{y} + 12\sqrt{y}$
$15\sqrt{y}$

27. $\sqrt{8} - \sqrt{32}$
$-2\sqrt{2}$

28. $\sqrt{27} - \sqrt{75}$
$-2\sqrt{3}$

29. $\sqrt{128x} - \sqrt{98x}$
$\sqrt{2x}$

30. $\sqrt{48x} + \sqrt{147x}$
$11\sqrt{3x}$

31. $\sqrt{27a} - \sqrt{8a}$
$3\sqrt{3a} - 2\sqrt{2a}$

32. $\sqrt{18b} + \sqrt{75b}$
$3\sqrt{2b} + 5\sqrt{3b}$

33. $2\sqrt{2x^3} + 4x\sqrt{8x}$
$10x\sqrt{2x}$

Simplify:

34. $5y\sqrt{8y} + 2\sqrt{50y^3}$
$20y\sqrt{2y}$

35. $x\sqrt{75xy} - \sqrt{27x^3y}$
$2x\sqrt{3xy}$

36. $3\sqrt{8x^2y^3} - 2x\sqrt{32y^3}$
$-2xy\sqrt{2y}$

37. $2\sqrt{32x^2y^3} - xy\sqrt{98y}$
$xy\sqrt{2y}$

38. $6y\sqrt{x^3y} - 2\sqrt{x^3y^3}$
$4xy\sqrt{xy}$

39. $7b\sqrt{a^5b^3} - 2ab\sqrt{a^3b^3}$
$5a^2b^2\sqrt{ab}$

40. $2a\sqrt{27ab^5} + 3b\sqrt{3a^3b}$
$6ab^2\sqrt{3ab} + 3ab\sqrt{3ab}$

41. $\sqrt[3]{128} + \sqrt[3]{250}$
$9\sqrt[3]{2}$

42. $\sqrt[3]{16} - \sqrt[3]{54}$
$-\sqrt[3]{2}$

43. $2\sqrt[3]{3a^4} - 3a\sqrt[3]{81a}$
$-7a\sqrt[3]{3a}$

44. $2b\sqrt[3]{16b^2} + \sqrt[3]{128b^5}$
$8b\sqrt[3]{2b^2}$

45. $3\sqrt[3]{x^5y^7} - 8xy\sqrt[3]{x^2y^4}$
$-5xy^2\sqrt[3]{x^2y}$

46. $3\sqrt[4]{32a^5} - a\sqrt[4]{162a}$
$3a\sqrt[4]{2a}$

47. $2a\sqrt[4]{16ab^5} + 3b\sqrt[4]{256a^5b}$
$16ab\sqrt[4]{ab}$

48. $2\sqrt{50} - 3\sqrt{125} + \sqrt{98}$
$17\sqrt{2} - 15\sqrt{5}$

49. $3\sqrt{108} - 2\sqrt{18} - 3\sqrt{48}$
$6\sqrt{3} - 6\sqrt{2}$

50. $\sqrt{9b^3} - \sqrt{25b^3} + \sqrt{49b^3}$
$5b\sqrt{b}$

51. $\sqrt{4x^7y^5} + 9x^2\sqrt{x^3y^5} - 5xy\sqrt{x^5y^3}$
$6x^3y^2\sqrt{xy}$

52. $2x\sqrt{8xy^2} - 3y\sqrt{32x^3} + \sqrt{4x^3y^3}$
$-8xy\sqrt{2x} + 2xy\sqrt{xy}$

53. $5a\sqrt{3a^3b} + 2a^2\sqrt{27ab} - 4\sqrt{75a^5b}$
$-9a^2\sqrt{3ab}$

54. $\sqrt[3]{54xy^3} - 5\sqrt[3]{2xy^3} + y\sqrt[3]{128x}$
$2y\sqrt[3]{2x}$

55. $2\sqrt[3]{24x^3y^4} + 4x\sqrt[3]{81y^4} - 3y\sqrt[3]{24x^3y}$
$10xy\sqrt[3]{3y}$

56. $2a\sqrt[4]{32b^5} - 3b\sqrt[4]{162a^4b} + \sqrt[4]{2a^4b^5}$
$-4ab\sqrt[4]{2b}$

57. $6y\sqrt[4]{48x^5} - 2x\sqrt[4]{243xy^4} - 4\sqrt[4]{3x^5y^4}$
$2xy\sqrt[4]{3x}$

▶ **Objective C**

Simplify:

58. $\sqrt{8}\sqrt{32}$
16

59. $\sqrt{14}\sqrt{35}$
$7\sqrt{10}$

60. $\sqrt[3]{4}\sqrt[3]{8}$
$2\sqrt[3]{4}$

61. $\sqrt[3]{6}\sqrt[3]{36}$
6

62. $\sqrt{x^2y^5}\sqrt{xy}$
$xy^3\sqrt{x}$

63. $\sqrt{a^3b}\sqrt{ab^4}$
$a^2b^2\sqrt{b}$

64. $\sqrt{2x^2y}\sqrt{32xy}$
$8xy\sqrt{x}$

65. $\sqrt{5x^3y}\sqrt{10x^3y^4}$
$5x^3y^2\sqrt{2y}$

66. $\sqrt[3]{x^2y}\sqrt[3]{16x^4y^2}$
$2x^2y\sqrt[3]{2}$

67. $\sqrt[3]{4a^2b^3}\sqrt[3]{8ab^5}$
$2ab^2\sqrt[3]{4b^2}$

68. $\sqrt[4]{12ab^3}\sqrt[4]{4a^5b^2}$
$2ab\sqrt[4]{3a^2b}$

69. $\sqrt[4]{36a^2b^4}\sqrt[4]{12a^5b^3}$
$2ab\sqrt[4]{27a^3b^3}$

70. $\sqrt{3}(\sqrt{27} - \sqrt{3})$
6

71. $\sqrt{10}(\sqrt{10} - \sqrt{5})$
$10 - 5\sqrt{2}$

72. $\sqrt{x}(\sqrt{x} - \sqrt{2})$
$x - \sqrt{2x}$

73. $\sqrt{y}(\sqrt{y} - \sqrt{5})$
$y - \sqrt{5y}$

74. $\sqrt{2x}(\sqrt{8x} - \sqrt{32})$
$4x - 8\sqrt{x}$

75. $\sqrt{3a}(\sqrt{27a^2} - \sqrt{a})$
$9a\sqrt{a} - a\sqrt{3}$

76. $(\sqrt{x} - 3)^2$
$x - 6\sqrt{x} + 9$

77. $(\sqrt{2x} + 4)^2$
$2x + 8\sqrt{2x} + 16$

78. $(4\sqrt{5} + 2)^2$
$84 + 16\sqrt{5}$

79. $2\sqrt{3x^2} \cdot 3\sqrt{12xy^3} \cdot \sqrt{6x^3y}$
$36x^3y^2\sqrt{6}$

80. $2\sqrt{14xy} \cdot 4\sqrt{7x^2y} \cdot 3\sqrt{8xy^2}$
$672x^2y^2$

81. $\sqrt[3]{8ab}\sqrt[3]{4a^2b^3}\sqrt[3]{9ab^4}$
$2ab^2\sqrt[3]{36ab^2}$

82. $\sqrt[3]{2a^2b}\sqrt[3]{4a^3b^2}\sqrt[3]{8a^5b^6}$
$4a^3b^3\sqrt[3]{a}$

83. $(\sqrt{2} - 3)(\sqrt{2} + 4)$
$-10 + \sqrt{2}$

84. $(\sqrt{5} - 5)(2\sqrt{5} + 2)$
$-8\sqrt{5}$

85. $(\sqrt{y} - 2)(\sqrt{y} + 2)$
$y - 4$

86. $(\sqrt{x} - y)(\sqrt{x} + y)$
$x - y^2$

87. $(\sqrt{2x} - 3\sqrt{y})(\sqrt{2x} + 3\sqrt{y})$
$2x - 9y$

88. $(2\sqrt{3x} - \sqrt{y})(2\sqrt{3x} + \sqrt{y})$
$12x - y$

▶ **Objective D**

Simplify:

89. $\dfrac{\sqrt{32x^2}}{\sqrt{2x}}$

$4\sqrt{x}$

90. $\dfrac{\sqrt{60y^4}}{\sqrt{12y}}$

$y\sqrt{5y}$

91. $\dfrac{\sqrt{42a^3b^5}}{\sqrt{14a^2b}}$

$b^2\sqrt{3a}$

92. $\dfrac{\sqrt{65ab^4}}{\sqrt{5ab}}$

$b\sqrt{13b}$

93. $\dfrac{1}{\sqrt{5}}$

$\dfrac{\sqrt{5}}{5}$

94. $\dfrac{1}{\sqrt{2}}$

$\dfrac{\sqrt{2}}{2}$

95. $\dfrac{1}{\sqrt{2x}}$

$\dfrac{\sqrt{2x}}{2x}$

96. $\dfrac{2}{\sqrt{3y}}$

$\dfrac{2\sqrt{3y}}{3y}$

97. $\dfrac{5}{\sqrt{5x}}$

$\dfrac{\sqrt{5x}}{x}$

98. $\dfrac{9}{\sqrt{3a}}$

$\dfrac{3\sqrt{3a}}{a}$

99. $\sqrt{\dfrac{x}{5}}$

$\dfrac{\sqrt{5x}}{5}$

100. $\sqrt{\dfrac{y}{2}}$

$\dfrac{\sqrt{2y}}{2}$

101. $\dfrac{3}{\sqrt[3]{2}}$

$\dfrac{3\sqrt[3]{4}}{2}$

102. $\dfrac{5}{\sqrt[3]{9}}$

$\dfrac{5\sqrt[3]{3}}{3}$

103. $\dfrac{3}{\sqrt[3]{4x^2}}$

$\dfrac{3\sqrt[3]{2x}}{2x}$

104. $\dfrac{5}{\sqrt[3]{3y}}$

$\dfrac{5\sqrt[3]{9y^2}}{3y}$

105. $\dfrac{\sqrt{40x^3y^2}}{\sqrt{80x^2y^3}}$

$\dfrac{\sqrt{2xy}}{2y}$

106. $\dfrac{\sqrt{15a^2b^5}}{\sqrt{30a^5b^3}}$

$\dfrac{b\sqrt{2a}}{2a^2}$

107. $\dfrac{\sqrt{24a^2b}}{\sqrt{18ab^4}}$

$\dfrac{2\sqrt{3ab}}{3b^2}$

108. $\dfrac{\sqrt{12x^3y}}{\sqrt{20x^4y}}$

$\dfrac{\sqrt{15x}}{5x}$

109. $\dfrac{2}{\sqrt{5}+2}$

$2\sqrt{5}-4$

110. $\dfrac{5}{2-\sqrt{7}}$

$\dfrac{10+5\sqrt{7}}{3}$

111. $\dfrac{3}{\sqrt{y}-2}$

$\dfrac{3\sqrt{y}+6}{y-4}$

112. $\dfrac{-7}{\sqrt{x}-3}$

$\dfrac{7\sqrt{x}+21}{x-9}$

113. $\dfrac{\sqrt{2}-\sqrt{3}}{\sqrt{2}+\sqrt{3}}$

$-5+2\sqrt{6}$

114. $\dfrac{\sqrt{3}+\sqrt{4}}{\sqrt{2}+\sqrt{3}}$

$-\sqrt{6}+3-2\sqrt{2}+2\sqrt{3}$

115. $\dfrac{2+3\sqrt{7}}{5-2\sqrt{7}}$

$\dfrac{52-19\sqrt{7}}{3}$

116. $\dfrac{2+3\sqrt{5}}{1-\sqrt{5}}$

$\dfrac{17+5\sqrt{5}}{4}$

117. $\dfrac{2\sqrt{3}-1}{3\sqrt{3}+2}$

$\dfrac{20-7\sqrt{3}}{23}$

118. $\dfrac{2\sqrt{a}-\sqrt{b}}{4\sqrt{a}+3\sqrt{b}}$

$\dfrac{8a-10\sqrt{ab}+3b}{16a-9b}$

119. $\dfrac{2\sqrt{x}-4}{\sqrt{x}+2}$

$\dfrac{2x-8\sqrt{x}+8}{x-4}$

120. $\dfrac{3\sqrt{y}-y}{\sqrt{y}+2y}$

$\dfrac{3-7\sqrt{y}+2y}{1-4y}$

121. $\dfrac{3\sqrt{x}-4\sqrt{y}}{3\sqrt{x}-2\sqrt{y}}$

$\dfrac{9x-6\sqrt{xy}-8y}{9x-4y}$

SECTION 5.3 — Complex Numbers

| Objective A | **To simplify a complex number** |

The radical expression $\sqrt{-4}$ is not a real number, since there is no real number whose square is -4. However, the solution of an algebraic equation is sometimes the square root of a negative number.

For example, the equation $x^2 + 1 = 0$ does not have a real number solution, since there is no real number whose square is a negative number.

$$x^2 + 1 = 0$$
$$x^2 = -1$$

Around the 17th century, a new number, called an **imaginary number,** was defined so that a negative number would have a square root. The letter i was chosen to represent the number whose square is -1.

$$i^2 = -1$$

An imaginary number is defined in terms of i.

If a is a positive real number, then the principal square root of negative a is the imaginary number $i\sqrt{a}$.

$$\sqrt{-a} = i\sqrt{a}$$

Here are some examples.

$$\sqrt{-16} = i\sqrt{16} = 4i$$
$$\sqrt{-12} = i\sqrt{12} = 2i\sqrt{3}$$
$$\sqrt{-21} = i\sqrt{21}$$
$$\sqrt{-1} = i\sqrt{1} = i$$

It is customary to write i in front of a radical to avoid confusing $\sqrt{a}i$ with $\sqrt{ai}$.

The real numbers and imaginary numbers make up the complex numbers.

Complex Number

A **complex number** is a number of the form **$a + bi$,** where a and b are real numbers and $i = \sqrt{-1}$. The number a is the **real part** of $a + bi$ and b is the **imaginary part.**

Examples of complex numbers are shown at the right.

Real Part	Imaginary Part
a $+$	bi
3 $+$	$2i$
8 $-$	$10i$

Complex Numbers — $a + bi$

— Real Numbers — $a + 0i$

— Imaginary Numbers — $0 + bi$

A *real number* is a complex number in which $b = 0$.

An *imaginary number* is a complex number in which $a = 0$.

Simplify: $\sqrt{20} - \sqrt{-50}$

Write the complex number in the form $a + bi$.

$$\sqrt{20} - \sqrt{-50} = \sqrt{20} - i\sqrt{50}$$

Use the Product Property of Radicals to simplify each radical.

$$= \sqrt{2^2 \cdot 5} - i\sqrt{5^2 \cdot 2}$$
$$= 2\sqrt{5} - 5i\sqrt{2}$$

Example 1

Simplify: $\sqrt{-80}$

Solution

$$\sqrt{-80} = i\sqrt{80} = i\sqrt{2^4 \cdot 5} = 4i\sqrt{5}$$

Example 2

Simplify: $\sqrt{-45}$

Your solution

$3i\sqrt{5}$

Example 3

Simplify: $\sqrt{25} + \sqrt{-40}$

Solution

$$\sqrt{25} + \sqrt{-40} = \sqrt{25} + i\sqrt{40}$$
$$= \sqrt{5^2} + i\sqrt{2^2 \cdot 2 \cdot 5}$$
$$= 5 + 2i\sqrt{10}$$

Example 4

Simplify: $\sqrt{98} - \sqrt{-60}$

Your solution

$7\sqrt{2} - 2i\sqrt{15}$

Solutions on p. A25

| Objective B | **To add or subtract complex numbers** |

To add two complex numbers, add the real parts and add the imaginary parts. To subtract two complex numbers, subtract the real parts and subtract the imaginary parts.

$$(a + bi) + (c + di) = (a + c) + (b + d)i$$
$$(a + bi) - (c + di) = (a - c) + (b - d)i$$

Simplify: $(3 - 7i) - (4 - 2i)$

Subtract the real parts and subtract the imaginary parts of the complex numbers.

$$(3 - 7i) - (4 - 2i) = (3 - 4) + [-7 - (-2)]i$$ Do this step mentally.

$$= -1 - 5i$$

Simplify: $(3 + \sqrt{-12}) + (7 - \sqrt{-27})$

Write each complex number in the
form $a + bi$.

$$(3 + \sqrt{-12}) + (7 - \sqrt{-27})$$
$$= (3 + i\sqrt{12}) + (7 - i\sqrt{27})$$

Use the Product Property of Radicals
to simplify each radical.

$$= (3 + i\sqrt{2^2 \cdot 3}) + (7 - i\sqrt{3^2 \cdot 3})$$
$$= (3 + 2i\sqrt{3}) + (7 - 3i\sqrt{3})$$

Add the complex numbers.

$$= 10 - i\sqrt{3}$$

Example 5

Simplify: $(3 + 2i) + (6 - 5i)$

Solution

$(3 + 2i) + (6 - 5i) = 9 - 3i$

Example 6

Simplify: $(-4 + 2i) - (6 - 8i)$

Your solution

$-10 + 10i$

Example 7

Simplify: $(9 - \sqrt{-8}) - (5 + \sqrt{-32})$

Solution

$(9 - \sqrt{-8}) - (5 + \sqrt{-32})$
$= (9 - i\sqrt{8}) - (5 + i\sqrt{32})$
$= (9 - i\sqrt{2^2 \cdot 2}) - (5 + i\sqrt{2^4 \cdot 2})$
$= (9 - 2i\sqrt{2}) - (5 + 4i\sqrt{2})$
$= 4 - 6i\sqrt{2}$

Example 8

Simplify: $(16 - \sqrt{-45}) - (3 + \sqrt{-20})$

Your solution

$13 - 5i\sqrt{5}$

Example 9

Simplify: $(6 + 4i) + (-6 - 4i)$

Solution

$(6 + 4i) + (-6 - 4i) = 0 + 0i = 0$

This illustrates that the additive inverse
of $a + bi$ is $-a - bi$.

Example 10

Simplify: $(3 - 2i) + (-3 + 2i)$

Your solution

0

Solutions on p. A25

Objective C	**To multiply complex numbers**

When multiplying complex numbers, the term i^2 is frequently a part of the
product. Recall that $i^2 = -1$.

Simplify: $2i \cdot 3i$

Multiply the imaginary numbers. $2i \cdot 3i = 6i^2$

Replace i^2 by -1. $= 6(-1)$

Simplify. $= -6$

Simplify: $\sqrt{-6} \cdot \sqrt{-24}$

Write each radical as the product of a real number and i.

$$\sqrt{-6} \cdot \sqrt{-24} = i\sqrt{6} \cdot i\sqrt{24}$$

Multiply the imaginary numbers.

$$= i^2\sqrt{144}$$

Replace i^2 by -1.

$$= -\sqrt{144}$$

Simplify the radical expression.

$$= -12$$

Note from the last example that it would have been incorrect to multiply the radicands of the two radical expressions. To illustrate,

$$\sqrt{-6} \cdot \sqrt{-24} = \sqrt{(-6)(-24)} = \sqrt{144} = 12, \text{ not } -12.$$

The Product Property of Radicals does not hold true when both radicands are negative and the index is an even number.

Simplify: $4i(3 - 2i)$

Use the Distributive Property to remove parentheses.

$$4i(3 - 2i) = 12i - 8i^2$$

Replace i^2 by -1.

$$= 12i - 8(-1)$$

Write the answer in the form $a + bi$.

$$= 8 + 12i$$

The product of two complex numbers is defined as follows.

Product of Complex Numbers

$$(a + bi)(c + di) = (ac - bd) + (ad + bc)i$$

One way to remember this rule is to use the FOIL method.

Simplify: $(2 + 4i)(3 - 5i)$

Use the FOIL method to find the product.

$$(2 + 4i)(3 - 5i) = 6 - 10i + 12i - 20i^2$$
$$= 6 + 2i - 20i^2$$

Replace i^2 by -1.

$$= 6 + 2i - 20(-1)$$

Write the answer in the form $a + bi$.

$$= 26 + 2i$$

Simplify: $(a + bi)(a - bi)$

$$(a + bi)(a - bi) = a^2 - b^2i^2$$
$$= a^2 - b^2(-1)$$
$$= a^2 + b^2$$

Conjugate of $a + bi$

The conjugate of $a + bi$ is $a - bi$. The product of conjugates $(a + bi)(a - bi)$ is the real number $a^2 + b^2$.

Simplify: $(2 + 3i)(2 - 3i)$

The product of conjugates is $a^2 + b^2$.

$$(2 + 3i)(2 - 3i) = 2^2 + 3^2$$
$$= 4 + 9$$
$$= 13$$

Note that the product of a complex number and its conjugate is a real number.

Example 11

Simplify: $(2i)(-5i)$

Solution

$(2i)(-5i) = -10i^2 = (-10)(-1) = 10$

Example 12

Simplify: $(-3i)(-10i)$

Your solution

-30

Example 13

Simplify: $\sqrt{-10} \cdot \sqrt{-5}$

Solution

$\sqrt{-10} \cdot \sqrt{-5} = i\sqrt{10} \cdot i\sqrt{5}$
$= i^2\sqrt{50} = -\sqrt{5^2 \cdot 2} = -5\sqrt{2}$

Example 14

Simplify: $-\sqrt{-8} \cdot \sqrt{-5}$

Your solution

$2\sqrt{10}$

Example 15

Simplify: $3i(2 - 4i)$

Solution

$3i(2 - 4i) = 6i - 12i^2 = 6i - 12(-1) = 12 + 6i$

Example 16

Simplify: $-6i(3 + 4i)$

Your solution

$24 - 18i$

Example 17

Simplify: $\sqrt{-8}(\sqrt{6} - \sqrt{-2})$

Solution

$\sqrt{-8}(\sqrt{6} - \sqrt{-2}) = i\sqrt{8}(\sqrt{6} - i\sqrt{2})$
$= i\sqrt{48} - i^2\sqrt{16}$
$= i\sqrt{2^4 \cdot 3} - (-1)\sqrt{2^4}$
$= 4i\sqrt{3} + 4 = 4 + 4i\sqrt{3}$

Example 18

Simplify: $\sqrt{-3}(\sqrt{27} - \sqrt{-6})$

Your solution

$3\sqrt{2} + 9i$

Example 19

Simplify: $(3 - 4i)(2 + 5i)$

Solution

$(3 - 4i)(2 + 5i) = 6 + 15i - 8i - 20i^2$
$= 6 + 7i - 20i^2$
$= 6 + 7i - 20(-1) = 26 + 7i$

Example 20

Simplify: $(4 - 3i)(2 - i)$

Your solution

$5 - 10i$

Example 21

Simplify: $(4 + 5i)(4 - 5i)$

Solution

$(4 + 5i)(4 - 5i) = 4^2 + 5^2 = 16 + 25 = 41$

Example 22

Simplify: $(3 + 6i)(3 - 6i)$

Your solution

45

Example 23

Simplify: $\left(\frac{9}{10} + \frac{3}{10}i\right)\left(1 - \frac{1}{3}i\right)$

Solution

$\left(\frac{9}{10} + \frac{3}{10}i\right)\left(1 - \frac{1}{3}i\right) = \frac{9}{10} - \frac{3}{10}i + \frac{3}{10}i - \frac{1}{10}i^2$
$= \frac{9}{10} - \frac{1}{10}i^2 = \frac{9}{10} - \frac{1}{10}(-1)$
$= \frac{9}{10} + \frac{1}{10} = 1$

Example 24

Simplify: $(3 - i)\left(\frac{3}{10} + \frac{1}{10}i\right)$

Your solution

1

Solutions on p. A25

| Objective D | **To divide complex numbers** |

A rational expression containing one or more complex numbers is in simplest form when no imaginary number remains in the denominator.

Simplify: $\dfrac{2-3i}{2i}$

Multiply the numerator and denominator by $\frac{i}{i}$.
$$\dfrac{2-3i}{2i} = \dfrac{2-3i}{2i} \cdot \dfrac{i}{i}$$
$$= \dfrac{2i - 3i^2}{2i^2}$$

Replace i^2 by -1.
$$= \dfrac{2i - 3(-1)}{2(-1)}$$

Simplify.
$$= \dfrac{3 + 2i}{-2}$$

Write the answer in the form $a + bi$.
$$= -\dfrac{3}{2} - i$$

Simplify: $\dfrac{3+2i}{1+i}$

Multiply the numerator and denominator by the conjugate of $1 + i$.
$$\dfrac{3+2i}{1+i} = \dfrac{3+2i}{1+i} \cdot \dfrac{1-i}{1-i}$$
$$= \dfrac{3 - 3i + 2i - 2i^2}{1^2 - i^2}$$

Replace i^2 by -1 and simplify.
$$= \dfrac{3 - i - 2(-1)}{2}$$
$$= \dfrac{5 - i}{2}$$

Write the answer in the form $a + bi$.
$$= \dfrac{5}{2} - \dfrac{1}{2}i$$

Example 25
Simplify: $\dfrac{5+4i}{3i}$

Solution
$$\dfrac{5+4i}{3i} = \dfrac{5+4i}{3i} \cdot \dfrac{i}{i} = \dfrac{5i + 4i^2}{3i^2}$$
$$= \dfrac{5i + 4(-1)}{3(-1)} = \dfrac{-4 + 5i}{-3} = \dfrac{4}{3} - \dfrac{5}{3}i$$

Example 26
Simplify: $\dfrac{2-3i}{4i}$

Your solution
$-\dfrac{3}{4} - \dfrac{1}{2}i$

Example 27
Simplify: $\dfrac{5-3i}{4+2i}$

Solution
$$\dfrac{5-3i}{4+2i} = \dfrac{5-3i}{4+2i} \cdot \dfrac{4-2i}{4-2i}$$
$$= \dfrac{20 - 10i - 12i + 6i^2}{4^2 + 2^2}$$
$$= \dfrac{20 - 22i + 6(-1)}{20}$$
$$= \dfrac{14 - 22i}{20} = \dfrac{7 - 11i}{10}$$
$$= \dfrac{7}{10} - \dfrac{11}{10}i$$

Example 28
Simplify: $\dfrac{2+5i}{3-2i}$

Your solution
$-\dfrac{4}{13} + \dfrac{19}{13}i$

Solutions on p. A25

5.3   EXERCISES

▶ **Objective A**

Simplify:

1. $\sqrt{-4}$
$2i$

2. $\sqrt{-64}$
$8i$

3. $\sqrt{-98}$
$7i\sqrt{2}$

4. $\sqrt{-72}$
$6i\sqrt{2}$

5. $\sqrt{-27}$
$3i\sqrt{3}$

6. $\sqrt{-75}$
$5i\sqrt{3}$

7. $\sqrt{16} + \sqrt{-4}$
$4 + 2i$

8. $\sqrt{25} + \sqrt{-9}$
$5 + 3i$

9. $\sqrt{12} - \sqrt{-18}$
$2\sqrt{3} - 3i\sqrt{2}$

10. $\sqrt{60} - \sqrt{-48}$
$2\sqrt{15} - 4i\sqrt{3}$

11. $\sqrt{160} - \sqrt{-147}$
$4\sqrt{10} - 7i\sqrt{3}$

12. $\sqrt{96} - \sqrt{-125}$
$4\sqrt{6} - 5i\sqrt{5}$

▶ **Objective B**

Simplify:

13. $(2 + 4i) + (6 - 5i)$
$8 - i$

14. $(6 - 9i) + (4 + 2i)$
$10 - 7i$

15. $(-2 - 4i) - (6 - 8i)$
$-8 + 4i$

16. $(3 - 5i) + (8 - 2i)$
$11 - 7i$

17. $(8 - \sqrt{-4}) - (2 + \sqrt{-16})$
$6 - 6i$

18. $(5 - \sqrt{-25}) - (11 - \sqrt{-36})$
$-6 + i$

19. $(12 - \sqrt{-50}) + (7 - \sqrt{-8})$
$19 - 7i\sqrt{2}$

20. $(5 - \sqrt{-12}) - (9 + \sqrt{-108})$
$-4 - 8i\sqrt{3}$

21. $(\sqrt{8} + \sqrt{-18}) + (\sqrt{32} - \sqrt{-72})$
$6\sqrt{2} - 3i\sqrt{2}$

22. $(\sqrt{40} - \sqrt{-98}) - (\sqrt{90} + \sqrt{-32})$
$-\sqrt{10} - 11i\sqrt{2}$

23. $(5 - 3i) + 2i$
$5 - i$

24. $(6 - 8i) + 4i$
$6 - 4i$

25. $(7 + 2i) + (-7 - 2i)$
0

26. $(8 - 3i) + (-8 + 3i)$
0

▶ Objective C

Simplify:

27. $(7i)(-9i)$
63

28. $(-6i)(-4i)$
-24

29. $\sqrt{-2}\sqrt{-8}$
-4

30. $\sqrt{-5}\sqrt{-45}$
-15

31. $\sqrt{-3}\sqrt{-6}$
$-3\sqrt{2}$

32. $\sqrt{-5}\sqrt{-10}$
$-5\sqrt{2}$

33. $2i(6 + 2i)$
$-4 + 12i$

34. $-3i(4 - 5i)$
$-15 - 12i$

35. $\sqrt{-2}(\sqrt{8} + \sqrt{-2})$
$-2 + 4i$

36. $\sqrt{-3}(\sqrt{12} - \sqrt{-6})$
$3\sqrt{2} + 6i$

37. $(5 - 2i)(3 + i)$
$17 - i$

38. $(2 - 4i)(2 - i)$
$-10i$

39. $(6 + 5i)(3 + 2i)$
$8 + 27i$

40. $(4 - 7i)(2 + 3i)$
$29 - 2i$

41. $(1 - i)\left(\frac{1}{2} + \frac{1}{2}i\right)$
1

42. $\left(\frac{4}{5} - \frac{2}{5}i\right)\left(1 + \frac{1}{2}i\right)$
1

43. $\left(\frac{6}{5} + \frac{3}{5}i\right)\left(\frac{2}{3} - \frac{1}{3}i\right)$
1

44. $(2 - i)\left(\frac{2}{5} + \frac{1}{5}i\right)$
1

▶ Objective D

Simplify:

45. $\frac{3}{i}$
$-3i$

46. $\frac{4}{5i}$
$-\frac{4}{5}i$

47. $\frac{2 - 3i}{-4i}$
$\frac{3}{4} + \frac{1}{2}i$

48. $\frac{16 + 5i}{-3i}$
$-\frac{5}{3} + \frac{16}{3}i$

49. $\frac{4}{5 + i}$
$\frac{10}{13} - \frac{2}{13}i$

50. $\frac{6}{5 + 2i}$
$\frac{30}{29} - \frac{12}{29}i$

51. $\frac{2}{2 - i}$
$\frac{4}{5} + \frac{2}{5}i$

52. $\frac{5}{4 - i}$
$\frac{20}{17} + \frac{5}{17}i$

53. $\frac{1 - 3i}{3 + i}$
$-i$

54. $\frac{2 + 12i}{5 + i}$
$\frac{11}{13} + \frac{29}{13}i$

55. $\frac{\sqrt{-10}}{\sqrt{8} - \sqrt{-2}}$
$-\frac{\sqrt{5}}{5} + \frac{2\sqrt{5}}{5}i$

56. $\frac{\sqrt{-2}}{\sqrt{12} - \sqrt{-8}}$
$-\frac{1}{5} + \frac{\sqrt{6}}{10}i$

57. $\frac{2 - 3i}{3 + i}$
$\frac{3}{10} - \frac{11}{10}i$

58. $\frac{3 + 5i}{1 - i}$
$-1 + 4i$

59. $\frac{5 + 3i}{3 - i}$
$\frac{6}{5} + \frac{7}{5}i$

SECTION 5.4

Solving Equations Containing Radical Expressions

Objective A

To solve an equation containing one or more radical expressions

An equation that contains a variable expression in a radicand is a **radical equation.**

$$\left.\begin{array}{l}\sqrt{x+2}=\sqrt{3x-4}\\ \sqrt[3]{x-4}=2\end{array}\right\}\begin{array}{l}\text{Radical}\\ \text{Equations}\end{array}$$

The following property of equality is used to solve radical equations.

The Property of Raising Both Sides of an Equation to a Power

If two numbers are equal, then the same powers of the numbers are equal. That is:

If a and b are real numbers and $a = b$, then $a^n = b^n$.

Solve: $\sqrt{x-2}-6=0$

Rewrite the equation with the radical on one side of the equation and the constant on the other side.

$$\sqrt{x-2}-6=0$$
$$\sqrt{x-2}=6$$

Square each side of the equation.

$$(\sqrt{x-2})^2=6^2$$

Solve the resulting equation.

$$x-2=36$$
$$x=38$$

Check the solution.
When raising both sides of an equation to an even power, the resulting equation may have a solution which is not a solution of the original equation.

Check:
$$\begin{array}{r}\sqrt{x-2}-6=0\\ \hline \sqrt{38-2}-6 \quad | \quad 0\\ \sqrt{36}-6 \quad | \quad 0\\ 6-6 \quad | \quad 0\\ 0=0\end{array}$$

38 checks as a solution.

The solution is 38.

Solve: $\sqrt[3]{x+2}=-3$

$$\sqrt[3]{x+2}=-3$$

Cube each side of the equation.

$$(\sqrt[3]{x+2})^3=(-3)^3$$

Solve the resulting equation.

$$x+2=-27$$
$$x=-29$$

Check the solution.

Check:
$$\begin{array}{r}\sqrt[3]{x+2}=-3\\ \hline \sqrt[3]{-29+2} \quad | \quad -3\\ \sqrt[3]{-27} \quad | \quad -3\\ -3=-3\end{array}$$

-29 checks as a solution.

The solution is -29.

When a radical equation contains two radical expressions, solve for one radical expression, square each side. Now solve the resulting radical equation.

Solve: $\sqrt{2x - 1} + \sqrt{x} = 2$

Solve for one of the radical expressions.

$$\sqrt{2x - 1} + \sqrt{x} = 2$$
$$\sqrt{2x - 1} = 2 - \sqrt{x}$$

Square each side. Recall that $(a - b)^2 = a^2 - 2ab + b^2$.

$$(\sqrt{2x - 1})^2 = (2 - \sqrt{x})^2$$
$$2x - 1 = 4 - 4\sqrt{x} + x$$

Solve the new radical equation.

$$x - 5 = -4\sqrt{x}$$
$$(x - 5)^2 = (-4\sqrt{x})^2$$

Solve the quadratic equation by factoring.

$$x^2 - 10x + 25 = 16x$$
$$x^2 - 26x + 25 = 0$$
$$(x - 25)(x - 1) = 0$$
$$x = 25 \quad \text{or} \quad x = 1$$

Check.

$$
\begin{array}{r|l}
\sqrt{2x - 1} + \sqrt{x} = 2 & \\
\hline
\sqrt{2(25) - 1} + \sqrt{25} & 2 \\
7 + 5 & 2 \\
12 \neq 2 &
\end{array}
\qquad
\begin{array}{r|l}
\sqrt{2x - 1} + \sqrt{x} = 2 & \\
\hline
\sqrt{2(1) - 1} + \sqrt{1} & 2 \\
1 + 1 & 2 \\
2 = 2 &
\end{array}
$$

25 does not check as a solution. 1 checks as a solution. The solution is 1.

Example 1

Solve: $\sqrt{x - 1} + \sqrt{x + 4} = 5$

Solution

$$\sqrt{x - 1} + \sqrt{x + 4} = 5$$
$$\sqrt{x + 4} = 5 - \sqrt{x - 1}$$
$$(\sqrt{x + 4})^2 = (5 - \sqrt{x - 1})^2$$
$$x + 4 = 25 - 10\sqrt{x - 1} + x - 1$$
$$2 = \sqrt{x - 1}$$
$$2^2 = (\sqrt{x - 1})^2$$
$$4 = x - 1$$
$$5 = x$$

The solution is 5.

Example 2

Solve: $\sqrt{x} - \sqrt{x + 5} = 1$

Your solution

The equation has no solution.

Example 3

Solve: $\sqrt[3]{3x - 1} = -4$

Solution

$$\sqrt[3]{3x - 1} = -4$$
$$(\sqrt[3]{3x - 1})^3 = (-4)^3$$
$$3x - 1 = -64$$
$$3x = -63$$
$$x = -21$$

Check:
$$
\begin{array}{r|l}
\sqrt[3]{3x - 1} = -4 & \\
\hline
\sqrt[3]{3(-21) - 1} & -4 \\
\sqrt[3]{-63 - 1} & -4 \\
\sqrt[3]{-64} & -4 \\
-4 = -4 &
\end{array}
$$

The solution is −21.

Example 4

Solve: $\sqrt[4]{x - 8} = 3$

Your solution

89

Solutions on p. A26

Content and Format © 1991 HMCo.

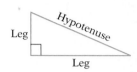

| Objective B | **To solve application problems** |

A right triangle contains one 90° angle. The side opposite the 90° angle is called the **hypotenuse.** The other two sides are called **legs.**

Pythagoras, a Greek mathematician, discovered that the square of the hypotenuse of a right triangle is equal to the sum of the squares of the two legs. This is called the **Pythagorean Theorem.**

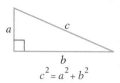

$$c^2 = a^2 + b^2$$

Example 5

A ladder 20 ft long is leaning against a building. How high on the building will the ladder reach when the bottom of the ladder is 8 ft from the building? Round to the nearest tenth.

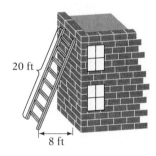

Example 6

Find the diagonal of a rectangle that is 6 cm long and 3 cm wide. Round to the nearest tenth.

Strategy

To find the distance, use the Pythagorean Theorem. The hypotenuse is the length of the ladder. One leg is the distance from the bottom of the ladder to the base of the building. The distance along the building from the ground to the top of the ladder is the unknown leg.

Your strategy

Solution

$$c^2 = a^2 + b^2$$
$$20^2 = 8^2 + b^2$$
$$400 = 64 + b^2$$
$$336 = b^2$$
$$(336)^{1/2} = (b^2)^{1/2}$$
$$\sqrt{336} = b$$
$$18.3 \approx b$$

The distance is 18.3 ft.

Your solution

6.7 cm

Solution on p. A26

Example 7

An object is dropped from a high building. Find the distance the object has fallen when the speed reaches 96 ft/s. Use the equation $v = \sqrt{64d}$, where v is the speed of the object and d is the distance.

Strategy

To find the distance the object has fallen, replace v in the equation with the given value and solve for d.

Solution

$$v = \sqrt{64d}$$
$$96 = \sqrt{64d}$$
$$(96)^2 = (\sqrt{64d})^2$$
$$9216 = 64d$$
$$144 = d$$

The object has fallen 144 ft.

Example 8

How far would a submarine periscope have to be above the water to locate a ship 5.5 mi away? The equation for the distance in miles that the lookout can see is $d = 1.4\sqrt{h}$, where h is the height in feet above the surface of the water. Round to the nearest thousandth.

Your strategy

Your solution

15.434 ft

Example 9

Find the length of a pendulum that makes one swing in 1.5 s. The equation for the time of one swing is given by $T = 2\pi\sqrt{\dfrac{L}{32}}$, where T is the time in seconds and L is the length in feet. Use 3.14 for π. Round to the nearest hundredth.

Strategy

To find the length of the pendulum, replace T in the equation with the given value and solve for L.

Solution

$$T = 2\pi\sqrt{\frac{L}{32}}$$

$$1.5 = 2(3.14)\sqrt{\frac{L}{32}}$$

$$\frac{1.5}{2(3.14)} = \sqrt{\frac{L}{32}}$$

$$\left[\frac{1.5}{2(3.14)}\right]^2 = \left(\sqrt{\frac{L}{32}}\right)^2$$

$$\left(\frac{1.5}{6.28}\right)^2 = \frac{L}{32}$$

$$1.83 \approx L$$

The length of the pendulum is 1.83 ft.

Example 10

Find the distance required for a car to reach a velocity of 88 ft/s when the acceleration is 22 ft/s². Use the equation $v = \sqrt{2as}$, where v is the velocity, a is the acceleration, and s is the distance.

Your strategy

Your solution

176 ft

Solutions on p. A26

5.4   EXERCISES

▶ **Objective A**

Solve:

1. $\sqrt[3]{4x} = -2$
-2

2. $\sqrt[3]{6x} = -3$
$-\dfrac{9}{2}$

3. $\sqrt{3x - 2} = 5$
9

4. $\sqrt{5x - 4} = 9$
17

5. $\sqrt{2x} = -4$
No solution

6. $\sqrt{5x} = -5$
No solution

7. $\sqrt{3x + 9} - 12 = 0$
45

8. $\sqrt{4x - 3} - 5 = 0$
7

9. $\sqrt{4x - 2} = \sqrt{3x + 9}$
11

10. $\sqrt{2x + 4} = \sqrt{5x - 9}$
$\dfrac{13}{3}$

11. $\sqrt[3]{x - 2} = 3$
29

12. $\sqrt[3]{2x - 6} = 4$
35

13. $\sqrt[3]{3x - 9} = \sqrt[3]{2x + 12}$
21

14. $\sqrt[3]{x - 12} = \sqrt[3]{5x + 16}$
-7

15. $\sqrt[4]{4x + 1} = 2$
$\dfrac{15}{4}$

16. $\sqrt[4]{2x - 9} = 3$
45

17. $\sqrt{2x - 3} - 2 = 1$
6

18. $\sqrt{3x - 5} - 5 = 3$
23

19. $\sqrt[3]{2x - 3} + 5 = 2$
-12

20. $\sqrt[3]{x - 4} + 7 = 5$
-4

21. $\sqrt{x} + \sqrt{x - 5} = 5$
9

22. $\sqrt{x + 3} + \sqrt{x - 1} = 2$
1

23. $\sqrt{2x + 5} - \sqrt{2x} = 1$
2

24. $\sqrt{3x} - \sqrt{3x - 5} = 1$
3

25. $\sqrt{2x} - \sqrt{x - 1} = 1$
2

26. $\sqrt{2x - 5} + \sqrt{x + 1} = 3$
3

27. $\sqrt{2x + 2} + \sqrt{x} = 3$
1

▶ **Objective B   *Application Problems***

Solve:

28. An object is dropped from an airplane. Find the distance the object has fallen when the speed reaches 400 ft/s. Use the equation $v = \sqrt{64d}$, where v is the speed of the object and d is the distance.
2500 ft

29. An object is dropped from a bridge. Find the distance the object has fallen when the speed reaches 100 ft/s. Use the equation $v = \sqrt{64d}$, where v is the speed of the object and d is the distance.
156.25 ft

30. A 16-ft ladder is leaning against a building. How high on the building will the ladder reach when the bottom of the ladder is 5 ft from the building? Round to the nearest tenth.
15.2 ft

31. Find the width of a rectangle that has a diagonal of 10 ft and a length of 8 ft.
6 ft

32. How far would a submarine periscope have to be above the water to locate a ship 3.6 mi away? The equation for the distance in miles that the lookout can see is $d = 1.4\sqrt{h}$, where h is the height in feet above the surface of the water. Round to the nearest hundredth.
6.61 ft

33. How far would a submarine periscope have to be above the water to locate a ship 4.2 mi away? The equation for the distance in miles that the lookout can see is $d = 1.4\sqrt{h}$, where h is the height in feet above the surface of the water.
9 ft

34. Find the distance required for a car to reach a velocity of 60 ft/s when the acceleration is 15 ft/s². Use the equation $v = \sqrt{2as}$, where v is the velocity, a is the acceleration, and s is the distance.
120 ft

35. Find the distance required for a car to reach a velocity of 40 m/s when the acceleration is 10 m/s². Use the equation $v = \sqrt{2as}$, where v is the velocity, a is the acceleration, and s is the distance.
80 m

36. Find the length of a pendulum that makes one swing in 2.4 s. The equation for the time of one swing of a pendulum is given by $T = 2\pi\sqrt{\frac{L}{32}}$, where T is the time in seconds and L is the length in feet. Use 3.14 for π. Round to the nearest hundredth.
4.67 ft

37. Find the length of a pendulum that makes one swing in 3 s. The equation for the time of one swing of a pendulum is given by $T = 2\pi\sqrt{\frac{L}{32}}$, where T is the time in seconds and L is the length in feet. Use 3.14 for π. Round to the nearest hundredth.
7.30 ft

Calculators and Computers

The $\boxed{y^x}$ **Key on a Calculator**

The $\boxed{y^x}$ key on a calculator is used to find powers of a number. For example, to find 4^7, enter the following key strokes:

$$4 \boxed{y^x} 7 \boxed{=}$$

The number 1 6 3 8 4 should be in the display of your calculator.

Fractional powers can also be calculated by using the $\boxed{y^x}$ key. The memory key on your calculator is useful for this calculation. For example, to find $17^{2/7}$, first find the decimal equivalent for $\frac{2}{7}$. Store this result in the calculator's memory. Now the $\boxed{y^x}$ key is used to find the power. Here are the key strokes.

$$2 \boxed{\div} 7 \boxed{=} \boxed{M+} 17 \boxed{y^x} \boxed{MR} \boxed{=}$$

The number 2 .2 4 6 7 6 0 8 should be in the display.

The symbols "M+" and "MR" are used here to mean "store in memory" and "recall from memory." These symbols may be different on your calculator.

The calculation of an installment loan payment uses the $\boxed{y^x}$ key. Here is the formula and a sample calculation.

$$PMT = PRIN\left(\frac{i}{1 - (1 + i)^{-n}}\right)$$

In this formula, *PMT* is the monthly payment, *PRIN* is the amount borrowed, and *i* is the monthly interest rate as a decimal. The monthly interest rate is the annual rate divided by 12. The number of months of the loan is given by *n*.

A car is purchased and a loan of $8000 is secured at an annual interest rate of 8.5% for 5 years. Find the monthly payment.

First calculate $(1 + i)^{-n}$; $n = 5 \cdot 12 = 60$; $i = \frac{0.085}{12} = 0.0070833$

Enter: 1.0070833 $\boxed{y^x}$ 60 $\boxed{+/-}$ $\boxed{=}$ $\boxed{M+}$

The number 0 .6 5 4 7 5 1 3 should be displayed. This number is also stored in memory.

Now the final calculation.

Enter: 8000 $\boxed{\times}$ 0.0070833 $\boxed{\div}$ $\boxed{(}$ 1 $\boxed{-}$ $\boxed{MR}$ $\boxed{)}$ $\boxed{=}$

The number 1 6 4. 1 3 2 1 should be in the display.

The monthly payment is $164.13.

Chapter Summary

Key Words   The *nth root of a* is $a^{1/n}$. (The expression $\sqrt[n]{a}$ is another symbol for the *n*th root of *a*.)

In the expression $\sqrt[n]{a}$, the symbol $\sqrt{}$ is called a *radical, n* is the *index* of the radical, and *a* is the *radicand*.

The symbol $\sqrt{}$ is used to indicate the positive or *principal square root* of a number.

A *complex number* is a number of the form $a + bi$, where *a* and *b* are real numbers and $i = \sqrt{-1}$.

For the complex number $a + bi$, *a* is called the *real part* of the complex number and *b* is called the *imaginary part* of the complex number.

Essential Rules

The Product Property of Radicals
If $\sqrt[n]{a}$ and $\sqrt[n]{b}$ are positive real numbers, then $\sqrt[n]{a}\sqrt[n]{b} = \sqrt[n]{ab}$ and $\sqrt[n]{ab} = \sqrt[n]{a} \cdot \sqrt[n]{b}$.

The Quotient Property of Radicals
If $\sqrt[n]{a}$ and $\sqrt[n]{b}$ are positive real numbers and $b \neq 0$, then $\dfrac{\sqrt[n]{a}}{\sqrt[n]{b}} = \sqrt[n]{\dfrac{a}{b}}$ and $\sqrt[n]{\dfrac{a}{b}} = \dfrac{\sqrt[n]{a}}{\sqrt[n]{b}}$.

Addition of Complex Numbers
If $a + bi$ and $c + di$ are complex numbers, then $(a + bi) + (c + di) = (a + c) + (b + d)i$.

Multiplication of Complex Numbers
If $a + bi$ and $c + di$ are complex numbers, then $(a + bi)(c + di) = (ac - bd) + (ad + bc)i$.

Property of Raising Both Sides of an Equation to a Power
If *a* and *b* are real numbers and $a = b$, then $a^n = b^n$.

Pythagorean Theorem
$c^2 = a^2 + b^2$

Chapter Review

SECTION 1

1. Simplify: $\dfrac{x^{-3/2}}{x^{7/2}}$

    $\dfrac{1}{x^5}$

2. Simplify: $\dfrac{(16x^{-4}y^{12})^{1/4}(100x^6y^{-2})^{1/2}}{20x^2y^2}$

3. Rewrite $3x^{3/4}$ as a radical expression.

    $3\sqrt[4]{x^3}$

4. Rewrite $7y\sqrt[3]{x^2}$ as an exponential expression.

    $7yx^{2/3}$

5. Simplify: $\sqrt[4]{81a^8b^{12}}$

    $3a^2b^3$

6. Simplify: $\sqrt[3]{-8a^6b^{12}}$

    $-2a^2b^4$

SECTION 2

7. Simplify: $\sqrt{18a^3b^6}$

    $3ab^3\sqrt{2a}$

8. Simplify: $\sqrt[5]{-64a^8b^{12}}$

    $-2ab^2\sqrt[5]{2a^3b^2}$

9. Simplify: $\sqrt{50a^4b^3} - ab\sqrt{18a^2b}$

    $2a^2b\sqrt{2b}$

10. Simplify: $3x\sqrt[3]{54x^8y^{10}} - 2x^2y\sqrt[3]{16x^5y^7}$

    $5x^3y^3\sqrt[3]{2x^2y}$

11. Simplify: $4x\sqrt{12x^2y} + \sqrt{3x^4y} - x^2\sqrt{27y}$

    $6x^2\sqrt{3y}$

12. Simplify: $\sqrt[3]{16x^4y}\sqrt[3]{4xy^5}$

    $4xy^2\sqrt[3]{x^2}$

13. Simplify: $(5 - \sqrt{6})^2$

    $31 - 10\sqrt{6}$

14. Simplify: $(\sqrt{3} + 8)(\sqrt{3} - 2)$

    $6\sqrt{3} - 13$

15. Simplify: $\dfrac{8}{\sqrt{3y}}$

    $\dfrac{8\sqrt{3y}}{3y}$

16. Simplify: $\dfrac{x + 2}{\sqrt{x} + \sqrt{2}}$

    $\dfrac{x\sqrt{x} - x\sqrt{2} + 2\sqrt{x} - 2\sqrt{2}}{x - 2}$

SECTION 3

17. Simplify: $\sqrt{-50}$

$5i\sqrt{2}$

18. Simplify: $(-8 + 3i) - (4 - 7i)$

$-12 + 10i$

19. Simplify: $(\sqrt{50} + \sqrt{-72}) - (\sqrt{162} - \sqrt{-8})$

$-4\sqrt{2} + 8i\sqrt{2}$

20. Simplify: $i(3 - 7i)$

$7 + 3i$

21. Simplify: $\sqrt{-12}\sqrt{-6}$

$-6\sqrt{2}$

22. Simplify: $(6 - 5i)(4 + 3i)$

$39 - 2i$

23. Simplify: $\dfrac{5 + 2i}{3i}$

$\dfrac{2}{3} - \dfrac{5}{3}i$

24. Simplify: $\dfrac{5 + 9i}{1 - i}$

$-2 + 7i$

SECTION 4

25. Solve: $\sqrt[4]{3x - 5} = 2$

7

26. Solve: $\sqrt{4x + 9} + 10 = 11$

-2

27. Solve: $\sqrt{x - 5} + \sqrt{x + 6} = 11$

30

28. The velocity of the wind determines the amount of power generated by a windmill. A typical equation for this relationship is $v = 4.05\sqrt[3]{P}$, where v is the velocity in mph and P is the power in watts. Find the amount of power generated by a 20 mph wind. Round to the nearest whole number.

120 watts

29. Find the distance required for a car to reach a velocity of 88 ft/s when the acceleration is 16 ft/s^2. Use the equation $v = \sqrt{2as}$, where v is the velocity, a is the acceleration, and s is the distance.

242 ft

30. A 12-ft ladder is leaning against a building. How far from the building is the bottom of the ladder when the top of the ladder touches the building 10 ft above the ground? Round to the nearest hundredth.

6.63 ft

Chapter Test

1. Simplify: $\dfrac{r^{2/3}r^{-1}}{r^{-1/2}}$

   $r^{1/6}$ [5.1A]

2. Simplify: $\dfrac{(2x^{1/3}y^{-2/3})^6}{(x^{-4}y^8)^{1/4}}$

   $\dfrac{64x^3}{y^6}$ [5.1A]

3. Simplify: $\left(\dfrac{4a^4}{b^2}\right)^{-3/2}$

   $\dfrac{b^3}{8a^6}$ [5.1A]

4. Write $3y^{2/5}$ as a radical expression.

   $3\sqrt[5]{y^2}$ [5.1B]

5. Write $\frac{1}{2}\sqrt[4]{x^3}$ as an exponential expression.

   $\frac{1}{2}x^{3/4}$ [5.1B]

6. Simplify: $\sqrt[3]{8x^3y^6}$

   $2xy^2$ [5.1C]

7. Simplify: $\sqrt{32x^4y^7}$

   $4x^2y^3\sqrt{2y}$ [5.2A]

8. Simplify: $\sqrt[3]{27a^4b^3c^7}$

   $3abc^2\sqrt[3]{ac}$ [5.2A]

9. Simplify: $\sqrt{18a^3} + a\sqrt{50a}$

   $8a\sqrt{2a}$ [5.2B]

10. Simplify: $\sqrt[3]{54x^7y^3} - x\sqrt[3]{128x^4y^3} - x^2\sqrt[3]{2xy^3}$

    $-2x^2y\sqrt[3]{2x}$ [5.2B]

11. Simplify: $\sqrt{3x}(\sqrt{x} - \sqrt{25x})$

    $-4x\sqrt{3}$ [5.2C]

12. Simplify: $(2\sqrt{3} + 4)(3\sqrt{3} - 1)$

    $14 + 10\sqrt{3}$ [5.2C]

13. Simplify: $(\sqrt{a} - 3\sqrt{b})(2\sqrt{a} + 5\sqrt{b})$

    $2a - \sqrt{ab} - 15b$ [5.2C]

14. Simplify: $(2\sqrt{x} + \sqrt{y})^2$

    $4x + 4\sqrt{xy} + y$ [5.2C]

15. Simplify: $\dfrac{\sqrt{32x^5y}}{\sqrt{2xy^3}}$

    $\dfrac{4x^2}{y}$ [5.2D]

16. Simplify: $\dfrac{4 - 2\sqrt{5}}{2 - \sqrt{5}}$

    2 [5.2D]

17. Simplify: $\dfrac{\sqrt{x}}{\sqrt{x} - \sqrt{y}}$

    $\dfrac{x + \sqrt{xy}}{x - y}$ [5.2D]

18. Simplify: $(\sqrt{-8})(\sqrt{-2})$

    -4 [5.3C]

19. Simplify: $(5 - 2i) - (8 - 4i)$

    $-3 + 2i$ [5.3B]

20. Simplify: $(2 + 5i)(4 - 2i)$

    $18 + 16i$ [5.3C]

21. Simplify: $\dfrac{2 + 3i}{1 - 2i}$

    $-\dfrac{4}{5} + \dfrac{7}{5}i$ [5.3D]

22. $(2 + i) + (2 - i)(3 + 2i)$

    $10 + 2i$ [5.3C]

23. Solve: $\sqrt{x + 12} - \sqrt{x} = 2$

    4 [5.4A]

24. Solve: $\sqrt[3]{2x - 2} + 4 = 2$

    -3 [5.4A]

25. An object is dropped from a high building. Find the distance the object has fallen when the speed reaches 192 ft/s. Use the equation $v = \sqrt{64d}$, where v is the speed of the object and d is the distance.

    576 ft [5.4B]

Cumulative Review

1. Identify the property that justifies the statement.
   $(a + 2)b = ab + 2b$
   Distributive Property   [1.3B]

2. Simplify: $2x - 3[x - 2(x - 4) + 2x]$
   $-x - 24$ [1.3C]

3. Solve: $5 - \frac{2}{3}x = 4$
   $x = \frac{3}{2}$ [2.1B]

4. Solve: $2[4 - 2(3 - 2x)] = 4(1 - x)$
   $x = \frac{2}{3}$ [2.1C]

5. Solve: $2 + |4 - 3x| = 5$
   $\frac{1}{3}$ and $\frac{7}{3}$ [2.3A]

6. Solve: $6x - 3(2x + 2) > 3 - 3(x + 2)$
   $\{x \mid x > 1\}$ [2.2A]

7. Solve: $|2x + 3| \leq 9$
   $\{x \mid -6 \leq x \leq 3\}$ [2.3B]

8. Factor: $81x^2 - y^2$
   $(9x + y)(9x - y)$ [3.4A]

9. Factor: $x^5 + 2x^3 - 3x$
   $x(x^2 + 3)(x + 1)(x - 1)$ [3.4D]

10. Simplify: $\dfrac{4a^2 + 8a}{a^3 + a^2 - 2a}$
    $\dfrac{4}{a - 1}$ [4.1A]

11. Simplify: $\dfrac{1 - \frac{1}{a^2}}{\frac{1}{a} + \frac{1}{a^2}}$
    $a - 1$ [4.4A]

12. Solve $P = \dfrac{R - C}{n}$ for C.
    $C = R - nP$ [4.6B]

13. Simplify: $(2^{-1}x^2y^{-6})(2^{-1}y^{-4})^{-2}$
    $2x^2y^2$ [5.1A]

14. Simplify: $\left(\dfrac{x^{-2/3}y^{1/2}}{y^{-1/3}}\right)^6$
    $\dfrac{y^5}{x^4}$ [5.1A]

15. Simplify: $\sqrt{40x^3} - x\sqrt{90x}$
    $-x\sqrt{10x}$ [5.2A]

16. Simplify: $(\sqrt{3} - 2)(\sqrt{3} - 5)$
    $13 - 7\sqrt{3}$ [5.2C]

17. Simplify: $\dfrac{7}{\sqrt{10} - \sqrt{3}}$
    $\sqrt{10} + \sqrt{3}$ [5.2D]

18. Simplify: $(3 - \sqrt{-4}) + (4 + \sqrt{-9})$
    $7 + i$ [5.3A]

19. Simplify: $\frac{2i}{3-i}$

$-\frac{1}{5} + \frac{3}{5}i$ [5.3D]

20. Solve: $\sqrt[3]{3x-4} + 5 = 1$

$x = -20$ [5.4A]

21. Simplify: $\frac{4}{\sqrt{6} - \sqrt{2}}$

$\sqrt{6} + \sqrt{2}$ [5.2D]

22. A collection of thirty stamps consists of 13¢ stamps and 18¢ stamps. The total value of the stamps is $4.85. Find the number of 18¢ stamps.

19 [2.4B]

23. An investment of $2500 is made at an annual simple interest rate of 7.2%. How much additional money must be invested at an annual simple interest rate of 8.4% so that the total interest earned is $516?

$4000 [2.6A]

24. A sales executive traveled 25 mi by car and then an additional 625 mi by plane. The rate of the plane was five times faster than the rate of the car. The total time of the trip was 3 h. Find the rate of the plane.

250 mph [4.6D]

25. How long does it take light to travel to the earth from the moon when the moon is 232,500 mi from the earth? Light travels 1.86×10^5 mi/s.

1.25 s [3.1E]

26. How far would a submarine periscope have to be above the water to locate a ship 7 mi away? The equation for the distance in miles that the lookout can see is $d = 1.4\sqrt{h}$, where h is the height in feet above the surface of the water.

25 ft [5.4B]

6

Linear Equations in Two Variables

OBJECTIVES

▶ To graph points on a rectangular coordinate system

▶ To determine a solution of a linear equation in two variables

▶ To graph an equation of the form $y = mx + b$

▶ To graph an equation of the form $Ax + By = C$

▶ To find the slope of a line given two points

▶ To find the *x*- and the *y*-intercept of a straight line

▶ To graph a line given a point and the slope

▶ To find the equation of a line given a point and the slope

▶ To find the equation of a line given two points

▶ To find parallel and perpendicular lines

▶ To obtain data from a graph

▶ To graph the solution set of an inequality in two variables

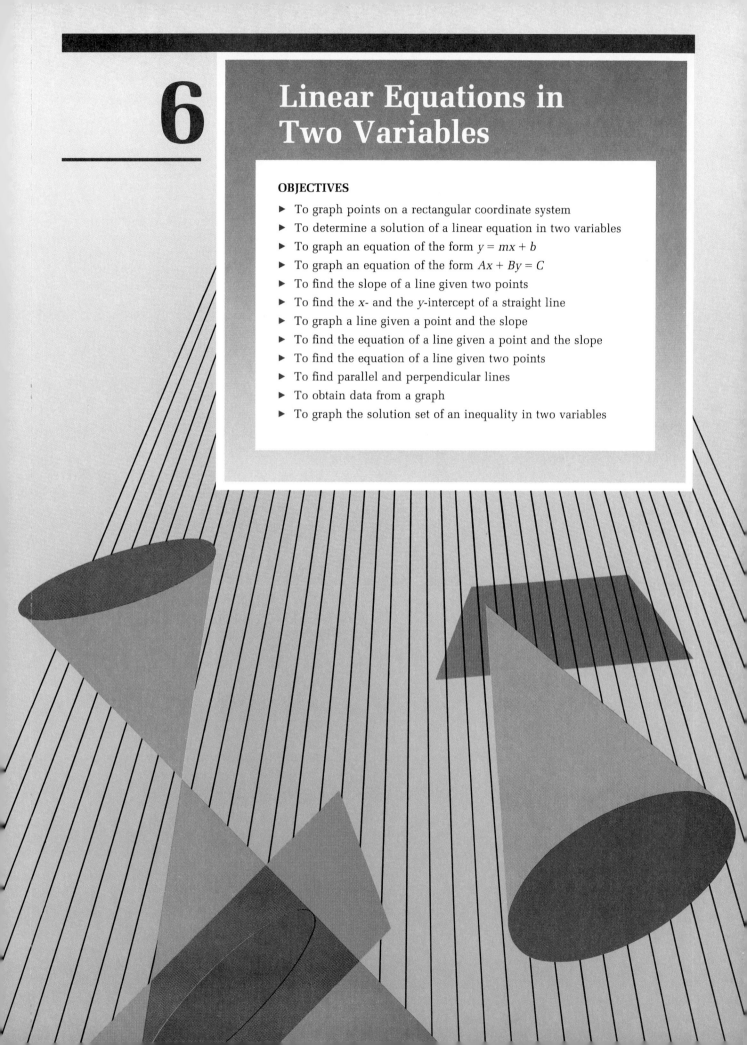

Brachistochrone Problem

Consider the diagram at the right. What curve should be drawn so that a ball allowed to roll along the curve will travel from *A* to *B* in the shortest time?

At first thought, one might conjecture that a straight line should connect the two points, since that shape is the shortest *distance* between the two points. Actually, however, the answer is half of one arch of an inverted cycloid.

A cycloid is shown below as the graph in bold. One way to draw this curve is to think of a wheel rolling along a straight line without slipping. Then a point on the rim of the wheel traces a cycloid.

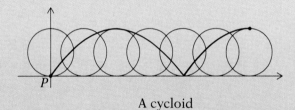

A cycloid

There are many applications of the idea of finding the shortest time between two points. As the above problem illustrates, the path of shortest time is not necessarily the path of shortest distance. Problems involving paths of shortest time are called *brachistochrone* problems.

SECTION 6.1   The Rectangular Coordinate System

Objective A   To graph points on a rectangular coordinate system

A **rectangular coordinate system** is formed by two number lines, one horizontal and one vertical, that intersect at the zero point of each line. The point of intersection is called the **origin**. The two lines are called the **coordinate axes**, or simply **axes**.

The axes determine a plane and divide the plane into four regions, called **quadrants**. The quadrants are numbered counterclockwise from I to IV.

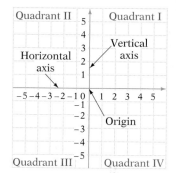

Each point in the plane can be identified by a pair of numbers called an **ordered pair.** The first number of the pair measures a horizontal distance and is called the **abscissa.** The second number of the pair measures a vertical distance and is called the **ordinate.** The **coordinates** of a point are the numbers in the ordered pair associated with the point.

horizontal distance ⎯⎯⎤   ⎡⎯ vertical distance
                        ↓   ↓

ordered pair → (1 , 2)

abscissa ⎯⎯⎤   ⎡⎯ ordinate

The **graph of an ordered pair** is a point in the plane. The graphs of the points $(-2, 3)$ and $(3, -2)$ are shown at the right. Notice that they are different points. The order in which the numbers in an ordered pair appear *is* important.

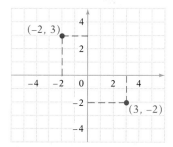

Example 1   Graph the ordered pairs $(2, -1)$ and $(-3, -4)$. Draw a line between the two points.

Solution

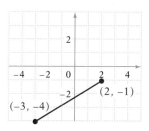

Example 2   Graph the ordered pairs $(5, -2)$ and $(-2, 3)$. Draw a line between the two points.

Your solution

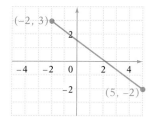

Solution on p. A27

Example 3   Find the coordinates of each
of the points.

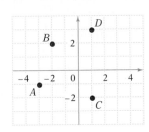

Example 4   Find the coordinates of
each of the points.

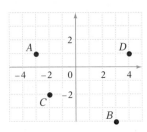

Solution $A(-3, -1)$
$B(-2, 2)$
$C(1, -2)$
$D(1, 3)$

Your solution $A(-3, 1)$
$B(3, -4)$
$C(-2, -2)$
$D(4, 1)$

Example 5   Draw a line through all
points with an abscissa of 2.

Solution

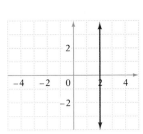

Example 6   Draw a line through all
points with an ordinate
of -1.

Your solution

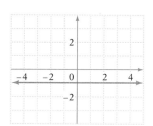

Solutions on p. A27

| Objective B |

**To determine a solution of a linear equation in two
variables**

An equation of the form $y = mx + b$, where m and b are constants, is a **linear
equation in two variables.**

Examples of linear equations in two variables are shown at the right.

$y = 5x - 7$ $(m = 5, \quad b = -7)$

$y = \frac{2}{3}x - 8$ $(m = \frac{2}{3}, \quad b = -8)$

$y = -\frac{1}{2}x + 2$ $(m = -\frac{1}{2}, b = 2)$

A **solution of an equation in two variables** is an ordered pair of numbers (x, y)
which makes the equation a true statement.

Is $(2, 3)$ a solution of $y = \frac{1}{2}x + 2$?

Replace x with 2, the abscissa.
Replace y with 3, the ordinate.

Compare the results. If the results are equal, the given ordered pair is a solution. If the results are not equal, the given ordered pair is not a solution.

$$y = \frac{1}{2}x + 2$$

| 3 | $\frac{1}{2}(2) + 2$ |
| | $1 + 2$ |

$3 = 3$

Yes, $(2, 3)$ is a solution of the equation $y = \frac{1}{2}x + 2$.

Besides the ordered pair $(2, 3)$, there are many other ordered pair solutions of the equation $y = \frac{1}{2}x + 2$. For example, the method used above can be used to show that $(-4, 0)$, $(-2, 1)$, and $(4, 4)$ are also solutions.

In general, a linear equation in two variables has an infinite number of solutions. By choosing any value for x and substituting that value into the linear equation, a corresponding value of y can be found.

Find the ordered pair solution of $y = \frac{3}{2}x - 2$ corresponding to $x = 2$.

Substitute 2 for x. Solve for y. $y = \frac{3}{2}x - 2 = \frac{3}{2}(2) - 2 = 3 - 2 = 1$

The ordered pair solution is $(2, 1)$.

Example 7 Find the ordered pair solution of $y = \frac{2}{3}x + 3$ corresponding to $x = -5$.

Solution $y = \frac{2}{3}x + 3$

$y = \frac{2}{3}(-5) + 3 = -\frac{10}{3} + 3 = -\frac{1}{3}$

The ordered pair solution is $\left(-5, -\frac{1}{3}\right)$.

Example 8 Find the ordered pair solution of $y = -2x + 5$ corresponding to $x = \frac{1}{3}$.

Your solution $\left(\frac{1}{3}, \frac{13}{3}\right)$

Solution on p. A27

Objective C   **To graph an equation of the form $y = mx + b$**

6

The **graph of an equation in two variables** is a drawing of the ordered pair solutions of the equation. For a linear equation in two variables, the graph is a straight line.

To graph a linear equation, find ordered pair solutions of the equation. Do this by choosing any value of x and finding the corresponding value of y. Repeat this procedure, choosing different values for x, until you have found the number of solutions desired. Since the graph of a linear equation in two variables is a straight line, and a straight line is determined by two points, it is necessary to find only two solutions. However, it is recommended that at least three solutions be used to insure accuracy.

Graph: $y = -2x + 1$

Choose any values of x and find the corresponding values of y. It is convenient to record these solutions in a table.

x	$y = -2x$	$+ 1$	y
0	$-2(0)$	$+ 1$	1
2	$-2(2)$	$+ 1$	-3
-2	$-2(-2)$	$+ 1$	5

The horizontal axis is the x-axis. The vertical axis is the y-axis. Graph the ordered pair solutions $(-2, 5)$, $(0, 1)$, and $(2, -3)$. Draw a line through the ordered pair solutions.

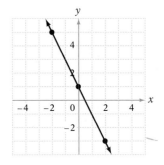

Remember that a graph is a drawing of the ordered pair solutions of the equation. Therefore, every point on the graph is a solution of the equation and every solution of the equation is a point on the graph.

Graph: $y = -\frac{3}{4}x + 2$

Find at least three solutions.

When m is a fraction ($m = -\frac{3}{4}$), choose values of x that will simplify the evaluations. Display the ordered pairs in a table.

x	y
0	2
4	-1
-4	5

Graph the ordered pairs on a rectangular coordinate system and draw a straight line through the points.

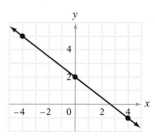

Example 9   Graph: $y = -\frac{3}{2}x - 3$

Solution

x	y
0	-3
-2	0
-4	3

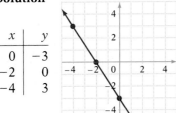

Example 10   Graph: $y = \frac{3}{5}x - 4$

Your solution

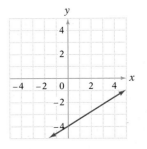

Solution on p. A27

Example 11   Graph: $y = \frac{2}{3}x$

Solution

x	y
0	0
3	2
−3	−2

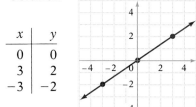

Example 12   Graph: $y = -\frac{3}{4}x$

Your solution

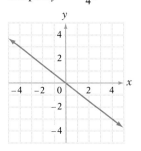

Solution on p. A27

Objective D   To graph an equation of the form $Ax + By = C$

An equation of the form $Ax + By = C$, where A, B, and C are constants, is also a linear equation.

Examples of equations of the form $Ax + By = C$ are shown at the right.

$$3x - 2y = -5 \qquad (A = 3, \quad B = -2, \ C = -5)$$
$$-\tfrac{1}{2}x + 2y = 4 \qquad (A = -\tfrac{1}{2}, \ B = 2, \quad C = 4)$$

An equation of the form $Ax + By = C$ can be written in the form $y = mx + b$.

Write the equation $3x - 2y = -5$ in the form $y = mx + b$.

Add the additive inverse of $3x$ to both sides of the equation.

$$3x - 2y = -5$$
$$-2y = -3x - 5$$

Divide each side of the equation by the coefficient of y.

$$\frac{-2y}{-2} = \frac{-3x - 5}{-2}$$

$$y = \tfrac{3}{2}x + \tfrac{5}{2}$$

To graph an equation of the form $Ax + By = C$, first solve the equation for y. Then follow the same procedure used for graphing an equation of the form $y = mx + b$.

Graph: $-3x + 4y = -4$

Solve the equation for y.

$$-3x + 4y = -4$$
$$4y = 3x - 4$$
$$y = \tfrac{3}{4}x - 1$$

Find at least three solutions.

Graph the ordered pairs on a rectangular coordinate system.

x	y
0	−1
4	2
−4	−4

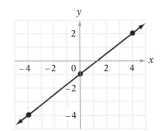

Draw a straight line through the points.

The graph of an equation in which one of the variables is missing is either a horizontal or a vertical line.

The equation $y = -2$ could be written:

$$0 \cdot x + y = -2$$

No matter what value of x is chosen, y is always -2. Some solutions of the equation are $(3, -2)$, $(0, -2)$, and $(-2, -2)$. The graph is shown at the right.

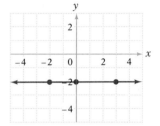

The **graph of $y = b$** is a horizontal line passing through point $(0, b)$.

The equation $x = 2$ could be written:

$$x + 0 \cdot y = 2$$

No matter what value of y is chosen, x is always 2. Some solutions of the equation are $(2, 2)$, $(2, 0)$, and $(2, -3)$. The graph is shown at the right.

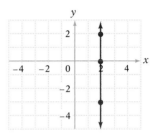

The **graph of $x = a$** is a vertical line passing through point $(a, 0)$.

Example 13   Graph: $3x + 2y = 6$

Solution $3x + 2y = 6$
$$2y = -3x + 6$$
$$y = -\tfrac{3}{2}x + 3$$

x	y
0	3
2	0
4	-3

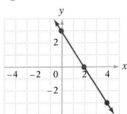

Example 14   Graph: $-3x + 2y = 4$

Your solution

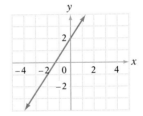

Example 15   Graph: $x = -4$

Solution   The graph of an equation of the form $x = a$ is a vertical line passing through point $(a, 0)$.

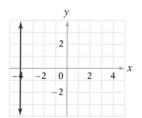

Example 16   Graph: $y = 3$

Your solution

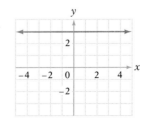

Solutions on p. A27

6.1   EXERCISES

▶ **Objective A**

1. Graph the ordered pairs (3, 2) and (−1, 4). Draw a line between the two points.

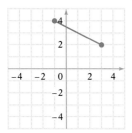

2. Graph the ordered pairs (−1, −3) and (3, −2). Draw a line between the two points.

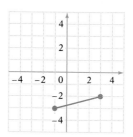

3. Graph the ordered pairs (−3, −3) and (2, −2). Draw a line between the two points.

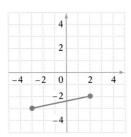

4. Graph the ordered pairs (−3, 2) and (4, 2). Draw a line between the two points.

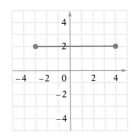

5. Find the coordinates of each of the points.

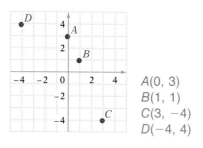

A(0, 3)
B(1, 1)
C(3, −4)
D(−4, 4)

6. Find the coordinates of each of the points.

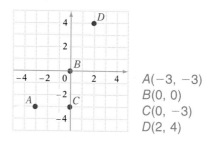

A(−3, −3)
B(0, 0)
C(0, −3)
D(2, 4)

7. Find the coordinates of each of the points.

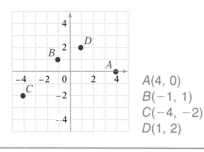

A(4, 0)
B(−1, 1)
C(−4, −2)
D(1, 2)

8. Find the coordinates of each of the points.

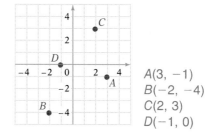

A(3, −1)
B(−2, −4)
C(2, 3)
D(−1, 0)

9. Draw a line through all points with an abscissa of 2.

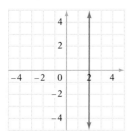

10. Draw a line through all points with an abscissa of -3.

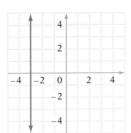

11. Draw a line through all points with an ordinate of -3.

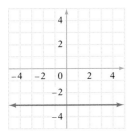

12. Draw a line through all points with an ordinate of 4.

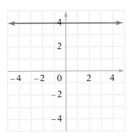

▶ **Objective B**

13. Find the ordered pair solution of $y = \frac{2}{3}x - 4$ corresponding to $x = -3$.
$(-3, -6)$

14. Find the ordered pair solution of $y = \frac{1}{2}x - 5$ corresponding to $x = 4$.
$(4, -3)$

15. Find the ordered pair solution of $y = -\frac{2}{3}x + 4$ corresponding to $x = -6$.
$(-6, 8)$

16. Find the ordered pair solution of $y = -\frac{3}{4}x + 5$ corresponding to $x = -4$.
$(-4, 8)$

17. Find the ordered pair solution of $y = \frac{3}{2}x + 3$ corresponding to $x = -4$.
$(-4, -3)$

18. Find the ordered pair solution of $y = \frac{4}{3}x - 5$ corresponding to $x = 3$.
$(3, -1)$

19. Find the ordered pair solution of $y = -2x - 3$ corresponding to $x = -2$.
$(-2, 1)$

20. Find the ordered pair solution of $y = 3x + 5$ corresponding to $x = -3$.
$(-3, -4)$

21. Find the ordered pair solution of $y = -\frac{4}{3}x - 3$ corresponding to $x = -2$.
$(-2, -\frac{1}{3})$

22. Find the ordered pair solution of $y = \frac{3}{2}x - 4$ corresponding to $x = 3$.
$(3, \frac{1}{2})$

▶ **Objective C**

Graph:

23. $y = 3x - 4$

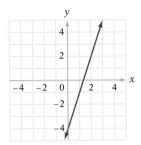

24. $y = -2x + 3$

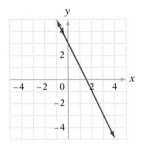

25. $y = -\frac{2}{3}x$

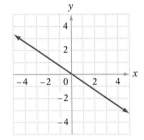

26. $y = \frac{3}{2}x$

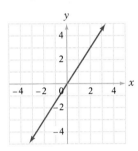

27. $y = \frac{2}{3}x - 4$

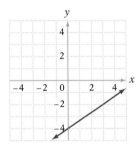

28. $y = \frac{3}{4}x + 2$

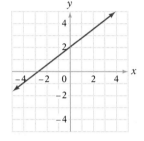

29. $y = -\frac{1}{3}x + 2$

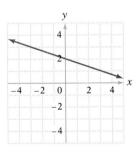

30. $y = -\frac{3}{2}x - 3$

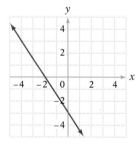

31. $y = \frac{3}{5}x - 1$

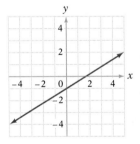

32. $y = -\frac{2}{3}x + 4$

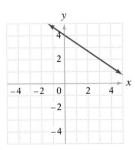

33. $y = -\frac{4}{3}x + 2$

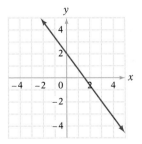

34. $y = \frac{3}{2}x - 3$

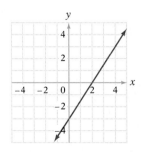

▶ **Objective D**

Graph:

35. $2x - y = 3$

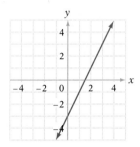

36. $2x + y = -3$

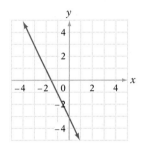

37. $2x + 5y = 10$

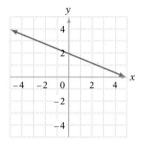

38. $x - 4y = 8$

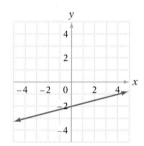

39. $y = -2$

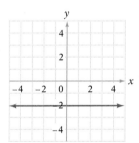

40. $y = \frac{1}{3}x$

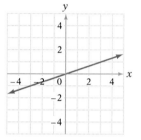

41. $2x - 3y = 12$

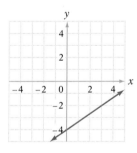

42. $3x - y = -2$

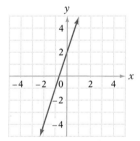

43. $3x - 2y = 8$

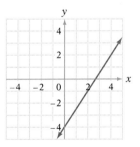

44. $x + 4y = 12$

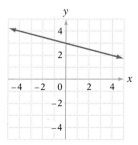

45. $2x - 3y = -9$

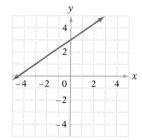

46. $3x + 4y = 8$

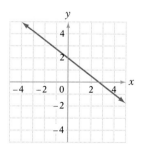

SECTION 6.2   Slopes and Intercepts of Straight Lines

Objective A   **To find the slope of a line given two points**

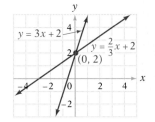

The graphs of $y = 3x + 2$ and $y = \frac{2}{3}x + 2$ are shown at the right. Each graph crosses the y-axis at the point $(0, 2)$, but the graphs have different slants. The **slope** of a line is a measure of the slant of a line. The symbol for slope is m.

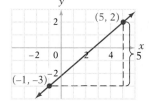

The slope of a line containing two points is the ratio of the change in the y values of the two points to the change in the x values. The line containing the points $(-1, -3)$ and $(5, 2)$ is graphed at the right.

The change in the y values is the difference between the two ordinates.

Change in $y = 2 - (-3) = 5$.

The change in the x values is the difference between the two abscissas.

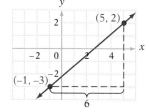

Change in $x = 5 - (-1) = 6$.

Slope $= m = \dfrac{\text{change in } y}{\text{change in } x} = \dfrac{5}{6}$.

Slope Formula

The slope of a line containing the two points, P_1 and P_2, whose coordinates are (x_1, y_1) and (x_2, y_2), is given by:

Slope $= m = \dfrac{y_2 - y_1}{x_2 - x_1},\ x_1 \neq x_2$

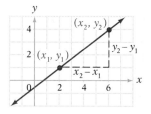

Find the slope of the line containing the points $(-2, 0)$ and $(4, 5)$.

Let $P_1 = (-2, 0)$ and $P_2 = (4, 5)$.
(It does not matter which point is named P_1 or P_2; the slope will be the same.)

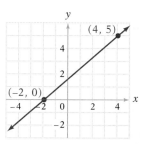

$$m = \frac{y_2 - y_1}{x_2 - x_1} = \frac{5 - 0}{4 - (-2)} = \frac{5}{6}$$

A line which slants upward to the right always has a **positive slope.**

Positive slope

Find the slope of the line containing the points $(-3, 4)$ and $(4, 2)$.

Let $P_1 = (-3, 4)$ and $P_2 = (4, 2)$.

$$m = \frac{y_2 - y_1}{x_2 - x_1} = \frac{2 - 4}{4 - (-3)} = \frac{-2}{7} = -\frac{2}{7}$$

A line that slants downward to the right always has a **negative slope.**

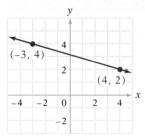

Negative slope

Find the slope of the line containing the points $(-2, 2)$ and $(4, 2)$.

Let $P_1 = (-2, 2)$ and $P_2 = (4, 2)$.

$$m = \frac{y_2 - y_1}{x_2 - x_1} = \frac{2 - 2}{4 - (-2)} = \frac{0}{6} = 0$$

A horizontal line has **zero slope.**

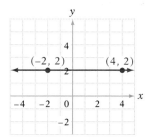

Zero slope

Find the slope of the line containing the points $(1, -2)$ and $(1, 3)$.

Let $P_1 = (1, -2)$ and $P_2 = (1, 3)$.

$$m = \frac{y_2 - y_1}{x_2 - x_1} = \frac{3 - (-2)}{1 - 1} = \frac{5}{0} \quad \text{Not a real number}$$

The slope of a vertical line is **undefined.**

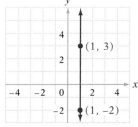

Undefined

Example 1

Find the slope of the line containing the points $(2, -5)$ and $(-4, 2)$.

Solution

Let $P_1 = (2, -5)$ and $P_2 = (-4, 2)$.

$$m = \frac{y_2 - y_1}{x_2 - x_1} = \frac{2 - (-5)}{-4 - 2} = \frac{7}{-6}$$

The slope is $-\frac{7}{6}$.

Example 2

Find the slope of the line containing the points $(4, -3)$ and $(2, 7)$.

Your solution

-5

Solution on p. A27

| Objective B | **To find the x- and the y-intercept of a straight line** |

The graph of the equation $x - 2y = 4$ is shown at the right. The graph crosses the x-axis at the point $(4, 0)$. This point is called the **x-intercept**. The graph also crosses the y-axis at the point $(0, -2)$. This point is called the **y-intercept**.

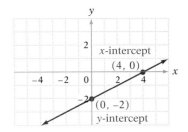

Find the x- and y-intercepts of the graph of the equation $3x + 4y = -12$.

To find the x-intercept, let $y = 0$. (Any point on the x-axis has y-coordinate 0.)

$$3x + 4y = -12$$
$$3x + 4(0) = -12$$
$$3x = -12$$
$$x = -4$$

The x-intercept is $(-4, 0)$.

To find the y-intercept, let $x = 0$. (Any point on the y-axis has x-coordinate 0.)

$$3x + 4y = -12$$
$$3(0) + 4y = -12$$
$$4y = -12$$
$$y = -3$$

The y-intercept is $(0, -3)$.

Find the y-intercept of $y = \frac{2}{3}x + 7$.

To find the y-intercept, let $x = 0$.

$$y = \frac{2}{3}x + 7$$
$$y = \frac{2}{3}(0) + 7$$
$$y = 7$$

The y-intercept is $(0, 7)$.

For any equation of the form $y = mx + b$, the y-intercept is $(0, b)$.

A linear equation can be graphed by finding the x- and y-intercepts and then drawing a line through the two points.

Graph $3x - 2y = 6$ by using the x- and y-intercepts.

To find the x-intercept, let $y = 0$.

$$3x - 2y = 6$$
$$3x - 2(0) = 6$$
$$3x = 6$$
$$x = 2$$

To find the y-intercept, let $x = 0$.

$$3x - 2y = 6$$
$$3(0) - 2y = 6$$
$$-2y = 6$$
$$y = -3$$

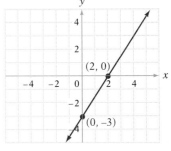

The x-intercept is $(2, 0)$.
The y-intercept is $(0, -3)$.

Example 3   Graph $4x - y = 4$ by using the x- and y-intercepts.

Solution x-intercept: y-intercept:
$4x - y = 4$ $4x - y = 4$
$4x - 0 = 4$ $4(0) - y = 4$
$\quad\ \ 4x = 4$ $\qquad -y = 4$
$\qquad x = 1$ $\qquad\ \ y = -4$
$(1, 0)$ $(0, -4)$

Example 4   Graph $3x - y = 2$ by using the x- and y-intercepts.

Your solution x-intercept: $(\frac{2}{3}, 0)$

y-intercept: $(0, -2)$

Example 5   Graph $y = \frac{2}{3}x - 2$ by using the x- and y-intercepts.

Solution x-intercept: y-intercept:
$\quad y = \frac{2}{3}x - 2$ $(0, b)$
$\qquad\qquad\qquad b = -2$
$\quad 0 = \frac{2}{3}x - 2$
$\qquad\qquad\quad (0, -2)$
$-\frac{2}{3}x = -2$

$\qquad x = 3$

$(3, 0)$

Example 6   Graph $y = \frac{1}{4}x + 1$ by using the x- and y-intercepts.

Your solution x-intercept: $(-4, 0)$
y-intercept: $(0, 1)$

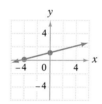

Solutions on p. A28

Objective C   **To graph a line given a point and the slope**

The graph of the equation $y = -\frac{3}{4}x + 4$ is shown at the right. The points $(-4, 7)$ and $(4, 1)$ are on the graph. The slope of the line is:

$$m = \frac{7 - 1}{-4 - 4} = \frac{6}{-8} = -\frac{3}{4}$$

Note that the slope of the line has the same value as the coefficient of x. The y-intercept is $(0, 4)$.

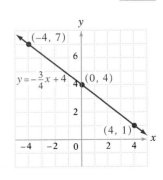

Slope-Intercept Formula

The equation $y = mx + b$ is called the slope-intercept form of a straight line. The slope of the line is m, the coefficient of x. The y-intercept is $(0, b)$.

When the equation of a straight line is in the form $y = mx + b$, the graph can be drawn using the slope and y-intercept. First locate the y-intercept. Use the slope to find a second point on the line. Then draw a line through the two points.

Graph $y = \frac{5}{3}x - 4$ by using the slope and y-intercept.

$m = \frac{5}{3} = \dfrac{\text{change in } y}{\text{change in } x}$

y-intercept $= (0, b) = (0, -4)$

Beginning at the y-intercept $(0, -4)$, move right 3 units (change in x) and then up 5 units (change in y).

(3, 1) is a second point on the graph.

Draw a line through the points $(0, -4)$ and $(3, 1)$.

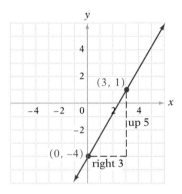

Graph $x + 2y = 4$ by using the slope and y-intercept.

Solve the equation for y.
$$x + 2y = 4$$
$$2y = -x + 4$$
$$y = -\frac{1}{2}x + 2$$

$m = -\frac{1}{2} = \frac{-1}{2} = \dfrac{\text{change in } y}{\text{change in } x}$

y-intercept $= (0, b) = (0, 2)$

Beginning at the y-intercept $(0, 2)$, move right 2 units and then down 1 unit.

(2, 1) is a second point on the graph.

Draw a line through the points (0, 2) and (2, 1).

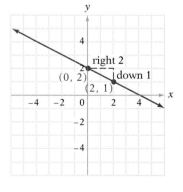

The graph of a line can be drawn when a point on the line and the slope of the line is given.

Graph the line that passes through point (2, 1) and has slope $\frac{2}{3}$.

Locate the point (2, 1) on the graph.

$m = \frac{2}{3} = \frac{\text{change in } y}{\text{change in } x}$

Beginning at the point (2, 1), move right 3 units and then up 2 units.

(5, 3) is a second point on the line.

Draw a line through the points (2, 1) and (5, 3).

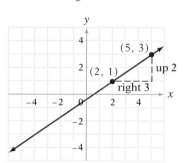

Example 7   Graph $y = -\frac{3}{2}x + 4$ by using the slope and y-intercept.

Solution $m = -\frac{3}{2} = \frac{-3}{2}$
y-intercept $= (0, 4)$

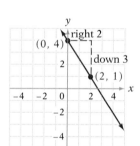

Example 8   Graph $2x + 3y = 6$ by using the slope and y-intercept.

Your solution $m = -\frac{2}{3}$
y-intercept: (0, 2)

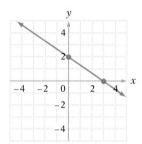

Example 9   Graph the line that passes through point $(-2, 3)$ and has slope $-\frac{4}{3}$.

Solution $(x_1, y_1) = (-2, 3)$
$m = -\frac{4}{3} = \frac{-4}{3}$

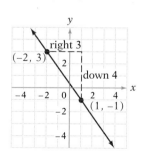

Example 10   Graph the line that passes through point $(-3, -2)$ and has slope 3.

Your solution

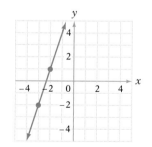

Solutions on p. A28

6.2   EXERCISES

▶ **Objective A**

Find the slope of the line containing the points.

1. $P_1(1, 3)$, $P_2(3, 1)$
-1

2. $P_1(2, 3)$, $P_2(5, 1)$
$-\frac{2}{3}$

3. $P_1(-1, 4)$, $P_2(2, 5)$
$\frac{1}{3}$

4. $P_1(3, -2)$, $P_2(1, 4)$
-3

5. $P_1(-1, 3)$, $P_2(-4, 5)$
$-\frac{2}{3}$

6. $P_1(-1, -2)$, $P_2(-3, 2)$
-2

7. $P_1(0, 3)$, $P_2(4, 0)$
$-\frac{3}{4}$

8. $P_1(-2, 0)$, $P_2(0, 3)$
$\frac{3}{2}$

9. $P_1(2, 4)$, $P_2(2, -2)$
undefined

10. $P_1(4, 1)$, $P_2(4, -3)$
undefined

11. $P_1(2, 5)$, $P_2(-3, -2)$
$\frac{7}{5}$

12. $P_1(4, 1)$, $P_2(-1, -2)$
$\frac{3}{5}$

13. $P_1(2, 3)$, $P_2(-1, 3)$
0

14. $P_1(3, 4)$, $P_2(0, 4)$
0

15. $P_1(0, 4)$, $P_2(-2, 5)$
$-\frac{1}{2}$

16. $P_1(3, 0)$, $P_2(-1, -4)$
1

17. $P_1(-3, 4)$, $P_2(-2, 1)$
-3

18. $P_1(4, -2)$, $P_2(2, -4)$
1

19. $P_1(-2, 3)$, $P_2(-2, 5)$
undefined

20. $P_1(-3, -1)$, $P_2(-3, 4)$
undefined

21. $P_1(-2, -5)$, $P_2(-4, -1)$
-2

22. $P_1(-3, -2)$, $P_2(0, -5)$
-1

23. $P_1(3, -1)$, $P_2(-2, -1)$
0

24. $P_1(0, -3)$, $P_2(-2, -3)$
0

25. $P_1(0, -3)$, $P_2(4, -2)$
$\frac{1}{4}$

26. $P_1(1, 0)$, $P_2(0, -1)$
1

27. $P_1(2, 5)$, $P_2(-2, -5)$
$\frac{5}{2}$

▶ **Objective B**

Find the *x*- and *y*-intercepts and graph.

28. $x - 2y = -4$

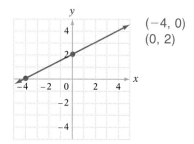

$(-4, 0)$
$(0, 2)$

29. $3x + y = 3$

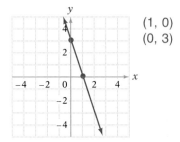

$(1, 0)$
$(0, 3)$

30. $4x - 2y = 5$

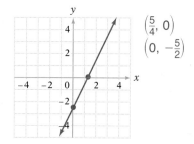

$\left(\dfrac{5}{4}, 0\right)$
$\left(0, -\dfrac{5}{2}\right)$

31. $2x - y = 4$

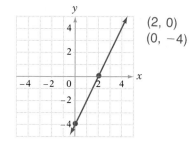

$(2, 0)$
$(0, -4)$

32. $3x + 2y = 5$

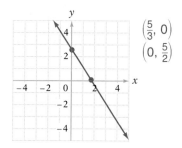

$\left(\dfrac{5}{3}, 0\right)$
$\left(0, \dfrac{5}{2}\right)$

33. $4x - 3y = 8$

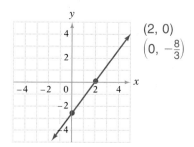

$(2, 0)$
$\left(0, -\dfrac{8}{3}\right)$

34. $2x - 3y = 4$

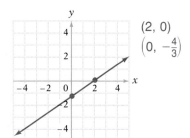

$(2, 0)$
$\left(0, -\dfrac{4}{3}\right)$

35. $3x - 5y = 9$

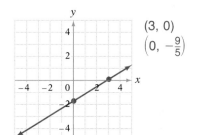

$(3, 0)$
$\left(0, -\dfrac{9}{5}\right)$

Find the *x*- and *y*-intercepts and graph.

36. $2x - 3y = 9$

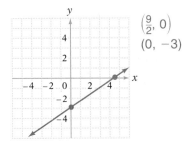

$\left(\frac{9}{2}, 0\right)$
$(0, -3)$

37. $3x - 4y = 4$

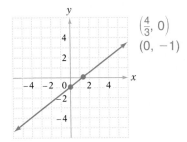

$\left(\frac{4}{3}, 0\right)$
$(0, -1)$

38. $2x + y = 3$

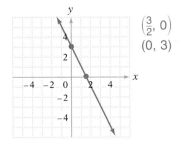

$\left(\frac{3}{2}, 0\right)$
$(0, 3)$

39. $3x + y = -5$

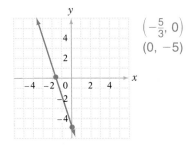

$\left(-\frac{5}{3}, 0\right)$
$(0, -5)$

40. $3x + 2y = 4$

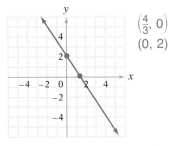

$\left(\frac{4}{3}, 0\right)$
$(0, 2)$

41. $3x + 4y = -12$

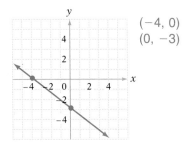

$(-4, 0)$
$(0, -3)$

42. $2x - 3y = -6$

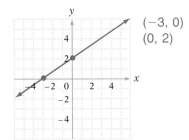

$(-3, 0)$
$(0, 2)$

43. $4x - 3y = 6$

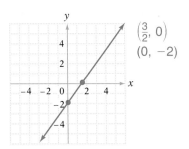

$\left(\frac{3}{2}, 0\right)$
$(0, -2)$

▶ **Objective C**

Graph by using the slope and the *y*-intercept.

44. $y = \frac{1}{2}x + 2$

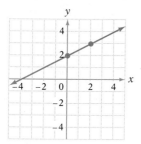

45. $y = \frac{2}{3}x - 3$

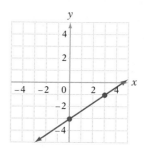

46. $y = -\frac{2}{3}x + 4$

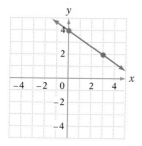

47. $y = -\frac{1}{2}x + 2$

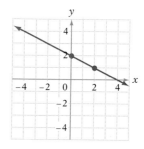

48. $y = -\frac{3}{2}x$

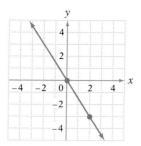

49. $y = \frac{3}{4}x$

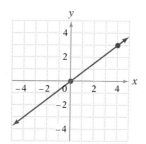

50. $y = \frac{2}{3}x - 1$

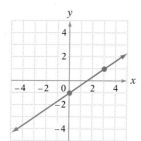

51. $y = -\frac{1}{4}x - 3$

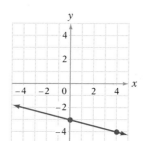

Graph by using the slope and the *y*-intercept.

52. $2x - 3y = 6$

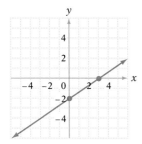

53. $3x - y = 2$

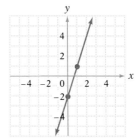

54. $4x + y = 2$

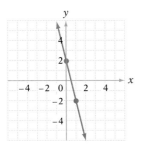

55. $3x + 2y = 8$

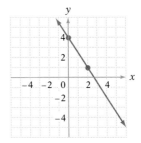

56. $4x - 5y = 5$

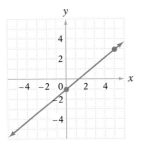

57. $3x - 2y = 6$

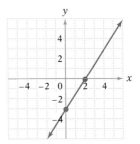

58. $x - 3y = 3$

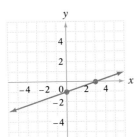

59. $x + 2y = 4$

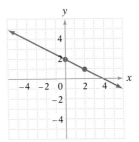

60. Graph the line that passes through point (2, 3) and has slope $\frac{1}{2}$.

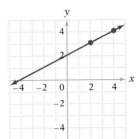

61. Graph the line that passes through point (−4, 1) and has slope $\frac{2}{3}$.

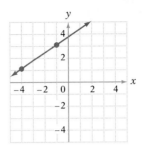

62. Graph the line that passes through point (1, 4) and has slope $-\frac{2}{3}$.

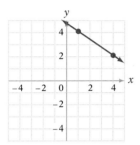

63. Graph the line that passes through point (0, −2) and has slope $-\frac{1}{3}$.

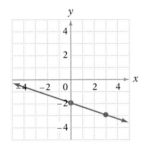

64. Graph the line that passes through point (−1, −3) and has slope $\frac{4}{3}$.

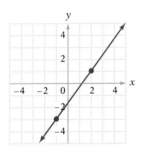

65. Graph the line that passes through point (−2, −3) and has slope $\frac{5}{4}$.

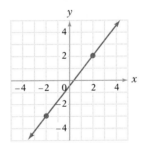

66. Graph the line that passes through point (−3, 0) and has slope −3.

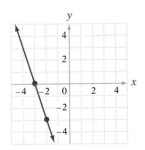

67. Graph the line that passes through point (2, 0) and has slope −1.

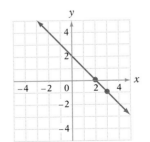

SECTION 6.3 **Finding Equations of Lines**

Objective A   **To find the equation of a line given a point and the slope**

When the slope of a line and a point on the line are known, the equation of the line can be determined.

Find the equation of the line that contains the point (0, 3) and has slope $\frac{1}{2}$.

The given point, (0, 3), is the y-intercept.
Use the slope-intercept form of the equation. $y = mx + b$

Replace m with $\frac{1}{2}$, the given slope. Replace b $y = \frac{1}{2}x + 3$
with 3.

The equation of the line is $y = \frac{1}{2}x + 3$.

One method for finding the equation of a line, given the slope and *any* point on the line, involves use of the point-slope formula. The point-slope formula is derived from the formula for slope.

Let (x_1, y_1) be the given point on the line and (x, y) be any other point on the line.

Formula for slope $\frac{y - y_1}{x - x_1} = m$

Multiply both sides of the equation by $(x - x_1)$. $\frac{y - y_1}{x - x_1}(x - x_1) = m(x - x_1)$

Simplify. $y - y_1 = m(x - x_1)$

Point-Slope Formula

> Let m be the slope of a line and (x_1, y_1) the coordinates of a point on the line. The equation of the line can be found from the point-slope formula.
> $$y - y_1 = m(x - x_1).$$

Find the equation of the line that contains the point (2, −3) and has slope $-\frac{3}{4}$.

$m = -\frac{3}{4}$ $(x_1, y_1) = (2, -3)$ $y - y_1 = m(x - x_1)$

Substitute the slope, $-\frac{3}{4}$, and the coordinates of the $y - (-3) = -\frac{3}{4}(x - 2)$
given point, (2, −3), into the point-slope formula.
Then simplify. $y + 3 = -\frac{3}{4}x + \frac{3}{2}$

$y = -\frac{3}{4}x + \frac{3}{2} - 3$

$y = -\frac{3}{4}x - \frac{3}{2}$

The equation of the line is $y = -\frac{3}{4}x - \frac{3}{2}$.

Example 1 Find the equation of the line that contains the point (0, −2) and has slope $-\frac{2}{3}$.

Solution $m = -\frac{2}{3}$ $b = -2$

$y = mx + b$

$y = -\frac{2}{3}x - 2$

The equation of the line is

$y = -\frac{2}{3}x - 2$.

Example 2 Find the equation of the line that contains the point (0, 3) and has slope $-\frac{5}{4}$.

Your solution $y = -\frac{5}{4}x + 3$

Example 3 Find the equation of the line that contains the point (3, 0) and has slope −4.

Solution $m = -4$ $(x_1, y_1) = (3, 0)$

$y - y_1 = m(x - x_1)$
$y - 0 = -4(x - 3)$
$y = -4x + 12$

The equation of the line is
$y = -4x + 12$.

Example 4 Find the equation of the line that contains the point (−3, −2) and has slope $-\frac{1}{3}$.

Your solution $y = -\frac{1}{3}x - 3$

Example 5 Find the equation of the line that contains the point (−2, 4) and has slope 2.

Solution $m = 2$ $(x_1, y_1) = (-2, 4)$

$y - y_1 = m(x - x_1)$
$y - 4 = 2[x - (-2)]$
$y - 4 = 2(x + 2)$
$y - 4 = 2x + 4$
$y = 2x + 8$

The equation of the line is
$y = 2x + 8$.

Example 6 Find the equation of the line that contains the point (4, −3) and has slope −3.

Your solution $y = -3x + 9$

Solutions on pp. A28–A29

| **Objective B** | **To find the equation of a line given two points** |

The point-slope formula and the formula for slope are used to find the equation of a line when two points are known.

Find the equation of the line containing the points $(3, 2)$ and $(-5, 6)$.

Find the slope. Let $(x_1, y_1) = (3, 2)$ and $(x_2, y_2) = (-5, 6)$.

$$m = \frac{y_2 - y_1}{x_2 - x_1} = \frac{6 - 2}{-5 - 3} = \frac{4}{-8} = -\frac{1}{2}$$

Substitute the slope and the coordinates of either one of the known points in the point-slope formula. Point $(3, 2)$ is used at the right.

$$y - y_1 = m(x - x_1)$$
$$y - 2 = -\frac{1}{2}(x - 3)$$
$$y - 2 = -\frac{1}{2}x + \frac{3}{2}$$
$$y = -\frac{1}{2}x + \frac{7}{2}$$

The equation of the line is $y = -\frac{1}{2}x + \frac{7}{2}$.

Example 7 Find the equation of the line containing the points $(2, 3)$ and $(4, 1)$.

Solution Let $(x_1, y_1) = (2, 3)$ and $(x_2, y_2) = (4, 1)$.

$$m = \frac{y_2 - y_1}{x_2 - x_1} = \frac{1 - 3}{4 - 2} = \frac{-2}{2} = -1$$

$$y - y_1 = m(x - x_1)$$
$$y - 3 = -1(x - 2)$$
$$y - 3 = -x + 2$$
$$y = -x + 5$$

The equation of the line is $y = -x + 5$.

Example 8 Find the equation of the line containing the points $(2, 0)$ and $(5, 3)$.

Your solution $y = x - 2$

Solution on p. A29

Example 9   Find the equation of the line containing the points $(-2, 5)$ and $(-4, -1)$.

Example 10   Find the equation of the line containing the points $(4, -2)$ and $(-1, -7)$.

Solution   Let $(x_1, y_1) = (-2, 5)$ and $(x_2, y_2) = (-4, -1)$.

$$m = \frac{y_2 - y_1}{x_2 - x_1}$$
$$= \frac{-1 - 5}{-4 - (-2)}$$
$$= \frac{-6}{-2} = 3$$

$y - y_1 = m(x - x_1)$
$y - 5 = 3[x - (-2)]$
$y - 5 = 3(x + 2)$
$y - 5 = 3x + 6$
$\quad\ \ y = 3x + 11$

The equation of the line is $y = 3x + 11$.

Your solution $y = x - 6$

Example 11   Find the equation of the line containing the points $(2, -3)$ and $(2, 5)$.

Example 12   Find the equation of the line containing the points $(2, 3)$ and $(-5, 3)$.

Solution   Let $(x_1, y_1) = (2, -3)$ and $(x_2, y_2) = (2, 5)$.

$$m = \frac{y_2 - y_1}{x_2 - x_1}$$
$$= \frac{5 - (-3)}{2 - 2}$$
$$= \frac{5 + 3}{0} = \frac{8}{0}$$

The slope of the line is undefined, therefore the line is a vertical line. All points on the line have an abscissa of 2.

The equation of the line is $x = 2$.

Your solution $y = 3$

Solutions on p. A29

tent and Format © 1991 HMCo.

6.3   EXERCISES

▶ Objective A

Find the equation of the line that contains the given point and has the given slope.

1. Point $(0, 5)$, $m = 2$
$y = 2x + 5$

2. Point $(0, 3)$, $m = 1$
$y = x + 3$

3. Point $(2, 3)$, $m = \frac{1}{2}$
$y = \frac{1}{2}x + 2$

4. Point $(5, 1)$, $m = \frac{2}{3}$
$y = \frac{2}{3}x - \frac{7}{3}$

5. Point $(-1, 4)$, $m = \frac{5}{4}$
$y = \frac{5}{4}x + \frac{21}{4}$

6. Point $(-2, 1)$, $m = \frac{3}{2}$
$y = \frac{3}{2}x + 4$

7. Point $(3, 0)$, $m = -\frac{5}{3}$
$y = -\frac{5}{3}x + 5$

8. Point $(-2, 0)$, $m = \frac{3}{2}$
$y = \frac{3}{2}x + 3$

9. Point $(2, 3)$, $m = -3$
$y = -3x + 9$

10. Point $(1, 5)$, $m = -\frac{4}{5}$
$y = -\frac{4}{5}x + \frac{29}{5}$

11. Point $(-1, 7)$, $m = -3$
$y = -3x + 4$

12. Point $(-2, 4)$, $m = -4$
$y = -4x - 4$

13. Point $(-1, -3)$, $m = \frac{2}{3}$
$y = \frac{2}{3}x - \frac{7}{3}$

14. Point $(-2, -4)$, $m = \frac{1}{4}$
$y = \frac{1}{4}x - \frac{7}{2}$

15. Point $(0, 0)$, $m = \frac{1}{2}$
$y = \frac{1}{2}x$

16. Point $(0, 0)$, $m = \frac{3}{4}$
$y = \frac{3}{4}x$

17. Point $(2, -3)$, $m = 3$
$y = 3x - 9$

18. Point $(4, -5)$, $m = 2$
$y = 2x - 13$

19. Point $(3, 5)$, $m = -\frac{2}{3}$
$y = -\frac{2}{3}x + 7$

20. Point $(5, 1)$, $m = -\frac{4}{5}$
$y = -\frac{4}{5}x + 5$

21. Point $(0, -3)$, $m = -1$
$y = -x - 3$

22. Point $(2, 0)$, $m = \frac{5}{6}$
$y = \frac{5}{6}x - \frac{5}{3}$

23. Point $(1, -4)$, $m = \frac{7}{5}$
$y = \frac{7}{5}x - \frac{27}{5}$

24. Point $(3, 5)$, $m = -\frac{3}{7}$
$y = -\frac{3}{7}x + \frac{44}{7}$

Find the equation of the line that contains the given point and has the given slope.

25. Point $(-2, 0)$, $m = 0$
$y = 0$

26. Point $(4, 0)$, $m = 0$
$y = 0$

27. Point $(-2, -3)$, $m = \frac{2}{3}$
$y = \frac{2}{3}x - \frac{5}{3}$

28. Point $(-3, -1)$, $m = \frac{4}{3}$
$y = \frac{4}{3}x + 3$

29. Point $(0, 2)$, slope is undefined
$x = 0$

30. Point $(0, 5)$, slope is undefined
$x = 0$

31. Point $(0, -3)$, $m = 3$
$y = 3x - 3$

32. Point $(0, -2)$, $m = 0$
$y = -2$

33. Point $(-5, 0)$, $m = -\frac{5}{2}$
$y = -\frac{5}{2}x - \frac{25}{2}$

34. Point $(-2, 0)$, $m = \frac{2}{5}$
$y = \frac{2}{5}x + \frac{4}{5}$

35. Point $(0, 5)$, $m = \frac{4}{3}$
$y = \frac{4}{3}x + 5$

36. Point $(0, -5)$, $m = \frac{6}{5}$
$y = \frac{6}{5}x - 5$

37. Point $(4, -1)$, $m = -\frac{2}{5}$
$y = -\frac{2}{5}x + \frac{3}{5}$

38. Point $(-3, 5)$, $m = -\frac{1}{4}$
$y = -\frac{1}{4}x + \frac{17}{4}$

39. Point $(3, -4)$, slope is undefined
$x = 3$

40. Point $(-2, 5)$, slope is undefined
$x = -2$

41. Point $(-2, -5)$, $m = -\frac{5}{4}$
$y = -\frac{5}{4}x - \frac{15}{2}$

42. Point $(-3, -2)$, $m = -\frac{2}{3}$
$y = -\frac{2}{3}x - 4$

43. Point $(-2, -3)$, $m = 0$
$y = -3$

44. Point $(-3, -2)$, $m = 0$
$y = -2$

45. Point $(4, -5)$, $m = -2$
$y = -2x + 3$

46. Point $(-3, 5)$, $m = 3$
$y = 3x + 14$

47. Point $(-5, -1)$, slope is undefined
$x = -5$

48. Point $(0, 4)$, slope is undefined
$x = 0$

▶ **Objective B**

Find the equation of the line containing the given points.

49. $P_1(0, 2)$, $P_2(3, 5)$
$y = x + 2$

50. $P_1(0, 4)$, $P_2(1, 5)$
$y = x + 4$

51. $P_1(0, -3)$, $P_2(-4, 5)$
$y = -2x - 3$

52. $P_1(0, -2)$, $P_2(-3, 4)$
$y = -2x - 2$

53. $P_1(2, 3)$, $P_2(5, 5)$
$y = \frac{2}{3}x + \frac{5}{3}$

54. $P_1(4, 1)$, $P_2(6, 3)$
$y = x - 3$

55. $P_1(-1, 3)$, $P_2(2, 4)$
$y = \frac{1}{3}x + \frac{10}{3}$

56. $P_1(-1, 1)$, $P_2(4, 4)$
$y = \frac{3}{5}x + \frac{8}{5}$

57. $P_1(-1, -2)$, $P_2(3, 4)$
$y = \frac{3}{2}x - \frac{1}{2}$

58. $P_1(-3, -1)$, $P_2(2, 4)$
$y = x + 2$

59. $P_1(0, 3)$, $P_2(2, 0)$
$y = -\frac{3}{2}x + 3$

60. $P_1(0, 4)$, $P_2(2, 0)$
$y = -2x + 4$

61. $P_1(-3, -1)$, $P_2(2, -1)$
$y = -1$

62. $P_1(-3, -5)$, $P_2(4, -5)$
$y = -5$

63. $P_1(-2, -3)$, $P_2(-1, -2)$
$y = x - 1$

64. $P_1(-4, -1)$, $P_2(-5, -2)$
$y = x + 3$

65. $P_1(-2, 3)$, $P_2(-1, 1)$
$y = -2x - 1$

66. $P_1(-3, 5)$, $P_2(-2, 3)$
$y = -2x - 1$

67. $P_1(-2, 5)$, $P_2(-2, 4)$
$x = -2$

68. $P_1(3, 6)$, $P_2(3, -2)$
$x = 3$

69. $P_1(0, -2)$, $P_2(-2, 5)$
$y = -\frac{7}{2}x - 2$

70. $P_1(3, 3)$, $P_2(-5, 4)$
$y = -\frac{1}{8}x + \frac{27}{8}$

71. $P_1(5, 0)$, $P_2(2, 4)$
$y = -\frac{4}{3}x + \frac{20}{3}$

72. $P_1(-4, -5)$, $P_2(1, -2)$
$y = \frac{3}{5}x - \frac{13}{5}$

Find the equation of the line containing the given points.

73. $P_1(3, 2)$, $P_2(-1, 5)$

$y = -\frac{3}{4}x + \frac{17}{4}$

74. $P_1(4, 1)$, $P_2(-2, 4)$

$y = -\frac{1}{2}x + 3$

75. $P_1(3, -1)$, $P_2(2, -4)$

$y = 3x - 10$

76. $P_1(4, 1)$, $P_2(3, -2)$

$y = 3x - 11$

77. $P_1(-2, 3)$, $P_2(2, -1)$

$y = -x + 1$

78. $P_1(3, 1)$, $P_2(-3, -2)$

$y = \frac{1}{2}x - \frac{1}{2}$

79. $P_1(2, 3)$, $P_2(5, 5)$

$y = \frac{2}{3}x + \frac{5}{3}$

80. $P_1(7, 2)$, $P_2(4, 4)$

$y = -\frac{2}{3}x + \frac{20}{3}$

81. $P_1(2, 0)$, $P_2(0, -1)$

$y = \frac{1}{2}x - 1$

82. $P_1(0, 4)$, $P_2(-2, 0)$

$y = 2x + 4$

83. $P_1(3, -4)$, $P_2(-2, -4)$

$y = -4$

84. $P_1(-3, 3)$, $P_2(-2, 3)$

$y = 3$

85. $P_1(0, 0)$, $P_2(4, 3)$

$y = \frac{3}{4}x$

86. $P_1(2, -5)$, $P_2(0, 0)$

$y = -\frac{5}{2}x$

87. $P_1(2, -1)$, $P_2(-1, 3)$

$y = -\frac{4}{3}x + \frac{5}{3}$

88. $P_1(3, -5)$, $P_2(-2, 1)$

$y = -\frac{6}{5}x - \frac{7}{5}$

89. $P_1(-2, 5)$, $P_2(-2, -5)$

$x = -2$

90. $P_1(3, 2)$, $P_2(3, -4)$

$x = 3$

91. $P_1(2, 1)$, $P_2(-2, -3)$

$y = x - 1$

92. $P_1(-3, -2)$, $P_2(1, -4)$

$y = -\frac{1}{2}x - \frac{7}{2}$

93. $P_1(-4, -3)$, $P_2(2, 5)$

$y = \frac{4}{3}x + \frac{7}{3}$

94. $P_1(4, 5)$, $P_2(-4, 3)$

$y = \frac{1}{4}x + 4$

95. $P_1(0, 3)$, $P_2(3, 0)$

$y = -x + 3$

96. $P_1(1, -3)$, $P_2(-2, 4)$

$y = -\frac{7}{3}x - \frac{2}{3}$

SECTION 6.4	# Parallel and Perpendicular Lines

Objective A	## To find parallel and perpendicular lines

Two lines that have the same slope do not intersect and are called **parallel lines.**

Two vertical lines are parallel lines. Two horizontal lines are parallel lines.

The slope of each of the lines at the right is $\frac{2}{3}$.

The lines are parallel.

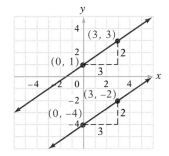

For two nonvertical parallel lines, **$m_1 = m_2$.**

Is the line that contains the points $(-2, 1)$ and $(-5, -1)$ parallel to the line that contains the points $(1, 0)$ and $(4, 2)$?

Find the slope of each line. $m_1 = \frac{-1 - 1}{-5 - (-2)} = \frac{-2}{-3} = \frac{2}{3}$

$m_2 = \frac{2 - 0}{4 - 1} = \frac{2}{3}$

$m_1 = m_2 = \frac{2}{3}$ The lines are parallel.

Are the lines $3x - 4y = 8$ and $6x - 8y = 2$ parallel?

Solve each equation for y.

$3x - 4y = 8$	$6x - 8y = 2$
$-4y = -3x + 8$	$-8y = -6x + 2$
$y = \frac{3}{4}x - 2$	$y = \frac{3}{4}x - \frac{1}{4}$

Find the slope of each line. $m_1 = \frac{3}{4}$ $m_2 = \frac{3}{4}$

$m_1 = m_2 = \frac{3}{4}$. Therefore the lines are parallel.

Find the equation of the line containing the point $(2, 3)$ and parallel to the line $y = \frac{1}{2}x - 4$.

Find the slope of the given line. $m = \frac{1}{2}$

Parallel lines have the same slope. Substitute the slope of the given line and the coordinates of the given point in the point-slope formula.

$y - y_1 = m(x - x_1)$

$y - 3 = \frac{1}{2}(x - 2)$

$y - 3 = \frac{1}{2}x - 1$

$y = \frac{1}{2}x + 2$

The equation of the line is $y = \frac{1}{2}x + 2$.

Find the equation of the line containing the point $(-1, 4)$ and parallel to the line $2x - 3y = 2$.

Solve the given equation for y.

$$2x - 3y = 2$$
$$-3y = -2x + 2$$
$$y = \tfrac{2}{3}x - \tfrac{2}{3}$$

Find the slope of the given line.

$$m = \tfrac{2}{3}$$

Parallel lines have the same slope. Substitute the slope of the given line and the coordinates of the given point in the point-slope formula.

$$y - y_1 = m(x - x_1)$$
$$y - 4 = \tfrac{2}{3}[x - (-1)]$$
$$y - 4 = \tfrac{2}{3}x + \tfrac{2}{3}$$
$$y = \tfrac{2}{3}x + \tfrac{14}{3}$$

The equation of the line is $y = \tfrac{2}{3}x + \tfrac{14}{3}$.

Two lines that intersect at right angles are **perpendicular lines.**

Any horizontal line is perpendicular to any vertical line. For example, $x = 3$ is perpendicular to $y = -2$.

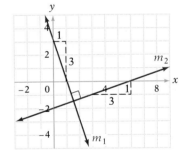

Slopes of Perpendicular Lines

If m_1 and m_2 are the slopes of two lines, neither of which is a vertical line, then the lines are perpendicular if and only if

$$m_1 \cdot m_2 = -1.$$

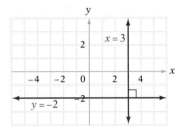

Solving $m_1 \cdot m_2 = -1$ for m_1 gives $m_1 = -\dfrac{1}{m_2}$. This last equation states that the slopes of perpendicular lines are *negative reciprocals of each other.*

Is the line that contains the points $(4, 2)$ and $(-2, 5)$ perpendicular to the line that contains the points $(-4, 3)$ and $(-3, 5)$?

Find the slope of each line.

$$m_1 = \frac{5 - 2}{-2 - 4} = \frac{3}{-6} = -\frac{1}{2}$$
$$m_2 = \frac{5 - 3}{-3 - (-4)} = \frac{2}{1} = 2$$

Find the product of the slopes of the two lines. If the product is -1, the lines are perpendicular. If the product is not -1, the lines are not perpendicular.

$$m_1 \cdot m_2 = -\tfrac{1}{2}(2) = -1$$

The lines are perpendicular.

Are the lines $3x + 4y = -8$ and $4x - 3y = -9$ perpendicular?

Solve each equation for y.

$$3x + 4y = -8 \qquad\qquad 4x - 3y = -9$$
$$4y = -3x - 8 \qquad\qquad -3y = -4x - 9$$
$$y = -\tfrac{3}{4}x - 2 \qquad\qquad y = \tfrac{4}{3}x + 3$$

Find the slope of each line. $m_1 = -\tfrac{3}{4} \qquad\qquad m_2 = \tfrac{4}{3}$

Find the product of the slopes of $m_1 \cdot m_2 = -\tfrac{3}{4}\left(\tfrac{4}{3}\right) = -1$
the two lines.

The lines are perpendicular.

Find the equation of the line containing the point $(-2, 1)$ and perpendicular to the line $y = -\tfrac{2}{3}x + 2$.

Find the slope of the given line. $m_1 = -\tfrac{2}{3}$

For perpendicular lines, $m_1 \cdot m_2 = -1$. $m_1 \cdot m_2 = -1$

Substitute $-\tfrac{2}{3}$ for m_1 and solve for m_2. $-\tfrac{2}{3}m_2 = -1$

$$m_2 = \tfrac{3}{2}$$

Substitute the slope and the coordinates of the given point in the point-slope formula.

$$y - y_1 = m(x - x_1)$$
$$y - 1 = \tfrac{3}{2}[x - (-2)]$$
$$y - 1 = \tfrac{3}{2}x + 3$$
$$y = \tfrac{3}{2}x + 4$$

The equation of the line is

$y = \tfrac{3}{2}x + 4$.

Find the equation of the line containing the point $(3, -4)$ and perpendicular to the line $2x - y = -3$.

Solve the given equation for y.

$$2x - y = -3$$
$$-y = -2x - 3$$
$$y = 2x + 3$$

Find the slope of the given line. $m_1 = 2$

For perpendicular lines, $m_1 \cdot m_2 = -1$. $m_1 \cdot m_2 = -1$

Substitute 2 for m_1 and solve for m_2. $2m_2 = -1$

$$m_2 = -\tfrac{1}{2}$$

Substitute the slope and the coordinates of the given point in the point-slope formula.

$$y - y_1 = m(x - x_1)$$
$$y - (-4) = -\tfrac{1}{2}(x - 3)$$
$$y + 4 = -\tfrac{1}{2}x + \tfrac{3}{2}$$
$$y = -\tfrac{1}{2}x - \tfrac{5}{2}$$

The equation of the line is

$y = -\tfrac{1}{2}x - \tfrac{5}{2}$.

Example 1 Is the line that contains the points $(-4, 2)$ and $(1, 6)$ parallel to the line that contains the points $(2, -4)$ and $(7, 0)$?

Solution $m_1 = \frac{6 - 2}{1 - (-4)} = \frac{4}{5}$

$m_2 = \frac{0 - (-4)}{7 - 2} = \frac{4}{5}$

$m_1 = m_2 = \frac{4}{5}$

The lines are parallel.

Example 2 Is the line that contains the points $(-2, -3)$ and $(7, 1)$ perpendicular to the line that contains the points $(4, 1)$ and $(6, -5)$?

Your solution No

Example 3 Are the lines $4x - y = -2$ and $x + 4y = -12$ perpendicular?

Solution $4x - y = -2$
$\quad\quad -y = -4x - 2$
$\quad\quad\quad y = 4x + 2$

$m_1 = 4$
$x + 4y = -12$
$\quad 4y = -x - 12$

$\quad\quad y = -\frac{1}{4}x - 3$

$m_2 = -\frac{1}{4}$

$m_1 \cdot m_2 = 4\left(-\frac{1}{4}\right) = -1$

The lines are perpendicular.

Example 4 Are the lines $5x + 2y = 2$ and $5x + 2y = -6$ parallel?

Your solution Yes

Example 5 Find the equation of the line containing the point $(3, -1)$ and parallel to the line $3x - 2y = 4$.

Solution $3x - 2y = 4$
$\quad\quad -2y = -3x + 4$

$\quad\quad\quad y = \frac{3}{2}x - 2$

$m = \frac{3}{2}$

$y - y_1 = m(x - x_1)$

$y - (-1) = \frac{3}{2}(x - 3)$

$\quad y + 1 = \frac{3}{2}x - \frac{9}{2}$

$\quad\quad\quad y = \frac{3}{2}x - \frac{11}{2}$

The equation of the line is
$y = \frac{3}{2}x - \frac{11}{2}$.

Example 6 Find the equation of the line containing the point $(-2, 2)$ and perpendicular to the line $x - 4y = 3$.

Your solution $y = -4x - 6$

Solutions on p. A29

6.4 EXERCISES

▶ **Objective A**

1. Is the line $x = -2$ perpendicular to the line $y = 3$?
Yes

2. Is the line $y = \frac{1}{2}$ perpendicular to the line $y = -4$?
No

3. Is the line $x = -3$ parallel to the line $y = \frac{1}{3}$?
No

4. Is the line $x = 4$ parallel to the line $x = -4$?
Yes

5. Is the line $y = \frac{2}{3}x - 4$ parallel to the line $y = -\frac{3}{2}x - 4$?
No

6. Is the line $y = -2x + \frac{2}{3}$ parallel to the line $y = -2x + 3$?
Yes

7. Is the line $y = \frac{4}{3}x - 2$ perpendicular to the line $y = -\frac{3}{4}x + 2$?
Yes

8. Is the line $y = \frac{1}{2}x + \frac{3}{2}$ perpendicular to the line $y = -\frac{1}{2}x + \frac{3}{2}$?
No

9. Are the lines $2x + 3y = 2$ and $2x + 3y = -4$ parallel?
Yes

10. Are the lines $2x - 4y = 3$ and $2x + 4y = -3$ parallel?
No

11. Are the lines $x - 4y = 2$ and $4x + y = 8$ perpendicular?
Yes

12. Are the lines $4x - 3y = 2$ and $4x + 3y = -7$ perpendicular?
No

13. Is the line that contains the points $(3, 2)$ and $(1, 6)$ parallel to the line that contains the points $(-1, 3)$ and $(-1, -1)$?
No

14. Is the line that contains the points $(4, -3)$ and $(2, 5)$ parallel to the line that contains the points $(-2, -3)$ and $(-4, 1)$?
No

15. Is the line that contains the points $(-3, 2)$ and $(4, -1)$ perpendicular to the line that contains the points $(1, 3)$ and $(-2, -4)$?
Yes

16. Is the line that contains the points $(-1, 2)$ and $(3, 4)$ perpendicular to the line that contains the points $(-1, 3)$ and $(-4, 1)$?
No

17. Is the line that contains the points $(-5, 0)$ and $(0, 2)$ parallel to the line that contains the points $(5, 1)$ and $(0, -1)$?
Yes

18. Is the line that contains the points $(3, 5)$ and $(-3, 3)$ perpendicular to the line that contains the points $(2, -5)$ and $(-4, 4)$?
No

19. Find the equation of the line containing the point $(-2, -4)$ and parallel to the line $2x - 3y = 2$.

$y = \frac{2}{3}x - \frac{8}{3}$

20. Find the equation of the line containing the point $(3, 2)$ and parallel to the line $3x + y = -3$.

$y = -3x + 11$

21. Find the equation of the line containing the point $(4, 1)$ and perpendicular to the line $y = -3x + 4$.

$y = \frac{1}{3}x - \frac{1}{3}$

22. Find the equation of the line containing the point $(2, -5)$ and perpendicular to the line $y = \frac{5}{2}x - 4$.

$y = -\frac{2}{5}x - \frac{21}{5}$

23. Find the equation of the line containing the point $(-1, -3)$ and perpendicular to the line $3x - 5y = 2$.

$y = -\frac{5}{3}x - \frac{14}{3}$

24. Find the equation of the line containing the point $(-1, 3)$ and perpendicular to the line $2x + 4y = -1$.

$y = 2x + 5$

25. Find the equation of the line containing the point $(-3, 1)$ and parallel to the line $y = \frac{2}{3}x - 1$.

$y = \frac{2}{3}x + 3$

26. Find the equation of the line containing the point $(4, -3)$ and parallel to the line $y = -\frac{4}{3}x + 2$.

$y = -\frac{4}{3}x + \frac{7}{3}$

27. Find the equation of the line containing the point $(-5, 4)$ and parallel to the line $y = -\frac{5}{3}x - 7$.

$y = -\frac{5}{3}x - \frac{13}{3}$

28. Find the equation of the line containing the point $(3, -4)$ and parallel to the line $y = \frac{3}{2}x$.

$y = \frac{3}{2}x - \frac{17}{2}$

29. Find the equation of the line containing the point $(-4, -2)$ and parallel to the line $5x - 2y = -4$.

$y = \frac{5}{2}x + 8$

30. Find the equation of the line containing the point $(-2, 0)$ and parallel to the line $3x - 4y = 6$.

$y = \frac{3}{4}x + \frac{3}{2}$

31. Find the equation of the line containing the point $(4, -3)$ and perpendicular to the line $y = 5x$.

$y = -\frac{1}{5}x - \frac{11}{5}$

32. Find the equation of the line containing the point $(-2, 5)$ and perpendicular to the line $y = \frac{2}{3}x - 5$.

$y = -\frac{3}{2}x + 2$

33. Find the equation of the line containing the point $(-3, -3)$ and perpendicular to the line $3x - 2y = 3$.

$y = -\frac{2}{3}x - 5$

34. Find the equation of the line containing the point $(-1, 6)$ and perpendicular to the line $4x + y = -3$.

$y = \frac{1}{4}x + \frac{25}{4}$

SECTION 6.5   Applications of Linear Equations

Objective A   To obtain data from a graph

The rectangular coordinate system is used in business, science, and mathematics to show a relationship between two variables. One variable is represented along the horizontal axis and the other variable is represented along the vertical axis. A linear relationship between the variables is represented on the coordinate system as a straight line.

A company purchases a computer system for $10,000. The graph at the right shows the depreciation of the computer system over a five-year period.

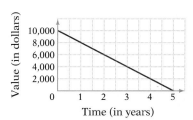

Information can be obtained from the graph. For example, the depreciated value of the computer after two years is $6000; the depreciated value of the computer after four years is $2000.

From the graph, an equation of the line that represents the depreciation can be written.

Use any two points shown on the graph to find the slope of the line. Note that the points (0, 10,000), (1, 8000), (2, 6000), (3, 4000), (4, 2000), and (5, 0) are all points on the graph. Points (0, 10,000) and (5, 0) are used here.

$(x_1, y_1) = (0, 10,000)$
$(x_2, y_2) = (5, 0)$
$m = \frac{y_2 - y_1}{x_2 - x_1} = \frac{0 - 10,000}{5 - 0} = \frac{-10,000}{5} = -2000$

Locate the y-intercept of the line on the graph.

The y-intercept is (0, 10,000).

Use the slope-intercept form of an equation to write the equation of the line.

$y = mx + b$
$y = -2000x + 10,000$

The equation of the line that represents the depreciation is $y = -2000x + 10,000$.

In the equation, the slope represents the annual depreciation of the computer system. The y-intercept represents the computers' value at the time of purchase.

Once the equation of the line has been found, the equation can be used to determine the depreciated value of the computer system after any given number of years.

To find the depreciated value of the computer after two and one-half years, substitute 2.5 for x in the equation and solve for y.

$y = -2000x + 10,000$
$y = -2000(2.5) + 10,000$
$y = -5000 + 10,000$
$y = 5000$

The depreciated value of the computer system after two and one-half years is $5000.

Example 1

The graph below shows the relationship between the total cost of manufacturing toasters and the number of toasters manufactured. Write the equation of the line that represents the total cost of manufacturing the toasters. Use the equation to find the total cost of manufacturing 300 toasters.

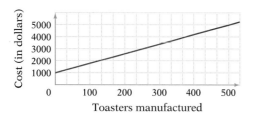

Example 2

The relationship between Fahrenheit and Celsius temperature is shown in the graph below. Write the equation for the Fahrenheit temperature in terms of the Celsius temperature. Use the equation to find the Fahrenheit temperature when the Celsius temperature is 40°.

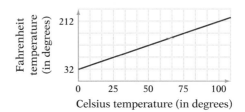

Strategy

To write the equation:

- Use two points on the graph to find the slope of the line.
- Locate the y-intercept of the line on the graph.
- Use the slope-intercept form of an equation to write the equation of the line.

To find the cost of manufacturing 300 toasters, substitute 300 for x in the equation and solve for y.

Your strategy

Solution

$(x_1, y_1) = (0, 1000)$ $(x_2, y_2) = (500, 5000)$

$$m = \frac{y_2 - y_1}{x_2 - x_1} = \frac{5000 - 1000}{500 - 0} = \frac{4000}{500} = 8$$

The y-intercept is (0, 1000).

In the equation, the slope represents the unit cost, or the cost to manufacture one toaster. The y-intercept represents the fixed costs of operating the plant.

$y = mx + b$
$y = 8x + 1000$

The equation of the line is $y = 8x + 1000$.

$y = 8x + 1000$
$y = 8(300) + 1000 = 2400 + 1000 = 3400$

The cost of manufacturing 300 toasters is $3400.

Your solution

$y = \frac{9}{5}x + 32$

104°F

Solution on p. A30

6.5   EXERCISES

▶ Objective A   *Application Problems*

1. The graph below represents the relationship between the amount invested and the annual income from an investment. Write the equation for the annual income in terms of the amount of the investment. Use the equation to find the annual income from a $4000 investment. *y = 0.08x*; $320

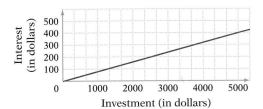

2. The graph below represents the relationship between the distance traveled by a motorist and the time of travel. Write the equation for the distance traveled in terms of the time of travel. Use the equation to find the distance traveled in three and one-half hours. *y = 50x*; 175 mi

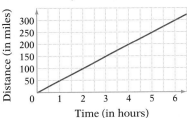

3. A milk tank has a capacity of 1000 cubic feet. The graph below shows the relationship between the amount of milk in the tank and the time it takes to fill the tank. Write the equation for the amount of milk in terms of the time. Use the equation to find the amount of milk in the tank after one hour. *y = 10x*; 600 gal

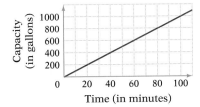

4. The graph below shows the relationship between the monthly income and the sales of an account executive. Write an equation for the line that represents the income of the account executive. Use the equation to find the monthly income when the monthly sales are $60,000. *y = 0.04x + 1000*; $3400

5. The relationship between the cost of a building and the depreciation allowed for income tax purposes is shown in the graph below. Write the equation for the line that represents the depreciated value of the building. Use the equation to find the value of the building after 12 years.

$y = -\frac{25{,}000}{3}x + 250{,}000$; $150,000

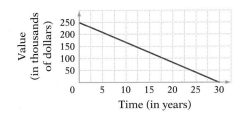

6. The graph below represents the relationship between a company's cost for new office equipment and the depreciation allowed for income tax purposes. Write the equation for the line that represents the depreciated value of the office equipment. Use the equation to find the depreciated value of the office equipment after two and one-half years. *y = −3125x + 25,000*; $17,187.50

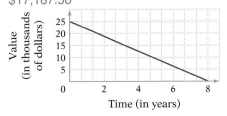

7. The relationship between the cost of manufacturing cameras and the number of cameras manufactured is shown in the graph below. Write the equation that represents the cost of manufacturing the cameras. Use the equation to find the cost of manufacturing 35 cameras.

$y = 120x + 3000$; $7200

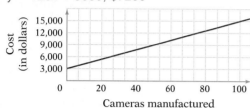

8. The graph below shows the relationship between the cost of skis and the number of skis manufactured. Write the equation that represents the cost of manufacturing the skis. Use the equation to find the cost of manufacturing 150 skis.

$y = 80x + 4000$; $16,000

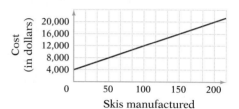

Economists frequently express the relationship between the price of a product and consumer demand for that product in a graph, called a demand curve. Demand curves are ofter linear.

9. Write the equation of the demand curve shown below. Use the equation to find the demand for the product when the price is $8 per unit.

$y = -\frac{1}{2}x + 20$; 24 units

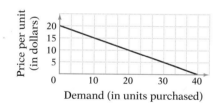

10. Write the equation of the demand curve shown below. Use the equation to find the demand for the product when the price is $25 per unit.

$y = -\frac{1}{5}x + 100$; 375 units

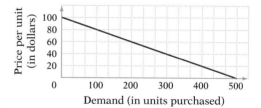

11. Write the equation of the demand curve shown below. Use the equation to find the demand for the product when the price is $3 per unit.

$y = -\frac{1}{4}x + 5$; 8 units

12. Write the equation of the demand curve shown below. Use the equation to find the demand for the product when the price is $10 per unit.

$y = -\frac{1}{8}x + 15$; 40 units

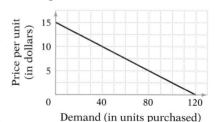

SECTION 6.6

Inequalities in Two Variables

Objective A

To graph the solution set of an inequality in two variables

The graph of the linear equation $y = x - 1$ separates the plane into three sets:

the set of points on the line,

the set of points above the line,

the set of points below the line.

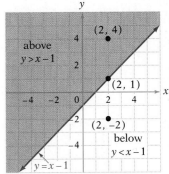

The point $(2, 1)$ is a solution of $y = x - 1$.

The point $(2, 4)$ is a solution of $y > x - 1$.

The point $(2, -2)$ is a solution of $y < x - 1$.

The solution set of $y = x - 1$ is all points on the line. The solution set of the linear inequality $y > x - 1$ is all points above the line. The solution set of the linear inequality $y < x - 1$ is all points below the line.

The solution set of an inequality in two variables is a **half-plane**.

The following illustrates the procedure for graphing a linear inequality.

Graph the solution set of $3x - 4y < 12$.

Solve the inequality for y.

$$3x - 4y < 12$$
$$-4y < -3x + 12$$
$$y > \tfrac{3}{4}x - 3$$
$$y = \tfrac{3}{4}x - 3$$

Change the inequality to an equality and graph the line. If the inequality is $\leq$ or $\geq$, the line is in the solution set and is shown by a **solid line**. If the inequality is $<$ or $>$, the line is not a part of the solution set and is shown by a **dotted line**.

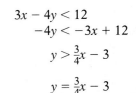

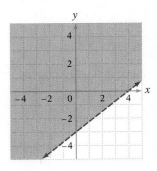

If the inequality is $>$ or $\geq$, shade the **upper half-plane**. If the inequality is $<$ or $\leq$, shade the **lower half-plane**.

As a check, the point $(0, 0)$ can be used to determine if the correct region of the plane has been shaded. If $(0, 0)$ is a solution of the inequality, then $(0, 0)$ should be in the shaded region. If $(0, 0)$ is not a solution of the inequality, then $(0, 0)$ should not be in the shaded region. Note in the above example that $(0, 0)$ is in the shaded region and that $(0, 0)$ is a solution of the inequality.

Example 1

Graph the solution set of $x + 2y \leq 4$.

Solution

$x + 2y \leq 4$

$\quad 2y \leq -x + 4$

$\quad\quad y \leq -\frac{1}{2}x + 2$

Graph $y = -\frac{1}{2}x + 2$
as a solid line.

Shade the lower half-plane.

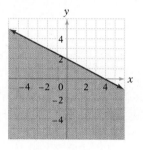

Example 2

Graph the solution set of $x + 3y > 6$.

Your solution

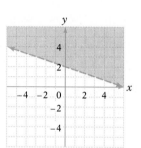

Example 3

Graph the solution set of $x \geq -1$.

Solution

Graph $x = -1$ as a solid line.

Shade the half-plane to the right of the line.

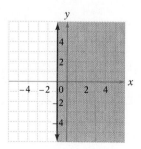

Example 4

Graph the solution set of $y < 2$.

Your solution

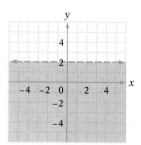

Solutions on p. A30

Content and Format © 1991 HMCo.

6.6   EXERCISES

▶ Objective A

Graph the solution set.

1. $3x - 2y \geq 6$

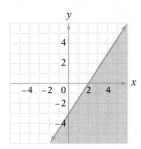

2. $4x - 3y \leq 12$

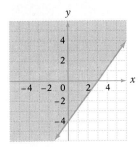

3. $x + 3y < 4$

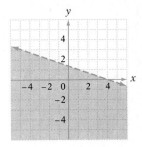

4. $x + 3y < 6$

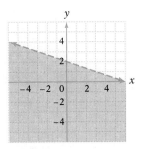

5. $2x - 5y \leq 10$

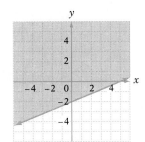

6. $2x + 3y \geq 6$

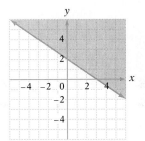

7. $3x - 5y > 15$

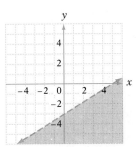

8. $4x - 5y > 10$

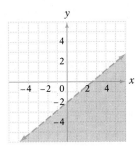

9. $4x + 3y < 9$

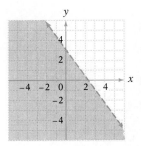

Graph the solution set.

10. $3x + 2y < 4$

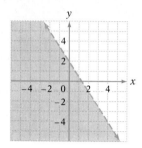

11. $-x + 2y > -8$

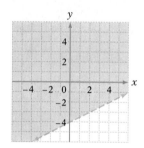

12. $-3x + 2y > 2$

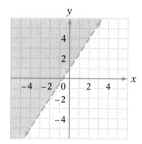

13. $y - 4 < 0$

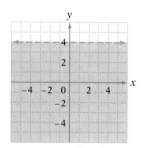

14. $x + 2 \geq 0$

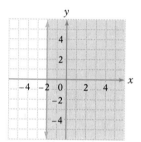

15. $6x + 5y < 15$

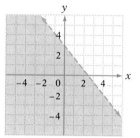

16. $3x - 5y < 10$

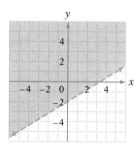

17. $-5x + 3y \geq -12$

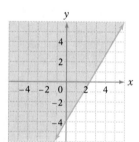

18. $3x + 4y \geq 12$

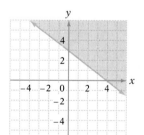

Calculators and Computers

Linear Equations in Two Variables

The program EQUATION OF A STRAIGHT LINE on the Math ACE Disk can be used to practice finding the equation of a line. Two types of problems are presented. One type asks you to find the equation of a line given the slope and a point on the line. The second type of problem is to find the equation of a line given two points on the line.

In each case, solve the problem using paper and pencil. When you are ready, press the RETURN key and the answer will be displayed. The equation of the line is always given in the form $Ax + By = C$.

You may practice as long as you like. When you wish to quit, press the letter 'Q' on the keyboard.

Chapter Summary

Key Words

A *rectangular coordinate system* is formed by two number lines, one horizontal and one vertical, that intersect at the zero point of each line.

The number lines that make up a coordinate system are called the *coordinate axes* or simply *axes*.

The *origin* is the point of intersection of the two coordinate axes.

A rectangular coordinate system divides the plane into four regions called *quadrants*.

An *ordered pair* (a, b) is used to locate a point in a plane.

The first number in an ordered pair is called the *abscissa*.

The second number in an ordered pair is called the *ordinate*.

The point at which a graph crosses the *x*-axis is called the *x-intercept*.

The point at which a graph crosses the *y*-axis is called the *y-intercept*.

An equation of the form $Ax + By = C$ is a *linear equation in two variables*.

The *graph* of a linear equation in two variables is a straight line.

The *slope* of a line is a measure of the slant or tilt of a line. The symbol for slope is *m*.

A line that slants upward to the right has a *positive slope*.

A line that slants downward to the right has a *negative slope*.

A horizontal line has *zero slope*.

The slope of a vertical line is *undefined*.

Two lines that have the same slope do not intersect and are called *parallel lines*.

Two lines that intersect at right angles are called *perpendicular lines*.

The solution set of an inequality in two variables is a *half-plane*.

Essential Rules   ***Slope of a straight line*** $\text{slope} = m = \frac{y_2 - y_1}{x_2 - x_1}$

Slope-intercept form of a straight line $y = mx + b$

Point-slope form of a straight line $y - y_1 = m(x - x_1)$

Chapter Review

SECTION 1

1. Graph the ordered pairs $(-2, 4)$ and $(4, -2)$. Draw a line between the two points.

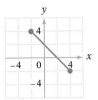

2. Graph the ordered pairs $(3, 5)$ and $(-3, 1)$. Draw a line between the two points.

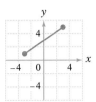

3. Find the ordered pair solution of $y = 3x - 5$ corresponding to $x = 2$.
$(2, 1)$

4. Find the ordered pair solution of $y = \frac{2x}{3} + 4$ corresponding to $x = -6$.
$(-6, 0)$

5. Graph: $y = -2x + 2$

6. Graph: $2x - 3y = -6$

SECTION 2

7. Find the slope of the line containing the points $(3, -2)$ and $(-1, 2)$.
-1

8. Find the slope of the line containing the points $(3, -2)$ and $(3, 5)$.
undefined

9. Graph $3x + 2y = -4$ by using the x-and the y-intercepts.

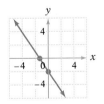

10. Graph the line that passes through the point $(-1, 4)$ and has slope $-\frac{1}{3}$.

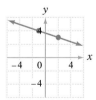

SECTION 3

11. Find the equation of the line that contains the point $(-3, 4)$ and has slope $\frac{5}{2}$.

$y = \frac{5}{2}x + \frac{23}{2}$

12. Find the equation of the line that contains the point $(3, 5)$ and has slope $-\frac{4}{3}$.

$y = -\frac{4}{3}x + 9$

13. Find the equation of the line that contains the points $(-2, 4)$ and $(4, -3)$.

$y = -\frac{7}{6}x + \frac{5}{3}$

14. Find the equation of the line that contains the points $(3, 4)$ and $(0, 4)$.

$y = 4$

SECTION 4

15. Find the equation of the line containing the point $(3, -2)$ and parallel to the line $y = -3x + 4$.

$y = -3x + 7$

16. Find the equation of the line containing the point $(-2, -4)$ and parallel to the line $2x - 3y = 4$.

$y = \frac{2}{3}x - \frac{8}{3}$

17. Find the equation of the line containing the point $(2, 5)$ and perpendicular to the line $y = -\frac{2}{3}x + 6$.

$y = \frac{3}{2}x + 2$

18. Find the equation of the line containing the point $(-3, -1)$ and perpendicular to the line $4x - 2y = 7$.

$y = -\frac{1}{2}x - \frac{5}{2}$

SECTION 5

19. The relationship between the cost of manufacturing calculators and the number of calculators manufactured is shown in the graph below. Write the equation that represents the cost of manufacturing the calculators. Use the equation to find the cost of manufacturing 125 calculators.

$y = 20x + 2000;\ \$4500$

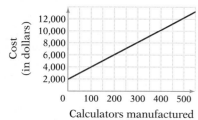

Calculators manufactured

20. The graph below shows the relationship between the distance traveled by a plane and the time of travel. Write an equation for the distance traveled in terms of the time of travel. Use the equation to find the distance traveled in two and one-half hours.

$y = 400x;\ 1000$ mi

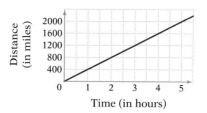

Time (in hours)

SECTION 6

21. Graph the solution set of $y \geq 2x - 3$.

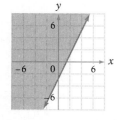

22. Graph the solution set of $3x - 2y < 6$.

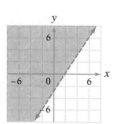

Chapter Test

1. Graph the ordered pairs (3, −4) and (4, −1). Draw a line between the two points.

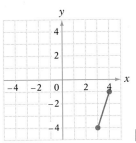

[6.1A]

2. Find the ordered pair solution of $y = 2x + 6$ corresponding to $x = -3$.

(−3, 0) [6.1B]

3. Graph: $y = \frac{2}{3}x - 4$

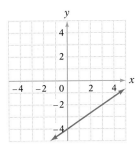

[6.1C]

4. Graph: $2x + 3y = -3$

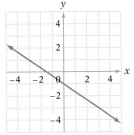

[6.1D]

5. Find the equation of the vertical line that contains the point (−2, 3).

$x = -2$ [6.3A]

6. Find the equation of the horizontal line that contains the point (4, −3).

$y = -3$ [6.3A]

7. Find the slope of the line containing the points (−2, 3) and (4, 2).

$-\frac{1}{6}$ [6.2A]

8. Find the slope of the line containing the points (2, 5) and (−2, 5).

0 [6.2A]

9. Graph $2x - 3y = 6$ by using the x- and the y-intercepts.

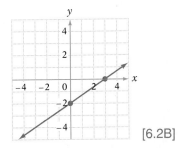

[6.2B]

10. Graph the line that passes through point (−2, 3) and has slope $-\frac{3}{2}$.

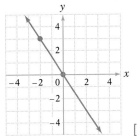

[6.2C]

11. Find the equation of the line that contains the point $(-5, 2)$ and has slope $\frac{2}{5}$.

$y = \frac{2}{5}x + 4$ [6.3A]

12. Find the equation of the line that contains the point $(0, -2)$ and has slope $-\frac{3}{4}$.

$y = -\frac{3}{4}x - 2$ [6.3A]

13. Find the equation of the line containing the points $(3, -4)$ and $(-2, 3)$.

$y = -\frac{7}{5}x + \frac{1}{5}$ [6.3B]

14. Find the equation of the line containing the points $(2, -3)$ and $(2, -5)$.

$x = 2$ [6.3B]

15. Find the equation of the line passing through $(2, 3)$ that is perpendicular to the vertical line $x = 5$.

$y = 3$ [6.4A]

16. Find the equation of the line containing the point $(1, 2)$ and parallel to the line $y = -\frac{3}{2}x - 6$.

$y = -\frac{3}{2}x + \frac{7}{2}$ [6.4A]

17. Find the equation of the line containing the point $(-2, -3)$ and perpendicular to the line $y = -\frac{1}{2}x - 3$.

$y = 2x + 1$ [6.4A]

18. Find the equation of the line containing the point $(-3, 4)$ and parallel to the line $2x + 3y = 9$.

$y = -\frac{2}{3}x + 2$ [6.4A]

19. The graph below shows the relationship between the cost of a rental house and the depreciation allowed for income tax purposes. Write the equation for the depreciation.

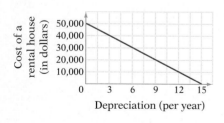

$y = -\frac{10,000}{3}x + 50,000$ [6.5A]

20. Graph the solution set of $3x - 4y > 8$.

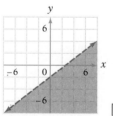

[6.6A]

Cumulative Review

1. Identify the property that justifies the statement $(x + y) \cdot 2 = 2 \cdot (x + y)$.
   Commutative Property of Multiplication
   [1.3B]

2. Solve: $3 - \frac{x}{2} = \frac{3}{4}$
   $x = \frac{9}{2}$ [2.1B]

3. Solve:
   $2[y - 2(3 - y) + 4] = 4 - 3y$
   $y = \frac{8}{9}$ [2.1C]

4. Solve:
   $\frac{1 - 3x}{2} + \frac{7x - 2}{6} = \frac{4x + 2}{9}$
   $x = -\frac{1}{14}$ [2.1C]

5. Solve:
   $x - 3 < -4$ or $2x + 2 > 3$
   $\left\{x \mid x < -1 \quad \text{or} \quad x > \frac{1}{2}\right\}$ [2.2B]

6. Solve:
   $8 - |2x - 1| = 4$
   $\frac{5}{2}$ and $-\frac{3}{2}$ [2.3A]

7. Solve: $|3x - 5| < 5$
   $\left\{x \mid 0 < x < \frac{10}{3}\right\}$ [2.3B]

8. Factor: $x^3 - 64y^3$
   $(x - 4y)(x^2 + 4xy + 16y^2)$ [3.4B]

9. Factor:
   $ab^2 - b^3 - a^3 + a^2b$
   $(a - b)(b - a)(b + a)$ [3.4D]

10. Simplify:
    $\frac{x^3 - y^3}{x^4 - x^2y^2} \div \frac{x^3 + x^2y + xy^2}{x^3 + x^2}$
    $\frac{x + 1}{x(x + y)}$ [4.1C]

11. Solve $P = \frac{R - C}{n}$ for C.
    $C = R - Pn$ [4.6B]

12. Simplify: $\frac{x - 2}{\sqrt{x} + 2}$
    $\frac{x\sqrt{x} - 2x - 2\sqrt{x} + 4}{x - 4}$ [5.2D]

13. Simplify:
    $y\sqrt{18x^5y^4} - x\sqrt{98x^3y^6}$
    $-4x^2y^3\sqrt{2x}$ [5.2B]

14. Simplify: $\frac{i}{2 - i}$
    $-\frac{1}{5} + \frac{2}{5}i$ [5.3D]

15. Find the ordered pair solution of $y = -\frac{5}{4}x + 3$ corresponding to $x = -8$.
    $(-8, 13)$ [6.1B]

16. Find the slope of the line containing the points $(-1, 3)$ and $(3, -4)$.
    $-\frac{7}{4}$ [6.2A]

17. Find the equation of the line that contains the point $(-1, 5)$ and has slope $\frac{3}{2}$.
    $y = \frac{3}{2}x + \frac{13}{2}$ [6.3A]

18. Find the equation of the line containing the points $(4, -2)$ and $(0, 3)$.
    $y = -\frac{5}{4}x + 3$ [6.3B]

19. Find the equation of the line containing the point (2, 4) and parallel to the line $y = -\frac{3}{2}x + 2$.

$y = -\frac{3}{2}x + 7$ [6.4A]

20. Find the equation of the line containing the point (4, 0) and perpendicular to the line $3x - 2y = 5$.

$y = -\frac{2}{3}x + \frac{8}{3}$ [6.4A]

21. A coin purse contains 17 coins with a value of $1.60. The purse contains nickels, dimes, and quarters. There are four times as many nickels as quarters. Find the number of dimes in the purse.

7 dimes [2.4B]

22. Two planes are 1800 mi apart and traveling toward each other. One plane is traveling twice as fast as the other plane. The planes meet in 3 h. Find the speed of each plane.

First plane: 200 mph
Second plane: 400 mph [2.5B]

23. A grocer combines coffee costing $8.00 per pound with coffee costing $3.00 per pound. How many pounds of each should be used to make 80 pounds of a blend costing $5.00 per pound?

32 lb of $8 coffee
48 lb of $3 coffee [2.5A]

24. Graph $3x - 5y = 15$ by using the x- and y-intercepts.

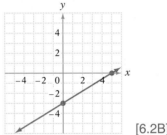

[6.2B]

25. Graph the line that passes through the point (−3, 1) and has slope $-\frac{3}{2}$.

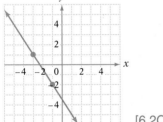

[6.2C]

26. Graph the inequality $3x - 2y \geq 6$.

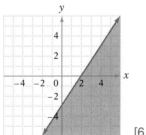

[6.6A]

27. The relationship between the cost of a truck and the depreciation allowed for income tax purposes is shown in the graph at the right. Write the equation for the line that represents the depreciated value of the truck. Use the equation to find the value of the truck after three and one-half years.

$y = -2500x + 15,000$
6250 [6.5A]

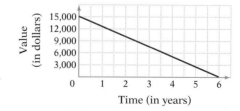

7

Quadratic Equations

OBJECTIVES

▶ To solve a quadratic equation by factoring

▶ To write a quadratic equation given its solutions

▶ To solve a quadratic equation by taking square roots

▶ To solve a quadratic equation by completing the square

▶ To solve a quadratic equation by using the quadratic formula

▶ To solve an equation that is quadratic in form

▶ To solve a radical equation that is reducible to a quadratic equation

▶ To solve a fractional equation that is reducible to a quadratic equation

▶ To solve a non-linear inequality

▶ To solve application problems

Complex Numbers

Negative numbers were not universally accepted in the mathematical community until well into the 14th century. It is no wonder then that *imaginary numbers* took an even longer time to gain acceptance.

Beginning in the mid-sixteenth century, mathematicians were beginning to integrate imaginary numbers into their writings. One notation for 3i was R̲ (0 m 3). Literally this was interpreted as $\sqrt{0-3}$.

By the mid-eighteenth century, the symbol i was introduced. Still later it was shown that complex numbers could be thought of as points in the plane. The complex number $3 + 4i$ was associated with the point (3, 4).

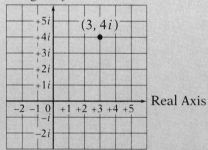

By the end of the nineteenth century, complex numbers were fully integrated into mathematics. This was due in large part to some eminent mathematicians who used complex numbers to prove theorems that had previously eluded proof.

SECTION 7.1	# Solving Quadratic Equations by Factoring or by Taking Square Roots

Objective A	**To solve a quadratic equation by factoring**

Recall from Chapter 3, a **quadratic equation** is an equation of the form $ax^2 + bx + c = 0$, where a, b, and c are constants and $a \neq 0$.

Quadratic Equations
$$\begin{cases} 3x^2 - x + 2 = 0, & a = 3, \quad b = -1, \quad c = 2 \\ -x^2 + 4 = 0, & a = -1, \quad b = 0, \quad c = 4 \\ 6x^2 - 5x = 0, & a = 6, \quad b = -5, \quad c = 0 \end{cases}$$

A quadratic equation is in **standard form** when the polynomial is in descending order and equal to zero. Since the degree of the polynomial $ax^2 + bx + c$ is 2, a quadratic equation is also called a **second-degree equation.**

As shown in Chapter 3, quadratic equations sometimes can be solved by using the Principle of Zero Products. This method is reviewed here.

The Principle of Zero Products

> If a and b are real numbers and $ab = 0$, then $a = 0$ or $b = 0$.

The Principle of Zero Products states that if the product of two factors is zero, then at least one of the factors must be zero.

Solve by factoring: $3x^2 = 2 - 5x$

$$3x^2 = 2 - 5x$$

Write the equation in standard form. $3x^2 + 5x - 2 = 0$

Factor. $(3x - 1)(x + 2) = 0$

Use the Principle of Zero Products to write two equations. $3x - 1 = 0 \qquad x + 2 = 0$

Solve each equation.
$$3x = 1 \qquad\qquad x = -2$$
$$x = \tfrac{1}{3}$$

Write the solutions. $\tfrac{1}{3}$ and -2 check as solutions.

The solutions are $\tfrac{1}{3}$ and -2.

Solve by factoring: $x^2 - 6x = -9$

$$x^2 - 6x = -9$$

Write the equation in standard form. $x^2 - 6x + 9 = 0$

Factor. $(x - 3)(x - 3) = 0$

Use the Principle of Zero Products. $x - 3 = 0 \qquad x - 3 = 0$

Solve each equation. $x = 3 \qquad\qquad x = 3$

Write the solutions.   3 checks as a solution.

The solution is 3.

When a quadratic equation has two solutions that are the same number, the solution is called a **double root** of the equation. The solution 3 is a double root of the equation $x^2 - 6x = -9$.

Example 1 Solve by factoring:
$2x(x - 3) = x + 4$

Solution

$$2x(x - 3) = x + 4$$
$$2x^2 - 6x = x + 4$$
$$2x^2 - 7x - 4 = 0$$
$$(2x + 1)(x - 4) = 0$$

$$2x + 1 = 0 \qquad x - 4 = 0$$
$$2x = -1 \qquad x = 4$$
$$x = -\frac{1}{2}$$

The solutions are $-\frac{1}{2}$ and 4.

Example 2 Solve by factoring:
$2x^2 = 7x - 3$

Your solution $\frac{1}{2}$ and 3

Example 3 Solve for x by factoring:
$x^2 - 4ax - 5a^2 = 0$

Solution This is a literal equation. Solve for x in terms of a.
$$x^2 - 4ax - 5a^2 = 0$$
$$(x + a)(x - 5a) = 0$$

$$x + a = 0 \qquad x - 5a = 0$$
$$x = -a \qquad x = 5a$$

The solutions are $-a$ and $5a$.

Example 4 Solve for x by factoring:
$x^2 - 3ax - 4a^2 = 0$

Your solution $-a$ and $4a$

Solutions on p. A31

Objective B   **To write a quadratic equation given its solutions**

7

As shown below, the solutions of the equation $(x - r_1)(x - r_2) = 0$ are r_1 and r_2.

$$(x - r_1)(x - r_2) = 0$$

$$x - r_1 = 0 \qquad\qquad x - r_2 = 0$$
$$x = r_1 \qquad\qquad x = r_2$$

Check:
$$\begin{array}{c|c} (x - r_1)(x - r_2) = 0 & (x - r_1)(x - r_2) = 0 \\ \hline (r_1 - r_1)(r_1 - r_2) & (r_2 - r_1)(r_2 - r_2) \\ 0 \cdot (r_1 - r_2) & (r_2 - r_1) \cdot 0 \\ \qquad 0 = 0 & \qquad 0 = 0 \end{array}$$

Using the equation $(x - r_1)(x - r_2) = 0$ and the fact that r_1 and r_2 are solutions of this equation, it is possible to write a quadratic equation given its solutions.

Write a quadratic equation that has solutions 4 and -5.

$$(x - r_1)(x - r_2) = 0$$
Replace r_1 by 4 and r_2 by -5. $\quad (x - 4)[x - (-5)] = 0$

Simplify. $\qquad\qquad\qquad\qquad\qquad (x - 4)(x + 5) = 0$

Multiply. $\qquad\qquad\qquad\qquad\qquad x^2 + x - 20 = 0$

Write a quadratic equation that has integer coefficients and has solutions $\frac{2}{3}$ and $\frac{1}{2}$.

Replace r_1 by $\frac{2}{3}$ and r_2 by $\frac{1}{2}$.

$$(x - r_1)(x - r_2) = 0$$
$$\left(x - \tfrac{2}{3}\right)\left(x - \tfrac{1}{2}\right) = 0$$

Multiply.

$$x^2 - \tfrac{7}{6}x + \tfrac{1}{3} = 0$$

Multiply each side of the equation by the LCM of the denominators.

$$6\left(x^2 - \tfrac{7}{6}x + \tfrac{1}{3}\right) = 6 \cdot 0$$
$$6x^2 - 7x + 2 = 0$$

Example 5   Write a quadratic equation that has integer coefficients and has solutions $\frac{1}{2}$ and -4.

Solution

$$(x - r_1)(x - r_2) = 0$$
$$\left(x - \tfrac{1}{2}\right)[x - (-4)] = 0$$
$$\left(x - \tfrac{1}{2}\right)(x + 4) = 0$$
$$x^2 + \tfrac{7}{2}x - 2 = 0$$
$$2\left(x^2 + \tfrac{7}{2}x - 2\right) = 2 \cdot 0$$
$$2x^2 + 7x - 4 = 0$$

Example 6   Write a quadratic equation that has integer coefficients and has solutions 3 and $-\frac{1}{2}$.

Your solution $2x^2 - 5x - 3 = 0$

Solution on p. A31

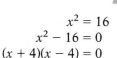

Objective C   **To solve a quadratic equation by taking square roots**

The solution of the quadratic equation $x^2 = 16$ is shown at the right.

$$x^2 = 16$$
$$x^2 - 16 = 0$$
$$(x + 4)(x - 4) = 0$$

Note that the solution is the positive or the negative square root of 16, 4 or -4.

$$x + 4 = 0 \qquad x - 4 = 0$$
$$x = -4 \qquad\quad x = 4$$

The solution can also be found by taking the square root of each side of the equation and writing the positive and the negative square roots of the number. The notation $x = \pm 4$ means $x = 4$ or $x = -4$.

$$x^2 = 16$$
$$\sqrt{x^2} = \sqrt{16}$$
$$x = \pm\sqrt{16} = \pm 4$$

The solutions are 4 and -4.

Solve by taking square roots: $3x^2 = 54$

Solve for x^2.

$$3x^2 = 54$$
$$x^2 = 18$$

Take the square root of each side of the equation.

$$\sqrt{x^2} = \sqrt{18}$$

Simplify.

$$x = \pm\sqrt{18} = \pm 3\sqrt{2}$$

Write the solutions.

The solutions are $3\sqrt{2}$ and $-3\sqrt{2}$.

$3\sqrt{2}$ and $-3\sqrt{2}$ check as solutions.

Solving a quadratic equation by taking the square root of each side of the equation can lead to solutions which are complex numbers.

Solve by taking square roots: $2x^2 + 18 = 0$

Solve for x^2.

$$2x^2 + 18 = 0$$
$$2x^2 = -18$$
$$x^2 = -9$$

Take the square root of each side of the equation.

$$\sqrt{x^2} = \sqrt{-9}$$

Simplify.

$$x = \pm\sqrt{-9} = \pm 3i$$

Write the solutions.

The solutions are $3i$ and $-3i$.

$3i$ and $-3i$ check as solutions.

An equation containing the square of a binomial can be solved by taking square roots.

Solve by taking square roots: $(x + 2)^2 - 24 = 0$

Solve for $(x + 2)^2$.

$$(x + 2)^2 - 24 = 0$$
$$(x + 2)^2 = 24$$

Take the square root of each side of the equation.

$$\sqrt{(x + 2)^2} = \sqrt{24}$$

Simplify.

$$x + 2 = \pm\sqrt{24} = \pm 2\sqrt{6}$$

Solve for x.

$$x + 2 = 2\sqrt{6} \qquad\qquad x + 2 = -2\sqrt{6}$$
$$x = -2 + 2\sqrt{6} \qquad\qquad x = -2 - 2\sqrt{6}$$

Write the solutions.

The solutions are $-2 + 2\sqrt{6}$ and $-2 - 2\sqrt{6}$.

$-2 + 2\sqrt{6}$ and $-2 - 2\sqrt{6}$ check as solutions.

Example 7

Solve by taking square roots:
$3(x - 2)^2 + 12 = 0$

Solution

$$3(x - 2)^2 + 12 = 0$$
$$3(x - 2)^2 = -12$$
$$(x - 2)^2 = -4$$
$$\sqrt{(x - 2)^2} = \sqrt{-4}$$
$$x - 2 = \pm\sqrt{-4} = \pm 2i$$

$$x - 2 = 2i \qquad\qquad x - 2 = -2i$$
$$x = 2 + 2i \qquad\qquad x = 2 - 2i$$

The solutions are $2 + 2i$ and $2 - 2i$.

Example 8

Solve by taking square roots:
$2(x + 1)^2 - 24 = 0$

Your solution

$-1 + 2\sqrt{3}$ and $-1 - 2\sqrt{3}$

Solution on p. A31

7.1 EXERCISES

▶ **Objective A**

Solve by factoring.

1. $x^2 - 4x = 0$
0 and 4

2. $y^2 + 6y = 0$
0 and −6

3. $t^2 - 25 = 0$
5 and −5

4. $p^2 - 81 = 0$
9 and −9

5. $s^2 - s - 6 = 0$
3 and −2

6. $v^2 + 4v - 5 = 0$
1 and −5

7. $y^2 - 6y + 9 = 0$
3

8. $x^2 + 10x + 25 = 0$
−5

9. $9z^2 - 18z = 0$
0 and 2

10. $4y^2 + 20y = 0$
0 and −5

11. $r^2 - 3r = 10$
5 and −2

12. $p^2 + 5p = 6$
1 and −6

13. $v^2 + 10 = 7v$
2 and 5

14. $t^2 - 16 = 15t$
16 and −1

15. $2x^2 - 9x - 18 = 0$
6 and $-\frac{3}{2}$

16. $3y^2 - 4y - 4 = 0$
$-\frac{2}{3}$ and 2

17. $4z^2 - 9z + 2 = 0$
2 and $\frac{1}{4}$

18. $2s^2 - 9s + 9 = 0$
$\frac{3}{2}$ and 3

19. $3w^2 + 11w = 4$
$\frac{1}{3}$ and −4

20. $2r^2 + r = 6$
$\frac{3}{2}$ and −2

21. $6x^2 = 23x + 18$
$\frac{9}{2}$ and $-\frac{2}{3}$

22. $6x^2 = 7x - 2$
$\frac{2}{3}$ and $\frac{1}{2}$

23. $4 - 15u - 4u^2 = 0$
$\frac{1}{4}$ and −4

24. $3 - 2y - 8y^2 = 0$
$-\frac{3}{4}$ and $\frac{1}{2}$

25. $x + 18 = x(x - 6)$
9 and −2

26. $t + 24 = t(t + 6)$
3 and −8

27. $4s(s + 3) = s - 6$
−2 and $-\frac{3}{4}$

28. $3v(v - 2) = 11v + 6$
$-\frac{1}{3}$ and 6

29. $u^2 - 2u + 4 = (2u - 3)(u + 2)$
2 and −5

30. $(3v - 2)(2v + 1) = 3v^2 - 11v - 10$
−2 and $-\frac{4}{3}$

31. $(3x - 4)(x + 4) = x^2 - 3x - 28$
−4 and $-\frac{3}{2}$

Solve for x by factoring.

32. $x^2 + 14ax + 48a^2 = 0$
  $-6a$ and $-8a$

33. $x^2 - 9bx + 14b^2 = 0$
  $2b$ and $7b$

34. $x^2 + 9xy - 36y^2 = 0$
  $-12y$ and $3y$

35. $x^2 - 6cx - 7c^2 = 0$
  $7c$ and $-c$

36. $x^2 - ax - 20a^2 = 0$
  $-4a$ and $5a$

37. $2x^2 + 3bx + b^2 = 0$
  $-\frac{b}{2}$ and $-b$

38. $3x^2 - 4cx + c^2 = 0$
  $\frac{c}{3}$ and c

39. $3x^2 - 14ax + 8a^2 = 0$
  $4a$ and $\frac{2a}{3}$

40. $3x^2 - 11xy + 6y^2 = 0$
  $\frac{2y}{3}$ and $3y$

41. $3x^2 - 8ax - 3a^2 = 0$
  $-\frac{a}{3}$ and $3a$

42. $3x^2 - 4bx - 4b^2 = 0$
  $-\frac{2b}{3}$ and $2b$

43. $4x^2 + 8xy + 3y^2 = 0$
  $-\frac{3y}{2}$ and $-\frac{y}{2}$

44. $6x^2 - 11cx + 3c^2 = 0$
  $\frac{c}{3}$ and $\frac{3c}{2}$

45. $6x^2 + 11ax + 4a^2 = 0$
  $-\frac{a}{2}$ and $-\frac{4a}{3}$

46. $12x^2 - 5xy - 2y^2 = 0$
  $-\frac{y}{4}$ and $\frac{2y}{3}$

▶ Objective B

Write a quadratic equation that has integer coefficients and has as solutions the given pair of numbers.

47. 2 and 5
  $x^2 - 7x + 10 = 0$

48. 3 and 1
  $x^2 - 4x + 3 = 0$

49. -2 and -4
  $x^2 + 6x + 8 = 0$

50. -1 and -3
  $x^2 + 4x + 3 = 0$

51. 6 and -1
  $x^2 - 5x - 6 = 0$

52. -2 and 5
  $x^2 - 3x - 10 = 0$

53. 3 and -3
  $x^2 - 9 = 0$

54. 5 and -5
  $x^2 - 25 = 0$

55. 4 and 4
  $x^2 - 8x + 16 = 0$

56. 2 and 2
  $x^2 - 4x + 4 = 0$

57. 0 and 5
  $x^2 - 5x = 0$

58. 0 and -2
  $x^2 + 2x = 0$

59. 0 and 3
  $x^2 - 3x = 0$

60. 0 and -1
  $x^2 + x = 0$

61. 3 and $\frac{1}{2}$
  $2x^2 - 7x + 3 = 0$

Write a quadratic equation that has integer coefficients and has as solutions the given pair of numbers.

62. 2 and $\frac{2}{3}$

$3x^2 - 8x + 4 = 0$

63. $-\frac{3}{4}$ and 2

$4x^2 - 5x - 6 = 0$

64. $-\frac{1}{2}$ and 5

$2x^2 - 9x - 5 = 0$

65. $-\frac{5}{3}$ and -2

$3x^2 + 11x + 10 = 0$

66. $-\frac{3}{2}$ and -1

$2x^2 + 5x + 3 = 0$

67. $-\frac{2}{3}$ and $\frac{2}{3}$

$9x^2 - 4 = 0$

68. $-\frac{1}{2}$ and $\frac{1}{2}$

$4x^2 - 1 = 0$

69. $\frac{1}{2}$ and $\frac{1}{3}$

$6x^2 - 5x + 1 = 0$

70. $\frac{3}{4}$ and $\frac{2}{3}$

$12x^2 - 17x + 6 = 0$

71. $\frac{6}{5}$ and $-\frac{1}{2}$

$10x^2 - 7x - 6 = 0$

72. $\frac{3}{4}$ and $-\frac{3}{2}$

$8x^2 + 6x - 9 = 0$

73. $-\frac{1}{4}$ and $-\frac{1}{2}$

$8x^2 + 6x + 1 = 0$

74. $-\frac{5}{6}$ and $-\frac{2}{3}$

$18x^2 + 27x + 10 = 0$

75. $\frac{3}{5}$ and $-\frac{1}{10}$

$50x^2 - 25x - 3 = 0$

76. $\frac{7}{2}$ and $-\frac{1}{4}$

$8x^2 - 26x - 7 = 0$

▶ **Objective C**

Solve by taking square roots.

77. $y^2 = 49$

7 and -7

78. $x^2 = 64$

8 and -8

79. $z^2 = -4$

$2i$ and $-2i$

80. $v^2 = -16$

$4i$ and $-4i$

81. $s^2 - 4 = 0$

2 and -2

82. $r^2 - 36 = 0$

6 and -6

83. $4x^2 - 81 = 0$

$\frac{9}{2}$ and $-\frac{9}{2}$

84. $9x^2 - 16 = 0$

$\frac{4}{3}$ and $-\frac{4}{3}$

85. $y^2 + 49 = 0$

$7i$ and $-7i$

86. $z^2 + 16 = 0$

$4i$ and $-4i$

87. $v^2 - 48 = 0$

$4\sqrt{3}$ and $-4\sqrt{3}$

88. $s^2 - 32 = 0$

$4\sqrt{2}$ and $-4\sqrt{2}$

89. $r^2 - 75 = 0$

$5\sqrt{3}$ and $-5\sqrt{3}$

90. $u^2 - 54 = 0$

$3\sqrt{6}$ and $-3\sqrt{6}$

91. $z^2 + 18 = 0$

$3i\sqrt{2}$ and $-3i\sqrt{2}$

Solve by taking square roots.

92. $t^2 + 27 = 0$
$3i\sqrt{3}$ and $-3i\sqrt{3}$

93. $(x - 1)^2 = 36$
7 and -5

94. $(x + 2)^2 = 25$
3 and -7

95. $3(y + 3)^2 = 27$
0 and -6

96. $4(s - 2)^2 = 36$
5 and -1

97. $5(z + 2)^2 = 125$
-7 and 3

98. $2(y - 3)^2 = 18$
6 and 0

99. $\left(v - \frac{1}{2}\right)^2 = \frac{1}{4}$
1 and 0

100. $\left(r + \frac{2}{3}\right)^2 = \frac{1}{9}$
$-\frac{1}{3}$ and -1

101. $(x + 5)^2 - 6 = 0$
$-5 + \sqrt{6}$ and $-5 - \sqrt{6}$

102. $(t - 1)^2 - 15 = 0$
$1 + \sqrt{15}$ and $1 - \sqrt{15}$

103. $(s - 2)^2 - 24 = 0$
$2 + 2\sqrt{6}$ and $2 - 2\sqrt{6}$

104. $(y + 3)^2 - 18 = 0$
$-3 + 3\sqrt{2}$ and $-3 - 3\sqrt{2}$

105. $(z + 1)^2 + 12 = 0$
$-1 + 2i\sqrt{3}$ and $-1 - 2i\sqrt{3}$

106. $(r - 2)^2 + 28 = 0$
$2 + 2i\sqrt{7}$ and $2 - 2i\sqrt{7}$

107. $(v - 3)^2 + 45 = 0$
$3 + 3i\sqrt{5}$ and $3 - 3i\sqrt{5}$

108. $(x + 5)^2 + 32 = 0$
$-5 + 4i\sqrt{2}$ and $-5 - 4i\sqrt{2}$

109. $\left(u + \frac{2}{3}\right)^2 - 18 = 0$
$-\frac{2 - 9\sqrt{2}}{3}$ and $-\frac{2 + 9\sqrt{2}}{3}$

110. $\left(z - \frac{1}{2}\right)^2 - 20 = 0$
$\frac{1 + 4\sqrt{5}}{2}$ and $\frac{1 - 4\sqrt{5}}{2}$

111. $\left(t - \frac{3}{4}\right)^2 - 27 = 0$
$\frac{3 + 12\sqrt{3}}{4}$ and $\frac{3 - 12\sqrt{3}}{4}$

112. $\left(y + \frac{2}{5}\right)^2 - 72 = 0$
$-\frac{2 - 30\sqrt{2}}{5}$ and $-\frac{2 + 30\sqrt{2}}{5}$

113. $\left(x + \frac{1}{2}\right)^2 + 40 = 0$
$-\frac{1}{2} + 2i\sqrt{10}$ and $-\frac{1}{2} - 2i\sqrt{10}$

114. $\left(r - \frac{3}{2}\right)^2 + 48 = 0$
$\frac{3}{2} + 4i\sqrt{3}$ and $\frac{3}{2} - 4i\sqrt{3}$

| **SECTION 7.2** | # Solving Quadratic Equations by Completing the Square |

| **Objective A** | **To solve a quadratic equation by completing the square** |

Recall that a perfect square trinomial is the square of a binomial.

Perfect Square Trinomial		**Square of a Binomial**
$x^2 + 8x + 16$	$=$	$(x + 4)^2$
$x^2 - 10x + 25$	$=$	$(x - 5)^2$
$x^2 + 2ax + a^2$	$=$	$(x + a)^2$

For each perfect square trinomial, the square of $\frac{1}{2}$ of the coefficient of x equals the constant term.

$$x^2 + 8x + 16, \quad \left(\frac{1}{2} \cdot 8\right)^2 = 16$$

$$x^2 - 10x + 25, \quad \left[\frac{1}{2}(-10)\right]^2 = 25$$

$$x^2 + 2ax + a^2, \quad \left(\frac{1}{2} \cdot 2a\right)^2 = a^2$$

$$\left(\tfrac{1}{2} \text{ coefficient of x}\right)^2 = \text{constant term}$$

This relationship can be used to write the constant term for a perfect square trinomial. Adding to a binomial the constant term which makes it a perfect square trinomial is called **completing the square.**

Complete the square on $x^2 - 12x$. Write the resulting perfect square trinomial as the square of a binomial.

Find the constant term.

$$\left[\tfrac{1}{2}(-12)\right]^2 = (-6)^2 = 36$$

Complete the square on $x^2 - 12x$ by adding the constant term.

$$x^2 - 12x + 36$$

Write the resulting perfect square trinomial as the square of a binomial.

$$x^2 - 12x + 36 = (x - 6)^2$$

Complete the square on $z^2 + 3z$. Write the resulting perfect square trinomial as the square of a binomial.

Find the constant term.

$$\left(\tfrac{1}{2} \cdot 3\right)^2 = \left(\tfrac{3}{2}\right)^2 = \tfrac{9}{4}$$

Complete the square on $z^2 + 3z$ by adding the constant term.

$$z^2 + 3z + \tfrac{9}{4}$$

Write the resulting perfect square trinomial as the square of a binomial.

$$z^2 + 3z + \tfrac{9}{4} = \left(z + \tfrac{3}{2}\right)^2$$

Any quadratic equation can be solved by completing the square. Add to each side of the equation the term which completes the square. Rewrite the equation in the form $(x + a)^2 = b$. Take the square root of each side of the equation.

Solve by completing the square: $x^2 - 4x - 14 = 0$

$$x^2 - 4x - 14 = 0$$

Add the opposite of the constant term to each side of the equation.

$$x^2 - 4x = 14$$

Find the constant term which completes the square on $x^2 - 4x$.

$$\left[\tfrac{1}{2}(-4)\right]^2 = (-2)^2 = 4$$ Do this step mentally.

Add this term to each side of the equation.

$$x^2 - 4x + 4 = 14 + 4$$

Factor the perfect square trinomial.

$$(x - 2)^2 = 18$$

Take the square root of each side of the equation.

$$\sqrt{(x - 2)^2} = \sqrt{18}$$

Simplify.

$$x - 2 = \pm\sqrt{18} = \pm 3\sqrt{2}$$

Solve for x.

$$x - 2 = 3\sqrt{2} \qquad x - 2 = -3\sqrt{2}$$
$$x = 2 + 3\sqrt{2} \qquad x = 2 - 3\sqrt{2}$$

Write the solutions.

The solutions are $2 + 3\sqrt{2}$ and $2 - 3\sqrt{2}$.

Check:

$$x^2 - 4x - 14 = 0$$
$$\overline{(2 + 3\sqrt{2})^2 - 4(2 + 3\sqrt{2}) - 14 \;\Big|}$$
$$4 + 12\sqrt{2} + 18 - 8 - 12\sqrt{2} - 14 \;\Big|$$
$$0 = 0$$

$$x^2 - 4x - 14 = 0$$
$$\overline{(2 - 3\sqrt{2})^2 - 4(2 - 3\sqrt{2}) - 14 \;\Big|}$$
$$4 - 12\sqrt{2} + 18 - 8 + 12\sqrt{2} - 14 \;\Big|$$
$$0 = 0$$

Solve by completing the square: $2x^2 - x - 2 = 0$

$$2x^2 - x - 2 = 0$$

Add the opposite of the constant term to each side of the equation.

$$2x^2 - x = 2$$

To complete the square, the coefficient of the x^2 term must be 1. Multiply each side of the equation by the reciprocal of the coefficient of x^2.

$$\tfrac{1}{2}(2x^2 - x) = \tfrac{1}{2} \cdot 2$$
$$x^2 - \tfrac{1}{2}x = 1$$

Content and Format © 1991 HMCo.

Add the term which completes the square on $x^2 - \frac{1}{2}x$ to each side of the equation.

$$x^2 - \frac{1}{2}x + \frac{1}{16} = 1 + \frac{1}{16}$$

Factor the perfect square trinomial.

$$\left(x - \frac{1}{4}\right)^2 = \frac{17}{16}$$

Take the square root of each side of the equation.

$$\sqrt{\left(x - \frac{1}{4}\right)^2} = \sqrt{\frac{17}{16}}$$

Simplify.

$$x - \frac{1}{4} = \pm\frac{\sqrt{17}}{4}$$

Solve for x.

$$x - \frac{1}{4} = \frac{\sqrt{17}}{4} \qquad x - \frac{1}{4} = -\frac{\sqrt{17}}{4}$$

$$x = \frac{1}{4} + \frac{\sqrt{17}}{4} \qquad x = \frac{1}{4} - \frac{\sqrt{17}}{4}$$

Write the solutions.

The solutions are $\frac{1 + \sqrt{17}}{4}$ and $\frac{1 - \sqrt{17}}{4}$.

$\frac{1 + \sqrt{17}}{4}$ and $\frac{1 - \sqrt{17}}{4}$ check as solutions.

Completing the square can also be used to find complex number solutions of a quadratic equation.

Solve by completing the square: $x^2 - 3x + 5 = 0$

$$x^2 - 3x + 5 = 0$$

Add the opposite of the constant term to each side of the equation.

$$x^2 - 3x = -5$$

Add the term which completes the square on $x^2 - 3x$ to each side of the equation.

$$x^2 - 3x + \frac{9}{4} = -5 + \frac{9}{4}$$

Factor the perfect square trinomial.

$$\left(x - \frac{3}{2}\right)^2 = -\frac{11}{4}$$

Take the square root of each side of the equation.

$$\sqrt{\left(x - \frac{3}{2}\right)^2} = \sqrt{-\frac{11}{4}}$$

Simplify.

$$x - \frac{3}{2} = \pm\frac{i\sqrt{11}}{2}$$

Solve for x.

$$x - \frac{3}{2} = \frac{i\sqrt{11}}{2} \qquad x - \frac{3}{2} = -\frac{i\sqrt{11}}{2}$$

$$x = \frac{3}{2} + \frac{i\sqrt{11}}{2} \qquad x = \frac{3}{2} - \frac{i\sqrt{11}}{2}$$

Write the solutions.

The solutions are $\frac{3}{2} + \frac{\sqrt{11}}{2}i$ and $\frac{3}{2} - \frac{\sqrt{11}}{2}i$.

$\frac{3}{2} + \frac{\sqrt{11}}{2}i$ and $\frac{3}{2} - \frac{\sqrt{11}}{2}i$ check as solutions.

Example 1 Solve by completing the square: $4x^2 - 8x + 1 = 0$

Solution $4x^2 - 8x + 1 = 0$

$4x^2 - 8x = -1$

$\frac{1}{4}(4x^2 - 8x) = \frac{1}{4}(-1)$

$x^2 - 2x = -\frac{1}{4}$

Complete the square.

$x^2 - 2x + 1 = -\frac{1}{4} + 1$

$(x - 1)^2 = \frac{3}{4}$

$\sqrt{(x - 1)^2} = \sqrt{\frac{3}{4}}$

$x - 1 = \pm \frac{\sqrt{3}}{2}$

$x - 1 = \frac{\sqrt{3}}{2} \qquad x - 1 = -\frac{\sqrt{3}}{2}$

$x = 1 + \frac{\sqrt{3}}{2} \qquad x = 1 - \frac{\sqrt{3}}{2}$

The solutions are $\frac{2 + \sqrt{3}}{2}$ and $\frac{2 - \sqrt{3}}{2}$.

Example 2 Solve by completing the square: $4x^2 - 4x - 1 = 0$

Your solution $\frac{1 + \sqrt{2}}{2}$ and $\frac{1 - \sqrt{2}}{2}$

Example 3 Solve by completing the square: $x^2 + 4x + 5 = 0$

Solution $x^2 + 4x + 5 = 0$

$x^2 + 4x = -5$

Complete the square.

$x^2 + 4x + 4 = -5 + 4$

$(x + 2)^2 = -1$

$\sqrt{(x + 2)^2} = \sqrt{-1}$

$x + 2 = \pm i$

$x + 2 = i \qquad x + 2 = -i$

$x = -2 + i \qquad x = -2 - i$

The solutions are $-2 + i$ and $-2 - i$.

Example 4 Solve by completing the square: $2x^2 + x - 5 = 0$

Your solution $\frac{-1 + \sqrt{41}}{4}$ and $\frac{-1 - \sqrt{41}}{4}$

Solutions on p. A31

7.2   EXERCISES

▶ **Objective A**

Solve by completing the square.

1. $x^2 - 4x - 5 = 0$
   5 and -1

2. $y^2 + 6y + 5 = 0$
   -1 and -5

3. $v^2 + 8v - 9 = 0$
   -9 and 1

4. $w^2 - 2w - 24 = 0$
   6 and -4

5. $z^2 - 6z + 9 = 0$
   3

6. $u^2 + 10u + 25 = 0$
   -5

7. $r^2 + 4r - 7 = 0$
   $-2 + \sqrt{11}$ and $-2 - \sqrt{11}$

8. $s^2 + 6s - 1 = 0$
   $-3 + \sqrt{10}$ and $-3 - \sqrt{10}$

9. $x^2 - 6x + 7 = 0$
   $3 + \sqrt{2}$ and $3 - \sqrt{2}$

10. $y^2 + 8y + 13 = 0$
    $-4 + \sqrt{3}$ and $-4 - \sqrt{3}$

11. $z^2 - 2z + 2 = 0$
    $1 + i$ and $1 - i$

12. $t^2 - 4t + 8 = 0$
    $2 + 2i$ and $2 - 2i$

13. $s^2 - 5s - 24 = 0$
    8 and -3

14. $v^2 + 7v - 44 = 0$
    4 and -11

15. $x^2 + 5x - 36 = 0$
    4 and -9

16. $y^2 - 9y + 20 = 0$
    5 and 4

17. $p^2 - 3p + 1 = 0$
    $\frac{3 + \sqrt{5}}{2}$ and $\frac{3 - \sqrt{5}}{2}$

18. $r^2 - 5r - 2 = 0$
    $\frac{5 + \sqrt{33}}{2}$ and $\frac{5 - \sqrt{33}}{2}$

19. $t^2 - t - 1 = 0$
    $\frac{1 + \sqrt{5}}{2}$ and $\frac{1 - \sqrt{5}}{2}$

20. $u^2 - u - 7 = 0$
    $\frac{1 + \sqrt{29}}{2}$ and $\frac{1 - \sqrt{29}}{2}$

21. $y^2 - 6y = 4$
    $3 + \sqrt{13}$ and $3 - \sqrt{13}$

22. $w^2 + 4w = 2$
    $-2 + \sqrt{6}$ and $-2 - \sqrt{6}$

23. $x^2 = 8x - 15$
    5 and 3

24. $z^2 = 4z - 3$
    3 and 1

25. $v^2 = 4v - 13$
    $2 + 3i$ and $2 - 3i$

26. $x^2 = 2x - 17$
    $1 + 4i$ and $1 - 4i$

27. $p^2 + 6p = -13$
    $-3 + 2i$ and $-3 - 2i$

28. $x^2 + 4x = -20$
    $-2 + 4i$ and $-2 - 4i$

29. $y^2 - 2y = 17$
    $1 + 3\sqrt{2}$ and $1 - 3\sqrt{2}$

30. $x^2 + 10x = 7$
    $-5 + 4\sqrt{2}$ and $-5 - 4\sqrt{2}$

31. $z^2 = z + 4$
    $\frac{1 + \sqrt{17}}{2}$ and $\frac{1 - \sqrt{17}}{2}$

32. $r^2 = 3r - 1$
    $\frac{3 + \sqrt{5}}{2}$ and $\frac{3 - \sqrt{5}}{2}$

33. $x^2 + 13 = 2x$
    $1 + 2i\sqrt{3}$ and $1 - 2i\sqrt{3}$

Solve by completing the square.

34. $x^2 + 27 = 6x$
$3 + 3i\sqrt{2}$ and $3 - 3i\sqrt{2}$

35. $2y^2 + 3y + 1 = 0$
-1 and $-\frac{1}{2}$

36. $2t^2 + 5t - 3 = 0$
$\frac{1}{2}$ and -3

37. $4r^2 - 8r = -3$
$\frac{1}{2}$ and $\frac{3}{2}$

38. $4u^2 - 20u = -9$
$\frac{9}{2}$ and $\frac{1}{2}$

39. $6y^2 - 5y = 4$
$\frac{4}{3}$ and $-\frac{1}{2}$

40. $6v^2 - 7v = 3$
$\frac{3}{2}$ and $-\frac{1}{3}$

41. $4x^2 - 4x + 5 = 0$
$\frac{1}{2} + i$ and $\frac{1}{2} - i$

42. $4t^2 - 4t + 17 = 0$
$\frac{1}{2} + 2i$ and $\frac{1}{2} - 2i$

43. $9x^2 - 6x + 2 = 0$
$\frac{1}{3} + \frac{1}{3}i$ and $\frac{1}{3} - \frac{1}{3}i$

44. $9y^2 - 12y + 13 = 0$
$\frac{2}{3} + i$ and $\frac{2}{3} - i$

45. $2s^2 = 4s + 5$
$\frac{2 + \sqrt{14}}{2}$ and $\frac{2 - \sqrt{14}}{2}$

46. $3u^2 = 6u + 1$
$\frac{3 + 2\sqrt{3}}{3}$ and $\frac{3 - 2\sqrt{3}}{3}$

47. $2r^2 = 3 - r$
1 and $-\frac{3}{2}$

48. $2x^2 = 12 - 5x$
$\frac{3}{2}$ and -4

49. $y - 2 = (y - 3)(y + 2)$
$1 + \sqrt{5}$ and $1 - \sqrt{5}$

50. $8s - 11 = (s - 4)(s - 2)$
$7 + \sqrt{30}$ and $7 - \sqrt{30}$

51. $6t - 2 = (2t - 3)(t - 1)$
$\frac{1}{2}$ and 5

52. $2z + 9 = (2z + 3)(z + 2)$
$\frac{1}{2}$ and -3

53. $(x - 4)(x + 1) = x - 3$
$2 + \sqrt{5}$ and $2 - \sqrt{5}$

54. $(y - 3)^2 = 2y + 10$
$4 + \sqrt{17}$ and $4 - \sqrt{17}$

55. $11x - 4 = (x + 3)(x + 2)$
$3 + i$ and $3 - i$

56. $3p + 17 = (p - 3)(p + 4)$
$1 + \sqrt{30}$ and $1 - \sqrt{30}$

Solve by completing the square. Approximate the solutions to the nearest thousandth.

57. $z^2 + 2z = 4$
1.236 and -3.236

58. $t^2 - 4t = 7$
5.317 and -1.317

59. $2x^2 = 4x - 1$
1.707 and 0.293

60. $3y^2 = 5y - 1$
0.232 and 1.434

61. $4z^2 + 2z - 1 = 0$
0.309 and -0.809

62. $4w^2 - 8w = 3$
2.323 and -0.323

SECTION 7.3

Solving Quadratic Equations by Using the Quadratic Formula

Objective A

To solve a quadratic equation by using the quadratic formula

A general formula known as the **quadratic formula** can be derived by applying the method of completing the square to the standard form of a quadratic equation. This formula can be used to solve any quadratic equation.

Solve $ax^2 + bx + c = 0$ by completing the square.

$$ax^2 + bx + c = 0$$

Add the opposite of the constant term to each side of the equation.

$$ax^2 + bx + c + (-c) = 0 + (-c)$$

$$ax + bx = -c$$

Multiply each side of the equation by the reciprocal of a, the coefficient of x^2.

$$\frac{1}{a}(ax^2 + bx) = \frac{1}{a}(-c)$$

$$x^2 + \frac{b}{a}x = -\frac{c}{a}$$

Complete the square by adding $\left(\frac{1}{2} \cdot \frac{b}{a}\right)^2 = \frac{b^2}{4a^2}$ to each side of the equation.

$$x^2 + \frac{b}{a}x + \frac{b^2}{4a^2} = \frac{b^2}{4a^2} - \frac{c}{a}$$

Simplify the right side of the equation.

$$x^2 + \frac{b}{a}x + \frac{b^2}{4a^2} = \frac{b^2}{4a^2} - \left(\frac{c}{a} \cdot \frac{4a}{4a}\right)$$

$$x^2 + \frac{b}{a}x + \frac{b^2}{4a^2} = \frac{b^2}{4a^2} - \frac{4ac}{4a^2}$$

$$x^2 + \frac{b}{a}x + \frac{b^2}{4a^2} = \frac{b^2 - 4ac}{4a^2}$$

Factor the perfect square trinomial on the left side of the equation.

$$\left(x + \frac{b}{2a}\right)^2 = \frac{b^2 - 4ac}{4a^2}$$

Take the square root of each side of the equation.

$$\sqrt{\left(x + \frac{b}{2a}\right)^2} = \sqrt{\frac{b^2 - 4ac}{4a^2}}$$

$$\left(x + \frac{b}{2a}\right) = \pm\frac{\sqrt{b^2 - 4ac}}{2a}$$

Solve for x.

$$x + \frac{b}{2a} = \frac{\sqrt{b^2 - 4ac}}{2a} \qquad x + \frac{b}{2a} = -\frac{\sqrt{b^2 - 4ac}}{2a}$$

$$x = -\frac{b}{2a} + \frac{\sqrt{b^2 - 4ac}}{2a} \qquad x = -\frac{b}{2a} - \frac{\sqrt{b^2 - 4ac}}{2a}$$

$$= \frac{-b + \sqrt{b^2 - 4ac}}{2a} \qquad = \frac{-b - \sqrt{b^2 - 4ac}}{2a}$$

The Quadratic Formula

The solutions of $ax^2 + bx + c = 0$, $a \neq 0$, are
$$\frac{-b + \sqrt{b^2 - 4ac}}{2a} \quad \text{and} \quad \frac{-b - \sqrt{b^2 - 4ac}}{2a}.$$

The quadratic formula is frequently written in the form $x = \dfrac{-b \pm \sqrt{b^2 - 4ac}}{2a}$.

Solve by using the quadratic formula: $4x^2 = 8x - 13$

Write the equation in standard form.

$$4x^2 = 8x - 13$$
$$4x^2 - 8x + 13 = 0$$
$$a = 4,\ b = -8,\ c = 13$$

Replace a, b, and c in the quadratic formula by their values.

$$x = \frac{-b \pm \sqrt{b^2 - 4ac}}{2a}$$
$$= \frac{-(-8) \pm \sqrt{(-8)^2 - 4 \cdot 4 \cdot 13}}{2 \cdot 4}$$

Simplify.

$$= \frac{8 \pm \sqrt{64 - 208}}{8} = \frac{8 \pm \sqrt{-144}}{8}$$
$$= \frac{8 \pm 12i}{8} = \frac{2 \pm 3i}{2}$$

Write the solutions.

The solutions are $1 + \frac{3}{2}i$ and $1 - \frac{3}{2}i$.

Check:

$4x^2 = 8x - 13$	
$4\left(1 + \frac{3}{2}i\right)^2$	$8\left(1 + \frac{3}{2}i\right) - 13$
$4\left(1 + 3i - \frac{9}{4}\right)$	$8 + 12i - 13$
$4\left(-\frac{5}{4} + 3i\right)$	$-5 + 12i$
$-5 + 12i = -5 + 12i$	

$4x^2 = 8x - 13$	
$4\left(1 - \frac{3}{2}i\right)^2$	$8\left(1 - \frac{3}{2}i\right) - 13$
$4\left(1 - 3i - \frac{9}{4}\right)$	$8 - 12i - 13$
$4\left(-\frac{5}{4} - 3i\right)$	$-5 - 12i$
$-5 - 12i = -5 - 12i$	

Solve by using the quadratic formula: $4x^2 + 12x + 9 = 0$

The equation is in standard form.

$$4x^2 + 12x + 9 = 0$$
$$a = 4,\ b = 12,\ c = 9$$

Replace a, b, and c in the quadratic formula by their values.

$$x = \frac{-b \pm \sqrt{b^2 - 4ac}}{2a}$$
$$= \frac{-12 \pm \sqrt{12^2 - 4 \cdot 4 \cdot 9}}{2 \cdot 4}$$

Simplify.

$$= \frac{-12 \pm \sqrt{0}}{8} = \frac{-12}{8} = -\frac{3}{2}$$

The equation has a double root. Write the solution.

The solution is $-\frac{3}{2}$.

$-\frac{3}{2}$ checks as a solution.

Note in the first example above that the solutions of the equation are complex numbers, and in the second example, the solution is a real number that is a double root.

In the quadratic formula, the quantity $b^2 - 4ac$ is called the **discriminant**. When a, b, and c are real numbers, the discriminant determines whether a quadratic equation will have a double root, two real number solutions which are not equal, or two complex number solutions.

The Effect of the Discriminant on the Solutions of a Quadratic Equation

1. If $b^2 - 4ac = 0$, the equation has one real number solution, a double root.
2. If $b^2 - 4ac > 0$, the equation has two unequal real number solutions.
3. If $b^2 - 4ac < 0$, the equation has two complex number solutions.

Use the discriminant to determine whether $x^2 - 4x - 5 = 0$ has one real number solution, two real number solutions, or two complex number solutions.

Evaluate the discriminant.

$$a = 1, b = -4, c = -5$$
$$b^2 - 4ac$$
$$(-4)^2 - 4(1)(-5) = 16 + 20 = 36$$
$$36 > 0$$

Since the discriminant is greater than zero, the equation has two real number solutions.

Example 1   Solve by using the quadratic formula:
$x^2 - 4x - 4 = 0$

Solution $x^2 - 4x - 4 = 0$
$a = 1, b = -4, c = -4$

$$x = \frac{-b \pm \sqrt{b^2 - 4ac}}{2a}$$

$$= \frac{-(-4) \pm \sqrt{(-4)^2 - 4(1)(-4)}}{2 \cdot 1}$$

$$= \frac{4 \pm \sqrt{16 + 16}}{2} = \frac{4 \pm \sqrt{32}}{2}$$

$$= \frac{4 \pm 4\sqrt{2}}{2} = 2 \pm 2\sqrt{2}$$

The solutions are $2 + 2\sqrt{2}$ and $2 - 2\sqrt{2}$.

Example 2   Solve by using the quadratic formula:
$x^2 + 6x - 9 = 0$

Your solution $-3 + 3\sqrt{2}$ and $-3 - 3\sqrt{2}$

Example 3   Solve by using the quadratic formula:
$2x^2 - x + 5 = 0$

Solution $2x^2 - x + 5 = 0$
$a = 2, b = -1, c = 5$

$$x = \frac{-b \pm \sqrt{b^2 - 4ac}}{2a}$$

$$= \frac{-(-1) \pm \sqrt{(-1)^2 - 4(2)(5)}}{2 \cdot 2}$$

$$= \frac{1 \pm \sqrt{1 - 40}}{4} = \frac{1 \pm \sqrt{-39}}{4}$$

$$= \frac{1 \pm i\sqrt{39}}{4}$$

The solutions are $\frac{1}{4} + \frac{\sqrt{39}}{4}i$ and $\frac{1}{4} - \frac{\sqrt{39}}{4}i$.

Example 4   Solve by using the quadratic formula:
$x^2 - 2x + 10 = 0$

Your solution $1 + 3i$ and $1 - 3i$

Solutions on p. A32

Example 5

Solve by using the quadratic formula:
$2x^2 = (x - 2)(x - 3)$

Solution

$2x^2 = (x - 2)(x - 3)$
$2x^2 = x^2 - 5x + 6$

$x^2 + 5x - 6 = 0$

$a = 1, b = 5, c = -6$

$x = \dfrac{-b \pm \sqrt{b^2 - 4ac}}{2a}$

$\quad = \dfrac{-5 \pm \sqrt{5^2 - 4(1)(-6)}}{2 \cdot 1}$

$\quad = \dfrac{-5 \pm \sqrt{25 + 24}}{2} = \dfrac{-5 \pm \sqrt{49}}{2}$

$\quad = \dfrac{-5 \pm 7}{2}$

$x = \dfrac{-5 + 7}{2} \qquad x = \dfrac{-5 - 7}{2}$

$\quad = \dfrac{2}{2} = 1 \qquad \quad = \dfrac{-12}{2} = -6$

The solutions are 1 and -6.

Example 6

Solve by using the quadratic formula:
$4x^2 = 4x - 1$

Your solution

$\dfrac{1}{2}$

Example 7

Use the discriminant to determine whether $4x^2 - 2x + 5 = 0$ has one real number solution, two real number solutions, or two complex number solutions.

Solution

$a = 4, b = -2, c = 5$

$b^2 - 4ac = (-2)^2 - 4(4)(5)$
$\qquad\qquad = 4 - 80$
$\qquad\qquad = -76$

$-76 < 0$

Since the discriminant is less than zero, the equation has two complex number solutions.

Example 8

Use the discriminant to determine whether $3x^2 - x - 1 = 0$ has one real number solution, two real number solutions, or two complex number solutions.

Your solution

Two real number solutions

Solutions on p. A32

7.3 EXERCISES

▶ **Objective A**

Solve by using the quadratic formula.

1. $x^2 - 3x - 10 = 0$
5 and −2

2. $z^2 - 4z - 8 = 0$
$2 + 2\sqrt{3}$ and $2 - 2\sqrt{3}$

3. $y^2 + 5y - 36 = 0$
4 and −9

4. $z^2 - 3z - 40 = 0$
8 and −5

5. $w^2 = 8w + 72$
$4 + 2\sqrt{22}$ and $4 - 2\sqrt{22}$

6. $t^2 = 2t + 35$
7 and −5

7. $v^2 = 24 - 5v$
3 and −8

8. $x^2 = 18 - 7x$
2 and −9

9. $2y^2 + 5y - 3 = 0$
−3 and $\frac{1}{2}$

10. $4p^2 - 7p + 3 = 0$
1 and $\frac{3}{4}$

11. $8s^2 = 10s + 3$
$\frac{3}{2}$ and $-\frac{1}{4}$

12. $12t^2 = 5t + 2$
$\frac{2}{3}$ and $-\frac{1}{4}$

13. $6x^2 = 6 - 5x$
$-\frac{3}{2}$ and $\frac{2}{3}$

14. $6r^2 = 7r - 2$
$\frac{2}{3}$ and $\frac{1}{2}$

15. $z^2 - 4z + 2 = 0$
$2 + \sqrt{2}$ and $2 - \sqrt{2}$

16. $v^2 - 2v - 7 = 0$
$1 + 2\sqrt{2}$ and $1 - 2\sqrt{2}$

17. $t^2 - 2t - 11 = 0$
$1 + 2\sqrt{3}$ and $1 - 2\sqrt{3}$

18. $y^2 - 8y - 20 = 0$
10 and −2

19. $x^2 = 14x - 24$
2 and 12

20. $v^2 = 12v - 24$
$6 + 2\sqrt{3}$ and $6 - 2\sqrt{3}$

21. $2z^2 - 2z - 1 = 0$
$\frac{1 + \sqrt{3}}{2}$ and $\frac{1 - \sqrt{3}}{2}$

22. $4x^2 - 4x - 7 = 0$
$\frac{1 + 2\sqrt{2}}{2}$ and $\frac{1 - 2\sqrt{2}}{2}$

23. $2p^2 - 8p + 5 = 0$
$\frac{4 + \sqrt{6}}{2}$ and $\frac{4 - \sqrt{6}}{2}$

24. $2s^2 - 3s + 1 = 0$
1 and $\frac{1}{2}$

25. $4w^2 - 4w - 1 = 0$
$\frac{1 + \sqrt{2}}{2}$ and $\frac{1 - \sqrt{2}}{2}$

26. $3x^2 + 10x + 6 = 0$
$\frac{-5 - \sqrt{7}}{3}$ and $\frac{-5 + \sqrt{7}}{3}$

27. $3v^2 = 6v - 2$
$\frac{3 + \sqrt{3}}{3}$ and $\frac{3 - \sqrt{3}}{3}$

28. $6w^2 = 19w - 10$
$\frac{5}{2}$ and $\frac{2}{3}$

29. $z^2 + 2z + 2 = 0$
$-1 + i$ and $-1 - i$

30. $p^2 - 4p + 5 = 0$
$2 + i$ and $2 - i$

31. $y^2 - 2y + 5 = 0$
$1 + 2i$ and $1 - 2i$

32. $x^2 + 6x + 13 = 0$
$-3 + 2i$ and $-3 - 2i$

33. $s^2 - 4s + 13 = 0$
$2 + 3i$ and $2 - 3i$

Solve by using the quadratic formula.

34. $t^2 - 6t + 10 = 0$
$3 + i$ and $3 - i$

35. $2w^2 - 2w + 5 = 0$
$\frac{1}{2} + \frac{3}{2}i$ and $\frac{1}{2} - \frac{3}{2}i$

36. $4v^2 + 8v + 3 = 0$
$-\frac{1}{2}$ and $-\frac{3}{2}$

37. $2x^2 + 6x + 5 = 0$
$-\frac{3}{2} + \frac{1}{2}i$ and $-\frac{3}{2} - \frac{1}{2}i$

38. $2y^2 + 2y + 13 = 0$
$-\frac{1}{2} + \frac{5}{2}i$ and $-\frac{1}{2} - \frac{5}{2}i$

39. $4t^2 - 6t + 9 = 0$
$\frac{3}{4} + \frac{3\sqrt{3}}{4}i$ and $\frac{3}{4} - \frac{3\sqrt{3}}{4}i$

40. $3v^2 + 6v + 1 = 0$
$\frac{-3 - \sqrt{6}}{3}$ and $\frac{-3 + \sqrt{6}}{3}$

41. $2r^2 = 4r - 11$
$1 + \frac{3\sqrt{2}}{2}i$ and $1 - \frac{3\sqrt{2}}{2}i$

42. $3y^2 = 6y - 5$
$1 + \frac{\sqrt{6}}{3}i$ and $1 - \frac{\sqrt{6}}{3}i$

43. $2x(x - 2) = x + 12$
4 and $-\frac{3}{2}$

44. $10y(y + 4) = 15y - 15$
-1 and $-\frac{3}{2}$

45. $(3s - 2)(s + 1) = 2$
$-\frac{4}{3}$ and 1

46. $(2t + 1)(t - 3) = 9$
4 and $-\frac{3}{2}$

Use the discriminant to determine whether the quadratic equation has one real number solution, two real number solutions, or two complex number solutions.

47. $2z^2 - z + 5 = 0$
two complex

48. $3y^2 + y + 1 = 0$
two complex

49. $9x^2 - 12x + 4 = 0$
one real

50. $4x^2 + 20x + 25 = 0$
one real

51. $2v^2 - 3v - 1 = 0$
two real

52. $3w^2 + 3w - 2 = 0$
two real

53. $2p^2 + 5p + 1 = 0$
two real

54. $2t^2 + 9t + 3 = 0$
two real

55. $5z^2 + 2 = 0$
two complex

Solve by using the quadratic formula. Approximate the solutions to the nearest thousandth.

56. $x^2 + 6x - 6 = 0$
0.873 and -6.873

57. $p^2 - 8p + 3 = 0$
7.606 and 0.394

58. $r^2 - 2r - 4 = 0$
3.236 and -1.236

59. $w^2 + 4w - 1 = 0$
0.236 and -4.236

60. $3t^2 = 7t + 1$
2.468 and -0.135

61. $2y^2 = y + 5$
1.851 and -1.351

Content and Format © 1991 HMCo.

SECTION 7.4   Solving Equations That are Reducible to Quadratic Equations

Objective A   **To solve an equation that is quadratic in form**

Certain equations that are not quadratic can be expressed in quadratic form by making suitable substitutions. An equation is quadratic in form if it can be written as $au^2 + bu + c = 0$.

The equation $x^4 - 4x^2 - 5 = 0$ is quadratic in form.
Let $x^2 = u$.

$$x^4 - 4x^2 - 5 = 0$$
$$(x^2)^2 - 4(x^2) - 5 = 0$$
$$u^2 - 4u - 5 = 0$$

The equation $y - y^{1/2} - 6 = 0$ is quadratic in form.
Let $y^{1/2} = u$.

$$y - y^{1/2} - 6 = 0$$
$$(y^{1/2})^2 - (y^{1/2}) - 6 = 0$$
$$u^2 - u - 6 = 0$$

The key to recognizing equations that are quadratic in form: when the equation is written in standard form, the exponent on one variable term is $\frac{1}{2}$ the exponent on the other variable term.

Solve: $z + 7z^{1/2} - 18 = 0$

The equation is quadratic in form. Let $z^{1/2} = u$.

$$z + 7z^{1/2} - 18 = 0$$
$$(z^{1/2})^2 + 7(z^{1/2}) - 18 = 0$$
$$u^2 + 7u - 18 = 0$$

Solve for u by factoring.

$$(u - 2)(u + 9) = 0$$

$u - 2 = 0$	$u + 9 = 0$
$u = 2$	$u = -9$

Replace u by $z^{1/2}$.

$$z^{1/2} = 2 \qquad z^{1/2} = -9$$
$$\sqrt{z} = 2 \qquad \sqrt{z} = -9$$

Solve for z by squaring each side of the equation.

$$(\sqrt{z})^2 = 2^2 \qquad (\sqrt{z})^2 = (-9)^2$$
$$z = 4 \qquad z = 81$$

Check each solution. When squaring each side of an equation, the resulting equation may have a solution which is not a solution of the original equation.

Check:

$$\begin{array}{c|c} z + 7z^{1/2} - 18 = 0 & \\ \hline 4 + 7(4)^{1/2} - 18 & 0 \\ 4 + 7 \cdot 2 - 18 & \\ 4 + 14 - 18 & \end{array}$$
$$0 = 0$$

$$\begin{array}{c|c} z + 7z^{1/2} - 18 = 0 & \\ \hline 81 + 7(81)^{1/2} - 18 & 0 \\ 81 + 7 \cdot 9 - 18 & \\ 81 + 63 - 18 & \end{array}$$
$$126 \neq 0$$

4 checks as a solution, but 81 does not check as a solution.

The solution is 4.

Example 1   Solve: $x^4 + x^2 - 12 = 0$

Solution

$$x^4 + x^2 - 12 = 0$$
$$(x^2)^2 + (x^2) - 12 = 0$$
$$u^2 + u - 12 = 0$$
$$(u - 3)(u + 4) = 0$$

$$u - 3 = 0 \qquad u + 4 = 0$$
$$u = 3 \qquad\quad u = -4$$

Replace u by x^2.

$$x^2 = 3 \qquad\qquad x^2 = -4$$
$$\sqrt{x^2} = \sqrt{3} \qquad \sqrt{x^2} = \sqrt{-4}$$
$$x = \pm\sqrt{3} \qquad x = \pm 2i$$

The solutions are $\sqrt{3}$, $-\sqrt{3}$, $2i$, and $-2i$.

Example 2   Solve: $x - 5x^{1/2} + 6 = 0$

Your solution   4 and 9

Solution on p. A32

Objective B

7

To solve a radical equation that is reducible to a quadratic equation

Certain equations containing radicals can be expressed in quadratic form.

Solve: $\sqrt{x + 2} + 4 = x$

$$\sqrt{x + 2} + 4 = x$$

Solve for the radical expression.

$$\sqrt{x + 2} = x - 4$$

Square each side of the equation.

$$(\sqrt{x + 2})^2 = (x - 4)^2$$

Simplify.

$$x + 2 = x^2 - 8x + 16$$

Write the equation in standard form.

$$0 = x^2 - 9x + 14$$

Solve for x by factoring.

$$0 = (x - 7)(x - 2)$$

$$x - 7 = 0 \qquad x - 2 = 0$$
$$x = 7 \qquad\quad x = 2$$

Check each solution.

Check:

$$\begin{array}{c|c} \sqrt{x + 2} + 4 = x & \\ \hline \sqrt{7 + 2} + 4 & 7 \\ \sqrt{9} + 4 & \\ 3 + 4 & \\ \end{array}$$

$$7 = 7$$

$$\begin{array}{c|c} \sqrt{x + 2} + 4 = x & \\ \hline \sqrt{2 + 2} + 4 & 2 \\ \sqrt{4} + 4 & \\ 2 + 4 & \\ \end{array}$$

$$6 \neq 2$$

7 checks as a solution, but 2 does not check as a solution.

The solution is 7.

Example 3   Solve: $\sqrt{7y - 3} + 3 = 2y$

Solution $\sqrt{7y - 3} + 3 = 2y$
$\sqrt{7y - 3} = 2y - 3$
$(\sqrt{7y - 3})^2 = (2y - 3)^2$
$7y - 3 = 4y^2 - 12y + 9$
$0 = 4y^2 - 19y + 12$
$0 = (4y - 3)(y - 4)$

$4y - 3 = 0 \qquad y - 4 = 0$
$4y = 3 \qquad\quad y = 4$
$y = \frac{3}{4}$

4 checks as a solution.

$\frac{3}{4}$ does not check as a solution.

The solution is 4.

Example 4   Solve: $\sqrt{2x + 1} + x = 7$

Your solution   4

Example 5   Solve: $\sqrt{2y + 1} - \sqrt{y} = 1$

Solution $\sqrt{2y + 1} - \sqrt{y} = 1$

Solve for one of the radical expressions.

$\sqrt{2y + 1} = \sqrt{y} + 1$
$(\sqrt{2y + 1})^2 = (\sqrt{y} + 1)^2$
$2y + 1 = y + 2\sqrt{y} + 1$
$y = 2\sqrt{y}$

Square each side of the equation.

$y^2 = (2\sqrt{y})^2$
$y^2 = 4y$
$y^2 - 4y = 0$
$y(y - 4) = 0$

$y = 0 \qquad y - 4 = 0$
$\qquad\qquad y = 4$

0 and 4 check as solutions.

The solutions are 0 and 4.

Example 6   Solve: $\sqrt{2x - 1} + \sqrt{x} = 2$

Your solution   1

Content and Format © 1991 HMCo.

Solutions on pp. A32–A33

| Objective C | To solve a fractional equation that is reducible to a quadratic equation |

After each side of a fractional equation has been multiplied by the LCM of the denominators, the resulting equation may be a quadratic equation.

Solve: $\frac{1}{r} + \frac{1}{r+1} = \frac{3}{2}$

$$\frac{1}{r} + \frac{1}{r+1} = \frac{3}{2}$$

Multiply each side of the equation by the LCM of the denominators.

$$2r(r+1)\left(\frac{1}{r} + \frac{1}{r+1}\right) = 2r(r+1) \cdot \frac{3}{2}$$
$$2(r+1) + 2r = r(r+1) \cdot 3$$
$$2r + 2 + 2r = 3r(r+1)$$
$$4r + 2 = 3r^2 + 3r$$

Write the equation in standard form.

$$0 = 3r^2 - r - 2$$

Solve for r by factoring.

$$0 = (3r+2)(r-1)$$

$$3r + 2 = 0 \qquad r - 1 = 0$$
$$3r = -2 \qquad r = 1$$
$$r = -\frac{2}{3}$$

Write the solutions.

The solutions are $-\frac{2}{3}$ and 1.

$-\frac{2}{3}$ and 1 check as solutions.

Example 7   Solve: $\frac{9}{x-3} = 2x + 1$

Solution
$$\frac{9}{x-3} = 2x + 1$$

$$(x-3)\frac{9}{x-3} = (x-3)(2x+1)$$
$$9 = 2x^2 - 5x - 3$$
$$0 = 2x^2 - 5x - 12$$
$$0 = (2x+3)(x-4)$$

$$2x + 3 = 0 \qquad x - 4 = 0$$
$$2x = -3 \qquad x = 4$$
$$x = -\frac{3}{2}$$

The solutions are $-\frac{3}{2}$ and 4.

Example 8   Solve: $3y + \frac{25}{3y-2} = -8$

Your solution -1

Solution on p. A33

7.4 EXERCISES

▶ **Objective A**

Solve:

1. $x^4 - 13x^2 + 36 = 0$
3, −3, 2, −2

2. $y^4 - 5y^2 + 4 = 0$
1, −1, 2, −2

3. $z^4 - 6z^2 + 8 = 0$
$\sqrt{2}, -\sqrt{2}, 2, -2$

4. $t^4 - 12t^2 + 27 = 0$
$\sqrt{3}, -\sqrt{3}, 3, -3$

5. $p - 3p^{1/2} + 2 = 0$
1 and 4

6. $v - 7v^{1/2} + 12 = 0$
9 and 16

7. $x - x^{1/2} - 12 = 0$
16

8. $w - 2w^{1/2} - 15 = 0$
25

9. $z^4 + 3z^2 - 4 = 0$
2i, −2i, 1, −1

10. $y^4 + 5y^2 - 36 = 0$
2, −2, 3i, −3i

11. $x^4 + 12x^2 - 64 = 0$
4i, −4i, 2, −2

12. $x^4 - 81 = 0$
3, −3, 3i, −3i

13. $p + 2p^{1/2} - 24 = 0$
16

14. $v + 3v^{1/2} - 4 = 0$
1

15. $y^{2/3} - 9y^{1/3} + 8 = 0$
1 and 512

16. $z^{2/3} - z^{1/3} - 6 = 0$
−8 and 27

17. $9w^4 - 13w^2 + 4 = 0$
$\frac{2}{3}, -\frac{2}{3}, 1, -1$

18. $4y^4 - 7y^2 - 36 = 0$
$\frac{3}{2}i, -\frac{3}{2}i, 2, -2$

▶ **Objective B**

Solve:

19. $\sqrt{x + 1} + x = 5$
3

20. $\sqrt{x - 4} + x = 6$
5

21. $x = \sqrt{x + 6}$
9

22. $\sqrt{2y - 1} = y - 2$
5

23. $\sqrt{3w + 3} = w + 1$
2 and −1

24. $\sqrt{2s + 1} = s - 1$
4

25. $\sqrt{4y + 1} - y = 1$
0 and 2

26. $\sqrt{3s + 4} + 2s = 12$
4

27. $\sqrt{10x + 5} - 2x = 1$
2 and $-\frac{1}{2}$

Solve:

28. $\sqrt{t + 8} = 2t + 1$
1

29. $\sqrt{p + 11} = 1 - p$
−2

30. $x - 7 = \sqrt{x - 5}$
9

31. $\sqrt{x - 1} - \sqrt{x} = -1$
1

32. $\sqrt{y} + 1 = \sqrt{y + 5}$
4

33. $\sqrt{2x - 1} = 1 - \sqrt{x - 1}$
1

34. $\sqrt{x + 6} + \sqrt{x + 2} = 2$
−2

35. $\sqrt{t + 3} + \sqrt{2t + 7} = 1$
−3

36. $\sqrt{5 - 2x} = \sqrt{2 - x} + 1$
2 and −2

▶ **Objective C**

Solve:

37. $x = \dfrac{10}{x - 9}$
10 and −1

38. $z = \dfrac{5}{z - 4}$
5 and −1

39. $\dfrac{t}{t + 1} = \dfrac{-2}{t - 1}$
$-\dfrac{1}{2} + \dfrac{\sqrt{7}}{2}i$ and $-\dfrac{1}{2} - \dfrac{\sqrt{7}}{2}i$

40. $\dfrac{2v}{v - 1} = \dfrac{5}{v + 2}$
$\dfrac{1}{4} + \dfrac{\sqrt{39}}{4}i$ and $\dfrac{1}{4} - \dfrac{\sqrt{39}}{4}i$

41. $\dfrac{y - 1}{y + 2} + y = 1$
1 and −3

42. $\dfrac{2p - 1}{p - 2} + p = 8$
5 and 3

43. $\dfrac{3r + 2}{r + 2} - 2r = 1$
0 and −1

44. $\dfrac{2v + 3}{v + 4} + 3v = 4$
$-\dfrac{13}{3}$ and 1

45. $\dfrac{2}{2x + 1} + \dfrac{1}{x} = 3$
$\dfrac{1}{2}$ and $-\dfrac{1}{3}$

46. $\dfrac{3}{s} - \dfrac{2}{2s - 1} = 1$
$\dfrac{3}{2}$ and 1

47. $\dfrac{16}{z - 2} + \dfrac{16}{z + 2} = 6$
$-\dfrac{2}{3}$ and 6

48. $\dfrac{2}{y + 1} + \dfrac{1}{y - 1} = 1$
0 and 3

49. $\dfrac{t}{t - 2} + \dfrac{2}{t - 1} = 4$
$\dfrac{4}{3}$ and 3

50. $\dfrac{4t + 1}{t + 4} + \dfrac{3t - 1}{t + 1} = 2$
$-\dfrac{11}{5}$ and 1

51. $\dfrac{5}{2p - 1} + \dfrac{4}{p + 1} = 2$
$-\dfrac{1}{4}$ and 3

52. $\dfrac{3w}{2w + 3} + \dfrac{2}{w + 2} = 1$
0 and −3

53. $\dfrac{2v}{v + 2} + \dfrac{3}{v + 4} = 1$
$\dfrac{-5 - \sqrt{33}}{2}$ and $\dfrac{-5 + \sqrt{33}}{2}$

54. $\dfrac{x + 3}{x + 1} - \dfrac{x - 2}{x + 3} = 5$
$\dfrac{-13 - \sqrt{89}}{10}$ and $\dfrac{-13 + \sqrt{89}}{10}$

Quadratic Inequalities and Rational Inequalities

| Objective A | **To solve a non-linear inequality** |

A **quadratic inequality** is one that can be written in the form $ax^2 + bx + c < 0$ or $ax^2 + bx + c > 0$, where $a \neq 0$. The symbols $\leq$ or $\geq$ can also be used. The solution set of a quadratic inequality can be found by solving a compound inequality.

To solve $x^2 - 3x - 10 > 0$, first factor the trinomial.

$$x^2 - 3x - 10 > 0$$
$$(x + 2)(x - 5) > 0$$

There are two cases for which the product of the factors will be positive:
(1) both factors are positive, or
(2) both factors are negative.

(1) $x + 2 > 0$ and $x - 5 > 0$
(2) $x + 2 < 0$ and $x - 5 < 0$

Solve each pair of compound inequalities.

(1) $x + 2 > 0$ and $x - 5 > 0$
     $x > -2$ $x > 5$
   $\{x \mid x > -2\} \cap \{x \mid x > 5\} = \{x \mid x > 5\}$

(2) $x + 2 < 0$ and $x - 5 < 0$
     $x < -2$ $x < 5$
   $\{x \mid x < -2\} \cap \{x \mid x < 5\} = \{x \mid x < -2\}$

Since the two cases for which the product will be positive are connected by or, the solution set is the union of the solution sets of each pair of inequalities.

$\{x \mid x > 5\} \cup \{x \mid x < -2\} =$
$\{x \mid x > 5$ or $x < -2\}$

Although the solution set of any quadratic inequality can be found by using the method outlined above, frequently a graphical method is easier to use.

Solve and graph the solution set of $x^2 - x - 6 < 0$.

Factor the trinomial.

$$x^2 - x - 6 < 0$$
$$(x - 3)(x + 2) < 0$$

On a number line, draw lines indicating the numbers that make each factor equal to zero.
$x - 3 = 0$ $x + 2 = 0$
    $x = 3$ $x = -2$

For each factor, place plus signs above the number line for those regions where the factor is positive and negative signs where the factor is negative.
Since $x^2 - x - 6 < 0$, the solution set will be the regions where one factor is positive and the other factor is negative.

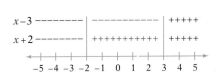

Write the solution set.

$\{x|-2 < x < 3\}$

The graph of the solution set of
$x^2 - x - 6 < 0$

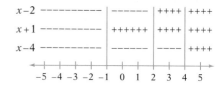

Solve and graph the solution set of $(x - 2)(x + 1)(x - 4) > 0$.

On a number line, identify for each factor the regions where the factor is positive and where the factor is negative.

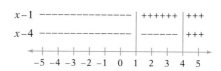

There are two regions where the product of the three factors is positive.

Write the solution set.

$\{x|-1 < x < 2 \text{ or } x > 4\}$

The graph of the solution set of
$(x - 2)(x + 1)(x - 4) > 0$

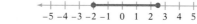

Solve: $\frac{2x - 5}{x - 4} \le 1$

$$\frac{2x - 5}{x - 4} \le 1$$

Rewrite the inequality so that 0 appears on the right side of the inequality.
Simplify.

$$\frac{2x - 5}{x - 4} - 1 \le 0$$

$$\frac{2x - 5}{x - 4} - \frac{x - 4}{x - 4} \le 0$$

$$\frac{x - 1}{x - 4} \le 0$$

On a number line, identify for each factor of the numerator and each factor of the denominator the regions where the factor is positive and where the factor is negative.

The region where the quotient of the two factors is negative is between 1 and 4.

Write the solution set.

$\{x|1 \le x < 4\}$

Note that 1 is part of the solution set, but 4 is not part of the solution set, since the denominator of the rational expression is zero when $x = 4$.

Example 1

Solve and graph the solution set of
$2x^2 - x - 3 \ge 0$.

Solution

$2x^2 - x - 3 \ge 0$
$(2x - 3)(x + 1) \ge 0$

$\left\{x|x \le -1 \text{ or } x \ge \frac{3}{2}\right\}$

Example 2

Solve and graph the solution set of
$2x^2 - x - 10 \le 0$.

Your solution

$\left\{x|-2 \le x \le \frac{5}{2}\right\}$

Solution on p. A33

7.5   EXERCISES

▶ **Objective A**

Solve and graph the solution set.

1. $(x - 4)(x + 2) > 0$

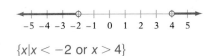

$\{x | x < -2 \text{ or } x > 4\}$

2. $(x + 1)(x - 3) > 0$

$\{x | x < -1 \text{ or } x > 3\}$

3. $x^2 - 3x + 2 \geq 0$

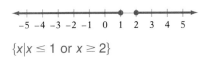

$\{x | x \leq 1 \text{ or } x \geq 2\}$

4. $x^2 + 5x + 6 > 0$

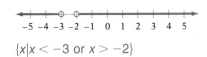

$\{x | x < -3 \text{ or } x > -2\}$

5. $x^2 - x - 12 < 0$

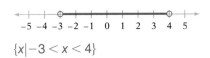

$\{x | -3 < x < 4\}$

6. $x^2 + x - 20 < 0$

$\{x | -5 < x < 4\}$

7. $(x - 1)(x + 2)(x - 3) < 0$

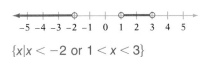

$\{x | x < -2 \text{ or } 1 < x < 3\}$

8. $(x + 4)(x - 2)(x + 1) > 0$

$\{x | -4 < x < -1 \text{ or } x > 2\}$

9. $(x + 4)(x - 2)(x - 1) \geq 0$

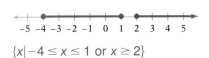

$\{x | -4 \leq x \leq 1 \text{ or } x \geq 2\}$

10. $(x - 1)(x + 5)(x - 2) \leq 0$

$\{x | x \leq -5 \text{ or } 1 \leq x \leq 2\}$

11. $\dfrac{x - 4}{x + 2} > 0$

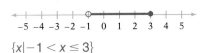

$\{x | x < -2 \text{ or } x > 4\}$

12. $\dfrac{x + 2}{x - 3} > 0$

$\{x | x < -2 \text{ or } x > 3\}$

13. $\dfrac{x - 3}{x + 1} \leq 0$

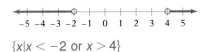

$\{x | -1 < x \leq 3\}$

14. $\dfrac{x - 1}{x} > 0$

$\{x | x < 0 \text{ or } x > 1\}$

15. $\dfrac{(x - 1)(x + 2)}{x - 3} \leq 0$

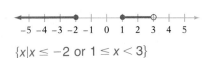

$\{x | x \leq -2 \text{ or } 1 \leq x < 3\}$

16. $\dfrac{(x + 3)(x - 1)}{x - 2} \geq 0$

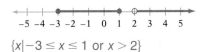

$\{x | -3 \leq x \leq 1 \text{ or } x > 2\}$

Solve:

17. $x^2 - 16 > 0$
$\{x | x > 4 \text{ or } x < -4\}$

18. $x^2 - 4 \geq 0$
$\{x | x \leq -2 \text{ or } x \geq 2\}$

19. $x^2 - 9x \leq 36$
$\{x | -3 \leq x \leq 12\}$

20. $x^2 + 4x > 21$
$\{x | x < -7 \text{ or } x > 3\}$

21. $2x^2 - 5x + 2 \geq 0$
$\left\{x | x \geq 2 \text{ or } x \leq \frac{1}{2}\right\}$

22. $4x^2 - 9x + 2 < 0$
$\left\{x | \frac{1}{4} < x < 2\right\}$

23. $4x^2 - 8x + 3 < 0$
$\left\{x | \frac{1}{2} < x < \frac{3}{2}\right\}$

24. $2x^2 + 11x + 12 \geq 0$
$\left\{x | x \leq -4 \text{ or } x \geq -\frac{3}{2}\right\}$

25. $(x - 6)(x + 3)(x - 2) \leq 0$
$\{x | x \leq -3 \text{ or } 2 \leq x \leq 6\}$

26. $(x + 5)(x - 2)(x - 3) > 0$
$\{x | -5 < x < 2 \text{ or } x > 3\}$

27. $(2x - 1)(x - 4)(2x + 3) > 0$
$\left\{x | -\frac{3}{2} < x < \frac{1}{2} \text{ or } x > 4\right\}$

28. $(x - 2)(3x - 1)(x + 2) \leq 0$
$\left\{x | x \leq -2 \text{ or } \frac{1}{3} \leq x \leq 2\right\}$

29. $(x - 5)(2x - 7)(x + 1) < 0$
$\left\{x | x < -1 \text{ or } \frac{7}{2} < x < 5\right\}$

30. $(x + 4)(2x + 7)(x - 2) > 0$
$\left\{x | -4 < x < -\frac{7}{2} \text{ or } x > 2\right\}$

31. $\frac{3x}{x - 2} > 1$
$\{x | x < -1 \text{ or } x > 2\}$

32. $\frac{2x}{x + 1} < 1$
$\{x | -1 < x < 1\}$

33. $\frac{2}{x + 1} \geq 2$
$\{x | -1 < x \leq 0\}$

34. $\frac{3}{x - 1} < 2$
$\left\{x | x < 1 \text{ or } x > \frac{5}{2}\right\}$

35. $\frac{x}{(x - 1)(x + 2)} \geq 0$
$\{x | x > 1 \text{ or } -2 < x \leq 0\}$

36. $\frac{x - 2}{(x + 1)(x - 1)} \leq 0$
$\{x | x < -1 \text{ or } 1 < x \leq 2\}$

37. $\frac{1}{x} < 2$
$\left\{x | x > \frac{1}{2} \text{ or } x < 0\right\}$

38. $\frac{x}{2x - 1} \geq 1$
$\left\{x | \frac{1}{2} < x \leq 1\right\}$

39. $\frac{x}{2x - 3} \leq 1$
$\left\{x | x < \frac{3}{2} \text{ or } x \geq 3\right\}$

40. $\frac{x}{2 - x} \leq -3$
$\{x | 2 < x \leq 3\}$

41. $\frac{3}{x - 2} > \frac{2}{x + 2}$
$\{x | x > 2 \text{ or } -10 < x < -2\}$

42. $\frac{3}{x - 5} > \frac{1}{x + 1}$
$\{x | x > 5 \text{ or } -4 < x < -1\}$

SECTION 7.6   Applications of Quadratic Equations

Objective A   **To solve application problems**

The application problems in this section are similar to problems solved earlier in the text. Each of the strategies for the problems in this section will result in a quadratic equation.

A small pipe takes 16 min longer to empty a tank than does a larger pipe. Working together, the pipes can empty the tank in 6 min. How long would it take each pipe working alone to empty the tank?

Strategy for Solving an Application Problem

> Determine the type of problem. Is it a uniform-motion problem, a geometry problem, an integer problem, or a work problem?

The problem is a work problem.

> Choose a variable to represent the unknown quantity. Write numerical or variable expressions for all the remaining quantities. These results can be recorded in a table.

The unknown time of the larger pipe: t
The unknown time of the smaller pipe: $t + 16$

	Rate of work	$\cdot$	*Time worked*	$=$	*Part of task completed*
Large pipe	$\dfrac{1}{t}$	$\cdot$	6	$=$	$\dfrac{6}{t}$
Smaller pipe	$\dfrac{1}{t + 16}$	$\cdot$	6	$=$	$\dfrac{6}{t + 16}$

> Determine how the quantities are related. If necessary, review the strategies presented in Chapter 2.

The sum of the parts of the task completed must equal 1.

$$\frac{6}{t} + \frac{6}{t + 16} = 1$$

$$t(t + 16)\left(\frac{6}{t} + \frac{6}{t + 16}\right) = t(t + 16) \cdot 1$$

$$(t + 16)6 + 6t = t^2 + 16t$$

$$6t + 96 + 6t = t^2 + 16t$$

$$0 = t^2 + 4t - 96$$

$$0 = (t + 12)(t - 8)$$

$$t + 12 = 0 \qquad\qquad t - 8 = 0$$

$$t = -12 \qquad\qquad t = 8$$

The solution $t = -12$ is not possible, since time cannot be a negative number.

The time for the smaller pipe is $t + 16$.
Replace t by 8 and evaluate. $t + 16 = 8 + 16 = 24$

The larger pipe requires 8 min to empty the tank.
The smaller pipe requires 24 min to empty the tank.

Example 1

In 8 h, two canoers rowed 15 mi down a river and then rowed back to their campsite. The rate of the river's current was 1 mph. Find the rate at which the campers rowed.

Example 2

The length of a rectangle is 3 m more than the width. The area is 54 m². Find the length of the rectangle.

Strategy

- This is a uniform-motion problem.
- Unknown rowing rate of the campers: r

	Distance	Rate	Time
Down river	15	$r + 1$	$\dfrac{15}{r+1}$
Up river	15	$r - 1$	$\dfrac{15}{r-1}$

- The total time of the trip was 8 h.

$$\frac{15}{r+1} + \frac{15}{r-1} = 8$$

Your strategy

Solution

$$\frac{15}{r+1} + \frac{15}{r-1} = 8$$

$$(r+1)(r-1)\left(\frac{15}{r+1} + \frac{15}{r-1}\right) = (r+1)(r-1)8$$

$$(r-1)15 + (r+1)15 = (r^2 - 1)8$$

$$15r - 15 + 15r + 15 = 8r^2 - 8$$

$$30r = 8r^2 - 8$$

$$0 = 8r^2 - 30r - 8$$

$$0 = 2(4r^2 - 15r - 4)$$

$$0 = 2(4r + 1)(r - 4)$$

$$4r + 1 = 0 \qquad r - 4 = 0$$

$$4r = -1 \qquad\quad r = 4$$

$$r = -\frac{1}{4}$$

The solution $r = -\frac{1}{4}$ is not possible, since the rate cannot be a negative number.

The rowing rate was 4 mph.

Your solution

9 m

Solution on p. A33

7.6 EXERCISES

▶ **Objective A** *Application Problems*

1. The base of a triangle is one less than five times the height of the triangle. The area of the triangle is 21 cm². Find the height and the length of the base of the triangle.
height: 3 cm; base: 14 cm

2. The height of a triangle is 3 in. less than the base of a triangle. The area of the triangle is 90 in.². Find the height and the length of the base of the triangle.
height: 12 in.; base: 15 in.

3. The length of a rectangle is two feet less than three times the width of a rectangle. The area of the rectangle is 65 ft². Find the length and width of the rectangle.
length: 13 ft; width: 5 ft

4. The length of a rectangle is 2 cm less than twice the width. The area of the rectangle is 180 cm². Find the length and width of the rectangle.
length: 18 cm; width: 10 cm

5. The sum of the squares of two consecutive odd integers is thirty-four. Find the two integers.
3 and 5 or −5 and −3

6. The sum of the squares of three consecutive odd integers is eighty-three. Find the three integers.
3, 5, and 7 or −7, −5, and −3

7. Five times an integer plus the square of the integer is twenty-four. Find the integer.
3 or 8

8. The sum of five times an integer and twice the square of the integer is three. Find the integer.
−3

9. The height of a projectile fired upward is given by the formula $s = v_0 t - 16t^2$, where s is the height, v_0 is the initial velocity, and t is the time. Find the time for a projectile to return to earth if it has an initial velocity of 200 feet per second.
12.5 s

10. The height of a projectile fired upward is given by the formula $s = v_0 t - 16t^2$, where s is the height, v_0 is the initial velocity, and t is the time. Find the time for a projectile to reach a height of 64 ft if it has an initial velocity of 128 ft/sec. Round to the nearest hundredth of a second.
7.46 s or 0.54 s

11. A chemistry experiment requires that a vacuum be created in a chamber. A small vacuum pump requires 15 s longer than does a second, larger pump to evacuate the chamber. Working together, the pumps can evacuate the chamber in 4 s. Find the time required for the larger vacuum pump working alone to evacuate the chamber.
5 s

12. An old computer requires 3 h longer to print the payroll than does a new computer. With both computers running, the payroll can be completed in 2 h. Find the time required for the new computer running alone to complete the payroll.
3 h

13. A small air conditioner requires 16 min longer to cool a room 3° than does a larger air conditioner. Working together, the two air conditioners can cool the room 3° in 6 min. How long would it take each air conditioner working alone to cool the room 3°?
larger air conditioner: 8 min; smaller air conditioner: 24 min

14. A small pipe can fill a tank in 6 min more time than it takes a larger pipe to fill the same tank. Working together, both pipes can fill the tank in 4 min. How long would it take each pipe working alone to fill the tank?
smaller pipe: 12 min; larger pipe: 6 min

15. An old mechanical sorter takes 21 min longer to sort a batch of mail than does a second, newer model. With both sorters working, a batch of mail can be sorted in 10 min. How long would it take each sorter working alone to sort the batch of mail?
new sorter: 14 min; old sorter: 35 min

16. A small heating unit takes 8 h longer to melt a piece of iron than does a larger unit. Working together, the heating units can melt the iron in 3 h. How long would it take each heating unit working alone to melt the iron?
smaller unit: 12 h; larger unit: 4 h

17. A cruise ship made a trip of 100 mi in 8 h. The ship traveled the first 40 mi at a constant rate before increasing its speed by 5 mph. Another 60 mi was traveled at the increased speed. Find the rate of the cruise ship for the first 40 mi.
10 mph

18. A cyclist traveled 60 mi at a constant rate before reducing the speed by 2 mph. Another 40 mi was traveled at the reduced speed. The total time for the 100-mile trip was 9 h. Find the rate during the first 60 mi.
12 mph

19. The rate of a single-engine plane in calm air is 100 mph. Flying with the wind, the plane can fly 240 mi in one hour less than the time required to make the return trip of 240 mi. Find the rate of the wind.
20 mph

20. A car travels 120 mi. A second car, traveling 10 mph faster than the first car, makes the same trip in 1 h less time. Find the speed of each car.
1st car: 30 mph; 2nd car: 40 mph

21. The rate of a river's current is 2 mph. A rowing crew can row 16 mi down this river and back in 6 h. Find the rowing rate of the crew in calm water.
6 mph

22. A boat traveled 30 mi down a river and then returned. The total time for the round trip was 4 h, and the rate of the river's current was 4 mph. Find the rate of the boat in still water.
16 mph

Calculators and Computers

Solving Quadratic Equations   The solutions of all quadratic equations can be found by using the quadratic formula. Although this formula is available, sometimes the coefficients of the equation make the use of the formula difficult because the computations are very tedious.

Consider trying to solve the equation

$$2.984x^2 + 9834.1x - 509.0023 = 0$$

by using the quadratic formula. The computations would take a long time even with a calculator.

On the Math ACE Disk there is a program QUADRATIC EQUATIONS that will solve any quadratic equation with real number coefficients, including those whose solutions are complex numbers. You can use the program to test your ability to solve equations by comparing your answer to that of the program or to experiment with different equations.

Remember when entering the coefficients that any fraction must be converted to a decimal before entering the number.

Chapter Summary

Key Words   A *quadratic equation* is an equation of the form $ax^2 + bx + c = 0$, where a, b, and c are constants and $a \neq 0$. A quadratic equation is also called a *second-degree equation*.

A quadratic equation is in *standard form* when the polynomial is in descending order and equal to zero.

When a quadratic equation has two solutions that are the same number, the solution is called a *double root* of the equation.

For an equation of the form $ax^2 + bx + c = 0$, the quantity $b^2 - 4ac$ is called the *discriminant*.

A *quadratic inequality* is one that can be written in the form $ax^2 + bx + c > 0$ or $ax^2 + bx + c < 0$, where $a \neq 0$. The symbols $\leq$ or $\geq$ can also be used.

Essential Rules   ***The Principle of Zero Products***   If $ab = 0$, then $a = 0$ or $b = 0$.

The Quadratic Formula $x = \dfrac{-b \pm \sqrt{b^2 - 4ac}}{2a}$

The Effect of the Discriminant on the Solutions of a Quadratic Equation

1. If $b^2 - 4ac = 0$, then the equation has one real number solution, a double root.
2. If $b^2 - 4ac > 0$, then the equation has two real number solutions that are not equal.
3. If $b^2 - 4ac < 0$, then the equation has two complex number solutions.

Chapter Review

SECTION 1

1. Solve by factoring:
$2x^2 - 3x = 0$
0 and $\frac{3}{2}$

2. Solve by factoring:
$6x^2 + 9cx = 6c^2$
$\frac{c}{2}$ and $-2c$

3. Write a quadratic equation that has integer coefficients and has solutions 0 and -3.
$x^2 + 3x = 0$

4. Write a quadratic equation that has integer coefficients and has solutions $\frac{3}{4}$ and $-\frac{2}{3}$.
$12x^2 - x - 6 = 0$

5. Solve by taking square roots:
$x^2 = 48$
$4\sqrt{3}$ and $-4\sqrt{3}$

6. Solve by taking square roots:
$\left(x + \frac{1}{2}\right)^2 + 4 = 0$
$-\frac{1}{2} + 2i$ and $-\frac{1}{2} - 2i$

SECTION 2

7. Solve by completing the square:
$x^2 + 4x + 3 = 0$
-3 and -1

8. Solve by completing the square:
$7x^2 - 14x + 3 = 0$
$\frac{7 + 2\sqrt{7}}{7}$ and $\frac{7 - 2\sqrt{7}}{7}$

9. Solve by completing the square:
$x^2 - 2x + 8 = 0$
$1 + i\sqrt{7}$ and $1 - i\sqrt{7}$

10. Solve by completing the square:
$(x - 2)(x + 3) = x - 10$
$2i$ and $-2i$

SECTION 3

11. Solve by using the quadratic formula:
$12x^2 - 25x + 12 = 0$
$\frac{3}{4}$ and $\frac{4}{3}$

12. Solve by using the quadratic formula:
$x^2 - x + 8 = 0$
$\frac{1}{2} + \frac{\sqrt{31}}{2}i$ and $\frac{1}{2} - \frac{\sqrt{31}}{2}i$

13. Solve by using the quadratic formula:
$3x(x - 3) = 2x - 4$
$\frac{11 + \sqrt{73}}{6}$ and $\frac{11 - \sqrt{73}}{6}$

14. Use the discriminant to determine whether $3x^2 - 5x + 1 = 0$ has one real number solution, two real number solutions, or two complex number solutions.
Two real number solutions

SECTION 4

15. Solve: $x^{2/3} + x^{1/3} - 12 = 0$
27 and -64

16. Solve: $2(x - 1) + 3\sqrt{x - 1} - 2 = 0$
$\frac{5}{4}$

17. Solve: $x = \sqrt{x} + 2$
4

18. Solve: $2x = \sqrt{5x + 24} + 3$
5

19. Solve: $3x = \frac{9}{x - 2}$
3 and -1

20. Solve: $\frac{3x + 7}{x + 2} + x = 3$
-1

21. Solve: $\frac{x - 2}{2x + 3} - \frac{x - 4}{x} = 2$
$\frac{-3 + \sqrt{249}}{10}$ and $\frac{-3 - \sqrt{249}}{10}$

22. Solve: $1 - \frac{x + 4}{2 - x} = \frac{x - 3}{x + 2}$
$\frac{-11 + \sqrt{129}}{2}$ and $\frac{-11 - \sqrt{129}}{2}$

SECTION 5

23. Solve: $(x + 3)(2x - 5) < 0$
$\left\{ x \mid -3 < x < \frac{5}{2} \right\}$

24. Solve: $(x - 2)(x + 4)(2x + 3) \le 0$
$\left\{ x \mid x \le -4 \text{ or } -\frac{3}{2} \le x \le 2 \right\}$

25. Solve and graph the solution set.
$\frac{x - 2}{2x - 3} \ge 0$

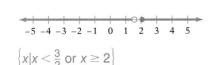

$\left\{ x \mid x < \frac{3}{2} \text{ or } x \ge 2 \right\}$

26. Solve and graph the solution set.
$\frac{(2x - 1)(x + 3)}{x - 4} \le 0$

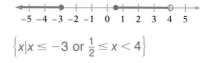

$\left\{ x \mid x \le -3 \text{ or } \frac{1}{2} \le x < 4 \right\}$

27. The length of a rectangle is two more than twice the width. The area of the rectangle is 60 cm². Find the length and width of the rectangle.
length: 12 cm; width: 5 cm

28. The sum of the squares of three consecutive even integers is fifty-six. Find the three integers.
2, 4, and 6 or -6, -4, and -2

29. An older computer requires 12 minutes longer to print the payroll than does a newer computer. Together the computers can print the payroll in 8 minutes. Find the time for the new computer working alone to complete the payroll.
12 min

30. A car travels 200 mi. A second car, traveling 10 mph faster than the first car, makes the same trip in 1 h less time. Find the speed of each car.
first car: 40 mph; second car: 50 mph

Chapter Test

1. Solve by factoring: $3x^2 + 10x = 8$

   $\frac{2}{3}$ and -4 [7.1A]

2. Solve by factoring: $6x^2 - 5x - 6 = 0$

   $\frac{3}{2}$ and $-\frac{2}{3}$ [7.1A]

3. Write a quadratic equation that has integer coefficients and has solutions 3 and -3.

   $x^2 - 9 = 0$ [7.1B]

4. Write a quadratic equation that has integer coefficients and has solutions $\frac{1}{2}$ and -4.

   $2x^2 + 7x - 4 = 0$ [7.1B]

5. Solve by taking square roots:
   $3(x - 2)^2 - 24 = 0$
   $2 + 2\sqrt{2}$ and $2 - 2\sqrt{2}$ [7.1C]

6. Solve by completing the square:
   $x^2 - 6x - 2 = 0$
   $3 + \sqrt{11}$ and $3 - \sqrt{11}$ [7.2A]

7. Solve by completing the square:
   $3x^2 - 6x = 2$

   $\frac{3 + \sqrt{15}}{3}$ and $\frac{3 - \sqrt{15}}{3}$ [7.2A]

8. Solve by using the quadratic formula:
   $2x^2 - 2x = 1$

   $\frac{1 + \sqrt{3}}{2}$ and $\frac{1 - \sqrt{3}}{2}$ [7.3A]

9. Solve by using the quadratic formula:
   $x^2 + 4x + 12 = 0$
   $-2 + 2i\sqrt{2}$ and $-2 - 2i\sqrt{2}$ [7.3A]

10. Use the discriminant to determine whether $3x^2 - 4x = 1$ has one real number solution, two real number solutions, or two complex number solutions.
    Two real number solutions [7.3A]

11. Use the discriminant to determine whether $x^2 - 6x = -15$ has one real number solution, two real number solutions, or two complex number solutions.
    Two complex number solutions [7.3A]

12. Solve: $2x + 7x^{1/2} - 4 = 0$

    $\frac{1}{4}$ [7.4A]

13. Solve: $x^4 - 4x^2 + 3 = 0$
1, −1, $\sqrt{3}$, −$\sqrt{3}$ [7.4A]

14. Solve: $\sqrt{2x + 1} + 5 = 2x$
4   [7.4B]

15. Solve: $\sqrt{x - 2} = \sqrt{x} - 2$
No solution   [7.4B]

16. Solve: $\frac{2x}{x - 3} + \frac{5}{x - 1} = 1$
2 and −9   [7.4C]

17. Solve and graph the solution set of $(x - 2)(x + 4)(x - 4) < 0$.

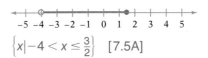

$\{x | x < -4 \text{ or } 2 < x < 4\}$ [7.5A]

18. Solve and graph the solution set of $\frac{2x - 3}{x + 4} \le 0$.

$\left\{x \middle| -4 < x \le \frac{3}{2}\right\}$ [7.5A]

19. The base of a triangle is three feet more than three times the height. The area of the triangle is 30 ft². Find the base and height of the triangle.
base: 15 ft; height: 4 ft   [7.6A]

20. The rate of a river's current is 2 mph. A canoe was rowed 6 mi down the river and back in 4 h. Find the rowing rate in calm water.
4 mph   [7.6A]

Cumulative Review

1. Evaluate $2a^2 - b^2 \div c^2$ when $a = 3$, $b = -4$, and $c = -2$.
   14   [1.3A]

2. Solve: $\frac{2x - 3}{4} - \frac{x + 4}{6} = \frac{3x - 2}{8}$
   $x = -28$ [2.1C]

3. Find the slope of the line containing the points $(3, -4)$ and $(-1, 2)$.
   $-\frac{3}{2}$ [6.2A]

4. Find the equation of the line containing the point $(1, 2)$ and parallel to the line $x - y = 1$.
   $y = x + 1$ [6.4A]

5. Factor: $-3x^3y + 6x^2y^2 - 9xy^3$
   $-3xy(x^2 - 2xy + 3y^2)$ [3.3A]

6. Factor: $6x^2 - 7x - 20$
   $(2x - 5)(3x + 4)$ [3.3D]

7. Factor:
   $a^nx + a^ny - 2x - 2y$
   $(x + y)(a^n - 2)$ [3.3B]

8. Simplify:
   $(3x^3 - 13x^2 + 10) \div (3x - 4)$
   $x^2 - 3x - 4 - \frac{6}{3x - 4}$ [4.2A]

9. Simplify: $\frac{x^2 + 2x + 1}{8x^2 + 8x} \cdot \frac{4x^3 - 4x^2}{x^2 - 1}$
   $\frac{x}{2}$ [4.1B]

10. Solve: $\frac{x}{2x + 3} - \frac{3}{4x^2 - 9} = \frac{x}{2x - 3}$
    $x = -\frac{1}{2}$ [4.6A]

11. Solve $S = \frac{n}{2}(a + b)$ for b.
    $b = \frac{2S - an}{n}$ [4.6B]

12. Simplify: $-2i(7 - 4i)$
    $-8 - 14i$ [5.3C]

13. Simplify: $a^{-1/2}(a^{1/2} - a^{3/2})$
    $1 - a$ [5.1A]

14. Simplify: $\frac{\sqrt[3]{8x^4y^5}}{\sqrt[3]{16xy^6}}$
    $\frac{x\sqrt[3]{4y^2}}{2y}$ [5.2D]

15. Solve by factoring:
    $3x^2 + 7x = 6$
    $\frac{2}{3}$ and -3 [7.1A]

16. Solve by using the quadratic formula:
    $x^2 + 6x + 10 = 0$
    $-3 + i$ and $-3 - i$ [7.3A]

17. Solve: $x^4 - 6x^2 + 8 = 0$
    $2, -2, \sqrt{2},$ and $-\sqrt{2}$ [7.4A]

18. Solve: $\sqrt{3x + 1} - 1 = x$
    0 and 1   [7.4B]

19. Solve: $\frac{x}{x+2} - \frac{4x}{x+3} = 1$

$-\frac{3}{2}$ and -1 [7.4C]

20. Find the x- and y-intercepts of the graph of $6x - 5y = 15$.

$\left(\frac{5}{2}, 0\right)$ and $(0, -3)$ [6.2B]

21. A piston rod for an automobile is $9\frac{3}{8}$ in. with a tolerance of $\frac{1}{64}$ in. Find the lower and upper limits of the length of the piston rod.

$9\frac{23}{64}$ in. lower limit; $9\frac{25}{64}$ in. upper limit   [2.3C]

22. The base of a triangle is $(x + 8)$ ft. The height is $(2x - 4)$ ft. Find the area of the triangle in terms of the variable x.

$(x^2 + 6x - 16)$ ft^2 [3.2D]

23. Solve and graph the solution set of $x^3 + x^2 - 6x < 0$.

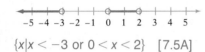

$\{x|x < -3 \text{ or } 0 < x < 2\}$ [7.5A]

24. Solve and graph the solution set of $\frac{(x - 1)(x - 5)}{x + 3} \geq 0$.

$\{x|-3 < x \leq 1 \text{ or } x \geq 5\}$ [7.5A]

25. Use the discriminant to determine whether $2x^2 + 4x + 3 = 0$ has one real number solution, two real number solutions, or two complex number solutions.

Two complex number solutions   [7.3A]

26. The rate of a plane in calm air is 450 mph. Flying with the wind, the plane can fly 2000 mi in one hour less time than is required to make the return trip of 2000 mi. Find the rate of the wind.

50 mph   [7.6A]

8

Functions and Relations

OBJECTIVES

▶ To evaluate a function
▶ To find the domain and range of a function
▶ To graph a function
▶ To determine if a relation is a function
▶ To determine from a graph whether a function is one-to-one
▶ To find the composition of two functions
▶ To find the inverse of a function
▶ To solve variation problems

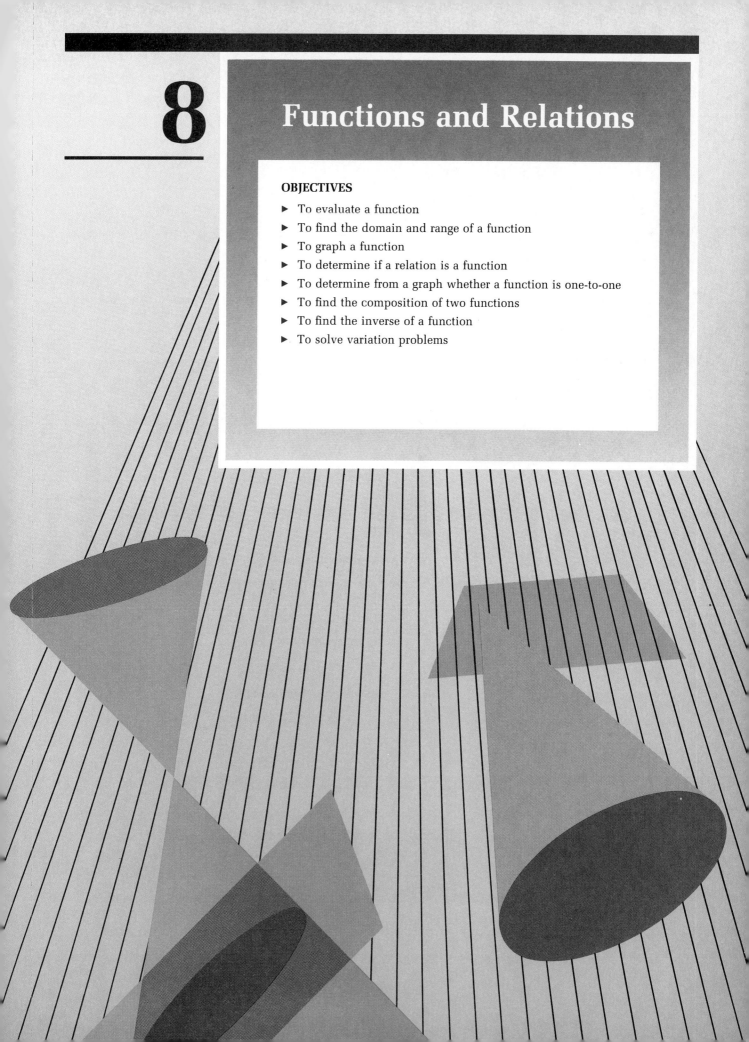

History of Equal Signs

A portion of a page of the first book that used an equals sign, =, is shown below. This book was written in 1557 by Robert Recorde and was titled *The Whetstone of Witte.*

Notice in the illustration the words "bicause noe 2 thyngs can be more equalle." Recorde decided that two things could not be more equal than two parallel lines of the same length. Therefore, it made sense to use this symbol to show equality.

This page also illustrates the use of the plus sign, +, and the minus sign, −. These symbols had been widely used only for about 100 years when this book was written.

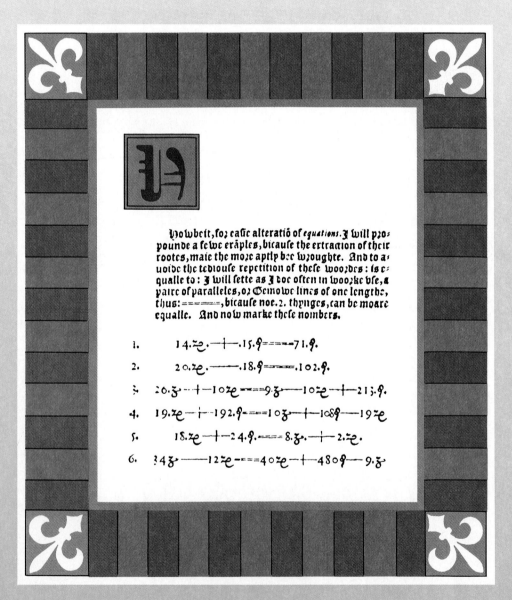

SECTION 8.1   Functions and Relations

Objective A   **To evaluate a function**

There are many situations in science, business, and mathematics where a correspondence exists between two quantities. The correspondence can be described by a formula, in a table, or in a graph, and then recorded as ordered pairs.

The formula $d = 16t^2$ describes a correspondence between the distance (d) a rock will fall and the time (t) of its fall. For each value of t, the formula assigns *only one* value for the distance. The ordered pair (2, 64) indicates that in 2 seconds a rock will fall 64 feet. Some of the other ordered pairs determined by the correspondence are shown at the right.

Time in seconds	Distance in feet
↓	↓
(1,	16)
(3,	144)
(4,	256)
(5,	400)

The table at the right describes a grading scale which defines a correspondence between a percent score and a letter grade. For any percent score, the table assigns *only one* letter grade. The ordered pair (86, B) indicates that a score of 86% receives a letter grade of B.

Score	*Grade*
90–100	A
80–89	B
70–79	C
60–69	D
0–59	F

The graph at the right defines a correspondence, used by an archeologist, between the percent of carbon-14 originally contained in an object and the age of the object. The order pair (50, 5000) indicates that an object with 50% of its original carbon-14 is approximately 5000 years old.

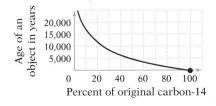

In each of these examples, a correspondence or a rule determines a set of ordered pairs. The set of ordered pairs determined by the correspondence is called a *function*.

> A **function** is a set of ordered pairs in which no two ordered pairs that have the same first component have different second components.

For example, the ordered pair (2, 64) was determined by the formula $d = 16t^2$. The first component, 2, can only have a second component of 64; no other second component can be paired with this first component.

Note that the ordered pairs of a function can have *different* first components paired with the *same* second component. For example, the ordered pairs determined by the grading scale shown above include (80, B), (83, B), and (87, B).

Not every correspondence between two sets is a function. Consider the correspondence which assigns to a postitive real number a square root of that number.

This is not a function since each positive real number can be paired with the positive or the negative square root of that number. For example, 9 can be paired with 3 or with -3. The same first component can be paired with different second components.

$(9, 3)$
$(9, -3)$

A **relation** assigns to each member of a first set one or more members of a second set. The correspondence that pairs a positive real number with a square root of that number is a relation. In general, a relation is any set of ordered pairs. A function is a special kind of relation.

Although a function can always be described in terms of ordered pairs, frequently it is more convenient to describe a function by an equation. The letter f is commonly used to represent a function, but any letter or combination of letters can be used.

The function that assigns to each real number its square is given by the equation $f(x) = x^2$. Read "$f(x)$" as "f of x" or "the value of f at x." The symbol $f(x)$ stands for the number that is paired with x. In terms of ordered pairs, this is written $(x, f(x))$. It is important to remember that $f(x)$ does not mean f times x. The letter f is the name of the function while $f(x)$ is the number that is paired with x.

The function that assigns to each real number its absolute value is described by $ABS(x) = |x|$ (taken from a computer programming language); it is a combination of letters used to name the absolute value function. The letters used to name this or any function are somewhat arbitrary. The symbols $g(x) = |x|$ or $v(t) = |t|$ also name the absolute value function.

To **evaluate a function** means to find the number that is paired with a given number.

Evaluate the function $f(x) = 2x + 1$ at $x = 3$.

$f(3)$ is the number that is paired with 3. $f(x) = 2x + 1$

To find $f(3)$, replace x by 3 and simplify. $f(3) = 2(3) + 1$
$f(3) = 7$

An ordered pair of this function is
$(3, 7)$.

Evaluate the function $s(t) = 2t^2 + 3t - 4$ at $t = -3$.

Replace t by its value. $s(t) = 2t^2 + 3t - 4$

Then simplify. $s(-3) = 2(-3)^2 + 3(-3) - 4 = 18 - 9 - 4$

An ordered pair of this function is $s(-3) = 5$
$(-3, 5)$.

For $f(x) = x^2$ and $g(x) = 3x$, evaluate $f(x) - g(x)$ at $x = 5$.

Replace x in each function by its value. $f(x) - g(x) = x^2 - 3x$
Then simplify. $f(5) - g(5) = 5^2 - 3(5) = 25 - 15$
$f(5) - g(5) = 10$

Evaluate the function $f(x) = x^2$ at $x = a + h$.

Replace x by its value.

Then simplify.

$$f(x) = x^2$$
$$f(a + h) = (a + h)^2$$
$$f(a + h) = a^2 + 2ah + h^2$$

Example 1

Evaluate the function $p(v) = \frac{v}{v-1}$ at $v = 3$.

Solution

$$p(v) = \frac{v}{v-1}$$

$$p(3) = \frac{3}{3-1}$$

$$p(3) = \frac{3}{2}$$

Example 2

Evaluate the function $s(t) = \frac{t}{t^2+1}$ at $t = -1$.

Your solution

$$s(-1) = -\frac{1}{2}$$

Example 3

For $f(x) = 2x + 1$ and $g(x) = x - 3$, evaluate $f(x) - g(x)$ at $x = 2$.

Solution

$$f(x) - g(x) = (2x + 1) - (x - 3)$$
$$f(2) - g(2) = (2 \cdot 2 + 1) - (2 - 3) = 5 - (-1)$$
$$f(2) - g(2) = 6$$

Example 4

For $f(x) = 2x^2$ and $g(x) = 4x - 1$, evaluate $f(x) - g(x)$ at $x = -1$.

Your solution

$$f(-1) - g(-1) = 7$$

Example 5

Find $f(3 + h) - f(3)$ for $f(x) = x^2 - 1$.

Solution

$$f(3 + h) - f(3) = [(3 + h)^2 - 1] - (3^2 - 1)$$
$$= (9 + 6h + h^2 - 1) - (9 - 1)$$
$$f(3 + h) - f(3) = h^2 + 6h$$

Example 6

Find $f(a + h) - f(a)$ for $f(x) = 2x^2 + 3$.

Your solution

$$f(a + h) - f(a) = 2h^2 + 4ah$$

Solutions on p. A34

Objective B	**To find the domain and range of a function**

The basic concept of a function is a set of ordered pairs. The ordered pairs may be formed by using a rule that pairs a member of a first set with only one member of a second set. The first set is the **domain** of the function. The second set is the **range** of the function.

In the grading scale shown at the right, a percent score (rounded to the nearest percent) is paired with one of the letters A, B, C, D, or F.

Score	Grade
90–100	A
80–89	B
70–79	C
60–69	D
0–59	F

The domain of the grading scale is the percent scores 0 to 100.

The range of the grading scale is the set of letters A, B, C, D, and F.

When a function is described as a set of ordered pairs, the domain of the function is the set of the first components in the ordered pairs. The range is the set of the second components in the ordered pairs.

Find the domain and range of the function $\{(-4, 3), (-2, 0), (0, -3), (2, -6)\}$.

The domain of the function is the set of the first components in the ordered pairs.

The domain is $\{-4, -2, 0, 2\}$.

The range of the function is the set of the second components in the ordered pairs.

The range is $\{3, 0, -3, -6\}$.

When a function is described by an equation, the range of the function can be found by evaluating the function at each number in the domain of the function.

Find the range of the function given by $f(x) = x^2 - x$ if the domain is $\{-2, -1, 0, 1, 2\}$.

Evaluate the function for each member of the domain. The range includes $f(-2)$, $f(-1)$, $f(0)$, $f(1)$, and $f(2)$.

$f(x) = x^2 - x$
$f(-2) = (-2)^2 - (-2) = 4 + 2 = 6$
$f(-1) = (-1)^2 - (-1) = 1 + 1 = 2$
$f(0) = 0^2 - 0 = 0 - 0 = 0$
$f(1) = 1^2 - 1 = 1 - 1 = 0$
$f(2) = 2^2 - 2 = 4 - 2 = 2$

The range is $\{0, 2, 6\}$.

Note from this example that the range does not include repetitions of elements. The numbers 0 and 2 are listed only once in the range.

Given a number c in the range of a function, it is possible to find an element in the domain that corresponds to c. For example, the number 3 is in the range of the function $f(x) = x^2 - 1$. Because 3 is in the range of the function, there must be at least one element c in the domain of the function for which $f(c) = 3$.

To find c, solve an equation.	$f(c) = 3$
Since $f(x) = x^2 - 1$, $f(c) = c^2 - 1$.	$c^2 - 1 = 3$
Solve for c.	$c^2 = 4$
	$c = \pm 2$

In this case there are two values of c, -2 and 2, that can be paired with 3. Two ordered pairs that belong to the function are $(-2, 3)$ and $(2, 3)$. Remember: A function can have different first elements paired with the same second element. A function cannot have the same first element paired with different second elements.

When an equation is used to define a function and the domain is not stated, the domain is the set of real numbers for which the equation produces real numbers. For example:

For all real numbers x, $f(x) = x^2 + 5$ is a real number. The domain of f is the set of real numbers.

The domain of $g(x) = \frac{1}{x + 2}$ is all real numbers except -2. When $x = -2$, $g(x) = \frac{1}{-2 + 2} = \frac{1}{0}$, which is not defined.

The domain of $h(x) = \sqrt{x - 3}$ is all real numbers greater than or equal to 3. When x is less than 3, $h(x)$ is not a real number. For example,

$h(-1) = \sqrt{-1 - 3} = \sqrt{-4}$ and $\sqrt{-4}$ is not a real number.

Domain of a Function Agreement

Unless otherwise stated, the domain of a function is all real numbers except:

1. those values for which the denominator of the function is zero; or
2. those values for which the value of the function is not a real number.

Example 7

Find the domain and range of the function $\{(-3, 3), (-2, 5), (0, 9), (1, 11)\}$.

Solution

The domain is $\{-3, -2, 0, 1\}$.
The range is $\{3, 5, 9, 11\}$.

Example 8

Find the domain and range of the function $\{(0, 2), (1, 0), (2, 0), (3, 2)\}$.

Your solution

The domain is $\{0, 1, 2, 3\}$.
The range is $\{0, 2\}$.

Solution on p. A34

Example 9

Find the range of the function defined by $f(x) = 3x - 1$ if the domain is $\{-1, 0, 1, 2\}$.

Solution

$$f(x) = 3x - 1$$
$$f(-1) = 3(-1) - 1 = -3 - 1 = -4$$
$$f(0) = 3(0) - 1 = 0 - 1 = -1$$
$$f(1) = 3(1) - 1 = 3 - 1 = 2$$
$$f(2) = 3(2) - 1 = 6 - 1 = 5$$

The range is $\{-4, -1, 2, 5\}$.

Example 10

Find the range of the function defined by $g(x) = x^2 - 1$ if the domain is $\{1, 2, 3, 4\}$.

Your solution

The range is $\{0, 3, 8, 15\}$.

Example 11

Find the domain of the function defined by $f(x) = \frac{x}{x^2 - 4}$.

Solution

By the Domain of a Function Agreement, the domain of the function must exclude values of x for which the denominator is zero. Equate the denominator to zero and solve for x.

$$x^2 - 4 = 0$$
$$x^2 = 4$$
$$x = \pm 2$$

The numbers -2 and 2 must be excluded from the domain. Therefore the domain is all real numbers except -2 and 2.

Example 12

Find the domain of the function defined by $g(x) = \sqrt{x^2 + 4}$.

Your solution

The domain of the function is all real numbers.

8.1   EXERCISES

▶ **Objective A**

For the function $f(x) = 3x^2$, find:

1. $f(2)$
12

2. $f(1)$
3

3. $f(-1)$
3

4. $f(-2)$
12

5. $f(a)$
$3a^2$

6. $f(w)$
$3w^2$

For the function $g(x) = x^2 - x + 1$, find:

7. $g(0)$
1

8. $g(1)$
1

9. $g(-2)$
7

10. $g(-1)$
3

11. $g(t)$
$t^2 - t + 1$

12. $g(s)$
$s^2 - s + 1$

For the function $f(x) = 4x - 3$, find:

13. $f(-2)$
-11

14. $f(3)$
9

15. $f(2 + h)$
$4h + 5$

16. $f(3 + h)$
$4h + 9$

17. $f(1 + h) - f(1)$
$4h$

18. $f(-1 + h) - f(-1)$
$4h$

For the function $g(x) = x^2 - 1$, find:

19. $g(2)$
3

20. $g(-3)$
8

21. $g(1 + h)$
$h^2 + 2h$

22. $g(2 + h)$
$h^2 + 4h + 3$

23. $g(3 + h) - g(3)$
$h^2 + 6h$

24. $g(-1 + h) - g(-1)$
$h^2 - 2h$

For $f(x) = 2x^2 - 3$ and $g(x) = -2x + 3$, find:

25. $f(2) - g(2)$
6

26. $f(3) - g(3)$
18

27. $f(0) + g(0)$
0

28. $f(1) + g(1)$
0

29. $f(1 + h) - f(1)$
$2h^2 + 4h$

30. $f(2 + h) - f(2)$
$2h^2 + 8h$

31. $\dfrac{g(1 + h) - g(1)}{h}$
-2

32. $\dfrac{g(-2 + h) - g(-2)}{h}$
-2

33. $\dfrac{f(3 + h) - f(3)}{h}$
$2h + 12$

34. $\dfrac{f(-1 + h) - f(-1)}{h}$
$2h - 4$

35. $\dfrac{g(a + h) - g(a)}{h}$
-2

36. $\dfrac{f(a + h) - f(a)}{h}$
$2h + 4a$

▶ **Objective B**

Find the domain and range of the function.

37. $\{(1, 1), (2, 4), (3, 7), (4, 10), (5, 13)\}$
D: $\{1, 2, 3, 4, 5\}$; R: $\{1, 4, 7, 10, 13\}$

38. $\{(2, 6), (4, 18), (6, 38), (8, 66), (10, 102)\}$
D: $\{2, 4, 6, 8, 10\}$; R: $\{6, 18, 38, 66, 102\}$

39. $\{(0, 1), (2, 2), (4, 3), (6, 4)\}$
D: $\{0, 2, 4, 6,\}$; R: $\{1, 2, 3, 4\}$

40. $\{(0, 1), (1, 2), (4, 3), (9, 4)\}$
D: $\{0, 1, 4, 9\}$; R: $\{1, 2, 3, 4\}$

41. $\{(1, 0), (3, 0), (5, 0), (7, 0), (9, 0)\}$
D: $\{1, 3, 5, 7, 9\}$; R: $\{0\}$

42. $\{(-2, -4), (2, 4), (-1, 1), (1, 1), (-3, 9), (3, 9)\}$
D: $\{-3, -2, -1, 1, 2, 3\}$; R: $\{-4, 1, 4, 9\}$

43. $\{(0, 0), (1, 1), (-1, 1), (2, 2), (-2, 2)\}$
D: $\{-2, -1, 0, 1, 2\}$; R: $\{0, 1, 2\}$

44. $\{(0, -5), (5, 0), (10, 5), (15, 10)\}$
D: $\{0, 5, 10, 15\}$; R: $\{-5, 0, 5, 10\}$

Find the range of the function defined by each equation.

45. $f(x) = 4x - 3$; domain = {0, 1, 2, 3, 4}
R: {−3, 1, 5, 9, 13}

46. $g(x) = x^2 + 2x - 1$; domain = {−2, −1, 0, 1, 2}
R: {−2, −1, 2, 7}

47. $h(x) = \frac{x}{2} + 3$; domain = {−4, −2, 0, 2, 4}
R: {1, 2, 3, 4, 5}

48. $F(x) = \sqrt{x + 1}$; domain = {−1, 0, 3, 8, 15}
R: {0, 1, 2, 3, 4}

49. $G(x) = \frac{2}{x + 3}$; domain = {−2, −1, 0, 1, 2}
R: $\left\{\frac{2}{5}, \frac{1}{2}, \frac{2}{3}, 1, 2\right\}$

50. $f(x) = |5x - 2|$; domain = {−10, −5, 0, 5, 10}
R: {2, 23, 27, 48, 52}

51. $g(a) = \frac{a^2 + 1}{3a - 1}$; domain = {−1, 0, 1, 2}
R: $\left\{-1, -\frac{1}{2}, 1\right\}$

52. $h(a) = (a^2 + 3a)^2$; domain = {−2, −1, 0, 1, 2}
R: {0, 4, 16, 100}

In Exercises 53–62, a function and a number in the range of that function are given. Find an element in the domain that corresponds to the number, and write an ordered pair that belongs to the function.

53. $f(x) = x + 5$; −3
−8; (−8, −3)

54. $g(x) = x - 4$; 6
10; (10, 6)

55. $h(a) = 3a + 2$; −1
−1; (−1, −1)

56. $f(a) = 2a - 5$; 0
$\frac{5}{2}$; $\left(\frac{5}{2}, 0\right)$

57. $g(x) = \frac{2}{3}x + \frac{1}{3}$; 1
1; (1, 1)

58. $h(x) = \frac{3}{2}x - 1$; −4
−2; (−2, −4)

59. $h(x) = x^2 + 3$; 7
± 2; (2, 7) or (−2, 7)

60. $g(x) = x^2 − 3$; −2
± 1; (1, −2) or (−1, −2)

61. $f(x) = \frac{x + 1}{5}$; 7
34; (34, 7)

62. $g(a) = \frac{a − 3}{4}$; 5
23; (23, 5)

What values of x are excluded from the domain of the function?

63. $f(x) = x^2 + 3x − 4$
none

64. $g(x) = 2x^2 − x + 5$
none

65. $h(x) = \frac{2x}{3x^2 − x}$

0, $\frac{1}{3}$

66. $F(x) = \frac{6}{5x^2 − 2x}$

0, $\frac{2}{5}$

67. $G(x) = \frac{x + 1}{x^2 + x − 6}$

−3, 2

68. $f(x) = \frac{x − 2}{x^2 − 3x − 4}$

−1, 4

69. $g(x) = \sqrt{6x − 2}$
$\left\{ x \mid x < \frac{1}{3} \right\}$

70. $h(x) = \sqrt{3x − 9}$
$\{ x \mid x < 3 \}$

71. $F(x) = |3x − 7|$
none

72. $G(x) = \frac{x}{|x|}$

0

SECTION 8.2	**Graphs of Functions**

Objective A	**To graph a function**

A function is a set of ordered pairs in which no two ordered pairs that have the same first component have different second components. The ordered pairs of a function can be written as $(x, f(x))$. However, often the value of the function, $f(x)$, is labeled y, and the ordered pairs are written (x, y), where $y = f(x)$. The **graph of a function** is a graph of the ordered pairs (x, y) of the function.

Since the graph of the equation $y = mx + b$ is a straight line, a function of the form $f(x) = mx + b$ is a **linear function.**

To graph the function $f(x) = 2x + 1$, think of the function as the equation $y = mx + b$.

This is the equation of a straight line. The slope is 2, and the y-intercept is 1.

Graph the straight line.

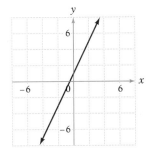

In the equation $y = 2x + 1$, the value of the variable y *depends* on the value of x. Thus y is called the **dependent variable**, and x is called the **independent variable.** For $f(x) = 2x + 1$, $f(x)$ is a symbol for the dependent variable.

In graphing functions, it is important to remember that $f(x)$ is the y-coordinate of an ordered pair.

The graph of the linear function given by $f(x) = x + 2$ is shown at the right.

When $x = -6$, $y = f(-6) = -4$.

When $x = 5$, $y = f(5) = 7$.

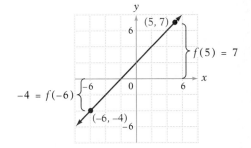

A **quadratic function** is given by $f(x) = ax^2 + bx + c$, $a \neq 0$. To graph a function of this type, consider the equation $y = ax^2 + bx + c$. As with a linear function, solutions of this quadratic equation in two variables are ordered pairs. The graph of a quadratic function can be drawn by finding ordered pair solutions of the quadratic equation in two variables. The graph of a quadratic function is called a **parabola.**

Graph: $f(x) = x^2 - 3$

Think of this as the equation $y = x^2 - 3$. Find enough ordered pairs (by evaluating the function for various values of x) to determine the shape of the graph. For convenience, the ordered pairs can be recorded in a table.

x	$f(x) = x^2 - 3$	$f(x)$	$(x, f(x))$
-3	$(-3)^2 - 3$	6	$(-3, 6)$
-2	$(-2)^2 - 3$	1	$(-2, 1)$
-1	$(-1)^2 - 3$	-2	$(-1, -2)$
0	$0^2 - 3$	-3	$(0, -3)$
1	$1^2 - 3$	-2	$(1, -2)$
2	$2^2 - 3$	1	$(2, 1)$
3	$3^2 - 3$	6	$(3, 6)$

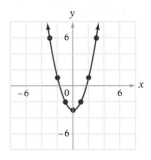

Draw a smooth curve through the points.

Because $f(x) = x^2 - 3$ is a real number for all values of x, the *domain* of the *function* is the *set of real numbers*. But note from the graph of this function that no portion of the parabola is below a y value of -3. For any value of x, $y \geq -3$. The range of this function is $\{y | y \geq -3\}$.

In general, the graph of any quadratic will resemble a "cup" shape as shown to the right. The cup will open up when $a > 0$ and open down when $a < 0$.

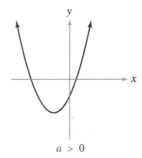

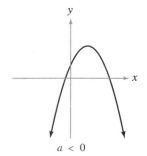

Graph: $f(x) = -x^2 - 2x + 2$

Because $a = -1$, the graph of this quadratic function will open down. Find enough ordered pairs to determine the position of the graph.

x	$f(x) = -x^2 - 2x + 2$	$f(x)$	$(x, f(x))$
-4	$-(-4)^2 - 2(-4) + 2$	-6	$(-4, -6)$
-3	$-(-3)^2 - 2(-3) + 2$	-1	$(-3, -1)$
-2	$-(-2)^2 - 2(-2) + 2$	2	$(-2, 2)$
-1	$-(-1)^2 - 2(-1) + 2$	3	$(-1, 3)$
0	$-(0)^2 - 2(0) + 2$	2	$(0, 2)$
1	$-(1)^2 - 2(1) + 2$	-1	$(1, -1)$
2	$-(2)^2 - 2(2) + 2$	-6	$(2, -6)$

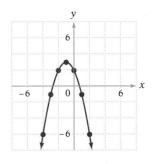

Content and Format © 1991 HMCo.

The domain of the function given by $f(x) = -x^2 - 2x + 2$ is the set of real numbers. To find the range of the function, complete the square of $-x^2 - 2x + 2$.

$$f(x) = -x^2 - 2x + 2$$
$$= -(x^2 + 2x) + 2$$
$$= -(x^2 + 2x + 1) + 1 + 2$$
$$= -(x + 1)^2 + 3$$

Because $-(x + 1)^2 \leq 0$ for all x, $-(x + 1)^2 + 3 \leq 3$ for all values of x. The range of the function is $\{y | y \leq 3\}$.

To graph $f(x) = x^3 - 3x + 3$, think of the function as the equation $y = x^3 - 3x + 3$.

Find enough ordered pairs to draw the graph. For more complicated curves, you may need to graph many points before an accurate graph can be drawn.

x	y
-3	-15
-2	1
-1	5
0	3
1	1
2	5

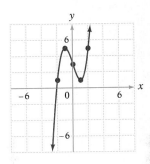

This function is an example of a cubic function. The domain of any cubic function is the set of real numbers. The range of any cubic function is the set of real numbers.

Example 1

Graph $f(x) = \frac{2}{3}x - 1$. Find the domain and range.

Solution

This is a linear function.

The slope is $\frac{2}{3}$ and the y-intercept is $(0, -1)$.

The domain of the function is the set of real numbers. The range is the set of real numbers.

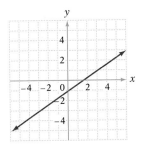

Example 2

Graph $f(x) = -2x + 1$. Find the domain and range.

Your solution

This is a linear function. The slope is -2 and the y-intercept is $(0, 1)$. The domain of the function is the set of real numbers. The range is the set of real numbers.

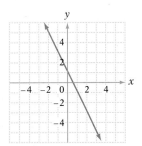

Solution on p. A34

Example 3

Graph: $f(x) = |x| - 1$
State the domain and range of the function.

Solution

x	y
−3	2
−2	1
−1	0
0	−1
1	0
2	1
3	2

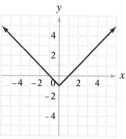

The domain is the set of real numbers.
The range is $\{y | y \geq -1\}$.

Example 4

Graph: $f(x) = |x - 1|$
State the domain and range of the function.

Your solution

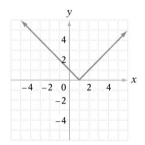

The domain is the set of real numbers.
The range is $\{y | y \geq 0\}$.

Example 5

Graph: $f(x) = 2x^2 + 6x$
State the domain and range of the function.

Solution

This is a quadratic function with $a > 0$.
The graph is a parabola that opens up.

x	y
−4	8
−3	0
−2	−4
−1	−4
0	0
1	8

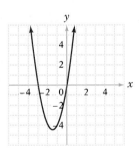

The domain is the set of real numbers.

$2x^2 + 6x = 2(x^2 + 3x)$

$\qquad = 2\left(x^2 + 3x + \frac{9}{4}\right) - \frac{9}{2}$

$\qquad = 2\left(x + \frac{3}{2}\right)^2 - \frac{9}{2}$

The range is $\left\{y \middle| y \geq -\frac{9}{2}\right\}$.

Example 6

Graph: $f(x) = -x^3 + 2$
State the domain and range of the function.

Your solution

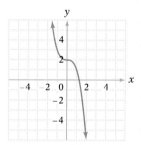

The domain is the set of real numbers.
The range is the set of real numbers.

| Objective B | **To determine if a relation is a function** |

As stated previously, not every correspondence between two sets is a function. A function assigns to each member of the domain one and only one member of the range. A relation assigns to each member of the domain one or more members of the range.

For example, the correspondence that pairs a positive real number with a square root is not a function because each positive real number can be paired with a positive or negative square root of that number. For example, since 16 can be paired with 4 or -4, the ordered pairs (16, 4) and (16, -4) would belong to the correspondence. But by the definition of a function, no two ordered pairs with the same first component can have different second components.

The ordered pairs (1, 1), (1, -1), (4, 2), (4, -2), (9, 3), and (9, -3) belong to the relation that pairs a positive real number with its square root. The graph of the relation is shown below.

Because each positive real number x has two square roots, a vertical line intersects the graph of the relation more than once.

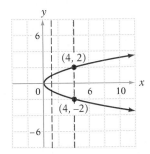

A function is a set of ordered pairs in which no two ordered pairs with the same first component have different second components. Consider the function $f(x) = x^3$. The ordered pairs (-2, -8), (-1, -1), (0, 0), (1, 1), (2, 8) belong to the function.

Because every real number has only one cube, there is only one y coordinate for each x. A vertical line intersects the graph of a function no more than once.

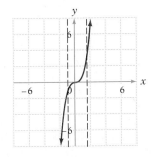

Vertical Line Test

> A graph is the graph of a function if any vertical line intersects the graph at no more than one point.

For example, the graph of a circle is not the graph of a function because a vertical line can intersect the graph more than once. The graph of a quadratic function is the graph of a function. Any vertical line intersects the graph at most once.

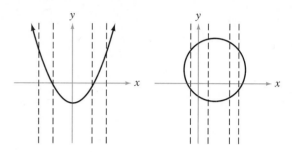

Example 7

Determine if the set of ordered pairs {(−1, 1), (0, 2), (1, 3), (2, 4), (1, 5)} is a function.

Solution

Because (1, 3) and (1, 5) belong to the set, there are two members of the set with the same first component and different second components. The set of ordered pairs is not a function.

Example 8

Determine if the set of ordered pairs {(−2, 3), (−1, 3), (0, 3), (1, 3), (2, 3)} is a function.

Your solution

No member of the set has the same first component and a different second component. The set of ordered pairs is a function.

Example 9

Determine if the graph is the graph of a function.

Solution

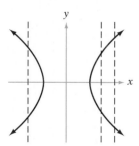

A vertical line intersects the graph more than once. The graph is not the graph of a function.

Example 10

Determine if the graph is the graph of a function.

Your solution

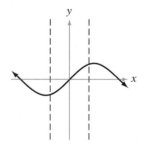

A vertical line intersects the graph at most once. The graph is the graph of a function.

Solutions on pp. A34–A35

Objective C	**To determine from a graph whether a function is one-to-one**

Recall that a function is a set of ordered pairs in which no two ordered pairs that have the same first component have different second components. This means that given any x there is only one y that can be paired with that x. A **one-to-one function** satisfies the additional condition that given any y, there is only one x which can be paired with the given y. One-to-one functions are commonly expressed by writing "1–1."

One-to-One Function

A function is a one-to-one function if, given any b in the range of f, there is exactly one c in the domain of f for which $f(c) = b$.

The function defined by $f(x) = x^2$ is *not* a 1–1 function since, given 4 in the range of f, there are two possible values in the domain of f, -2 and 2, for which $f(-2) = 4$ and $f(2) = 4$.

Given $f(x) = 2x - 1$ with domain $\{-1, 0, 1, 2, 3\}$, determine if f is a 1–1 function.

Evaluate the function at each element of the domain.

$$f(-1) = 2(-1) - 1 = -3$$
$$f(0) = 2(0) - 1 = -1$$
$$f(1) = 2(1) - 1 = 1$$
$$f(2) = 2(2) - 1 = 3$$
$$f(3) = 2(3) - 1 = 5$$

Because no two elements of the range $\{-3, -1, 1, 3, 5\}$ are equal, the function is a 1–1 function.

Given $f(x) = x^2 - 4x$ with domain $\{-1, 0, 1, 2, 3\}$, determine if f is a 1–1 function.

Evaluate the function at each element of the domain.

$$f(-1) = (-1)^2 - 4(-1) = 5$$
$$f(0) = (0)^2 - 4(0) = 0$$
$$f(1) = (1)^2 - 4(1) = -3$$
$$f(2) = (2)^2 - 4(2) = -4$$
$$f(3) = (3)^2 - 4(3) = -3$$

Because $f(1) = f(3)$ and $1 \neq 3$, f is not a 1–1 function.

When the domain of a function is an infinite set, it is not possible to decide if the function is a 1–1 function by evaluating the function at each element of the domain. In these cases, the graph of the function is helpful to determine a 1–1 function.

The ordered pairs $(-3, 9)$, $(-2, 4)$, $(-1, 1)$, $(0, 0)$, $(1, 1)$, $(2, 4)$, and $(3, 9)$ belong to the function defined by $f(x) = x^2$. The graph of this function is shown below.

Note, because the square of a positive or negative real number is positive, a horizontal line intersects the graph of f more than once.

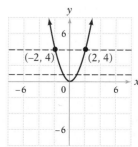

The function defined by $g(x) = x^3$ is a 1–1 function. Given an element b in the range of g, there is exactly one element of the domain of g that is the cube root of b. For example, given 8 in the range of g, 2 is the only member of the domain of g for which $g(2) = 8$. Some of the ordered pairs of g are $(-2, -8)$, $(-1, -1)$, $(0, 0)$, $(1, 1)$, and $(2, 8)$. The graph of g is shown below.

Note, because every real number has only one cube, a horizontal line intersects the graph of g once.

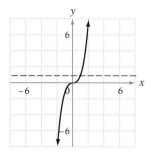

Horizontal Line Test

The graph of a function represents a 1–1 function if any horizontal line intersects the graph at no more than one point.

Example 11

Determine if the graph represents the graph of a 1–1 function.

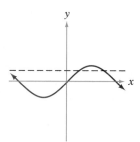

Solution

Since a horizontal line intersects the graph more than once, the graph is not the graph of a 1–1 function.

Example 12

Determine if the graph represents the graph of a 1–1 function.

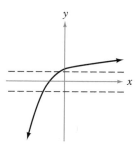

Your solution

Since any horizontal line will intersect the graph no more than once, the graph is the graph of a 1–1 function.

Solution on p. **A35**

8.2   EXERCISES

▶ **Objective A**

Graph:

1. $f(x) = 2x - 1$

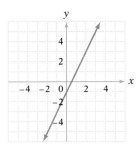

2. $f(x) = 2x + 4$

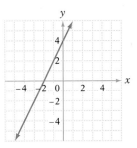

3. $f(x) = -3x - 1$

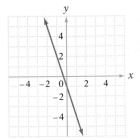

4. $f(x) = -\frac{1}{2}x + 1$

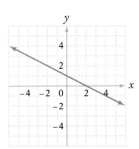

5. $f(x) = \frac{2}{3}x + 4$

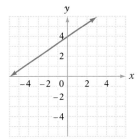

6. $f(x) = \frac{3}{4}x + 1$

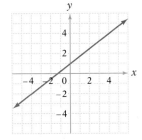

7. $f(x) = -\frac{4}{3}x + 3$

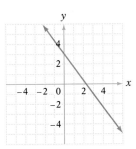

8. $f(x) = \frac{2}{3}x - 5$

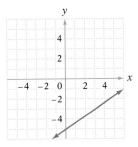

9. $f(x) = 3x - 6$

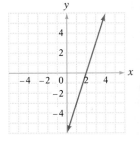

10. $f(x) = -x + 4$

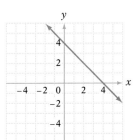

11. $f(x) = 2x^2 - 1$

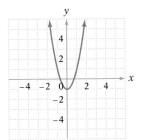

12. $f(x) = 2x^2 - 3$

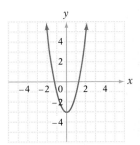

Graph:

13. $f(x) = \frac{1}{2}x^2$

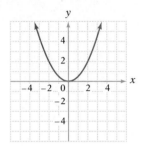

14. $f(x) = \frac{1}{3}x^2$

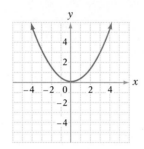

15. $f(x) = -x^2 + 2x - 1$

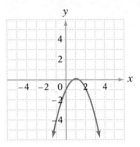

16. $f(x) = -x^2 + 4x - 4$

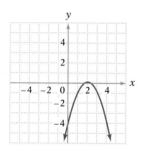

17. $f(x) = -\frac{1}{2}x^2 + 1$

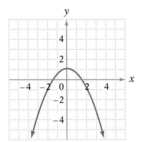

18. $f(x) = -2x^2 + 2$

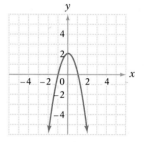

19. $f(x) = x^2 + 4x + 4$

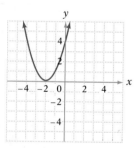

20. $f(x) = x^2 + 5x + 6$

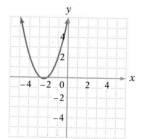

21. $f(x) = |x - 3|$

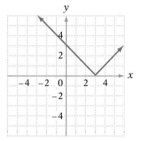

22. $f(x) = |x| + 1$

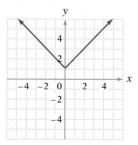

23. $f(x) = 2|x| - 1$

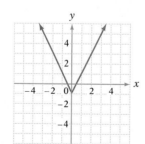

24. $f(x) = 2|x| + 2$

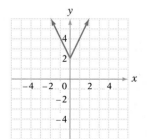

Graph:

25. $f(x) = |2x - 1|$

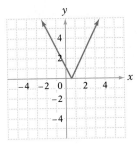

26. $f(x) = |2x + 2|$

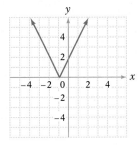

27. $f(x) = 2|x + 1|$

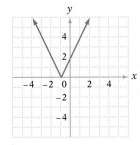

28. $f(x) = 3|2 - x|$

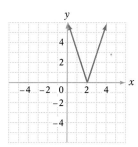

29. $f(x) = 2|2x| - 3$

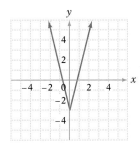

30. $f(x) = \left|\frac{1}{2}x\right| - 2$

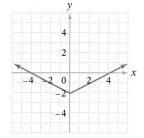

31. $f(x) = x^3 - 1$

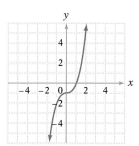

32. $f(x) = 1 - x^3$

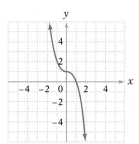

33. $f(x) = x^3 - 3x + 2$

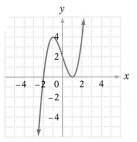

34. $f(x) = x^3 + 4x^2 + 4x$

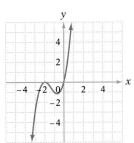

35. $f(x) = x^3 - x^2 - x + 1$

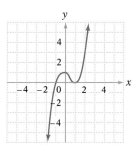

36. $f(x) = x^3 - x^2 - 4x + 4$

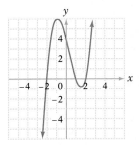

▶ **Objective B**

Determine if the given set of ordered pairs is a function. If it is a function, determine if it is a 1–1 function.

37. {(−2, 3), (−1, 3), (0, 3), (1, 3)}
function, not 1–1 function

38. {(2, 6), (3, 5), (4, 4), (5, 3)}
function, 1–1 function

39. {(−4, 4), (−2, 3), (4, 2), (4, 1)}
not a function

40. {(3, 1,) (3, 5), (4, 7) (5, 8)}
not a function

41. {(0, −2), (−1, 5), (3, 3), (−4, 6)}
function, 1–1 function

42. {(−2, −2), (0, 0), (2, 2), (4, 4)}
function, 1–1 function

43. {(0, 2), (0, 1), (0, 3), (0, 4)}
not a function

44. {(−5, −1), (−4, 0), (−3, −1), (−2, 2)}
function, not 1–1 function

Determine if the graph represents the graph of a function.

45.

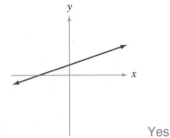

Yes

46.

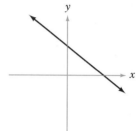

Yes

47.

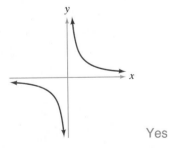

Yes

48.

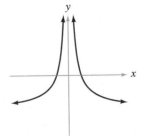

Yes

Determine if the graph represents the graph of a function.

49.

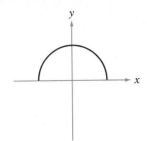

Yes

50.

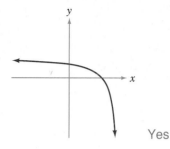

Yes

51.

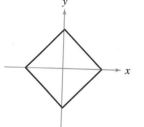

No

52.

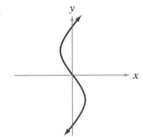

No

▶ Objective C

Determine if the graph represents the graph of a 1–1 function.

53.

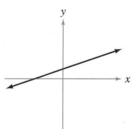

Yes

54.

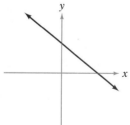

Yes

Determine if the graph represents the graph of a 1–1 function.

55.

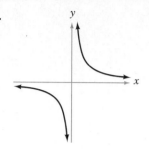

Yes

56.

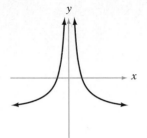

No

57.

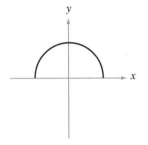

No

58.

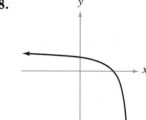

Yes

59.

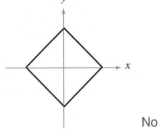

No

60.

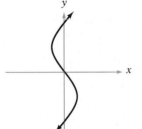

No

SECTION 8.3	**Composite Functions and Inverse Functions**

Objective A	**To find the composition of two functions**

Recall from Section 8.1 that to evaluate a function, replace the variable by the given value and then simplify.

For example, to evaluate the function

$$f(x) = -2x^2 + x - 3 \text{ at } x = 2$$

replace x by 2 and then simplify.

$$f(x) = -2x^2 + x - 3$$
$$f(2) = -2(2)^2 + 2 - 3 = -8 + 2 - 3 = -9$$

The value of the function at 2 is -9.

A function can be evaluated at the value of another function. For example, consider the functions

$$f(x) = 2x + 7 \text{ and } g(x) = x^2 + 1$$

The expression $f(g(-2))$ means to evaluate the function f at $g(-2)$.

$$g(-2) = (-2)^2 + 1 = 4 + 1 = 5$$
$$f(g(-2)) = f(5) = 2(5) + 7 = 10 + 7 = 17$$

In general, a function whose value is $f(g(x))$ can be defined as long as $g(x)$ is in the domain of f.

Composition of Two Functions

Let f and g be two functions such that $g(x)$ is in the domain of f for all x in the domain of g. Then the **composition** of the two functions, denoted by $f \circ g$, is the function whose value at x is given by $(f \circ g)(x) = f(g(x))$.

The function defined by $f(g(x))$ is called the **composite** of the two functions f and g.

Consider the two functions given by $f(x) = 3x - 2$ and $g(x) = x^2 - 2x$.

$$(f \circ g)(x) = f(g(x)) = 3(x^2 - 2x) - 2$$
$$= 3x^2 - 6x - 2$$

On the other hand,

$$(g \circ f)(x) = g(f(x)) = (3x - 2)^2 - 2(3x - 2)$$
$$= 9x^2 - 12x + 4 - 6x + 4$$
$$= 9x^2 - 18x + 8$$

This example illustrates that $f(g(x))$ can be different from $g(f(x))$.

Example 1

Given $f(x) = x^2 - x$ and $g(x) = 3x - 2$, find $f(g(3))$ and $g(f(1))$.

Solution

$g(3) = 3(3) - 2 = 9 - 2 = 7$
$f(g(3)) = f(7) = 7^2 - 7 = 49 - 7 = 42$

$f(1) = 1^2 - 1 = 1 - 1 = 0$
$g(f(1)) = g(0) = 3(0) - 2 = 0 - 2 = -2$

Example 2

Given $f(x) = 1 - 2x$ and $g(x) = x^2$, find $f(g(-1))$ and $g(f(-1))$.

Your solution

$f(g(-1)) = -1$
$g(f(-1)) = 9$

Example 3

Given $f(x) = 3x + 4$ and $g(x) = 1 - x^2$, find $f(g(x))$.

Solution

$f(g(x)) = 3(1 - x^2) + 4$
$= 3 - 3x^2 + 4$
$= 7 - 3x^2$

Example 4

Given $f(x) = 2x$ and $g(x) = \dfrac{1}{x^2 - 2}$, find $g(f(x))$.

Your solution

$\dfrac{1}{4x^2 - 2}$

Solutions on p. A35

Objective B	**To find the inverse of a function**

The **inverse of a function** is the set of ordered pairs formed by reversing the components of each ordered pair of the function.

For example, some of the ordered pairs of the function defined by $f(x) = 2x$ are $(0, 0)$, $(-1, -2)$, $(3, 6)$, and $\left(\frac{1}{2}, 1\right)$.

The inverse of this function would contain the ordered pairs $(0, 0)$, $(-2, -1)$, $(6, 3)$, and $\left(1, \frac{1}{2}\right)$.

Now consider the function defined by $g(x) = x^2$. Some of the ordered pairs of this function are $(0, 0)$, $(-1, 1)$, $(1, 1)$, $(-3, 9)$, and $(3, 9)$. Reversing the ordered pairs gives $(0, 0)$, $(1, -1)$, $(1, 1)$, $(9, -3)$, and $(9, 3)$. These ordered pairs do not satisfy the definition of a function because there are ordered pairs with the same first element and different second elements. This example illustrates that not all functions have an inverse function.

The graphs of the two functions $f(x) = 2x$ and $g(x) = x^2$ are shown below.

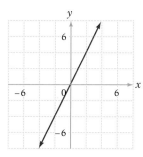
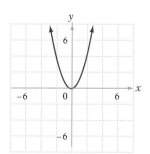

By the horizontal line test, the function f is a 1–1 function while the function g is not a 1–1 function.

Condition for an Inverse Function

A function f has an inverse function if and only if f is a 1–1 function.

The symbol f^{-1} is used to denote the inverse of a function. $f^{-1}(x)$ is read "f inverse of x." Note that this is not the reciprocal of $f(x)$ but is the notation used for the inverse of a 1–1 function.

Because the ordered pairs of an inverse function are reversed from those of the function, the domain of f^{-1} is the range of f, and the range of f^{-1} is the domain of f.

To find the inverse function of a linear function, interchange x and y. Then solve for y.

To find the inverse function of $f(x) = 3x + 6$, think of the function as the equation $y = 3x + 6$.

$$f(x) = 3x + 6$$
$$y = 3x + 6$$

Interchange x and y.

$$x = 3y + 6$$

Solve for y.

$$3y = x - 6$$
$$y = \tfrac{1}{3}x - 2$$

Replace y with $f^{-1}(x)$.

$$f^{-1}(x) = \tfrac{1}{3}x - 2$$

The inverse of the linear function $f(x) = 3x + 6$ is the linear function $f^{-1}(x) = \tfrac{1}{3}x - 2$.

The inverse of a linear function with $m \neq 0$ will always be another linear function.

The graph of $f(x) = 2x - 4$ and its inverse, $f^{-1}(x) = \tfrac{1}{2}x + 2$, is shown at the right.

The graph of $f^{-1}(x) = \tfrac{1}{2}x + 2$ is the mirror image of the graph of $f(x) = 2x - 4$ with respect to the line $y = x$.

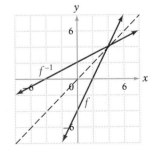

If two functions are inverses, then their graphs are mirror images of each other with respect to the line $y = x$.

The function given by $f(x) = \frac{1}{2}x^2$ does not have an inverse function. Two of the ordered pair solutions of this function are $(4, 8)$ and $(-4, 8)$.

The graph of $f(x) = \frac{1}{2}x^2$ is shown at the right. This graph does not pass the horizontal line test for the graph of a 1–1 function. The mirror image of the graph with respect to the line $y = x$ is also shown. This graph does not pass the vertical line test for the graph of a function.

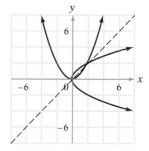

A quadratic function with domain the real numbers does not have an inverse function.

The composite functions $f(f^{-1}(x))$ and $f^{-1}(f(x))$ have the following property:

$$f(f^{-1}(x)) = f^{-1}(f(x)) = x$$

This property can be used to determine if one function is the inverse of a second function. For the functions $f(x) = 3x + 6$ and $f^{-1}(x) = \frac{1}{3}x - 2$,

$$f(f^{-1}(x)) = f\left(\frac{1}{3}x - 2\right) \qquad f^{-1}(f(x)) = f^{-1}(3x + 6)$$
$$= 3\left(\frac{1}{3}x - 2\right) + 6 \qquad = \frac{1}{3}(3x + 6) - 2$$
$$= x - 6 + 6 \qquad = x + 2 - 2$$
$$= x \qquad = x$$

This concept of inverse functions is similar to the additive inverse and multiplicative inverse used in arithmetic operations. For example, adding the number a and its additive inverse $(-a)$ to an expression results in the original expression.

$$x = x + a + (-a) = x$$

Inverse functions operate in a similar manner; one undoes the other.

For the functions $f(x) = 2x - 4$ and $f^{-1}(x) = \frac{1}{2}x + 2$,

$$f^{-1}(f(2)) = f^{-1}(0) \qquad f(f^{-1}(2)) = f(3)$$
$$= 2 \qquad = 2$$

Example 5

Find the inverse of the function defined by $f(x) = 4x - 8$.

Solution

$$f(x) = 4x - 8$$
$$y = 4x - 8$$
$$x = 4y - 8 \qquad \text{Interchange } x \text{ and } y.$$
$$x + 8 = 4y \qquad \text{Solve for } y.$$
$$\tfrac{1}{4}x + 2 = y$$

The inverse function is $f^{-1}(x) = \tfrac{1}{4}x + 2$.

Example 6

Find the inverse of the function defined by $f(x) = \tfrac{1}{2}x + 1$.

Your solution

$f^{-1}(x) = 2x - 2$

Example 7

Are the functions $f(x) = 4 - 2x$ and $g(x) = -\tfrac{1}{2}x + 2$ inverses of each other?

Solution

$$f(g(x)) = f\left[-\tfrac{1}{2}x + 2\right]$$
$$= 4 - 2\left[-\tfrac{1}{2}x + 2\right]$$
$$= 4 + x - 4 = x$$

$$g(f(x)) = g(4 - 2x)$$
$$= -\tfrac{1}{2}(4 - 2x) + 2$$
$$= -2 + x + 2 = x$$

$f(g(x)) = g(f(x)) = x$. Therefore the functions are inverses of each other.

Example 8

Are the functions $f(x) = 3x + 12$ and $g(x) = \tfrac{1}{3}x - 4$ inverses of each other?

Your solution

$f(g(x)) = g(f(x)) = x$. Therefore the functions are inverses of each other.

Solutions on p. A35

Content and Format © 1991 HMCo.

8.3   **EXERCISES**

▶ Objective A

Given $f(x) = 2x - 3$ and $g(x) = 4x - 1$, evaluate the composite function.

1. $f(g(0))$
    -5

2. $g(f(0))$
    -13

3. $f(g(2))$
    11

4. $g(f(-2))$
    -29

5. $f(g(x))$
    $8x - 5$

6. $g(f(x))$
    $8x - 13$

Given $h(x) = 2x + 4$ and $f(x) = \frac{1}{2}x + 2$, evaluate the composite function.

7. $h(f(0))$
    8

8. $f(h(0))$
    4

9. $h(f(2))$
    10

10. $f(h(-1))$
    3

11. $h(f(x))$
    $x + 8$

12. $f(h(x))$
    $x + 4$

Given $g(x) = x^2 + 3$ and $h(x) = x - 2$, evaluate the composite function.

13. $g(h(0))$
    7

14. $h(g(0))$
    1

15. $g(h(4))$
    7

16. $h(g(-2))$
    5

17. $g(h(x))$
    $x^2 - 4x + 7$

18. $h(g(x))$
    $x^2 + 1$

Given $f(x) = x^2 + x + 1$ and $h(x) = 3x + 2$, evaluate the composite function.

19. $f(h(0))$
    7

20. $h(f(0))$
    5

21. $f(h(-1))$
    1

22. $h(f(-2))$
    11

23. $f(h(x))$
    $9x^2 + 15x + 7$

24. $h(f(x))$
    $3x^2 + 3x + 5$

▶ Objective B

Find the inverse of the function. If the function does not have an inverse function, write "no inverse."

25. $\{(1, 0), (2, 3), (3, 8), (4, 15)\}$
    $\{(0, 1), (3, 2), (8, 3), (15, 4)\}$

26. $\{(1, 0), (2, 1), (-1, 0), (-2, 1)\}$
    no inverse

27. $\{(3, 5), (-3, -5), (2, 5), (-2, -5)\}$
    no inverse

28. $\{(-5, -5), (-3, -1), (-1, 3), (1, 7)\}$
    $\{(-5, -5), (-1, -3), (3, -1), (7, 1)\}$

Find the inverse of the function. If the function does not have an inverse function, write "no inverse."

29. $f(x) = 4x - 8$
$f^{-1}(x) = \frac{1}{4}x + 2$

30. $f(x) = 3x + 6$
$f^{-1}(x) = \frac{1}{3}x - 2$

31. $f(x) = x^2 - 1$
no inverse

32. $f(x) = 2x + 4$
$f^{-1}(x) = \frac{1}{2}x - 2$

33. $f(x) = x - 5$
$f^{-1}(x) = x + 5$

34. $f(x) = \frac{1}{2}x - 1$
$f^{-1}(x) = 2x + 2$

35. $f(x) = \frac{1}{3}x + 2$
$f^{-1}(x) = 3x - 6$

36. $f(x) = -2x + 2$
$f^{-1}(x) = -\frac{1}{2}x + 1$

37. $f(x) = -3x - 9$
$f^{-1}(x) = -\frac{1}{3}x - 3$

38. $f(x) = 2x^2 + 2$
no inverse

39. $f(x) = \frac{2}{3}x + 4$
$f^{-1}(x) = \frac{3}{2}x - 6$

40. $f(x) = \frac{3}{4}x - 4$
$f^{-1}(x) = \frac{4}{3}x + \frac{16}{3}$

41. $f(x) = -\frac{1}{3}x + 1$
$f^{-1}(x) = -3x + 3$

42. $f(x) = -\frac{1}{2}x + 2$
$f^{-1}(x) = -2x + 4$

43. $f(x) = 2x - 5$
$f^{-1}(x) = \frac{1}{2}x + \frac{5}{2}$

44. $f(x) = 3x + 4$
$f^{-1}(x) = \frac{1}{3}x - \frac{4}{3}$

45. $f(x) = x^2 + 3$
no inverse

46. $f(x) = 5x - 2$
$f^{-1}(x) = \frac{1}{5}x + \frac{2}{5}$

47. $f(x) = 4x - 2$
$f^{-1}(x) = \frac{1}{4}x + \frac{1}{2}$

48. $f(x) = 6x - 3$
$f^{-1}(x) = \frac{1}{6}x + \frac{1}{2}$

49. $f(x) = -8x + 4$
$f^{-1}(x) = -\frac{1}{8}x + \frac{1}{2}$

50. $f(x) = -6x + 2$
$f^{-1}(x) = -\frac{1}{6}x + \frac{1}{3}$

51. $f(x) = 8x + 6$
$f^{-1}(x) = \frac{1}{8}x - \frac{3}{4}$

52. $f(x) = \frac{1}{2}x^2 - 4$
no inverse

Are the functions inverses of each other?

53. $f(x) = 4x$; $g(x) = \frac{x}{4}$
yes

54. $g(x) = x + 5$; $h(x) = x - 5$
yes

55. $f(x) = 3x$; $h(x) = \frac{1}{3x}$
no

56. $h(x) = x + 2$; $g(x) = 2 - x$
no

57. $g(x) = 3x + 2$; $f(x) = \frac{1}{3}x - \frac{2}{3}$
yes

58. $h(x) = 4x - 1$; $f(x) = \frac{1}{4}x + \frac{1}{4}$
yes

59. $f(x) = \frac{1}{2}x - \frac{3}{2}$; $g(x) = 2x + 3$
yes

60. $g(x) = -\frac{1}{2}x - \frac{1}{2}$; $h(x) = -2x + 1$
no

Complete:

61. The domain of the inverse function f^{-1} is the _____ of f.
range

62. The range of the inverse function f^{-1} is the _____ of f.
domain

SECTION 8.4	# Variation

Objective A | **To solve variation problems**

Direct variation is a special function which can be expressed as the equation $y = kx$, where k is a constant. The equation $y = kx$ is read "y varies directly as x" or "y is proportional to x." The constant k is called the **constant of variation** or the **constant of proportionality**.

The circumference (C) of a circle varies directly as the diameter (d). The direct variation equation is written $C = \pi d$. The constant of variation is π.

A nurse makes \$20 per hour. The total wage (w) of the nurse is directly proportional to the number of hours (h) worked. The equation of variation is $w = 20h$. The constant of proportionality is 20.

A direct variation equation can be written in the form $y = kx^n$, where n is a positive number. For example, the equation $y = kx^2$ is read "y varies directly as the square of x."

The area (A) of a circle varies directly as the square of the radius (r) of the circle. The direct variation equation is $A = \pi r^2$. The constant of variation is π.

Given that V varies directly as r and that $V = 20$ when $r = 4$, find the constant of variation and the equation of variation.

Write the basic direct variation equation.	$V = kr$
Replace V and r by the given values.	$20 = k \cdot 4$
Solve for the constant of variation.	$5 = k$

The constant of variation is 5.

Write the direct variation equation by substituting the value of k into the basic direct variation equation.	$V = 5r$

The tension (T) in a spring varies directly as the distance (x) it is stretched. If $T = 8$ lb when $x = 2$ in., find T when $x = 4$ in.

Write the basic direct variation equation.	$T = kx$
Replace T and x by the given values.	$8 = k \cdot 2$
Solve for the constant of variation.	$4 = k$
Write the direct variation equation by substituting the value of k into the basic direct variation equation.	$T = 4x$
To find T when $x = 4$, substitute 4 for x in the equation and solve for T.	$T = 4 \cdot 4 = 16$

The tension is 16 lb.

Inverse variation is a function which can be expressed as the equation $y = \frac{k}{x}$, where k is a constant. The equation $y = \frac{k}{x}$ is read "y varies inversely as x" or "y is inversely proportional to x."

In general, an inverse variation equation can be written $y = \frac{k}{x^n}$, where n is a positive number. For example, the equation $y = \frac{k}{x^2}$ is read "y varies inversely as the square of x."

Given that P varies inversely as the square of x and that $P = 5$ when $x = 2$, find the variation constant and the equation of variation.

Write the basic inverse variation equation. $P = \frac{k}{x^2}$

Replace P and x by the given values. $5 = \frac{k}{2^2}$

Solve for the constant of variation. $5 = \frac{k}{4}$

$20 = k$

The constant of variation is 20.

Write the inverse variation equation by substituting the value of k into the basic inverse variation equation. $P = \frac{20}{x^2}$

The length (L) of a rectangle with fixed area is inversely proportional to the width (w). If $L = 6$ ft when $w = 2$ ft, find L when $w = 3$ ft.

Write the basic inverse variation equation. $L = \frac{k}{w}$

Replace L and w by the given values. $6 = \frac{k}{2}$

Solve for the constant of variation. $12 = k$

Write the inverse variation equation by substituting the value of k into the basic inverse variation equation. $L = \frac{12}{w}$

To find L when $w = 3$ ft, substitute 3 for w in the equation and solve for L. $L = \frac{12}{3} = 4$

The length is 4 ft.

Joint variation is a variation in which a variable varies directly as the product of two or more other variables. A joint variation can be expressed as the equation $z = kxy$, where k is a constant. The equation $z = kxy$ is read "z varies jointly as x and y."

The area (A) of a triangle varies jointly as the base (b) and the height (h). The joint variation equation is written $A = \frac{1}{2}bh$. The constant of variation is $\frac{1}{2}$.

A **combined variation** is a variation in which two or more types of variation occur at the same time. For example, in physics, the volume (V) of a gas varies directly as the temperature (T) and inversely as the pressure (P). This combined variation is written $V = \frac{kT}{P}$.

A ball is being twirled on the end of a string. The tension (T) in the string is directly proportional to the square of the speed (v) of the ball and inversely proportional to the length (r) of the string. If the tension is 96 lb when the length of the string is 0.5 ft and the speed is 4 ft/s, find the tension when the length of the string is 1 ft and the speed is 5 ft/s.

Write the basic combined variation equation.

$$T = \frac{kv^2}{r}$$

Replace T, v, and r by the given values.

$$96 = \frac{k \cdot 4^2}{0.5}$$

Solve for the constant of variation.

$$96 = \frac{k \cdot 16}{0.5}$$
$$96 = k \cdot 32$$
$$3 = k$$

Write the combined variation equation by substituting the value of k into the basic combined variation equation.

$$T = \frac{3v^2}{r}$$

To find T when $r = 1$ ft and $v = 5$ ft/s, substitute 1 for r and 5 for v and solve for T.

$$T = \frac{3 \cdot 5^2}{1} = 3 \cdot 25 = 75$$

The tension is 75 lb.

Example 1

The amount (A) of medication prescribed for a person is directly related to the person's weight (W). For a 50-kilogram person, 2 ml of medication are prescribed. How many milliliters of medication are required for a person who weighs 75 kg?

Strategy

To find the required amount of medication:

- Write the basic direct variation equation, replace the variables by the given values, and solve for k.
- Write the direct variation equation, replacing k by its value. Substitute 75 for W and solve for A.

Solution

$$A = kW$$
$$2 = k \cdot 50$$
$$\frac{1}{25} = k$$
$$A = \frac{1}{25}W = \frac{1}{25} \cdot 75 = 3$$

The required amount of medication is 3 ml.

Example 2

The distance (s) a body falls from rest varies directly as the square of the time (t) of the fall. An object falls 64 ft in 2 s. How far will it fall in 5 s?

Your strategy

Your solution

400 ft

Solution on p. A35

Example 3

A company which produces personal computers has determined that the number of computers it can sell (s) is inversely proportional to the price (P) of the computer. Two thousand computers can be sold when the price is $2500. How many computers can be sold if the price of a computer is $2000?

Strategy

To find the number of computers:

■ Write the basic inverse variation equation, replace the variables by the given values, and solve for k.
■ Write the inverse variation equation, replacing k by its value. Substitute 2000 for P and solve for s.

Solution

$$s = \frac{k}{P}$$

$$2000 = \frac{k}{2500}$$

$$5,000,000 = k$$

$$s = \frac{5,000,000}{P} = \frac{5,000,000}{2000} = 2500$$

At $2000 each, 2500 computers can be sold.

Example 4

The resistance (R) to the flow of electric current in a wire of fixed length is inversely proportional to the square of the diameter (d) of a wire. If a wire of diameter 0.01 cm has a resistance of 0.5 ohms, what is the resistance in a wire which is 0.02 cm in diameter?

Your strategy

Your solution

0.125 ohms

Example 5

The pressure (P) of a gas varies directly as the temperature (T) and inversely as the volume (V). When $T = 50°$ and $V = 275$ in^3, $P = 20$ lb/in^2. Find the pressure of a gas when $T = 60°$ and $V = 250$ in^3.

Strategy

To find the pressure:

■ Write the basic combined variation equation, replace the variables by the given values, and solve for k.
■ Write the combined variation equation, replacing k by its value. Substitute 60 for T and 250 for V, and solve for P.

Solution

$$P = \frac{kT}{V}$$

$$20 = \frac{k \cdot 50}{275}$$

$$110 = k$$

$$P = \frac{110T}{V} = \frac{110 \cdot 60}{250} = 26.4$$

The pressure is 26.4 lb/in^2.

Example 6

The strength (s) of a rectangular beam varies directly as its width (w) and inversely as the square of its depth (d). If the strength of a beam 2 in. wide and 12 in. deep is 1200 lb, find the strength of a beam 4 in. wide and 8 in. deep.

Your strategy

Your solution

5400 lb

Solutions on p. A36

Content and Format © 1991 HMCo.

8.4   EXERCISES

▶ Objective A   *Application Problems*

Solve:

1. The profit (*P*) realized by a company varies directly as the number of products it sells (*s*). If a company makes a profit of $4000 on the sale of 250 products, what is the profit when the company sells 5000 products?
$80,000

2. The number of bushels of wheat (*b*) produced by a farm is directly proportional to the number of acres (*A*) planted in wheat. If a 20-acre farm yields 450 bushels of wheat, what is the yield of a farm which has 30 acres of wheat?
675 bushels

3. The pressure (*p*) on a diver in the water varies directly as the depth (*d*). If the pressure is 4.5 lb/in.^{2} when the depth is 10 ft, what is the pressure when the depth is 15 ft?
6.75 lb/in.^{2}

4. The distance (*d*) a spring will stretch varies directly as the force (*f*) applied to the spring. If a force of 6 lb is required to stretch a spring 3 in., what force is required to stretch the spring 4 in.?
8 lb

5. The distance (*d*) a person can see to the horizon from a point above the surface of the earth varies directly as the square root of the height (*H*). If, for a height of 500 ft, the horizon is 19 mi away, how far is the horizon from a point that is 800 ft high? Round to the nearest hundredth.
24.04 mi

6. The period (*p*) of a pendulum, or the time it takes for a pendulum to make one complete swing, varies directly as the square root of the length (*L*) of the pendulum. If the period of a pendulum is 1.5 s when the length is 2 ft, find the period when the length is 5 ft. Round to the nearest hundredth.
2.37 s

7. The distance (*s*) a ball will roll down an inclined plane is directly proportional to the square of the time (*t*). If the ball rolls 6 ft in one second, how far will it roll in 3 s?
54 ft

8. The stopping distance (*s*) of a car varies directly as the square of its speed (*v*). If a car traveling 50 mph requires 170 ft to stop, find the stopping distance for a car traveling 60 mph.
244.8 ft

9. The length (*L*) of a rectangle of fixed area varies inversely as the width (*w*). If the length of a rectangle is 8 ft when the width is 5 ft, find the length of the rectangle when the width is 4 ft.
10 ft

10. The number of items (*n*) that can be purchased for a given amount of money is inversely proportional to the cost (*C*) of an item. If 60 items can be purchased when the cost per item is $.25, how many items can be purchased when the cost per item is $.20?
75 items

Solve:

11. For a constant temperature, the pressure (P) of a gas varies inversely as the volume (V). If the pressure is 30 lb/in.^{2} when the volume is 500 ft^3, find the pressure when the volume is 200 ft^3.
    75 lb/in.^{2}

12. The speed (v) of a gear varies inversely as the number of teeth (t). If a gear which has 45 teeth makes 24 revolutions per minute, how many revolutions per minute will a gear which has 36 teeth make?
    30 revolutions/min

13. The pressure (p) in a liquid varies directly as the product of the depth (d) and the density (D) of the liquid. If the pressure is 150 lb/in.^{2} when the depth is 100 in. and the density is 1.2, find the pressure when the density remains the same and the depth is 75 in.
    112.5 lb/in.^{2}

14. The current (l) in a wire varies directly as the voltage (v) and inversely as the resistance (r). If the current is 22 amps when the voltage is 110 volts and the resistance is 5 ohms, find the current when the voltage is 195 volts and the resistance is 15 ohms.
    13 amps

15. The repulsive force (f) between two north poles of a magnet is inversely proportional to the square of the distance (d) between them. If the repulsive force is 20 lb when the distance is 4 in., find the repulsive force when the distance is 2 in.
    80 lb

16. The intensity (l) of a light source is inversely proportional to the square of the distance (d) from the source. If the intensity is 12 lumens at a distance of 10 ft, what is the intensity when the distance is 5 ft?
    48 lumens

17. The resistance (R) of a wire varies directly as the length (L) of the wire and inversely as the square of the diameter (d). If the resistance is 9 ohms in 50 ft of wire which has a diameter of 0.05 in., find the resistance in 50 ft of a similar wire which has a diameter of 0.02 in.
    56.25 ohms

18. The frequency of vibration (f) of a string varies directly as the square root of the tension (T) and inversely as the length (L) of the string. If the frequency is 40 vibrations per second when the tension is 25 lb and the length of the string is 3 ft, find the frequency when the tension is 36 lb and the string is 4 ft.
    36 vibrations/s

19. The wind force (w) on a vertical surface varies directly as the product of the area (A) of the surface and the square of the wind velocity (v). When the wind is blowing at 30 mph, the force on a 10-square-foot area is 45 lb. Find the force on this area when the wind is blowing at 60 mph.
    180 lb

20. The power (P) in an electric circuit is directly proportional to the product of the current (l) and the square of the resistance (R). If the power is 100 watts when the current is 4 amps and the resistance is 5 ohms, find the power when the current is 2 amps and the resistance is 10 ohms.
    200 watts

Calculators and Computers

The program FUNCTION PLOTTER on the Math ACE Disk allows you to use a computer to graph the functions discussed in this chapter. Once the program has been loaded, use the arrow keys to highlight EQUATIONS and then press enter. You will enter the function following the notation "$f(x) = $".

You may now enter up to three equations to graph. (If you only want to graph one equation at a time, type in the equation and then press ENTER or RETURN.) The operational signs used to enter a function are +, −, * (times), / (divided by), and ^ (exponent). To graph an absolute value use "ABS" for the absolute value function. You may enter a function using upper or lower case letters.

Some examples of functions and the corresponding keystrokes are shown below.

Function	Keystroking
$f(x) = x^3 + 2x - 1$	x ^ $3 + 2$ * $x - 1$
$f(x) = 3x^2 + 2x - 4$	3 * x ^ $2 + 2$ * $x - 4$
$f(x) = |x - 1|$	abs($x - 1$)
$f(x) = |x| - 1$	abs(x) − 1
$f(x) = \frac{2}{3}x - \frac{7}{2}$	(2/3) * $x - $ (7/2)

Once the function has been entered, use the arrow keys to select DRAW and a graph of the function will appear on the screen. If you wish to graph another equation, select EQUATIONS by using the arrow keys and begin again. If you want to leave the program, select QUIT.

Chapter Summary

Key Words   A *relation* is a set of ordered pairs.

A *function* is a relation in which each element of the first set is assigned to one and only one member of the second set. No two ordered pairs that have the same first component have different second components.

The *domain* of a function is the set of first components of the ordered pairs of the function. The *range* of a function is the set of second components of the ordered pairs of the function.

A function given by $f(x) = mx + b$ is a *linear function*.

A function given by $f(x) = ax^2 + bx + c$, $a \neq 0$ is a *quadratic function*.

A *one-to-one function* is a function that satisfies the additional condition that given any y, there is only one x that can be paired with the given y.

The *vertical line test* is used to determine whether or not a graph is the graph of a function. The *horizontal line test* is used to determine whether or not the graph of a function is the graph of a 1–1 function.

A *composite function* is a function that involves more than one function.

The *inverse of a one-to-one function* is a function in which the components of each ordered pair are reversed.

Direct variation is a special function that can be expressed as the equation $y = kx^n$, where k is a constant called the *constant of variation* or the *constant of proportionality*.

Joint variation is a variation in which a variable varies directly as the product of two or more variables. A joint variation can be expressed as the equation $z = kxy$, where k is a constant.

Inverse variation is a function that can be expressed as the equation $y = \frac{k}{x^n}$, where k is a constant.

Combined variation is a variation in which two or more types of variation occur at the same time.

Essential Rule   *For the function f and its inverse f^{-1}* $f(f^{-1}(x)) = f^{-1}(f(x)) = x$

Chapter Review

SECTION 1

1. For the function $f(x) = x^3 - 8$, find $f(-2)$.
    −16

2. For the function $f(x) = x^2 + x + 2$, find $f(4 + h) - f(4)$.
    $h^2 + 9h$

3. For $f(x) = 3x^2 + 8$, find $\frac{f(2 + h) - f(2)}{h}$.
    $3h + 12$

4. Find the domain and range of the function $\{(3, 8), (5, 8), (7, 8)\}$.
    domain: $\{3, 5, 7\}$
    range: $\{8\}$

5. Find the range of the function $f(x) = 3x$; domain = $\{0, 2, 4, 6\}$.
    $\{0, 6, 12, 18\}$

6. The number 8 is in the range of the function $f(x) = x^2 + 4$. Find an element in the domain that corresponds to 8.
    −2 or 2

7. What values of x are excluded from the domain of the function $f(x) = \frac{3x + 1}{x^2 - 2x - 3}$?
    −1 and 3

8. What values of x are excluded from the domain of the function $f(x) = \sqrt{4x - 12}$.
    $\{x \mid x < 3\}$

SECTION 2

9. Graph $f(x) = -3x + 6$. State the domain and range.

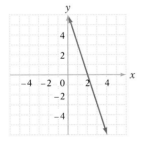

domain: all real numbers
range: all real numbers

10. Graph $f(x) = x^2 + 2x - 4$. State the domain and range.

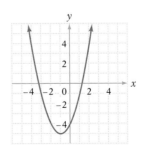

domain: all real numbers
range: $\{y \mid y \geq -5\}$

11. Graph $f(x) = |x| - 3$. State the domain and range.

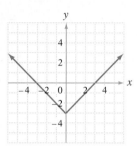

domain: all real numbers
range: $\{y | y \geq -3\}$

12. Determine if the graph represents the graph of a function.

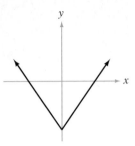

yes

13. Is the set of ordered pairs a function?
$\{(3, 2), (4, 8), (4, 7), (1, 1)\}$
no

14. Is the set of ordered pairs a 1–1 function?
$\{(1, 2), (3, 8), (8, 3), (2, 1)\}$
yes

15. Determine if the graph represents the graph of a 1–1 function.

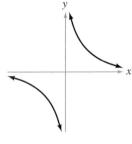

yes

16. Determine if the graph represents the graph of a 1–1 function.

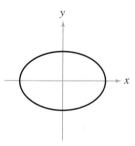

no

17. Determine if the graph represents the graph of a 1–1 function.

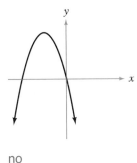

no

18. Determine if the graph represents the graph of a function.

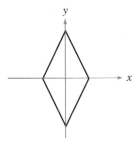

no

SECTION 3

19. Given $f(x) = 3x^2 - 4$ and $g(x) = 2x + 1$, find $f(g(x))$.
$12x^2 + 12x - 1$

20. Given $f(x) = x^2 + 4$ and $g(x) = 4x - 1$, find $f(g(0))$.
5

21. Given $f(x) = 6x + 8$ and $g(x) = 4x + 2$, find $g(f(-1))$.
10

22. Given $f(x) = 2x^2 + x - 5$ and $g(x) = 3x - 1$, find $g(f(x))$.
$6x^2 + 3x - 16$

23. Find the inverse of the function $f(x) = \frac{1}{2}x + 8$.
$f^{-1}(x) = 2x - 16$

24. Find the inverse of the function $f(x) = -6x + 4$.
$f^{-1}(x) = -\frac{1}{6}x + \frac{2}{3}$

25. Find the inverse of the function $f(x) = \frac{2}{3}x - 12$.
$f^{-1}(x) = \frac{3}{2}x + 18$

26. Are the functions $f(x) = -\frac{1}{4}x + \frac{5}{4}$ and $g(x) = -4x + 5$ inverses of each other?
yes

SECTION 4

27. The pressure (p) of wind on a flat surface varies jointly as the area (A) of the surface and the square of the wind's velocity (v). If the pressure on 22 ft^2 is 10 lb when the wind's velocity is 10 mph, find the pressure on the same surface when the wind's velocity is 20 mph.
40 lb

28. The illumination (i) produced by a light varies inversely as the square of the distance (d) from the light. If the illumination produced 10 ft from a light is 12 footcandles, find the illumination 2 ft from the light.
300 footcandles

29. The electrical resistance (r) of a cable varies directly as its length (l) and inversely as the square of its diameter (d). If a cable 16,000 ft long and $\frac{1}{4}$ in. in diameter has a resistance of 3.2 ohms, what is the resistance of a cable which is 8000 ft long and $\frac{1}{2}$ in. in diameter?
0.4 ohm

30. For a constant temperature, the pressure (P) of a gas varies inversely as the volume (V). If the pressure is 16 lb/in.^{2} when the volume is 200 ft^3, find the pressure when the volume is 64 ft^3.
50 lb/in.^{2}

Chapter Test

1. For $f(x) = x^2 - 1$, find $f(4)$.
   $f(4) = 15$ [8.1A]

2. Evaluate $g(x) = -x^2 + 3x - 2$ at $x = -2$.
   $g(-2) = -12$ [8.1A]

3. Is the set of ordered pairs a function?
   {(1, 3), (3, 3), (2, 4), (4, 4)}
   Yes [8.1B]

4. What values of x are excluded from the domain of the function $f(x) = \frac{3}{x^2 - 8x}$?
   0 and 8 [8.1B]

5. Find the range of the function $f(x) = 4x - 3$ if the domain is $\{-2, -1, 0, 1, 2,\}$.
   $\{-11, -7, -3, 1, 5\}$ [8.1B]

6. Find the domain of the function {(3, −1), (5, −3), (7, −5), (9, −7)}.
   {3, 5, 7, 9} [8.1B]

7. For $g(x) = -2x + 3$, find $g(1 + h)$.
   $g(1 + h) = 1 - 2h$ [8.1A]

8. What values of x are excluded from the domain of $f(x) = \sqrt{x - 7}$?
   $\{x | x < 7\}$ [8.1B]

9. The number -2 is in the range of $f(x) = 3x + 2$. Find an element in the domain that corresponds to -2.
   $-\frac{4}{3}$ [8.1B]

10. For $g(x) = 2x^2 + 1$ and $h(x) = 3x + 4$, find $g(0) + h(0)$.
    $g(0) + h(0) = 5$ [8.1A]

11. Graph $f(x) = \frac{1}{2}x - 2$.

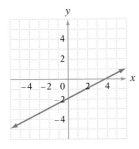

[8.2A]

12. Graph $f(x) = -x^2 + 1$.

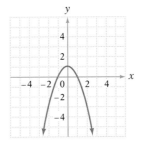

[8.2A]

13. Graph $f(x) = |x + 1|$. State the domain and range of the function.

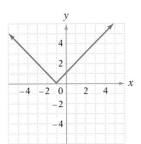

Domain: all real numbers
Range: $\{y | y \geq 0\}$
[8.2A]

14. Graph $f(x) = -x^2 + 2x$. State the domain and range of the function.

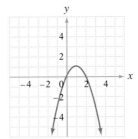

Domain: all real numbers
Range: $\{y | y \leq 1\}$
[8.2A]

15. Graph $f(x) = |x| - 2$. State the domain and range of the function.

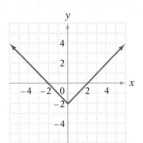

Domain: all real numbers
Range: $\{y \mid y \geq -2\}$
[8.2A]

16. Graph $f(x) = x^3 + 2$. State the domain and range of the function.

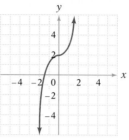

Domain: all real numbers
Range: all real numbers
[8.2A]

17. Determine if the graph represents the graph of a function.

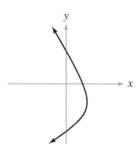

No   [8.2B]

18. Determine if the graph represents the graph of a 1–1 function.

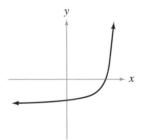

Yes   [8.2C]

19. Given $f(x) = 2x^2 - 3$ and $g(x) = 4x + 1$, evaluate $f(g(0))$.
$f(g(0)) = -1$ [8.3A]

20. Find the inverse of the function $f(x) = \frac{1}{2}x - 1$.
$f^{-1}(x) = 2x + 2$ [8.3B]

21. Given $g(x) = -3x + 2$ and $h(x) = x - 4$, evaluate $g(h(x))$.
$g(h(x)) = -3x + 14$ [8.3A]

22. Are the functions given by $f(x) = \frac{1}{2}x - 2$ and $g(x) = 2x - 4$ inverses of each other?
No   [8.3B]

23. Find the inverse of the function given by $f(x) = 3x + 6$.
$f^{-1}(x) = \frac{1}{3}x - 2$ [8.3B]

24. The stopping distance (s) of a car varies directly as the square of the speed (v) of the car. For a car traveling at 50 mph, the stopping distance is 170 ft. Find the stopping distance of a car that is traveling at 30 mph.
61.2 ft   [8.4A]

25. The current (I) in an electric circuit varies inversely as the resistance (R). If the current in the circuit is 4 amps when the resistance is 50 ohms, find the current in the circuit when the resistance is 100 ohms.
2 amps   [8.4A]

Cumulative Review

1. Evaluate $-3a + \left| \dfrac{3b - ab}{3b - c} \right|$ when $a = 2$, $b = 2$, and $c = -2$.

   $-\dfrac{23}{4}$ [1.3A]

2. Graph the solution set of $\{x \mid x < -3\} \cap \{x \mid x > -4\}$.

   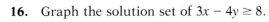
   [1.4A]

3. Solve: $\dfrac{3x - 1}{6} - \dfrac{5 - x}{4} = \dfrac{5}{6}$

   3 [2.1C]

4. Solve: $4x - 2 < -10$ or $3x - 1 > 8$

   $\{x \mid x < -2$ or $x > 3\}$ [2.2B]

5. Solve: $|8 - 2x| \geq 0$

   all real numbers [2.3B]

6. Simplify: $\left(\dfrac{3a^3 b}{2a} \right)^2 \left(\dfrac{a^2}{-3b^2} \right)^3$

   $-\dfrac{a^{10}}{12b^4}$ [3.1C]

7. Simplify: $(x - 4)(2x^2 + 4x - 1)$

   $2x^3 - 4x^2 - 17x + 4$ [3.2B]

8. Factor: $a^4 - 2a^2 - 8$

   $(a + 2)(a - 2)(a^2 + 2)$ [3.4D]

9. Factor: $x^3 y + x^2 y^2 - 6xy^3$

   $xy(x - 2y)(x + 3y)$ [3.3D]

10. Solve: $(b + 2)(b - 5) = 2b + 14$

    -3 and 8 [3.5A]

11. Solve: $x^2 - 2x > 15$

    $\{x \mid x < -3$ or $x > 5\}$ [7.5A]

12. Simplify: $\dfrac{x^2 + 4x - 5}{2x^2 - 3x + 1} - \dfrac{x}{2x - 1}$

    $\dfrac{5}{2x - 1}$ [4.3B]

13. Solve: $\dfrac{5}{x^2 + 7x + 12} = \dfrac{9}{x + 4} - \dfrac{2}{x + 3}$

    -2 [4.6A]

14. Simplify: $\dfrac{4 - 6i}{2i}$

    $-3 - 2i$ [5.3D]

15. Graph $f(x) = \frac{1}{4}x^2$.

    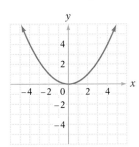
    [8.2A]

16. Graph the solution set of $3x - 4y \geq 8$.

    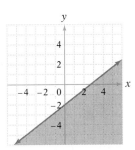
    [6.6A]

17. Find the equation of the line containing the points $(-3, 4)$ and $(2, -6)$.
$y = -2x - 2$ [6.3B]

18. Find the equation of the line containing the point $(-3, 1)$ and perpendicular to the line $2x - 3y = 6$.
$y = -\frac{3}{2}x - \frac{7}{2}$ [6.4A]

19. Solve: $3x^2 = 3x - 1$
$\frac{1}{2} + \frac{\sqrt{3}}{6}i$ and $\frac{1}{2} - \frac{\sqrt{3}}{6}i$ [7.3A]

20. Solve: $\sqrt{8x + 1} = 2x - 1$
3 [7.4B]

21. Evaluate $f(x) = 2x^2 - 3$ at $x = -2$.
$f(-2) = 5$ [8.1A]

22. Find the range of $f(x) = |3x - 4|$ if the domain is $\{0, 1, 2, 3\}$.
$\{1, 2, 4, 5\}$ [8.1B]

23. Is the set of ordered pairs a function?
$\{(-3, 0), (-2, 0), (-1, 1), (0, 1)\}$
Yes [8.2B]

24. Solve: $\sqrt[3]{5x - 2} = 2$
2 [5.4A]

25. Given $g(x) = 3x - 5$ and $h(x) = \frac{1}{2}x + 4$, find $g(h(2))$.
$g(h(2)) = 10$ [8.3A]

26. Find the inverse of the function given by $f(x) = -3x + 9$.
$f^{-1}(x) = -\frac{1}{3}x + 3$ [8.3B]

27. Find the cost per pound of a tea mixture made from 30 lb of tea costing $4.50 per pound and 45 lb of tea costing $3.60 per pound.
$3.96 [2.5A]

28. How many pounds of an 80% copper alloy must be mixed with 50 lb of a 20% copper alloy to make an alloy that is 40% copper?
25 lb [2.6B]

29. Six ounces of an insecticide are mixed with 16 gal of water to make a spray for spraying an orange grove. How much additional insecticide is required to be mixed with 28 gal of water?
4.5 oz [4.5B]

30. A large pipe can fill a tank in 8 min less time than it takes a smaller pipe to fill the same tank. Working together, both pipes can fill the tank in 3 min. How long would it take the larger pipe working alone to fill the tank?
4 min [7.6A]

31. The distance (d) a spring stretches varies directly as the force (f) used to stretch the spring. If a force of 50 lb can stretch the spring 30 in., how far will a force of 40 lb stretch the spring?
24 in. [8.4A]

32. The frequency of vibration (f) in an open pipe organ varies inversely as the length (L) of the pipe. If the air in a pipe 2 m long vibrates 60 times per minute, find the frequency in a pipe that is 1.5 m long.
80 vibrations/min [8.4A]

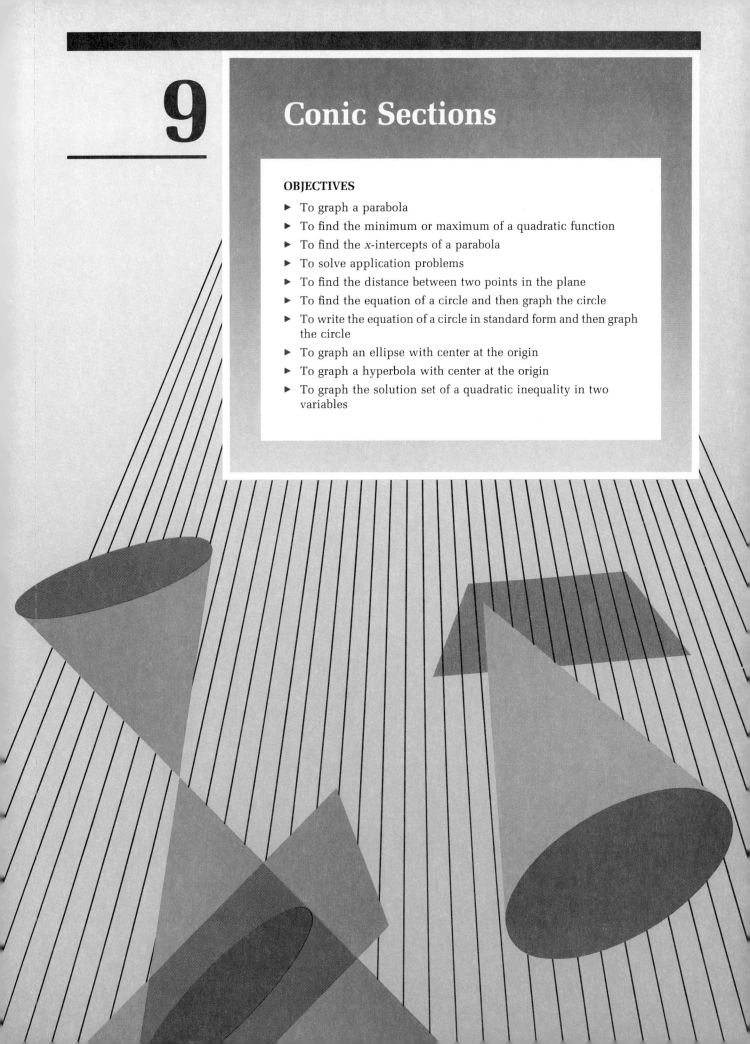

9

Conic Sections

OBJECTIVES

▶ To graph a parabola
▶ To find the minimum or maximum of a quadratic function
▶ To find the x-intercepts of a parabola
▶ To solve application problems
▶ To find the distance between two points in the plane
▶ To find the equation of a circle and then graph the circle
▶ To write the equation of a circle in standard form and then graph the circle
▶ To graph an ellipse with center at the origin
▶ To graph a hyperbola with center at the origin
▶ To graph the solution set of a quadratic inequality in two variables

Conic Sections

The graphs of three curves—the ellipse, the parabola, and the hyperbola—are discussed in this chapter. These curves were studied by the Greeks and were known prior to 400 B.C. The names of these curves were first used by Apollonius around 250 B.C. in *Conic Sections*, the most authoritative Greek discussion of these curves. Apollonius borrowed the names from a school founded by Pythagoras.

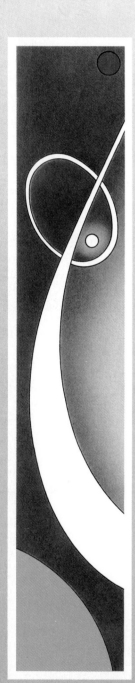

The diagram at the right shows the path of a planet around the sun. The curve traced out by the planet is an ellipse. The **aphelion** is the position of the planet when it is farthest from the sun. The **perihelion** is the position when the planet is nearest to the sun.

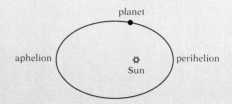

A telescope, like the one at the Palomar Observatory, has a cross section that is in the shape of a parabola. A parabolic mirror has the unusual property that all light rays parallel to the axis of symmetry that hit the mirror are reflected to the same point. This point is called the **focus of the parabola.**

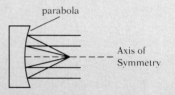

Some comets, unlike Halley's Comet, travel with such speed that they are not captured by the sun's gravitational field. The path of the comet as it comes around the sun is in the shape of a hyperbola.

SECTION 9.1   The Parabola

Objective A   **To graph a parabola**

A parabola is one of a number of curves called conic sections. The graph of a **conic section** can be represented by the intersection of a plane and a cone.

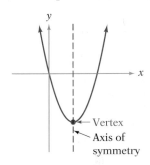

Every parabola has an axis of symmetry and a vertex which is on the axis of symmetry. To understand the axis of symmetry, think of folding the paper along that axis. The two halves of the curve will match up.

The graph of the equation $y = ax^2 + bx + c$, $a \neq 0$, is a parabola with the axis of symmetry parallel to the y-axis. The parabola opens up when $a > 0$ and opens down when $a < 0$. When the parabola opens up, the vertex is the lowest point on the parabola. When the parabola opens down, the vertex is the highest point on the parabola.

The coordinates of the vertex can be found by completing the square.

Find the vertex of the parabola whose equation is $y = x^2 - 4x + 5$.

$$y = x^2 - 4x + 5$$

Group the variable terms.

$$y = (x^2 - 4x) + 5$$

Complete the square on $x^2 - 4x$. Note that 4 is added and subtracted. Because $4 - 4 = 0$, the equation is not changed.

$$y = (x^2 - 4x + 4) - 4 + 5$$

Factor the trinomial and combine like terms.

$$y = (x - 2)^2 + 1$$

Since the coefficient of x^2 is positive, the parabola opens up. The vertex is the lowest point on the parabola, or that point that has the least y-coordinate.

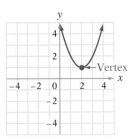

Since $(x - 2)^2 \geq 0$ for any x, the least y-coordinate occurs when $(x - 2)^2 = 0$. $(x - 2)^2 = 0$ when $x = 2$.

To find the y-coordinate of the vertex, replace x by 2 and solve for y.

$$y = (x - 2)^2 + 1$$
$$= (2 - 2)^2 + 1$$
$$= 1$$

The vertex is the point (2, 1).

By following the procedure of the last example and completing the square on the equation $y = ax^2 + bx + c$, the **x-coordinate of the vertex** is $-\frac{b}{2a}$. The y-coordinate of the vertex can then be determined by substituting this value of x into $y = ax^2 + bx + c$ and solving for y.

Since the axis of symmetry is parallel to the y-axis and passes through the vertex, the equation of the **axis of symmetry** is $x = -\frac{b}{2a}$.

Find the vertex and axis of symmetry of the parabola whose equation is $y = -3x^2 + 6x + 1$. Then sketch its graph.

Find the x-coordinate of the vertex and the axis of symmetry.
$a = -3, b = 6$

x-coordinate: $-\frac{b}{2a} = -\frac{6}{2(-3)} = 1$

The x-coordinate of the vertex is 1.
The axis of symmetry is the line $x = 1$.

To find the y-coordinate of the vertex, replace x by 1 and solve for y.

$$y = -3x^2 + 6x + 1$$
$$= -3(1)^2 + 6(1) + 1$$
$$= 4$$

The vertex is $(1, 4)$.

Since a is negative, the parabola opens down.

Find a few ordered pairs and use symmetry to sketch the graph.

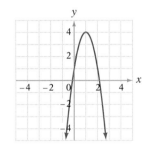

Find the vertex and axis of symmetry of the parabola whose equation is $y = x^2 - 2$. Then sketch its graph.

Find the x-coordinate of the vertex and the axis of symmetry.
$a = 1, b = 0$

x-coordinate: $-\frac{b}{2a} = -\frac{0}{2(1)} = 0$

The x-coordinate of the vertex is 0.
The axis of symmetry is the line $x = 0$.

To find the y-coordinate of the vertex, replace x by 0 and solve for y.

$$y = x^2 - 2$$
$$= 0^2 - 2$$
$$= -2$$

The vertex is $(0, -2)$.

Since a is positive, the parabola opens up.

Find a few ordered pairs and use symmetry to sketch the graph.

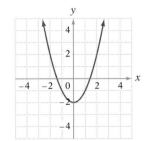

Content and Format © 1991 HMCo.

The graph of an equation of the form $x = ay^2 + by + c$, $a \neq 0$, is also a parabola. In this case, the parabola opens to the right when a is positive and opens to the left when a is negative.

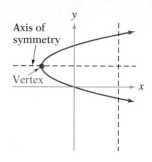

For a parabola of this form, the **y-coordinate of the vertex is $-\frac{b}{2a}$.** The **axis of symmetry** is the line $y = -\frac{b}{2a}$.

Using the vertical line test, the graph of a parabola of this form is not the graph of a function. The graph of $x = ay^2 + by + c$ is a relation.

Find the vertex and axis of symmetry of the parabola whose equation is $x = 2y^2 - 8y + 5$. Then sketch its graph.

Find the y-coordinate of the vertex and the axis of symmetry.
$a = 2, b = -8$

y-coordinate: $-\frac{b}{2a} = -\frac{-8}{2(2)} = 2$

The y-coordinate of the vertex is 2.
The axis of symmetry is the line $y = 2$.

To find the x-coordinate of the vertex, replace y by 2 and solve for x.

$x = 2y^2 - 8y + 5$
$ = 2(2)^2 - 8(2) + 5 = -3$
The vertex is $(-3, 2)$.

Since a is positive, the parabola opens to the right.

Find a few ordered pairs and use symmetry to sketch the graph.

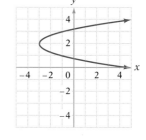

Find the vertex and axis of symmetry of the parabola whose equation is $x = -2y^2 - 4y - 3$. Then sketch its graph.

Find the y-coordinate of the vertex and the axis of symmetry.
$a = -2, b = -4$

y-coordinate: $-\frac{b}{2a} = -\frac{-4}{2(-2)} = -1$

The y-coordinate of the vertex is -1.
The axis of symmetry is the line $y = -1$.

To find the x-coordinate of the vertex, replace y by -1 and solve for x.

$x = -2y^2 - 4y - 3$
$ = -2(-1)^2 - 4(-1) - 3 = -1$

The vertex is $(-1, -1)$.

Since a is negative, the parabola opens to the left.

Find a few ordered pairs and use symmetry to sketch the graph.

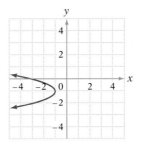

Example 1

Find the vertex and axis of symmetry of the parabola whose equation is $y = x^2 - 4x + 3$. Then sketch its graph.

Solution

$-\dfrac{b}{2a} = -\dfrac{-4}{2(1)} = 2$

Axis of symmetry:
$\quad x = 2$

$y = 2^2 - 4(2) + 3$
$\quad = -1$

Vertex: $(2, -1)$

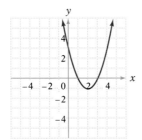

Example 2

Find the vertex and axis of symmetry of the parabola whose equation is $y = x^2 + 2x + 1$. Then sketch its graph.

Your solution

Vertex: $(-1, 0)$

Axis of symmetry:
$\quad x = -1$

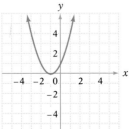

Example 3

Find the vertex and axis of symmetry of the parabola whose equation is $x = 2y^2 - 4y + 1$. Then sketch its graph.

Solution

$-\dfrac{b}{2a} = -\dfrac{-4}{2(2)} = 1$

Axis of symmetry:
$\quad y = 1$

$x = 2(1)^2 - 4(1) + 1$
$\quad = -1$

Vertex: $(-1, 1)$

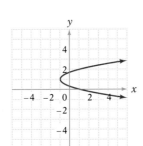

Example 4

Find the vertex and axis of symmetry of the parabola whose equation is $x = -y^2 - 2y + 2$. Then sketch its graph.

Your solution

Vertex: $(3, -1)$

Axis of symmetry:
$\quad y = -1$

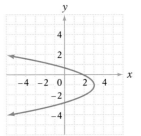

Example 5

Find the vertex and axis of symmetry of the parabola whose equation is $y = x^2 + 1$. Then sketch its graph.

Solution

$-\dfrac{b}{2a} = -\dfrac{0}{2(1)} = 0$

Axis of symmetry
$\quad x = 0$

$y = 0^2 + 1 = 1$

Vertex: $(0, 1)$

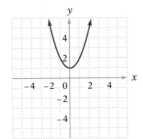

Example 6

Find the vertex and axis of symmetry of the parabola whose equation is $y = x^2 - 2x - 1$. Then sketch its graph.

Your solution

Vertex: $(1, -2)$

Axis of symmetry:
$\quad x = 1$

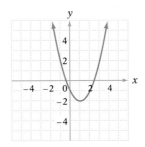

Solutions on pp. A36–A37

Content and Format © 1991 HMCo.

| Objective B | **To find the minimum or maximum of a quadratic function** |

The graph of the function $f(x) = x^2 - 2x + 3$ is shown at the right. Since a is positive, the parabola opens up. The vertex of the parabola is the lowest point on the parabola. It is the point that has the minimum y-coordinate. Therefore, the value of the function at this point is a **minimum.**

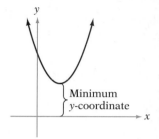

The graph of the function $f(x) = -x^2 + 2x + 1$ is shown at the right. Since a is negative, the parabola opens down. The vertex of the parabola is the highest point on the parabola. It is the point that has the maximum y-coordinate. Therefore, the value of the function at this point is a **maximum.**

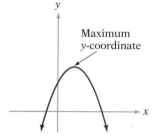

To find the minimum or maximum value of a quadratic function, first find the x-coordinate of the vertex. Then evaluate the function at that value.

Find the maximum value of the function $f(x) = -2x^2 + 4x + 3$.

Find the x-coordinate of the vertex.
$a = -2, b = 4$

$$x = -\frac{b}{2a} = -\frac{4}{2(-2)} = 1$$

Evaluate the function at $x = 1$.

$$\begin{aligned} f(x) &= -2x^2 + 4x + 3 \\ f(1) &= -2(1)^2 + 4(1) + 3 \\ &= -2 + 4 + 3 \\ &= 5 \end{aligned}$$

Because $a < 0$, the function has a maximum value.

The maximum value of the function is 5.

Example 7

Find the minimum value of $f(x) = 2x^2 - 3x + 1$.

Solution

$$x = -\frac{b}{2a} = -\frac{-3}{2(2)} = \frac{3}{4}$$

$$f(x) = 2x^2 - 3x + 1$$

$$f\left(\frac{3}{4}\right) = 2\left(\frac{3}{4}\right)^2 - 3\left(\frac{3}{4}\right) + 1 = \frac{9}{8} - \frac{9}{4} + 1 = -\frac{1}{8}$$

Because a is positive, the function has a minimum value.

The minimum value of the function is $-\frac{1}{8}$.

Example 8

Find the maximum value of $f(x) = -3x^2 + 4x - 1$.

Your solution

$\frac{1}{3}$

Solution on p. A37

Objective C   To find the x-intercepts of a parabola

The points at which a graph crosses a coordinate axis are called the **intercepts** of the graph.

When the graph of a parabola crosses the x-axis, the y-coordinate is zero.

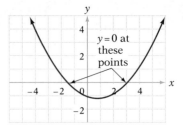

To find the x-intercepts for the parabola whose equation is $y = ax^2 + bx + c$, let $y = 0$. Then solve for x.

Find the x-intercepts for the parabola whose equation is $y = x^2 - 2x - 1$.

Let $y = 0$.

$$y = x^2 - 2x - 1$$
$$0 = x^2 - 2x - 1$$

Since the equation is difficult to factor, use the quadratic formula to solve for x.

$$x = \frac{-b \pm \sqrt{b^2 - 4ac}}{2a}$$

$$= \frac{-(-2) \pm \sqrt{(-2)^2 - 4(1)(-1)}}{2 \cdot 1}$$

$$= \frac{2 \pm \sqrt{4 + 4}}{2} = \frac{2 \pm \sqrt{8}}{2}$$

$$= \frac{2 \pm 2\sqrt{2}}{2} = 1 \pm \sqrt{2}$$

The x-intercepts are $(1 + \sqrt{2}, 0)$ and $(1 - \sqrt{2}, 0)$.

Find the x-intercepts for the parabola whose equation is $y = 4x^2 - 4x + 1$.

Let $y = 0$.

$$y = 4x^2 - 4x + 1$$
$$0 = 4x^2 - 4x + 1$$

Solve for x by factoring.

$$0 = (2x - 1)(2x - 1)$$

$$2x - 1 = 0 \qquad 2x - 1 = 0$$
$$2x = 1 \qquad 2x = 1$$
$$x = \frac{1}{2} \qquad x = \frac{1}{2}$$

The x-intercept is $\left(\frac{1}{2}, 0\right)$.

In this last example, the parabola has only one x-intercept. In this case the parabola is said to be **tangent** to the x-axis at $x = \frac{1}{2}$.

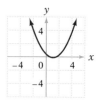

A parabola can have one x-intercept, as in the last example; two x-intercepts, as in the graph at the top of the page; or no x-intercepts, as shown in the graph at the right.

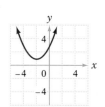

The parabola whose equation is $y = ax^2 + bx + c$ will have one x-intercept if the equation $0 = ax^2 + bx + c$ has one real number solution, two x-intercepts if the equation has two real number solutions, and no x-intercepts if the equation has no real number solutions.

Since the discriminant determines whether there are one, two, or no real number solutions of the equation $0 = ax^2 + bx + c$, it can also be used to determine whether there are one, two, or no x-intercepts of the parabola $y = ax^2 + bx + c$.

The Effect of the Discriminant on the Number of x-intercepts of a Parabola

1. If $b^2 - 4ac = 0$, the parabola has one x-intercept.

2. If $b^2 - 4ac > 0$, the parabola has two x-intercepts.

3. If $b^2 - 4ac < 0$, the parabola has no x-intercepts.

Use the discriminant to determine the number of x-intercepts of the parabola whose equation is $y = 2x^2 - x + 2$.

Evaluate the discriminant.

$a = 2, b = -1, c = 2$
$b^2 - 4ac$
$(-1)^2 - 4(2)(2) = 1 - 16 = -15$
$-15 < 0$

Since the discriminant is less than zero, the parabola has no x-intercepts.

Example 9   Find the x-intercepts of $y = 2x^2 - 5x + 2$.

Solution $y = 2x^2 - 5x + 2$
$0 = 2x^2 - 5x + 2$
$0 = (2x - 1)(x - 2)$

$2x - 1 = 0 \qquad x - 2 = 0$
$2x = 1 \qquad\qquad x = 2$
$x = \frac{1}{2}$

The x-intercepts are
$\left(\frac{1}{2}, 0\right)$ and $(2, 0)$.

Example 10   Find the x-intercepts of $y = x^2 + 4x + 5$.

Your solution   There are no x-intercepts.

Example 11   Use the discriminant to determine the number of x-intercepts of $y = x^2 - 6x + 9$.

Solution $a = 1, b = -6, c = 9$
$b^2 - 4ac$
$(-6)^2 - 4(1)(9) = 36 - 36 = 0$

Since the discriminant is equal to zero, the parabola has one x-intercept.

Example 12   Use the discriminant to determine the number of x-intercepts of $y = x^2 - x - 6$.

Your solution   two x-intercepts

Solutions on p. A37

Objective D   To solve application problems

Example 13

A mining company has determined that the cost in dollars (C) per ton of mining a mineral is given by the function $C(x) = 0.2x^2 - 2x + 12$, where x is the number of tons of the mineral that are mined. Find the number of tons of the mineral that should be mined to minimize the cost. What is the minimum cost?

Strategy

- To find the number of tons which will minimize the cost, find the x-coordinate of the vertex.
- To find the minimum cost, evaluate the function at the x-coordinate of the vertex.

Solution

$$x = -\frac{b}{2a} = -\frac{-2}{2(0.2)} = 5$$

To minimize cost, 5 tons should be mined.

$C(x) = 0.2x^2 - 2x + 12$
$C(5) = 0.2(5)^2 - 2(5) + 12 = 5 - 10 + 12 = 7$

The minimum cost per ton is $7.

Example 14

The height in feet (s) of a ball thrown straight up is given by the function $s(t) = -16t^2 + 64t$, where t is the time in seconds. Find the time it takes the ball to reach its maximum height. What is the maximum height?

Your strategy

Your solution

The ball reaches its maximum height in 2 s. The maximum height is 64 ft.

Example 15

Find two numbers whose difference is 10 and whose product is a minimum.

Strategy

Let x represent one number. Since the difference between the two numbers is 10, $x + 10$ represents the other number. Then their product is represented by $x^2 + 10x$.

- To find the first of the two numbers, find the x-coordinate of the vertex of the function $f(x) = x^2 + 10x$.
- To find the other number, replace x in $x + 10$ by the x-coordinate of the vertex and evaluate.

Solution

$$x = -\frac{b}{2a} = -\frac{10}{2(1)} = -5$$

$x + 10 = -5 + 10 = 5$

The numbers are -5 and 5.

Example 16

A mason is forming a rectangular floor for a storage shed. The perimeter of the rectangle is 44 ft. What dimensions would give the floor a maximum area?

Your strategy

Your solution

length: 11 ft
width: 11 ft

Solutions on p. A37

9.1 EXERCISES

▶ **Objective A**

Find the vertex and axis of symmetry of the parabola given by each equation.
Then sketch its graph.

1. $y = x^2 - 2x - 4$

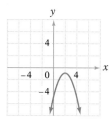

Vertex: $(1, -5)$
Axis of symmetry: $x = 1$

2. $y = x^2 + 4x - 4$

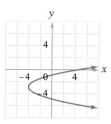

Vertex: $(-2, -8)$
Axis of symmetry: $x = -2$

3. $y = -x^2 + 2x - 3$

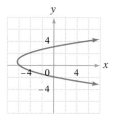

Vertex: $(1, -2)$
Axis of symmetry: $x = 1$

4. $y = -x^2 + 4x - 5$

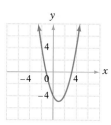

Vertex: $(2, -1)$
Axis of symmetry: $x = 2$

5. $x = y^2 + 6y + 5$

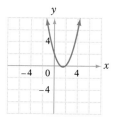

Vertex: $(-4, -3)$
Axis of symmetry: $y = -3$

6. $x = y^2 - y - 6$

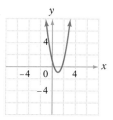

Vertex: $\left(-\frac{25}{4}, \frac{1}{2}\right)$
Axis of symmetry: $y = \frac{1}{2}$

7. $y = x^2 - x - 2$

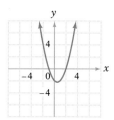

Vertex: $\left(\frac{1}{2}, -\frac{9}{4}\right)$
Axis of symmetry: $x = \frac{1}{2}$

8. $y = x^2 - 3x + 2$

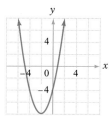

Vertex: $\left(\frac{3}{2}, -\frac{1}{4}\right)$
Axis of symmetry: $x = \frac{3}{2}$

9. $y = 2x^2 - 4x + 1$

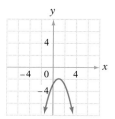

Vertex: $(1, -1)$
Axis of symmetry: $x = 1$

Find the vertex and axis of symmetry of the parabola given by each equation.
Then sketch its graph.

10. $y = 2x^2 + 4x - 5$

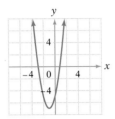

Vertex: $(-1, -7)$
Axis of symmetry: $x = -1$

11. $y = 2x^2 - x - 5$

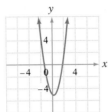

Vertex: $\left(\frac{1}{4}, -\frac{41}{8}\right)$
Axis of symmetry: $x = \frac{1}{4}$

12. $y = 2x^2 - x - 3$

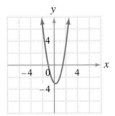

Vertex: $\left(\frac{1}{4}, -\frac{25}{8}\right)$
Axis of symmetry: $x = \frac{1}{4}$

13. $y = x^2 - 5x + 4$

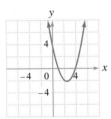

Vertex: $\left(\frac{5}{2}, -\frac{9}{4}\right)$
Axis of symmetry: $x = \frac{5}{2}$

14. $y = x^2 + 5x + 6$

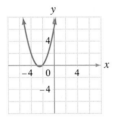

Vertex: $\left(-\frac{5}{2}, -\frac{1}{4}\right)$
Axis of symmetry: $x = -\frac{5}{2}$

15. $x = y^2 - 2y - 5$

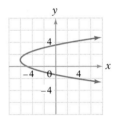

Vertex: $(-6, 1)$
Axis of symmetry: $y = 1$

16. $x = y^2 - 3y - 4$

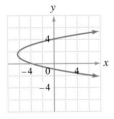

Vertex: $\left(-\frac{25}{4}, \frac{3}{2}\right)$
Axis of symmetry: $y = \frac{3}{2}$

17. $y = x^2 - 2$

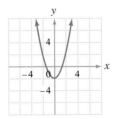

Vertex: $(0, -2)$
Axis of symmetry: $x = 0$

18. $y = x^2 + 2$

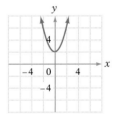

Vertex: $(0, 2)$
Axis of symmetry: $x = 0$

19. $y = x^2 - 2x$

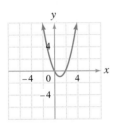

Vertex: $(1, -1)$
Axis of symmetry: $x = 1$

20. $y = x^2 + 2x$

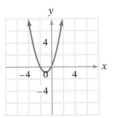

Vertex: $(-1, -1)$
Axis of symmetry: $x = -1$

21. $y = -3x^2 - 9x$

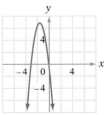

Vertex: $\left(-\frac{3}{2}, \frac{27}{4}\right)$
Axis of symmetry: $x = -\frac{3}{2}$

Find the vertex and axis of symmetry of the parabola given by each equation. Then sketch its graph.

22. $y = -2x^2 + 6x$

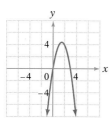

Vertex: $\left(\frac{3}{2}, \frac{9}{2}\right)$

Axis of symmetry: $x = \frac{3}{2}$

23. $y = x^2 + 4x + 5$

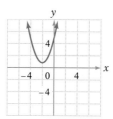

Vertex: $(-2, 1)$
Axis of symmetry: $x = -2$

24. $y = -x^2 + 2x - 1$

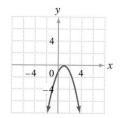

Vertex: $(1, 0)$
Axis of symmetry: $x = 1$

25. $x = -\frac{1}{2}y^2 + 4$

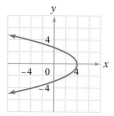

Vertex: $(4, 0)$
Axis of symmetry: $y = 0$

26. $x = -\frac{1}{4}y^2 - 1$

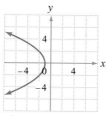

Vertex: $(-1, 0)$
Axis of symmetry: $y = 0$

27. $x = \frac{1}{2}y^2 - y + 1$

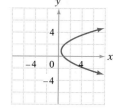

Vertex: $\left(\frac{1}{2}, 1\right)$

Axis of symmetry: $y = 1$

28. $x = -\frac{1}{2}y^2 + 2y - 3$

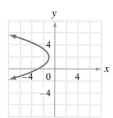

Vertex: $(-1, 2)$
Axis of symmetry: $y = 2$

29. $y = \frac{1}{2}x^2 + 2x - 6$

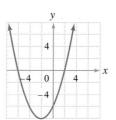

Vertex: $(-2, -8)$
Axis of symmetry: $x = -2$

30. $y = -\frac{1}{2}x^2 + x - 3$

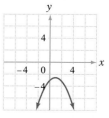

Vertex: $\left(1, -\frac{5}{2}\right)$

Axis of symmetry: $x = 1$

▶ **Objective B**

Find the minimum or maximum value of each quadratic function.

31. $f(x) = x^2 - 2x + 3$

minimum: 2

32. $f(x) = x^2 + 3x - 4$

minimum: $-\frac{25}{4}$

33. $f(x) = -2x^2 + 4x - 3$

maximum: -1

Find the minimum or maximum value of each quadratic function.

34. $f(x) = -x^2 - x + 2$

maximum: $\frac{9}{4}$

35. $f(x) = 3x^2 + 3x - 2$

minimum: $-\frac{11}{4}$

36. $f(x) = x^2 - 5x + 3$

minimum: $-\frac{13}{4}$

37. $f(x) = -3x^2 + 4x - 2$

maximum: $-\frac{2}{3}$

38. $f(x) = -2x^2 - 5x + 1$

maximum: $\frac{33}{8}$

39. $f(x) = 3x^2 + 5x + 2$

minimum: $-\frac{1}{12}$

▶ **Objective C**

Find the *x*-intercepts of the graph of the parabola given by each equation.

40. $y = x^2 - 4$
(2, 0)
(−2, 0)

41. $y = x^2 - 9$
(3, 0)
(−3, 0)

42. $y = 2x^2 - 4x$
(0, 0)
(2, 0)

43. $y = 3x^2 + 6x$
(0, 0)
(−2, 0)

44. $y = x^2 - x - 2$
(2, 0)
(−1, 0)

45. $y = x^2 - 2x - 8$
(4, 0)
(−2, 0)

46. $y = 2x^2 - x - 1$
$\left(-\frac{1}{2}, 0\right)$
(1, 0)

47. $y = 2x^2 - 5x - 3$
$\left(-\frac{1}{2}, 0\right)$
(3, 0)

48. $y = x^2 + 2x - 1$
$(-1 + \sqrt{2}, 0)$
$(-1 - \sqrt{2}, 0)$

49. $y = x^2 + 4x - 3$
$(-2 + \sqrt{7}, 0)$
$(-2 - \sqrt{7}, 0)$

50. $y = x^2 + 6x + 10$
No *x*-intercepts

51. $y = -x^2 - 4x - 5$
No *x*-intercepts

52. $y = x^2 - 2x - 2$
$(1 + \sqrt{3}, 0)$
$(1 - \sqrt{3}, 0)$

53. $y = -x^2 - 2x + 1$
$(-1 + \sqrt{2}, 0)$
$(-1 - \sqrt{2}, 0)$

54. $y = -x^2 + 4x + 1$
$(2 + \sqrt{5}, 0)$
$(2 - \sqrt{5}, 0)$

Use the discriminant to determine the number of x-intercepts.

55. $y = 2x^2 + x + 1$
No x-intercepts

56. $y = 2x^2 + 2x - 1$
Two

57. $y = -x^2 - x + 3$
Two

58. $y = -2x^2 + x + 1$
Two

59. $y = x^2 - 8x + 16$
One

60. $y = x^2 - 10x + 25$
One

61. $y = -3x^2 - x - 2$
No x-intercepts

62. $y = -2x^2 + x - 1$
No x-intercepts

63. $y = 4x^2 - x - 2$
Two

64. $y = 2x^2 + x + 4$
No x-intercepts

65. $y = -2x^2 - x - 5$
No x-intercepts

66. $y = -3x^2 + 4x - 5$
No x-intercepts

▶ Objective D   *Application Problems*

Solve:

67. The height in feet (s) of a rock thrown upward at an initial speed of 64 ft/s from a cliff 50 ft high is given by the function $s(t) = -16t^2 + 64t + 50$, where t is the time in seconds. Find the maximum height above the ground that the rock will attain.
114 ft

68. The height in feet (s) of a ball thrown upward at an initial speed of 80 ft/s from a platform 50 ft high is given by the function $s(t) = -16t^2 + 80t + 50$, where t is the time in seconds. Find the maximum height above the ground that the ball will attain.
150 ft

69. A pool is treated with a chemical to reduce the amount of algae. The amount of algae in the pool t days after the treatment can be approximated by the function $A(t) = 40t^2 - 400t + 500$. How many days after treatment will the pool have the least amount of algae?
5 days

Solve:

70. The suspension cable which supports a small foot-bridge hangs in the shape of a parabola. The height in feet of the cable above the bridge is given by the function $h(x) = 0.25x^2 - 0.8x + 25$, where x is the distance from one end of the bridge. What is the minimum height of the cable above the bridge?
24.36 ft

71. A manufacturer of microwave ovens believes that the revenue the company receives is related to the price (P) of an oven by the function $R(P) = 125P - \frac{1}{4}P^2$. What price will give the maximum revenue?
$250

72. A manufacturer of camera lenses estimated that the average monthly cost of a lens can be given by the function $C(x) = 0.1x^2 - 20x + 2000$, where x is the number of lenses produced each month. Find the number of lenses to produce in order to minimize the average cost.
100 lenses

73. Find two numbers whose sum is 20 and whose product is a maximum.
10 and 10

74. Find two numbers whose sum is 50 and whose product is a maximum.
25 and 25

75. Find two numbers whose difference is 24 and whose product is a minimum.
12 and −12

76. Find two numbers whose difference is 14 and whose product is a minimum.
7 and −7

77. A rectangle has a perimeter of 60 ft. Find the dimensions of the rectangle that will have the maximum area. What is the maximum area?
15 ft by 15 ft; 225 ft^2

78. The perimeter of a rectangular window is 24 ft. Find the dimensions of the window that will enclose the largest area. What is the maximum area?
6 ft by 6 ft; 36 ft^2

SECTION 9.2   The Circle

Objective A   **To find the distance between two points in the plane**

The distance between two points on a horizontal or vertical number line is the absolute value of the difference between the coordinates of the two points.

Find the distance between the points -2 and 4 on the vertical number line shown at the right.

The distance is the absolute value of the difference between the coordinates.

distance $= |4 - (-2)| = |6| = 6$

Absolute value is used so that the coordinates can be subtracted in either order.

distance $= |-2 - 4| = |-6| = 6$

Now consider the points and the right triangle in the coordinate plane shown at the right. The vertical distance between the two points (x_1, y_1) and (x_2, y_2), is $|y_1 - y_2|$.

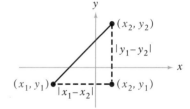

The horizontal distance between these two points is $|x_1 - x_2|$.

The Pythagorean Theorem is used to find the distance between two points (x_1, y_1) and (x_2, y_2).

$$d^2 = |x_1 - x_2|^2 + |y_1 - y_2|^2$$

Since the square of a number is always non-negative, the absolute value signs are not necessary.

$$d = \sqrt{(x_1 - x_2)^2 + (y_1 - y_2)^2}$$

The Distance Formula

> If (x_1, y_1) and (x_2, y_2) are two points in the plane, then the distance between the two points is given by $d = \sqrt{(x_1 - x_2)^2 + (y_1 - y_2)^2}$.

Example 1

Find the distance between the points $(2, -1)$ and $(-4, 3)$.

Solution

$(x_1, y_1) = (2, -1)$ $(x_2, y_2) = (-4, 3)$

$d = \sqrt{(x_1 - x_2)^2 + (y_1 - y_2)^2}$

$\quad = \sqrt{[2 - (-4)]^2 + (-1 - 3)^2} = \sqrt{6^2 + (-4)^2}$

$\quad = \sqrt{36 + 16} = \sqrt{52} = 2\sqrt{13}$

Example 2

Find the distance between the points $(3, -2)$ and $(-1, -5)$.

Your solution

5

Solution on p. A38

Objective B	To find the equation of a circle and then graph the circle

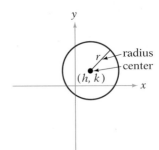

A **circle** is a conic section formed by the intersection of a cone and a plane parallel to the base of the cone.

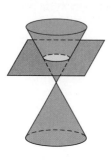

A **circle** can be defined as all the points (x, y) in the plane that are at a fixed distance from a given point (h, k) called the center. The fixed distance is the **radius** of the circle.

Using the distance formula, the equation of a circle can be determined.

Let (h, k) represent the center of the circle, r the radius, and (x, y) any point on the circle. Then $r = \sqrt{(x - h)^2 + (y - k)^2}$.

Squaring each side of the equation gives the equation of a circle.

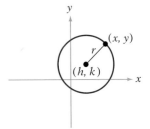

The Standard Form of the Equation of a Circle

Let r be the radius of a circle and (h, k) the center of the circle. Then the equation of the circle is given by

$$(x - h)^2 + (y - k)^2 = r^2.$$

Find the equation of the circle with radius 4 and center $(-1, 2)$. Then sketch its graph.

Use the standard form of the equation of a circle.

$$(x - h)^2 + (y - k)^2 = r^2$$

Replace r by 4, h by -1, and k by 2.

$$[x - (-1)]^2 + (y - 2)^2 = 4^2$$
$$(x + 1)^2 + (y - 2)^2 = 16$$

Sketch the graph by drawing a circle with center $(-1, 2)$ and radius 4.

Recall that the graph of a circle is not the graph of a function. The graph of a circle is the graph of a relation.

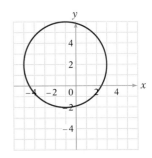

Content and Format © 1991 HMCo.

Find the equation of the circle that passes through point (3, −4) and whose center is (2, 1).

Since a point on the circle is known and the center is known, use the distance formula to find the radius of the circle.

$$\sqrt{(x_1 - x_2)^2 + (y_1 - y_2)^2} = r$$
$$\sqrt{(3 - 2)^2 + (-4 - 1)^2} = r$$
$$\sqrt{1^2 + (-5)^2} = r$$
$$\sqrt{1 + 25} = r$$
$$\sqrt{26} = r$$

Write the equation. The radius is $\sqrt{26}$. The center is (2, 1).

$$(x - 2)^2 + (y - 1)^2 = 26$$

Example 3

Sketch a graph of $(x + 2)^2 + (y - 1)^2 = 4$.

Solution

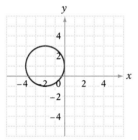

Example 4

Sketch a graph of $(x - 2)^2 + (y + 3)^2 = 9$.

Your solution

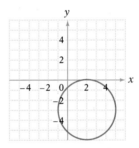

Example 5

Find the equation of the circle with radius 5 and center (−1, 3). Then sketch its graph.

Solution

$$(x - h)^2 + (y - k)^2 = r^2$$
$$[x - (-1)]^2 + (y - 3)^2 = 5^2$$
$$(x + 1)^2 + (y - 3)^2 = 25$$

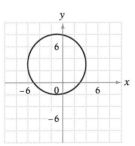

Example 6

Find the equation of the circle with radius 4 and center (2, −3). Then sketch its graph.

Your solution

$$(x - 2)^2 + (y + 3)^2 = 16$$

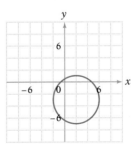

Solutions on p. A38

Objective C

9

To write the equation of a circle in standard form and then graph the circle

The equation of a circle can also be expressed as $x^2 + y^2 + ax + by + c = 0$. To rewrite this equation in standard form, it is necessary to complete the square on the x and y terms.

Write the equation of the circle $x^2 + y^2 + 4x + 2y + 1 = 0$ in standard form. Then sketch its graph.

Subtract the constant term from each side of the equation.

$$x^2 + y^2 + 4x + 2y + 1 = 0$$
$$x^2 + y^2 + 4x + 2y = -1$$

Rewrite the equation by grouping terms involving x and terms involving y.

$$(x^2 + 4x) + (y^2 + 2y) = -1$$

Complete the square on $x^2 + 4x$ and $y^2 + 2y$.

$$(x^2 + 4x + 4) + (y^2 + 2y + 1) = -1 + 4 + 1$$
$$(x^2 + 4x + 4) + (y^2 + 2y + 1) = 4$$

Factor each trinomial.

$$(x + 2)^2 + (y + 1)^2 = 4$$

Sketch the graph by drawing a circle with center $(-2, -1)$ and radius 2.

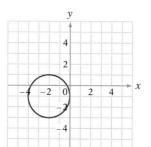

Example 7

Write the equation of the circle $x^2 + y^2 + 3x - 2y = 1$ in standard form. Then sketch its graph.

Solution

$$x^2 + y^2 + 3x - 2y = 1$$
$$(x^2 + 3x) + (y^2 - 2y) = 1$$
$$\left(x^2 + 3x + \tfrac{9}{4}\right) + (y^2 - 2y + 1) = 1 + \tfrac{9}{4} + 1$$
$$\left(x + \tfrac{3}{2}\right)^2 + (y - 1)^2 = \tfrac{17}{4}$$

Center: $\left(-\tfrac{3}{2}, 1\right)$ Radius: $\sqrt{\tfrac{17}{4}} = \tfrac{\sqrt{17}}{2}$

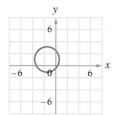

Example 8

Write the equation of the circle $x^2 + y^2 - 4x + 8y + 15 = 0$ in standard form. Then sketch its graph.

Your solution

$$(x - 2)^2 + (y + 4)^2 = 5$$

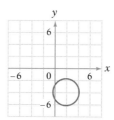

Solution on p. A38

9.2   EXERCISES

▶ **Objective A**

Find the distance between the two points.

1. $P_1 = (3, 2); P_2 = (4, 5)$
$\sqrt{10}$

2. $P_1 = (4, 5); P_2 = (1, 1)$
5

3. $P_1 = (-4, 2); P_2 = (1, -1)$
$\sqrt{34}$

4. $P_1 = (3, -2); P_2 = (-1, 4)$
$2\sqrt{13}$

5. $P_1 = (-1, -2); P_2 = (1, 5)$
$\sqrt{53}$

6. $P_1 = (-2, -3); P_2 = (3, 1)$
$\sqrt{41}$

7. $P_1 = (-1, -1); P_2 = (-4, -5)$
5

8. $P_1 = (-4, 1); P_2 = (3, -2)$
$\sqrt{58}$

▶ **Objective B**

Sketch a graph of the circle given by each equation.

9. $(x - 2)^2 + (y + 2)^2 = 9$

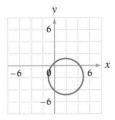

10. $(x + 2)^2 + (y - 3)^2 = 16$

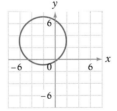

11. $(x + 3)^2 + (y - 1)^2 = 25$

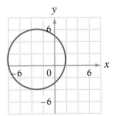

12. $(x - 2)^2 + (y + 3)^2 = 4$

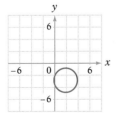

13. $(x + 2)^2 + (y + 2)^2 = 4$

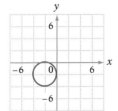

14. $(x - 1)^2 + (y - 2)^2 = 25$

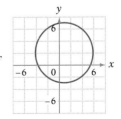

15. Find the equation of the circle with radius 2 and center (2, −1). Then sketch its graph.

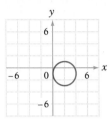

$(x − 2)^2 + (y + 1)^2 = 4$

16. Find the equation of the circle with radius 3 and center (−1, −2). Then sketch its graph.

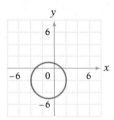

$(x + 1)^2 + (y + 2)^2 = 9$

17. Find the equation of the circle that passes through point (1, 2) and whose center is (−1, 1). Then sketch its graph.

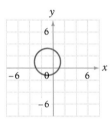

$(x + 1)^2 + (y − 1)^2 = 5$

18. Find the equation of the circle that passes through point (−1, 3) and whose center is (−2, 1). Then sketch its graph.

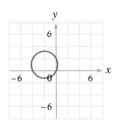

$(x + 2)^2 + (y − 1)^2 = 5$

▶ **Objective C**

Write the equation of the circle in standard form. Then sketch its graph.

19. $x^2 + y^2 − 2x + 4y − 20 = 0$

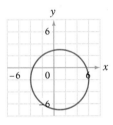

$(x − 1)^2 + (y + 2)^2 = 25$

20. $x^2 + y^2 − 4x + 8y + 4 = 0$

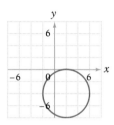

$(x − 2)^2 + (y + 4)^2 = 16$

21. $x^2 + y^2 + 6x + 8y + 9 = 0$

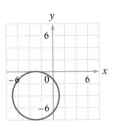

$(x + 3)^2 + (y + 4)^2 = 16$

22. $x^2 + y^2 − 6x + 10y + 25 = 0$

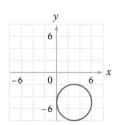

$(x − 3)^2 + (y + 5)^2 = 9$

| SECTION 9.3 | The Ellipse and the Hyperbola |

| Objective A | **To graph an ellipse with center at the origin** |

The orbits of the planets around the sun are "oval" shaped. This oval shape can be described as an **ellipse**, which is another of the conic sections.

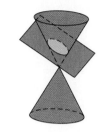

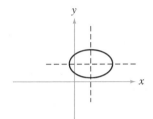

There are two **axes of symmetry** for an ellipse. The intersection of the two axes is the **center** of the ellipse.

An ellipse with center at the origin is shown at the right. Note that there are two x-intercepts and two y-intercepts.

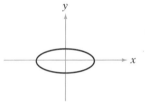

Using the vertical line test, the graph of an ellipse is not the graph of a function. The graph of an ellipse is the graph of a relation.

The Standard Form of the Equation of an Ellipse with Center at the Origin

The equation of an ellipse with center at the origin is $\frac{x^2}{a^2} + \frac{y^2}{b^2} = 1$.

The x-intercepts are $(a, 0)$ and $(-a, 0)$. The y-intercepts are $(0, b)$ and $(0, -b)$.

By finding the x- and y-intercepts for an ellipse and using the fact that the ellipse is "oval" shaped, a graph of an ellipse can be sketched.

Sketch a graph of the ellipse whose equation is $\frac{x^2}{9} + \frac{y^2}{4} = 1$.

$a^2 = 9, b^2 = 4$

The x-intercepts are $(a, 0)$ and $(-a, 0)$. x-intercepts: $(3, 0)$ and $(-3, 0)$

The y-intercepts are $(0, b)$ and $(0, -b)$. y-intercepts: $(0, 2)$ and $(0, -2)$

Use the intercepts and symmetry to sketch the graph of the ellipse.

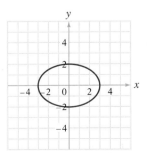

Sketch a graph of the ellipse whose equation is $\frac{x^2}{16} + \frac{y^2}{16} = 1$.

$a^2 = 16, b^2 = 16$

The x-intercepts are $(a, 0)$ and $(-a, 0)$. x-intercepts: $(4, 0)$ and $(-4, 0)$

The y-intercepts are $(0, b)$ and $(0, -b)$. y-intercepts: $(0, 4)$ and $(0, -4)$

Use the intercepts and symmetry to sketch the graph of the ellipse.

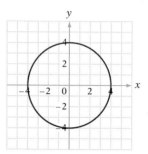

The graph in this last example is the graph of a circle. A circle is a special case of an ellipse and occurs when $a^2 = b^2$ in the equation $\frac{x^2}{a^2} + \frac{y^2}{b^2} = 1$.

Example 1

Sketch a graph of the ellipse whose equation is $\frac{x^2}{9} + \frac{y^2}{16} = 1$.

Solution

x-intercepts: $(3, 0)$ and $(-3, 0)$

y-intercepts: $(0, 4)$ and $(0, -4)$

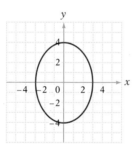

Example 2

Sketch a graph of the ellipse whose equation is $\frac{x^2}{4} + \frac{y^2}{25} = 1$.

Your solution

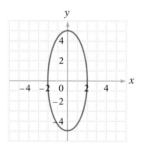

Example 3

Sketch a graph of the ellipse whose equation is $\frac{x^2}{16} + \frac{y^2}{12} = 1$.

Solution

x-intercepts: $(4, 0)$ and $(-4, 0)$

y-intercepts: $(0, 2\sqrt{3})$ and $(0, -2\sqrt{3})$

$(2\sqrt{3} \approx 3.5)$

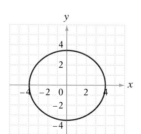

Example 4

Sketch a graph of the ellipse whose equation is $\frac{x^2}{18} + \frac{y^2}{9} = 1$.

Your solution

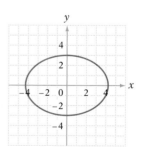

Solutions on p. A38

| Objective B | **To graph a hyperbola with center at the origin** |

A **hyperbola** is a conic section which is formed by the intersection of a cone by a plane perpendicular to the base of the cone.

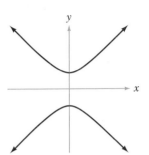

The hyperbola has two **vertices** and an **axis of symmetry** which passes through the vertices. The **center** of a hyperbola is the point halfway between the two vertices.

The graphs at the right show two possible graphs of a hyperbola with center at the origin.

In the first graph, an axis of symmetry is the *x*-axis and the vertices are *x*-intercepts.

In the second graph, an axis of symmetry is the *y*-axis and the vertices are *y*-intercepts.

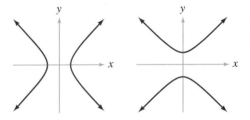

Note, in either case, the graph of a hyperbola is not the graph of a function. The graph of a hyperbola is the graph of a relation.

The Standard Form of the Equation of a Hyperbola with Center at the Origin

The equation of a hyperbola for which an axis of symmetry is the *x*-axis is $\frac{x^2}{a^2} - \frac{y^2}{b^2} = 1$. The vertices are $(a, 0)$ and $(-a, 0)$.

The equation of a hyperbola for which an axis of symmetry is the *y*-axis is $\frac{y^2}{b^2} - \frac{x^2}{a^2} = 1$. The vertices are $(0, b)$ and $(0, -b)$.

To sketch a hyperbola, it is helpful to draw two lines that are "approached" by the hyperbola. These two lines are called **asymptotes**. As the hyperbola gets farther from the origin, the hyperbola "gets closer to" the asymptotes.

Since the asymptotes are straight lines, their equations are linear equations. The equations of the asymptotes for a hyperbola with center at the origin are $y = \frac{b}{a}x$ and $y = -\frac{b}{a}x$.

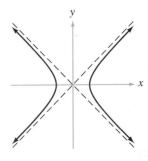

Sketch a graph of the hyperbola whose equation is $\frac{y^2}{9} - \frac{x^2}{4} = 1$.

$b^2 = 9$, $a^2 = 4$

The vertices are $(0, b)$ and $(0, -b)$.

The asymptotes are $y = \frac{b}{a}x$ and $y = -\frac{b}{a}x$.

Sketch the asymptotes.

Use symmetry and the fact that the hyperbola will approach the asymptotes to sketch its graph.

Axis of symmetry: y-axis

Vertices: $(0, 3)$ and $(0, -3)$

Asymptotes: $y = \frac{3}{2}x$ and $y = -\frac{3}{2}x$

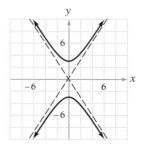

Example 5

Sketch a graph of the hyperbola whose equation is $\frac{x^2}{16} - \frac{y^2}{4} = 1$.

Solution

Axis of symmetry: x-axis

Vertices: $(4, 0)$ and $(-4, 0)$

Asymptotes: $y = \frac{1}{2}x$ and $y = -\frac{1}{2}x$

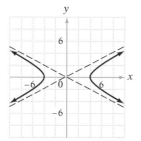

Example 6

Sketch a graph of the hyperbola whose equation is $\frac{x^2}{9} - \frac{y^2}{25} = 1$.

Your solution

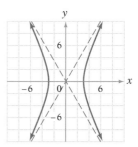

Example 7

Sketch a graph of the hyperbola whose equation is $\frac{y^2}{16} - \frac{x^2}{25} = 1$.

Solution

Axis of symmetry: y-axis

Vertices: $(0, 4)$ and $(0, -4)$

Asymptotes: $y = \frac{4}{5}x$ and $y = -\frac{4}{5}x$

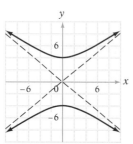

Example 8

Sketch a graph of the hyperbola whose equation is $\frac{y^2}{9} - \frac{x^2}{9} = 1$.

Your solution

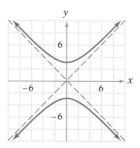

Solutions on p. A38

9.3   **EXERCISES**

▶ **Objective A**

Sketch a graph of the ellipse given by each equation.

1. $\dfrac{x^2}{4} + \dfrac{y^2}{9} = 1$

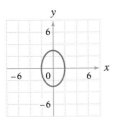

2. $\dfrac{x^2}{25} + \dfrac{y^2}{16} = 1$

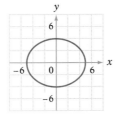

3. $\dfrac{x^2}{25} + \dfrac{y^2}{9} = 1$

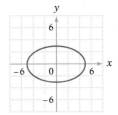

4. $\dfrac{x^2}{16} + \dfrac{y^2}{9} = 1$

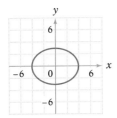

5. $\dfrac{x^2}{36} + \dfrac{y^2}{16} = 1$

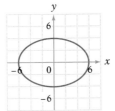

6. $\dfrac{x^2}{49} + \dfrac{y^2}{64} = 1$

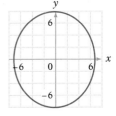

7. $\dfrac{x^2}{16} + \dfrac{y^2}{49} = 1$

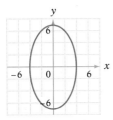

8. $\dfrac{x^2}{25} + \dfrac{y^2}{36} = 1$

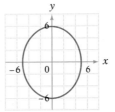

9. $\dfrac{x^2}{4} + \dfrac{y^2}{25} = 1$

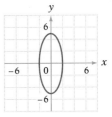

10. $\dfrac{x^2}{1} + \dfrac{y^2}{16} = 1$

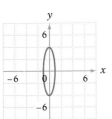

11. $\dfrac{x^2}{36} + \dfrac{y^2}{4} = 1$

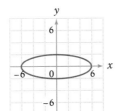

12. $\dfrac{x^2}{25} + \dfrac{y^2}{4} = 1$

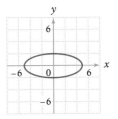

▶ **Objective B**

Sketch a graph of the hyperbola given by each equation.

13. $\dfrac{x^2}{9} - \dfrac{y^2}{16} = 1$

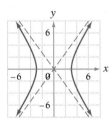

14. $\dfrac{x^2}{25} - \dfrac{y^2}{4} = 1$

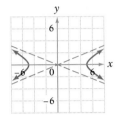

15. $\dfrac{y^2}{16} - \dfrac{x^2}{9} = 1$

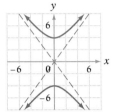

16. $\dfrac{y^2}{16} - \dfrac{x^2}{25} = 1$

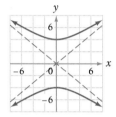

17. $\dfrac{x^2}{16} - \dfrac{y^2}{4} = 1$

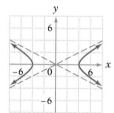

18. $\dfrac{x^2}{9} - \dfrac{y^2}{49} = 1$

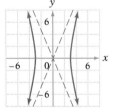

19. $\dfrac{y^2}{25} - \dfrac{x^2}{9} = 1$

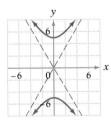

20. $\dfrac{y^2}{4} - \dfrac{x^2}{16} = 1$

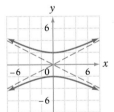

21. $\dfrac{x^2}{4} - \dfrac{y^2}{25} = 1$

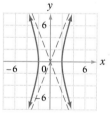

22. $\dfrac{x^2}{36} - \dfrac{y^2}{9} = 1$

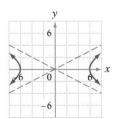

23. $\dfrac{y^2}{9} - \dfrac{x^2}{36} = 1$

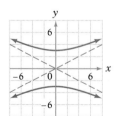

24. $\dfrac{y^2}{25} - \dfrac{x^2}{4} = 1$

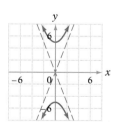

SECTION 9.4   Quadratic Inequalities

Objective A   **To graph the solution set of a quadratic inequality in two variables**

The **graph of a quadratic inequality** in two variables is a region of the plane that is bounded by one of the conic sections (parabola, circle, ellipse, or hyperbola). When graphing an inequality of this type, first replace the inequality symbol with an equal sign. Graph the resulting conic using a dashed curve when the original inequality is less than ($<$) or greater than ($>$). Use a solid curve if the original inequality is $\leq$ or $\geq$. Use the point $(0, 0)$ to determine which region of the plane to shade. If $(0, 0)$ is a solution of the inequality, then shade the region of the plane containing $(0, 0)$. If not, shade the other portion of the plane.

Graph the solution set of $x^2 + y^2 > 9$.

Change the inequality to an equality.

$$x^2 + y^2 > 9$$
$$x^2 + y^2 = 9$$

This is the equation of a circle with center $(0, 0)$ and radius 3.

Since the inequality is $>$, the graph is drawn as a dotted circle.

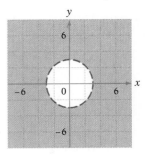

Substitute the point $(0, 0)$ in the inequality. Since $0^2 + 0^2 > 9$ is not true, the point $(0, 0)$ should not be in the shaded region.

Not all inequalities will result in shading a region of the plane. For example, the inequality

$$\frac{x^2}{9} + \frac{y^2}{4} < -1$$

has no ordered pair solutions. Therefore the solution set is the empty set and no region of the plane would be shaded.

Graph the solution set of $y \leq x^2 + 2x + 2$.

Change the inequality to an equality.

$$y \leq x^2 + 2x + 2$$
$$y = x^2 + 2x + 2$$

This is the equation of a parabola that opens up. The vertex is $(-1, 1)$ and the axis of symmetry is the line $x = -1$.

Since the inequality is $\leq$, the graph is drawn as a solid line.

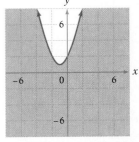

Substitute the point $(0, 0)$ into the inequality. Since $0 < 0^2 + 2(0) + 2$ is true, the point $(0, 0)$ should be in the shaded region.

Graph the solution set of $\frac{y^2}{9} - \frac{x^2}{4} \geq 1$.

Write the inequality as an equality.

$$\frac{y^2}{9} - \frac{x^2}{4} \geq 1$$
$$\frac{y^2}{9} - \frac{x^2}{4} = 1$$

This is the equation of a hyperbola. The vertices are $(0, -3)$ and $(0, 3)$. The equations of the asymptotes are $y = \frac{3}{2}x$ and $y = -\frac{3}{2}x$.

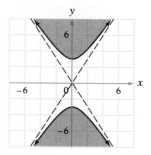

Since the inequality is $\geq$, the graph is drawn as a solid line.

Substitute the point $(0, 0)$ into the inequality. Since $\frac{0^2}{9} - \frac{0^2}{4} \geq 1$ is not true, the point $(0, 0)$ should not be in the shaded region.

Example 1

Graph the solution set of $\frac{x^2}{9} + \frac{y^2}{16} \leq 1$.

Solution

Graph the ellipse $\frac{x^2}{9} + \frac{y^2}{16} = 1$ as a solid line.

Shade the region of the plane that includes $(0, 0)$.

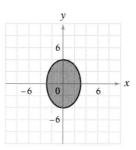

Example 2

Graph the solution set of $\frac{x^2}{9} - \frac{y^2}{16} < 1$.

Your solution

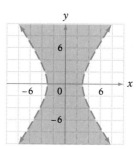

Example 3

Graph of the solution set of $x > y^2 - 4y + 2$.

Solution

Graph the parabola $x = y^2 - 4y + 2$ as a dashed curve.

Shade the region that does not include $(0, 0)$.

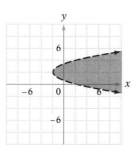

Example 4

Graph the solution set of $\frac{x^2}{16} + \frac{y^2}{4} > 1$.

Your solution

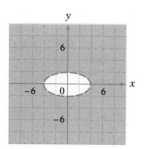

Solutions on p. A39

9.4 EXERCISES

▶ Objective A

Graph the solution set.

1. $y < x^2 - 4x + 3$

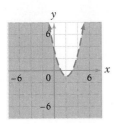

2. $y < x^2 - 2x - 3$

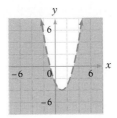

3. $(x - 1)^2 + (y + 2)^2 \leq 9$

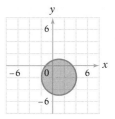

4. $(x + 2)^2 + (y - 3)^2 > 4$

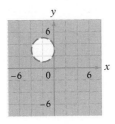

5. $(x + 3)^2 + (y - 2)^2 \geq 9$

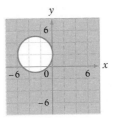

6. $(x - 2)^2 + (y + 1)^2 \leq 16$

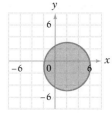

7. $\dfrac{x^2}{16} + \dfrac{y^2}{25} < 1$

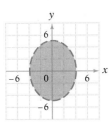

8. $\dfrac{x^2}{9} + \dfrac{y^2}{4} \geq 1$

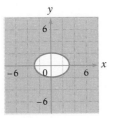

9. $\dfrac{x^2}{25} + \dfrac{y^2}{9} \leq 1$

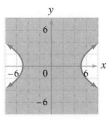

10. $\dfrac{y^2}{25} - \dfrac{x^2}{36} > 1$

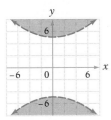

11. $\dfrac{x^2}{4} + \dfrac{y^2}{16} \geq 1$

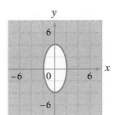

12. $\dfrac{x^2}{4} - \dfrac{y^2}{16} \leq 1$

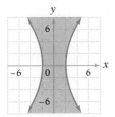

Graph the solution set.

13. $y \le x^2 - 2x + 3$

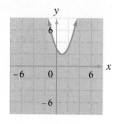

14. $x < y^2 + 2y + 1$

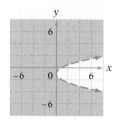

15. $\frac{y^2}{9} - \frac{x^2}{16} \le 1$

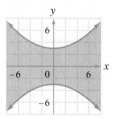

16. $\frac{x^2}{16} - \frac{y^2}{4} > 1$

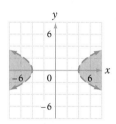

17. $\frac{x^2}{9} + \frac{y^2}{1} \le 1$

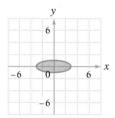

18. $\frac{x^2}{16} + \frac{y^2}{4} > 1$

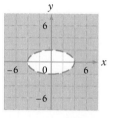

19. $(x - 1)^2 + (x + 3)^2 \le 25$

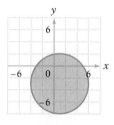

20. $(x + 1)^2 + (y - 2)^2 \ge 16$

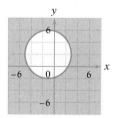

21. $\frac{y^2}{25} - \frac{x^2}{4} \le 1$

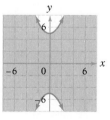

22. $\frac{x^2}{9} - \frac{y^2}{25} \ge 1$

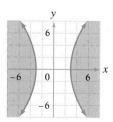

23. $\frac{x^2}{25} + \frac{y^2}{9} < 1$

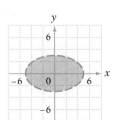

24. $\frac{x^2}{36} + \frac{y^2}{4} \le 1$

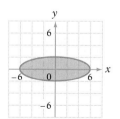

Calculators and Computers

Graph of an Ellipse or Hyperbola

The program GRAPH OF AN ELLIPSE OR HYPERBOLA on the Math ACE Disk will graph an ellipse or hyperbola with the center at the origin. The program asks you to select from one of the following.

$$1. \quad \frac{x^2}{a^2} + \frac{y^2}{b^2} = 1$$

$$2. \quad \frac{x^2}{b^2} + \frac{y^2}{a^2} = 1$$

$$3. \quad \frac{x^2}{a^2} - \frac{y^2}{b^2} = 1$$

$$4. \quad \frac{y^2}{b^2} - \frac{x^2}{a^2} = 1$$

You provide the values for a and b and the program will draw the graph. The values of a and b should be kept between 1 and 8. After the graph has been drawn, you may draw another graph without erasing the first graph, erase the graph and then draw another graph, or quit the program.

By choosing various values for a and b and viewing the resulting graphs, you can better understand the effect these numbers have on the graph.

Chapter Summary

Key Words

The graph of a *conic section* can be represented by the intersection of a plane and a cone.

A *circle* is the set of all points (x, y) in the plane that are a fixed distance from a given point (h, k) called the center. The fixed distance is the *radius* of the circle.

The *asymptotes* of a hyperbola are two straight lines that are "approached" by the hyperbola. As the graph of the hyperbola gets farther from the origin, the hyperbola gets "closer to" the asymptotes.

The *graph of a quadratic inequality* in two variables is a region of the plane that is bounded by one of the conic sections.

Essential Rules

Equation of a Parabola

$y = ax^2 + bx + c$
When $a > 0$, the parabola opens up.
When $a < 0$, the parabola opens down.
The x-coordinate of the vertex is $-\frac{b}{2a}$.
The axis of symmetry is the line $x = -\frac{b}{2a}$.

$x = ay^2 + by + c$
When $a > 0$, the parabola opens to the right.
When $a < 0$, the parabola opens to the left.
The y-coordinate of the vertex is $-\frac{b}{2a}$.
The axis of symmetry is the line $y = -\frac{b}{2a}$.

The Effect of the Discriminant on the Number of x-intercepts of a Parabola

1. If $b^2 - 4ac = 0$, the parabola has one x-intercept.
2. If $b^2 - 4ac > 0$, the parabola has two x-intercepts.
3. If $b^2 - 4ac < 0$, the parabola has no x-intercepts.

Distance Formula

If (x_1, y_1) and (x_2, y_2) are two points in the plane, then the distance between the two points is given by $d = \sqrt{(x_1 - x_2)^2 + (y_1 - y_2)^2}$.

Equation of a Circle

$(x - h)^2 + (y - k)^2 = r^2$
The center is (h, k) and the radius is r.

Equation of an Ellipse

$\dfrac{x^2}{a^2} + \dfrac{y^2}{b^2} = 1$

The x-intercepts are $(a, 0)$ and $(-a, 0)$.
The y-intercepts are $(0, b)$ and $(0, -b)$.

Equation of a Hyperbola

$\dfrac{x^2}{a^2} - \dfrac{y^2}{b^2} = 1$
An axis of symmetry is the x-axis.
The vertices are $(a, 0)$ and $(-a, 0)$.
The equations of the asymptotes are
$y = \pm \dfrac{b}{a}x$.

$\dfrac{y^2}{b^2} - \dfrac{x^2}{a^2} = 1$
An axis of symmetry is the y-axis.
The vertices are $(0, b)$ and $(0, -b)$.
The equations of the asymptotes are
$y = \pm \dfrac{b}{a}x$.

Chapter Review

SECTION 1

1. Find the vertex and axis of symmetry of the parabola $y = x^2 - 4x + 8$.

Vertex (2, 4)
Axis of symmetry: $x = 2$

2. Find the vertex and axis of symmetry of the parabola $y = -x^2 + 7x - 8$.

Vertex: $\left(\frac{7}{2}, \frac{17}{4}\right)$
Axis of symmetry: $x = \frac{7}{2}$

3. Sketch a graph of $y = x^2 - 2x + 3$.

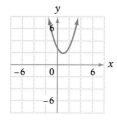

4. Sketch a graph of $y = -2x^2 + x - 2$.

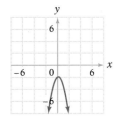

5. Sketch a graph of $x = y^2 - 3y + 5$.

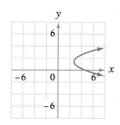

6. Sketch a graph of $x = 2y^2 - 6y + 5$.

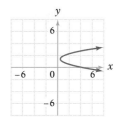

7. Find the minimum value of the function $f(x) = x^2 - 7x + 8$.

minimum: $-\frac{17}{4}$

8. Find the maximum value of the function $f(x) = -x^2 + 8x - 7$.

maximum: 9

9. Find the x-intercepts of the parabola $y = 3x^2 + 9x$.

(0, 0) (−3, 0)

10. Find the x-intercepts of the parabola $y = -2x^2 + 2x - 3$.

no x-intercepts

11. Use the discriminant to determine the number of x-intercepts of the parabola $y = -3x^2 + 4x + 6$.

two

12. Use the discriminant to determine the number of x-intercepts of the parabola $y = 2x^2 - 2x + 5$.

no x-intercepts

13. Find the maximum product of two numbers whose sum is 20.

100

14. The perimeter of a rectangle is 28 ft. What dimensions would give the rectangle a maximum area?

7 ft × 7 ft

SECTION 2

15. Find the distance between the points (2, −3) and (3, 4).

$5\sqrt{2}$

16. Find the distance between the points (0, −1) and (−3, 5).

$3\sqrt{5}$

17. Sketch the graph of
$(x + 3)^2 + (y + 1)^2 = 1$.

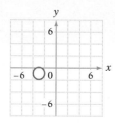

18. Sketch the graph of
$x^2 + (y - 2)^2 = 9$.

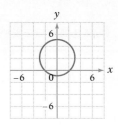

19. Find the equation of the circle with radius 6 and center $(-1, 5)$.
$(x + 1)^2 + (y - 5)^2 = 36$

20. Find the equation of the circle with radius 4 and center $(-3, -3)$.
$(x + 3)^2 + (y + 3)^2 = 16$

21. Find the equation of the circle that passes through $(4, 6)$ and whose center is $(0, -3)$.
$x^2 + (y + 3)^2 = 97$

22. Write the equation
$x^2 + y^2 + 4x - 2y = 4$
in standard form,
then sketch its graph.
$(x + 2)^2 + (y - 1)^2 = 9$

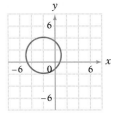

SECTION 3

23. Sketch a graph of
$\dfrac{x^2}{1} + \dfrac{y^2}{9} = 1$.

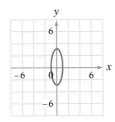

24. Sketch a graph of
$\dfrac{x^2}{25} + \dfrac{y^2}{9} = 1$.

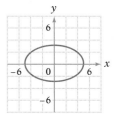

25. Sketch a graph of
$\dfrac{x^2}{25} - \dfrac{y^2}{1} = 1$.

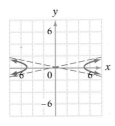

26. Sketch a graph of
$\dfrac{y^2}{16} - \dfrac{x^2}{9} = 1$.

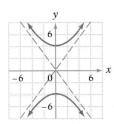

SECTION 4

27. Graph the solution
set of
$(x - 2)^2 + (y + 1)^2 \leq 16$.

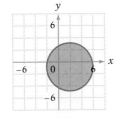

28. Graph the solution set of
$\dfrac{x^2}{9} - \dfrac{y^2}{16} < 1$.

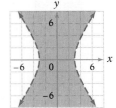

29. Graph the solution
set of $y \geq -x^2 - 2x + 3$.

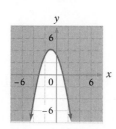

30. Graph the solution
set of $\dfrac{x^2}{16} + \dfrac{y^2}{4} > 1$.

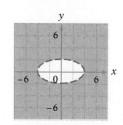

Chapter Test

1. Find the axis of symmetry of the parabola
   $y = -x^2 + 6x - 5$.
   $x = 3$ [9.1A]

2. Find the vertex of the parabola
   $y = -x^2 + 3x - 2$.
   $\left(\frac{3}{2}, \frac{1}{4}\right)$ [9.1A]

3. Sketch a graph of
   $y = \frac{1}{2}x^2 + x - 4$.

   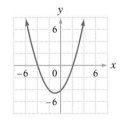

4. Sketch a graph of
   $x = y^2 - y - 2$.

   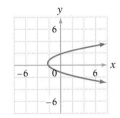

5. Find the maximum value of the function
   $f(x) = -3x^2 + 6x + 1$.
   4 [9.1B]

6. Find the minimum value of the function
   $f(x) = 2x^2 - 6x + 1$.
   $-\frac{7}{2}$ [9.1B]

7. Use the discriminant to determine the number of x-intercepts of $y = -2x^2 + 2x - 3$.
   no x-intercepts [9.1C]

8. Use the discriminant to determine the number of x-intercepts of $y = 3x^2 - 2x - 4$.
   two x-intercepts [9.1C]

9. Find two numbers whose difference is 30 and whose product is a minimum.
   15 and -15 [9.1D]

10. The perimeter of a rectangle is 48 cm. What dimensions would give the rectangle a maximum area?
    12 cm × 12 cm [9.1D]

11. Find the distance between the points $(3, -4)$ and $(0, 3)$.
    $\sqrt{58}$ [9.2A]

12. Sketch the graph of
    $(x - 2)^2 + (y + 1)^2 = 9$.
    [9.2B]

    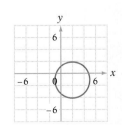

13. Find the equation of the circle with radius 3 and center $(-2, 4)$.
$(x + 2)^2 + (y - 4)^2 = 9$ [9.2B]

14. Find the equation of the circle that passes through point $(2, 5)$ and whose center is $(-2, 1)$.
$(x + 2)^2 + (y - 1)^2 = 32$ [9.2B]

15. Write the equation $x^2 + y^2 - 4x + 2y + 1 = 0$ in standard form then sketch its graph.
$(x - 2)^2 + (y + 1)^2 = 4$

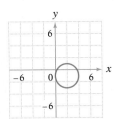

[9.2C]

16. Sketch a graph of $\frac{y^2}{25} - \frac{x^2}{16} = 1$.

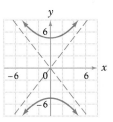

[9.3B]

17. Sketch a graph of $\frac{x^2}{9} - \frac{y^2}{4} = 1$.

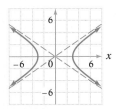

[9.3B]

18. Sketch a graph of $\frac{x^2}{16} + \frac{y^2}{4} = 1$.

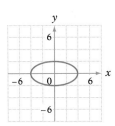

[9.3A]

19. Graph the solution set of $(x + 3)^2 + (y - 1)^2 \geq 16$.

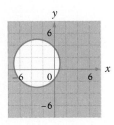

[9.4A]

20. Graph the solution set of $\frac{x^2}{16} - \frac{y^2}{25} < 1$.

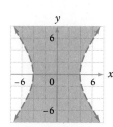

[9.4A]

Cumulative Review

1. Graph the solution set of
$\{x|x < 4\} \cap \{x|x > 2\}$.

$[1.4B]$

2. Solve: $\frac{5x-2}{3} - \frac{1-x}{5} = \frac{x+4}{10}$

$\frac{38}{53}$ $[2.1C]$

3. Solve: $4 + |3x + 2| < 6$
$\left\{x\left|-\frac{4}{3} < x < 0\right.\right\}$ $[2.3A]$

4. Find the equation of the line that contains the point $(2, -3)$ and has slope $-\frac{3}{2}$.
$y = -\frac{3}{2}x$ $[6.3A]$

5. Find the equation of the line containing the point $(4, -2)$ and perpendicular to the line $y = -x + 5$.
$y = x - 6$ $[6.4A]$

6. Simplify: $x^{2n}(x^{2n} + 2x^n - 3x)$
$x^{4n} + 2x^{3n} - 3x^{2n+1}$ $[3.2A]$

7. Factor: $(x - 1)^3 - y^3$
$(x - y - 1)(x^2 - 2x + 1 + xy - y + y^2)$ $[3.4B]$

8. Solve: $\frac{3x-2}{x+4} \le 1$
$\{x|-4 < x \le 3\}$ $[7.5A]$

9. Simplify: $\frac{ax - bx}{ax + ay - bx - by}$
$\frac{x}{x+y}$ $[4.1A]$

10. Simplify: $\frac{x-4}{3x-2} - \frac{1+x}{3x^2+x-2}$
$\frac{x-5}{3x-2}$ $[4.3B]$

11. Solve: $\frac{6x}{2x-3} - \frac{1}{2x-3} = 7$
$\frac{5}{2}$ $[4.6A]$

12. Graph the solution set of $5x + 2y > 10$.

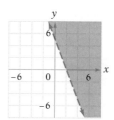

$[6.6A]$

13. Simplify: $\left(\frac{12a^2b^{-2}}{a^{-3}b^{-4}}\right)^{-1}\left(\frac{ab}{4^{-1}a^{-2}b^4}\right)^2$
$\frac{4a}{3b^8}$ $[3.1C]$

14. Write $2\sqrt[4]{x^3}$ as an exponential expression.
$2x^{3/4}$ $[5.1B]$

15. Simplify: $\sqrt{18} - \sqrt{-25}$
$3\sqrt{2} - 5i$ $[5.3A]$

16. Solve: $2x^2 + 2x - 3 = 0$
$\frac{-1+\sqrt{7}}{2}$ and $\frac{-1-\sqrt{7}}{2}$ $[7.3A]$

17. Solve: $a^{2/3} - 2a^{1/3} - 3 = 0$
27 and -1 $[7.4A]$

18. Solve: $x - \sqrt{2x-3} = 3$
6 $[7.4B]$

19. Evaluate the function $f(x) = -x^2 + 3x - 2$ at $x = -3$.
−20   [8.1A]

20. Find the inverse of the function $f(x) = 4x + 8$.
$f^{-1}(x) = \frac{1}{4}x - 2$ [8.3B]

21. Find the maximum value of the function $f(x) = -2x^2 + 4x - 2$.
0   [9.1B]

22. Find the maximum product of two numbers whose sum is 40.
400   [9.1D]

23. Find the distance between the points $(2, 4)$ and $(-1, 0)$.
5   [9.2A]

24. Find the equation of the circle that passes through $(3, 1)$ and whose center is $(-1, 2)$.
$(x + 1)^2 + (y - 2)^2 = 17$ [9.2B]

25. Graph: $x = y^2 - 2y + 3$
[9.1A]

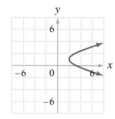

26. Graph: $\frac{x^2}{25} + \frac{y^2}{4} = 1$
[9.3A]

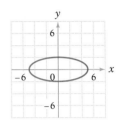

27. Graph: $\frac{y^2}{4} - \frac{x^2}{25} = 1$
[9.3B]

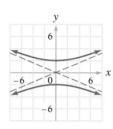

28. Graph: $\frac{x^2}{9} - \frac{y^2}{36} = 1$
[9.3B]

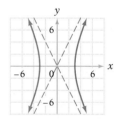

29. Tickets for a school play sold for $4.00 for each adult and $1.50 for each child. The total receipt for the 192 tickets sold was $493. Find the number of adult tickets sold.
82 tickets   [2.5A]

30. A motorcycle travels 180 mi in the same amount of time that a car travels 144 mi. The rate of the motorcycle is 12 mph faster than the rate of the car. Find the rate of the motorcycle.
60 mph   [4.6D]

31. The rate of a river's current is 1.5 mph. A rowing crew can row 12 mi down this river and 12 mi back in 6 h. Find the rowing rate of the crew in calm water.
4.5 mph   [4.6D]

32. The speed (v) of a gear varies inversely as the number of teeth (t). If a gear that has 36 teeth makes 30 revolutions per minute, how many revolutions per minute will a gear that has 60 teeth make?
18 revolutions/min   [8.4A]

10

Systems of Equations and Inequalities

OBJECTIVES

▶ To solve a system of linear equations by graphing

▶ To solve a system of linear equations by the substitution method

▶ To solve a system of two linear equations in two variables by the addition method

▶ To solve a system of three linear equations in three variables by the addition method

▶ To evaluate a determinant

▶ To solve a system of equations by using Cramer's Rule

▶ To solve rate-of-wind or current problems

▶ To solve application problems using two variables

▶ To solve a nonlinear system of equations

▶ To graph the solution set of a system of inequalities

Analytic Geometry

Euclid's Geometry (plane geometry is still taught in high school) was developed around 300 B.C. Euclid depends on a synthetic proof—a proof based on pure logic.

Algebra is the science of solving equations such as $ax^2 + bx + c = 0$. As early as the third century, equations were solved by trial and error. However, the appropriate terminology and symbols of operation were not fully developed until the seventeenth century.

Analytic geometry is the combining of algebraic methods with geometric concepts. The development of analytic geometry in the seventeenth century is attributed to Descartes.

The development of analytic geometry depended on the invention of a system of coordinates in which any point in the plane could be represented by an ordered pair of numbers. This invention allowed lines and curves (such as the conic sections) to be represented and described by algebraic equations.

Some equations are quite simple and have a simple graph. For example, the graph of the equation $y = 2x - 3$ is a line.

Other graphs are quite complicated and have complicated equations. The graph of the Mandelbrot set is very complicated. The graph is called a fractal.

SECTION 10.1

Solving Systems of Linear Equations by Graphing and by the Substitution Method

| Objective A | **To solve a system of linear equations by graphing** |

A **system of equations** is two or more equations considered together. The system at the right is a system of two linear equations in two variables. The graphs of the equations are straight lines.

$$3x + 4y = 7$$
$$2x - 3y = 6$$

A **solution of a system of equations in two variables** is an ordered pair that is a solution of each equation of the system.

Is $(3, -2)$ a solution of the system
$$2x - 3y = 12$$
$$5x + 2y = 11?$$

$2x - 3y = 12$		$5x + 2y = 11$	
$2(3) - 3(-2)$	12	$5(3) + 2(-2)$	11
$6 - (-6)$		$15 + (-4)$	
	$12 = 12$		$11 = 11$

Yes, since $(3, -2)$ is a solution of each equation, it is a solution of the system of equations.

A solution of a system of linear equations can be found by graphing the lines of the system on the same coordinate axes. The point of intersection of the lines is the ordered pair that lies on both lines. It is the solution of the system of equations.

Solve by graphing: $x + 2y = 4$
$2x + y = -1$

Graph each line.

Find the point of intersection.

The ordered pair $(-2, 3)$ lies on each line.

The solution is $(-2, 3)$.

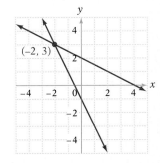

When the graphs of a system of equations intersect at only one point, the equations are called **independent equations.**

Solve by graphing: $2x + 3y = 6$
$4x + 6y = -12$

Graph each line.

The lines are parallel and therefore do not intersect. The system of equations has no solution.

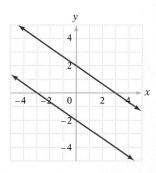

When a system of equations has no solution, it is called an **inconsistent system of equations.**

Solve by graphing: $x - 2y = 4$
$\qquad\qquad\qquad\quad 2x - 4y = 8$

Graph each line.

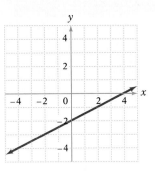

The two equations represent the same line. **The equations in this system are dependent equations.** When a system of two equations is dependent, solve one of the equations of the system for y. The solutions of the dependent system are $(x, mx + b)$, where $y = mx + b$. For this system,

$$x - 2y = 4$$
$$-2y = -x + 4$$
$$y = \tfrac{1}{2}x - 2$$

The solutions are the ordered pairs $\left(x, \tfrac{1}{2}x - 2\right)$.

Example 1   Solve by graphing:
$2x - y = 3$
$3x + y = 2$

Solution

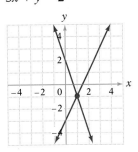

The solution is $(1, -1)$.

Example 2   Solve by graphing:
$x + y = 1$
$2x + y = 0$

Your solution

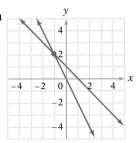

$(-1, 2)$

Example 3   Solve by graphing:
$2x + 3y = 6$

$\qquad y = -\tfrac{2}{3}x + 1$

Solution

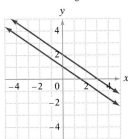

The lines are parallel and therefore do not intersect. The system of equations has no solution.

Example 4   Solve by graphing:
$3x - 4y = 12$

$\qquad y = \tfrac{3}{4}x - 3$

Your solution

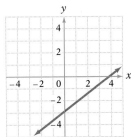

The two equations represent the same line. The equations are dependent. The solutions are the ordered pairs $\left(x, \tfrac{3}{4}x - 3\right)$.

Solutions on p. A39

| Objective B | **To solve a system of linear equations by the substitution method** |

Graphical methods of solving systems of equations are often unsatisfactory, since it can be difficult to read exact values of a point of intersection. For example, the point $\left(\frac{2}{3}, -\frac{4}{7}\right)$ would be difficult to read from a graph.

An algebraic method, called the **substitution method,** can be used to find an exact solution of a system of equations.

In the system of equations at the right, equation (2) states that $y = 3x - 4$.
Substitute $3x - 4$ for y in equation (1).

(1) $\qquad 2x - 3y = 5$
(2) $\qquad y = 3x - 4$

$$2x - 3(3x - 4) = 5$$

Solve for x.

$$2x - 9x + 12 = 5$$
$$-7x + 12 = 5$$
$$-7x = -7$$
$$x = 1$$

Substitute the value of x into equation (2) and solve for y.

(2) $\qquad y = 3x - 4$
$$y = 3(1) - 4$$
$$y = 3 - 4$$
$$y = -1$$

The solution is $(1, -1)$.

Solve by substitution: $\qquad 3x - y = 5$
$\qquad\qquad\qquad\qquad\qquad 2x + 5y = 9$

(1) $\qquad 3x - y = 5$
(2) $\qquad 2x + 5y = 9$

Solve equation (1) for y. Equation (1) is chosen because it is the easier equation to solve for one variable in terms of the other.

$$3x - y = 5$$
$$-y = -3x + 5$$
$$y = 3x - 5$$

Substitute $3x - 5$ for y in equation (2).

$$2x + 5y = 9$$
$$2x + 5(3x - 5) = 9$$
$$2x + 15x - 25 = 9$$
$$17x - 25 = 9$$
$$17x = 34$$
$$x = 2$$

Substitute the value of x into the equation $y = 3x - 5$ and solve for y.

$$y = 3x - 5$$
$$y = 3(2) - 5$$
$$y = 6 - 5$$
$$y = 1$$

The solution is $(2, 1)$.

Check: $\qquad 3x - y = 5 \qquad\qquad 2x + 5y = 9$

$$
\begin{array}{c|c}
3(2) - 1 & 5 \\
6 - 1 & \\
5 = 5 &
\end{array}
\qquad
\begin{array}{c|c}
2(2) + 5(1) & 9 \\
4 + 5 & \\
9 = 9 &
\end{array}
$$

Example 5 Solve by substitution:
(1) $3x - 2y = 4$
(2) $-x + 4y = -3$

Solution Solve equation (2) for x.

$-x + 4y = -3$
$\quad -x = -4y - 3$
$\quad\quad x = 4y + 3$

Substitute into equation (1).

$3x - 2y = 4$
$3(4y + 3) - 2y = 4$
$12y + 9 - 2y = 4$
$10y + 9 = 4$
$10y = -5$
$y = -\dfrac{5}{10} = -\dfrac{1}{2}$

Substitute into equation (2).

$-x + 4y = -3$
$-x + 4\left(-\dfrac{1}{2}\right) = -3$
$-x - 2 = -3$
$-x = -1$
$x = 1$

The solution is $\left(1, -\dfrac{1}{2}\right)$.

Example 6 Solve by substitution:
$3x - y = 3$
$6x + 3y = -4$

Your solution $\left(\dfrac{1}{3}, -2\right)$

Example 7 Solve by substitution:
$3x - 3y = 2$
$\quad y = x + 2$

Solution $3x - 3y = 2$
$3x - 3(x + 2) = 2$
$3x - 3x - 6 = 2$
$-6 = 2$

This is not a true equation. The lines are parallel and the system is inconsistent. The system does not have a solution.

Example 8 Solve by substitution:
$\quad y = 2x - 3$
$2x - 4y = 5$

Your solution $\left(\dfrac{7}{6}, -\dfrac{2}{3}\right)$

Example 9 Solve by substitution:
$9x + 3y = 12$
$\quad y = -3x + 4$

Solution $9x + 3y = 12$
$9x + 3(-3x + 4) = 12$
$9x - 9x + 12 = 12$
$12 = 12$

This is a true equation. The equations are dependent. The solutions are the ordered pairs $(x, -3x + 4)$.

Example 10 Solve by substitution:
$6x - 3y = 6$
$2x - y = 2$

Your solution dependent equations
The solutions are the ordered pairs $(x, 2x - 2)$.

Solutions on pp. A39–A40

10.1 EXERCISES

▶ **Objective A**

Solve by graphing:

1. $x + y = 2$
$x - y = 4$

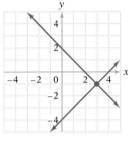

$(3, -1)$

2. $x + y = 1$
$3x - y = -5$

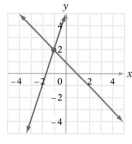

$(-1, 2)$

3. $x - y = -2$
$x + 2y = 10$

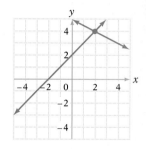

$(2, 4)$

4. $2x - y = 5$
$3x + y = 5$

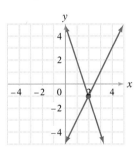

$(2, -1)$

5. $3x - 2y = 6$
$y = 3$

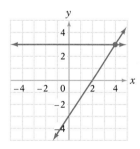

$(4, 3)$

6. $x = 4$
$3x - 2y = 4$

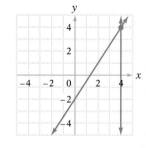

$(4, 4)$

7. $x = 4$
$y = -1$

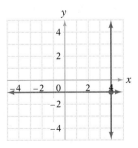

$(4, -1)$

8. $x + 2 = 0$
$y - 1 = 0$

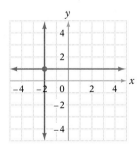

$(-2, 1)$

9. $2x + y = 3$
$x - 2 = 0$

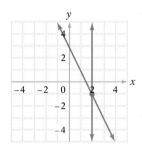

$(2, -1)$

Solve by graphing:

10. $x - 3y = 6$
$y + 3 = 0$

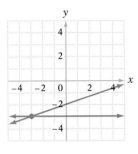

$(-3, -3)$

11. $x - y = 4$
$x + y = 2$

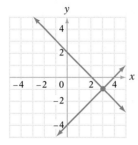

$(3, -1)$

12. $2x + y = 2$
$-x + y = 5$

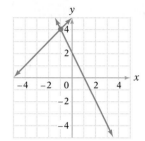

$(-1, 4)$

13. $y = x - 5$
$2x + y = 4$

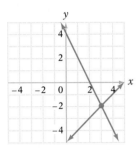

$(3, -2)$

14. $2x - 5y = 4$
$y = x + 1$

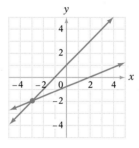

$(-3, -2)$

15. $y = \frac{1}{2}x - 2$
$x - 2y = 8$

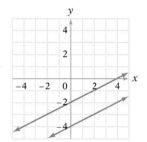

no solution

16. $2x + 3y = 6$
$y = -\frac{2}{3}x + 1$

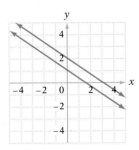

no solution

17. $2x - 5y = 10$
$y = \frac{2}{5}x - 2$

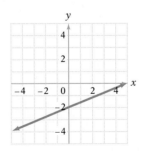

$\left(x, \frac{2}{5}x - 2\right)$

18. $3x - 2y = 6$
$y = \frac{3}{2}x - 3$

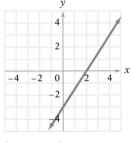

$\left(x, \frac{3}{2}x - 3\right)$

Solve by graphing:

19. $3x - 4y = 12$
$5x + 4y = -12$

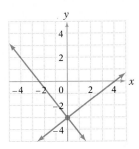

(0, −3)

20. $2x - 3y = 6$
$2x - 5y = 10$

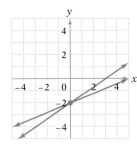

(0, −2)

21. $2x - 3y = 2$
$5x + 4y = 5$

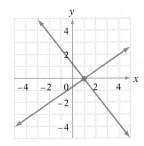

(1, 0)

22. $2x - 3y = 0$
$x - 2y = 1$

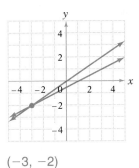

(−3, −2)

23. $2x - 3y = -1$
$x + 4y = 5$

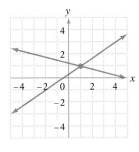

(1, 1)

24. $x + 3y = 6$
$-2x + y = 2$

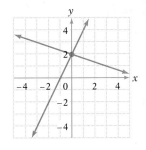

(0, 2)

▶ **Objective B**

Solve by substitution:

25. $3x - 2y = 4$
$x = 2$
(2, 1)

26. $y = -2$
$2x + 3y = 4$
(5, −2)

27. $y = 2x - 1$
$x + 2y = 3$
(1, 1)

28. $y = -x + 1$
$2x - y = 5$
(2, −1)

29. $x = 3y + 1$
$x - 2y = 6$
(16, 5)

30. $x = 2y - 3$
$3x + y = 5$
(1, 2)

31. $4x - 3y = 5$
$y = 2x - 3$
(2, 1)

32. $3x + 5y = -1$
$y = 2x - 8$
(3, −2)

33. $5x - 2y = 9$
$y = 3x - 4$
(−1, −7)

34. $4x - 3y = 2$
$y = 3x + 1$
(−1, −2)

35. $x = 2y + 4$
$4x + 3y = -17$
(−2, −3)

36. $3x - 2y = -11$
$x = 2y - 9$
(−1, 4)

Solve by substitution:

37. $5x + 4y = -1$
       $y = 2 - 2x$
  $(3, -4)$

38. $3x + 2y = 4$
       $y = 1 - 2x$
  $(-2, 5)$

39. $2x - 5y = -9$
       $y = 9 - 2x$
  $(3, 3)$

40. $5x + 2y = 15$
       $x = 6 - y$
  $(1, 5)$

41. $7x - 3y = 3$
       $x = 2y + 2$
  $(0, -1)$

42. $3x - 4y = 6$
       $x = 3y + 2$
  $(2, 0)$

43. $2x + 2y = 7$
       $y = 4x + 1$
  $\left(\frac{1}{2}, 3\right)$

44. $3x + 7y = -5$
       $y = 6x - 5$
  $\left(\frac{2}{3}, -1\right)$

45. $3x + y = 5$
   $2x + 3y = 8$
  $(1, 2)$

46. $4x + y = 9$
   $3x - 4y = 2$
  $(2, 1)$

47. $x + 3y = 5$
   $2x + 3y = 4$
  $(-1, 2)$

48. $x - 4y = 2$
   $2x - 5y = 1$
  $(-2, -1)$

49. $3x - y = 10$
   $6x - 2y = 5$
  inconsistent

50. $6x - 4y = 3$
   $3x - 2y = 9$
  inconsistent

51. $3x + 4y = 14$
   $2x + y = 1$
  $(-2, 5)$

52. $5x + 3y = 8$
   $3x + y = 8$
  $(4, -4)$

53. $3x + 5y = 0$
   $x - 4y = 0$
  $(0, 0)$

54. $2x - 7y = 0$
   $3x + y = 0$
  $(0, 0)$

55. $2x - 4y = 16$
   $-x + 2y = -8$
  $\left(x, \frac{1}{2}x - 4\right)$

56. $3x - 12y = -24$
   $-x + 4y = 8$
  $\left(x, \frac{1}{4}x + 2\right)$

57. $y = 3x + 2$
   $y = 2x + 3$
  $(1, 5)$

58. $y = 3x - 7$
   $y = 2x - 5$
  $(2, -1)$

59. $y = 3x + 1$
   $y = 6x - 1$
  $\left(\frac{2}{3}, 3\right)$

60. $y = 2x - 3$
   $y = 4x - 4$
  $\left(\frac{1}{2}, -2\right)$

61. $x = 2y + 1$
   $x = 3y - 1$
  $(5, 2)$

62. $x = 4y + 1$
   $x = -2y - 5$
  $(-3, -1)$

63. $y = 5x - 1$
   $y = 5 - x$
  $(1, 4)$

64. $y = 3 - 2x$
   $y = 2 - 3x$
  $(-1, 5)$

65. $-x + 2y = 13$
   $5x + 3y = 13$
  $(-1, 6)$

66. $3x - y = 8$
   $4x - 7y = 5$
  $(3, 1)$

SECTION 10.2

Solving Systems of Linear Equations by the Addition Method

Objective A

To solve a system of two linear equations in two variables by the addition method

The **addition method** is an alternative method for solving a system of equations. This method is based upon an Addition Property of Equations. It is appropriate when it is not convenient to solve one equation for one variable in terms of another variable.

Note, for the system of equations at the right, the effect of adding equation (2) to equation (1). Since $-3y$ and $3y$ are additive inverses, adding the equations results in an equation with only one variable.

$$(1) \quad 5x - 3y = 14$$
$$(2) \quad 2x + 3y = -7$$
$$7x + 0y = 7$$
$$7x = 7$$

The solution of the resulting equation is the first component of the ordered pair solution of the system.

$$7x = 7$$
$$x = 1$$

The second component is found by substituting the value of x into equation (1) or (2) and then solving for y. Equation (1) is used here.

$$(1) \quad 5x - 3y = 14$$
$$5(1) - 3y = 14$$
$$5 - 3y = 14$$
$$-3y = 9$$
$$y = -3$$

The solution is $(1, -3)$.

Sometimes adding the two equations does not eliminate one of the variables. In this case, use the Multiplication Property of Equations to rewrite one or both of the equations, so that when the equations are added, one of the variables is eliminated. To do this, first choose which variable to eliminate. The coefficients of that variable must be additive inverses. Multiply each equation by a constant that will produce coefficients that are additive inverses.

Solve by the addition method: $3x + 4y = 2$
$\qquad\qquad\qquad\qquad\qquad\quad 2x + 5y = -1$

$$(1) \quad 3x + 4y = 2$$
$$(2) \quad 2x + 5y = -1$$

Eliminate x. Multiply equation (1) by 2 and equation (2) by -3. Note how the constants are selected.

$$2 \diagdown (3x + 4y) = 2(2)$$
$$-3 \diagdown (2x + 5y) = -3(-1)$$
The negative sign is used so that the coefficients will be additive inverses.

The coefficients of the x terms are additive inverses.

$$6x + 8y = 4$$
$$-6x - 15y = 3$$

Add the equations.
Solve for y.

$$-7y = 7$$
$$y = -1$$

Substitute the value of y into one of the equations and solve for x. Equation (1) is used here.

$$(1) \quad 3x + 4y = 2$$
$$3x + 4(-1) = 2$$
$$3x - 4 = 2$$
$$3x = 6$$
$$x = 2$$

The solution is $(2, -1)$.

Solve by the addition method: $\frac{2}{3}x + \frac{1}{2}y = 4$ (1) $\frac{2}{3}x + \frac{1}{2}y = 4$

$\frac{1}{4}x - \frac{3}{8}y = -\frac{3}{4}$ (2) $\frac{1}{4}x - \frac{3}{8}y = -\frac{3}{4}$

Clear fractions. Multiply each equation by the LCM of the denominators.

$$6\left(\frac{2}{3}x + \frac{1}{2}y\right) = 6(4)$$

$$8\left(\frac{1}{4}x - \frac{3}{8}y\right) = 8\left(-\frac{3}{4}\right)$$

$$4x + 3y = 24$$
$$2x - 3y = -6$$

Eliminate y. Add the equations.
Solve for x.

$$6x = 18$$
$$x = 3$$

Substitute the value of x into equation (1) and solve for y.

$$\frac{2}{3}x + \frac{1}{2}y = 4$$

$$\frac{2}{3}(3) + \frac{1}{2}y = 4$$

$$2 + \frac{1}{2}y = 4$$

$$\frac{1}{2}y = 2$$

$$y = 4$$

The solution is (3, 4).

Solve by the addition method: $3x - 2y = 5$ (1) $3x - 2y = 5$

$6x - 4y = 1$ (2) $6x - 4y = 1$

Eliminate x. Multiply equation (1) by -2 and add to equation (2).

$$-6x + 4y = -10$$
$$6x - 4y = 1$$

This is not a true equation. The system is inconsistent and therefore has no solution.

$$0 = -9$$

The graph of the equations of the system is shown at the right. Note that the lines are parallel and therefore do not intersect.

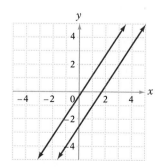

Example 1 Solve by the addition method:
(1) $3x - 2y = 2x + 5$
(2) $2x + 3y = -4$

Solution Write equation (1) in the form $Ax + By = C$.

$$3x - 2y = 2x + 5$$
$$x - 2y = 5$$

Solve the system:

$$x - 2y = 5$$
$$2x + 3y = -4$$

Eliminate x.

$$-2(x - 2y) = -2(5)$$
$$2x + 3y = -4$$

$$-2x + 4y = -10$$
$$2x + 3y = -4$$

Add the equations.

$$7y = -14$$
$$y = -2$$

Replace y in equation (2).

$$2x + 3y = -4$$
$$2x + 3(-2) = -4$$
$$2x + (-6) = -4$$
$$2x = 2$$
$$x = 1$$

The solution is $(1, -2)$.

Example 2 Solve by the addition method:
$2x + 5y = 6$
$3x - 2y = 6x + 2$

Your solution $(-2, 2)$

Example 3 Solve by the addition method:
(1) $4x - 8y = 36$
(2) $3x - 6y = 27$

Solution Eliminate x.

$$3(4x - 8y) = 3(36)$$
$$-4(3x - 6y) = -4(27)$$

$$12x - 24y = 108$$
$$-12x + 24y = -108$$

Add the equations.
$$0 = 0$$

The system of equations is dependent. The solutions are the ordered pairs $\left(x, \frac{1}{2}x - \frac{9}{2}\right)$.

Example 4 Solve by the addition method:
$2x + y = 5$
$4x + 2y = 6$

Your solution inconsistent

Solutions on p. A40

| Objective B | To solve a system of three linear equations in three variables by the addition method | 10 |

An equation of the form $Ax + By + Cz = D$, where A, B, C, and D are constants, is a **linear equation in three variables.** Examples of these equations are shown at the right. The graph of a linear equation in three variables is a plane.

$$2x + 4y - 3z = 7$$
$$x - 6y + z = -8$$

Just as a solution of an equation in two variables is an ordered pair (x, y), a **solution of an equation in three variables** is an ordered triple (x, y, z). For example, $(2, 1, -3)$ is a solution of the equation $2x - y - 2z = 9$. The ordered triple $(1, 3, 2)$ is not a solution.

A **system of linear equations in three variables** is shown at the right. A **solution of a system of equations in three variables** is an ordered triple that is a solution of each equation of the system.

$$x - 2y + z = 6$$
$$3x + y - 2z = 2$$
$$2x - 3y + 5z = 1$$

A system of linear equations in three variables can be solved by using the addition method. First, eliminate one variable from any two of the given equations. Then eliminate the same variable from any other two equations. The result will be a system of two equations in two variables. Solve this system by the addition method.

Solve:
$$x + 4y - z = 10$$
$$3x + 2y + z = 4$$
$$2x - 3y + 2z = -7$$

(1)	$x + 4y - z = 10$
(2)	$3x + 2y + z = 4$
(3)	$2x - 3y + 2z = -7$

Eliminate z from equations (1) and (2) by adding the two equations.

$$x + 4y - z = 10$$
$$3x + 2y + z = 4$$
$$(4) \qquad 4x + 6y = 14$$

Eliminate z from equations (1) and (3). Multiply equation (1) by 2 and add to equation (3).

$$2x + 8y - 2z = 20$$
$$2x - 3y + 2z = -7$$
$$(5) \qquad 4x + 5y = 13$$

Solve the system of two equations in two variables.

$$(4) \qquad 4x + 6y = 14$$
$$(5) \qquad 4x + 5y = 13$$

Eliminate x. Multiply equation (5) by -1 and add to equation (4).
Solve for y.

$$4x + 6y = 14$$
$$-4x - 5y = -13$$
$$y = 1$$

Substitute the value of y into equation (4) or (5) and solve for x. Equation (4) is used here.

$$4x + 6y = 14$$
$$4x + 6(1) = 14$$
$$4x = 8$$
$$x = 2$$

Substitute the value of y and the value of x into one of the equations in the original system. Equation (2) is used here.

$$3x + 2y + z = 4$$
$$3(2) + 2(1) + z = 4$$
$$6 + 2 + z = 4$$
$$8 + z = 4$$
$$z = -4$$

The solution is $(2, 1, -4)$.

Solve:
$$2x - 3y - z = 1$$
$$x + 4y + 3z = 2$$
$$4x - 6y - 2z = 5$$

(1) $2x - 3y - z = 1$
(2) $x + 4y + 3z = 2$
(3) $4x - 6y - 2z = 5$

Eliminate x from equations (1) and (2). Multiply equation (2) by -2 and add to equation (1).

$$2x - 3y - z = 1$$
$$-2x - 8y - 6z = -4$$
$$-11y - 7z = -3$$

Eliminate x from equations (1) and (3). Multiply equation (1) by -2 and add to equation (3).

$$-4x + 6y + 2z = -2$$
$$4x - 6y - 2z = 5$$
$$0 = 3$$

The equation $0 = 3$ is not a true equation. The system of equations is inconsistent and therefore has no solution.

Solve:
$$3x - z = -1$$
$$2y - 3z = 10$$
$$x + 3y - z = 7$$

(1) $3x - z = -1$
(2) $2y - 3z = 10$
(3) $x + 3y - z = 7$

Eliminate x from equations (1) and (3). Multiply equation (3) by -3 and add to equation (1).

$$3x - z = -1$$
$$-3x - 9y + 3z = -21$$

(4) $-9y + 2z = -22$

Use equations (2) and (4) to form a system of equations in two variables.

(2) $2y - 3z = 10$
(4) $-9y + 2z = -22$

Eliminate z. Multiply equation (2) by 2 and equation (4) by 3.

$$4y - 6z = 20$$
$$-27y + 6z = -66$$

Add the equations.
Solve for y.

$$-23y = -46$$
$$y = 2$$

Substitute the value of y into equation (2) or (4) and solve for z. Equation (2) is used here.

(2)
$$2y - 3z = 10$$
$$2(2) - 3z = 10$$
$$4 - 3z = 10$$
$$-3z = 6$$
$$z = -2$$

Substitute the value of z into equation (1) and solve for x.

(1)
$$3x - z = -1$$
$$3x - (-2) = -1$$
$$3x + 2 = -1$$
$$3x = -3$$
$$x = -1$$

The solution is $(-1, 2, -2)$.

Example 5 Solve: (1) $3x - y + 2z = 1$
(2) $2x + 3y + 3z = 4$
(3) $x + y - 4z = -9$

Solution Eliminate y. Add equations
(1) and (3).

$3x - y + 2z = 1$
$x + y - 4z = -9$
$4x - 2z = -8$

Multiply each side of the
equation by $\frac{1}{2}$.

(4) $2x - z = -4$

Multiply equation (1) by 3
and add to equation (2).

$9x - 3y + 6z = 3$
$2x + 3y + 3z = 4$
(5) $11x + 9z = 7$

Solve the system of two
equations.

(4) $2x - z = -4$
(5) $11x + 9z = 7$

Multiply equation (4) by 9
and add to equation (5).

$18x - 9z = -36$
$11x + 9z = 7$
$29x = -29$
$x = -1$

Replace x by -1 in equation
(4).

$2x - z = -4$
$2(-1) - z = -4$
$-2 - z = -4$
$-z = -2$
$z = 2$

Replace x by -1 and z by 2
in equation (3).

$x + y - 4z = -9$
$-1 + y - 4(2) = -9$
$-9 + y = -9$
$y = 0$

The solution is $(-1, 0, 2)$.

Example 6 Solve: $x - y + z = 6$
$2x + 3y - z = 1$
$x + 2y + 2z = 5$

Your solution $(3, -1, 2)$

Solution on p. A41

10.2   EXERCISES

▶ Objective A

Solve by the addition method:

1. $x - y = 5$
$x + y = 7$
(6, 1)

2. $x + y = 1$
$2x - y = 5$
(2, −1)

3. $3x + y = 4$
$x + y = 2$
(1, 1)

4. $x - 3y = 4$
$x + 5y = -4$
(1, −1)

5. $3x + y = 7$
$x + 2y = 4$
(2, 1)

6. $x - 2y = 7$
$3x - 2y = 9$
(1, −3)

7. $2x + 3y = -1$
$x + 5y = 3$
(−2, 1)

8. $x + 5y = 7$
$2x + 7y = 8$
(−3, 2)

9. $3x - y = 4$
$6x - 2y = 8$
(x, 3x − 4)

10. $x - 2y = -3$
$-2x + 4y = 6$
$\left(x, \frac{1}{2}x + \frac{3}{2}\right)$

11. $2x + 5y = 9$
$4x - 7y = -16$
$\left(-\frac{1}{2}, 2\right)$

12. $8x - 3y = 21$
$4x + 5y = -9$
$\left(\frac{3}{2}, -3\right)$

13. $4x - 6y = 5$
$2x - 3y = 7$
inconsistent

14. $3x + 6y = 7$
$2x + 4y = 5$
inconsistent

15. $3x - 5y = 7$
$x - 2y = 3$
(−1, −2)

16. $3x + 4y = 25$
$2x + y = 10$
(3, 4)

17. $x + 3y = 7$
$-2x + 3y = 22$
(−5, 4)

18. $2x - 3y = 14$
$5x - 6y = 32$
(4, −2)

19. $3x + 2y = 16$
$2x - 3y = -11$
(2, 5)

20. $2x - 5y = 13$
$5x + 3y = 17$
(4, −1)

21. $4x + 4y = 5$
$2x - 8y = -5$
$\left(\frac{1}{2}, \frac{3}{4}\right)$

Solve by the addition method:

22. $3x + 7y = 16$
$4x - 3y = 9$
$(3, 1)$

23. $5x + 4y = 0$
$3x + 7y = 0$
$(0, 0)$

24. $3x - 4y = 0$
$4x - 7y = 0$
$(0, 0)$

25. $5x + 2y = 1$
$2x + 3y = 7$
$(-1, 3)$

26. $3x + 5y = 16$
$5x - 7y = -4$
$(2, 2)$

27. $3x - 6y = 6$
$9x - 3y = 8$
$\left(\frac{2}{3}, -\frac{2}{3}\right)$

28. $4x - 8y = 5$
$8x + 2y = 1$
$\left(\frac{1}{4}, -\frac{1}{2}\right)$

29. $5x + 2y = 2x + 1$
$2x - 3y = 3x + 2$
$(1, -1)$

30. $3x + 3y = y + 1$
$x + 3y = 9 - x$
$(-3, 5)$

31. $\frac{2}{3}x - \frac{1}{2}y = 3$
$\frac{1}{3}x - \frac{1}{4}y = \frac{3}{2}$
$\left(x, \frac{4}{3}x - 6\right)$

32. $\frac{3}{4}x + \frac{1}{3}y = -\frac{1}{2}$
$\frac{1}{2}x - \frac{5}{6}y = -\frac{7}{2}$
$(-2, 3)$

33. $\frac{2}{5}x - \frac{1}{3}y = 1$
$\frac{3}{5}x + \frac{2}{3}y = 5$
$(5, 3)$

34. $\frac{5x}{6} + \frac{y}{3} = \frac{4}{3}$
$\frac{2x}{3} - \frac{y}{2} = \frac{11}{6}$
$(2, -1)$

35. $\frac{3x}{4} + \frac{2y}{5} = -\frac{3}{20}$
$\frac{3x}{2} - \frac{y}{4} = \frac{3}{4}$
$\left(\frac{1}{3}, -1\right)$

36. $\frac{2x}{5} - \frac{y}{2} = \frac{13}{2}$
$\frac{3x}{4} - \frac{y}{5} = \frac{17}{2}$
$(10, -5)$

37. $4x - 5y = 3y + 4$
$2x + 3y = 2x + 1$
$\left(\frac{5}{3}, \frac{1}{3}\right)$

38. $5x - 2y = 8x - 1$
$2x + 7y = 4y + 9$
$(-3, 5)$

39. $2x + 5y = 5x + 1$
$3x - 2y = 3y + 3$
inconsistent

40. $\frac{1}{x} + \frac{2}{y} = 3$
$\frac{1}{x} - \frac{3}{y} = 4$
$\left(\frac{5}{17}, -5\right)$

41. $\frac{3}{x} + \frac{2}{y} = 1$
$\frac{2}{x} + \frac{4}{y} = -2$
$(1, -1)$

42. $\frac{3}{x} - \frac{5}{y} = -\frac{3}{2}$
$\frac{1}{x} - \frac{2}{y} = -\frac{2}{3}$
$(3, 2)$

▶ **Objective B**

Solve by the addition method:

43. $x + 2y - z = 1$
$2x - y + z = 6$
$x + 3y - z = 2$
$(2, 1, 3)$

44. $x + 3y + z = 6$
$3x + y - z = -2$
$2x + 2y - z = 1$
$(-1, 2, 1)$

45. $2x - y + 2z = 7$
$x + y + z = 2$
$3x - y + z = 6$
$(1, -1, 2)$

46. $x - 2y + z = 6$
$x + 3y + z = 16$
$3x - y - z = 12$
$(6, 2, 4)$

47. $3x + y = 5$
$3y - z = 2$
$x + z = 5$
$(1, 2, 4)$

48. $2y + z = 7$
$2x - z = 3$
$x - y = 3$
$(4, 1, 5)$

49. $x - y + z = 1$
$2x + 3y - z = 3$
$-x + 2y - 4z = 4$
$(2, -1, -2)$

50. $2x + y - 3z = 7$
$x - 2y + 3z = 1$
$3x + 4y - 3z = 13$
$(3, 1, 0)$

51. $2x + 3z = 5$
$3y + 2z = 3$
$3x + 4y = -10$
$(-2, -1, 3)$

52. $3x + 4z = 5$
$2y + 3z = 2$
$2x - 5y = 8$
$(-1, -2, 2)$

53. $2x + 4y - 2z = 3$
$x + 3y + 4z = 1$
$x + 2y - z = 4$
inconsistent

54. $x - 3y + 2z = 1$
$x - 2y + 3z = 5$
$2x - 6y + 4z = 3$
inconsistent

Solve by the addition method:

55. $2x + y - z = 5$
$x + 3y + z = 14$
$3x - y + 2z = 1$
$(1, 4, 1)$

56. $3x - y - 2z = 11$
$2x + y - 2z = 11$
$x + 3y - z = 8$
$(2, 1, -3)$

57. $3x + y - 2z = 2$
$x + 2y + 3z = 13$
$2x - 2y + 5z = 6$
$(1, 3, 2)$

58. $4x + 5y + z = 6$
$2x - y + 2z = 11$
$x + 2y + 2z = 6$
$(2, -1, 3)$

59. $2x - y + z = 6$
$3x + 2y + z = 4$
$x - 2y + 3z = 12$
$(1, -1, 3)$

60. $3x + 2y - 3z = 8$
$2x + 3y + 2z = 10$
$x + y - z = 2$
$(6, -2, 2)$

61. $3x - 2y + 3z = -4$
$2x + y - 3z = 2$
$3x + 4y + 5z = 8$
$(0, 2, 0)$

62. $3x - 3y + 4z = 6$
$4x - 5y + 2z = 10$
$x - 2y + 3z = 4$
$(0, -2, 0)$

63. $3x - y + 2z = 2$
$4x + 2y - 7z = 0$
$2x + 3y - 5z = 7$
$(1, 5, 2)$

64. $2x + 2y + 3z = 13$
$-3x + 4y - z = 5$
$5x - 3y + z = 2$
$(2, 3, 1)$

65. $2x - 3y + 7z = 0$
$x + 4y - 4z = -2$
$3x + 2y + 5z = 1$
$(-2, 1, 1)$

66. $5x + 3y - z = 5$
$3x - 2y + 4z = 13$
$4x + 3y + 5z = 22$
$(1, 1, 3)$

Solving Systems of Equations by Using Determinants

Objective A   ### To evaluate a determinant

A **matrix** is a rectangular array of numbers. Each number in the matrix is called an **element** of the matrix. The matrix at the right, with three rows and four columns, is called a 3×4 (read 3 by 4) matrix.

$$A = \begin{pmatrix} 1 & -3 & 2 & 4 \\ 0 & 4 & -3 & 2 \\ 6 & -5 & 4 & -1 \end{pmatrix}$$

A matrix of m rows and n columns is said to be of **order $m \times n$**. The matrix above has order 3×4. The notation a_{ij} refers to the element of a matrix in the ith row and jth column. For matrix A, $a_{23} = -3$, $a_{31} = 6$, and $a_{13} = 2$.

A **square matrix** is one that has the same number of rows as columns. An example of a 2×2 and a 3×3 matrix is shown at the right.

$$\begin{pmatrix} -1 & 3 \\ 5 & 2 \end{pmatrix} \qquad \begin{pmatrix} 4 & 0 & 1 \\ 5 & -3 & 7 \\ 2 & 1 & 4 \end{pmatrix}$$

Associated with every square matrix is a number called its **determinant.**

> The determinant of a 2×2 matrix $\begin{pmatrix} a_{11} & a_{12} \\ a_{21} & a_{22} \end{pmatrix}$ is written $\begin{vmatrix} a_{11} & a_{12} \\ a_{21} & a_{22} \end{vmatrix}$. The value of this determinant is given by the formula
>
> $$\begin{vmatrix} a_{11} & a_{12} \\ a_{21} & a_{22} \end{vmatrix} = a_{11}a_{22} - a_{12}a_{21}.$$

Note that vertical bars are used to represent the determinant and curved lines are used to represent the matrix.

Find the value of the determinant $\begin{vmatrix} 3 & 4 \\ -1 & 2 \end{vmatrix}$.

Use the formula. $\begin{vmatrix} 3 & 4 \\ -1 & 2 \end{vmatrix} = 3 \cdot 2 - 4(-1) = 6 - (-4) = 10$

The value of the determinant is 10.

For a square matrix whose order is 3×3 or greater, the value of the determinant of that matrix is found by using 2×2 determinants.

The **minor of an element** in a 3×3 determinant is the 2×2 determinant that is obtained by eliminating the row and column that contain that element.

The minor of -3 for the determinant $\begin{vmatrix} 2 & -3 & 4 \\ 0 & 4 & 8 \\ -1 & 3 & 6 \end{vmatrix}$ $\begin{vmatrix} 2 & -3 & 4 \\ 0 & 4 & 8 \\ -1 & 3 & 6 \end{vmatrix}$

is the 2×2 determinant created by eliminating the row and column that contain -3.

The minor of -3 is $\begin{vmatrix} 0 & 8 \\ -1 & 6 \end{vmatrix}$.

> The **cofactor** of an element of a matrix is $(-1)^{i+j}$ times the minor of that element, where i is the row number of the element and j is the column number of the element.

For the determinant $\begin{vmatrix} 3 & -2 & 1 \\ 2 & -5 & -4 \\ 0 & 3 & 1 \end{vmatrix}$, find the cofactor of -2 and -5.

Because -2 is in the first row and second column, $i = 1$ and $j = 2$. Therefore $i + j = 1 + 2 = 3$, and $(-1)^{i+j} = (-1)^3 = -1$.

cofactor of -2 is $(-1)\begin{vmatrix} 2 & -4 \\ 0 & 1 \end{vmatrix}$

Because -5 is in the second row and second column, $i = 2$ and $j = 2$. Therefore $i + j = 2 + 2 = 4$, and $(-1)^{i+j} = (-1)^4 = 1$.

cofactor of -5 is $1 \cdot \begin{vmatrix} 3 & 1 \\ 0 & 1 \end{vmatrix}$

Note from this example that the cofactor of an element is -1 or 1 times the minor of that element, depending on the sum $i + j$.

The value of a 3×3 or larger determinant can be found by **expanding by cofactors** of any row or any column. The result of expanding by cofactors using the first row of a 3×3 matrix is shown below.

$$\begin{vmatrix} a_{11} & a_{12} & a_{13} \\ a_{21} & a_{22} & a_{23} \\ a_{31} & a_{32} & a_{33} \end{vmatrix} = a_{11}(-1)^{1+1}\begin{vmatrix} a_{22} & a_{23} \\ a_{32} & a_{33} \end{vmatrix} + a_{12}(-1)^{1+2}\begin{vmatrix} a_{21} & a_{23} \\ a_{31} & a_{33} \end{vmatrix} + a_{13}(-1)^{1+3}\begin{vmatrix} a_{21} & a_{22} \\ a_{31} & a_{32} \end{vmatrix}$$

$$= a_{11}\begin{vmatrix} a_{22} & a_{23} \\ a_{32} & a_{33} \end{vmatrix} - a_{12}\begin{vmatrix} a_{21} & a_{23} \\ a_{31} & a_{33} \end{vmatrix} + a_{13}\begin{vmatrix} a_{21} & a_{22} \\ a_{31} & a_{32} \end{vmatrix}$$

Find the value of the determinant $\begin{vmatrix} 2 & -3 & 2 \\ 1 & 3 & -1 \\ 0 & -2 & 2 \end{vmatrix}$.

Expand by cofactors of the first row.

$$\begin{vmatrix} 2 & -3 & 2 \\ 1 & 3 & -1 \\ 0 & -2 & 2 \end{vmatrix} = 2\begin{vmatrix} 3 & -1 \\ -2 & 2 \end{vmatrix} - (-3)\begin{vmatrix} 1 & -1 \\ 0 & 2 \end{vmatrix} + 2\begin{vmatrix} 1 & 3 \\ 0 & -2 \end{vmatrix}$$

$$= 2(6 - 2) - (-3)(2 - 0) + 2(-2 - 0)$$
$$= 2(4) - (-3)(2) + 2(-2)$$
$$= 8 - (-6) + (-4)$$
$$= 10$$

To illustrate a statement made earlier, the value of this determinant will now be found by expanding by cofactors using the second column.

$$\begin{vmatrix} 2 & -3 & 2 \\ 1 & 3 & -1 \\ 0 & -2 & 2 \end{vmatrix} = -3 \cdot (-1)^{1+2}\begin{vmatrix} 1 & -1 \\ 0 & 2 \end{vmatrix} + 3 \cdot (-1)^{2+2}\begin{vmatrix} 2 & 2 \\ 0 & 2 \end{vmatrix} + (-2)(-1)^{3+2}\begin{vmatrix} 2 & 2 \\ 1 & -1 \end{vmatrix}$$

$$= -3 \cdot (-1)\begin{vmatrix} 1 & -1 \\ 0 & 2 \end{vmatrix} + 3 \cdot 1\begin{vmatrix} 2 & 2 \\ 0 & 2 \end{vmatrix} + (-2)(-1)\begin{vmatrix} 2 & 2 \\ 1 & -1 \end{vmatrix}$$

$$= 3(2 - 0) + 3(4 - 0) + 2(-2 - 2)$$
$$= 3(2) + 3(4) + 2(-4)$$
$$= 6 + 12 + (-8)$$
$$= 10$$

Note that the value of the determinant is the same whether the first row or the second column is used to expand by cofactors. Any row or column can be used to evaluate a determinant by expanding by cofactors.

Example 1

Find the value of $\begin{vmatrix} 3 & -2 \\ 6 & -4 \end{vmatrix}$.

Solution

$\begin{vmatrix} 3 & -2 \\ 6 & -4 \end{vmatrix} = 3(-4) - (-2)(6) = -12 + 12 = 0$

The value of the determinant is 0.

Example 2

Find the value of $\begin{vmatrix} -1 & -4 \\ 3 & -5 \end{vmatrix}$.

Your solution

17

Example 3

Find the value of $\begin{vmatrix} -2 & 3 & 1 \\ 4 & -2 & 0 \\ 1 & -2 & 3 \end{vmatrix}$.

Solution

Expand by cofactors of the first row.

$\begin{vmatrix} -2 & 3 & 1 \\ 4 & -2 & 0 \\ 1 & -2 & 3 \end{vmatrix} = -2\begin{vmatrix} -2 & 0 \\ -2 & 3 \end{vmatrix} - 3\begin{vmatrix} 4 & 0 \\ 1 & 3 \end{vmatrix} + 1\begin{vmatrix} 4 & -2 \\ 1 & -2 \end{vmatrix}$

$= -2(-6 - 0) - 3(12 - 0) + 1(-8 + 2)$
$= -2(-6) - 3(12) + 1(-6)$
$= 12 - 36 - 6$
$= -30$

The value of the determinant is -30.

Example 4

Find the value of $\begin{vmatrix} 1 & 4 & -2 \\ 3 & 1 & 1 \\ 0 & -2 & 2 \end{vmatrix}$.

Your solution

-8

Example 5

Find the value of $\begin{vmatrix} 0 & -2 & 1 \\ 1 & 4 & 1 \\ 2 & -3 & 4 \end{vmatrix}$.

Solution

$\begin{vmatrix} 0 & -2 & 1 \\ 1 & 4 & 1 \\ 2 & -3 & 4 \end{vmatrix} = 0\begin{vmatrix} 4 & 1 \\ -3 & 4 \end{vmatrix} - (-2)\begin{vmatrix} 1 & 1 \\ 2 & 4 \end{vmatrix} + 1\begin{vmatrix} 1 & 4 \\ 2 & -3 \end{vmatrix}$

$= 0 - (-2)(4 - 2) + 1(-3 - 8)$
$= 2(2) + 1(-11)$
$= 4 - 11$
$= -7$

The value of the determinant is -7.

Example 6

Find the value of $\begin{vmatrix} 3 & -2 & 0 \\ 1 & 4 & 2 \\ -2 & 1 & 3 \end{vmatrix}$.

Your solution

44

Objective B   To solve a system of equations by using Cramer's Rule

The connection between determinants and systems of equations can be understood by solving a general system of linear equations.

Solve: $a_1x + b_1y = c_1$ (1) $a_1x + b_1y = c_1$
             $a_2x + b_2y = c_2$ (2) $a_2x + b_2y = c_2$

Eliminate y. Multiply equation (1) by b_2 $a_1b_2x + b_1b_2y = c_1b_2$
and equation (2) by $-b_1$. $-a_2b_1x - b_1b_2y = -c_2b_1$

Add the equations. $a_1b_2x - a_2b_1x = c_1b_2 - c_2b_1$

Assuming $a_1b_2 - a_2b_1 \neq 0$, solve for x. $(a_1b_2 - a_2b_1)x = c_1b_2 - c_2b_1$

$$x = \frac{c_1b_2 - c_2b_1}{a_1b_2 - a_2b_1}$$

The denominator $a_1b_2 - a_2b_1$ is the determinant of the coefficients of x and y. This is called the **coefficient determinant.**

$$a_1b_2 - a_2b_1 = \begin{vmatrix} a_1 & b_1 \\ a_2 & b_2 \end{vmatrix}$$

coefficients of x ——————
coefficients of y ——————

The numerator $c_1b_2 - c_2b_1$ is the determinant obtained by replacing the first column in the coefficient determinant by the constants c_1 and c_2. This is called a **numerator determinant.**

$$c_1b_2 - c_2b_1 = \begin{vmatrix} c_1 & b_1 \\ c_2 & b_2 \end{vmatrix}$$

constants of ——————
the equations

Following a similar procedure and eliminating x, the y-component of the solution can also be expressed in determinant form. These results are summarized in Cramer's Rule.

Cramer's Rule

The solution of the system of equations $\begin{array}{l} a_1x + b_1y = c_1 \\ a_2x + b_2y = c_2 \end{array}$ is given by $x = \dfrac{D_x}{D}$

and $y = \dfrac{D_y}{D}$ where $D = \begin{vmatrix} a_1 & b_1 \\ a_2 & b_2 \end{vmatrix}$, $D_x = \begin{vmatrix} c_1 & b_1 \\ c_2 & b_2 \end{vmatrix}$, $D_y = \begin{vmatrix} a_1 & c_1 \\ a_2 & c_2 \end{vmatrix}$, and $D \neq 0$.

Solve by using Cramer's Rule: $3x - 2y = 1$
                                            $2x + 5y = 3$

Find the value of the coefficient determinant.

$$D = \begin{vmatrix} 3 & -2 \\ 2 & 5 \end{vmatrix} = 19$$

Find the value of each of the numerator determinants.

$$D_x = \begin{vmatrix} 1 & -2 \\ 3 & 5 \end{vmatrix} = 11, \quad D_y = \begin{vmatrix} 3 & 1 \\ 2 & 3 \end{vmatrix} = 7$$

Use Cramer's Rule to write the solution.

$$x = \frac{D_x}{D} = \frac{11}{19}, \quad y = \frac{D_y}{D} = \frac{7}{19}$$

The solution is $\left(\frac{11}{19}, \frac{7}{19}\right)$.

A procedure similar to that followed for two equations in two variables can be used to extend Cramer's Rule to three equations in three variables.

Cramer's Rule for a System of Three Equations in Three Variables

The solution of the system of equations $\begin{aligned} a_1x + b_1y + c_1z &= d_1 \\ a_2x + b_2y + c_2z &= d_2 \\ a_3x + b_3y + c_3z &= d_3 \end{aligned}$ is given by

$x = \dfrac{D_x}{D}, y = \dfrac{D_y}{D},$ and $z = \dfrac{D_z}{D}$ where

$$D = \begin{vmatrix} a_1 & b_1 & c_1 \\ a_2 & b_2 & c_2 \\ a_3 & b_3 & c_3 \end{vmatrix}, D_x = \begin{vmatrix} d_1 & b_1 & c_1 \\ d_2 & b_2 & c_2 \\ d_3 & b_3 & c_3 \end{vmatrix}, D_y = \begin{vmatrix} a_1 & d_1 & c_1 \\ a_2 & d_2 & c_2 \\ a_3 & d_3 & c_3 \end{vmatrix}, D_z = \begin{vmatrix} a_1 & b_1 & d_1 \\ a_2 & b_2 & d_2 \\ a_3 & b_3 & d_3 \end{vmatrix},$$

and $D \neq 0$.

Solve by using Cramer's Rule: $\begin{aligned} 2x - y + z &= 1 \\ x + 3y - 2z &= -2 \\ 3x + y + 3z &= 4 \end{aligned}$

Find the value of the coefficient determinant.

$$D = \begin{vmatrix} 2 & -1 & 1 \\ 1 & 3 & -2 \\ 3 & 1 & 3 \end{vmatrix} = 2\begin{vmatrix} 3 & -2 \\ 1 & 3 \end{vmatrix} - (-1)\begin{vmatrix} 1 & -2 \\ 3 & 3 \end{vmatrix} + 1\begin{vmatrix} 1 & 3 \\ 3 & 1 \end{vmatrix}$$

$$= 2(11) + 1(9) + 1(-8)$$

$$= 23$$

Find the value of each of the numerator determinants.

$$D_x = \begin{vmatrix} 1 & -1 & 1 \\ -2 & 3 & -2 \\ 4 & 1 & 3 \end{vmatrix} = 1\begin{vmatrix} 3 & -2 \\ 1 & 3 \end{vmatrix} - (-1)\begin{vmatrix} -2 & -2 \\ 4 & 3 \end{vmatrix} + 1\begin{vmatrix} -2 & 3 \\ 4 & 1 \end{vmatrix}$$

$$= 1(11) + 1(2) + 1(-14)$$

$$= -1$$

$$D_y = \begin{vmatrix} 2 & 1 & 1 \\ 1 & -2 & -2 \\ 3 & 4 & 3 \end{vmatrix} = 2\begin{vmatrix} -2 & -2 \\ 4 & 3 \end{vmatrix} - 1\begin{vmatrix} 1 & -2 \\ 3 & 3 \end{vmatrix} + 1\begin{vmatrix} 1 & -2 \\ 3 & 4 \end{vmatrix}$$

$$= 2(2) - 1(9) + 1(10)$$

$$= 5$$

$$D_z = \begin{vmatrix} 2 & -1 & 1 \\ 1 & 3 & -2 \\ 3 & 1 & 4 \end{vmatrix} = 2\begin{vmatrix} 3 & -2 \\ 1 & 4 \end{vmatrix} - (-1)\begin{vmatrix} 1 & -2 \\ 3 & 4 \end{vmatrix} + 1\begin{vmatrix} 1 & 3 \\ 3 & 1 \end{vmatrix}$$

$$= 2(14) + 1(10) + 1(-8)$$

$$= 30$$

Use Cramer's Rule to write the solution.

$$x = \frac{D_x}{D} = \frac{-1}{23}, \qquad y = \frac{D_y}{D} = \frac{5}{23}, \qquad z = \frac{D_z}{D} = \frac{30}{23}$$

The solution is $\left(-\frac{1}{23}, \frac{5}{23}, \frac{30}{23}\right)$.

Example 7

Solve by using Cramer's Rule:
$2x - 3y = 8$
$5x + 6y = 11$

Solution

$$D = \begin{vmatrix} 2 & -3 \\ 5 & 6 \end{vmatrix} = 27, \qquad D_x = \begin{vmatrix} 8 & -3 \\ 11 & 6 \end{vmatrix} = 81,$$

$$D_y = \begin{vmatrix} 2 & 8 \\ 5 & 11 \end{vmatrix} = -18$$

$$x = \frac{D_x}{D} = \frac{81}{27} = 3 \qquad y = \frac{D_y}{D} = \frac{-18}{27} = -\frac{2}{3}$$

The solution is $\left(3, -\frac{2}{3}\right)$.

Example 8

Solve by using Cramer's Rule:
$6x - 6y = 5$
$2x - 10y = -1$

Your solution

$\left(\frac{7}{6}, \frac{1}{3}\right)$

Example 9

Solve by using Cramer's Rule:
$6x - 9y = 5$
$4x - 6y = 4$

Solution

$$D = \begin{vmatrix} 6 & -9 \\ 4 & -6 \end{vmatrix} = 0$$

Since $D = 0$, $\frac{D_x}{D}$ is undefined. Therefore, the equations are dependent or inconsistent.

Example 10

Solve by using Cramer's Rule:
$3x - y = 4$
$6x - 2y = 5$

Your solution

not independent

Example 11

Solve by using Cramer's Rule:
$3x - y + z = 5$
$x + 2y - 2z = -3$
$2x + 3y + z = 4$

Solution

$$D = \begin{vmatrix} 3 & -1 & 1 \\ 1 & 2 & -2 \\ 2 & 3 & 1 \end{vmatrix} = 28,$$

$$D_x = \begin{vmatrix} 5 & -1 & 1 \\ -3 & 2 & -2 \\ 4 & 3 & 1 \end{vmatrix} = 28,$$

$$D_y = \begin{vmatrix} 3 & 5 & 1 \\ 1 & -3 & -2 \\ 2 & 4 & 1 \end{vmatrix} = 0,$$

$$D_z = \begin{vmatrix} 3 & -1 & 5 \\ 1 & 2 & -3 \\ 2 & 3 & 4 \end{vmatrix} = 56$$

$$x = \frac{D_x}{D} = \frac{28}{28} = 1, \qquad y = \frac{D_y}{D} = \frac{0}{28} = 0,$$
$$z = \frac{D_z}{D} = \frac{56}{28} = 2$$

The solution is $(1, 0, 2)$.

Example 12

Solve by using Cramer's Rule:
$2x - y + z = -1$
$3x + 2y - z = 3$
$x + 3y + z = -2$

Your solution

$\left(\frac{3}{7}, -\frac{1}{7}, -2\right)$

Solutions on p. A42

10.3   EXERCISES

▶ **Objective A**

Evaluate the determinant:

1. $\begin{vmatrix} 2 & -1 \\ 3 & 4 \end{vmatrix}$

11

2. $\begin{vmatrix} 5 & 1 \\ -1 & 2 \end{vmatrix}$

11

3. $\begin{vmatrix} 6 & -2 \\ -3 & 4 \end{vmatrix}$

18

4. $\begin{vmatrix} -3 & 5 \\ 1 & 7 \end{vmatrix}$

−26

5. $\begin{vmatrix} 3 & 6 \\ 2 & 4 \end{vmatrix}$

0

6. $\begin{vmatrix} 5 & -10 \\ 1 & -2 \end{vmatrix}$

0

7. $\begin{vmatrix} 1 & -1 & 2 \\ 3 & 2 & 1 \\ 1 & 0 & 4 \end{vmatrix}$

15

8. $\begin{vmatrix} 4 & 1 & 3 \\ 2 & -2 & 1 \\ 3 & 1 & 2 \end{vmatrix}$

3

9. $\begin{vmatrix} 3 & -1 & 2 \\ 0 & 1 & 2 \\ 3 & 2 & -2 \end{vmatrix}$

−30

10. $\begin{vmatrix} 4 & 5 & -2 \\ 3 & -1 & 5 \\ 2 & 1 & 4 \end{vmatrix}$

−56

11. $\begin{vmatrix} 4 & 2 & 6 \\ -2 & 1 & 1 \\ 2 & 1 & 3 \end{vmatrix}$

0

12. $\begin{vmatrix} 3 & 6 & -3 \\ 4 & -1 & 6 \\ -1 & -2 & 3 \end{vmatrix}$

−54

▶ **Objective B**

Solve by using Cramer's Rule:

13. $2x - 5y = 26$
$5x + 3y = 3$
$(3, -4)$

14. $3x + 7y = 15$
$2x + 5y = 11$
$(-2, 3)$

15. $x - 4y = 8$
$3x + 7y = 5$
$(4, -1)$

16. $5x + 2y = -5$
$3x + 4y = 11$
$(-3, 5)$

17. $2x + 3y = 4$
$6x - 12y = -5$
$\left(\frac{11}{14}, \frac{17}{21} \right)$

18. $5x + 4y = 3$
$15x - 8y = -21$
$\left(-\frac{3}{5}, \frac{3}{2} \right)$

Solve by using Cramer's Rule:

19. $2x + 5y = 6$
$6x - 2y = 1$
$\left(\frac{1}{2}, 1\right)$

20. $7x + 3y = 4$
$5x - 4y = 9$
$(1, -1)$

21. $-2x + 3y = 7$
$4x - 6y = 9$
not independent

22. $9x + 6y = 7$
$3x + 2y = 4$
not independent

23. $2x - 5y = -2$
$3x - 7y = -3$
$(-1, 0)$

24. $8x + 7y = -3$
$2x + 2y = 5$
$\left(-\frac{41}{2}, 23\right)$

25. $2x - y + 3z = 9$
$x + 4y + 4z = 5$
$3x + 2y + 2z = 5$
$(1, -1, 2)$

26. $3x - 2y + z = 2$
$2x + 3y + 2z = -6$
$3x - y + z = 0$
$(-1, -2, 1)$

27. $3x - y + z = 11$
$x + 4y - 2z = -12$
$2x + 2y - z = -3$
$(2, -2, 3)$

28. $x + 2y + 3z = 8$
$2x - 3y + z = 5$
$3x - 4y + 2z = 9$
$(-3, -2, 5)$

29. $4x - 2y + 6z = 1$
$3x + 4y + 2z = 1$
$2x - y + 3z = 2$
not independent

30. $x - 3y + 2z = 1$
$2x + y - 2z = 3$
$3x - 9y + 6z = -3$
not independent

31. $5x - 4y + 2z = 4$
$3x - 5y + 3z = -4$
$3x + y - 5z = 12$
$\left(\frac{68}{25}, \frac{56}{25}, -\frac{8}{25}\right)$

32. $2x + 4y + z = 7$
$x + 3y - z = 1$
$3x + 2y - 2z = 5$
$\left(\frac{53}{19}, -\frac{1}{19}, \frac{31}{19}\right)$

SECTION 10.4   Application Problems in Two Variables

Objective A   To solve rate-of-wind or current problems

Motion problems that involve an object moving with or against a wind or current normally require two variables to solve.

A motorboat traveling with the current can go 24 mi in 2 h. Against the current it takes 3 h to go the same distance. Find the rate of the motorboat in calm water and the rate of the current.

Strategy for Solving Rate-of-Wind or Current Problems

> Choose one variable to represent the rate of the object in calm conditions and a second variable to represent the rate of the wind or current. Using these variables, express the rate of the object with and against the wind or current. Use the equation $rt = d$ to write expressions for the distance traveled by the object. The results can be recorded in a table.

Rate of the boat in calm water: x
Rate of the current: y

	Rate	·	*Time*	=	*Distance*
With the current	$x + y$	·	2	=	$2(x + y)$
Against the current	$x - y$	·	3	=	$3(x - y)$

> Determine how the expressions for distance are related.

The distance traveled with the current is 24 mi. $2(x + y) = 24$
The distance traveled against the current is 24 mi. $3(x - y) = 24$

Solve the system of equations.

$2(x + y) = 24$ $\frac{1}{2} \cdot 2(x + y) = \frac{1}{2} \cdot 24$ $x + y = 12$

$3(x - y) = 24$ $\frac{1}{3} \cdot 3(x - y) = \frac{1}{3} \cdot 24$ $x - y = 8$

$2x = 20$
$x = 10$

Replace x by 10 in the equation $x + y = 12$. $x + y = 12$
Solve for y. $10 + y = 12$
$y = 2$

The rate of the boat in calm water is 10 mph.
The rate of the current is 2 mph.

Example 1

Flying with the wind, a plane flew 1000 mi in 5 h. Flying against the wind, the plane could fly only 500 mi in the same amount of time. Find the rate of the plane in calm air and the rate of the wind.

Strategy

- Rate of the plane in still air: p
  Rate of the wind: w

	Rate	*Time*	*Distance*
With wind	$p + w$	5	$5(p + w)$
Against wind	$p - w$	5	$5(p - w)$

- The distance traveled with the wind is 1000 mi.
  The distance traveled against the wind is 500 mi.

$5(p + w) = 1000$
$5(p - w) = 500$

Solution

$5(p + w) = 1000$ $\qquad \frac{1}{5} \cdot 5(p + w) = \frac{1}{5} \cdot 1000$

$5(p - w) = 500$ $\qquad \frac{1}{5} \cdot 5(p - w) = \frac{1}{5} \cdot 500$

$$p + w = 200$$
$$p - w = 100$$

$$2p = 300$$
$$p = 150$$

$p + w = 200$
$150 + w = 200$
$w = 50$

The rate of the plane in calm air is 150 mph.
The rate of the wind is 50 mph.

Example 2

A rowing team rowing with the current traveled 18 mi in 2 h. Against the current, the team rowed 10 mi in 2 h. Find the rate of the rowing team in calm water and the rate of the current.

Your strategy

Your solution

rate of rowing team in calm water: 7 mph
rate of the current: 2 mph

Solution on p. A43

| Objective B | **To solve application problems using two variables** | 10 |

The application problems in this section are varieties of those problems solved earlier in the text. Each of the strategies for the problems in this section will result in a system of equations.

A store owner purchased twenty 60-watt light bulbs and 30 fluorescent lights for a total cost of $40. A second purchase, at the same prices, included thirty 60-watt bulbs and 10 fluorescent lights for a total cost of $25. Find the cost of a 60-watt bulb and a fluorescent light.

Strategy for Solving an Application Problem in Two Variables

Choose one variable to represent one of the unknown quantities and a second variable to represent the other unknown quantity. Write numerical or variable expressions for all the remaining quantities. These results can be recorded in two tables, one for each of the conditions.

Cost of a 60-watt bulb: b
Cost of a fluorescent light: f

First purchase

	Amount	·	*Unit cost*	=	*Value*
60-watt	20	·	b	=	$20b$
Fluorescent	30	·	f	=	$30f$

Second purchase

	Amount	·	*Unit cost*	=	*Value*
60-watt	30	·	b	=	$30b$
Fluorescent	10	·	f	=	$10f$

Determine a system of equations. The strategies presented in Chapter 2 can be used to determine the relationships between the expressions in the tables. Each table will give one equation of the system.

The total of the first purchase was $40. $20b + 30f = 40$
The total of the second purchase was $25. $30b + 10f = 25$

Solve the system of equations.

$20b + 30f = 40$ $3(20b + 30f) = 3 \cdot 40$ $60b + 90f = 120$
$30b + 10f = 25$ $-2(30b + 10f) = -2 \cdot 25$ $-60b - 20f = -50$

$$70f = 70$$
$$f = 1$$

Replace f by 1 in the equation $20b + 30f = 40$. $20b + 30f = 40$
Solve for b. $20b + 30(1) = 40$
      $20b + 30 = 40$
      $20b = 10$
      $b = 0.5$

The cost of a 60-watt bulb was $.50.
The cost of a fluorescent light was $1.00.

Example 3

The total value of the nickels and dimes in a coin bank is $2.50. If the nickels were dimes and the dimes were nickels, the total value of the coins would be $3.05. Find the number of dimes and the number of nickels in the bank.

Example 4

A citrus fruit grower purchased 25 orange trees and 20 grapefruit trees for $290. The next week, at the same prices, the grower bought 20 orange trees and 30 grapefruit trees for $330. Find the cost of an orange tree and a grapefruit tree.

Strategy

■ Number of nickels in the bank: n
  Number of dimes in the bank: d

Coins in the bank now:

Coin	Number	Value	Total Value
Nickels	n	5	$5n$
Dimes	d	10	$10d$

Coins in the bank if the nickels were dimes and the dimes were nickels:

Coin	Number	Value	Total Value
Nickels	d	5	$5d$
Dimes	n	10	$10n$

■ The value of the nickels and dimes in the bank is $2.50.
  The value of the nickels and dimes in the bank would be $3.05.

$5n + 10d = 250$
$10n + 5d = 305$

Your strategy

Solution

$5n + 10d = 250$ 　　$2(5n + 10d) = 2 \cdot 250$
$10n + 5d = 305$ 　　$-1(10n + 5d) = -1 \cdot 305$

$$10n + 20d = 500$$
$$-10n - 5d = -305$$
$$15d = 195$$
$$d = 13$$

$5n + 10d = 250$
$5n + 10(13) = 250$
$5n + 130 = 250$
$5n = 120$
$n = 24$

There are 24 nickels and 13 dimes in the bank.

Your solution

orange tree: $6
grapefruit tree: $7

Solution on p. A43

10.4   EXERCISES

▶ Objective A   *Application Problems*

Solve:

1. A motorboat traveling with the current went 36 mi in 2 h. Against the current it took 3 h to travel the same distance. Find the rate of the boat in calm water and the rate of the current.
    boat: 15 mph; current: 3 mph

2. A cabin cruiser traveling with the current went 45 mi in 3 h. Against the current it took 5 h to travel the same distance. Find the rate of the cabin cruiser in calm water and the rate of the current.
    cabin cruiser: 12 mph; current: 3 mph

3. A jet plane flying with the wind went 2200 mi in 4 h. Against the wind, the plane could fly only 1820 mi in the same amount of time. Find the rate of the plane in calm air and the rate of the wind.
    plane: 502.5 mph; wind: 47.5 mph

4. Flying with the wind, a small plane flew 300 mi in 2 h. Against the wind, the plane could fly only 270 mi in the same amount of time. Find the rate of the plane in calm air and the rate of the wind.
    plane: 142.5 mph; wind: 7.5 mph

5. A rowing team rowing with the current traveled 20 km in 2 h. Rowing against the current, the team rowed 12 km in the same amount of time. Find the rate of the team in calm water and the rate of the current.
    team: 8 km/h; current: 2 km/h

Content and Format © 1991 HMCo.

Solve:

6. A motorboat traveling with the current went 72 km in 3 h. Against the current, the boat could go only 48 km in the same amount of time. Find the rate of the boat in calm water and the rate of the current.

boat: 20 km/h; current: 4 km/h

7. A turbo-prop plane flying with the wind flew 800 mi in 4 h. Flying against the wind, the plane required 5 h to travel the same distance. Find the rate of the wind and the rate of the plane in calm air.

plane: 180 mph; wind: 20 mph

8. Flying with the wind, a pilot flew 600 mi between two cities in 4 h. The return trip against the wind took 5 h. Find the rate of the plane in calm air and the rate of the wind.

plane: 135 mph; wind: 15 mph

9. A merchant mixed 10 lb of a cinnamon tea with 5 lb of spice tea. The 15-pound mixture sells for $40. A second mixture included 12 lb of the cinnamon tea and 8 lb of the spice tea. The 20-pound mixture sells for $54. Find the cost per pound of the cinnamon tea and the spice tea.

cinnamon: $2.50/lb; spice: $3/lb

10. A carpenter purchased 60 ft of redwood and 80 ft of pine for a total cost of $27. A second purchase, at the same prices, included 100 ft of redwood and 60 ft of pine for a total cost of $34. Find the cost per foot of redwood and pine.

pine: $.15/ft; redwood: $.25/ft

▶ Objective B   *Application Problems*

Solve:

11. A coin bank contains only nickels and dimes. The total value of the coins in the bank is $2.50. If the nickels were dimes and the dimes were nickels, the total value of the coins would be $3.50. Find the number of nickels in the bank.
    30 nickels

12. A plane flying with a tailwind flew 600 mi in 5 h. Against the wind the plane required 6 h to fly the same distance. Find the rate of the plane in calm air and the rate of the wind.
    plane: 110 mph; wind: 10 mph

13. Flying with the wind, a plane flew 720 mi in 3 h. Against the wind, the plane required 4 h to fly the same distance. Find the rate of the plane in calm air and the rate of the wind.
    plane: 210 mph; wind: 30 mph

14. The total value of the quarters and dimes in a coin bank is $5.75. If the quarters were dimes and the dimes were quarters, the total value of the coins would be $6.50. Find the number of quarters in the bank.
    15 quarters

15. A contractor buys 16 yd of nylon carpet and 20 yd of wool carpet for $920. A second purchase, at the same prices, includes 18 yd of nylon carpet and 25 yd of wool carpet for $1100. Find the cost per yard of the wool carpet.
    $26/yd

16. During one month, a homeowner used 500 units of electricity and 100 units of gas for a total cost of $88. The next month, 400 units of electricity and 150 units of gas were used for a total cost of $76. Find the cost per unit of gas.
    $.08

Solve:

17. A company manufactures both 10-speed and standard model bicycles. The cost of materials for a 10-speed bicycle is $35, while the cost of materials for a standard bicycle is $25. The cost of labor to manufacture a 10-speed bicycle is $40, while the cost of labor to manufacture a standard bicycle is $20. During a week when the company has budgeted $1250 for materials and $1300 for labor, how many 10-speed bicycles does the company plan to manufacture?
25 10-speed bicycles

18. A company manufactures both color and black-and-white television sets. The cost of materials for a black-and-white TV is $20, while the cost of materials for a color TV is $80. The cost of labor to manufacture a black-and-white TV is $30, while the cost of labor to manufacture a color TV is $50. During a week when the company has budgeted $4200 for materials and $2800 for labor, how many color TVs does the company plan to manufacture?
50 color TVs

19. A chemist has two alloys, one of which is 10% gold and 15% lead, the other of which is 30% gold and 40% lead. How many grams of each of the two alloys should be used to make an alloy that contains 60 g of gold and 88 g of lead?
1st alloy: 480 g; 2nd alloy: 40 g

20. A pharmacist has two vitamin-supplement powders. The first powder is 20% vitamin B_1 and 10% vitamin B_2. The second is 15% vitamin B_1 and 20% vitamin B_2. How many milligrams of each of the two powders should the pharmacist use to make a mixture that contains 130 mg of vitamin B_1 and 80 mg of vitamin B_2?
1st powder: 560 mg; 2nd powder: 120 mg

SECTION 10.5   Solving Systems of Nonlinear Equations and Systems of Inequalities

Objective A   **To solve a nonlinear system of equations**

A nonlinear system of equations is one in which one or more of the equations are not linear equations. A nonlinear system of equations can be solved by using either a substitution method or an addition method.

Solve: $2x - y = 4$ (1)
$\qquad\quad y^2 = 4x$ (2)

$$2x - y = 4$$
$$y^2 = 4x$$

When a nonlinear system of equations contains a linear equation, the substitution method is used.

Solve equation (1) for y.

$$2x - y = 4$$
$$-y = -2x + 4$$
$$y = 2x - 4$$

Substitute $2x - 4$ for y into equation (2).

$$y^2 = 4x$$
$$(2x - 4)^2 = 4x$$

Write the equation in standard form.

$$4x^2 - 16x + 16 = 4x$$
$$4x^2 - 20x + 16 = 0$$

Solve for x by factoring.

$$4(x^2 - 5x + 4) = 0$$
$$4(x - 4)(x - 1) = 0$$

$$x - 4 = 0 \qquad x - 1 = 0$$
$$x = 4 \qquad\quad x = 1$$

Substitute the values of x into the equation $y = 2x - 4$ and solve for y.

$$y = 2x - 4 \qquad y = 2x - 4$$
$$y = 2(4) - 4 \qquad y = 2(1) - 4$$
$$y = 4 \qquad\qquad y = -2$$

The solutions are $(4, 4)$ and $(1, -2)$.

The graph of the system solved above is shown at the right. Note that the line intersects the parabola at two points. These points correspond to the solutions.

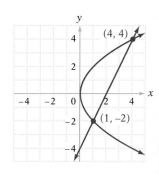

Solve: $x^2 + y^2 = 4$ (1) $x^2 + y^2 = 4$
                 $y = x + 3$ (2) $y = x + 3$

Use the substitution method.

Substitute the expression for y into equation (1).

$$x^2 + y^2 = 4$$
$$x^2 + (x + 3)^2 = 4$$

Write the equation in standard form.

$$x^2 + x^2 + 6x + 9 = 4$$
$$2x^2 + 6x + 5 = 0$$

Since the discriminant of the quadratic equation is less than zero, the equation has two complex number solutions. Therefore, the system of equations has no real number solution.

$$b^2 - 4ac = 6^2 - 4(2)(5)$$
$$= 36 - 40$$
$$= -4$$

The graphs of the equations of this system are shown at the right. Note that the two graphs do not intersect.

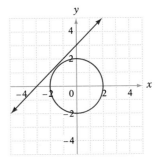

Solve: $4x^2 + y^2 = 16$ (1) $4x^2 + y^2 = 16$
            $x^2 + y^2 = 4$ (2) $x^2 + y^2 = 4$

Use the addition method.

Multiply equation (2) by -1 and add to equation (1).

$$-x^2 - y^2 = -4$$
$$4x^2 + y^2 = 16$$
$$3x^2 = 12$$

Solve for x.

$$x^2 = 4$$
$$x = \pm 2$$

Substitute the values of x into equation (2) and solve for y.

$$x^2 + y^2 = 4$$
$$2^2 + y^2 = 4$$
$$y^2 = 0$$
$$y = 0$$

$$x^2 + y^2 = 4$$
$$(-2)^2 + y^2 = 4$$
$$y^2 = 0$$
$$y = 0$$

The solutions are $(2, 0)$ and $(-2, 0)$.

The graphs of the equations in this system are shown at the right. Note that the graphs intersect at two points.

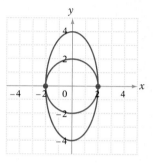

Solve: $2x^2 - y^2 = 1$
         $x^2 + 2y^2 = 18$

(1) $2x^2 - y^2 = 1$
(2) $x^2 + 2y^2 = 18$

Use the addition method.

Multiply equation (1) by 2 and add to equation (2).

$$4x^2 - 2y^2 = 2$$
$$x^2 + 2y^2 = 18$$
$$5x^2 = 20$$

Solve for x.

$$x^2 = 4$$
$$x = \pm 2$$

Substitute the values of x into one of the equations and solve for y. Equation (2) is used here.

$$x^2 + 2y^2 = 18$$
$$2^2 + 2y^2 = 18$$
$$2y^2 = 14$$
$$y^2 = 7$$
$$y = \pm\sqrt{7}$$

$$x^2 + 2y^2 = 18$$
$$(-2)^2 + 2y^2 = 18$$
$$2y^2 = 14$$
$$y^2 = 7$$
$$y = \pm\sqrt{7}$$

The solutions are $(2, \sqrt{7})$, $(2, -\sqrt{7})$, $(-2, \sqrt{7})$, and $(-2, -\sqrt{7})$.

The graphs of the equations in this system are shown at the right. Note that there are four points of intersection.

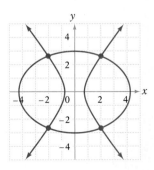

Example 1   Solve: (1) $y = 2x^2 - 3x - 1$
                          (2) $y = x^2 - 2x + 5$

Solution   Use the substitution method.

$$y = x^2 - 2x + 5$$
$$2x^2 - 3x - 1 = x^2 - 2x + 5$$
$$x^2 - x - 6 = 0$$
$$(x - 3)(x + 2) = 0$$

$$x - 3 = 0 \qquad x + 2 = 0$$
$$x = 3 \qquad\quad x = -2$$

Substitute each value of x into equation (1).

$$y = 2x^2 - 3x - 1$$
$$y = 2(3)^2 - 3(3) - 1$$
$$y = 18 - 9 - 1$$
$$y = 8$$

$$y = 2x^2 - 3x - 1$$
$$y = 2(-2)^2 - 3(-2) - 1$$
$$y = 8 + 6 - 1$$
$$y = 13$$

The solutions are (3, 8) and (−2, 13).

Example 2   Solve: $y = 2x^2 + x - 3$
                         $y = 2x^2 - 2x + 9$

Your solution   (4, 33)

Example 3   Solve: (1) $3x^2 - 2y^2 = 26$
                          (2) $x^2 - y^2 = 5$

Solution   Use the addition method. Multiply equation (2) by −2.

$$-2x^2 + 2y^2 = -10$$
$$3x^2 - 2y^2 = 26$$
$$x^2 = 16$$
$$x = \pm\sqrt{16} = \pm 4$$

Substitute each value of x into equation (2).

$$x^2 - y^2 = 5 \qquad\qquad x^2 - y^2 = 5$$
$$4^2 - y^2 = 5 \qquad\quad (-4)^2 - y^2 = 5$$
$$16 - y^2 = 5 \qquad\quad 16 - y^2 = 5$$
$$-y^2 = -11 \qquad\qquad -y^2 = -11$$
$$y^2 = 11 \qquad\qquad\quad y^2 = 11$$
$$y = \pm\sqrt{11} \qquad\qquad y = \pm\sqrt{11}$$

The solutions are $(4, \sqrt{11})$, $(4, -\sqrt{11})$, $(-4, \sqrt{11})$, and $(-4, -\sqrt{11})$.

Example 4   Solve: $x^2 - y^2 = 10$
                         $x^2 + y^2 = 8$

Your solution   no real solution

Solutions on p. A44

Example 5   Solve: (1) $y = x + 2$
                         (2) $x^2 + y^2 = 10$

Solution   Use the substitution method.

$$x^2 + y^2 = 10$$
$$x^2 + (x + 2)^2 = 10$$
$$x^2 + x^2 + 4x + 4 = 10$$
$$2x^2 + 4x + 4 = 10$$
$$2x^2 + 4x - 6 = 0$$
$$2(x^2 + 2x - 3) = 0$$
$$2(x + 3)(x - 1) = 0$$

$$x + 3 = 0 \qquad x - 1 = 0$$
$$x = -3 \qquad x = 1$$

Substitute each value of x into equation (1).

$$y = x + 2 \qquad y = x + 2$$
$$y = (-3) + 2 \qquad y = 1 + 2$$
$$y = -1 \qquad y = 3$$

The solutions are $(-3, -1)$ and $(1, 3)$.

Example 6   Solve: $2x^2 - y^2 = 14$
                          $y = x - 1$

Your solution $(-5, -6)$ and $(3, 2)$

Solution on p. A44

Objective B   **To graph the solution set of a system of inequalities**

10

The **solution set of a system of inequalities** is the intersection of the solution sets of each inequality. To graph the solution set of a system of inequalities, first graph the solution set for each inequality. The solution set of the system of inequalities is the region of the plane represented by the intersection of the two shaded regions.

Graph the solution set of $2x - y \leq 3$
$3x + 2y \geq 8$.

Graph the solution set of each inequality.

The solution set is the region of the plane represented by the intersection of the solution sets of each inequality.

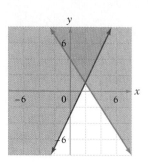

Example 7

Graph the solution set of $\dfrac{x^2}{9} + \dfrac{y^2}{4} \geq 1$
$\dfrac{x^2}{4} - \dfrac{y^2}{9} > 1$.

Solution

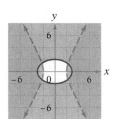

Example 8

Graph the solution set of $x^2 + y^2 < 16$
$y^2 > x$.

Your solution

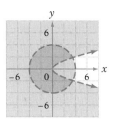

Example 9

Graph the solution set of $y > x^2$
$y < x + 2$.

Solution

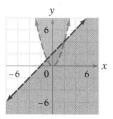

Example 10

Graph the solution set of $y \geq x - 1$
$y < -2x$.

Your solution

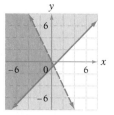

Solutions on p. A44

Content and Format © 1991 HMCo.

10.5

▶ **Objective A**

Solve:

1. $y = x^2 - x - 1$
$y = 2x + 9$
$(-2, 5)$
$(5, 19)$

2. $y = x^2 - 3x + 1$
$y = x + 6$
$(5, 11)$
$(-1, 5)$

3. $\quad y^2 = -x + 3$
$x - y = 1$
$(-1, -2)$
$(2, 1)$

4. $\quad y^2 = 4x$
$x - y = -1$
$(1, 2)$

5. $\quad y^2 = 2x$
$x + 2y = -2$
$(2, -2)$

6. $\quad y^2 = 2x$
$x - y = 4$
$(2, -2)$
$(8, 4)$

7. $x^2 + 2y^2 = 12$
$2x - y = 2$
$(2, 2)$
$\left(-\dfrac{2}{9}, -\dfrac{22}{9}\right)$

8. $x^2 + 4y^2 = 37$
$x - y = -4$
$\left(\dfrac{27}{5}, -\dfrac{7}{5}\right)$
$(-1, 3)$

9. $x^2 + y^2 = 13$
$x + y = 5$
$(3, 2)$
$(2, 3)$

10. $x^2 + y^2 = 16$
$x - 2y = -4$
$(-4, 0)$
$\left(\dfrac{12}{5}, \dfrac{16}{5}\right)$

11. $4x^2 + y^2 = 12$
$y = 4x^2$
$\left(\dfrac{\sqrt{3}}{2}, 3\right)$
$\left(-\dfrac{\sqrt{3}}{2}, 3\right)$

12. $2x^2 + y^2 = 6$
$y = 2x^2$
$(1, 2)$
$(-1, 2)$

13. $y = x^2 - 2x - 3$
$y = x - 6$
no real solution

14. $y = x^2 + 4x + 5$
$y = -x - 3$
no real solution

15. $3x^2 - y^2 = -1$
$x^2 + 4y^2 = 17$
$(1, 2)\ (1, -2)$
$(-1, 2)\ (-1, -2)$

Solve:

16. $x^2 + y^2 = 10$
$x^2 + 9y^2 = 18$
(3, 1) (−3, 1)
(3, −1) (−3, −1)

17. $2x^2 + 3y^2 = 30$
$x^2 + y^2 = 13$
(3, 2) (3, −2)
(−3, 2) (−3, −2)

18. $x^2 + y^2 = 61$
$x^2 − y^2 = 11$
(6, 5) (6, −5)
(−6, 5) (−6, −5)

19. $y = 2x^2 − x + 1$
$y = x^2 − x + 5$
(2, 7) (−2, 11)

20. $y = −x^2 + x − 1$
$y = x^2 + 2x − 2$
$\left(\frac{1}{2}, -\frac{3}{4}\right)$
(−1, −3)

21. $2x^2 + 3y^2 = 24$
$x^2 − y^2 = 7$
(3, $\sqrt{2}$) (3, −$\sqrt{2}$)
(−3, $\sqrt{2}$) (−3, −$\sqrt{2}$)

22. $2x^2 + 3y^2 = 21$
$x^2 + 2y^2 = 12$
($\sqrt{6}$, $\sqrt{3}$) ($\sqrt{6}$, −$\sqrt{3}$)
(−$\sqrt{6}$, $\sqrt{3}$) (−$\sqrt{6}$, −$\sqrt{3}$)

23. $x^2 + y^2 = 36$
$4x^2 + 9y^2 = 36$
no real solution

24. $2x^2 + 3y^2 = 12$
$x^2 − y^2 = 25$
no real solution

25. $11x^2 − 2y^2 = 4$
$3x^2 + y^2 = 15$
($\sqrt{2}$, 3) ($\sqrt{2}$, −3)
(−$\sqrt{2}$, 3) (−$\sqrt{2}$, −3)

26. $x^2 + 4y^2 = 25$
$x^2 − y^2 = 5$
(3, 2) (3, −2)
(−3, 2) (−3, −2)

27. $2x^2 − y^2 = 7$
$2x − y = 5$
(2, −1) (8, 11)

28. $3x^2 + 4y^2 = 7$
$x − 2y = −3$
$\left(-\frac{1}{2}, \frac{5}{4}\right)$ (−1, 1)

29. $y = 3x^2 + x − 4$
$y = 3x^2 − 8x + 5$
(1, 0)

30. $y = 2x^2 + 3x + 1$
$y = 2x^2 + 9x + 7$
(−1, 0)

Solve:

31. $x = y + 3$
$x^2 + y^2 = 5$
$(2, -1)\ (1, -2)$

32. $x - y = -6$
$x^2 + y^2 = 4$
no real solution

33. $y = x^2 + 4x + 4$
$x + 2y = 4$
$\left(-\frac{1}{2}, \frac{9}{4}\right)\ (-4, 4)$

34. $x = 2y^2 - 3y + 1$
$3x - 2y = 0$
$\left(\frac{2}{9}, \frac{1}{3}\right)\ \left(1, \frac{3}{2}\right)$

35. $x = y^2 - 2y + 1$
$x = 2y^2 - 3y - 5$
$(9, -2)\ (4, 3)$

36. $x = 3y^2 + 2y - 4$
$x = y^2 - 5y$
$\left(-\frac{9}{4}, \frac{1}{2}\right)\ (36, -4)$

▶ **Objective B**

Graph the solution set:

37. $2x + y \geq 4$
$3x - 2y < 6$

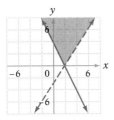

38. $3x - 4y < 12$
$x + 2y < 6$

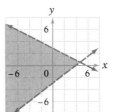

39. $x - 2y \leq 6$
$2x + 3y \leq 6$

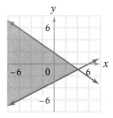

40. $x - 3y > 6$
$2x + y > 5$

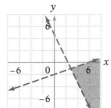

41. $x - 2y \leq 4$
$3x + 2y \leq 8$
$x > -1$

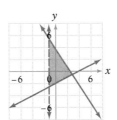

42. $3x - 2y < 0$
$5x + 3y > 9$
$y < 4$

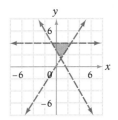

Graph the solution set:

43. $2x + 3y \leq 15$
$3x - y \leq 6$
$y \geq 0$

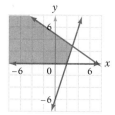

44. $x + y \leq 6$
$x - y \leq 2$
$x \geq 0$

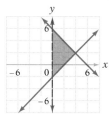

45. $x^2 + y^2 < 16$
$y > x + 1$

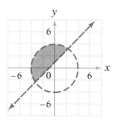

46. $y > x^2 - 4$
$y < x - 2$

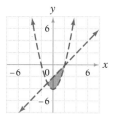

47. $\dfrac{x^2}{4} + \dfrac{y^2}{16} \leq 1$
$\phantom{\dfrac{x^2}{4} + }y \leq -\dfrac{1}{2}x + 2$

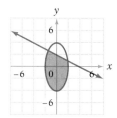

48. $\dfrac{y^2}{4} - \dfrac{x^2}{25} \geq 1$
$\phantom{\dfrac{y^2}{4} - }y \leq \dfrac{2}{3}x + 4$

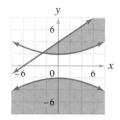

49. $x \geq y^2 - 3y + 2$
$y \geq 2x - 2$

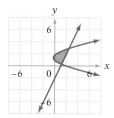

50. $x^2 + y^2 \leq 25$
$y \leq -\dfrac{1}{3}x + 2$

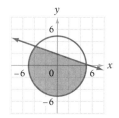

51. $x^2 + y^2 < 25$
$\dfrac{x^2}{9} + \dfrac{y^2}{36} < 1$

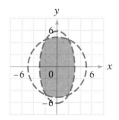

52. $\dfrac{x^2}{9} - \dfrac{y^2}{4} < 1$
$\dfrac{x^2}{25} + \dfrac{y^2}{9} < 1$

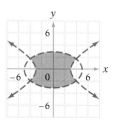

53. $x^2 + y^2 > 4$
$x^2 + y^2 < 25$

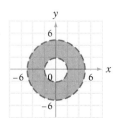

54. $\dfrac{x^2}{25} + \dfrac{y^2}{16} \leq 1$
$\dfrac{x^2}{4} + \dfrac{y^2}{4} \geq 1$

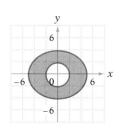

Calculators and Computers

Cramer's Rule   In this chapter, determinants were used as one method of solving a system of linear equations. By evaluating some determinants and then using Cramer's Rule, the solution to a system of equations could be found.

The difficulty with using Cramer's Rule when there are more than three variables in the system of equations is the number of computations necessary to evaluate a single determinant. There are two terms in the evaluation of a 2×2 determinant, 6 terms in a 3×3, 24 terms in a 4×4, 120 in a 5×5, 720 in a 6×6, and 5040 in a 7×7.

To solve a linear system of eight equations with eight unknowns would require evaluating more than 41,000 terms. And that does not count all the multiplications within each term. Happily, however, computer programs can be written to do the evaluations of determinants. One such program can be found on the Math ACE Disk, CRAMER'S RULE.

This program gives the solution to equations in terms of the variables x_1, x_2, x_3, and so on instead of the usual x, y, z. This is more convenient since it is impossible to run out of variables. Variables written in this way are called subscripted variables.

You are encouraged to use this program on some of the exercises in your text or make up some systems and try the program. The coefficients you use can be any real numbers, but remember to convert any fraction to a decimal before you enter that number.

Chapter Summary

Key Words   Equations considered together are called a *system of equations*.

Inequalities considered together are called a *system of inequalities*.

A *solution of a system of equations* in two variables is an ordered pair which is a solution of each equation of the system.

The *solution set of a system of inequalities* is the intersection of the solution sets of each inequality.

When the graphs of a system of equations intersect at only one point, the equations are called *independent equations*.

When the graphs of a system of equations coincide, the equations are called *dependent equations*.

When a system of equations has no solution, it is called an *inconsistent system of equations*.

An equation of the form $Ax + By + Cz = D$, where A, B, C, and D are constants, is called a *linear equation in three variables*.

A *solution of a system of equations in three variables* is an ordered triple which is a solution of each equation of the system.

A *matrix* is a rectangular array of numbers.

A *square matrix* has the same number of rows as columns.

A *determinant* is a number associated with a square matrix.

The *minor of an element* in a 3 × 3 determinant is the 2 × 2 determinant obtained by eliminating the row and column that contain that element.

The *cofactor* of an element in a matrix is $(-1)^{i+j}$ times the minor of that element, where i is the row number of the element and j is the column number.

The evaluation of the determinant of a 3 × 3 or larger matrix is accomplished by expanding the cofactors.

Essential Rules   A system of equations can be solved by the graphing method, the substitution method, or the addition method.

Cramer's Rule is a method of solving a system of equations by using determinants.

Chapter Review

SECTION 1

1. Solve by graphing: $x + y = 3$
$3x - 2y = -6$

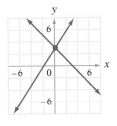

$(0, 3)$

2. Solve by graphing: $2x - y = 4$
$y = 2x - 4$

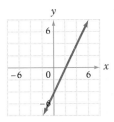

$(x, 2x - 4)$

3. Solve by substitution: $2x - 6y = 15$
$x = 3y + 8$

inconsistent

4. Solve by substitution: $3x + 12y = 18$
$x + 4y = 6$

$\left(x, -\frac{1}{4}x + \frac{3}{2}\right)$

SECTION 2

5. Solve by the addition method:
$3x + 2y = 2$
$x + y = 3$
$(-4, 7)$

6. Solve by the addition method:
$5x - 15y = 30$
$x - 3y = 6$
$\left(x, \frac{1}{3}x - 2\right)$

7. Solve by the addition method:
$\dfrac{3}{x} - \dfrac{6}{y} = 14$
$\dfrac{1}{x} + \dfrac{9}{y} = 1$
$\left(\frac{1}{4}, -3\right)$

8. Solve by the addition method:
$x + y + z = 4$
$2x - y + z = -1$
$2x + y + z = 3$
$(-1, 2, 3)$

9. Solve by the addition method:
$$3x + y = 13$$
$$2y + 3z = 5$$
$$x + 2z = 11$$
$(5, -2, 3)$

10. Solve by the addition method:
$$3x - 4y - 2z = 17$$
$$4x - 3y + 5z = 5$$
$$5x - 5y + 3z = 14$$
$(3, -1, -2)$

SECTION 3

11. Evaluate the determinant.
$$\begin{vmatrix} 6 & 1 \\ 2 & 5 \end{vmatrix}$$
28

12. Evaluate the determinant.
$$\begin{vmatrix} 1 & 5 & -2 \\ -2 & 1 & 4 \\ 4 & 3 & -8 \end{vmatrix}$$
0

13. Solve by using Cramer's Rule.
$$2x - y = 7$$
$$3x + 2y = 7$$
$(3, -1)$

14. Solve by using Cramer's Rule.
$$3x - 4y = 10$$
$$2x + 5y = 15$$
$\left(\dfrac{110}{23}, \dfrac{25}{23} \right)$

15. Solve by using Cramer's Rule.
$$x + y + z = 0$$
$$x + 2y + 3z = 5$$
$$2x + y + 2z = 3$$
$(-1, -3, 4)$

16. Solve by using Cramer's Rule.
$$x + 3y + z = 6$$
$$2x + y - z = 12$$
$$x + 2y - z = 13$$
$(2, 3, -5)$

SECTION 4

17. A cabin cruiser traveling with the current went 60 mi in 3 h. Against the current it took 5 h to travel the same distance. Find the rate of the cabin cruiser in calm water and the rate of the current.
cabin cruiser: 16 mph; current: 4 mph

18. A pilot flying with the wind flew 600 mi in 3 h. Flying against the wind, the pilot required 4 h to travel the same distance. Find the rate of the plane in calm air and the rate of the wind.
plane: 175 mph; wind: 25 mph

19. At a movie theater admission tickets are $5 for children and $8 for adults. The receipts for one Friday evening were $2500. The next day there were three times as many children as the preceding evening but only half the number of adults as the night before, yet the receipts were still $2500. Find the number of children who attended the movie Friday evening.
100 children

20. A confectioner mixed 3 lb of milk chocolate candy with 3 lb of semi-sweet chocolate candy. The 6-lb mixture sells for $30. A second mixture included 6 lb of the milk chocolate candy and 2 lb of the semi-sweet chocolate candy. The 8-lb mixture sells for $42. Find the cost per pound of the milk chocolate candy and the semi-sweet chocolate candy.
milk chocolate: $5.50/lb; semi-sweet chocolate: $4.50/lb

21. If either 3 pencils and 6 pens or 21 pencils and 2 pens can be bought for $6, find the price per pencil.
$.20

22. A wallet contains $44 in one-dollar bills and five-dollar bills. If the one-dollar bills were five-dollar bills and the five-dollar bills were ten-dollar bills, the wallet would contain $130. Find the number of one-dollar bills.
14 one-dollar bills

SECTION 5

23. Solve: $y = x^2 + 5x - 6$
          $y = x - 10$
   $(-2, -12)$

24. Solve: $x^2 + y^2 = 20$
          $x^2 - y^2 = 12$
   $(4, 2)\ (-4, 2)$
   $(4, -2)\ (-4, -2)$

25. Solve: $x^2 - y^2 = 24$
          $2x^2 + 5y^2 = 55$
   $(5, 1)\ (-5, 1)$
   $(5, -1)\ (-5, -1)$

26. Solve: $2x^2 + y^2 = 19$
          $3x^2 - y^2 = 6$
   $(\sqrt{5}, 3)\ (-\sqrt{5}, 3)$
   $(\sqrt{5}\ -3)\ (-\sqrt{5}, -3)$

27. Graph the solution set:
   $x + 3y \leq 6$
   $2x - y \geq 4$

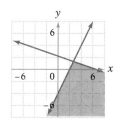

28. Graph the solution set.
   $x^2 + y^2 < 36$
   $x + y > 4$

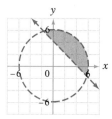

29. Graph the solution set:
   $y \geq x^2 - 4x + 2$
   $y \leq \frac{1}{3}x - 1$

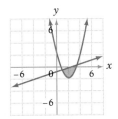

30. Graph the solution set:
   $\dfrac{x^2}{25} + \dfrac{y^2}{16} \leq 1$
   $\dfrac{y^2}{4} - \dfrac{x^2}{4} \geq 1$

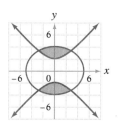

Chapter Test

1. Solve by graphing: $2x - 3y = -6$
$\qquad\qquad\qquad\quad 2x - y = 2$

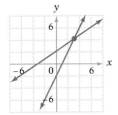

$(3, 4)$ [10.1A]

2. Solve by graphing: $x - 2y = -5$
$\qquad\qquad\qquad\quad 3x + 4y = -15$

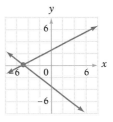

$(-5, 0)$ [10.1A]

3. Graph the solution set: $2x - y < 3$
$\qquad\qquad\qquad\qquad 4x + 3y < 11$

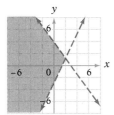

[10.5B]

4. Graph the solution set: $(x - 1)^2 + y^2 \leq 25$
$\qquad\qquad\qquad\qquad\qquad y^2 < x$

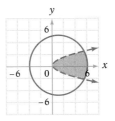

[10.5B]

5. Solve by substitution: $3x + 2y = 4$
$\qquad\qquad\qquad\qquad\quad x = 2y - 1$

$\left(\dfrac{3}{4}, \dfrac{7}{8}\right)$ [10.1B]

6. Solve by substitution: $5x + 2y = -23$
$\qquad\qquad\qquad\qquad\quad 2x + y = -10$

$(-3, -4)$ [10.1B]

7. Solve by substitution: $y = 3x - 7$
$\qquad\qquad\qquad\qquad\quad y = -2x + 3$

$(2, -1)$ [10.1B]

8. Solve by the addition method:
$3x + 4y = -2$
$2x + 5y = 1$
$(-2, 1)$ [10.2A]

9. Solve by the addition method:
$4x - 6y = 5$
$6x - 9y = 4$
inconsistent [10.2A]

10. Solve by the addition method:
$3x - y = 2x + y - 1$
$5x + 2y = y + 6$
$(1, 1)$ [10.2A]

11. Solve by the addition method:
$2x + 4y - z = 3$
$x + 2y + z = 5$
$4x + 8y - 2z = 7$
inconsistent [10.2B]

12. Solve by the addition method:
$x - y - z = 5$
$2x + z = 2$
$3y - 2z = 1$
$(2, -1, -2)$ [10.2B]

13. Evaluate the determinant: $\begin{vmatrix} 3 & -1 \\ -2 & 4 \end{vmatrix}$

10 [10.3A]

14. Evaluate the determinant: $\begin{vmatrix} 1 & -2 & 3 \\ 3 & 1 & 1 \\ 2 & -1 & -2 \end{vmatrix}$

-32 [10.3A]

15. Solve by using Cramer's Rule:

$5x + 2y = 9$
$3x + 5y = -7$

$\left(\frac{59}{19}, -\frac{62}{19}\right)$ [10.3B]

16. Solve by using Cramer's Rule:

$3x + 2y + 2z = 2$
$x - 2y - z = 1$
$2x - 3y - 3z = -3$

$(0, -2, 3)$ [10.3B]

17. Solve: $x - 2y = 6$
$\quad\quad\quad\quad y = x^2 - 17$

$(4, -1)\ \left(-\frac{7}{2}, -\frac{19}{4}\right)$ [10.5A]

18. Solve: $2x^2 + 3y^2 = 20$
$\quad\quad\quad\quad 3x^2 - y^2 = 8$

$(2, 2)\ (2, -2)$
$(-2, 2)\ (-2, -2)$ [10.5A]

19. A plane flying with the wind went 350 mi in two hours. The return trip, flying against the wind, took 2.8 h. Find the rate of the plane in calm air and the rate of the wind.

plane, 150 mph
wind, 25 mph [10.4A]

20. A clothing manufacturer purchased 60 yd of cotton and 90 yd of wool for a total cost of $900. Another purchase, at the same prices, included 80 yd of cotton and 20 yd of wool for a total cost of $500. Find the cost per yard of the cotton and the wool.

cotton, $4.50
wool, $7 [10.4B]

Cumulative Review

1. Solve: $\frac{3}{2}x - \frac{3}{8} + \frac{1}{4}x = \frac{7}{12}x - \frac{5}{6}$

$x = -\frac{11}{28}$ [2.1C]

2. Find the equation of the line containing the points $(2, -1)$ and $(3, 4)$.

$y = 5x - 11$ [6.3B]

3. Factor: $6x^2 - 19x + 10$

$(2x - 5)(3x - 2)$ [3.3D]

4. Simplify: $\frac{2x}{x^2 - 5x + 6} - \frac{3}{x^2 - 2x - 3}$

$\frac{2x^2 - x + 6}{(x - 3)(x - 2)(x + 1)}$ [4.3B]

5. Solve:
$\frac{3}{x^2 - 5x + 6} - \frac{x}{x - 3} = \frac{2}{x - 2}$

$x = -3$ [4.6A]

6. Simplify: $a^{-1} + a^{-1}b$

$\frac{1 + b}{a}$ [3.1C]

7. Simplify: $\sqrt[3]{-4ab^4} \ \sqrt[3]{2a^3b^4}$

$-2ab^2 \ \sqrt[3]{ab^2}$ [5.2C]

8. Solve: $\sqrt{3x + 10} = 1$

$x = -3$ [5.4A]

9. Solve by factoring: $2x^2 + 9x = 5$

$\frac{1}{2}$ and -5 [7.1A]

10. Solve by using the quadratic formula:
$3x^2 = 2x + 2$

$\frac{1 + \sqrt{7}}{3}$ and $\frac{1 - \sqrt{7}}{3}$ [7.3A]

11. Graph $f(x) = |x| - 2$.

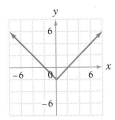

[8.2A]

12. Write the equation $x^2 + y^2 - 4x + 2y - 4 = 0$ in standard form. Then sketch the graph.

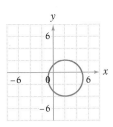

$(x - 2)^2 + (y + 1)^2 = 9$ [9.2C]

13. Find the distance between the points $(-5, 1)$ and $(2, 0)$.

$5\sqrt{2}$ [9.2A]

14. Use the discriminant to determine the number of x-intercepts of the parabola $y = -2x^2 - x - 2$.

no x-intercepts   [9.1C]

15. Solve by graphing:
$5x - 2y = 10$
$3x + 2y = 6$

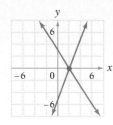

(2, 0) [10.1A]

16. Solve by substitution:
$3x - 2y = 7$
$y = 2x - 1$

$(-5, -11)$ [10.1B]

17. Solve by the addition method:
$3x + 2z = 1$
$2y - z = 1$
$x + 2y = 1$
$(1, 0, -1)$ [10.2A]

18. Evaluate the determinant.
$$\begin{vmatrix} 2 & -5 & 1 \\ 3 & 1 & 2 \\ 6 & -1 & 4 \end{vmatrix}$$
3 [10.3A]

19. Solve by using Cramer's Rule:
$4x - 3y = 17$
$3x - 2y = 12$
$(2, -3)$ [10.3B]

20. Solve: $x^2 + y^2 = 5$
$y^2 = 4x$
(1, 2) (1, -2) [10.5A]

21. A coin purse contains 40 coins in nickels, dimes, and quarters. There are three times as many dimes as quarters. The total value of the coins is $4.10. How many nickels are in the coin purse?
16 nickels [2.4B]

22. How many milliliters of pure water must be added to 100 ml of a 4% salt solution to make a 2.5% salt solution?
60 ml [2.6B]

23. The height of a triangle is twice the length of the base. The area of the triangle is 36 m². Find the length of the base of the triangle.
6 m [7.6A]

24. The distance (d) a spring stretches varies directly as the force (f) used to stretch the spring. If a force of 50 lb can stretch the spring 30 in., how far will a force of 40 lb stretch the spring?
24 in. [8.4A]

25. Flying with the wind, a small plane required 2 h to fly 150 mi. Against the wind, it took 3 h to fly the same distance. Find the rate of the wind.
12.5 mph [8.4A]

26. A restaurant manager buys 100 lb of hamburger and 50 lb of steak for a total cost of $270. A second purchase, at the same prices, includes 150 lb of hamburger and 100 lb of steak. The total cost is $480. Find the price of one pound of steak.
$3 [10.4B]

11

Exponential and Logarithmic Functions

OBJECTIVES

- ► To evaluate an exponential function
- ► To graph an exponential function
- ► To write equivalent exponential and logarithmic expressions
- ► To graph a logarithmic function
- ► To use the Properties of Logarithms to simplify expressions containing logarithms
- ► To find common logarithms
- ► To find common antilogarithms
- ► To use interpolation to find a common logarithm or antilogarithm
- ► To evaluate numerical expressions by using common logarithms
- ► To solve an exponential equation
- ► To solve a logarithmic equation
- ► To find the logarithm of a number other than base 10
- ► To solve application problems

Napier Rods

The labor which is involved in calculating the products of large numbers has led many people to devise ways to short-cut the procedure. One such way was first described in the early 1600's by John Napier and is based on Napier Rods.

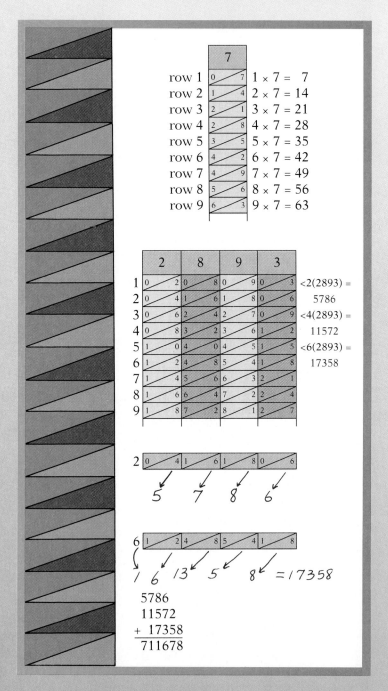

A Napier Rod consists of placing a number and the first 9 multiples of that number on a rectangular piece of paper (Napier's Rod). This is done for the first 9 positive integers. It is necessary to have more than one rod for each number. The rod for 7 is shown at the left.

To illustrate how these rods were used to multiply, an example will be used.

Multiply: 2893
            × 246

Place the rods for 2, 8, 9, and 3 next to one another. The products for 2, 4, and 6 are found by using the numbers along the 2nd, 4th, and 6th rows.

Each number is found by adding the digits diagonally downward on the diagonal and carrying to the next diagonal when necessary. The products for 2 and 6 are shown at the left.

Notice, that the sum of the 3rd diagonal is 13, thus carrying to the next diagonal is necessary.

The final product is then found by addition.

Before the invention of the electronic calculator, logarithms were used to ease the drudgery of lengthy calculations. John Napier is also credited with the invention of logarithms.

Content and Format © 1991 HMCo.

SECTION 11.1 — The Exponential and Logarithmic Functions

Objective A   **To evaluate an exponential function**

A function given by $f(x) = b^x$ is an **exponential function** where b is a positive real number not equal to 1. The number b is the **base** of the exponential function.

Evaluate $f(x) = 2^x$ at $x = 3$.

$$f(3) = 2^3 = 8$$

Evaluate $f(x) = 3^x$ at $x = -2$.

$$f(-2) = 3^{-2} = \frac{1}{3^2} = \frac{1}{9}$$

Evaluate $f(x) = (-4)^x$ at $x = \frac{1}{2}$.

$$f\left(\frac{1}{2}\right) = (-4)^{1/2} = \sqrt{-4} = 2i$$

The value is a complex number. For this reason, the base of an exponential function is required to be a positive real number.

Evaluate $f(x) = 4^x$ at $x = \sqrt{2}$.

Using a calculator, the value of this function can be found to the desired degree of accuracy by using approximations for $\sqrt{2}$.

$$f(\sqrt{2}) \approx f(1.41) \quad\approx 4^{1.41} \quad\approx 7.06$$
$$f(\sqrt{2}) \approx f(1.414) \quad\approx 4^{1.414} \quad\approx 7.101$$
$$f(\sqrt{2}) \approx f(1.4142) \quad\approx 4^{1.4142} \approx 7.1029$$
$$f(\sqrt{2}) \approx f(1.41421) \approx 4^{1.41421} \approx 7.10296$$

Example 1   Evaluate $f(x) = \left(\frac{1}{2}\right)^x$ at $x = 2$ and $x = -3$.

Solution $f(x) = \left(\frac{1}{2}\right)^x$

$$f(2) = \left(\frac{1}{2}\right)^2 = \frac{1}{4}$$

$$f(-3) = \left(\frac{1}{2}\right)^{-3} = 2^3 = 8$$

Example 2   Evaluate $f(x) = \left(\frac{2}{3}\right)^x$ at $x = 3$ and $x = -2$.

Your solution $f(3) = \frac{8}{27}$

$$f(-2) = \frac{9}{4}$$

Example 3   Evaluate $f(x) = 2^{3x-1}$ at $x = 1$ and $x = -1$.

Solution $f(x) = 2^{3x-1}$
$$f(1) = 2^{3(1)-1} = 2^2 = 4$$

$$f(-1) = 2^{3(-1)-1} = 2^{-4} = \frac{1}{2^4} = \frac{1}{16}$$

Example 4   Evaluate $f(x) = 2^{2x+1}$ at $x = 0$ and $x = -2$.

Your solution $f(0) = 2$

$$f(-2) = \frac{1}{8}$$

Solutions on p. A45

Objective B	**To graph an exponential function**

11

Some of the properties of an exponential function can be seen by considering its graph.

Graph $f(x) = 2^x$.

Think of this as the equation $y = 2^x$.

Choose values of x and find the corresponding values of y. The results can be recorded in a table.

Graph the ordered pairs on a rectangular coordinate system.

x	$f(x) = y$
-2	$2^{-2} = \frac{1}{4}$
-1	$2^{-1} = \frac{1}{2}$
0	$2^0 = 1$
1	$2^1 = 2$
2	$2^2 = 4$
3	$2^3 = 8$

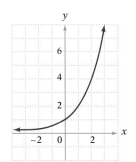

Connect the points with a smooth curve.

Note that a vertical line would intersect the graph at only one point. Therefore, by the vertical line test, the graph of $f(x) = 2^x$ is the graph of a function. Also notice that a horizontal line would intersect the graph at only one point. Therefore, the graph of $f(x) = 2^x$ is the graph of a one-to-one function.

Graph $f(x) = \left(\frac{1}{2}\right)^x$.

Think of this as the equation $y = \left(\frac{1}{2}\right)^x$.

Choose values of x and find the corresponding values of y.

Graph the ordered pairs on a rectangular coordinate system.

Connect the points with a smooth curve.

x	$f(x) = y$
-3	$\left(\frac{1}{2}\right)^{-3} = 8$
-2	$\left(\frac{1}{2}\right)^{-2} = 4$
-1	$\left(\frac{1}{2}\right)^{-1} = 2$
0	$\left(\frac{1}{2}\right)^0 = 1$
1	$\left(\frac{1}{2}\right)^1 = \frac{1}{2}$
2	$\left(\frac{1}{2}\right)^2 = \frac{1}{4}$

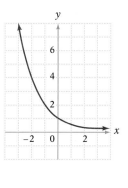

Applying the vertical and horizontal line tests, the graph of $f(x) = \left(\frac{1}{2}\right)^x$ is also the graph of a one-to-one function.

Graph $f(x) = 2^{-x}$.

Think of this as the equation $y = 2^{-x}$.

Choose values of x and find the corresponding values of y.

Graph the ordered pairs on a rectangular coordinate system.

Connect the points with a smooth curve.

x	y
-3	8
-2	4
-1	2
0	1
1	$\frac{1}{2}$
2	$\frac{1}{4}$

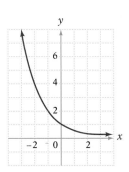

Note that since $2^{-x} = (2^{-1})^x = \left(\frac{1}{2}\right)^x$, the graphs of $f(x) = 2^{-x}$ and $f(x) = \left(\frac{1}{2}\right)^x$ are the same.

Example 5   Graph: $f(x) = 3^{\frac{1}{2}x - 1}$

Solution

x	y
-2	$\frac{1}{9}$
0	$\frac{1}{3}$
2	1
4	3

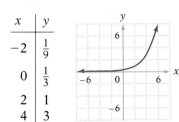

Example 6   Graph: $f(x) = 2^{-\frac{1}{2}x}$

Your solution

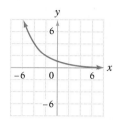

Example 7   Graph: $f(x) = 2^x - 1$

Solution

x	y
-2	$-\frac{3}{4}$
-1	$-\frac{1}{2}$
0	0
1	1
2	3
3	7

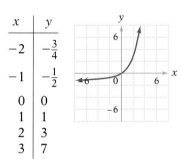

Example 8   Graph: $f(x) = 2^x + 1$

Your solution

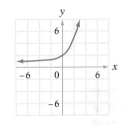

Solutions on p. A45

Objective C   **To write equivalent exponential and logarithmic expressions**

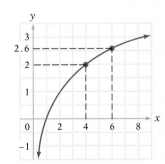

Recall that the inverse of a function is formed by interchanging x and y. There-fore, the **inverse of the exponential function $y = 2^x$ is $x = 2^y$.**

For the inverse function $x = 2^y$, the value of y is called the **logarithm** of x to the base 2 and is written $y = \log_2 x$, where the abbreviation "log" is used for "loga-rithm." Thus, $y = \log_2 x$ is equivalent to $x = 2^y$.

The graph of the inverse function $x = 2^y$ or $y = \log_2 x$ is shown at the right. Using this graph, it is possible to estimate a value of y when x is given.

When $x = 4$, $y = 2$.
This can be written as $4 = 2^2$ or $\log_2 4 = 2$.
Read $\log_2 4$ as "the logarithm of 4 to the base 2" or "the log base 2 of 4."

When $x = 6$, $y \approx 2.6$.
Therefore, $6 \approx 2^{2.6}$ or $\log_2 6 \approx 2.6$.

Given the equation $x = b^y$, the logarithm of x is y, the exponent on b. Basically, a logarithm is a special kind of exponent.

For $b > 0$, $b \neq 1$, $y = \log_b x$ is equivalent to $x = b^y$.

Read $\log_b x$ as "the logarithm of x to the base b" or "the log base b of x."

The table at the right shows equivalent statements written in both exponential and logarithmic form.

Exponential Form	*Logarithmic Form*
$2^4 = 16$	$\log_2 16 = 4$
$\left(\frac{2}{3}\right)^2 = \frac{4}{9}$	$\log_{\frac{2}{3}}\left(\frac{4}{9}\right) = 2$
$10^{-1} = 0.1$	$\log_{10} 0.1 = -1$

Write $\log_3 81 = 4$ in exponential form.

$\log_3 81 = 4$ is equivalent to $3^4 = 81$.

Write $10^{-2} = 0.01$ in logarithmic form.

$10^{-2} = 0.01$ is equivalent to $\log_{10} 0.01 = -2$.

The 1–1 property of an exponential function can be used to evaluate some logarithms.

1–1 Property of Exponential Functions

If $b^u = b^v$, then $u = v$.

Evaluate: $\log_2 8$

Write an equation.	$\log_2 8 = x$
Write the equation in its equivalent exponential form.	$8 = 2^x$
Write 8 in exponential form using 2 as the base.	$2^3 = 2^x$
Solve for x using the 1–1 property of exponential functions.	$3 = x$
	$\log_2 8 = 3$

Solve for x: $\log_4 x = -2$

	$\log_4 x = -2$
Write the equation in its equivalent exponential form.	$4^{-2} = x$
Solve for x.	$\frac{1}{16} = x$

The solution is $\frac{1}{16}$.

Example 9 Write $3^4 = 81$ in logarithmic form.

Solution $3^4 = 81$ is equivalent to $\log_3 81 = 4$.

Example 10 Write $10^3 = 1000$ in logarithmic form.

Your solution $\log_{10} 1000 = 3$

Solution on p. A45

Example 11   Evaluate: $\log_3\left(\frac{1}{9}\right)$

Solution $\log_3\left(\frac{1}{9}\right) = x$

$$\frac{1}{9} = 3^x$$
$$3^{-2} = 3^x$$
$$-2 = x$$

$$\log_3\left(\frac{1}{9}\right) = -2$$

Example 12   Evaluate: $\log_4 64$

Your solution   3

Example 13   Solve for x: $\log_5 x = 2$

Solution $\log_5 x = 2$
$$5^2 = x$$
$$25 = x$$

The solution is 25.

Example 14   Solve for x: $\log_2 x = -4$

Your solution $\frac{1}{16}$

Solutions on p. **A45**

| Objective D | **To graph a logarithmic function** | |

The graph of a logarithmic function can be found by using the relationship between the exponential and logarithmic functions.

Graph $f(x) = \log_2 x$.

Think of this as the equation $y = \log_2 x$.

$$f(x) = \log_2 x$$
$$y = \log_2 x$$

Write the equivalent exponential equation.

$$x = 2^y$$

Since the equation is solved for x in terms of y, it is easier to choose values of y and find the corresponding values of x. The results can be recorded in a table.

Graph the ordered pairs on a rectangular coordinate system.

x	y
$\frac{1}{4}$	-2
$\frac{1}{2}$	-1
1	0
2	1
4	2

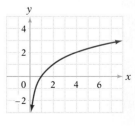

Connect the points with a smooth curve.

Applying the vertical and horizontal line tests, $f(x) = \log_2 x$ is a one-to-one function.

Graph $f(x) = \log_2 x + 1$.

Think of this as the equation $y = \log_2 x + 1$.

$$f(x) = \log_2 x + 1$$
$$y = \log_2 x + 1$$

Solve the equation for $\log_2 x$.

$$y - 1 = \log_2 x$$

Write the equivalent exponential equation.

$$2^{y-1} = x$$

Choose values of y and find the corresponding values of x.

Graph the ordered pairs on a rectangular coordinate system.

Connect the points with a smooth curve.

x	y
$\frac{1}{4}$	-1
$\frac{1}{2}$	0
1	1
2	2
4	3

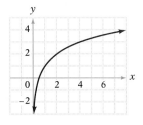

Example 15   Graph: $f(x) = \log_3 x$

Solution $f(x) = \log_3 x$
$\quad\quad\quad y = \log_3 x$

$y = \log_3 x$ is equivalent to $x = 3^y$.

x	y
$\frac{1}{9}$	-2
$\frac{1}{3}$	-1
1	0
3	1

Example 16   Graph: $f(x) = \log_2(x - 1)$

Your solution

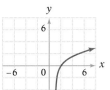

Example 17   Graph: $f(x) = 2\log_3 x$

Solution $f(x) = 2\log_3 x$
$\quad\quad\quad y = 2\log_3 x$

$\dfrac{y}{2} = \log_3 x$

$\dfrac{y}{2} = \log_3 x$ is equivalent to $x = 3^{y/2}$.

x	y
$\frac{1}{9}$	-4
$\frac{1}{3}$	-2
1	0
3	2

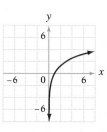

Example 18   Graph: $f(x) = \log_3 2x$

Your solution

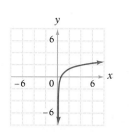

Solutions on p. A45

11.1   **EXERCISES**

▶ **Objective A**

Given the function $f(x) = 3^x$, find:

1. $f(2)$ 9	**2.** $f(3)$ 27	**3.** $f(-2)$ $\frac{1}{9}$	**4.** $f(-1)$ $\frac{1}{3}$	**5.** $f(0)$ 1	**6.** $f(1)$ 3

Given the function $f(x) = 2^{x+1}$, find:

7. $f(3)$ 16	**8.** $f(1)$ 4	**9.** $f(-3)$ $\frac{1}{4}$	**10.** $f(-4)$ $\frac{1}{8}$	**11.** $f(-1)$ 1	**12.** $f(0)$ 2

Given the function $f(x) = \left(\frac{1}{2}\right)^{2x}$, find:

13. $f(0)$ 1	**14.** $f(1)$ $\frac{1}{4}$	**15.** $f(-2)$ 16	**16.** $f(-1)$ 4	**17.** $f\left(\frac{3}{2}\right)$ $\frac{1}{8}$	**18.** $f\left(-\frac{1}{2}\right)$ 2

Given the function $f(x) = \left(\frac{1}{3}\right)^{x-1}$, find:

19. $f(2)$ $\frac{1}{3}$	**20.** $f(3)$ $\frac{1}{9}$	**21.** $f(-1)$ 9	**22.** $f(-2)$ 27	**23.** $f(0)$ 3	**24.** $f(1)$ 1

Given the function $f(x) = 2^{x^2}$, find:

25. $f(1)$ 2	**26.** $f(0)$ 1	**27.** $f(2)$ 16	**28.** $f(3)$ 512	**29.** $f(-1)$ 2	**30.** $f(-2)$ 16

▶ **Objective B**

Graph:

31. $f(x) = 3^x$

32. $f(x) = 3^{-x}$

33. $f(x) = 2^{x+1}$

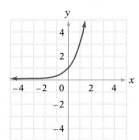

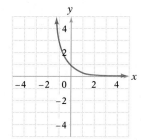

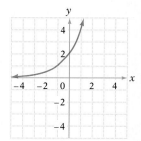

Graph:

34. $f(x) = 2^{x-1}$

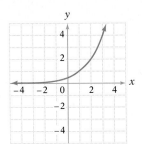

35. $f(x) = \left(\frac{1}{3}\right)^x$

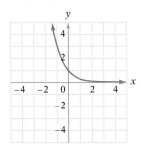

36. $f(x) = \left(\frac{2}{3}\right)^x$

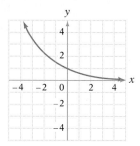

37. $f(x) = 2^{-x} + 1$

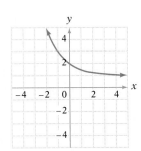

38. $f(x) = 2^x - 3$

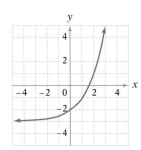

39. $f(x) = \left(\frac{1}{3}\right)^{-x}$

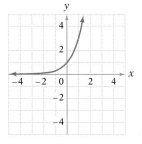

40. $f(x) = \left(\frac{3}{2}\right)^{-x}$

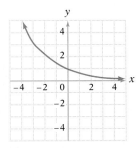

41. $f(x) = \left(\frac{1}{2}\right)^{-x} + 2$

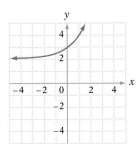

42. $f(x) = \left(\frac{1}{2}\right)^x - 1$

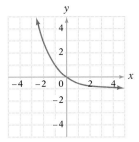

▶ **Objective C**

Write the exponential expression in logarithmic form.

43. $2^5 = 32$
$\log_2 32 = 5$

44. $3^4 = 81$
$\log_3 81 = 4$

45. $5^2 = 25$
$\log_5 25 = 2$

46. $10^3 = 1000$
$\log_{10} 1000 = 3$

47. $4^{-2} = \frac{1}{16}$
$\log_4 \frac{1}{16} = -2$

48. $3^{-3} = \frac{1}{27}$
$\log_3 \frac{1}{27} = -3$

49. $\left(\frac{1}{2}\right)^2 = \frac{1}{4}$
$\log_{\frac{1}{2}} \frac{1}{4} = 2$

50. $\left(\frac{1}{3}\right)^4 = \frac{1}{81}$
$\log_{\frac{1}{3}} \frac{1}{81} = 4$

Write the exponential expression in logarithmic form.

51. $3^0 = 1$
$\log_3 1 = 0$

52. $10^0 = 1$
$\log_{10} 1 = 0$

53. $a^x = w$
$\log_a w = x$

54. $b^y = c$
$\log_b c = y$

Write the logarithmic expression in exponential form.

55. $\log_3 9 = 2$
$3^2 = 9$

56. $\log_2 32 = 5$
$2^5 = 32$

57. $\log_4 4 = 1$
$4^1 = 4$

58. $\log_7 7 = 1$
$7^1 = 7$

59. $\log_{10} 1 = 0$
$10^0 = 1$

60. $\log_8 1 = 0$
$8^0 = 1$

61. $\log_{10} 0.01 = -2$
$10^{-2} = 0.01$

62. $\log_5 \frac{1}{5} = -1$
$5^{-1} = \frac{1}{5}$

63. $\log_{\frac{1}{3}}\left(\frac{1}{9}\right) = 2$
$\left(\frac{1}{3}\right)^2 = \frac{1}{9}$

64. $\log_{\frac{1}{4}}\left(\frac{1}{16}\right) = 2$
$\left(\frac{1}{4}\right)^2 = \frac{1}{16}$

65. $\log_b u = v$
$b^v = u$

66. $\log_c x = y$
$c^y = x$

Evaluate:

67. $\log_4 16$
2

68. $\log_3 27$
3

69. $\log_2 32$
5

70. $\log_{10} 1000$
3

71. $\log_{10} 100$
2

72. $\log_5 125$
3

73. $\log_6 216$
3

74. $\log_7 1$
0

75. $\log_8 1$
0

76. $\log_3 243$
5

77. $\log_5 625$
4

78. $\log_2 64$
6

Solve for *x:*

79. $\log_3 x = 2$
9

80. $\log_5 x = 1$
5

81. $\log_4 x = 3$
64

82. $\log_2 x = 6$
64

83. $\log_7 x = -1$
$\frac{1}{7}$

84. $\log_8 x = -2$
$\frac{1}{64}$

85. $\log_6 x = 3$
216

86. $\log_4 x = 0$
1

87. $\log_5 x = 0$
1

88. $\log_5 x = 3$
125

89. $\log_3 x = -3$
$\frac{1}{27}$

90. $\log_2 x = 3$
8

▶ **Objective D**

Graph:

91. $f(x) = \log_4 x$

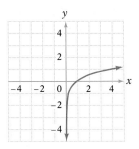

92. $f(x) = \log_2(x + 1)$

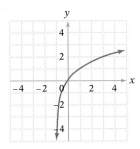

93. $f(x) = \log_3(2x - 1)$

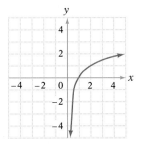

94. $f(x) = \log_2\left(\frac{1}{2}x\right)$

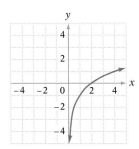

95. $f(x) = 3\log_2 x$

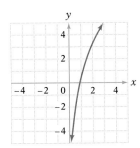

96. $f(x) = \frac{1}{2}\log_2 x$

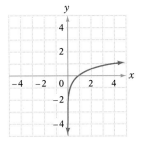

97. $f(x) = -\log_2 x$

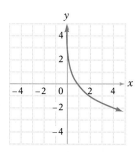

98. $f(x) = -\log_3 x$

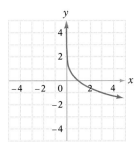

99. $f(x) = \log_2(x - 1)$

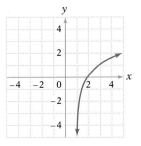

100. $f(x) = \log_3(2 - x)$

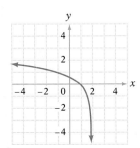

101. $f(x) = -\log_2(x - 1)$

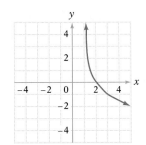

102. $f(x) = -\log_2(1 - x)$

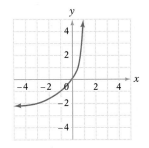

The Properties of Logarithms

Objective A

To use the Properties of Logarithms to simplify expressions containing logarithms

Since a logarithm is a special kind of exponent, the Properties of Logarithms are similar to the Properties of Exponents.

The table at the right shows some powers of 2 and the equivalent logarithmic form.

The table can be used to show that $\log_2 4 + \log_2 8$ equals $\log_2 32$.

$$2^0 = 1 \qquad \log_2 1 = 0$$
$$2^1 = 2 \qquad \log_2 2 = 1$$
$$2^2 = 4 \qquad \log_2 4 = 2$$
$$2^3 = 8 \qquad \log_2 8 = 3$$
$$2^4 = 16 \qquad \log_2 16 = 4$$
$$2^5 = 32 \qquad \log_2 32 = 5$$

$$\log_2 4 + \log_2 8 = 2 + 3 = 5$$
$$\log_2 32 = 5$$
$$\log_2 4 + \log_2 8 = \log_2 32$$

Note that $\log_2 32 = \log_2(4 \times 8) = \log_2 4 + \log_2 8$.

The property of logarithms that states that the logarithm of the product of two numbers equals the sum of the logarithms of the two numbers is similar to the property of exponents which states that to multiply two exponential expressions with the same base, add the exponents.

The Logarithm Property of the Product of Two Numbers

> For any positive real numbers x, y, and b, $b \neq 1$, $\log_b xy = \log_b x + \log_b y$.

A proof of this property can be found in the Appendix on page A7.

Write $\log_b 6z$ in expanded form.
Use the Logarithm Property of Products. $\qquad \log_b 6z = \log_b 6 + \log_b z$

Write $\log_b 12 + \log_b r$ as a single logarithm.
Use the Logarithm Property of Products. $\qquad \log_b 12 + \log_b r = \log_b 12r$

The Logarithm Property of Products can be extended to include the logarithm of the product of more than two factors. For example:

$$\log_b xyz = \log_b(xy)z = \log_b xy + \log_b z = \log_b x + \log_b y + \log_b z$$

Write $\log_b 5st$ in expanded form.
Use the Logarithm Property of Products. $\qquad \log_b 5st = \log_b 5 + \log_b s + \log_b t$

A second property of logarithms involves the logarithm of the quotient of two numbers. This property of logarithms is also based on the fact that a logarithm is an exponent and that to divide two exponential expressions with the same base, the exponents are subtracted.

The Logarithm Property of the Quotient of Two Numbers

For any positive real numbers x, y, and b, $b \neq 1$, $\log_b \frac{x}{y} = \log_b x - \log_b y$.

A proof of this property can be found in the Appendix on page A7.

Write $\log_b \frac{p}{8}$ in expanded form.

Use the Logarithm Property of Quotients. $\quad \log_b \frac{p}{8} = \log_b p - \log_b 8$

Write $\log_b y - \log_b v$ as a single logarithm.

Use the Logarithm Property of Quotients. $\quad \log_b y - \log_b v = \log_b \frac{y}{v}$

A third property of logarithms, which is especially useful in the computation of the power of a number, is based on the fact that a logarithm is an exponent and the power of an exponential expression is found by multiplying the exponents.

The table of the powers of 2 shown on the previous page can be used to show that $\log_2 2^3$ equals $3 \log_2 2$.

$$\log_2 2^3 = \log_2 8 = 3$$
$$3 \log_2 2 = 3 \cdot 1 = 3$$
$$\log_2 2^3 = 3 \log_2 2$$

The Logarithm Property of the Power of a Number

For any positive real numbers x and b, $b \neq 1$, and for any real number r, $\log_b x^r = r \log_b x$.

A proof of this property can be found in the Appendix on page A7.

Rewrite $\log_b x^3$ in terms of $\log_b x$.

Use the Logarithm Property of Powers. $\quad \log_b x^3 = 3 \log_b x$

Rewrite $\frac{2}{3} \log_4 z$ with a coefficient of 1.

Use the Logarithm Property of Powers. $\quad \frac{2}{3} \log_4 z = \log_4 z^{2/3}$

Two other properties of logarithmic functions can be established directly from the equivalence of the expressions $\log_b x = y$ and $x = b^y$.

For any positive real number b, $b \neq 1$, and any real number n, $\log_b b^n = n$.
For any positive real number b, $b \neq 1$, $\log_b 1 = 0$.

The Properties of Logarithms can be used in combination to simplify expressions containing logarithms.

Write $\log_b \frac{xy}{z}$ in expanded form.

Use the Logarithm Property of Quotients. $\log_b \frac{xy}{z} = \log_b(xy) - \log_b z$

Use the Logarithm Property of Products. $= \log_b x + \log_b y - \log_b z$

Write $\log_b \frac{x^2}{y^3}$ in expanded form.

Use the Logarithm Property of Quotients. $\log_b \frac{x^2}{y^3} = \log_b x^2 - \log_b y^3$

Use the Logarithm Property of Powers. $= 2 \log_b x - 3 \log_b y$

Write $2 \log_b x + 4 \log_b y$ as a single logarithm with a coefficient of 1.

Use the Logarithm Property of Powers. $2 \log_b x + 4 \log_b y = \log_b x^2 + \log_b y^4$

Use the Logarithm Property of Products. $= \log_b x^2 y^4$

Example 1

Write $\log_7 xy^2$ in expanded form.

Solution

$\log_7 xy^2 = \log_7 x + \log_7 y^2 = \log_7 x + 2 \log_7 y$

Example 2

Write $\log_6 y^{1/3} z^3$ in expanded form.

Your solution

$\frac{1}{3} \log_6 y + 3 \log_6 z$

Example 3

Write $\log_{10} \sqrt{x^3 y}$ in expanded form.

Solution

$\log_{10} \sqrt{x^3 y} = \log_{10}(x^3 y)^{1/2} = \frac{1}{2} \log_{10} x^3 y$

$= \frac{1}{2}(\log_{10} x^3 + \log_{10} y)$

$= \frac{1}{2}(3 \log_{10} x + \log_{10} y)$

$= \frac{3}{2} \log_{10} x + \frac{1}{2} \log_{10} y$

Example 4

Write $\log_8 \sqrt[3]{xy^2}$ in expanded form.

Your solution

$\frac{1}{3} \log_8 x + \frac{2}{3} \log_8 y$

Solutions on p. A45

Example 5

Write $2 \log_b x - 3 \log_b y - \log_b z$ as a single logarithm with a coefficient of 1.

Solution

$2 \log_b x - 3 \log_b y - \log_b z$
$= \log_b x^2 - \log_b y^3 - \log_b z$
$= \log_b x^2 - (\log_b y^3 + \log_b z)$
$= \log_b x^2 - \log_b y^3 z = \log_b \frac{x^2}{y^3 z}$

Example 6

Write $3 \log_b x + 2 \log_b y - \log_b z$ as a single logarithm with a coefficient of 1.

Your solution

$\log_b \frac{x^3 y^2}{z}$

Example 7

Write $3 \log_5 x + \log_5 y - 2 \log_5 z$ as a single logarithm with a coefficient of 1.

Solution

$3 \log_5 x + \log_5 y - 2 \log_5 z$
$= \log_5 x^3 + \log_5 y - \log_5 z^2$
$= \log_5 x^3 y - \log_5 z^2 = \log_5 \frac{x^3 y}{z^2}$

Example 8

Write $3 \log_4 x + 2 \log_4 y$ as a single logarithm with a coefficient of 1.

Your solution

$\log_4 x^3 y^2$

Example 9

Write $\frac{1}{2}(\log_3 x - 3 \log_3 y + \log_3 z)$ as a single logarithm with a coefficient of 1.

Solution

$\frac{1}{2}(\log_3 x - 3 \log_3 y + \log_3 z)$
$= \frac{1}{2}(\log_3 x - \log_3 y^3 + \log_3 z)$
$= \frac{1}{2}(\log_3 \frac{x}{y^3} + \log_3 z)$
$= \frac{1}{2}(\log_3 \frac{xz}{y^3}) = \log_3 (\frac{xz}{y^3})^{1/2} = \log_3 \sqrt{\frac{xz}{y^3}}$

Example 10

Write $\frac{1}{3}(\log_4 x - 2 \log_4 y + \log_4 z)$ as a single logarithm with a coefficient of 1.

Your solution

$\log_4 \sqrt[3]{\frac{xz}{y^2}}$

Example 11

Find $8 \log_4 4$.

Solution

$8 \log_4 4 = \log_4 4^8$
Since $\log_b b^n = n$, $\log_4 4^8 = 8$.

Example 12

Find $\log_9 1$.

Your solution

0

11.2   **EXERCISES**

▶ **Objective A**

Write the logarithm in expanded form.

1. $\log_8(xz)$
$\log_8 x + \log_8 z$

2. $\log_7(4y)$
$\log_7 4 + \log_7 y$

3. $\log_3 x^5$
$5 \log_3 x$

4. $\log_2 y^7$
$7 \log_2 y$

5. $\log_b\left(\dfrac{r}{s}\right)$
$\log_b r - \log_b s$

6. $\log_c\left(\dfrac{z}{4}\right)$
$\log_c z - \log_c 4$

7. $\log_3(x^2 y^6)$
$2 \log_3 x + 6 \log_3 y$

8. $\log_4(t^4 u^2)$
$4 \log_4 t + 2 \log_4 u$

9. $\log_7\left(\dfrac{u^3}{v^4}\right)$
$3 \log_7 u - 4 \log_7 v$

10. $\log_{10}\left(\dfrac{s^5}{t^2}\right)$
$5 \log_{10} s - 2 \log_{10} t$

11. $\log_2(rs)^2$
$2 \log_2 r + 2 \log_2 s$

12. $\log_3(x^2 y)^3$
$6 \log_3 x + 3 \log_3 y$

13. $\log_9 x^2 yz$
$2 \log_9 x + \log_9 y + \log_9 z$

14. $\log_6 xy^2 z^3$
$\log_6 x + 2 \log_6 y + 3 \log_6 z$

15. $\log_5\left(\dfrac{xy^2}{z^4}\right)$
$\log_5 x + 2 \log_5 y - 4 \log_5 z$

16. $\log_b\left(\dfrac{r^2 s}{t^3}\right)$
$2 \log_b r + \log_b s - 3 \log_b t$

Write the logarithm in expanded form.

17. $\log_8\left(\frac{x^2}{yz^2}\right)$

$2\log_8 x - \log_8 y - 2\log_8 z$

18. $\log_9\left(\frac{x}{y^2z^3}\right)$

$\log_9 x - 2\log_9 y - 3\log_9 z$

19. $\log_7\sqrt{xy}$

$\frac{1}{2}\log_7 x + \frac{1}{2}\log_7 y$

20. $\log_8\sqrt[3]{xz}$

$\frac{1}{3}\log_8 x + \frac{1}{3}\log_8 z$

21. $\log_2\sqrt{\frac{x}{y}}$

$\frac{1}{2}\log_2 x - \frac{1}{2}\log_2 y$

22. $\log_3\sqrt[3]{\frac{r}{s}}$

$\frac{1}{3}\log_3 r - \frac{1}{3}\log_3 s$

23. $\log_4\sqrt{x^3y}$

$\frac{3}{2}\log_4 x + \frac{1}{2}\log_4 y$

24. $\log_3\sqrt{x^5y^3}$

$\frac{5}{2}\log_3 x + \frac{3}{2}\log_3 y$

25. $\log_7\sqrt{\frac{x^3}{y}}$

$\frac{3}{2}\log_7 x - \frac{1}{2}\log_7 y$

26. $\log_b\sqrt[3]{\frac{r^2}{t}}$

$\frac{2}{3}\log_b r - \frac{1}{3}\log_b t$

27. $\log_b x\sqrt{\frac{y}{z}}$

$\log_b x + \frac{1}{2}\log_b y - \frac{1}{2}\log_b z$

28. $\log_4 y\sqrt[3]{\frac{r}{s}}$

$\log_4 y + \frac{1}{3}\log_4 r - \frac{1}{3}\log_4 s$

29. $\log_3\frac{t}{\sqrt{x}}$

$\log_3 t - \frac{1}{2}\log_3 x$

30. $\log_4\left(\frac{\sqrt{uv}}{x}\right)$

$\frac{1}{2}\log_4 u + \frac{1}{2}\log_4 v - \log_4 x$

Express as a single logarithm with a coefficient of 1.

31. $\log_3 x^3 - \log_3 y$

$\log_3 \frac{x^3}{y}$

32. $\log_7 t + \log_7 v^2$

$\log_7 tv^2$

33. $\log_8 x^4 + \log_8 y^2$

$\log_8 x^4 y^2$

34. $\log_2 r^2 + \log_2 s^3$

$\log_2 r^2 s^3$

35. $3 \log_7 x$

$\log_7 x^3$

36. $4 \log_8 y$

$\log_8 y^4$

37. $3 \log_5 x + 4 \log_5 y$

$\log_5 x^3 y^4$

38. $2 \log_6 x + 5 \log_6 y$

$\log_6 x^2 y^5$

39. $-2 \log_4 x$

$\log_4 \frac{1}{x^2}$

40. $-3 \log_2 y$

$\log_2 \frac{1}{y^3}$

41. $2 \log_3 x - \log_3 y + 2 \log_3 z$

$\log_3 \frac{x^2 z^2}{y}$

42. $4 \log_5 r - 3 \log_5 s + \log_5 t$

$\log_5 \frac{r^4 t}{s^3}$

43. $\log_b x - (2 \log_b y + \log_b z)$

$\log_b \frac{x}{y^2 z}$

44. $2 \log_2 x - (3 \log_2 y + \log_2 z)$

$\log_2 \frac{x^2}{y^3 z}$

Express as a single logarithm with a coefficient of 1.

45. $2(\log_4 x + \log_4 y)$

$\log_4 x^2 y^2$

46. $3(\log_5 r + \log_5 t)$

$\log_5 r^3 t^3$

47. $\frac{1}{2}(\log_6 x - \log_6 y)$

$\log_6 \sqrt{\dfrac{x}{y}}$

48. $\frac{1}{3}(\log_8 x - \log_8 y)$

$\log_8 \sqrt[3]{\dfrac{x}{y}}$

49. $2(\log_4 s - 2\log_4 t + \log_4 r)$

$\log_4 \dfrac{s^2 r^2}{t^4}$

50. $3(\log_9 x + 2\log_9 y - 2\log_9 z)$

$\log_9 \dfrac{x^3 y^6}{z^6}$

51. $\log_5 x - 2(\log_5 y + \log_5 z)$

$\log_5 \dfrac{x}{y^2 z^2}$

52. $\log_4 t - 3(\log_4 u + \log_4 v)$

$\log_4 \dfrac{t}{u^3 v^3}$

53. $3\log_2 t - 2(\log_2 r - \log_2 v)$

$\log_2 \dfrac{t^3 v^2}{r^2}$

54. $2\log_{10} x - 3(\log_{10} y - \log_{10} z)$

$\log_{10} \dfrac{x^2 z^3}{y^3}$

55. $\frac{1}{2}(3\log_4 x - 2\log_4 y + \log_4 z)$

$\log_4 \sqrt{\dfrac{x^3 z}{y^2}}$

56. $\frac{1}{3}(4\log_5 t - 3\log_5 u - 3\log_5 v)$

$\log_5 \sqrt[3]{\dfrac{t^4}{u^3 v^3}}$

| SECTION 11.3 | **Computations with Logarithms** |

| Objective A | **To find common logarithms** |

Logarithms to the base 10 are called **common logarithms.** Usually the base, 10, is omitted when writing the common logarithm of a number. For the remainder of this unit, $\log x$ will be used for $\log_{10} x$.

Using the definition of logarithms, the examples at the right show that the logarithm of a power of 10 is the exponent on 10.

$$\log 100 = \log 10^2 \quad = 2$$
$$\log 10 = \log 10^1 \quad = 1$$
$$\log 1 = \log 10^0 \quad = 0$$
$$\log 0.1 = \log 10^{-1} = -1$$
$$\log 0.01 = \log 10^{-2} = -2$$

To find the base 10 logarithm of a number other than a power of 10, it is necessary to use a calculator or a table of logarithms. Since most logarithms are irrational numbers, a calculator or table of logarithms gives approximate values of logarithms. Despite the fact that the logarithms are approximate values, it is customary to use the equals sign (=) rather than the approximately equal sign ($\approx$) when writing logarithms.

The base 10 logarithms of numbers from 1 to 9.99 can be found in the Table of Common Logarithms on pages A4–A5. The logarithms in this table have been rounded to the nearest ten-thousandth. A portion of this table is shown below.

x	0	1	2	3	4	5	6
2.0	.3010	.3032	.3054	.3075	.3096	.3118	.3139
2.1	.3222	.3243	.3263	.3284	.3304	.3324	.3345
2.2	.3424	.3444	.3464	.3483	.3502	.3522	.3541
2.3	.3617	.3636	.3655	.3674	.3692	.3711	.3729
2.4	.3802	.3820	.3838	.3856	.3874	.3892	.3909
2.5	.3979	.3997	.4014	.4031	.4048	.4065	.4082
2.6	.4150	.4166	.4183	.4200	.4216	.4232	.4249

Find log 2.35.

Locate 2.3 under x in the left-hand column of the table.
Move right to the column headed by 5. $\log 2.35 = 0.3711$

Remember that the common logarithm of a number, N, is the power to which 10 must be raised to equal N.

Note from the table above that $\log 2 = 0.3010$. Therefore, $2 = 10^{0.3010}$.

From the example above, $\log 2.35 = 0.3711$. Therefore, $2.35 = 10^{0.3711}$.

To find the common logarithm of a number not in the Table of Common Logarithms, first write the number in scientific notation. Then use the table and the Properties of logarithms to find the logarithm of the number.

Find log 247.

Write 247 in scientific notation.	$\log 247 = \log (2.47 \times 10^2)$
Use the Logarithm Property of Products.	$= \log 2.47 + \log 10^2$
Locate log 2.47 in the table.	$= 0.3927 + \log 10^2$
By the definition of logarithms, the logarithm of a power of 10 is the exponent on 10.	$= 0.3927 + 2$
Add.	$= 2.3927$

Find log 2470.

Write 2470 in scientific notation.	$\log 2470 = \log (2.47 \times 10^3)$
Use the Logarithm Property of Products.	$= \log 2.47 + \log 10^3$
Use the table and the definition of logarithms.	$= 0.3927 + 3$
Add.	$= 3.3927$

Note that for log 247 and log 2470, the decimal part of the logarithm, 0.3927, is the same. The integer part is the exponent on 10 when the number is written in scientific notation.

The decimal part of the logarithm is called the **mantissa.** The integer part is called the **characteristic.**

Find log 0.00247.

Write 0.00247 in scientific notation.	$\log 0.00247 = \log (2.47 \times 10^{-3})$
Use the Logarithm Property of Products.	$= \log 2.47 + 10^{-3}$
Use the table and the definition of logarithms.	$= 0.3927 + (-3)$

In this last example, adding the characteristic to the mantissa would result in a negative logarithm. This form of a logarithm is inconvenient to use with logarithm tables, since these tables contain only positive logarithms. Therefore, when the characteristic of a logarithm is negative, it is customary to leave the logarithm as written or to rewrite the logarithm so that the mantissa remains unchanged.

For example: $\log 0.00247 = 0.3927 + (-3)$
$\log 0.00247 = 0.3927 + (7 - 10) = 7.3927 - 10$
$\log 0.00247 = 0.3927 + (12 - 15) = 12.3927 - 15$

Of these forms, $\log 0.247 = 7.3927 - 10$ is most common.

Find log 0.756.

Write 0.756 in scientific notation.	$\log 0.756 = \log (7.56 \times 10^{-1})$
Use the Logarithm Property of Products.	$= \log 7.56 + \log 10^{-1}$
Use the table and the definition of logarithms.	$= 0.8785 + (-1)$
	$= 0.8785 + (9 - 10)$ Do this step mentally.
	$= 9.8785 - 10$

To find the common logarithm of a number, N, using a calculator, enter N. Then press the log key. The log key on a calculator computes base 10 logarithms.

Example 1

Find log 8360.

Solution

$$\log 8360 = \log (8.36 \times 10^3)$$
$$= \log 8.36 + \log 10^3 = 0.9222 + 3$$
$$= 3.9222$$

Example 2

Find log 93,000.

Your solution

4.9685

Example 3

Find log 0.0217.

Solution

$$\log 0.0217 = \log (2.17 \times 10^{-2})$$
$$= \log 2.17 + \log 10^{-2}$$
$$= 0.3365 + (-2)$$
$$= 8.3365 - 10$$

Example 4

Find log 0.0006.

Your solution

6.7782 − 10

Solutions on p. A46

Objective B   **To find common antilogarithms**

In the last objective, given a number, N, the logarithm of N was found. In this objective, the reverse process is examined. Given the logarithm of N, the number N will be found.

Given log N = 0.3636, find N.

Locate 0.3636 in the body of the Table of Common Logarithms.

Find the number which corresponds to this mantissa.

The number is 2.31.

$$\log 2.31 = 0.3636$$

x	0	1	2	3
2.0	.3010	.3032	.3054	.3075
2.1	.3222	.3243	.3263	.3284
2.2	.3424	.3444	.3464	.3483
2.3	.3617	.3636	.3655	.3674
2.4	.3802	.3820	.3838	.3856
2.5	.3979	.3997	.4014	.4031
2.6	.4150	.4166	.4183	.4200

In the equation log 2.31 = 0.3636, the number 2.31 is called the **antilogarithm** of 0.3636. This is written antilog 0.3636 = 2.31, where the abbreviation antilog is used for antilogarithm.

The common logarithm of a number, N, is the power to which 10 must be raised to equal N.

$\log 2 = 0.3010$ because $2 = 10^{0.3010}$.

The common antilogarithm of a number, N, is the Nth power of 10.

antilog 0.3010 = 2 because $10^{0.3010} = 2$.

Solve $\log N = 1.1206$.

Write 1.1206 as the sum of the mantissa and the characteristic.	$\log N = 1.1206$ $\log N = 0.1206 + 1$

Locate the mantissa in the Table of Common Logarithms and find the number that corresponds to this mantissa.

Since the characteristic is 1, the exponent on 10 is 1 when the number is written in scientific notation. $N = 1.32 \times 10^1$

Simplify. $N = 13.2$

$$\text{Check:} \quad \log 13.2 = \log (1.32 + 10^1)$$
$$= \log 1.32 + 10^1$$
$$= 0.1206 + 1$$
$$= 1.1206$$

Find antilog 3.6946.

antilog 3.6946 =

Write 3.6946 as the sum of the mantissa and the characteristic.   antilog (0.6946 + 3) =   Do this step mentally.

Use the table to find the number that corresponds to the mantissa. The characteristic is the exponent on 10. $4.95 \times 10^3 =$

Simplify.   4950

Find antilog $(7.2601 - 10)$.

antilog $(7.2601 - 10) =$

Write $(7.2601 - 10)$ as the sum of the mantissa and the characteristic.   antilog $[0.2601 + (-3)]$ Do this step mentally.

Use the table to find the number that corresponds to the mantissa. The characteristic is the exponent on 10. $1.82 \times 10^{-3} =$

Simplify.   0.00182

Given $\log N = x$, then antilog $x = N$. Since $\log N = x$ is equivalent to $N = 10^x$ and to antilog $x = N$, 10^x can be substituted for N in the equation antilog $x = N$. The resulting equation is:

antilog $x = 10^x$

This equation is especially useful when using a calculator to find antilogarithms. To find antilog x, enter x. Then press the 10^x key.

Example 5

Find antilog 1.9745.

Solution

antilog $1.9745 = 9.43 \times 10^1 = 94.3$

Example 6

Find antilog 2.3365.

Your solution

217

Example 7

Find antilog (8.7332 − 10).

Solution

antilog $(8.7332 - 10) = 5.41 \times 10^{-2}$
$= 0.0541$

Example 8

Find antilog (9.7846 − 10).

Your solution

0.609

Solutions on p. A46

Objective C   **To use interpolation to find a common logarithm or antilogarithm**

Since 3.257 is not included in the Table of Common Logarithms on pages A4–A5, it is not possible to find log 3.257 directly from the table. However, using the first number less than 3.257 which can be found in the table (3.250) and the first number greater than 3.257 (3.260), it is possible to obtain an approximation for log 3.257 by using linear interpolation. Linear interpolation is based on the fact that many functions can be approximated, over small intervals, by a straight line.

The graph of the equation $y = \log x$ is shown at the right. The units on the *x*- and *y*-axes have been distorted to illustrate linear interpolation.

When $x = 3.250$, $y = 0.5119$. When $x = 3.260$, $y = 0.5132$. When $x = 3.257$, the value of y is greater than 0.5119 and less than 0.5132. The actual value of log 3.257 is on the curve. An approximation for this value is on the straight line.

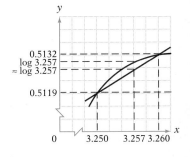

A proportion can be written using the fact that between any two points on a straight line, the slope is the same.

$$\frac{y_2 - y_1}{x_2 - x_1} = \frac{y - y_1}{x - x_1}$$

Let $(x_1, y_1) = (3.250, 0.5119)$
and $(x_2, y_2) = (3.260, 0.5132)$.
(x, y) is the point where $x = 3.257$.

Solve the proportion for *y*.

$$\frac{0.5132 - 0.5119}{3.260 - 3.250} = \frac{y - 0.5119}{3.257 - 3.250}$$

$$\frac{0.0013}{0.01} = \frac{y - 0.5119}{0.007}$$

$$0.00091 = y - 0.5119$$
$$0.51281 = y$$

$\log 3.257 \approx 0.5128$, rounded to the nearest ten-thousandth.

A more convenient method of using linear interpolation is illustrated in the following example.

Find log 32.57.

$$\log 32.57 = \log (3.257 \times 10^1)$$
$$= \log 3.257 + \log 10^1$$

Arrange the numbers 3.250, 3.257, and 3.260 and their corresponding mantissas in a table. Indicate the differences between the numbers and the mantissas, as shown at the right, using d to represent the unknown difference between the mantissa of 3.250 and the mantissa of 3.257.

	Number	Mantissa	
	3.250	0.5119	
0.01 ⌐ 0.007	3.257	0.5119 + d	d 0.0013
	3.260	0.5132	

Write a proportion.

$$\frac{0.007}{0.01} = \frac{d}{0.0013}$$

Solve for d, rounding to the nearest ten-thousandth.

$$0.0009 = d$$

Add the value of d to the mantissa 0.5119.

$$\log 3.257 + \log 10^1 = (0.5119 + d) + 1$$
$$= (0.5119 + 0.0009) + 1$$
$$= 0.5128 + 1$$
$$= 1.5128$$

Linear interpolation can also be used to approximate an antilogarithm.

Find antilog (7.3431 − 10).

Use the table to find the first mantissa less than 0.3431 and the first mantissa greater than 0.3431. Arrange the mantissas and their corresponding antilogs in a table. Indicate the differences between the mantissas and the antilogs, using d to represent the unknown difference.

	Mantissa	Antilog	
	0.3424	2.200	
0.002 ⌐ 0.0007	0.3431	2.200 + d	d 0.01
	0.3444	2.210	

Write a proportion.

$$\frac{0.0007}{0.002} = \frac{d}{0.01}$$

Solve for d, rounding to the nearest thousandth.

$$0.004 = d$$

Add the value of d to the antilog 2.20.

$$\text{antilog } (7.3431 - 10) = (2.20 + d) \times 10^{-3}$$
$$= (2.20 + 0.004) \times 10^{-3}$$
$$= 2.204 \times 10^{-3}$$
$$= 0.002204$$

Example 9

Find log 0.02903.

Solution

$$\log 0.02903 = \log (2.903 \times 10^{-2})$$
$$= \log 2.903 + \log 10^{-2}$$

Number   Mantissa

2.900   0.4624
0.003 d
0.01   2.903   0.4624 + d 0.0015
2.910   0.4639

$$\frac{0.003}{0.01} = \frac{d}{0.0015}$$
$$0.0005 = d$$

$$\log 2.903 + \log 10^{-2} = (0.4624 + d) + (-2)$$
$$= (0.4624 + 0.0005) + (-2)$$
$$= 0.4629 + (-2)$$
$$= 8.4629 - 10$$

Example 10

Find log 31,560.

Your solution

4.4991

Example 11

Find antilog (9.8465 − 10).

Solution

Mantissa   Antilog

0.8463   7.02
0.0002 d
0.0007   0.8465   7.02 + d 0.01
0.8470   7.03

$$\frac{0.0002}{0.0007} = \frac{d}{0.01}$$
$$0.003 = d$$

$$\text{antilog } (9.8465 - 10) = (7.02 + d) \times 10^{-1}$$
$$= (7.02 + 0.003) \times 10^{-1}$$
$$= 7.023 \times 10^{-1} = 0.7023$$

Example 12

Find antilog 1.4907.

Your solution

30.95

Solutions on p. A46

Objective D   **To evaluate numerical expressions by using common logarithms**

11

Logarithms were developed in the 17th century as an aid in scientific calculations. Today, with the widespread use of calculators, the necessity of using logarithms for computations has been minimized. However, logarithms do play an important role in the analysis of economic and scientific problems.

To solve a numerical problem by using logarithms, use the fact that if $x = y$, then $\log x = \log y$.

Evaluate $\frac{(48.6)(9.37)}{803}$ by using common logarithms.

Write an equation.

$$N = \frac{(48.6)(9.37)}{803}$$

Take the common logarithm of each side of the equation.

$$\log N = \log \frac{(48.6)(9.37)}{803}$$

Rewrite the equation using the Properties of Logarithms and the table on pages A4-A5.

$$\begin{aligned} &= \log 48.6 + \log 9.37 - \log 803 \\ &= 1.6866 + 0.9717 - 2.9047 \\ &= 9.7536 - 10 \end{aligned}$$

Find the antilogarithm. Use interpolation if necessary.

$$\begin{aligned} N &= \text{antilog } (9.7536 - 10) \\ &= 5.67 + 10^{-1} \\ &= 0.567 \end{aligned}$$

Evaluate $\sqrt[3]{0.179}$ by using common logarithms.

Write an equation.

$$N = \sqrt[3]{0.179} = 0.179^{1/3}$$

Take the common logarithm of each side of the equation.

$$\log N = \log 0.179^{1/3}$$
$$= \tfrac{1}{3}\log 0.179$$
$$= \tfrac{1}{3}[0.2529 + (-1)]$$

The characteristic is written so that the product results in an integer characteristic.

$$= \tfrac{1}{3}(2.2529 - 3)$$
$$= 0.7510 - 1$$

Find the antilogarithm. Use interpolation if necessary.

$$\begin{aligned} N &= \text{antilog } (0.7510 - 1) \\ &= 5.636 \times 10^{-1} \\ &= 0.5636 \end{aligned}$$

Example 13

Evaluate $\sqrt{(46.1)(52.3)}$ by using common logarithms.

Example 14

Evaluate $\sqrt[3]{\dfrac{14.1}{3.7}}$ by using common logarithms.

Solution

$N = \sqrt{(46.1)(52.3)} = [(46.1)(52.3)]^{1/2}$

$\log N = \log [(46.1)(52.3)]^{1/2}$

$ = \tfrac{1}{2} \log (46.1)(52.3)$

$ = \tfrac{1}{2} (\log 46.1 + \log 52.3)$

$ = \tfrac{1}{2} (1.6637 + 1.7185)$

$ = \tfrac{1}{2} (3.3822) = 1.6911$

$N = \text{antilog } 1.6911 = 4.91 \times 10^1 = 49.1$

Your solution

1.562

Solution on p. A46

11.3 EXERCISES

▶ **Objective A**

Find the logarithm.

1. 5.87
0.7686

2. 6.42
0.8075

3. 34.5
1.5378

4. 10.9
1.0374

5. 389
2.5899

6. 592
2.7723

7. 1030
3.0128

8. 9060
3.9571

9. 47,300
4.6749

10. 68,200
4.8338

11. 0.114
9.0569 − 10

12. 0.209
9.3201 − 10

13. 0.002
7.3010 − 10

14. 0.034
8.5315 − 10

15. 0.00175
7.2430 − 10

16. 0.0806
8.9063 − 10

17. 0.984
9.9930 − 10

18. 0.79
9.8976 − 10

▶ **Objective B**

Find the antilogarithm.

19. 0.7396
5.49

20. 0.6474
4.44

21. 1.5353
34.3

22. 1.4829
30.4

23. 3.9015
7970

24. 3.9557
9030

25. 2.8007
632

26. 2.0128
103

27. 9.6580 − 10
0.455

28. 9.7033 − 10
0.505

29. 8.3892 − 10
0.0245

30. 8.8344 − 10
0.0683

31. 7.8704 − 10
0.00742

32. 7.9791 − 10
0.00953

33. 1.9025
79.9

34. 2.6767
475

35. 0.1492 + (−1)
0.141

36. 0.6920 + (−1)
0.492

▶ Objective C

Find the logarithm.

37. 3.845 0.5849	**38.** 4.965 0.6960	**39.** 84.52 1.9270	**40.** 67.38 1.8286
41. 499.3 2.6984	**42.** 699.6 2.8449	**43.** 0.07071 8.8495 − 10	**44.** 0.01453 8.1623 − 10
45. 75,930 4.8804	**46.** 83,090 4.9196	**47.** 0.8407 9.9247 − 10	**48.** 0.9813 9.9918 − 10

Find the antilogarithm.

49. 0.9006 7.954	**50.** 0.8466 7.024	**51.** 2.7360 544.5	**52.** 2.5375 344.8
53. 9.7408 − 10 0.5505	**54.** 9.7320 − 10 0.5395	**55.** 2.6003 398.4	**56.** 2.7004 501.7
57. 8.3088 − 10 0.02036	**58.** 8.5372 − 10 0.03445	**59.** 1.6525 44.93	**60.** 1.6809 47.97

▶ Objective D

Evaluate the expression using common logarithms. Interpolate if necessary.

61. $(16)(19.8)$
316.8

62. $(5.47)(81.7)$
446.9

63. $(0.45)(0.611)$
0.2749

64. $(0.98)(0.0153)$
0.01499

65. $\frac{79.3}{26.9}$
2.948

66. $\frac{40.9}{36.3}$
1.127

67. $\frac{(57)(31.4)}{107}$
16.73

68. $\frac{(68)(12.3)}{217}$
3.854

69. $(8.7)^3$
658.4

70. $(9.03)^4$
6650

71. $\frac{\sqrt{19.8 \cdot 52}}{7.09}$
32.65

72. $\frac{\sqrt[3]{145} \cdot 7.62}{17.3}$
2.315

73. $\sqrt{(48.5)(3.13)}$
12.32

74. $\sqrt[3]{(70.7)(91.4)}$
18.63

75. $\left(\frac{83}{257}\right)^4$
0.01089

76. $\sqrt[3]{0.688}$
0.8829

77. $\sqrt[3]{0.719}$
0.8958

78. $\sqrt{\frac{76.3}{89.7}}$
0.9224

SECTION 11.4

Solving Exponential and Logarithmic Equations

Objective A

To solve an exponential equation

An **exponential equation** is one in which the variable occurs in the exponent. The examples at the right are exponential equations.

$$6^{2x+1} = 6^{3x-2}$$

$$4^x = 3$$

$$2^{x+1} = 7$$

An exponential equation in which each side of the equation can be expressed in terms of the same base can be solved by using the property that the exponential function is one-to-one.

Recall the one-to-one property of an exponential function. If $b^x = b^y$, then $x = y$.

Solve: $10^{3x+5} = 10^{x-3}$

Use the one-to-one property of the exponential function to equate the exponents.

$$10^{3x+5} = 10^{x-3}$$
$$3x + 5 = x - 3$$

Solve the resulting equation.

$$2x + 5 = -3$$
$$2x = -8$$
$$x = -4$$

The solution is -4.

Check:
$$10^{3x+5} = 10^{x-3}$$

$10^{3(-4)+5}$	10^{-4-3}
10^{-12+5}	10^{-7}

$$10^{-7} = 10^{-7}$$

Solve: $9^{x+1} = 27^{x-1}$

Rewrite each side of the equation using the same base.
$9 = 3^2$, $27 = 3^3$.

$$9^{x+1} = 27^{x-1}$$

$$(3^2)^{x+1} = (3^3)^{x-1}$$
$$3^{2x+2} = 3^{3x-3}$$

Use the one-to-one property of the exponential function to equate the exponents.

$$2x + 2 = 3x - 3$$
$$2 = x - 3$$

Solve the resulting equation.

$$5 = x$$

The solution is 5.

Check:
$$9^{x+1} = 27^{x-1}$$

9^{5+1}	27^{5-1}
9^6	27^4

$$531{,}441 = 531{,}441$$

When each side of an exponential equation cannot easily be expressed in terms of the same base, logarithms are used to solve the exponential equation.

Solve: $4^x = 7$

Take the common logarithm of each side of the equation.	$4^x = 7$ $\log 4^x = \log 7$
Rewrite the equation using the Properties of Logarithms.	$x \log 4 = \log 7$
Solve for x.	$x = \dfrac{\log 7}{\log 4} = \dfrac{0.8451}{0.6021} = 1.4036$

Note that $\dfrac{\log 7}{\log 4} \neq \log 7 - \log 4$.

The solution is 1.4036.

Solve: $3^{x+1} = 5$

Take the common logarithm of each side of the equation.	$3^{x+1} = 5$ $\log 3^{x+1} = \log 5$
Rewrite the equation using the Properties of Logarithms.	$(x + 1)\log 3 = \log 5$
Solve for x.	$x + 1 = \dfrac{\log 5}{\log 3} = \dfrac{0.6990}{0.4771}$ $x + 1 = 1.4651$ $x = 0.4651$

The solution is 0.4651.

Example 1   Solve for x: $3^{2x} = 4$

Solution
$$3^{2x} = 4$$
$$\log 3^{2x} = \log 4$$
$$2x \log 3 = \log 4$$
$$2x = \frac{\log 4}{\log 3} = \frac{0.6021}{0.4771}$$
$$2x = 1.2620$$
$$x = 0.6310$$

The solution is 0.6310.

Example 2   Solve for x: $4^{3x} = 25$

Your solution   0.7739

Example 3   Solve for n: $(1.1)^n = 2$

Solution
$$(1.1)^n = 2$$
$$\log (1.1)^n = \log 2$$
$$n \log 1.1 = \log 2$$
$$n = \frac{\log 2}{\log 1.1} = \frac{0.3010}{0.0414}$$
$$n = 7.2705$$

The solution is 7.2705.

Example 4   Solve for n: $(1.06)^n = 1.5$

Your solution   6.9605

Solutions on p. A47

| **Objective B** | **To solve a logarithmic equation** |

A logarithmic equation can be solved by using the Properties of Logarithms.

Solve: $\log_9 x + \log_9 (x - 8) = 1$

Use the Logarithm Property of Products to rewrite the left side of the equation.

$$\log_9 x + \log_9(x - 8) = 1$$
$$\log_9 x(x - 8) = 1$$

Write the equation in exponential form.

$$9^1 = x(x - 8)$$

Simplify.

$$9 = x^2 - 8x$$

Solve for x.

$$0 = x^2 - 8x - 9$$

Factor and use the Principle of Zero Products.

$$0 = (x - 9)(x + 1)$$

$$x - 9 = 0 \qquad x + 1 = 0$$
$$x = 9 \qquad x = -1$$

Replacing x by 9 in the original equation, 9 checks as a solution. Replacing x by -1 in the original equation results in the expression $\log_9(-1)$. Since the logarithm of a negative number is not a real number, -1 does not check as a solution. Therefore, the solution of the equation is 9.

Example 5 Solve for x: $\log_3(2x - 1) = 2$

Solution $\log_3(2x - 1) = 2$

Rewrite in exponential form.
$$3^2 = 2x - 1$$
$$9 = 2x - 1$$
$$10 = 2x$$
$$5 = x$$

The solution is 5.

Example 6 Solve for x: $\log_4(x^2 - 3x) = 1$

Your solution -1 and 4

Example 7 Solve for x:
$\log_2 x - \log_2(x - 1) = \log_2 2$

Solution $\log_2 x - \log_2(x - 1) = \log_2 2$

$$\log_2\left(\frac{x}{x - 1}\right) = \log_2 2$$

Use the fact that if $\log_b u = \log_b v$, then $u = v$.

$$\frac{x}{x - 1} = 2$$
$$(x - 1)\left(\frac{x}{x - 1}\right) = (x - 1)2$$
$$x = 2x - 2$$
$$-x = -2$$
$$x = 2$$

The solution is 2.

Example 8 Solve for x:
$\log_3 x + \log_3(x + 3) = \log_3 4$

Your solution 1

| Objective C | **To find the logarithm of a number other than base 10** |

Once a table of logarithms for a given base has been developed, it is possible to find the logarithm of a number to any other positive base other than 1.

Since the base 10 logarithms are given on pages A4-A5, this table will be used to find $\log_b x$ for any base b, $b > 0$, $b \ne 1$.

Find $\log_2 7$.

Write an equation. $\log_2 7 = x$

Write in exponential form. $2^x = 7$

Solve the exponential equation using common logarithms.

$$\log 2^x = \log 7$$
$$x\log 2 = \log 7$$
$$x = \frac{\log 7}{\log 2} = \frac{0.8451}{0.3010} = 2.8076$$
$$\log_2 7 = 2.8076$$

Using a procedure similar to the one shown above, a formula can be written that makes it possible to use any given table of logarithms to find the logarithm of a number with a different base.

The Change of Base Formula

$$\log_a N = \frac{\log_b N}{\log_b a}$$

Find $\log_e 5$, where $e = 2.72$.

Use the Change of Base Formula.
$N = 5$, $a = e$, and $b = 10$.

$$\log_e 5 = \frac{\log_{10} 5}{\log_{10} e} = \frac{\log_{10} 5}{\log_{10} 2.72} = \frac{0.6990}{0.4346} = 1.6084$$

The logarithm base e used in the last example is frequently used in scientific and economic applications of mathematics. When e is used as a base of the logarithm, the logarithm is referred to as the **natural logarithm** and is abbreviated ln x. The number e is an irrational number and is approximately equal to 2.718282183. To find the natural logarithm of a number, N, using a calculator, enter N. Then press the ln key.

Example 9   Find $\log_3 12$.

Solution $\log_3 12 = \frac{\log_{10} 12}{\log_{10} 3}$
$$= \frac{1.0792}{0.4771}$$
$$= 2.2620$$

Example 10   Find $\log_9 23$.

Your solution   1.4271

Solution on p. A47

11.4   EXERCISES

▶ **Objective A**

Solve for x. Round to the nearest ten-thousandth.

1. $5^{4x-1} = 5^{x-2}$

$-\dfrac{1}{3}$

2. $7^{4x-3} = 7^{2x+1}$

2

3. $8^{x-4} = 8^{5x+8}$

-3

4. $10^{4x-5} = 10^{x+4}$

3

5. $5^x = 6$
1.1133

6. $7^x = 10$
1.1833

7. $12^x = 6$
0.7211

8. $10^x = 5$
0.6990

9. $\left(\dfrac{1}{2}\right)^x = 3$

-1.5850

10. $\left(\dfrac{1}{3}\right)^x = 2$

-0.6309

11. $(1.5)^x = 2$
1.7093

12. $(2.7)^x = 3$
1.1059

13. $10^x = 21$
1.3222

14. $10^x = 37$
1.5682

15. $2^{-x} = 7$
-2.8076

16. $3^{-x} = 14$
-2.4022

17. $2^{x-1} = 6$
3.5854

18. $4^{x+1} = 9$
0.5848

19. $3^{2x-1} = 4$
1.1310

20. $4^{-x+2} = 12$
0.2076

21. $9^x = 3^{x+1}$
1

22. $2^{x-1} = 4^x$
-1

23. $8^{x+2} = 16^x$
6

24. $9^{3x} = 81^{x-4}$
-8

25. $16^{2-x} = 32^{2x}$

$\dfrac{4}{7}$

26. $27^{2x-3} = 81^{4-x}$

$\dfrac{5}{2}$

27. $25^{3-x} = 125^{2x-1}$

$\dfrac{9}{8}$

28. $8^{4x-7} = 64^{x-3}$

$\dfrac{1}{2}$

▶ **Objective B**

Solve for x.

29. $\log_3(x + 1) = 2$
8

30. $\log_5(x - 1) = 1$
6

31. $\log_2(2x - 3) = 3$

$\dfrac{11}{2}$

32. $\log_4(3x + 1) = 2$
5

33. $\log_2(x^2 + 2x) = 3$
2 and -4

34. $\log_3(x^2 + 6x) = 3$
-9 and 3

35. $\log_5\left(\dfrac{2x}{x-1}\right) = 1$

$\dfrac{5}{3}$

36. $\log_6\left(\dfrac{3x}{x+1}\right) = 1$

-2

37. $\log x = \log(1 - x)$

$\dfrac{1}{2}$

Solve for x.

38. $\log_3 x + \log_3(x - 1) = \log_3 6$
3

39. $\log_4 x + \log_4(x - 2) = \log_4 15$
5

40. $\log_2 8x - \log_2(x^2 - 1) = \log_2 3$
3

41. $\log_5 3x - \log_5(x^2 - 1) = \log_5 2$
2

42. $\log_9 x + \log_9(2x - 3) = \log_9 2$
2

43. $\log_6 x + \log_6(3x - 5) = \log_6 2$
2

44. $\log_8 6x = \log_8 2 + \log_8(x - 4)$
No solution

45. $\log_7 5x = \log_7 3 + \log_7(2x + 1)$
No solution

46. $\log_9 7x = \log_9 2 + \log_9(x^2 - 2)$
4

47. $\log_3 x = \log_3 2 + \log_3(x^2 - 3)$
2

▶ **Objective C**

Find the logarithm.

48. $\log_3 7$
1.7713

49. $\log_2 9$
3.1701

50. $\log_5 12$
1.5439

51. $\log_5 10$
1.4306

52. $\log_8 4$
0.6667

53. $\log_7 6$
0.9208

54. $\log_3 15$
2.4651

55. $\log_{12} 9$
0.8842

56. $\log_8 6$
0.8617

57. $\log_4 8$
1.4999

58. $\log_5 30$
2.1132

59. $\log_6 28$
1.8597

60. $\log_3 8.6$
1.9587

61. $\log_5 9.2$
1.3788

62. $\log_9 4$
0.6310

63. $\log_8 5$
0.7740

64. $\log_3(0.5)$
−0.6309

65. $\log_5(0.6)$
−0.3173

66. $\log_7(1.7)$
0.2726

67. $\log_6(2.3)$
0.4648

68. $\log_7(1.4)$
0.1729

| SECTION 11.5 | Applications of Exponential and Logarithmic Functions: A Calculator Approach |

| Objective A | **To solve application problems** | |

Solving problems that involve exponential or logarithmic equations can be quite tedious without a calculator. Each of the problems in this section was solved by using a calculator.

A biologist places one single-celled bacterium in a culture and each hour that particular species of bacteria divides into two bacteria. After one hour there will be two bacteria. After two hours, each of the two bacteria will divide and there will be four bacteria. After three hours, each of the four bacteria will divide and there will be eight bacteria.

The table at the right shows the number of bacteria in the culture after various intervals of time, t, in hours. Values in this table could also be found by using the exponential equation $N = 2^t$.

Time, t	Number of bacteria, N
0	1
1	2
2	4
3	8
4	16

The equation $N = 2^t$ is an example of an **exponential growth equation.** In general, any equation that can be written in the form $A = A_0 b^{kt}$, where A is the size at time t, A_0 is the initial size, $b > 1$, and k is a positive real number, is an exponential growth equation. These equations are important not only in population growth studies, but in physics, chemistry, psychology, and economics.

Recall from Chapter 2 that interest is the amount of money paid or received when borrowing or investing money. **Compound interest** is interest that is computed not only on the original principal, but also on the interest already earned. The compound interest formula is an exponential equation.

The **compound interest formula** is $P = A(1 + i)^n$, where A is the original value of an investment, i is the interest rate per compounding period, n is the total number of compounding periods, and P is the value of the investment after n periods.

An investment broker deposits \$1000 into an account which earns 12% annual interest compounded quarterly. What is the value of the investment after two years?

Find i, the interest rate per quarter. The quarterly rate is the annual rate divided by 4, the number of quarters in one year.

$$i = \frac{12\%}{4} = \frac{0.12}{4} = 0.03$$

Find n, the number of compounding periods. The investment is compounded quarterly, 4 times a year, for 2 years.

$$n = 4 \cdot 2 = 8$$

Use the compound interest formula.

$$P = A(1 + i)^n$$

Replace A, i, and n by their values.

$$P = 1000(1 + 0.03)^8$$

Solve for P.

$$P \approx 1267$$

The value of the investment after two years is \$1267.

Exponential decay is another important example of an exponential equation. One of the most common illustrations is the decay of a radioactive substance.

An isotope of cobalt has a half-life of approximately 5 years. This means that one half of any given amount of a cobalt isotope will disintegrate in 5 years.

The table at the right indicates the amount of the initial 10 mg of a cobalt isotope that remains after various intervals of time, t, in years. Values in this table could also be found by using the exponential equation $A = 10\left(\frac{1}{2}\right)^{t/5}$.

Time, t	Amount, A
0	10
5	5
10	2.5
15	1.25
20	0.625

The equation $A = 10\left(\frac{1}{2}\right)^{t/5}$ is an example of an **exponential decay equation.** Comparing this equation to the equation on exponential growth, note that for exponential growth, the base of the exponential equation is greater than 1, while for exponential decay, the base is between 0 and 1.

A method by which an archeologist can measure the age of a bone is called **carbon dating.** Carbon dating is based on a radioactive isotope of carbon called carbon 14, which has a half-life of approximately 5570 years. The exponential decay equation is given by $A = A_0\left(\frac{1}{2}\right)^{t/5570}$, where A_0 is the original amount of carbon 14 present in the bone, t is the age of the bone, and A is the amount present after t years.

A bone that originally contained 100 mg of carbon 14 now has 70 mg of carbon 14. What is the appproximate age of the bone?

Use the exponential decay equation.

$$A = A_0\left(\frac{1}{2}\right)^{t/5570}$$

Replace A_0 and A by their given values and solve for t.

$$70 = 100\left(\frac{1}{2}\right)^{t/5570}$$
$$70 = 100(0.5)^{t/5570}$$

Divide each side of the equation by 100.

$$0.7 = (0.5)^{t/5570}$$

$$\log 0.7 = \log (0.5)^{t/5570}$$

Take the common logarithm of each side of the equation. Then simplify.

$$\log 0.7 = \frac{t}{5570} \log 0.5$$

$$\frac{5570 \log 0.7}{\log 0.5} = t$$

$$2866 = t$$

The age of the bone is approximately 2866 years.

A chemist measures the acidity or alkalinity of a solution by the concentration of hydrogen ions, H^+, in the solution by the formula **pH $= -$log (H^+)**. A neutral solution such as distilled water has a pH of 7, acids have a pH of less than 7, and alkaline solutions (also called basic solutions) have a pH of greater than 7.

Content and Format © 1991 HMCo.

Find the pH of vinegar for which $H^+ = 1.26 \times 10^{-3}$. Round to the nearest tenth.

Use the pH equation. $H^+ = 1.26 \times 10^{-3}$.

$$pH = -\log (H^+)$$
$$= -\log (1.26 \times 10^{-3})$$
$$= -(\log 1.26 + \log 10^{-3})$$
$$= -[0.1004 + (-3)] = 2.8996$$

The pH of vinegar is 2.9.

The **Richter scale** measures the magnitude, M, of an earthquake in terms of the intensity, I, of its shock waves. This can be expressed as the logarithmic equation $M = \log \frac{I}{I_0}$, where I_0 is a constant.

How many times stronger is an earthquake that has magnitude 4 on the Richter scale than one that has magnitude 2 on the scale?

Let I_1 represent the intensity of the earthquake which has magnitude 4 and I_2 represent the intensity of the earthquake that has magnitude 2. The ratio of I_1 to I_2, $\frac{I_1}{I_2}$, measures how much stronger I_1 is than I_2.

$$4 = \log \frac{I_1}{I_0}$$
$$2 = \log \frac{I_2}{I_0}$$

Use the Richter equation to write a system of equations, one equation for magnitude 4 and one for magnitude 2. Then rewrite the system using the Properties of Logarithms.

$$4 = \log I_1 - \log I_0$$
$$2 = \log I_2 - \log I_0$$

Use the addition method to eliminate $\log I_0$.

$$2 = \log I_1 - \log I_2$$

Rewrite the equations using the Properties of Logarithms.

$$2 = \log \frac{I_1}{I_2}$$

Solve for the ratio using the relationship between logarithms and exponents.

$$\frac{I_1}{I_2} = 10^2 = 100$$

An earthquake that has magnitude 4 on the Richter scale is 100 times stronger than an earthquake that has magnitude 2.

The percent of light that will pass through a substance is given by the equation **$\log P = -kd$**, where P is the percent of light passing through the substance, k is a constant which depends upon the substance, and d is the thickness of the substance.

Find the percent of light that will pass through opaque glass for which $k = 0.4$ and d is 0.5 cm.

Replace k and d in the equation by their given values and solve for P.

$$\log P = -kd$$
$$\log P = -(0.4)(0.5)$$
$$\log P = -0.2$$

Use the relationship between the logarithmic and exponential function.

$$P = 10^{-0.2}$$
$$P = 0.6310$$

Approximately 63.1% of the light will pass through the glass.

Example 1

An investment of $3000 is placed into an account which earns 12% annual interest compounded monthly. In approximately how many years will the investment be worth twice the original amount?

Strategy

To find the time, solve the compound interest formula for n. Use $P = 6000$, $A = 3000$, and $i = \frac{12\%}{12} = \frac{0.12}{12} = 0.01$.

Solution

$$P = A(1 + i)^n$$
$$6000 = 3000(1 + 0.01)^n$$
$$6000 = 3000(1.01)^n$$
$$2 = (1.01)^n$$
$$\log 2 = \log (1.01)^n$$
$$\log 2 = n \log 1.01$$
$$\frac{\log 2}{\log 1.01} = n$$
$$70 = n$$

70 months ÷ 12 ≈ 5.8 years

In approximately 6 years, the investment will be worth $6000.

Example 2

Find the pH of sodium carbonate for which $H^+ = 2.51 \times 10^{-12}$. Round to the nearest tenth.

Your strategy

Your solution

11.6

Example 3

The number of words per minute a student can type will increase with practice and can be approximated by the equation $N = 100[1 - (0.9)^t]$, where N is the number of words typed per minute after t days of instruction. Find the number of words a student will type per minute after 8 days of instruction.

Strategy

To find the number of words per minute, replace t by its given value in the equation and solve for N.

Solution

$N = 100[1 - (0.9)^t] = 100[1 - (0.9)^8]$
$\quad = 56.95$

After 8 days of instruction, a student will type approximately 57 words per minute.

Example 4

An earthquake that measures 5 on the Richter scale can cause serious damage to buildings. The San Francisco earthquake of 1906 would have measured about 7.8 on this scale. How many times stronger was the San Francisco earthquake than one that can cause serious damage? Round to the nearest whole number.

Your strategy

Your solution

631 times stronger

Solutions on p. A47

11.5   EXERCISES

▶ Objective A   *Application Problems*

Solve. Round to the nearest whole number.

Use the compound interest formula $P = A(1 + i)^n$, where A is the original value of an investment, i is the interest rate per compounding period, n is the total number of compounding periods, and P is the value of the investment after n periods.

1. An investment club deposits $5000 into an account that earns 9% annual interest compounded monthly. What is the value of the investment after two years?
$5982

2. An investment advisor deposits $8000 into an account that earns 8% annual interest compounded daily. What is the value of the investment after one year? (1 year = 365 days.)
$8666

3. An investor deposits $12,000 into an account that earns 10% annual interest compounded semi-annually. In approximately how many years will the investment double?
7 years

4. An insurance broker deposits $4000 into an account that earns 7% annual interest compounded monthly. In approximately how many years will the investment be worth $8000?
10 years

5. The shop foreman for a company estimates that in 4 years the company will need to purchase a new bottling machine at a cost of $25,000. How much money must be deposited in an account that earns 10% annual interest compounded monthly so that the value of the account in four years will be $25,000?
$16,786

6. The comptroller of a company has determined that it will be necessary to purchase a new computer in three years. The estimated cost of the computer is $10,000. How much money must be deposited in an account that earns 9% annual interest compounded quarterly so that the value of the account in three years will be $10,000?
$7657

Use the exponential decay equation $A = A_0 \left(\frac{1}{2}\right)^{t/k}$, where A is the amount of a radioactive material present after a time t, k is the half-life, and A_0 is the original amount of radioactive material.

7. An isotope of carbon has a half-life of approximately 1600 years. How long will it take an original sample of 15 mg of this isotope to decay to 10 mg?

936 years

8. An isotope has a half-life of 80 days. How many days are required for a 10 mg sample of this isotope to decay to 1 mg?

266 days

9. A laboratory assistant measures the amount of a radioactive material as 15 mg. Five hours later a second measurement shows that there are 12 mg of the material remaining. Find the half-life of this radioactive material.

16 hours

10. A scientist measured the amount of a radioactive substance as 10 mg. Twenty-four hours later, another measurement showed that there were 8 mg remaining. What is the half-life of this substance?

75 hours

The percent of correct welds a student welder can make will increase with practice and can be approximated by the equation $P = 100[1 - (0.75)^t]$, where P is the percent of correct welds and t is the number of weeks of practice.

11. How many weeks of practice are necessary before a student will make 80% of the welds correctly?

    6 weeks

12. Find the percent of correct welds a student will make after four weeks of practice.

    68%

Use the pH equation $pH = -\log (H^+)$, where H^+ is the hydrogen ion concentration of a solution.

13. Find the pH of a sodium hydroxide solution for which the hydrogen ion concentration is 7.5×10^{-9}.

    8.1

14. Find the pH of a hydrogen chloride solution for which the hydrogen ion concentration is 2.4×10^{-3}.

    2.6

The percent of light that will pass through a material is given by the equation $\log P = -kd$, where P is the percent of light passing through the material, k is a constant which depends upon the material, and d is the thickness of the material in centimeters.

15. The constant k for a piece of opaque glass which is 0.5 cm thick is 0.2. Find the percent of light that will pass through the glass.

    79%

16. The constant k for a piece of tinted glass is 0.5. How thick a piece of this glass is needed so that 60% of the light incident to the glass will pass through it?

    0.44 cm

Use the Richter equation $M = \log \frac{I}{I_0}$, where M is the magnitude of an earthquake, I is the intensity of its shock waves, and I_0 is a constant.

17. An earthquake in China in 1976 measured 8.2 on the Richter scale and resulted in a death toll of 750,000. The greatest physical damage from an earthquake resulted from the 1988 earthquake in Armenia which measured 6.9 on the Richter scale. How much stronger was China's earthquake?
     20 times

18. How many times stronger is an earthquake of magnitude 7.2 on the Richter scale than one of magnitude 3.2 on the scale?
     10,000 times

The number of decibels, D, of a sound can be given by the equation $D = 10(\log I + 16)$, where I is the power of the sound measured in watts.

19. Find the number of decibels of normal conservation. The power of the sound of normal conservation is approximately 3.2×10^{-10} watts.
     65 decibels

20. The loudest sound of any animal is made by the blue whale and can be heard over 500 miles away. The power of the sound made by the blue whale is 630 watts. Find the number of decibels from the blue whale.
     188 decibels

Calculators and Computers

The $\boxed{10^x}$ and $\boxed{\log}$ Keys on a Calculator

The $\boxed{\log}$ key on a calculator gives the logarithm base 10 of a number. For this discussion, $\log x$ will mean $\log_{10} x$. The $\boxed{10^x}$ key gives the antilogarithm of x.

Here are some examples along with the key strokes.

Find log 18.   Enter: 18 $\boxed{\log}$
                    The answer is 1.2552725.

Find log 0.0235.   Enter: 0.0235 $\boxed{\log}$
                    The answer is -1.62893.

Find antilog 1.23514.   Enter: 1.23514 $\boxed{10^x}$
                        The answer is 17.18462.

Find antilog (-2.351).   Enter: 2.351 $\boxed{+/-}$ $\boxed{10^x}$
                           The answer is 0.0044566.

Note that in this last example a negative mantissa can be used.

For those calculators that do not have a $\boxed{10^x}$ key, the antilogarithm can be found by using the $\boxed{INV}$ or $\boxed{2nd}$ keys. The $\boxed{INV}$ key will be used here but the key-strokes for the $\boxed{2nd}$ key would be the same.

Find antilog 2.9132.   Enter: 2.9132 $\boxed{INV}$ $\boxed{\log}$
                       The answer is 818.8418.

Here is a further illustration of the use of the $\boxed{\log}$ key.

An investor wishes to double an investment of $4000 in 5 years. What interest rate, compounded quarterly, is necessary to reach this goal?

Use the compound interest equation $P = A(1 + i)^n$, where $P = 8000$, $A = 4000$, and $n = 20$.

$$P = A(1 + i)^n$$
$$8000 = 4000\,(1 + i)^{20}$$
$$2 = (1 + i)^{20}$$
$$\log 2 = 20 \log (1 + i)$$
$$\frac{\log 2}{20} = \log (1 + i)$$

Use your calculator to find $\frac{\log 2}{20}$.
Enter: 2 $\boxed{\log}$ $\boxed{\div}$ 20 $\boxed{=}$

$$0.01505 = \log (1 + i)$$
$$10^{0.01505} = 1 + i$$

Recall that $\log_{10} a = b$ means $10^b = a$.
Use your calculator to find $10^{0.01505}$
Enter: 0.01505 $\boxed{10^x}$

$$1.03526 = 1 + i$$
$$0.03526 = i$$

The quarterly interest rate is 3.53%.
The annual interest rate is $4 \times 3.53\% = 14.1\%$.

Chapter Summary

Key Words   A function of the form $f(x) = b^x$ is an *exponential function* where b is a positive real number not equal to one. The number b is the *base* of the exponential function.

For $b > 0$, $b \neq 1$, $y = \log_b x$ *is equivalent to* $x = b^y$.

$\log_b x$ is the *logarithm* of x to the base b.

Common logarithms are logarithms to the base 10. (Usually the base 10 is omitted when writing the common logarithm of a number.)

The *mantissa* is the decimal part of a logarithm.

The *characteristic* is the integer part of a logarithm.

The *common antilogarithm* of a number, N, is the Nth power of 10, written antilog $N = 10^N$.

An *exponential equation* is one in which the variable occurs in the exponent.

Essential Rules   ***The Logarithm Property of***   For any positive real numbers x, y, and
the Product of Two Numbers b, $b \neq 1$,
$\log_b xy = \log_b x + \log_b y$.

The Logarithm Property of   For any positive real numbers x, y, and
the Quotient of Two Numbers b, $b \neq 1$,
$\log_b \dfrac{x}{y} = \log_b x - \log_b y$.

The Logarithm Property   For any positive real numbers x and b,
of the Power of a Number $b \neq 1$, and any real number r,
$\log_b x^r = r \log_b x$.

Additional Properties   For any positive real number b, $b \neq 1$,
of Logarithms   and any real number n, $\log_b b^n = n$.

For any positive real number b, $b \neq 1$,
$\log_b 1 = 0$.

Change of Base Formula $\log_a N = \dfrac{\log_b N}{\log_b a}$

One-to-One Property of   If $b^x = b^y$, then $x = y$.
the Exponential Function

To Solve Numerical Problems   If $x = y$, then $\log_b x = \log_b y$.
Using Logarithms

Chapter Review

SECTION 1

1. Evaluate the function $f(x) = 3^{x-2}$ at $x = 2$.
1

2. Evaluate the function $f(x) = \left(\frac{2}{3}\right)^{x+2}$ at $x = -3$.
$\frac{3}{2}$

3. Graph: $f(x) = \left(\frac{2}{3}\right)^{x+1}$

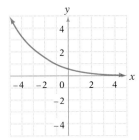

4. Graph: $f(x) = 3^{-x} + 2$

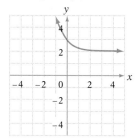

5. Write $2^5 = 32$ in logarithmic form.
$\log_2 32 = 5$

6. Write $\log_5 25 = 2$ in exponential form.
$5^2 = 25$

7. Solve for x: $\log_2 x = 5$
32

8. Solve for x: $\log_{10} x = 3$
1000

9. Graph: $f(x) = \log_3(x - 1)$

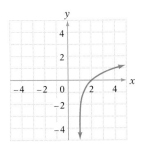

10. Graph: $f(x) = \log_2(2x - 1)$

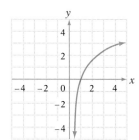

SECTION 2

11. Write $\log_3 \sqrt[5]{x^2 y^4}$ in expanded form.
$\frac{2}{5} \log_3 x + \frac{4}{5} \log_3 y$

12. Write $\log_8 \sqrt{\frac{x^5}{y^3}}$ in expanded form.
$\frac{1}{2}(5 \log_8 x - 3 \log_8 y)$

13. Write $2 \log_3 x - 5 \log_3 y$ as a single logarithm with a coefficient of 1.
$\log_3 \frac{x^2}{y^5}$

14. Write $\frac{1}{3}(\log_7 x + 4 \log_7 y)$ as a single logarithm with a coefficient of 1.
$\log_7 \sqrt[3]{xy^4}$

SECTION 3

15. Find log 567,000.
5.7536

16. Find antilog $(7.7110 - 10)$.
0.00514

17. Evaluate 0.067^5 by using common logarithms.
1.35×10^{-6}

18. Evaluate $\sqrt{\dfrac{23.5}{0.67}}$ by using common logarithms.
5.923

SECTION 4

19. Solve: $27^{2x+4} = 81^{x-3}$
-12

20. Solve $3^{x+2} = 5$. Round to the nearest thousandth.
-0.5350

21. Solve: $\log_3(x + 2) = 4$
79

22. Solve: $\log_5 \dfrac{7x + 2}{3x} = 1$
$\dfrac{1}{4}$

23. Solve: $\log_8(x + 2) - \log_8 x = \log_8 4$
$\dfrac{2}{3}$

24. Solve: $\log_6 2x = \log_6 2 + \log_6(3x - 4)$
2

25. Find $\log_6 22$.
1.7251

26. Find $\log_3 1.6$.
0.4278

SECTION 5

27. Use the compound interest formula $P = A(1 + i)^n$, where A is the original value of an investment, i is the interest rate per compounding period, and n is the number of compounding periods, to find the value of an investment after two years. The amount of the investment is $4000 and it is invested at 8% compounded monthly.
$4692

28. Use the Richter equation $M = \log\dfrac{I}{I_0}$, where M is the magnitude of an earthquake, I is the intensity of its shock, and I_0 is a constant, to find how many times stronger is an earthquake with magnitude 6 on the Richter scale than one with magnitude 3 on the scale.
1000 times

29. Use the exponential decay equation $A = A_0\left(\dfrac{1}{2}\right)^{t/k}$, where A is the amount of a radioactive material present after time t, k is the half-life, and A_0 is the original amount of radioactive material, to find the half-life of a material that decays from 25 mg to 15 mg in 20 days. Round to the nearest whole number.
27 days

30. The number of decibels, D, of a sound can be given by the equation $D = 10(\log I + 16)$, where I is the power of the sound measured in watts. Find the number of decibels from a busy street corner for which the power of the sound is 5×10^{-6} watts.
107 decibels

Chapter Test

1. Evaluate $f(x) = \left(\frac{2}{3}\right)^x$ at $x = 0$.

$f(0) = 1$ [11.1A]

2. Evaluate $f(x) = 3^{x+1}$ at $x = -2$.

$f(-2) = \frac{1}{3}$ [11.1A]

3. Graph: $f(x) = 2^x - 3$

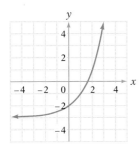

[11.1B]

4. Graph: $f(x) = 2^x + 2$

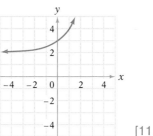

[11.1B]

5. Evaluate: $\log_4 16$

2   [11.1C]

6. Solve for x: $\log_3 x = -2$

$\frac{1}{9}$ [11.1C]

7. Graph: $f(x) = \log_2(2x)$

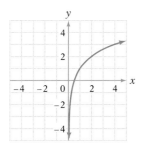

[11.1D]

8. Graph: $f(x) = \log_3(x + 1)$

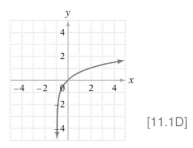

[11.1D]

9. Write $\log_6 \sqrt{xy^3}$ in expanded form.

$\frac{1}{2}(\log_6 x + 3 \log_6 y)$ [11.2A]

10. Write $\frac{1}{2}(\log_3 x - \log_3 y)$ as a single logarithm with a coefficient of 1.

$\log_3 \sqrt{\frac{x}{y}}$ [11.2A]

11. Find log 0.00367.
7.5647 − 10   [11.3A]

12. Find antilog (8.6064 − 10).
0.0404   [11.3B]

13. Find antilog 2.8523.
711.7   [11.3C]

14. Evaluate $\frac{2.74}{42.6}$ by using common logarithms.
0.06433   [11.3D]

15. Evaluate $\sqrt{0.638}$ by using common logarithms.
0.7988   [11.3D]

16. Solve for x: $3^{7x+1} = 3^{4x-5}$
−2   [11.4A]

17. Solve for x: $8^x = 2^{x-6}$
−3   [11.4A]

18. Solve for x:
$\log x + \log (x - 4) = \log 12$
6   [11.4B]

19. Find $\log_3 19$.
2.6804   [11.4C]

20. Use the exponential decay equation $A = A_0 \left(\frac{1}{2}\right)^{t/k}$, where A is the amount of a radioactive material present after time t, k is the half-life, and A_0 is the original amount of radioactive material, to find the half-life of a material which decays from 10 mg to 9 mg in 5 h. Round to the nearest whole number.
33 h   [11.5A]

Cumulative Review

1. Solve:
   $4 - 2[x - 3(2 - 3x) - 4x] = 2x$

   $x = \frac{8}{7}$ [2.1C]

2. Find the equation of the line containing the point $(2, -2)$ and parallel to the line $2x - y = 5$.
   $y = 2x - 6$ [6.4A]

3. Factor: $4x^{2n} + 7x^n + 3$
   $(4x^n + 3)(x^n + 1)$ [3.4C]

4. Simplify: $\dfrac{1 - \frac{5}{x} + \frac{6}{x^2}}{1 + \frac{1}{x} - \frac{6}{x^2}}$

   $\frac{x-3}{x+3}$ [4.4A]

5. Simplify: $\dfrac{\sqrt{xy}}{\sqrt{x} - \sqrt{y}}$

   $\frac{x\sqrt{y} + y\sqrt{x}}{x - y}$ [5.2D]

6. Solve by completing the square:
   $x^2 - 4x - 6 = 0$
   $2 + \sqrt{10}$ and $2 - \sqrt{10}$ [7.2A]

7. Write a quadratic equation that has integer coefficients and has solutions $\frac{1}{3}$ and -3.

   $3x^2 + 8x - 3 = 0$ [7.1B]

8. Graph: $y = -x^2 - 2x + 3$

   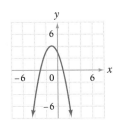

   [9.1A]

9. Solve by the addition method.
   $3x - y + z = 3$
   $x + y + 4z = 7$
   $3x - 2y + 3z = 8$
   $(0, -1, 2)$ [10.2B]

10. Solve: $y = x^2 + x - 3$
    $y = 2x - 1$
    $(2, 3), (-1, -3)$ [10.5A]

11. Solve: $x^2 + 4x - 5 \le 0$
    $\{x | -5 \le x \le 1\}$ [7.5A]

12. Solve: $|2x - 5| \le 3$
    $\{x | 1 \le x \le 4\}$ [2.3B]

13. Graph: $f(x) = \left(\frac{1}{2}\right)^x + 1$

    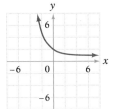

    [11.1C]

14. Graph: $f(x) = \log_2 x - 1$

    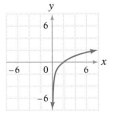

    [11.1D]

15. Evaluate the function
$f(x) = 2^{-x-1}$ at $x = -3$.
4   [11.1A]

16. Solve for x: $\log_5 x = 3$
$x = 125$ [11.1C]

17. Write $3 \log_b x - 5 \log_b y$ as a single logarithm with a coefficient of 1.
$\log_b \frac{x^3}{y^5}$ [11.2A]

18. Evaluate $\frac{49.1}{68.3}$ by using common logarithms.
0.719   [11.3D]

19. Solve for x: $4^{5x-2} = 4^{3x+2}$
$x = 2$ [11.4A]

20. Solve for x: $\log x + \log(2x + 3) = \log 2$
$x = \frac{1}{2}$ [11.4B]

21. A bank offers two types of checking accounts. One account has a charge of $5 per month plus 2 cents per check. The second account has a charge of $2 per month and 8 cents per check. How many checks can a customer who has the second type of account write if it is to cost the customer less than the first type of checking account?
49 checks   [2.2C]

22. Find the cost per pound of a mixture made from 16 lb of chocolate that cost $4.00 per pound and 24 lb of chocolate that cost $2.50 per pound.
$3.10   [2.5A]

23. A plane can fly at a rate of 225 mph in calm air. Traveling with the wind, the plane flew 1000 mi in the same amount of time that it flew 800 mi against the wind. Find the rate of the wind.
25 mph   [4.6D]

24. The distance (d) a spring stretches varies directly as the force (f) used to stretch the spring. If a force of 20 lb stretches a spring 6 in., how far will a force of 34 lb stretch the spring?
10.2 in.   [8.4A]

25. A carpenter purchased 80 ft of redwood and 140 ft of fir for a total cost of $67. A second purchase, at the same prices, included 140 ft of redwood and 100 ft of fir for a total cost of $81. Find the cost of redwood and fir.
redwood: 40¢ per foot
fir: 25¢ per foot   [10.4B]

26. The compound interest formula is $P = A(1 + i)^n$, where A is the original value of an investment, i is the interest rate per compounding period, n is the total number of compounding periods, and P is the value of the investment after n periods. Use the compound interest formula to find the number of years in which an investment of $5000 will double in value. The investment earns 9% annual interest and is compounded semiannually.
8 years   [11.5A]

12

Sequences and Series

OBJECTIVES

▶ To write the terms of a sequence
▶ To find the sum of a series
▶ To find the nth term of an arithmetic sequence
▶ To find the sum of an arithmetic series
▶ To solve application problems
▶ To find the nth term of a geometric sequence
▶ To find the sum of a finite geometric series
▶ To find the sum of an infinite geometric series
▶ To solve application problems
▶ To expand $(a + b)^n$

Tower of Hanoi

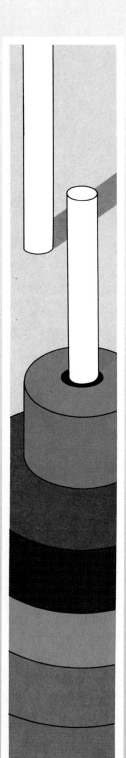

The Tower of Hanoi is a puzzle that has the following form.

Three pegs are placed in a board. A number of disks, graded in size, are stacked on one of the pegs with the largest disk on the bottom and the succeeding smaller disks placed on top.

The disks are moved according to the following rules:

1) Only one disk at a time may be moved.
2) A larger disk cannot be placed over a smaller disk.

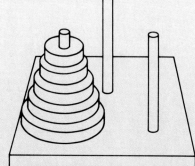

The object of the puzzle is to transfer all the disks from one peg to one of the other two pegs. If initially there is only one disk, then only one move would be required. With two disks initially, three moves are required, and with three disks, seven moves are required.

You can try this puzzle using playing cards. Select a number of playing cards, in sequence, from a deck. The number of cards corresponds to the number of disks and the number on the card corresponds to the size of the disk. For example, an "ace" would correspond to the smallest disk, a "two" would correspond to the next largest disk and so on for five cards. Now place three coins on a table to be used as the pegs. Pile the cards on one of the coins (in order) and try to move the pile to a second coin. Now a numerically larger card cannot be placed on a numerically smaller card. Try this for 5 cards. The minimum number of moves is 31.

Below is a chart that shows the minimum number of moves required for an initial number of disks. The difference between the number of moves for each succeeding disk is also given.

1	2	3	4	5	6	7	8
1	3	7	15	31	63	127	255
	2	4	8	16	32	64	128

For this last list of numbers, each succeeding number can be found by multiplying the preceding number by a constant (in this case 2). Such a list of numbers is called a *geometric sequence*.

The formula for the minimum number of moves is given by $M = 2^n - 1$, where M is the number of moves and n is the number of disks. This equation is an exponential equation.

Here's a hint for solving the Tower of Hanoi puzzle with 5 disks. First solve the puzzle for three disks. Now solve the puzzle for 4 disks by first moving the top three disks to one peg. (You just did this when you solved the puzzle for three disks.) Now move the fourth disk to a peg and then again use your solution for 3 disks to move the disks back to the peg with the fourth disk. Now use this solution for 4 disks to solve the 5-disk problem.

Introduction to Sequences and Series

Objective A   **To write the terms of a sequence**

An investor deposits $100 in an account that earns 10% interest compounded annually. The amount of interest earned each year can be determined by using the compound interest formula used in Chapter 11.

The amount of interest earned in each of the first four years of the investment is shown at the right.

Year	1	2	3	4
Interest Earned	$10	$11	$12.10	$13.31

The list of numbers 10, 11, 12.10, 13.31 is called a sequence. A **sequence** is an ordered list of numbers. The list 10, 11, 12.10, 13.31 is ordered because the position of a number in this list indicates the year in which that amount of interest was earned.

Examples of other sequences are shown at the right. These sequences are separated into two groups. A **finite sequence** contains a finite number of terms. An **infinite sequence** contains an infinite number of terms.

$$1, 1, 2, 3, 5, 8$$
$$1, 2, 3, 4, 5, 6, 7, 8$$ Finite
$$1, -1, 1, -1$$ Sequences

$$1, 3, 5, 7, \ldots$$
$$1, \frac{1}{2}, \frac{1}{4}, \frac{1}{8}, \ldots$$ Infinite
$$1, 1, 2, 3, 5, 8, \ldots$$ Sequences

Each of the numbers of a sequence is called a **term** of the sequence.

For the sequence at the right, the first term is 2, the second term is 4, the third term is 6, and the fourth term is 8.

$$2, 4, 6, 8, \ldots$$

A general sequence is shown at the right. The first term is a_1, the second term is a_2, the third term is a_3, and the nth term, also called the **general term** of the sequence, is a_n. Note that each term of the sequence is paired with a natural number.

$$a_1, a_2, a_3, \ldots a_n, \ldots$$

Frequently a sequence has a definite pattern which can be expressed by a formula.

Each term of the sequence shown at the right is paired with a natural number by the formula $a_n = 3n$. The first term, a_1, is 3. The second term, a_2, is 6. The third term, a_3, is 9. The nth term, a_n, is $3n$.

$$a_n = 3n$$
$$a_1, \quad a_2, \quad a_3, \ldots \quad a_n, \ldots$$
$$3(1), \quad 3(2), \quad 3(3), \ldots 3(n), \ldots$$
$$3, \quad 6, \quad 9, \ldots \quad 3n, \ldots$$

Write the first three terms of the sequence whose nth term is given by the formula $a_n = 4n$.

$$a_n = 4n$$

Replace n by 1.	$a_1 = 4(1) = 4$	The first term is 4.
Replace n by 2.	$a_2 = 4(2) = 8$	The second term is 8.
Replace n by 3.	$a_3 = 4(3) = 12$	The third term is 12.

Find the fifth and seventh terms of the sequence whose nth term is given by the formula $a_n = \frac{1}{n^2}$.

$$a_n = \frac{1}{n^2}$$

Replace n by 5. $\qquad a_5 = \frac{1}{5^2} = \frac{1}{25}$ $\qquad$ The fifth term is $\frac{1}{25}$.

Replace n by 7. $\qquad a_7 = \frac{1}{7^2} = \frac{1}{49}$ $\qquad$ The seventh term is $\frac{1}{49}$.

Example 1

Write the first three terms of the sequence whose nth term is given by the formula $a_n = 2n - 1$.

Solution

$a_n = 2n - 1$
$a_1 = 2(1) - 1 = 1$ $\quad$ The first term is 1.
$a_2 = 2(2) - 1 = 3$ $\quad$ The second term is 3.
$a_3 = 2(3) - 1 = 5$ $\quad$ The third term is 5.

Example 2

Write the first four terms of the sequence whose nth term is given by the formula $a_n = n(n + 1)$.

Your solution

2, 6, 12, 20

Example 3

Find the eighth and tenth terms of the sequence whose nth term is given by the formula $a_n = \frac{n}{n + 1}$.

Solution

$a_n = \frac{n}{n + 1}$

$a_8 = \frac{8}{8 + 1} = \frac{8}{9}$ $\qquad$ The eighth term is $\frac{8}{9}$.

$a_{10} = \frac{10}{10 + 1} = \frac{10}{11}$ $\qquad$ The tenth term is $\frac{10}{11}$.

Example 4

Find the sixth and ninth terms of the sequence whose nth term is given by the formula $a_n = \frac{1}{n(n + 2)}$.

Your solution

$a_6 = \frac{1}{48}$

$a_9 = \frac{1}{99}$

Solutions on p. A48

Objective B $\qquad$ **To find the sum of a series**

12

In the last objective, the sequence 10, 11, 12.10, 13.31 was shown to represent the amount of interest earned in each of 4 years of an investment.

10, 11, 12.10, 13.31

The sum of the terms of this sequence represents the total interest earned by the investment over the four-year period.

$10 + 11 + 12.10 + 13.31 = 46.41$

The total interest earned over the four-year period is $46.41.

The indicated sum of the terms of a sequence is called a **series.** Given the sequence 10, 11, 12.10, 13.31, the series $10 + 11 + 12.10 + 13.31$ can be written.

S_n is used to designate the sum of the first n terms of a sequence.

For the example above, the sums of the series S_1, S_2, S_3, and S_4 represent the total interest earned for 1, 2, 3, and 4 years, respectively.

$$S_1 = 10 = 10$$
$$S_2 = 10 + 11 = 21$$
$$S_3 = 10 + 11 + 12.10 = 33.10$$
$$S_4 = 10 + 11 + 12.10 + 13.31 = 46.41$$

For the general sequence $a_1, a_2, a_3, \ldots a_n$, the series S_1, S_2, S_3, and S_n are shown at the right.

$$S_1 = a_1$$
$$S_2 = a_1 + a_2$$
$$S_3 = a_1 + a_2 + a_3$$
$$S_n = a_1 + a_2 + a_3 + \cdots + a_n$$

It is convenient to represent a series in a compact form called **summation notation,** or **sigma notation.** The Greek letter sigma, Σ, is used to indicate a sum.

The first four terms of the sequence whose nth term is given by the formula $a_n = 2n$ are 2, 4, 6, 8. The corresponding series is shown at the right written in summation notation, read "the sum from 1 to 4 of $2n$." The letter n is called the **index** of the summation.

$$\sum_{n=1}^{4} 2n$$

To write the terms of the series, replace n by the consecutive integers from 1 to 4.

$$\sum_{n=1}^{4} 2n = 2(1) + 2(2) + 2(3) + 2(4)$$

The series is $2 + 4 + 6 + 8$.

$$= 2 + 4 + 6 + 8$$

The sum of the series is 20.

$$= 20$$

Find the sum of the series $\sum_{n=1}^{6} n^2$.

Replace n by 1, 2, 3, 4, 5, and 6.

$$\sum_{n=1}^{6} n^2 = 1^2 + 2^2 + 3^2 + 4^2 + 5^2 + 6^2$$

Write the series.

$$= 1 + 4 + 9 + 16 + 25 + 36$$

Find the sum of the series.

$$= 91$$

Find the sum of the series $\sum_{i=1}^{3} (2i - 1)$.

Replace i by 1, 2, and 3.

$$\sum_{i=1}^{3} (2i - 1) = [2(1) - 1] + [2(2) - 1] + [2(3) - 1]$$

Write the series.

$$= 1 + 3 + 5$$

Find the sum of the series.

$$= 9$$

Find the sum of the series $\displaystyle\sum_{n=3}^{6} \frac{1}{2}n$.

Replace n by 3, 4, 5, and 6. $\displaystyle\sum_{n=3}^{6} \frac{1}{2}n = \frac{1}{2}(3) + \frac{1}{2}(4) + \frac{1}{2}(5) + \frac{1}{2}(6)$

Write the series. $= \frac{3}{2} + 2 + \frac{5}{2} + 3$

Find the sum of the series. $= 9$

Write $\displaystyle\sum_{i=1}^{5} x^i$ in expanded form.

This is a variable series.

Replace i by 1, 2, 3, 4, and 5. $\displaystyle\sum_{i=1}^{5} x^i = x + x^2 + x^3 + x^4 + x^5$

Example 5

Find the sum of the series $\displaystyle\sum_{n=1}^{4} (7 - n)$.

Solution

$\displaystyle\sum_{n=1}^{4} (7 - n)$

$= (7 - 1) + (7 - 2) + (7 - 3) + (7 - 4)$
$= 6 + 5 + 4 + 3 = 18$

Example 6

Find the sum of the series $\displaystyle\sum_{i=3}^{6} (i^2 - 2)$.

Your solution

78

Example 7

Write $\displaystyle\sum_{i=1}^{5} ix$ in expanded form.

Solution

$\displaystyle\sum_{i=1}^{5} ix = 1 \cdot x + 2 \cdot x + 3 \cdot x + 4 \cdot x + 5 \cdot x$

$= x + 2x + 3x + 4x + 5x$

Example 8

Write $\displaystyle\sum_{n=1}^{4} \frac{x}{n}$ in expanded form.

Your solution

$x + \frac{x}{2} + \frac{x}{3} + \frac{x}{4}$

Solutions on p. A48

12.1   **EXERCISES**

▶ Objective A

Write the first four terms of the sequence whose nth term is given by the formula:

1. $a_n = n + 1$
2, 3, 4, 5

2. $a_n = n - 1$
0, 1, 2, 3

3. $a_n = 2n + 1$
3, 5, 7, 9

4. $a_n = 3n - 1$
2, 5, 8, 11

5. $a_n = 2 - 2n$
0, −2, −4, −6

6. $a_n = 1 - 2n$
−1, −3, −5, −7

7. $a_n = 2^n$
2, 4, 8, 16

8. $a_n = 3^n$
3, 9, 27, 81

9. $a_n = n^2 + 1$
2, 5, 10, 17

10. $a_n = n^2 - 1$
0, 3, 8, 15

11. $a_n = \dfrac{n}{n^2 + 1}$

$\dfrac{1}{2}, \dfrac{2}{5}, \dfrac{3}{10}, \dfrac{4}{17}$

12. $a_n = \dfrac{n^2 - 1}{n}$

$0, \dfrac{3}{2}, \dfrac{8}{3}, \dfrac{15}{4}$

13. $a_n = n - \dfrac{1}{n}$

$0, \dfrac{3}{2}, \dfrac{8}{3}, \dfrac{15}{4}$

14. $a_n = n^2 - \dfrac{1}{n}$

$0, \dfrac{7}{2}, \dfrac{26}{3}, \dfrac{63}{4}$

15. $a_n = (-1)^{n+1} n$
1, −2, 3, −4

Find the indicated term of the sequence whose nth term is given by the formula:

16. $a_n = 3n + 4; a_{12}$
40

17. $a_n = 2n - 5; a_{10}$
15

18. $a_n = n(n - 1); a_{11}$
110

19. $a_n = \dfrac{n}{n + 1}; a_{12}$

$\dfrac{12}{13}$

20. $a_n = (-1)^{n-1} n^2; a_{15}$
225

21. $a_n = (-1)^{n-1}(n - 1); a_{25}$
24

22. $a_n = \left(\dfrac{1}{2}\right)^n; a_8$

$\dfrac{1}{256}$

23. $a_n = \left(\dfrac{2}{3}\right)^n; a_5$

$\dfrac{32}{243}$

24. $a_n = (n + 2)(n + 3); a_{17}$
380

25. $a_n = (n + 4)(n + 1); a_7$
88

26. $a_n = \dfrac{(-1)^{2n-1}}{n^2}; a_6$

$-\dfrac{1}{36}$

27. $a_n = \dfrac{(-1)^{2n}}{n + 4}; a_{16}$

$\dfrac{1}{20}$

▶ **Objective B**

Find the sum of the series.

28. $\displaystyle\sum_{n=1}^{5} (2n + 3)$
45

29. $\displaystyle\sum_{i=1}^{7} (i + 2)$
42

30. $\displaystyle\sum_{i=1}^{4} 2i$
20

31. $\displaystyle\sum_{n=1}^{7} n$
28

32. $\displaystyle\sum_{i=1}^{6} i^2$
91

33. $\displaystyle\sum_{i=1}^{5} (i^2 + 1)$
60

34. $\displaystyle\sum_{n=1}^{6} (-1)^n$
0

35. $\displaystyle\sum_{n=1}^{4} \frac{1}{2n}$
$\frac{25}{24}$

36. $\displaystyle\sum_{i=3}^{6} i^3$
432

37. $\displaystyle\sum_{n=2}^{4} 2^n$
28

38. $\displaystyle\sum_{n=3}^{7} \frac{n}{n-1}$
$\frac{129}{20}$

39. $\displaystyle\sum_{i=3}^{6} \frac{i+1}{i}$
$\frac{99}{20}$

40. $\displaystyle\sum_{i=1}^{4} \frac{1}{2^i}$
$\frac{15}{16}$

41. $\displaystyle\sum_{i=1}^{5} \frac{1}{2i}$
$\frac{137}{120}$

42. $\displaystyle\sum_{n=1}^{4} (-1)^{n-1} n^2$
-10

43. $\displaystyle\sum_{i=1}^{4} (-1)^{i-1}(i + 1)$
-2

44. $\displaystyle\sum_{n=3}^{5} \frac{(-1)^{n-1}}{n-2}$
$\frac{5}{6}$

45. $\displaystyle\sum_{n=4}^{7} \frac{(-1)^{n-1}}{n-3}$
$-\frac{7}{12}$

Write the series in expanded form.

46. $\displaystyle\sum_{n=1}^{5} 2x^n$
$2x + 2x^2 + 2x^3 + 2x^4 + 2x^5$

47. $\displaystyle\sum_{n=1}^{4} \frac{2n}{x}$
$\frac{2}{x} + \frac{4}{x} + \frac{6}{x} + \frac{8}{x}$

48. $\displaystyle\sum_{i=1}^{5} \frac{x^i}{i}$
$x + \frac{x^2}{2} + \frac{x^3}{3} + \frac{x^4}{4} + \frac{x^5}{5}$

49. $\displaystyle\sum_{i=1}^{4} \frac{x^i}{i+1}$
$\frac{x}{2} + \frac{x^2}{3} + \frac{x^3}{4} + \frac{x^4}{5}$

50. $\displaystyle\sum_{i=3}^{5} \frac{x^i}{2i}$
$\frac{x^3}{6} + \frac{x^4}{8} + \frac{x^5}{10}$

51. $\displaystyle\sum_{i=2}^{4} \frac{x^i}{2i-1}$
$\frac{x^2}{3} + \frac{x^3}{5} + \frac{x^4}{7}$

52. $\displaystyle\sum_{n=1}^{5} x^{2n}$
$x^2 + x^4 + x^6 + x^8 + x^{10}$

53. $\displaystyle\sum_{n=1}^{4} x^{2n-1}$
$x + x^3 + x^5 + x^7$

54. $\displaystyle\sum_{i=1}^{4} \frac{x^i}{i^2}$
$x + \frac{x^2}{4} + \frac{x^3}{9} + \frac{x^4}{16}$

55. $\displaystyle\sum_{n=1}^{4} (2n)x^n$
$2x + 4x^2 + 6x^3 + 8x^4$

56. $\displaystyle\sum_{n=1}^{4} nx^{n-1}$
$1 + 2x + 3x^2 + 4x^3$

57. $\displaystyle\sum_{i=1}^{5} x^{-i}$
$\frac{1}{x} + \frac{1}{x^2} + \frac{1}{x^3} + \frac{1}{x^4} + \frac{1}{x^5}$

SECTION 12.2

Arithmetic Sequences and Series

Objective A

To find the nth term of an arithmetic sequence

A company's expenses for training a new employee are quite high. To encourage employees to continue their employment with the company, a company which has a six-month training program offers a starting salary of $800 a month and then a $100-per-month pay increase each month during the training period.

The sequence at the right shows the employee's monthly salaries during the training period. Each term of the sequence is found by adding $100 to the previous term.

Month	1	2	3	4	5	6
Salary	800	900	1000	1100	1200	1300

The sequence 800, 900, 1000, 1100, 1200, 1300 is called an arithmetic sequence. An **arithmetic sequence**, or **arithmetic progression**, is one in which the difference between any two consecutive terms is constant. The difference between consecutive terms is called the **common difference** of the sequence.

Each of the sequences shown at the right is an arithmetic sequence. To find the common difference of an arithmetic sequence, subtract the first term from the second term.

2, 7, 12, 17, 22, . . .   Common difference: 5

3, 1, -1, -3, -5, . . .   Common difference: -2

1, $\frac{3}{2}$, 2, $\frac{5}{2}$, 3, $\frac{7}{2}$, . . .   Common difference: $\frac{1}{2}$

Consider an arithmetic sequence in which the first term is a_1 and the common difference is d. By adding the common difference to each successive term of the arithmetic sequence, a formula for the nth term can be found.

The first term is a_1.
$$a_1 = a_1$$

To find the second term, add the common difference d to the first term.
$$a_2 = a_1 + d$$

To find the third term, add the common difference d to the second term.
$$a_3 = a_2 + d = (a_1 + d) + d$$
$$a_3 = a_1 + 2d$$

To find the fourth term, add the common difference d to the third term.
$$a_4 = a_3 + d = (a_1 + 2d) + d$$
$$a_4 = a_1 + 3d$$

Note the relationship between the term number and the number which multiplies d. The multiplier of d is one less than the term number.
$$a_n = a_1 + (n - 1)d$$

The Formula for the nth Term of an Arithmetic Sequence

The nth term of an arithmetic sequence with a common difference of d is given by $a_n = a_1 + (n - 1)d$.

Find the 27th term of the arithmetic sequence $-4, -1, 2, 5, 8, \ldots$.

Find the common difference.	$d = a_2 - a_1 = -1 - (-4) = 3$

Use the Formula for the nth Term of an Arithmetic Sequence to find the 27th term. $n = 27$, $a_1 = -4$, $d = 3$

$a_n = a_1 + (n - 1)d$
$a_{27} = -4 + (27 - 1)3 = -4 + (26)3 = -4 + 78$
$a_{27} = 74$

Find the formula for the nth term of the arithmetic sequence $-5, -2, 1, 4, \ldots$.

Find the common difference. $d = a_2 - a_1 = -2 - (-5) = 3$

Use the Formula for the nth Term of an Arithmetic Sequence. $a_1 = -5$, $d = 3$

$a_n = a_1 + (n - 1)d$
$a_n = -5 + (n - 1)3$
$a_n = -5 + 3n - 3$
$a_n = 3n - 8$

Find the number of terms in the finite arithmetic sequence $7, 10, 13, \ldots, 55$.

Find the common difference. $d = a_2 - a_1 = 10 - 7 = 3$

Use the Formula for the nth Term of an Arithmetic Sequence. $a_n = 55$, $a_1 = 7$, $d = 3$
Solve for n.

$a_n = a_1 + (n - 1)d$
$55 = 7 + (n - 1)3$
$55 = 7 + 3n - 3$
$55 = 4 + 3n$
$51 = 3n$
$17 = n$

There are 17 terms in the sequence.

Example 1

Find the 10th term of the arithmetic sequence $1, 6, 11, 16, \ldots$

Solution

$d = a_2 - a_1 = 6 - 1 = 5$

$a_n = a_1 + (n - 1)d$
$a_{10} = 1 + (10 - 1)5 = 1 + (9)5 = 1 + 45$
$a_{10} = 46$

Example 3

Find the number of terms in the finite arithmetic sequence $-3, 1, 5, \ldots, 41$.

Solution

$d = a_2 - a_1 = 1 - (-3) = 4$

$a_n = a_1 + (n - 1)d$
$41 = -3 + (n - 1)4$
$41 = -3 + 4n - 4$
$41 = -7 + 4n$
$48 = 4n$
$12 = n$

There are 12 terms in the sequence.

Example 2

Find the 15th term of the arithmetic sequence $9, 3, -3, -9, \ldots$

Your solution

-75

Example 4

Find the number of terms in the finite arithmetic sequence $7, 9, 11, \ldots, 59$.

Your solution

27

Solutions on p. A48

Content and Format © 1991 HMCo.

| Objective B | **To find the sum of an arithmetic series** |

The indicated sum of the terms of an arithmetic sequence is called an **arithmetic series.** The sum of an arithmetic series can be found by using a formula. A proof of this formula can be found on page A8.

The Formula for the Sum of *n* Terms of an Arithmetic Series

Let a_1 be the first term of a finite arithmetic sequence, n the number of terms, and a_n the last term of the sequence. Then the sum of the series S_n is given by $S_n = \frac{n}{2}(a_1 + a_n)$.

Find the sum of the first 10 terms of the arithmetic sequence 2, 4, 6, 8, ...

Find the common difference. $d = a_2 - a_1 = 4 - 2 = 2$

Find the 10th term. $a_n = a_1 + (n - 1)d$
                                   $a_{10} = 2 + (10 - 1)2 = 2 + (9)2 = 2 + 18 = 20$

Use the Formula for the Sum of n Terms of an Arithmetic Series. $S_n = \frac{n}{2}(a_1 + a_n)$

$n = 10, a_1 = 2, a_n = 20$ $S_{10} = \frac{10}{2}(2 + 20) = 5(22) = 110$

Find the sum of the arithmetic series $\displaystyle\sum_{n=1}^{25} (3n + 1)$.

                                   $a_n = 3n + 1$
Find the first term. $a_1 = 3(1) + 1 = 4$

Find the 25th term. $a_{25} = 3(25) + 1 = 76$

Use the Formula for the Sum of n Terms of an Arithmetic Series. $S_n = \frac{n}{2}(a_1 + a_n)$

$n = 25, a_1 = 4, a_n = 76$ $S_{25} = \frac{25}{2}(4 + 76) = \frac{25}{2}(80) = 1000$

Example 5

Find the sum of the first 20 terms of the arithmetic sequence 3, 8, 13, 18, ...

Solution

$d = a_2 - a_1 = 8 - 3 = 5$

$a_n = a_1 + (n - 1)d$
$a_{20} = 3 + (20 - 1)5 = 3 + (19)5$
$\quad\;\; = 3 + 95 = 98$

$S_n = \frac{n}{2}(a_1 + a_n)$

$S_{20} = \frac{20}{2}(3 + 98) = 10(101) = 1010$

Example 6

Find the sum of the first 25 terms of the arithmetic sequence $-4, -2, 0, 2, 4, ...$

Your solution

500

Solution on p. A48

Example 7

Find the sum of the arithmetic series

$$\sum_{n=1}^{15} (2n + 3).$$

Solution

$a_n = 2n + 3$
$a_1 = 2(1) + 3 = 5$
$a_{15} = 2(15) + 3 = 33$

$S_n = \frac{n}{2}(a_1 + a_n)$

$S_{15} = \frac{15}{2}(5 + 33) = \frac{15}{2}(38) = 285$

Example 8

Find the sum of the arithmetic series

$$\sum_{n=1}^{18} (3n - 2).$$

Your solution

477

Solution on p. A48

Objective C   To solve application problems

Example 9

The distance a ball rolls down a ramp each second is given by an arithmetic sequence. The distance in feet traveled by the ball during the nth second is given by $2n - 1$. Find the distance the ball will travel during the 10th second.

Strategy

To find the distance:
■ Find the common difference of the arithmetic sequence.
■ Use the Formula for the nth Term of an Arithmetic Sequence.

Solution

$a_n = 2n - 1$
$a_1 = 2(1) - 1 = 1$
$a_2 = 2(2) - 1 = 3$
$d = a_2 - a_1 = 3 - 1 = 2$

$a_n = a_1 + (n - 1)d$
$a_{10} = 1 + (10 - 1)2 = 1 + (9)2$
$\quad = 1 + 18 = 19$

The ball will travel 19 ft during the 10th second.

Example 10

A contest offers 20 prizes. The first prize is $10,000 and each successive prize is $300 less than the preceding prize. What is the value of the 20th-place prize? What is the total amount of prize money being awarded?

Your strategy

Your solution

20th-place prize: $4300
total amount being awarded: $143,000

Solution on p. A49

12.2   EXERCISES

▶ Objective A

Find the indicated term of the arithmetic sequence.

1. 1, 11, 21, . . . ; a_{15}
141

2. 3, 8, 13, . . . ; a_{20}
98

3. −6, −2, 2, . . . ; a_{15}
50

4. −7, −2, 3, . . . ; a_{14}
58

5. 2, $\frac{5}{2}$, 3, . . . ; a_{31}
17

6. 1, $\frac{5}{4}$, $\frac{3}{2}$, . . . ; a_{17}
5

7. −4, −$\frac{5}{2}$, −1, . . . ; a_{12}
$12\frac{1}{2}$

8. −$\frac{5}{3}$, −1, −$\frac{1}{3}$, . . . ; a_{22}
$12\frac{1}{3}$

9. 8, 5, 2, . . . ; a_{40}
−109

10. $\frac{4}{3}$, $\frac{1}{3}$, −$\frac{2}{3}$, . . . ; a_{30}
$-27\frac{2}{3}$

11. 6, 5.75, 5.50, . . . ; a_{10}
3.75

12. 4, 3.7, 3.4, . . . ; a_{12}
0.7

Find the formula for the *n*th term of the arithmetic sequence.

13. 1, 2, 3, . . .
$a_n = n$

14. 1, 4, 7, . . .
$a_n = 3n − 2$

15. 6, 2, −2, . . .
$a_n = −4n + 10$

16. 3, 0, −3, . . .
$a_n = −3n + 6$

17. 2, $\frac{7}{2}$, 5, . . .
$a_n = \frac{3n + 1}{2}$

18. 7, 4.5, 2, . . .
$a_n = −2.5n + 9.5$

19. −8, −13, −18, . . .
$a_n = −5n − 3$

20. 17, 30, 43, . . .
$a_n = 13n + 4$

21. 26, 16, 6, . . .
$a_n = −10n + 36$

Find the number of terms in the finite arithmetic sequence.

22. 1, 5, 9, . . . , 81
21

23. 3, 8, 13, . . . , 98
20

24. 2, 0, −2, . . . , −56
30

25. 1, −3, −7, . . . , −75
20

26. $\frac{5}{2}$, 3, $\frac{7}{2}$, . . . , 13
22

27. $\frac{7}{3}$, $\frac{13}{3}$, $\frac{19}{3}$, . . . , $\frac{79}{3}$
13

28. 1, 0.75, 0.50, . . . , −4
21

29. 3.5, 2, 0.5, . . . , −25
20

30. −3.4, −2.8, −2.2, . . . , 11
25

▶ **Objective B**

Find the sum of the indicated number of terms of the arithmetic sequence.

31. 1, 3, 5, . . . ; $n = 50$
2500

32. 2, 4, 6, . . . ; $n = 25$
650

33. 20, 18, 16, . . . ; $n = 40$
−760

34. 25, 20, 15, . . . ; $n = 22$
−605

35. $\frac{1}{2}$, 1, $\frac{3}{2}$, . . . ; $n = 27$
189

36. 2, $\frac{11}{4}$, $\frac{7}{2}$, . . . ; $n = 10$
$\frac{215}{4}$

Find the sum of the arithmetic series.

37. $\sum\limits_{i=1}^{15}(3i - 1)$
345

38. $\sum\limits_{i=1}^{15}(3i + 4)$
420

39. $\sum\limits_{n=1}^{17}\left(\frac{1}{2}n + 1\right)$
$\frac{187}{2}$

40. $\sum\limits_{n=1}^{10}(1 - 4n)$
−210

41. $\sum\limits_{i=1}^{15}(4 - 2i)$
−180

42. $\sum\limits_{n=1}^{10}(5 - n)$
−5

▶ **Objective C** *Application Problems*

Solve:

43. The distance that an object dropped from a cliff will fall is 16 ft the first second, 48 ft the next second, 80 ft the third second, and so on in an arithmetic sequence. What is the total distance the object will fall in 6 s?
576 ft

44. An exercise program calls for walking 10 min each day for a week. Each week thereafter, the amount of time spent walking increases by 5 min per day. In how many weeks will a person be walking 60 min each day?
11 weeks

45. A display of cans in a grocery store consists of 24 cans in the bottom row, 21 cans in the next row, and so on in an arithmetic sequence. The top row has 3 cans. Find the total number of cans in the display.
108 cans

46. The salary schedule for an engineering assistant is $700 for the first month and a $35-per-month salary increase for the next eight months. Find the monthly salary during the eighth month. Find the total salary for the eight-month period.
$945; $6580

47. The loge seating section in a concert hall consists of 27 rows of chairs. There are 73 seats in the first row, 79 seats in the second row, 85 seats in the third row, and so on in an arithmetic sequence. How many seats are in the loge seating section?
4077 seats

48. A "theater in the round" has 68 seats in the first row, 80 seats in the second row, 92 seats in the third row, and so on in an arithmetic sequence. Find the total number of seats in the theater if there are 22 rows of seats.
4268 seats

| SECTION 12.3 | # Geometric Sequences and Series |

Objective A   ### To find the *n*th term of a geometric sequence

An ore sample contains 20 mg of a radioactive material with a half-life of one week. The amount of the radioactive material which the sample contains at the beginning of each week can be determined by using the exponential decay equation used in Chapter 11.

The sequence at the right represents the amount in the sample at the beginning of each week. Each term of the sequence is found by multiplying the preceding term by $\frac{1}{2}$.

Week	1	2	3	4	5
Amount	20	10	5	2.5	1.25

The sequence 20, 10, 5, 2.5, 1.25 is called a geometric sequence. **A geometric sequence,** or **geometric progression,** is one in which each successive term of the sequence is the same non-zero constant multiple of the preceding term. The common multiple is called the **common ratio** of the sequence.

Each of the sequences shown at the right is a geometric sequence. To find the common ratio of a geometric sequence, divide the second term of the sequence by the first term.

3, 6, 12, 24, 48, . . .   Common ratio: 2

4, −12, 36, −108, 324, . . .   Common ratio: −3

6, 4, $\frac{8}{3}$, $\frac{16}{9}$, $\frac{32}{27}$, . . .   Common ratio: $\frac{2}{3}$

Consider a geometric sequence in which the first term is a_1 and the common ratio is r. By multiplying each successive term of the geometric sequence by the common ratio, a formula for the *n*th term can be found.

The first term is a_1.

$$a_1 = a_1$$

To find the second term, multiply the first term by the common ratio r.

$$a_2 = a_1 r$$

To find the third term, multiply the second term by the common ratio r.

$$a_3 = (a_1 r)r$$
$$a_3 = a_1 r^2$$

To find the fourth term, multiply the third term by the common ratio r.

$$a_4 = (a_1 r^2)r$$
$$a_4 = a_1 r^3$$

Note the relationship between the term number and the number that is the exponent on r. The exponent on r is one less than the term number.

$$a_n = a_1 r^{n-1}$$

The Formula for the nth Term of a Geometric Sequence

> The nth term of a geometric sequence with first term a_1 and common ratio r is given by $a_n = a_1 r^{n-1}$.

Find the 6th term of the geometric sequence 3, 6, 12, ...

Find the common ratio.

$$r = \frac{a_2}{a_1} = \frac{6}{3} = 2$$

Use the Formula for the nth Term of a Geometric Sequence to find the 6th term.
$n = 6$, $a_1 = 3$, $r = 2$

$$a_n = a_1 r^{n-1}$$
$$a_6 = 3(2)^{6-1} = 3(2)^5 = 3 \cdot 32$$
$$a_6 = 96$$

Find a_3 for the geometric sequence 8, a_2, a_3, 27, . . .

Find the common ratio.
$a_4 = 27$, $a_1 = 8$, $n = 4$

$$a_n = a_1 r^{n-1}$$
$$a_4 = a_1 r^{4-1}$$
$$27 = 8r^{4-1}$$
$$27 = 8r^3$$
$$\frac{27}{8} = r^3$$
$$\frac{3}{2} = r$$

Use the Formula for the nth Term of a Geometric Sequence.

$$a_3 = 8\left(\frac{3}{2}\right)^{3-1} = 8\left(\frac{3}{2}\right)^2 = 8\left(\frac{9}{4}\right)$$
$$a_3 = 18$$

Example 1

Find the 7th term of the geometric sequence 3, −6, 12, ...

Solution

$$r = \frac{a_2}{a_1} = \frac{-6}{3} = -2$$

$$a_n = a_1 r^{n-1}$$
$$a_7 = 3(-2)^{7-1} = 3(-2)^6 = 3(64)$$
$$a_7 = 192$$

Example 2

Find the 5th term of the geometric sequence 5, 2, $\frac{4}{5}$, ...

Your solution

$\frac{16}{125}$

Solution on p. A49

Example 3

Find a_2 for the geometric sequence
2, a_2, a_3, 54, ...

Solution

$a_n = a_1 r^{n-1}$
$a_4 = 2r^{4-1}$
$54 = 2r^{4-1}$
$54 = 2r^3$
$27 = r^3$
$3 = r$
$a_n = a_1 r^{n-1}$
$a_2 = 2(3)^{2-1} = 2(3)$
$a_2 = 6$

Example 4

Find a_3 for the geometric sequence
3, a_2, a_3, −192, ...

Your solution

48

Solution on p. A49

Objective B	**To find the sum of a finite geometric series**	

The indicated sum of the terms of a geometric sequence is called a **geometric series.** The sum of a geometric series can be found by a formula. A proof of this formula can be found on page A7.

The Formula for the Sum of n Terms of a Finite Geometric Series

Let a_1 be the first term of a finite geometric sequence, n the number of terms, and r the common ratio. Then the sum of the series S_n is given by
$S_n = \frac{a_1(1 - r^n)}{1 - r}$.

Find the sum of the geometric sequence 2, 8, 32, 128, 512.

Find the common ratio.

$r = \frac{a_2}{a_1} = \frac{8}{2} = 4$

Use the Formula for the Sum of n Terms of a Finite Geometric Series.
$n = 5$, $a_1 = 2$, $r = 4$

$S_n = \frac{a_1(1 - r^n)}{1 - r}$

$S_5 = \frac{2(1 - 4^5)}{1 - 4} = \frac{2(1 - 1024)}{-3} = \frac{2(-1023)}{-3}$

$= \frac{-2046}{-3} = 682$

Find the sum of the geometric series $\displaystyle\sum_{n=1}^{10}(-20)(-2)^{n-1}$.

Find the first term.

$a_n = (-20)(-2)^{n-1}$
$a_1 = (-20)(-2)^{1-1} = (-20)(-2)^0 = (-20)(1) = -20$

Find the second term.

$a_2 = (-20)(-2)^{2-1} = (-20)(-2)^1 = (-20)(-2) = 40$

Find the common ratio.

$r = \dfrac{a_2}{a_1} = \dfrac{40}{-20} = -2$

Use the Formula for the Sum of n Terms of a Finite Geometric Series.
$n = 10, a_1 = -20, r = -2$

$S_n = \dfrac{a_1(1-r^n)}{1-r}$

$S_{10} = \dfrac{-20[1-(-2)^{10}]}{1-(-2)} = \dfrac{-20(1-1024)}{3}$

$= \dfrac{-20(-1023)}{3} = \dfrac{20{,}460}{3} = 6820$

Example 5

Find the sum of the geometric sequence 3, 6, 12, 24, 48, 96.

Solution

$r = \dfrac{a_2}{a_1} = \dfrac{6}{3} = 2$

$S_n = \dfrac{a_1(1-r^n)}{1-r}$

$S_6 = \dfrac{3(1-2^6)}{1-2} = \dfrac{3(1-64)}{-1} = -3(-63) = 189$

Example 6

Find the sum of the geometric sequence $1, -\dfrac{1}{3}, \dfrac{1}{9}, -\dfrac{1}{27}$.

Your solution

$\dfrac{20}{27}$

Example 7

Find the sum of the geometric series

$\displaystyle\sum_{n=1}^{4}4^n$.

Solution

$a_n = 4^n$

$a_1 = 4^1 = 4$

$a_2 = 4^2 = 16$

$r = \dfrac{a_2}{a_1} = \dfrac{16}{4} = 4$

$S_n = \dfrac{a_1(1-r^n)}{1-r}$

$S_4 = \dfrac{4(1-4^4)}{1-4} = \dfrac{4(1-256)}{-3} = \dfrac{4(-255)}{-3} = \dfrac{-1020}{-3} = 340$

Example 8

Find the sum of the geometric series

$\displaystyle\sum_{n=1}^{5}\left(\dfrac{1}{2}\right)^n$.

Your solution

$\dfrac{31}{32}$

Solutions on p. A49

Content and Format © 1991 HMCo.

| Objective C | **To find the sum of an infinite geometric series** |

When the absolute value of the common ratio of a geometric sequence is less than 1, $|r| < 1$, then as n becomes larger, r^n becomes closer to zero.

Examples of geometric sequences for which $|r| < 1$ are shown at the right. Note that as the number of terms increases, the value of the last term listed is closer to zero.

$$1, \frac{1}{3}, \frac{1}{9}, \frac{1}{27}, \frac{1}{81}, \frac{1}{243}, \cdots$$

$$1, -\frac{1}{2}, \frac{1}{4}, -\frac{1}{8}, \frac{1}{16}, -\frac{1}{32}, \cdots$$

The indicated sum of the terms of an infinite geometric sequence is called an **infinite geometric series.**

An example of an infinite geometric series is shown at the right. The first term is 1. The common ratio is $\frac{1}{3}$.

$$1 + \frac{1}{3} + \frac{1}{9} + \frac{1}{27} + \frac{1}{81} + \frac{1}{243} + \cdots$$

The sum of the first, 5, 7, 12, and 15 terms, along with the values of r^n, are shown at the right. Note that as n increases, the sum of the terms is closer to 1.5, and the value of r^n is closer to zero.

n	S_n	r^n
5	1.4938272	0.0041152
7	1.4993141	0.0004572
12	1.4999972	0.0000019
15	1.4999999	0.0000001

Using the Formula for the Sum of n Terms of a Geometric Sequence and the fact that r^n approaches zero when $|r| < 1$ and n increases, a formula for an infinite geometric series can be found.

The sum of the first n terms of a geometric series is shown at the right. If $|r| < 1$, then r^n can be made very close to zero by using larger and larger values of n. Therefore, the sum of the first n terms is approximately $\frac{a_1}{1-r}$.

approximately zero

$$S_n = \frac{a_1(1 - r^n)}{1 - r}$$

$$S_n \approx \frac{a_1(1 - 0)}{1 - r} \approx \frac{a_1}{1 - r}$$

The Formula for the Sum of an Infinite Geometric Series

The sum of an infinite geometric series in which $|r| < 1$ is $S = \frac{a_1}{1-r}$.

When $|r| > 1$, the infinite geometric series does not have a sum. For example, the sum of the infinite geometric series $1 + 2 + 4 + 8 + \cdots$ increases without limit.

Find the sum of the infinite geometric sequence $1, -\frac{1}{2}, \frac{1}{4}, -\frac{1}{8}, \cdots$

The common ratio is $-\frac{1}{2}$. $\left|-\frac{1}{2}\right| < 1$.

Use the Formula for the Sum of an Infinite Geometric Series.

$$S = \frac{a_1}{1-r} = \frac{1}{1-\left(-\frac{1}{2}\right)} = \frac{1}{\frac{3}{2}} = \frac{2}{3}$$

The sum of an infinite geometric series can be applied to non-terminating, repeating decimals.

The repeating decimal shown at the right has been rewritten as an infinite geometric series, with first term $\frac{3}{10}$ and common ratio $\frac{1}{10}$.

$$0.33\overline{3} = 0.3 + 0.03 + 0.003 + \cdots$$
$$= \frac{3}{10} + \frac{3}{100} + \frac{3}{1000} + \cdots$$

Use the Formula for the Sum of an Infinite Geometric Series.

$$S = \frac{a_1}{1-r} = \frac{\frac{3}{10}}{1-\frac{1}{10}} = \frac{\frac{3}{10}}{\frac{9}{10}} = \frac{3}{9} = \frac{1}{3}$$

This method can be used to find an equivalent numerical fraction for any repeating decimal.

Find an equivalent fraction for $0.12\overline{2}$.

Write the decimal as an infinite geometric series. Note that the geometric series does not begin with the first term. The series begins with $\frac{2}{100}$. The common ratio is $\frac{1}{10}$.

$$0.12\overline{2} = 0.1 + 0.02 + 0.002 + 0.0002 + \cdots$$
$$= \frac{1}{10} + \frac{2}{100} + \frac{2}{1000} + \frac{2}{10,000} + \cdots$$

Use the Formula for the Sum of an Infinite Geometric Series.

$$S = \frac{a_1}{1-r} = \frac{\frac{2}{100}}{1-\frac{1}{10}} = \frac{\frac{2}{100}}{\frac{9}{10}} = \frac{2}{90}$$

Add $\frac{1}{10}$ to the sum of the geometric series.

$$0.12\overline{2} = \frac{1}{10} + \frac{2}{90} = \frac{11}{90}$$

An equivalent fraction is $\frac{11}{90}$.

Example 9

Find the sum of the infinite geometric sequence $2, \frac{1}{2}, \frac{1}{8}, \ldots$

Solution

$r = \dfrac{a_2}{a_1} = \dfrac{\frac{1}{2}}{2} = \dfrac{1}{4}$

$S = \dfrac{a_1}{1-r} = \dfrac{2}{1-\frac{1}{4}} = \dfrac{2}{\frac{3}{4}} = \dfrac{8}{3}$

Example 10

Find the sum of the infinite geometric sequence $3, -2, \frac{4}{3}, -\frac{8}{9}, \ldots$

Your solution

$\frac{9}{5}$

Example 11

Find an equivalent fraction for $0.36\overline{36}$.

Solution

$0.36\overline{36} = 0.36 + 0.0036 + 0.000036 + \cdots$

$\qquad = \dfrac{36}{100} + \dfrac{36}{10,000} + \dfrac{36}{1,000,000} + \cdots$

$S = \dfrac{a_1}{1-r} = \dfrac{\frac{36}{100}}{1-\frac{1}{100}} = \dfrac{\frac{36}{100}}{\frac{99}{100}} = \dfrac{36}{99} = \dfrac{4}{11}$

An equivalent fraction is $\frac{4}{11}$.

Example 12

Find an equivalent fraction for $0.6\overline{6}$.

Your solution

$\frac{2}{3}$

Objective D   To solve application problems

Example 13

On the first swing, the length of the arc through which a pendulum swings is 16 in. The length of each successive swing is $\frac{7}{8}$ of the preceding swing. Find the length of the arc on the fifth swing. Round to the nearest tenth.

Strategy

To find the length of the arc on the fifth swing, use the Formula for the nth Term of a Geometric Sequence.

Solution

$n = 5$, $a_1 = 16$, $r = \frac{7}{8}$

$a_n = a_1 r^{n-1}$

$a_5 = 16\left(\frac{7}{8}\right)^{5-1} = 16\left(\frac{7}{8}\right)^{4} = 16\left(\frac{2401}{4096}\right)$

$\quad = \frac{38,416}{4096} \approx 9.4$

The length of the arc on the fifth swing is 9.4 in.

Example 14

You start a chain letter and send it to three friends. Each of the three friends sends the letter to three other friends, and the sequence is repeated. Assuming no one breaks the chain, how many letters will have been mailed from the first through the sixth mailings?

Your strategy

Your solution

1092 letters

Solution on p. A50

12.3 EXERCISES

▶ **Objective A**

Find the indicated term of the geometric sequence.

1. $2, 8, 32, \ldots; a_9$
131,072

2. $4, 3, \frac{9}{4}, \ldots; a_8$
$\frac{2187}{4096}$

3. $6, -4, \frac{8}{3}, \ldots; a_7$
$\frac{128}{243}$

4. $-5, 15, -45, \ldots; a_7$
-3645

5. $-\frac{1}{16}, \frac{1}{8}, -\frac{1}{4}, \ldots; a_{10}$
32

6. $\frac{1}{27}, \frac{1}{9}, \frac{1}{3}, \ldots; a_7$
27

7. $1, \sqrt{2}, 2, \ldots; a_9$
16

8. $3, 3\sqrt{3}, 9, \ldots; a_8$
$81\sqrt{3}$

9. $-\frac{\sqrt{2}}{4}, \frac{1}{2}, -\frac{\sqrt{2}}{2}, \ldots; a_{11}$
$-8\sqrt{2}$

Find a_2 and a_3 for the geometric sequence.

10. $9, a_2, a_3, \frac{8}{3}, \ldots$
6, 4

11. $8, a_2, a_3, \frac{27}{8}, \ldots$
6, $\frac{9}{2}$

12. $3, a_2, a_3, -\frac{8}{9}, \ldots$
$-2, \frac{4}{3}$

13. $6, a_2, a_3, -48, \ldots$
$-12, 24$

14. $-3, a_2, a_3, 192, \ldots$
12, -48

15. $5, a_2, a_3, 625, \ldots$
25, 125

▶ **Objective B**

Find the sum of the indicated number of terms of the geometric sequence.

16. $2, 6, 18, \ldots; n = 7$
2186

17. $-4, 12, -36, \ldots; n = 7$
-2188

18. $12, 9, \frac{27}{4}, \ldots; n = 5$
$\frac{2343}{64}$

19. $3, 3\sqrt{2}, 6, \ldots; n = 12$
$189 + 189\sqrt{2}$

Find the sum of the geometric series.

20. $\sum_{i=1}^{5} (2)^i$
62

21. $\sum_{n=1}^{6} \left(\frac{3}{2}\right)^n$
$\frac{1995}{64}$

22. $\sum_{i=1}^{5} \left(\frac{1}{3}\right)^i$
$\frac{121}{243}$

23. $\sum_{n=1}^{6} \left(\frac{2}{3}\right)^n$
$\frac{1330}{729}$

▶ **Objective C**

Find the sum of the infinite geometric series.

24. $3 + 2 + \frac{4}{3} + \cdots$
9

25. $2 - \frac{1}{4} + \frac{1}{32} + \cdots$
$\frac{16}{9}$

Find the sum of the infinite geometric series.

26. $6 - 4 + \frac{8}{3} + \cdots$

$\frac{18}{5}$

27. $\frac{1}{10} + \frac{1}{100} + \frac{1}{1000} + \cdots$

$\frac{1}{9}$

28. $\frac{7}{10} + \frac{7}{100} + \frac{7}{1000} + \cdots$

$\frac{7}{9}$

29. $\frac{5}{100} + \frac{5}{10,000} + \frac{5}{1,000,000} + \cdots$

$\frac{5}{99}$

Find an equivalent fraction for the repeating decimal.

30. $0.8\overline{88}$

$\frac{8}{9}$

31. $0.5\overline{55}$

$\frac{5}{9}$

32. $0.2\overline{22}$

$\frac{2}{9}$

33. $0.9\overline{99}$

1

34. $0.45\overline{45}$

$\frac{5}{11}$

35. $0.18\overline{18}$

$\frac{2}{11}$

36. $0.16\overline{66}$

$\frac{1}{6}$

37. $0.83\overline{33}$

$\frac{5}{6}$

▶ Objective D   *Application Problems*

38. A laboratory ore sample contains 400 mg of a radioactive material with a half-life of 1 h. Find the amount of radioactive material in the sample at the beginning of the fourth hour.
50 mg

39. On the first swing, the length of the arc through which a pendulum swings is 20 in. The length of each successive swing is $\frac{4}{5}$ of the preceding swing. What is the total distance the pendulum has traveled during four swings? Round to the nearest tenth.
59 in.

40. To test the "bounce" of a tennis ball, the ball is dropped from a height of 10 ft. The ball bounces to 75% of its previous height with each bounce. How high does the ball bounce on the sixth bounce? Round to the nearest tenth.
1.8 ft

41. The temperature of a hot water spa is 80°F. Each hour the temperature is 5% higher than during the previous hour. Find the temperature of the spa after 5 h. Round to the nearest tenth.
102.1°F

42. A real estate broker estimates that a piece of land will increase in value at a rate of 15% each year. If the original value of the land is $25,000, what will be its value in 10 years?
$101,138.94

43. Suppose an employee receives a wage of 1¢ for the first day of work, 2¢ the second day, 4¢ the third day, and so on in a geometric sequence. Find the total amount of money earned for working 30 days.
$10,737,418.23

Content and Format © 1991 HMCo.

SECTION 12.4   Binomial Expansions

Objective A   To expand $(a + b)^n$

By carefully observing the series for each expansion of the binomial $(a + b)^n$ shown below, it is possible to identify some interesting patterns.

$(a + b)^1 = a + b$
$(a + b)^2 = a^2 + 2ab + b^2$
$(a + b)^3 = a^3 + 3a^2b + 3ab^2 + b^3$
$(a + b)^4 = a^4 + 4a^3b + 6a^2b^2 + 4ab^3 + b^4$
$(a + b)^5 = a^5 + 5a^4b + 10a^3b^2 + 10a^2b^3 + 5ab^4 + b^5$

Patterns for the Variable Part

1. The first term is a^n. The exponent on a decreases by 1 for each successive term.

2. The exponent on b increases by 1 for each successive term. The last term is b^n.

3. The degree of each term is n.

Write the variable parts of the terms of the expansion of $(a + b)^6$.

The first term is a^6. For each successive term, the exponent on a decreases by 1, and the exponent on b increases by 1. The last term is b^6.

$a^6, a^5b, a^4b^2, a^3b^3, a^2b^4, ab^5, b^6$

The variable parts of the general expansion of $(a + b)^n$ are:

$a^n, a^{n-1}b, a^{n-2}b^2, \ldots, a^{n-r}b^r, \ldots, ab^{n-1}, b^n$

A pattern for the coefficients of the terms of the expanded binomial can be found by writing the coefficients in a triangular array, which is known as **Pascal's Triangle.**

Each row begins and ends with the number 1. Any other number in a row is the sum of the two closest numbers above it. For example, $4 + 6 = 10$.

For $(a + b)^1$:   1   1
For $(a + b)^2$:   1   2   1
For $(a + b)^3$:   1   3   3   1
For $(a + b)^4$:   1   4   6   4   1
For $(a + b)^5$: 1   5   10   10   5   1

Write the sixth row of Pascal's Triangle.

To write the sixth row, first write the numbers of the fifth row. The first and last numbers of the sixth row are 1. Each of the other numbers of the sixth row can be obtained by finding the sum of the two closest numbers above it in the fifth row.

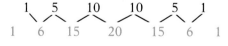

These numbers will be the coefficients of the terms of the expansion of $(a + b)^6$.

Using the numbers of the sixth row of Pascal's Triangle for the coefficients and the pattern for the variable part of each term, the expanded form of $(a + b)^6$ can be written:

$$(a + b)^6 = a^6 + 6a^5b + 15a^4b^2 + 20a^3b^3 + 15a^2b^4 + 6ab^5 + b^6$$

Although Pascal's Triangle can be used to find the coefficients for the expanded form of the power of any binomial, this method is inconvenient when the power of the binomial is large. An alternate method for determining these coefficients is based on the concept of **factorial.**

n Factorial

> $n!$ (read "n factorial") is the product of the first n consecutive natural numbers. $0!$ is defined to be 1.
>
> $$\mathbf{n! = n(n-1)(n-2) \ldots 3 \cdot 2 \cdot 1}$$

1!, 5!, and 7! are shown at the right.

$$1! = 1$$
$$5! = 5 \cdot 4 \cdot 3 \cdot 2 \cdot 1 = 120$$
$$7! = 7 \cdot 6 \cdot 5 \cdot 4 \cdot 3 \cdot 2 \cdot 1 = 5040$$

Evaluate: 6!

Write the factorial as a product.
Then simplify.

$$6! = 6 \cdot 5 \cdot 4 \cdot 3 \cdot 2 \cdot 1$$
$$= 720$$

Evaluate: $\frac{7!}{4!3!}$

Write each factorial as a product.

$$\frac{7!}{4!3!} = \frac{7 \cdot 6 \cdot 5 \cdot 4 \cdot 3 \cdot 2 \cdot 1}{(4 \cdot 3 \cdot 2 \cdot 1)(3 \cdot 2 \cdot 1)}$$

Then simplify.

$$= 35$$

The coefficients in a binomial expansion can be given in terms of factorials.

Note that in the expansion of $(a + b)^5$, the coefficient of a^2b^3 can be given by $\frac{5!}{2!3!}$. The numerator is the factorial of the power of the binomial. The denominator is the product of the factorials of the exponents on a and b.

$$(a + b)^5 =$$
$$a^5 + 5a^4b + 10a^3b^2 + 10a^2b^3 + 5ab^4 + b^5$$

$$\frac{5!}{2!3!} = \frac{5 \cdot 4 \cdot 3 \cdot 2 \cdot 1}{(2 \cdot 1)(3 \cdot 2 \cdot 1)} = 10$$

Binomial Coefficients

> The coefficients of $a^{n-r}b^r$ in the expression of $(a + b)^n$ are given by $\frac{n!}{(n-r)!r!}$. The symbol $\binom{n}{r}$ is used to express this quotient.
>
> $$\binom{n}{r} = \frac{n!}{(n-r)!r!}$$

Evaluate: $\binom{8}{5}$

Write the quotient of the factorials.
Then simplify.

$$\binom{8}{5} = \frac{8!}{(8-5)!5!} = \frac{8!}{3!5!}$$

$$= \frac{8 \cdot 7 \cdot 6 \cdot 5 \cdot 4 \cdot 3 \cdot 2 \cdot 1}{(3 \cdot 2 \cdot 1)(5 \cdot 4 \cdot 3 \cdot 2 \cdot 1)}$$

$$= 56$$

Evaluate: $\dbinom{7}{0}$

Write the quotient of the factorials. $\dbinom{7}{0} = \dfrac{7!}{(7-0)!0!} = \dfrac{7!}{7!0!}$

Recall that $0! = 1$. $= \dfrac{7 \cdot 6 \cdot 5 \cdot 4 \cdot 3 \cdot 2 \cdot 1}{(7 \cdot 6 \cdot 5 \cdot 4 \cdot 3 \cdot 2 \cdot 1)(1)}$

Simplify. $= 1$

Using factorials and the pattern for the variable part of each term, a formula for any natural number power of a binomial can be written.

The Binomial Expansion Formula

$$(a + b)^n = \dbinom{n}{0}a^n + \dbinom{n}{1}a^{n-1}b + \dbinom{n}{2}a^{n-2}b^2 + \cdots + \dbinom{n}{r}a^{n-r}b^r + \cdots + \dbinom{n}{n}b^n$$

The Binomial Expansion Formula is used below to expand $(a + b)^7$.

$(a + b)^7$

$= \dbinom{7}{0}a^7 + \dbinom{7}{1}a^6b + \dbinom{7}{2}a^5b^2 + \dbinom{7}{3}a^4b^3 + \dbinom{7}{4}a^3b^4 + \dbinom{7}{5}a^2b^5 + \dbinom{7}{6}ab^6 + \dbinom{7}{7}b^7$

$= a^7 + 7a^6b + 21a^5b^2 + 35a^4b^3 + 35a^3b^4 + 21a^2b^5 + 7ab^6 + b^7$

Write $(x - 2)^5$ in expanded form.

Use the Binomial Expansion Formula. Then simplify each term.

$(x - 2)^5$

$= \dbinom{5}{0}x^5 + \dbinom{5}{1}x^4(-2) + \dbinom{5}{2}x^3(-2)^2 + \dbinom{5}{3}x^2(-2)^3 + \dbinom{5}{4}x(-2)^4 + \dbinom{5}{5}(-2)^5$

$= x^5 + 5x^4(-2) + 10x^3(4) + 10x^2(-8) + 5x(16) + 1(-32)$

$= x^5 - 10x^4 + 40x^3 - 80x^2 + 80x - 32$

Write $(4x + 3y)^3$ in expanded form.

Use the Binomial Expansion Formula. Then simplify each term.

$(4x + 3y)^3 = \dbinom{3}{0}(4x)^3 + \dbinom{3}{1}(4x)^2(3y) + \dbinom{3}{2}(4x)(3y)^2 + \dbinom{3}{3}(3y)^3$

$= 1(64x^3) + 3(16x^2)(3y) + 3(4x)(9y^2) + 1(27y^3)$

$= 64x^3 + 144x^2y + 108xy^2 + 27y^3$

The Binomial Theorem can also be used to write any term of a binomial expansion.

Note that in the expansion of $(a + b)^5$, the exponent on b is one less than the term number.

$(a + b)^5$
$= a^5 + 5a^4b + 10a^3b^2 + 10a^2b^3 + 5ab^4 + b^5$

The Formula for the rth Term in a Binomial Expansion

The rth term of $(a + b)^n$ is $\dbinom{n}{r-1}a^{n-r+1}b^{r-1}$.

Find the 4th term in the expansion of $(x + 3)^7$.

Use the Formula for the rth Term in a Binomial Expansion. $r = 4$

$$\binom{n}{r-1}a^{n-r+1}b^{r-1}$$

$$\binom{7}{4-1}x^{7-4+1}(3)^{4-1} = \binom{7}{3}x^4(3)^3$$

Simplify.

$$= 35x^4(27)$$
$$= 945x^4$$

Find the 12th term in the expansion of $(2x - 1)^{14}$.

Use the Formula for the rth Term in a Binomial Expansion. $r = 12$
Simplify.

$$\binom{14}{12-1}(2x)^{14-12+1}(-1)^{12-1} = \binom{14}{11}(2x)^3(-1)^{11}$$

$$= 364(8x^3)(-1)$$
$$= -2912x^3$$

Example 1

Write $(2x + y)^3$ in expanded form.

Solution

$(2x + y)^3$

$$= \binom{3}{0}(2x)^3 + \binom{3}{1}(2x)^2(y) + \binom{3}{2}(2x)(y)^2 + \binom{3}{3}(y)^3$$
$$= 1(8x^3) + 3(4x^2)(y) + 3(2x)(y^2) + 1(y^3)$$
$$= 8x^3 + 12x^2y + 6xy^2 + y^3$$

Example 2

Write $(3m - n)^4$ in expanded form.

Your solution

$81m^4 - 108m^3n + 54m^2n^2 - 12mn^3 + n^4$

Example 3

Find the first 3 terms in the expansion of $(x + 3)^{15}$.

Solution

$(x + 3)^{15}$

$$= \binom{15}{0}x^{15} + \binom{15}{1}x^{14}(3) + \binom{15}{2}x^{13}(3)^2 + \cdots$$
$$= 1(x^{15}) + 15x^{14}(3) + 105x^{13}(9) + \cdots$$
$$= x^{15} + 45x^{14} + 945x^{13} + \cdots$$

Example 4

Find the first three terms in the expansion of $(y - 2)^{10}$.

Your solution

$y^{10} - 20y^9 + 180y^8$

Example 5

Find the 4th term in the expansion of $(5x - y)^6$.

Solution

$n = 6, a = 5x, b = -y, r = 4$

$$\binom{6}{4-1}(5x)^{6-4+1}(-y)^{4-1}$$
$$= \binom{6}{3}(5x)^3(-y)^3$$
$$= 20(125x^3)(-y^3)$$
$$= -2500x^3y^3$$

Example 6

Find the 3rd term in the expansion of $(r - 2s)^7$.

Your solution

$84r^5s^2$

Solutions on p. A50

12.4   EXERCISES

▶ **Objective A**

Evaluate:

1. 3!
6

2. 4!
24

3. 8!
40,320

4. 9!
362,880

5. 0!
1

6. 1!
1

7. $\frac{5!}{2!3!}$
10

8. $\frac{8!}{5!3!}$
56

9. $\frac{6!}{6!0!}$
1

10. $\frac{10!}{10!0!}$
1

11. $\frac{9!}{6!3!}$
84

12. $\frac{10!}{2!8!}$
45

Evaluate:

13. $\binom{7}{2}$
21

14. $\binom{8}{6}$
28

15. $\binom{10}{2}$
45

16. $\binom{9}{6}$
84

17. $\binom{9}{0}$
1

18. $\binom{10}{10}$
1

19. $\binom{6}{3}$
20

20. $\binom{7}{6}$
7

21. $\binom{11}{1}$
11

22. $\binom{13}{1}$
13

23. $\binom{4}{2}$
6

24. $\binom{8}{4}$
70

Write in expanded form.

25. $(x + y)^4$
$x^4 + 4x^3y + 6x^2y^2 + 4xy^3 + y^4$

26. $(r - s)^3$
$r^3 - 3r^2s + 3rs^2 - s^3$

27. $(x - y)^5$
$x^5 - 5x^4y + 10x^3y^2 - 10x^2y^3 + 5xy^4 - y^5$

28. $(y - 3)^4$
$y^4 - 12y^3 + 54y^2 - 108y + 81$

29. $(2m + 1)^4$
$16m^4 + 32m^3 + 24m^2 + 8m + 1$

30. $(2x + 3y)^3$
$8x^3 + 36x^2y + 54xy^2 + 27y^3$

31. $(2r - 3)^5$
$32r^5 - 240r^4 + 720r^3 - 1080r^2 + 810r - 243$

32. $(x + 3y)^4$
$x^4 + 12x^3y + 54x^2y^2 + 108xy^3 + 81y^4$

Find the first three terms in the expansion.

33. $(a + b)^{10}$
$a^{10} + 10a^9b + 45a^8b^2$

34. $(a + b)^9$
$a^9 + 9a^8b + 36a^7b^2$

35. $(a - b)^{11}$
$a^{11} - 11a^{10}b + 55a^9b^2$

36. $(a - b)^{12}$
$a^{12} - 12a^{11}b + 66a^{10}b^2$

37. $(2x + y)^8$
$256x^8 + 1024x^7y + 1792x^6y^2$

38. $(x + 3y)^9$
$x^9 + 27x^8y + 324x^7y^2$

39. $(4x - 3y)^8$
$65,536x^8 - 393,216x^7y + 1,032,192x^6y^2$

40. $(2x - 5)^7$
$128x^7 - 2240x^6 + 16,800x^5$

41. $\left(x + \dfrac{1}{x}\right)^7$
$x^7 + 7x^5 + 21x^3$

42. $\left(x - \dfrac{1}{x}\right)^8$
$x^8 - 8x^6 + 28x^4$

43. $(x^2 + 3)^5$
$x^{10} + 15x^8 + 90x^6$

44. $(x^2 - 2)^6$
$x^{12} - 12x^{10} + 60x^8$

Find the indicated term in the expansion.

45. $(2x - 1)^7$; 4th term
$-560x^4$

46. $(x + 4)^5$; 3rd term
$160x^3$

47. $(x^2 - y^2)^6$; 2nd term
$-6x^{10}y^2$

48. $(x^2 + y^2)^7$; 6th term
$21x^4y^{10}$

49. $(y - 1)^9$; 5th term
$126y^5$

50. $(x - 2)^8$; 8th term
$-1024x$

51. $\left(n + \dfrac{1}{n}\right)^5$; 2nd term
$5n^3$

52. $\left(x + \dfrac{1}{2}\right)^6$; 3rd term
$\dfrac{15}{4}x^4$

53. $\left(\dfrac{x}{2} + 2\right)^5$; 1st term
$\dfrac{x^5}{32}$

54. $\left(y - \dfrac{2}{3}\right)^6$; 3rd term
$\dfrac{20}{3}y^4$

Calculators and Computers

Continued Fractions   An example of a **continued fraction** is

$$1 + \cfrac{1}{2 + \cfrac{9}{2 + \cfrac{25}{2 + \cfrac{49}{2 + \cdots}}}}$$

The pattern continues with each numerator the square of an odd integer. Approximations to these fractions can be found by using a calculator.

1. To evaluate the continued fraction above, begin by dividing 49 by 2. $49\ \boxed{\div}\ 2$

2. Now add 2. Store the result in memory. $\boxed{+}\ 2\ \boxed{M+}$

3. Divide 25 by the result in memory. Then add 2 and store in memory. The MC clears the memory before the new result is stored. $25\ \boxed{\div}\ \boxed{MR}\ \boxed{+}\ 2\ \boxed{MC}\ \boxed{M+}$

4. Repeat step 3 but divide 9 by the result in memory.

5. Repeat step 3 but divide 1 by the result in memory and add 1 instead of 2.

The result is 1.197719. This value approximates $\frac{4}{\pi}$. In fact, as the fraction is continued, the approximation becomes closer and closer to $\frac{4}{\pi}$.

Here are two other continued fractions. Evaluate them by using your calculator.

$$1 + \cfrac{1}{2 + \cfrac{1}{2 + \cfrac{1}{2 + \cdots}}} \qquad 1 + \cfrac{1}{1 + \cfrac{1}{1 + \cfrac{1}{1 + \cdots}}}$$

The first continued fraction is approximately $\sqrt{2}$. As the fraction is continued, the approximation becomes closer to $\sqrt{2}$. The second continued fraction is approximately $\frac{1+\sqrt{5}}{2}$. This number is called the "golden ratio" and has played an important role in art and architecture.

Chapter Summary

Key Words   A *sequence* is an ordered list of numbers.

Each of the numbers of a sequence is called a *term* of the sequence.

A *finite sequence* contains a finite number of terms.

An *infinite sequence* contains an infinite number of terms.

The indicated sum of a sequence is a *series*.

An *arithmetic sequence*, or arithmetic progression, is one in which the difference between any two consecutive terms is constant. The difference between consecutive terms is called the *common difference* of the sequence.

A *geometric sequence*, or geometric progression, is one in which each successive term of the sequence is the same non-zero constant multiple of the preceding term. The common multiple is called the *common ratio* of the sequence.

n factorial, written *n!*, is the product of the first *n* positive integers.

Essential Rules

Formula for the nth Term of an Arithmetic Sequence

The *n*th term of an arithmetic sequence with a common difference of *d* is given by $a_n = a_1 + (n - 1)d$.

Formula for the Sum of n Terms of an Arithmetic Sequence

Let a_1 be the first term of a finite arithmetic sequence, *n* the number of terms, and a_n the last term of the sequence. Then the sum of the series S_n is given by $S_n = \frac{n(a_1 + a_n)}{2}$.

Formula for the nth Term of a Geometric Sequence

The *n*th term of a geometric sequence with first term a_1 and common ratio *r* is given by $a_n = a_1 r^{n-1}$.

Formula for the Sum of n Terms of a Geometric Series

Let a_1 be the first term of a finite geometric sequence, *n* the number of terms, and *r* the common ratio. Then the sum of the series S_n is given by $S_n = \frac{a_1(1 - r^n)}{1 - r}$.

Formula for the Sum of an Infinite Geometric Series

The sum of an infinite geometric series in which $|r| < 1$ and a_1 is the first term is given by $S = \frac{a_1}{1 - r}$.

The Binomial Formula

$$(a + b)^n = \binom{n}{0}a^n + \binom{n}{1}a^{n-1}b + \binom{n}{2}a^{n-2}b^2 + \cdots + \binom{n}{n-1}ab^{n-1} + \binom{n}{n}b^n$$

Formula for the rth Term in a Binomial Expansion

The *r*th term of $(a + b)^n$ is $\binom{n}{r-1}a^{n-r+1}b^{r-1}$.

Chapter Review

SECTION 1

1. Write the 10th term of the sequence whose nth term is given by the formula $a_n = 3n - 2$.

28

2. Write the 5th term of the sequence whose nth term is given by the formula

$$a_n = \frac{(-1)^{2n-1}n}{n^2 + 2}.$$

$-\frac{5}{27}$

3. Find the sum of the series $\sum_{n=1}^{5} (3n - 2)$.

35

4. Find the sum of the series $\sum_{n=1}^{4} \frac{(-1)^{n-1}n}{n + 1}$.

$-\frac{13}{60}$

5. Write $\sum_{i=1}^{4} 2x^{i-1}$ in expanded form.

$2 + 2x + 2x^2 + 2x^3$

6. Write $\sum_{i=1}^{5} \frac{(2x)^i}{i}$ in expanded form.

$2x + 2x^2 + \frac{8x^3}{3} + 4x^4 + \frac{32x^5}{5}$

SECTION 2

7. Find the 20th term of the arithmetic sequence $1, 5, 9, \ldots$.

77

8. Find the formula for the nth term of the arithmetic sequence $-7, -2, 3, \ldots$.

$5n - 12$

9. Find the number of terms in the finite arithmetic sequence $8, 2, -4, \ldots, -118$.

22 terms

10. Find the sum of the first 40 terms of the arithmetic sequence $11, 13, 15, \ldots$.

2000

11. Find the sum of the arithmetic series

$$\sum_{i=1}^{30} (4i - 1).$$

1830

12. The salary schedule for an apprentice electrician is \$1200 for the first month and a \$40-per-month salary increase for the next nine months. Find the total salary for the nine-month period.

\$12,240

SECTION 3

13. Find the 8th term of the geometric sequence $3, 1, \frac{1}{3}, \ldots$.

$\frac{1}{729}$

14. Find the 12th term of the geometric sequence $1, \sqrt{3}, 3, \ldots$.

$243\sqrt{3}$

15. Find the sum of the first seven terms of the geometric sequence $5, 5\sqrt{5}, 25, \ldots$. Round to the nearest whole number.
1127

16. Find the sum of the geometric series
$$\sum_{n=1}^{5} 2(3)^n.$$

726

17. Find the sum of the geometric series
$$\sum_{n=1}^{8} \left(\frac{1}{2}\right)^n.$$ Round to the nearest thousandth.

0.996

18. Find the sum of the infinite geometric series $2 + \frac{4}{3} + \frac{8}{9} + \cdots$.
6

19. Find the sum of the infinite geometric series $\frac{3}{10} + \frac{3}{100} + \frac{3}{1000} + \cdots$.
$\frac{1}{3}$

20. Find the sum of the infinite geometric series $4 - 1 + \frac{1}{4} - \cdots$.
$\frac{16}{5}$

21. Find an equivalent fraction for $0.2\overline{23}$.
$\frac{23}{99}$

22. Find an equivalent fraction for $0.63\overline{3}$.
$\frac{19}{30}$

23. The temperature of a hot water spa is 102°F. Each hour the temperature is 5% lower than during the previous hour. Find the temperature of the spa after 8 hours. Round to the nearest tenth.
67.7°F

24. A laboratory radioactive sample contains 200 mg of a material with a half-life of 1 h. Find the amount of radioactive material in the sample at the beginning of the seventh hour.
3.125 mg

SECTION 4

25. Evaluate: $\frac{12!}{5!8!}$
99

26. Evaluate: $\binom{12}{9}$
220

27. Evaluate: $\binom{9}{3}$
84

28. Write $(x - 3y^2)^5$ in expanded form.
$x^5 - 15x^4y^2 + 90x^3y^4 - 270x^2y^6 + 405xy^8 - 243y^{10}$

29. Find the fifth term in the expansion of $(3x - y)^8$.
$5670x^4y^4$

30. Find the eighth term in the expansion of $(x - 2y)^{11}$.
$-42{,}240x^4y^7$

Chapter Test

1. Write the 14th term of the sequence whose nth term is given by the formula $a_n = \frac{8}{n+2}$.

   $\frac{1}{2}$ [12.1A]

2. Write the 6th and 7th terms of the sequence whose nth term is given by the formula $a_n = \frac{n+1}{n}$.

   $\frac{7}{6}, \frac{8}{7}$ [12.1A]

3. Find the sum of the series

   $\sum_{n=1}^{4} (3n + 1)$.

   34 [12.1B]

4. Write in expanded form:

   $\sum_{i=1}^{4} 3x^i$

   $3x + 3x^2 + 3x^3 + 3x^4$ [12.1B]

5. Find the 35th term of the arithmetic sequence $-13, -16, -19, \ldots$.

   -115 [12.2A]

6. Find the formula for the nth term of the arithmetic sequence $12, 9, 6, \ldots$.

   $a_n = -3n + 15$ [12.2A]

7. Find the number of terms in the finite arithmetic sequence $-5, -8, -11, \ldots, -50$.

   16 [12.2A]

8. Find the sum of the first 18 terms of the arithmetic sequence $-25, -19, -13, \ldots$.

   468 [12.2B]

9. Find the sum of the first 21 terms of the arithmetic sequence $5, 12, 19, \ldots$.

   1575 [12.2B]

10. An inventory of supplies for a fabric manufacturer indicated that 7500 yd of material were in stock on January 1. On February 1, and on the first of the month for each successive month, the manufacturer sent 550 yd of material to retail outlets. How much material was in stock after the shipment on October 1?

    2550 yd [12.2C]

11. Find the 7th term of the geometric sequence
4, 4√2, 8, ...
32   [12.3A]

12. Find the 5th term of the geometric sequence
6, 2, $\frac{2}{3}$, ...
$\frac{2}{27}$ [12.3A]

13. Find the sum of the first six terms of the geometric sequence 1, $\frac{3}{2}$, $\frac{9}{4}$, ...
$\frac{665}{32}$ [12.3B]

14. Find the sum of the first five terms of the geometric sequence −6, 12, −24, ...
−66   [12.3B]

15. Find the sum of the infinite geometric sequence 4, 3, $\frac{9}{4}$, ...
16   [12.3B]

16. Find an equivalent fraction for 0.2333̄.
$\frac{7}{30}$ [12.3C]

17. An ore sample contains 320 mg of a radioactive substance with a half-life of one day. Find the amount of radioactive material in the sample at the beginning of the fifth day.
20 mg   [12.3D]

18. Evaluate: $\frac{8!}{4!4!}$
70   [12.4A]

19. Evaluate: $\binom{9}{3}$
84   [12.4A]

20. Find the 4th term in the expansion of $(x - 2y)^7$.
$-280x^4y^3$ [12.4A]

Cumulative Review

1. Graph: $3x - 2y = -4$
   [6.1D]

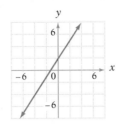

2. Factor: $2x^6 + 16$
   $2(x^2 + 2)(x^4 - 2x^2 + 4)$ [3.4D]

3. Simplify: $\frac{4x^2}{x^2 + x - 2} - \frac{3x - 2}{x + 2}$

   $\frac{x^2 + 5x - 2}{(x + 2)(x - 1)}$ [4.3B]

4. Simplify: $\left(\frac{x^{-3/4} \cdot x^{3/2}}{x^{-5/2}}\right)^{-8}$

   $\frac{1}{x^{26}}$ [5.1A]

5. Simplify: $\sqrt{2y}\,(\sqrt{8xy} - \sqrt{y})$

   $4y\sqrt{x} - y\sqrt{2}$ [5.2C]

6. Solve by using the quadratic formula:
   $2x^2 - x + 7 = 0$

   $\frac{1}{4} + \frac{\sqrt{55}}{4}i$ and $\frac{1}{4} - \frac{\sqrt{55}}{4}i$ [7.3A]

7. Solve: $5 - \sqrt{x} = \sqrt{x + 5}$

   $x = 4$ [5.4A]

8. Find the equation of the circle that passes through point $(4, 2)$ and whose center is $(-1, -1)$.

   $(x + 1)^2 + (y + 1)^2 = 34$ [9.2B]

9. Solve by the addition method:
   $3x - 3y = 2$
   $6x - 4y = 5$

   $\left(\frac{7}{6}, \frac{1}{2}\right)$ [10.2A]

10. Evaluate the determinant:
    $\begin{vmatrix} -3 & 1 \\ 4 & 2 \end{vmatrix}$

    -10 [10.3A]

11. Solve: $2x - 1 > 3$ or $1 - 3x > 7$
    $\{x \,|\, x < -2 \text{ or } x > 2\}$ [2.2B]

12. Graph: $2x - 3y < 9$
    [6.6A]

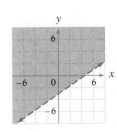

13. Write $\log_5 \sqrt{\dfrac{x}{y}}$ in expanded form.

    $\frac{1}{2}\log_5 x - \frac{1}{2}\log_5 y$ [11.2A]

14. Solve for x: $4^x = 8^{x-1}$
    $x = 3$ [11.4A]

15. Write the 5th and 6th terms of the sequence whose nth term is given by the formula $a_n = n(n - 1)$.
    $a_5 = 20$
    $a_6 = 30$ [12.1A]

16. Find the sum of the series

    $\displaystyle\sum_{n=1}^{7} (-1)^{n-1}(n + 2).$

    6 [12.1B]

17. Find the 33rd term of the arithmetic sequence $-7, -10, -13, \ldots$.
-103 [12.2A]

18. Find the sum of the infinite geometric sequence $3, -2, \frac{4}{3}, \ldots$.

$\frac{9}{5}$ [12.3C]

19. Find an equivalent fraction for $0.4\overline{6}$.
$\frac{7}{15}$ [12.3C]

20. Find the 6th term in the expansion of $(2x + y)^6$.
$12xy^5$ [12.4A]

21. How many ounces of pure water must be added to 200 oz of an 8% salt solution to make a 5% salt solution?
120 oz   [2.6B]

22. A new computer can complete a payroll in 16 minutes less time than it takes an older computer to complete the same payroll. Working together, both computers can complete the payroll in 15 minutes. How long would it take each computer working alone to complete the payroll?
new computer: 24 min;   old computer: 40 min   [4.6C]

23. A boat traveling with the current went 15 mi in 2 h. Against the current it took 3 h to travel the same distance. Find the rate of the boat in calm water and the rate of the current.
boat: 6.25 mph;   current: 1.25 mph   [4.6D]

24. An 80 mg sample of a radioactive material decays to 55 mg in 30 days. Use the exponential decay equation $A = A_0 \left(\frac{1}{2}\right)^{t/k}$ where A is the amount of radioactive material present after time t, k is the half-life, and A_0 is the original amount of radioactive material, to find the half-life of the 80-mg sample. Round to the nearest whole number.
55 days   [11.5A]

25. A "theatre in the round" has 62 seats in the first row, 74 seats in the second row, 86 seats in the third row, and so on in an arithmetic sequence. Find the total number of seats in the theatre if there are 12 rows of seats.
1536 seats   [12.2C]

26. To test the "bounce" of a ball, the ball is dropped from a height of 8 ft. The ball bounces to 80% of its previous height with each bounce. How high does the ball bounce on the fifth bounce? Round to the nearest tenth.
2.6 ft   [12.3D]

Final Exam

1. Simplify:
   $12 - 8[3 - (-2)]^2 \div 5 - 3$
   -31 [1.2B]

2. Evaluate $\frac{a^2 - b^2}{a - b}$ when $a = 3$ and
   $b = -4$.
   -1 [1.3A]

3. Simplify: $5 - 2[3x - 7(2 - x) - 5x]$
   $33 - 10x$ [1.3C]

4. Solve: $\frac{3}{4}x - 2 = 4$
   $x = 8$ [2.1B]

5. Solve: $\frac{2 - 4x}{3} - \frac{x - 6}{12} = \frac{5x - 2}{6}$
   $x = \frac{2}{3}$ [2.1C]

6. Solve: $8 - |5 - 3x| = 1$
   $4, -\frac{2}{3}$ [2.3A]

7. Graph $2x - 3y = 9$ by using the x- and y-intercepts.

   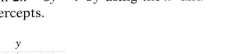

   x-intercept: $\left(\frac{9}{2}, 0\right)$

   y-intercept: $(0, -3)$ [6.2B]

8. Find the equation of the line containing the points $(3, -2)$ and $(1, 4)$.
   $y = -3x + 7$ [6.3B]

9. Find the equation of the line containing the point $(-2, 1)$ and perpendicular to the line $3x - 2y = 6$.
   $y = -\frac{2}{3}x - \frac{1}{3}$ [6.4A]

10. Simplify: $2a[5 - a(2 - 3a) - 2a] + 3a^2$
    $6a^3 - 5a^2 + 10a$ [3.2A]

11. Factor: $8 - x^3y^3$
$(2 - xy)(4 + 2xy + x^2y^2)$ [3.4B]

12. Factor: $x - y - x^3 + x^2y$
$(x - y)(1 + x)(1 - x)$ [3.4D]

13. Simplify: $(2x^3 - 7x^2 + 4) \div (2x - 3)$
$x^2 - 2x - 3 - \dfrac{5}{2x - 3}$ [4.2A]

14. Simplify: $\dfrac{x^2 - 3x}{2x^2 - 3x - 5} \div \dfrac{4x - 12}{4x^2 - 4}$
$\dfrac{x(x - 1)}{2x - 5}$ [4.1C]

15. Simplify: $\dfrac{x - 2}{x + 2} - \dfrac{x + 3}{x - 3}$
$\dfrac{-10x}{(x + 2)(x - 3)}$ [4.3B]

16. Simplify: $\dfrac{\dfrac{3}{x} + \dfrac{1}{x + 4}}{\dfrac{1}{x} + \dfrac{3}{x + 4}}$
$\dfrac{x + 3}{x + 1}$ [4.4A]

17. Solve: $\dfrac{5}{x - 2} - \dfrac{5}{x^2 - 4} = \dfrac{1}{x + 2}$
$x = -\dfrac{7}{4}$ [4.6A]

18. Solve $a_n = a_1 + (n - 1)d$ for d.
$d = \dfrac{a_n - a_1}{n - 1}$ [4.6B]

19. Simplify: $\left(\dfrac{4x^2y^{-1}}{3x^{-1}y}\right)^{-2}\left(\dfrac{2x^{-1}y^2}{9x^{-2}y^2}\right)^3$
$\dfrac{y^4}{162x^3}$ [3.1C]

20. Simplify: $\left(\dfrac{3x^{2/3}y^{1/2}}{6x^2y^{4/3}}\right)^6$
$\dfrac{1}{64x^8y^5}$ [5.1A]

21. Simplify: $x\sqrt{18x^2y^3} - y\sqrt{50x^4y}$
$-2x^2y\sqrt{2y}$ [5.2B]

22. Simplify: $\dfrac{\sqrt{16x^5y^4}}{\sqrt{32xy^7}}$
$\dfrac{x^2\sqrt{2y}}{2y^2}$ [5.2D]

Final Exam

23. Simplify: $\frac{3}{2+i}$

$\frac{6}{5} - \frac{3}{5} i$ [5.3D]

24. Write a quadratic equation that has integer coefficients and has solutions $-\frac{1}{2}$ and 2.

$2x^2 - 3x - 2 = 0$ [7.1B]

25. Solve by using the quadratic formula: $2x^2 - 3x - 1 = 0$

$\frac{3 + \sqrt{17}}{4}$ and $\frac{3 - \sqrt{17}}{4}$ [7.3A]

26. Solve: $x^{2/3} - x^{1/3} - 6 = 0$

27, -8 [7.4A]

27. Solve: $\frac{2}{x} - \frac{2}{2x + 3} = 1$

$\frac{3}{2}, -2$ [7.4C]

28. Graph $f(x) = -x^2 + 4$.

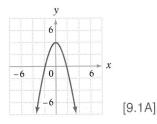

[9.1A]

29. Graph: $\frac{x^2}{16} + \frac{y^2}{4} = 1$

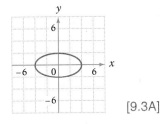

[9.3A]

30. Find the inverse of the function $f(x) = \frac{2}{3}x - 4$.

$f^{-1}(x) = \frac{3}{2}x + 6$ [8.3B]

31. Solve by the addition method:
$3x - 2y = 1$
$5x - 3y = 3$
(3, 4)   [10.2A]

32. Evaluate the determinant:
$\begin{vmatrix} 3 & 4 \\ -1 & 2 \end{vmatrix}$

10   [10.3A]

33. Solve: $x^2 - y^2 = 4$
$x + y = 1$
$\left(\frac{5}{2}, -\frac{3}{2}\right)$ [10.5A]

34. Solve:
$2 - 3x < 6$ and $2x + 1 > 4$
$\left\{x \mid x > \frac{3}{2}\right\}$ [2.2B]

35. Solve: $|2x + 5| < 3$
$\{x \mid -4 < x < -1\}$ [2.3B]

36. Graph the solution set of $3x + 2y > 6$.

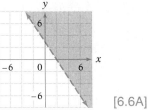

[6.6A]

37. Graph: $f(x) = \log_2(x + 1)$

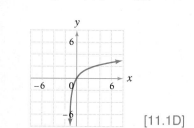

[11.1D]

38. Write $2(\log_2 a - \log_2 b)$ as a single logarithm with a coefficient of 1.

$\log_2 \frac{a^2}{b^2}$ [11.2A]

39. Solve for x: $\log_3 x - \log_3(x - 3) = \log_3 2$
$x = 6$ [11.4B]

40. Write $\sum\limits_{i=1}^{5} 2y^i$ in expanded form.

$2y + 2y^2 + 2y^3 + 2y^4 + 2y^5$ [12.1B]

41. Find an equivalent fraction for $0.5\overline{1}$.
$\frac{23}{45}$ [12.3C]

42. Find the third term in the expansion $(x - 2y)^9$.
$144x^7 y^2$ [12.4A]

Final Exam

43. An average score of 70–79 in a history class receives a C grade. A student has grades of 64, 58, 82, and 77 on four history tests. Find the range of scores on the fifth test that will give the student a C grade for the course.
$69 \leq x \leq 100$ [2.2C]

44. A jogger and a cyclist set out at 8 A.M. from the same point headed in the same direction. The average speed of the cyclist is two and a half times the speed of the jogger. In 2 h, the cyclist is 24 mi ahead of the jogger. How far did the cyclist ride?
40 mi   [2.5B]

45. You have a total of $12,000 invested in two simple interest accounts. On one account, a money market fund, the annual simple interest rate is 8.5%. On the other account, a tax free bond fund, the annual simple interest rate is 6.4%. The total annual interest earned by the two accounts is $936. How much do you have invested in each account?
$8000 at 8.5%
$4000 at 6.4%   [2.6A]

46. The length of a rectangle is one less than three times the width. The area of the rectangle is 140 ft². Find the length and width of the rectangle.
length: 20 ft
width: 7 ft   [7.6A]

47. Three hundred shares of a utility stock earns a yearly dividend of $486. How many additional shares of the utility stock would give a total dividend income of $810?
200 additional shares   [8.4A]

48. An account executive traveled 45 mi by car and then an additional 1050 mi by plane. The rate of the plane was seven times the rate of the car. The total time of the trip was $3\frac{1}{4}$ h. Find the rate of the plane.
420 mph   [10.4B]

49. An object is dropped from the top of a building. Find the distance the object has fallen when the speed reaches 75 ft/s. Use the equation $v = \sqrt{64d}$, where v is the speed of the object and d is the distance. Round to the nearest whole number.
88 ft   [5.4B]

50. A small plane made a trip of 660 mi in 5 h. The plane traveled the first 360 mi at a constant rate before increasing its speed by 30 mph. Another 300 mi was traveled at the increased speed. Find the rate of the plane for the first 360 mi.
120 mph   [7.6A]

51. The intensity (L) of a light source is inversely proportional to the square of the distance (d) from the source. If the intensity is 8 lumens at a distance of 20 ft, what is the intensity when the distance is 4 ft?
200 lumens   [8.4A]

52. A motorboat traveling with the current can go 30 mi in 2 h. Against the current, it takes 3 h to go the same distance. Find the rate of the motorboat in calm water and the rate of the current.
boat: 12.5 mph
current: 2.5 mph   [10.4A]

53. An investor deposits $4000 into an account that earns 9% annual interest compounded monthly. Use the compound interest formula $P = A(1 + i)^n$, where A is the original value of the investment, i is the interest rate per compounding period, n is the total number of compounding periods, and P is the value of the investment after n periods, to find the value of the investment after two years. Round to the nearest cent.
$4785.65   [11.5A]

54. Assume the average value of a home increases 6% per year. How much would a house costing $80,000 be worth in 20 years? Round to the nearest dollar.
$256,571   [12.3D]

APPENDIX

Table of Properties

Properties of Real Numbers

Associative Property of Addition

If a, b, and c are real numbers, then $(a + b) + c = a + (b + c)$.

Commutative Property of Addition

If a and b are real numbers, then $a + b = b + a$.

Addition Property of Zero

If a is a real number, then $a + 0 = 0 + a = a$.

Inverse Property of Addition

If a is a real number, then $a + (-a) = (-a) + a = 0$.

Associative Property of Multiplication

If a, b, and c are real numbers, then $(a \cdot b) \cdot c = a \cdot (b \cdot c)$.

Commutative Property of Multiplication

If a and b are real numbers, then $a \cdot b = b \cdot a$.

Multiplication Property of One

If a is a real number, then $a \cdot 1 = 1 \cdot a = a$.

Inverse Property of Multiplication

If a is a real number and $a \neq 0$, then $a \cdot \dfrac{1}{a} = \dfrac{1}{a} \cdot a = 1$.

Distributive Property

If a, b, and c are real numbers, then $a(b + c) = ab + ac$.

Properties of Equations

Addition Property of Equations

If $a = b$, then $a + c = b + c$.

Multiplication Property of Equations

If $a = b$ and $c \neq 0$, then $a \cdot c = b \cdot c$.

Properties of Exponents

If m and n are integers, then $x^m \cdot x^n = x^{m+n}$.

If m and n are integers, then $(x^m)^n = x^{m \cdot n}$.

If x is a real number and $x \neq 0$, then $x^0 = 1$.

If m and n are integers and $x \neq 0$, then $\dfrac{x^m}{x^n} = x^{m-n}$.

If m, n, and p are integers, then $(x^m \cdot y^n)^p = x^{m \cdot p} y^{n \cdot p}$.

If n is a positive integer and $x \neq 0$, then $x^{-n} = \dfrac{1}{x^n}$.

Property of Zero Products

If $a \cdot b = 0$, then $a = 0$ or $b = 0$.

Properties of Radical Expressions

If a and b are positive real numbers, then $\sqrt{ab} = \sqrt{a} \cdot \sqrt{b}$.

If a and b are positive real numbers and $b \neq 0$, then $\sqrt{\dfrac{a}{b}} = \dfrac{\sqrt{a}}{\sqrt{b}}$.

Properties of Inequalities

Addition Property of Inequalities

If $a > b$, then $a + c > b + c$.
If $a < b$, then $a + c < b + c$.

Multiplication Property of Inequalities

If $a > b$ and $c > 0$, then $ac > bc$.
If $a < b$ and $c > 0$, then $ac < bc$.
If $a > b$ and $c < 0$, then $ac < bc$.
If $a < b$ and $c < 0$, then $ac > bc$.

Property of Raising Both Sides of an Equation to a Power

If a and b are real numbers and $a = b$, then $a^n = b^n$.

Table of Square and Cube Roots

Decimal approximations have been rounded to the nearest thousandth.

Number	Square Root	Cube Root	Number	Square Root	Cube Root
1	1	1	51	7.141	3.708
2	1.414	1.260	52	7.211	3.733
3	1.732	1.442	53	7.280	3.756
4	2	1.587	54	7.348	3.780
5	2.236	1.710	55	7.416	3.803
6	2.449	1.817	56	7.483	3.826
7	2.646	1.913	57	7.550	3.849
8	2.828	2	58	7.616	3.871
9	3	2.080	59	7.681	3.893
10	3.162	2.154	60	7.746	3.915
11	3.317	2.224	61	7.810	3.936
12	3.464	2.289	62	7.874	3.958
13	3.606	2.351	63	7.937	3.979
14	3.742	2.410	64	8	4
15	3.873	2.466	65	8.062	4.021
16	4	2.520	66	8.124	4.041
17	4.123	2.571	67	8.185	4.062
18	4.243	2.621	68	8.246	4.082
19	4.359	2.668	69	8.307	4.102
20	4.472	2.714	70	8.367	4.121
21	4.583	2.759	71	8.426	4.141
22	4.690	2.802	72	8.485	4.160
23	4.796	2.844	73	8.544	4.179
24	4.899	2.884	74	8.602	4.198
25	5	2.924	75	8.660	4.217
26	5.099	2.962	76	8.718	4.236
27	5.196	3	77	8.775	4.254
28	5.292	3.037	78	8.832	4.273
29	5.385	3.072	79	8.888	4.291
30	5.477	3.107	80	8.944	4.309
31	5.568	3.141	81	9	4.327
32	5.657	3.175	82	9.055	4.344
33	5.745	3.208	83	9.110	4.362
34	5.831	3.240	84	9.165	4.380
35	5.916	3.271	85	9.220	4.397
36	6	3.302	86	9.274	4.414
37	6.083	3.332	87	9.327	4.431
38	6.164	3.362	88	9.381	4.448
39	6.245	3.391	89	9.434	4.465
40	6.325	3.420	90	9.487	4.481
41	6.403	3.448	91	9.539	4.498
42	6.481	3.476	92	9.592	4.514
43	6.557	3.503	93	9.644	4.531
44	6.633	3.530	94	9.695	4.547
45	6.708	3.557	95	9.747	4.563
46	6.782	3.583	96	9.798	4.579
47	6.856	3.609	97	9.849	4.595
48	6.928	3.634	98	9.899	4.610
49	7	3.659	99	9.950	4.626
50	7.071	3.684	100	10	4.642

Table of Common Logarithms

Decimal approximations have been rounded to the nearest ten-thousandth.

x	0	1	2	3	4	5	6	7	8	9
1.0	.0000	.0043	.0086	.0128	.0170	.0212	.0253	.0294	.0334	.0374
1.1	.0414	.0453	.0492	.0531	.0569	.0607	.0645	.0682	.0719	.0755
1.2	.0792	.0828	.0864	.0899	.0934	.0969	.1004	.1038	.1072	.1106
1.3	.1139	.1173	.1206	.1239	.1271	.1303	.1335	.1367	.1399	.1430
1.4	.1461	.1492	.1523	.1553	.1584	.1614	.1644	.1673	.1703	.1732
1.5	.1761	.1790	.1818	.1847	.1875	.1903	.1931	.1959	.1987	.2014
1.6	.2041	.2068	.2095	.2122	.2148	.2175	.2201	.2227	.2253	.2279
1.7	.2304	.2330	.2355	.2380	.2405	.2430	.2455	.2480	.2504	.2529
1.8	.2553	.2577	.2601	.2625	.2648	.2672	.2695	.2718	.2742	.2765
1.9	.2788	.2810	.2833	.2856	.2878	.2900	.2923	.2945	.2967	.2989
2.0	.3010	.3032	.3054	.3075	.3096	.3118	.3139	.3160	.3181	.3201
2.1	.3222	.3243	.3263	.3284	.3304	.3324	.3345	.3365	.3385	.3404
2.2	.3424	.3444	.3464	.3483	.3502	.3522	.3541	.3560	.3579	.3598
2.3	.3617	.3636	.3655	.3674	.3692	.3711	.3729	.3747	.3766	.3784
2.4	.3802	.3820	.3838	.3856	.3874	.3892	.3909	.3927	.3945	.3962
2.5	.3979	.3997	.4014	.4031	.4048	.4065	.4082	.4099	.4116	.4133
2.6	.4150	.4166	.4183	.4200	.4216	.4232	.4249	.4265	.4281	.4298
2.7	.4314	.4330	.4346	.4362	.4378	.4393	.4409	.4425	.4440	.4456
2.8	.4472	.4487	.4502	.4518	.4533	.4548	.4564	.4579	.4594	.4609
2.9	.4624	.4639	.4654	.4669	.4683	.4698	.4713	.4728	.4742	.4757
3.0	.4771	.4786	.4800	.4814	.4829	.4843	.4857	.4871	.4886	.4900
3.1	.4914	.4928	.4942	.4955	.4969	.4983	.4997	.5011	.5024	.5038
3.2	.5051	.5065	.5079	.5092	.5105	.5119	.5132	.5145	.5159	.5172
3.3	.5185	.5198	.5211	.5224	.5237	.5250	.5263	.5276	.5289	.5302
3.4	.5315	.5328	.5340	.5353	.5366	.5378	.5391	.5403	.5416	.5428
3.5	.5441	.5453	.5465	.5478	.5490	.5502	.5514	.5527	.5539	.5551
3.6	.5563	.5575	.5587	.5599	.5611	.5623	.5635	.5647	.5658	.5670
3.7	.5682	.5694	.5705	.5717	.5729	.5740	.5752	.5763	.5775	.5786
3.8	.5798	.5809	.5821	.5832	.5843	.5855	.5866	.5877	.5888	.5899
3.9	.5911	.5922	.5933	.5944	.5955	.5966	.5977	.5988	.5999	.6010
4.0	.6021	.6031	.6042	.6053	.6064	.6075	.6085	.6096	.6107	.6117
4.1	.6128	.6138	.6149	.6160	.6170	.6180	.6191	.6201	.6212	.6222
4.2	.6232	.6243	.6253	.6263	.6274	.6284	.6294	.6304	.6314	.6325
4.3	.6335	.6345	.6355	.6365	.6375	.6385	.6395	.6405	.6415	.6425
4.4	.6435	.6444	.6454	.6464	.6474	.6484	.6493	.6503	.6513	.6522
4.5	.6532	.6542	.6551	.6561	.6571	.6580	.6590	.6599	.6609	.6618
4.6	.6628	.6637	.6646	.6656	.6665	.6675	.6684	.6693	.6702	.6712
4.7	.6721	.6730	.6739	.6749	.6758	.6767	.6776	.6785	.6794	.6803
4.8	.6812	.6821	.6830	.6839	.6848	.6857	.6866	.6875	.6884	.6893
4.9	.6902	.6911	.6920	.6928	.6937	.6946	.6955	.6964	.6972	.6981
5.0	.6990	.6998	.7007	.7016	.7024	.7033	.7042	.7050	.7059	.7067
5.1	.7076	.7084	.7093	.7101	.7110	.7118	.7126	.7135	.7143	.7152
5.2	.7160	.7168	.7177	.7185	.7193	.7202	.7210	.7218	.7226	.7235
5.3	.7243	.7251	.7259	.7267	.7275	.7284	.7292	.7300	.7308	.7316
5.4	.7324	.7332	.7340	.7348	.7356	.7364	.7372	.7380	.7388	.7396

Table of Common Logarithms

x	0	1	2	3	4	5	6	7	8	9
5.5	.7404	.7412	.7419	.7427	.7435	.7443	.7451	.7459	.7466	.7474
5.6	.7482	.7490	.7497	.7505	.7513	.7520	.7528	.7536	.7543	.7551
5.7	.7559	.7566	.7574	.7582	.7489	.7597	.7604	.7612	.7619	.7627
5.8	.7634	.7642	.7649	.7657	.7664	.7672	.7679	.7686	.7694	.7701
5.9	.7709	.7716	.7723	.7731	.7738	.7745	.7752	.7760	.7767	.7774
6.0	.7782	.7789	.7796	.7803	.7810	.7818	.7825	.7832	.7839	.7846
6.1	.7853	.7860	.7868	.7875	.7882	.7889	.7896	.7903	.7910	.7917
6.2	.7924	.7931	.7938	.7945	.7952	.7959	.7966	.7973	.7980	.7987
6.3	.7993	.8000	.8007	.8014	.8021	.8028	.8035	.8041	.8048	.8055
6.4	.8062	.8069	.8075	.8082	.8089	.8096	.8102	.8109	.8116	.8122
6.5	.8129	.8136	.8142	.8149	.8156	.8162	.8169	.8176	.8182	.8189
6.6	.8195	.8202	.8209	.8215	.8222	.8228	.8235	.8241	.8248	.8254
6.7	.8261	.8267	.8274	.8280	.8287	.8293	.8299	.8306	.8312	.8319
6.8	.8325	.8331	.8338	.8344	.8351	.8357	.8363	.8370	.8376	.8382
6.9	.8388	.8395	.8401	.8407	.8414	.8420	.8426	.8432	.8439	.8445
7.0	.8451	.8457	.8463	.8470	.8476	.8482	.8488	.8494	.8500	.8506
7.1	.8513	.8519	.8525	.8531	.8537	.8543	.8549	.8555	.8561	.8567
7.2	.8573	.8579	.8585	.8591	.8597	.8603	.8609	.8615	.8621	.8627
7.3	.8633	.8639	.8645	.8651	.8657	.8663	.8669	.8675	.8681	.8686
7.4	.8692	.8698	.8704	.8710	.8716	.8722	.8727	.8733	.8739	.8745
7.5	.8751	.8756	.8762	.8768	.8774	.8779	.8785	.8791	.8797	.8802
7.6	.8808	.8814	.8820	.8825	.8831	.8837	.8842	.8848	.8854	.8859
7.7	.8865	.8871	.8876	.8882	.8887	.8893	.8899	.8904	.8910	.8915
7.8	.8921	.8927	.8932	.8938	.8943	.8949	.8954	.8960	.8965	.8971
7.9	.8976	.8982	.8987	.8993	.8998	.9004	.9009	.9015	.9020	.9025
8.0	.9031	.9036	.9042	.9047	.9053	.9058	.9063	.9069	.9074	.9079
8.1	.9085	.9090	.9096	.9101	.9106	.9112	.9117	.9122	.9128	.9133
8.2	.9138	.9143	.9149	.9154	.9159	.9165	.9170	.9175	.9180	.9186
8.3	.9191	.9196	.9201	.9206	.9212	.9217	.9222	.9227	.9232	.9238
8.4	.9243	.9248	.9253	.9258	.9263	.9269	.9274	.9279	.9284	.9289
8.5	.9294	.9299	.9304	.9309	.9315	.9320	.9325	.9330	.9335	.9340
8.6	.9345	.9350	.9355	.9360	.9365	.9370	.9375	.9380	.9385	.9390
8.7	.9395	.9400	.9405	.9410	.9415	.9420	.9425	.9430	.9435	.9440
8.8	.9445	.9450	.9455	.9460	.9465	.9469	.9474	.9479	.9484	.9489
8.9	.9494	.9499	.9504	.9509	.9513	.9518	.9523	.9528	.9533	.9538
9.0	.9542	.9547	.9552	.9557	.9562	.9566	.9571	.9576	.9581	.9586
9.1	.9590	.9595	.9600	.9605	.9609	.9614	.9619	.9624	.9628	.9633
9.2	.9638	.9643	.9647	.9652	.9657	.9661	.9666	.9671	.9675	.9680
9.3	.9685	.9689	.9694	.9699	.9703	.9708	.9713	.9717	.9722	.9727
9.4	.9731	.9736	.9741	.9745	.9750	.9754	.9759	.9763	.9768	.9773
9.5	.9777	.9782	.9786	.9791	.9795	.9800	.9805	.9809	.9814	.9818
9.6	.9823	.9827	.9832	.9836	.9841	.9845	.9850	.9854	.9859	.9863
9.7	.9868	.9872	.9877	.9881	.9886	.9890	.9894	.9899	.9903	.9908
9.8	.9912	.9917	.9921	.9926	.9930	.9934	.9939	.9943	.9948	.9952
9.9	.9956	.9961	.9965	.9969	.9974	.9978	.9983	.9987	.9991	.9996

Table of Geometric Formulas

Triangle

Perimeter: $P = a + b + c$

Area: $A = \frac{1}{2}bh$

Sum of the angles: $\angle A + \angle B + \angle C = 180°$

Right Triangle

Pythagorean Theorem: $c^2 = a^2 + b^2$

Square

Perimeter: $P = 4s$

Area: $A = s^2$

Rectangle

Perimeter: $P = 2L + 2w$

Area: $A = Lw$

Circle

Circumference: $C = \pi d$ or $C = 2\pi r$

Area: $A = \pi r^2$

Cube

Volume: $V = s^3$

Rectangular Solid

Volume: $V = Lwh$

Right Circular Cylinder

Surface area: $S = 2\pi r^2 + 2\pi rh$

Volume: $V = \pi r^2 h$

Sphere

Surface area: $S = 4\pi r^2$

Volume: $V = \frac{4}{3}\pi r^3$

Table of Measurement Abbreviations

U.S. Customary System

Length		**Capacity**		**Weight**		**Area**	
in.	inches	oz	ounces	oz	ounces	in²	square inches
ft	feet	c	cups	lb	pounds	ft²	square feet
yd	yards	qt	quarts				
mi	miles	gal	gallons				

Metric System

Length		**Capacity**		**Weight/Mass**		**Area**	
mm	millimeter (0.001 m)	ml	milliliter (0.001 L)	mg	milligram (0.001 g)	cm²	square centimeters
cm	centimeter (0.01 m)	cl	centiliter (0.01 L)	cg	centigram (0.01 g)	m²	square meters
dm	decimeter (0.1 m)	dl	deciliter (0.1 L)	dg	decigram (0.1 g)		
m	meter	L	liter	g	gram		
dam	decameter (10 m)	dal	decaliter (10 L)	dag	decagram (10 g)		
hm	hectometer (100 m)	hl	hectoliter (100 L)	hg	hectogram (100 g)		
km	kilometer (1000 m)	kl	kiloliter (1000 L)	kg	kilogram (1000 g)		

Time

h	hours	min	minutes	s	seconds

Proofs of Logarithmic Properties

In each of the following proofs of logarithmic properties, it is assumed that the Properties of Exponents are true for all real number exponents.

The Logarithm Property of the Product of Two Numbers

For any positive real numbers x, y, and b, $b \neq 1$, $\log_b xy = \log_b x + \log_b y$.

Proof: Let $\log_b x = m$ and $\log_b y = n$.

Write each equation in its equivalent exponential form. $\qquad x = b^m \qquad y = b^n$

Use substitution and the Properties of Exponents.
$$xy = b^m b^n$$
$$xy = b^{m+n}$$

Write the equation in its equivalent logarithmic form. $\qquad \log_b xy = m + n$

Substitute $\log_b x$ for m and $\log_b y$ for n. $\qquad \log_b xy = \log_b x + \log_b y$

The Logarithm Property of the Quotient of Two Numbers

For any positive real numbers x, y, and b, $b \neq 1$, $\log_b \dfrac{x}{y} = \log_b x - \log_b y$.

Proof: Let $\log_b x = m$ and $\log_b y = n$.

Write each equation in its equivalent exponential form. $\qquad x = b^m \qquad y = b^n$

Use substitution and the Properties of Exponents.
$$\frac{x}{y} = \frac{b^m}{b^n}$$
$$\frac{x}{y} = b^{m-n}$$

Write the equation in its equivalent logarithmic form. $\qquad \log_b \dfrac{x}{y} = m - n$

Substitute $\log_b x$ for m and $\log_b y$ for n. $\qquad \log_b \dfrac{x}{y} = \log_b x - \log_b y$

The Logarithm Property of the Power of a Number

For any real numbers x and b, $b \neq 1$, and for any real number r, $\log_b x^r = r\log_b x$.

Proof: Let $\log_b x = m$.

Write the equation in its equivalent exponential form. $\qquad x = b^m$

Raise both sides to the r power.
$$x^r = (b^m)^r$$
$$x^r = b^{mr}$$

Write the equation in its equivalent logarithm form. $\qquad \log_b x^r = mr$

Substitute $\log_b x$ for m. $\qquad \log_b x^r = r\log_b x$

Proof of the Formula for the Sum of n Terms of a Geometric Series

Theorem: The sum of the first n terms of a geometric sequence whose nth term is ar^{n-1} is given by $S_n = \dfrac{a(1-r^n)}{1-r}$.

Proof: Let S_n represent the sum of n terms of the sequence. $\qquad S_n = a + ar + ar^2 + \cdots + ar^{n-2} + ar^{n-1}$

Multiply each side of the equation by r. $\qquad rS_n = ar + ar^2 + ar^3 + \cdots + ar^{n-1} + ar^n$

Subtract the two equations. $\qquad S_n - rS_n = a - ar^n$

Assuming $r \neq 1$, solve for S_n. $\qquad (1-r)S_n = a(1-r^n)$

$$S_n = \frac{a(1-r^n)}{1-r}$$

Proof of the Formula for the Sum of *n* Terms of an Arithmetic Series

Each term of the arithmetic sequence shown at the right was found by adding 3 to the previous term.

$2, 5, 8, \ldots, 17, 20$

Each term of the reverse arithmetic sequence can be found by subtracting 3 from the previous term.

$20, 17, 14, \ldots, 5, 2$

This idea is used in the following proof.

Theorem: The sum of the first n terms of an arithmetic sequence for which a_1 is the first term, n is the number of terms, a_n is the last term, and d is the common difference is given by $S_n = \frac{n}{2}(a_1 + a_n)$.

    Proof: Let S_n represent the sum of the sequence.

$S_n = a_1 + (a_1 + d) + (a_1 + 2d) + \cdots + a_n$

Write the terms of the sum of the sequence in reverse order. The sum will be the same.

$S_n = a_n + (a_n - d) + (a_n - 2d) + \cdots + a_1$

Add the two equations.

$2S_n = (a_1 + a_n) + (a_1 + a_n) + (a_1 + a_n) + \cdots + (a_1 + a_n)$

Simplify the right side of the equation by using the fact that there are n terms in the sequence.

$2S_n = n(a_1 + a_n)$

Solve for S_n.

$S_n = \frac{n}{2}(a_1 + a_n)$

Table of Symbols

$+$	add	$<$	is less than
$-$	subtract	$\leq$	is less than or equal to
$\cdot, \times, (a)(b)$	multiply	$>$	is greater than
$\frac{a}{b}, \div$	divide	$\geq$	is greater than or equal to
$(\ )$	parentheses, a grouping symbol	(a, b)	an ordered pair whose first component is a and whose second component is b
$[\ ]$	brackets, a grouping symbol		
π	pi, a number approximately equal to $\frac{22}{7}$ or 3.14	$\circ$	degree (for angles)
		$\sqrt{a}$	the principal square root of a
$-a$	the opposite, or additive inverse, of a	$\emptyset, \{\ \}$	the empty set
$\frac{1}{a}$	the reciprocal, or multiplicative inverse, of a	$|a|$	the absolute value of a
		$\cup$	union of two sets
$=$	is equal to	$\cap$	intersection of two sets
$\approx$	is approximately equal to	$\in$	is an element of (for sets)
$\neq$	is not equal to	$\notin$	is not an element of (for sets)

SOLUTIONS to Chapter 1 Examples

SECTION 1.1   *pages 3–6*

Example 2   a. -18 b. 32

Example 4   a. 11 b. -51

Example 6
$-7 + (-13) - (-8)$
$-7 + (-13) + 8$
$-20 + 8$
-12

Example 8
$(-2)(3)(-4)(8)$
$-6(-4)(8)$
$24(8)$
192

Example 10
$-64 \div |-32|$
$-64 \div 32$
-2

Example 12
$-3^3 \cdot 2^2$
$-(3)(3)(3) \cdot (2)(2)$
$-27 \cdot 4$
-108

SECTION 1.2   *pages 9–14*

Example 2
$\dfrac{5}{6} - \dfrac{3}{8} + \dfrac{7}{9} = \dfrac{60}{72} - \dfrac{27}{72} + \dfrac{56}{72}$

$= \dfrac{60 - 27 + 56}{72} = \dfrac{89}{72}$

Example 4
$\begin{array}{r} 12.094 \\ -8.729 \\ \hline 3.365 \end{array}$

$-8.729 + 12.094 = 3.365$

Example 6
$\dfrac{5}{8} \div \left(-\dfrac{15}{40}\right) = \dfrac{5}{8} \cdot \left(-\dfrac{40}{15}\right)$

$= -\dfrac{5 \cdot 40}{8 \cdot 15}$

$= -\dfrac{\overset{1}{\cancel{5}} \cdot \overset{1}{\cancel{2}} \cdot \overset{1}{\cancel{2}} \cdot \overset{1}{\cancel{2}} \cdot 5}{\underset{1}{\cancel{2}} \cdot \underset{1}{\cancel{2}} \cdot \underset{1}{\cancel{2}} \cdot 3 \cdot \underset{1}{\cancel{5}}} = -\dfrac{5}{3}$

Example 8
$\begin{array}{r} 4.027 \\ \times \quad 0.49 \\ \hline 36243 \\ 16108 \quad \\ \hline 1.97323 \end{array} \approx 1.97$

$-4.027 \cdot 0.49 \approx -1.97$

Example 10
$(3.81 - 1.41)^2 \div 0.036 - 1.89$
$(2.40)^2 \div 0.036 - 1.89$
$5.76 \div 0.036 - 1.89$
$160 - 1.89$
158.11

Example 12
$\dfrac{1}{3} + \dfrac{5}{8} \div \dfrac{15}{16} - \dfrac{7}{12}$

$\dfrac{1}{3} + \dfrac{5}{8} \cdot \dfrac{16}{15} - \dfrac{7}{12}$

$\dfrac{1}{3} + \dfrac{2}{3} - \dfrac{7}{12}$

$1 - \dfrac{7}{12}$

$\dfrac{5}{12}$

Example 14
$\dfrac{11}{12} - \dfrac{\frac{5}{4}}{2 - \frac{7}{2}} \cdot \dfrac{3}{4}$

$\dfrac{11}{12} - \dfrac{\frac{5}{4}}{-\frac{3}{2}} \cdot \dfrac{3}{4}$

$\dfrac{11}{12} - \left[\dfrac{5}{4} \cdot \left(-\dfrac{2}{3}\right)\right] \cdot \dfrac{3}{4}$

$\dfrac{11}{12} - \left(-\dfrac{5}{6}\right) \cdot \dfrac{3}{4}$

$\dfrac{11}{12} - \left(-\dfrac{5}{8}\right)$

$\dfrac{37}{24}$

SECTION 1.3   *pages 17–22*

Example 2 $(b - c)^2 \div ab$
$[2 - (-4)]^2 \div (-3)(2)$
$[6]^2 \div (-3)(2)$
$36 \div (-3)(2)$
$-12(2)$
-24

Example 4 $|3 - 4x|$
$|3 - 4(-2)|$
$|3 + 8|$
$|11|$
11

Example 6 $3x + (-3x) = 0$

Example 8   The Associative Property of Addition

Example 10 $(2x + xy - y) - (5x - 7xy + y)$
$2x + xy - y - 5x + 7xy - y$
$-3x + 8xy - 2y$

Example 12 $2x - 3[y - 3(x - 2y + 4)]$
$2x - 3[y - 3x + 6y - 12]$
$2x - 3[7y - 3x - 12]$
$2x - 21y + 9x + 36$
$11x - 21y + 36$

Example 14   the first integer: n
the next consecutive integer:
$n + 1$
$\frac{1}{2}[n + (n + 1)]$
$\frac{1}{2}[2n + 1]$
$n + \frac{1}{2}$

Example 16   the unknown number: n
three eighths of the number: $\frac{3}{8}n$

five twelfths of the number: $\frac{5}{12}n$

$\frac{3}{8}n + \frac{5}{12}n$

$\frac{9}{24}n + \frac{10}{24}n$

$\frac{19}{24}n$

SECTION 1.4   *pages 27–30*

Example 2 $A = \{2, 4, 6, 8, 10\}$

Example 4 $A = \{1, 3, 5, 7, 9\}$

Example 6   0.35 is less than 2.

Yes, 0.35 is an element of the set.

Example 8 $A \cup C = \{-5, -2, -1, 0, 1, 2, 5\}$

Example 10 $D \cap C = \{3, 5, 7\}$

Example 12   There are no integers that are both odd and even.

$E \cap F = \varnothing$

Example 14   The solution set is $\{x | x > -3\}$.

Example 16   The solution set is $\{x | -3 < x < 4\}$.

Example 18   The solution set is $\{x | x \geq 1 \text{ or } x \leq -3\}$.

SOLUTIONS to Chapter 2 Examples

SECTION 2.1   *pages 43–48*

Example 2
$$x + 4 = -3$$
$$x + 4 - 4 = -3 - 4$$
$$x = -7$$

The solution is -7.

Example 4
$$-3x = 18$$
$$\frac{-3x}{-3} = \frac{18}{-3}$$
$$x = -6$$

The solution is -6.

Example 6
$$6x - 5 - 3x = 14 - 5x$$
$$3x - 5 = 14 - 5x$$
$$3x - 5 + 5x = 14 - 5x + 5x$$
$$8x - 5 = 14$$
$$8x - 5 + 5 = 14 + 5$$
$$8x = 19$$
$$\frac{8x}{8} = \frac{19}{8}$$
$$x = \frac{19}{8}$$

The solution is $\frac{19}{8}$.

Example 8
$$6(5 - x) - 12 = 2x - 3(4 + x)$$
$$30 - 6x - 12 = 2x - 12 - 3x$$
$$18 - 6x = -x - 12$$
$$18 - 6x + x = -x - 12 + x$$
$$18 - 5x = -12$$
$$18 - 5x - 18 = -12 - 18$$
$$-5x = -30$$
$$\frac{-5x}{-5} = \frac{-30}{-5}$$
$$x = 6$$

The solution is 6.

Example 10   The LCM of 3, 5, and 30 is 30.
$$\frac{2x - 7}{3} - \frac{5x + 4}{5} = \frac{-x - 4}{30}$$
$$30\left(\frac{2x - 7}{3} - \frac{5x + 4}{5}\right) = 30\left(\frac{-x - 4}{30}\right)$$
$$\frac{30(2x - 7)}{3} - \frac{30(5x + 4)}{5} = \frac{30(-x - 4)}{30}$$
$$10(2x - 7) - 6(5x + 4) = -x - 4$$
$$20x - 70 - 30x - 24 = -x - 4$$
$$-10x - 94 = -x - 4$$
$$-10x - 94 + x = -x - 4 + x$$
$$-9x - 94 = -4$$
$$-9x - 94 + 94 = -4 + 94$$
$$-9x = 90$$
$$\frac{-9x}{-9} = \frac{90}{-9}$$
$$x = -10$$

The solution is -10.

SECTION 2.2   *pages 53–58*

Example 2
$$2x - 1 < 6x + 7$$
$$-4x - 1 < 7$$
$$-4x < 8$$
$$\frac{-4x}{-4} > \frac{8}{-4}$$
$$x > -2$$
$$\{x \mid x > -2\}$$

Example 4
$$5x - 2 \leq 4 - 3(x - 2)$$
$$5x - 2 \leq 4 - 3x + 6$$
$$5x - 2 \leq 10 - 3x$$
$$8x - 2 \leq 10$$
$$8x \leq 12$$
$$\frac{8x}{8} \leq \frac{12}{8}$$
$$x \leq \frac{3}{2}$$
$$\left\{x \mid x \leq \frac{3}{2}\right\}$$

Example 6

$$-2 \le 5x + 3 \le 13$$
$$-2 - 3 \le 5x + 3 - 3 \le 13 - 3$$
$$-5 \le 5x \le 10$$
$$\frac{-5}{5} \le \frac{5x}{5} \le \frac{10}{5}$$
$$-1 \le x \le 2$$
$$\{x \mid -1 \le x \le 2\}$$

Example 8

$$2 - 3x > 11 \quad \text{or} \quad 5 + 2x > 7$$
$$-3x > 9 \qquad\qquad 2x > 2$$
$$x < -3 \qquad\qquad x > 1$$
$$\{x \mid x < -3\} \qquad \{x \mid x > 1\}$$
$$\{x \mid x < -3\} \cup \{x \mid x > 1\} =$$
$$\{x \mid x < -3 \text{ or } x > 1\}$$

Example 10

Strategy

To find the maximum height, substitute the given values in the inequality $\frac{1}{2}bh < 50$ and solve.

Solution

$$\frac{1}{2}bh < 50$$
$$\frac{1}{2}(12)(x + 2) < 50$$
$$6(x + 2) < 50$$
$$6x + 12 < 50$$
$$6x < 38$$
$$x < \frac{19}{3}$$

The largest integer less than $\frac{19}{3}$ is 6.

$$x + 2 = 6 + 2 = 8$$

The maximum height of the triangle is 8 in.

Example 12

Strategy

To find the range of scores, write and solve an inequality using N to represent the score on the last test.

Solution

$$80 \le \frac{72 + 94 + 83 + 70 + N}{5} \le 89$$
$$80 \le \frac{319 + N}{5} \le 89$$
$$5 \cdot 80 \le 5\left(\frac{319 + N}{5}\right) \le 5 \cdot 89$$
$$400 \le 319 + N \le 445$$
$$400 - 319 \le 319 + N - 319 \le 445 - 319$$
$$81 \le N \le 126$$

Since 100 is a maximum score, the range of scores to receive a B grade is $81 \le N \le 100$.

SECTION 2.3   *pages 65–70*

Example 2

$$|2x - 3| = 5$$
$$2x - 3 = 5 \qquad 2x - 3 = -5$$
$$2x = 8 \qquad\quad 2x = -2$$
$$x = 4 \qquad\qquad x = -1$$

The solutions are 4 and -1.

Example 4

$$|x - 3| = -2$$

There is no solution to this equation because the absolute value of a number must be non-negative.

Example 6

$$5 - |3x + 5| = 3$$
$$-|3x + 5| = -2$$
$$|3x + 5| = 2$$
$$3x + 5 = 2 \qquad 3x + 5 = -2$$
$$3x = -3 \qquad\quad 3x = -7$$
$$x = -1 \qquad\qquad x = -\frac{7}{3}$$

The solutions are -1 and $-\frac{7}{3}$.

Example 8

$$|3x + 2| < 8$$
$$-8 < 3x + 2 < 8$$
$$-8 - 2 < 3x + 2 - 2 < 8 - 2$$
$$-10 < 3x < 6$$
$$\frac{-10}{3} < \frac{3x}{3} < \frac{6}{3}$$
$$-\frac{10}{3} < x < 2$$
$$\left\{x \mid -\frac{10}{3} < x < 2\right\}$$

Example 10

$|3x - 7| < 0$

The absolute value of a number must be non-negative.

The solution set is the empty set.

Example 12

$|5x + 3| > 8$

$5x + 3 < -8$ or $5x + 3 > 8$
$5x < -11$ $5x > 5$
$x < -\frac{11}{5}$ $x > 1$

$\left\{x \middle| x < -\frac{11}{5}\right\}$ $\{x | x > 1\}$

$\left\{x \middle| x < -\frac{11}{5}\right\} \cup \{x | x > 1\} =$

$\left\{x \middle| x < -\frac{11}{5} \text{ or } x > 1\right\}$

Example 14

Strategy

Let b represent the diameter of the bushing, T the tolerance, and d the lower and upper limits of the diameter. Solve the absolute value inequality $|d - b| \leq T$ for d.

Solution

$|d - b| \leq T$
$|d - 2.55| \leq 0.003$

$-0.003 \leq d - 2.55 \leq 0.003$
$-0.003 + 2.55 \leq d - 2.55 + 2.55 \leq 0.003 + 2.55$
$2.547 \leq d \leq 2.553$

The lower and upper limits of the diameter of the bushing are 2.547 in. and 2.553 in.

SECTION 2.4   *pages 75–78*

Example 2

Strategy

- The first number: n
  The second number: $2n$
  The third number: $4n - 3$
- The sum of the numbers is 81.

$n + 2n + (4n - 3) = 81$

Solution

$n + 2n + (4n - 3) = 81$
$7n - 3 = 81$
$7n = 84$
$n = 12$

$2n = 2(12) = 24$
$4n - 3 = 4(12) - 3 = 48 - 3 = 45$

The numbers are 12, 24, and 45.

Example 4

Strategy

- First odd integer: n
  Second odd integer: $n + 2$
  Third odd integer: $n + 4$
- Three times the sum of the first two integers is ten more than the product of the third integer and four.

$3[n + (n + 2)] = (n + 4)4 + 10$

Solution

$3[n + (n + 2)] = (n + 4)4 + 10$
$3[2n + 2] = 4n + 16 + 10$
$6n + 6 = 4n + 26$
$2n + 6 = 26$
$2n = 20$
$n = 10$

Since 10 is not an odd integer, there is no solution.

Example 6

Strategy

- Number of 3¢ stamps: x
  Number of 10¢ stamps: $2x + 2$
  Number of 15¢ stamps: $3x$

Stamps	Number	Value	Total Value
3¢	x	3	$3x$
10¢	$2x + 2$	10	$10(2x + 2)$
15¢	$3x$	15	$45x$

- The sum of the total values of each type of stamp equals the total value of all the stamps (156 cents).

$3x + 10(2x + 2) + 45x = 156$

Solution

$$3x + 10(2x + 2) + 45x = 156$$
$$3x + 20x + 20 + 45x = 156$$
$$68x + 20 = 156$$
$$68x = 136$$
$$x = 2$$

$3x = 3(2) = 6$

There are six 15¢ stamps in the collection.

SECTION 2.5 *pages 81–84*

Example 2

Strategy

- Pounds of $3.00 hamburger: x
  Pounds of $1.80 hamburger: $75 - x$

	Amount	Cost	Value
$3.00 hamburger	x	3.00	$3.00x$
$1.80 hamburger	$75 - x$	1.80	$1.80(75 - x)$
Mixture	75	2.20	$75(2.20)$

- The sum of the values before mixing equals the value after mixing.

$3.00x + 1.80(75 - x) = 75(2.20)$

Solution

$$3.00x + 1.80(75 - x) = 75(2.20)$$
$$3x + 135 - 1.80x = 165$$
$$1.2x + 135 = 165$$
$$1.2x = 30$$
$$x = 25$$

$75 - x = 75 - 25 = 50$

The mixture must contain 25 lb of the $3.00 hamburger and 50 lb of the $1.80 hamburger.

Example 4

Strategy

- Rate of the second plane: r
  Rate of the first plane: $r + 30$

	Rate	Time	Distance
1st plane	$r + 30$	4	$4(r + 30)$
2nd plane	r	4	$4r$

- The total distance traveled by the two planes is 1160 mi.

$4(r + 30) + 4r = 1160$

Solution

$$4(r + 30) + 4r = 1160$$
$$4r + 120 + 4r = 1160$$
$$8r + 120 = 1160$$
$$8r = 1040$$
$$r = 130$$

$r + 30 = 130 + 30 = 160$

The first plane is traveling 160 mph.
The second plane is traveling 130 mph.

SECTION 2.6   *pages 87–90*

Example 2

Strategy

- Amount invested at 11.5%: x

	Principal	Rate	Interest
Amount at 13.2%	3500	0.132	0.132(3500)
Amount at 11.5%	x	0.115	0.115x

- The sum of the interest earned by the two investments equals the total annual interest earned ($1037).

$$0.132(3500) + 0.115x = 1037$$

Solution

$$0.132(3500) + 0.115x = 1037$$
$$462 + 0.115x = 1037$$
$$0.115x = 575$$
$$x = 5000$$

The amount invested at 11.5% is $5000.

Example 4

Strategy

- Pounds of 22% hamburger: x
  Pounds of 12% hamburger: $80 - x$

	Amount	Percent	Quantity
22%	x	0.22	0.22x
12%	$80 - x$	0.12	0.12(80 − x)
18%	80	0.18	0.18(80)

- The sum of the quantities before mixing is equal to the quantity after mixing.

$$0.22x + 0.12(80 - x) = 0.18(80)$$

Solution

$$0.22x + 0.12(80 - x) = 0.18(80)$$
$$0.22x + 9.6 - 0.12x = 14.4$$
$$0.10x + 9.6 = 14.4$$
$$0.10x = 4.8$$
$$x = 48$$

$$80 - x = 80 - 48 = 32$$

The butcher needs 48 lb of the hamburger that is 22% fat and 32 lb of the hamburger that is 12% fat.

SOLUTIONS to Chapter 3 Examples

SECTION 3.1   *pages 103–114*

Example 2

$$\begin{array}{r} -3x^2 - 4x + 9 \\ +\,-5x^2 - 7x + 1 \\ \hline -8x^2 - 11x + 10 \end{array}$$

Example 6

$(5x^{2n} - 3x^n - 7) - (-2x^{2n} - 5x^n + 8)$
$(5x^{2n} - 3x^n - 7) + (2x^{2n} + 5x^n - 8)$
$7x^{2n} + 2x^n - 15$

Example 10

$(-3a^2b^4)(-2ab^3)^4 = (-3a^2b^4)[(-2)^4 a^4 b^{12}]$
$= (-3a^2b^4)(16a^4b^{12})$
$= -48a^6b^{16}$

Example 4

$$\dfrac{-5x^2 + 2x - 3}{-\ 6x^2 + 3x - 7} = \begin{array}{r} -5x^2 + 2x - 3 \\ +\,-6x^2 - 3x + 7 \\ \hline -11x^2 - x + 4 \end{array}$$

Example 8

$(7xy^3)(-5x^2y^2)(-xy^2) = 35x^4y^7$

Example 12

$(y^{n-3})^2 = y^{(n-3)2} = y^{2n-6}$

Example 14

$[(ab^3)^3]^4 = (ab^3)^{3 \cdot 4} = (ab^3)^{12} = a^{12}b^{36}$

Example 16

$6a^2(2a) + 3a(2a^2) = 12a^3 + 6a^3 = 18a^3$

Example 18

$\dfrac{3^2}{3^{-3}} = 3^{2-(-3)} = 3^5 = 243$

Example 20

$(4r^{-2}t^{-1})^{-2}(r^4t^{-1})^3 = (4^{-2}r^4t^2)(r^{12}t^{-3})$

$$= 4^{-2}r^{16}t^{-1}$$

$$= \frac{r^{16}}{4^2t}$$

$$= \frac{r^{16}}{16t}$$

Example 22

$\dfrac{20r^{-2}t^{-5}}{-16r^{-3}s^{-2}} = -\dfrac{4 \cdot 5r^{-2-(-3)}s^2t^{-5}}{4 \cdot 4}$

$$= -\frac{5rs^2}{4t^5}$$

Example 24

$\dfrac{(9u^{-6}v^4)^{-1}}{(6u^{-3}v^{-2})^{-2}} = \dfrac{9^{-1}u^6v^{-4}}{6^{-2}u^6v^4}$

$$= 9^{-1} \cdot 6^2u^0v^{-8}$$

$$= \frac{36}{9v^8}$$

$$= \frac{4}{v^8}$$

Example 26

$\left[\dfrac{4a^4b^{-2}}{8a^5b^{-1}}\right]^{-2} = \left[\dfrac{a^{-1}b^{-1}}{2}\right]^{-2}$

$$= \frac{a^2b^2}{2^{-2}}$$

$$= 2^2a^2b^2$$

$$= 4a^2b^2$$

Example 28

$\left(\dfrac{4a^2}{5b^3}\right)^3\left(\dfrac{-10a^3}{2b^8}\right)^2 = \left(\dfrac{4a^2}{5b^3}\right)^3\left(\dfrac{-5a^3}{b^8}\right)^2$

$$= \left(\frac{4^3a^6}{5^3b^9}\right)\left(\frac{(-5)^2a^6}{b^{16}}\right) = \left(\frac{64a^6}{125b^9}\right)\left(\frac{25a^6}{b^{16}}\right)$$

$$= \frac{64 \cdot 25a^{12}}{125b^{25}} = \frac{64a^{12}}{5b^{25}}$$

Example 30

$\dfrac{y^{4n}}{y^{3n}} = y^{4n-3n} = y^n$

Example 32

$\dfrac{a^{2n+1}}{a^{n+3}} = a^{2n+1-(n+3)} = a^{2n+1-n-3} = a^{n-2}$

Example 34

$942,000,000 = 9.42 \times 10^8$

Example 36

$2.7 \times 10^{-5} = 0.000027$

Example 38

$\dfrac{5,600,000 \times 0.000000081}{900 \times 0.000000028}$

$$= \frac{5.6 \times 10^6 \times 8.1 \times 10^{-8}}{9 \times 10^2 \times 2.8 \times 10^{-8}}$$

$$= \frac{(5.6)(8.1) \times 10^{6+(-8)-2-(-8)}}{(9)(2.8)}$$

$$= 1.8 \times 10^4 = 18,000$$

Example 40

Strategy

To find the number of arithmetic operations:
- Find the reciprocal of 1×10^{-7}, which is the number of operations performed in one second.
- Write the number of seconds in one minute (60) in scientific notation.
- Multiply the number of arithmetic operations per second by the number of seconds in one minute.

Solution

$\dfrac{1}{1 \times 10^{-7}} = 10^7$

$60 = 6 \times 10$

$6 \times 10 \times 10^7$

6×10^8

The computer can perform 6×10^8 operations in one minute.

SECTION 3.2   *pages 121–126*

Example 2

$(2b^2 - 7b - 8)(-5b) = -10b^3 + 35b^2 + 40b$

Example 4

$x^2 - 2x[x - x(4x - 5) + x^2]$
$= x^2 - 2x[x - 4x^2 + 5x + x^2]$
$= x^2 - 2x[6x - 3x^2]$
$= x^2 - 12x^2 + 6x^3$
$= 6x^3 - 11x^2$

Example 6

$y^{n+3}(y^{n-2} - 3y^2 + 2)$
$= y^{n+3}(y^{n-2}) - (y^{n+3})(3y^2) + (y^{n+3})(2)$
$= y^{n+3+(n-2)} - 3y^{n+3+2} + 2y^{n+3}$
$= y^{2n+1} - 3y^{n+5} + 2y^{n+3}$

Example 8

$$\begin{array}{r} -2b^2 + 5b - 4 \\ -3b + 2 \\ \hline -\ \ 4b^2 + 10b - 8 \\ 6b^3 - 15b^2 + 12b \\ \hline 6b^3 - 19b^2 + 22b - 8 \end{array}$$

Example 10

$(3x - 4)(2x - 3) = 6x^2 - 9x - 8x + 12$
$\qquad\qquad\qquad = 6x^2 - 17x + 12$

Example 12

$(2x^n + y^n)(x^n - 4y^n) = 2x^{2n} - 8x^ny^n + x^ny^n - 4y^{2n}$
$\qquad\qquad\qquad\qquad = 2x^{2n} - 7x^ny^n - 4y^{2n}$

Example 14

$(3x - 7)(3x + 7) = 9x^2 - 49$

Example 16

$(2x^n + 3)(2x^n - 3) = 4x^{2n} - 9$

Example 18

$(3x - 4y)^2 = 9x^2 - 24xy + 16y^2$

Example 20

$(2x^n - 8)^2 = 4x^{2n} - 32x^n + 64$

Example 22

Strategy

To find the area, replace the variables b and h in the equation $A = \frac{1}{2}bh$ by the given values and solve for A.

Solution

$A = \frac{1}{2}bh$

$A = \frac{1}{2}(2x + 6)(x - 4)$

$A = (x + 3)(x - 4)$
$A = x^2 - 4x + 3x - 12$
$A = x^2 - x - 12$

The area is $(x^2 - x - 12)$ ft^2.

Example 24

Strategy

To find the volume, subtract the volume of the small rectangular solid from the volume of the large rectangular solid.

Large rectangular solid:
 Length = $L_1 = 12x$
 Width = $w_1 = 7x + 2$
 Height = $h_1 = 5x - 4$
Small rectangular solid:
 Length = $L_2 = 12x$
 Width = $w_2 = x$
 Height = $h_2 = 2x$

Solution

V = Volume of large rectangular solid − volume of
 small rectangular solid
$V = (L_1 \cdot w_1 \cdot h_1) - (L_2 \cdot w_2 \cdot h_2)$
$V = (12x)(7x + 2)(5x - 4) - (12x)(x)(2x)$
$V = (84x^2 + 24x)(5x - 4) - (12x^2)(2x)$
$V = (420x^3 - 336x^2 + 120x^2 - 96x) - (24x^3)$
$V = 396x^3 - 216x^2 - 96x$

The volume is $(396x^3 - 216x^2 - 96x)$ ft^3.

Example 26

Strategy

To find the area, replace the variable r in the equation $A = \pi r^2$ by the given value and solve for A.

Solution

$A = \pi r^2$
$A = 3.14(2x + 3)^2$
$A = 3.14(4x^2 + 12x + 9)$
$A = 12.56x^2 + 37.68x + 28.26$

The area is $(12.56x^2 + 37.68x + 28.26)$ cm^2.

SECTION 3.3   *pages 131–138*

Example 2

The GCF of $3x^3y$, $6x^2y^2$, and $3xy^3$ is $3xy$.

$3x^3y - 6x^2y^2 - 3xy^3 = 3xy(x^2 - 2xy - y^2)$

Example 6

$3(6x - 7y) - 2x^2(6x - 7y) = (6x - 7y)(3 - 2x^2)$

Example 10

$x^2 - x - 20 = (x + 4)(x - 5)$

Example 14

$4x^2 + 15x - 4 = (x + 4)(4x - 1)$

Example 18

The GCF of $3a^3b^3$, $3a^2b^2$, and $60ab$ is $3ab$.
$3a^3b^3 + 3a^2b^2 - 60ab = 3ab(a^2b^2 + ab - 20)$
$\qquad\qquad\qquad\qquad = 3ab(ab + 5)(ab - 4)$

Example 4

The GCF of $6t^{2n}$ and $9t^n$ is $3t^n$.

$6t^{2n} - 9t^n = 3t^n(2t^n - 3)$

Example 8

$4a^2 - 6a - 6ax + 9x = (4a^2 - 6a) - (6ax - 9x)$
$\qquad\qquad\qquad\quad = 2a(2a - 3) - 3x(2a - 3)$
$\qquad\qquad\qquad\quad = (2a - 3)(2a - 3x)$

Example 12

$x^2 + 5xy + 6y^2 = (x + 2y)(x + 3y)$

Example 16

$10x^2 + 39x + 14 = (2x + 7)(5x + 2)$

SECTION 3.4   *pages 143–148*

Example 2

$x^2 - 36y^4 = x^2 - (6y^2)^2$
$\qquad\quad\;\, = (x + 6y^2)(x - 6y^2)$

Example 6

$(a + b)^2 - (a - b)^2$
$\quad = [(a + b) + (a - b)][(a + b) - (a - b)]$
$\quad = (a + b + a - b)(a + b - a + b)$
$\quad = (2a)(2b) = 4ab$

Example 10

$(x - y)^3 + (x + y)^3$
$\;= [(x - y) + (x + y)][(x - y)^2 - (x - y)(x + y) + (x + y)^2]$
$\;= 2x[x^2 - 2xy + y^2 - (x^2 - y^2) + x^2 + 2xy + y^2]$
$\;= 2x(x^2 - 2xy + y^2 - x^2 + y^2 + x^2 + 2xy + y^2)$
$\;= 2x(x^2 + 3y^2)$

Example 4

$9x^2 + 12x + 4 = (3x + 2)^2$

Example 8

$8x^3 + y^3z^3 = (2x)^3 + (yz)^3$
$\qquad\qquad\; = (2x + yz)(4x^2 - 2xyz + y^2z^2)$

Example 12

Let $u = x^2$.

$3x^4 + 4x^2 - 4 = 3u^2 + 4u - 4$

$\qquad\qquad\quad = (u + 2)(3u - 2)$

$\qquad\qquad\quad = (x^2 + 2)(3x^2 - 2)$

Example 14

$4x - 4y - x^3 + x^2y = (4x - 4y) + (-x^3 + x^2y)$
$= 4(x - y) - x^2(x - y) = (x - y)(4 - x^2)$
$= (x - y)(2 + x)(2 - x)$

Example 16

$x^{4n} - x^{2n}y^{2n} = x^{2n+2n} - x^{2n}y^{2n}$
$= x^{2n}(x^{2n} - y^{2n})$
$= x^{2n}[(x^n)^2 - (y^n)^2]$
$= x^{2n}(x^n + y^n)(x^n - y^n)$

SECTION 3.5   *pages 153–154*

Example 2

$(x + 4)(x - 1) = 14$
$x^2 + 3x - 4 = 14$
$x^2 + 3x - 18 = 0$
$(x + 6)(x - 3) = 0$
$x + 6 = 0 \qquad x - 3 = 0$
$\qquad x = -6 \qquad\quad x = 3$

The solutions are -6 and 3.

Example 4

$$a^4 - 5a^2 + 4 = 0$$
$$(a^2 - 4)(a^2 - 1) = 0$$
$$(a - 2)(a + 2)(a - 1)(a + 1) = 0$$
$a - 2 = 0 \quad a + 2 = 0 \quad a - 1 = 0 \quad a + 1 = 0$
$\quad a = 2 \qquad a = -2 \qquad a = 1 \qquad a = -1$

The solutions are -2, -1, 1, and 2.

SOLUTIONS to Chapter 4 Examples

SECTION 4.1   *pages 167–170*

Example 2

$\dfrac{6x^4 - 24x^3}{12x^3 - 48x^2} = \dfrac{6x^3(x - 4)}{12x^2(x - 4)}$

$= \dfrac{6x^3\cancel{(x - 4)}}{12x^2\cancel{(x - 4)}} = \dfrac{x}{2}$

Example 4

$\dfrac{21a^3b - 14a^3b^2}{7a^2b} = \dfrac{7a^3b(3 - 2b)}{7a^2b}$
$= a(3 - 2b)$

Example 6

$\dfrac{20x - 15x^2}{15x^3 - 5x^2 - 20x} = \dfrac{5x(4 - 3x)}{5x(3x^2 - x - 4)}$

$= \dfrac{5x(4 - 3x)}{5x(3x - 4)(x + 1)}$

$= \dfrac{5x(4 - 3x)}{5x(3x - 4)(x + 1)}$

$= -\dfrac{1}{x + 1}$

Example 8

$\dfrac{x^{2n} + x^n - 12}{x^{2n} - 3x^n} = \dfrac{(x^n + 4)(x^n - 3)}{x^n(x^n - 3)}$

$= \dfrac{(x^n + 4)(x^n - 3)}{x^n(x^n - 3)}$

$= \dfrac{x^n + 4}{x^n}$

Example 10

$\dfrac{12 + 5x - 3x^2}{x^2 + 2x - 15} \cdot \dfrac{2x^2 + x - 45}{3x^2 + 4x}$

$= \dfrac{(4 + 3x)(3 - x)}{(x + 5)(x - 3)} \cdot \dfrac{(2x - 9)(x + 5)}{x(3x + 4)}$

$= \dfrac{(4 + 3x)(3 - x)(2x - 9)(x + 5)}{(x + 5)(x - 3) \cdot x(3x + 4)}$

$= \dfrac{(4 + 3x)(3 - x)(2x - 9)(x + 5)}{(x + 5)(x - 3) \cdot x(3x + 4)} = -\dfrac{2x - 9}{x}$

Example 12

$\dfrac{6x^2 - 3xy}{10ab^4} \div \dfrac{16x^2y^2 - 8xy^3}{15a^2b^2}$

$= \dfrac{6x^2 - 3xy}{10ab^4} \cdot \dfrac{15a^2b^2}{16x^2y^2 - 8xy^3}$

$= \dfrac{3x(2x - y)}{10ab^4} \cdot \dfrac{15a^2b^2}{8xy^2(2x - y)}$

$= \dfrac{3x(2x - y) \cdot 15a^2b^2}{10ab^4 \cdot 8xy^2(2x - y)}$

$= \dfrac{3x(2x - y) \cdot 15a^2b^2}{10ab^4 \cdot 8xy^2(2x - y)} = \dfrac{9a}{16b^2y^2}$

Example 14

$$\frac{6x^2 - 7x + 2}{3x^2 + x - 2} \div \frac{4x^2 - 8x + 3}{5x^2 + x - 4}$$

$$= \frac{6x^2 - 7x + 2}{3x^2 + x - 2} \cdot \frac{5x^2 + x - 4}{4x^2 - 8x + 3}$$

$$= \frac{(2x - 1)(3x - 2)}{(x + 1)(3x - 2)} \cdot \frac{(x + 1)(5x - 4)}{(2x - 1)(2x - 3)}$$

$$= \frac{(2x - 1)(3x - 2)(x + 1)(5x - 4)}{(x + 1)(3x - 2)(2x - 1)(2x - 3)}$$

$$= \frac{\overset{1}{\cancel{(2x - 1)}}\overset{1}{\cancel{(3x - 2)}}\overset{1}{\cancel{(x + 1)}}(5x - 4)}{\underset{1}{\cancel{(x + 1)}}\underset{1}{\cancel{(3x - 2)}}\underset{1}{\cancel{(2x - 1)}}(2x - 3)} = \frac{5x - 4}{2x - 3}$$

SECTION 4.2   *pages 175–178*

Example 2

$$
\require{enclose}
\begin{array}{r}
5x - 1 \\
3x + 4 \enclose{longdiv}{15x^2 + 17x - 20} \\
\underline{15x^2 + 20x } \\
-3x - 20 \\
\underline{-3x - 4} \\
-16
\end{array}
$$

$$\frac{15x^2 + 17x - 20}{3x + 4} = 5x - 1 - \frac{16}{3x + 4}$$

Example 4

$$
\require{enclose}
\begin{array}{r}
x^2 + 3x - 1 \\
3x - 1 \enclose{longdiv}{3x^3 + 8x^2 - 6x + 2} \\
\underline{3x^3 - x^2 } \\
9x^2 - 6x \\
\underline{9x^2 - 3x} \\
-3x + 2 \\
\underline{-3x + 1} \\
1
\end{array}
$$

$$\frac{3x^3 + 8x^2 - 6x + 2}{3x - 1} = x^2 + 3x - 1 + \frac{1}{3x - 1}$$

Example 6

$$
\require{enclose}
\begin{array}{r}
3x^2 - 2x + 4 \\
x^2 - 3x + 2 \enclose{longdiv}{3x^4 - 11x^3 + 16x^2 - 16x + 8} \\
\underline{3x^4 - 9x^3 + 6x^2 } \\
-2x^3 + 10x^2 - 16x \\
\underline{-2x^3 + 6x^2 - 4x } \\
4x^2 - 12x + 8 \\
\underline{4x^2 - 12x + 8} \\
0
\end{array}
$$

$$\frac{3x^4 - 11x^3 + 16x^2 - 16x + 8}{x^2 - 3x + 2} = 3x^2 - 2x + 4$$

Example 8

$$
\begin{array}{r|rrr}
-2 & 6 & 8 & -5 \\
   &   & -12 & 8 \\
\hline
   & 6 & -4 & 3
\end{array}
$$

$$(6x^2 + 8x - 5) \div (x + 2) = 6x - 4 + \frac{3}{x + 2}$$

Example 10

$$
\begin{array}{r|rrrr}
2 & 5 & -12 & -8 & 16 \\
  &   & 10 & -4 & -24 \\
\hline
  & 5 & -2 & -12 & -8
\end{array}
$$

$$(5x^3 - 12x^2 - 8x + 16) \div (x - 2)$$

$$= 5x^2 - 2x - 12 - \frac{8}{x - 2}$$

Example 12

$$
\begin{array}{r|rrrrr}
3 & 2 & -3 & -8 & 0 & -2 \\
  &   & 6 & 9 & 3 & 9 \\
\hline
  & 2 & 3 & 1 & 3 & 7
\end{array}
$$

$$(2x^4 - 3x^3 - 8x^2 - 2) \div (x - 3)$$

$$= 2x^3 + 3x^2 + x + 3 + \frac{7}{x - 3}$$

SECTION 4.3   *pages 181–184*

Example 2

The LCM is $(2x - 5)(x + 4)$.

$$\frac{2x}{2x - 5} = \frac{2x}{2x - 5} \cdot \frac{x + 4}{x + 4} = \frac{2x^2 + 8x}{(2x - 5)(x + 4)}$$

$$\frac{3}{x + 4} = \frac{3}{x + 4} \cdot \frac{2x - 5}{2x - 5} = \frac{6x - 15}{(2x - 5)(x + 4)}$$

Example 4

$2x^2 - 11x + 15 = (x - 3)(2x - 5); \ x^2 - 3x = x(x - 3)$
The LCM is $x(x - 3)(2x - 5)$.

$$\frac{3x}{2x^2 - 11x + 15} = \frac{3x}{(x - 3)(2x - 5)} \cdot \frac{x}{x} = \frac{3x^2}{x(x - 3)(2x - 5)}$$

$$\frac{x - 2}{x^2 - 3x} = \frac{x - 2}{x(x - 3)} \cdot \frac{2x - 5}{2x - 5} = \frac{2x^2 - 9x + 10}{x(x - 3)(2x - 5)}$$

Example 6

$2x - x^2 = x(2 - x) = -x(x - 2);$
$3x^2 - 5x - 2 = (x - 2)(3x + 1)$
The LCM is $x(x - 2)(3x + 1)$.

$$\frac{2x - 7}{2x - x^2} = -\frac{2x - 7}{x(x - 2)} \cdot \frac{3x + 1}{3x + 1} = -\frac{6x^2 - 19x - 7}{x(x - 2)(3x + 1)}$$

$$\frac{3x - 2}{3x^2 - 5x - 2} = \frac{3x - 2}{(x - 2)(3x + 1)} \cdot \frac{x}{x} = \frac{3x^2 - 2x}{x(x - 2)(3x + 1)}$$

Example 8

The LCM is ab.

$$\frac{2}{b} - \frac{1}{a} + \frac{4}{ab} = \frac{2}{b} \cdot \frac{a}{a} - \frac{1}{a} \cdot \frac{b}{b} + \frac{4}{ab} = \frac{2a}{ab} - \frac{b}{ab} + \frac{4}{ab}$$

$$= \frac{2a - b + 4}{ab}$$

Example 10

The LCM is $a(a - 5)(a + 5)$.

$$\frac{a - 3}{a^2 - 5a} + \frac{a - 9}{a^2 - 25}$$

$$= \frac{a - 3}{a(a - 5)} \cdot \frac{a + 5}{a + 5} + \frac{a - 9}{(a - 5)(a + 5)} \cdot \frac{a}{a}$$

$$= \frac{(a - 3)(a + 5) + a(a - 9)}{a(a - 5)(a + 5)}$$

$$= \frac{(a^2 + 2a - 15) + (a^2 - 9a)}{a(a - 5)(a + 5)}$$

$$= \frac{a^2 + 2a - 15 + a^2 - 9a}{a(a - 5)(a + 5)}$$

$$= \frac{2a^2 - 7a - 15}{a(a - 5)(a + 5)} = \frac{(2a + 3)(a - 5)}{a(a - 5)(a + 5)}$$

$$= \frac{(2a + 3)\overset{1}{(a - 5)}}{a\underset{1}{(a - 5)}(a + 5)} = \frac{2a + 3}{a(a + 5)}$$

Example 12

The LCM is $(x - 4)(x + 1)$.

$$\frac{2x}{x - 4} - \frac{x - 1}{x + 1} + \frac{2}{x^2 - 3x - 4}$$

$$= \frac{2x}{x - 4} \cdot \frac{x + 1}{x + 1} - \frac{x - 1}{x + 1} \cdot \frac{x - 4}{x - 4} + \frac{2}{(x - 4)(x + 1)}$$

$$= \frac{2x(x + 1) - (x - 1)(x - 4) + 2}{(x - 4)(x + 1)}$$

$$= \frac{(2x^2 + 2x) - (x^2 - 5x + 4) + 2}{(x - 4)(x + 1)}$$

$$= \frac{x^2 + 7x - 2}{(x - 4)(x + 1)}$$

SECTION 4.4   *pages 189–190*

Example 2

The LCM of $x + 3$ and $x + 2$ is $(x + 3)(x + 2)$.

$$\frac{1 - \dfrac{2}{x + 3}}{1 - \dfrac{1}{x + 2}} = \frac{1 - \dfrac{2}{x + 3}}{1 - \dfrac{1}{x + 2}} \cdot \frac{(x + 3)(x + 2)}{(x + 3)(x + 2)}$$

$$= \frac{(x + 3)(x + 2) - \dfrac{2}{x + 3}(x + 3)(x + 2)}{(x + 3)(x + 2) - \dfrac{1}{x + 2}(x + 3)(x + 2)}$$

$$= \frac{x^2 + 5x + 6 - 2x - 4}{x^2 + 5x + 6 - x - 3}$$

$$= \frac{x^2 + 3x + 2}{x^2 + 4x + 3} = \frac{(x + 2)(x + 1)}{(x + 3)(x + 1)} = \frac{x + 2}{x + 3}$$

Example 4

The LCM is $x - 3$.

$$\frac{2x + 5 + \dfrac{14}{x - 3}}{4x + 16 + \dfrac{49}{x - 3}} = \frac{2x + 5 + \dfrac{14}{x - 3}}{4x + 16 + \dfrac{49}{x - 3}} \cdot \frac{x - 3}{x - 3}$$

$$= \frac{(2x + 5)(x - 3) + \dfrac{14}{x - 3}(x - 3)}{(4x + 16)(x - 3) + \dfrac{49}{x - 3}(x - 3)}$$

$$= \frac{2x^2 - x - 15 + 14}{4x^2 + 4x - 48 + 49} = \frac{2x^2 - x - 1}{4x^2 + 4x + 1}$$

$$= \frac{(2x + 1)(x - 1)}{(2x + 1)(2x + 1)} = \frac{\overset{1}{(2x + 1)}(x - 1)}{\underset{1}{(2x + 1)}(2x + 1)} = \frac{x - 1}{2x + 1}$$

Example 6

The LCM of the complex fraction $\dfrac{1}{2 - \frac{1}{x}}$ is x.

$$2 - \frac{1}{2 - \frac{1}{x}} = 2 - \frac{1}{2 - \frac{1}{x}} \cdot \frac{\frac{x}{x}}{}$$

$$= 2 - \frac{1 \cdot x}{2 \cdot x - \frac{1}{x} \cdot x} = 2 - \frac{x}{2x - 1}$$

The LCM is $2x - 1$.

$$2 - \frac{x}{2x - 1} = 2 \cdot \frac{(2x - 1)}{(2x - 1)} - \frac{x}{2x - 1}$$

$$= \frac{4x - 2}{2x - 1} - \frac{x}{2x - 1}$$

$$= \frac{4x - 2 - x}{2x - 1} = \frac{3x - 2}{2x - 1}$$

SECTION 4.5   *pages 193–194*

Example 2

$$\frac{5}{x - 2} = \frac{3}{4}$$

$$\frac{5}{x - 2} \cdot 4(x - 2) = \frac{3}{4} \cdot 4(x - 2)$$

$$5 \cdot 4 = 3(x - 2)$$

$$20 = 3x - 6$$

$$26 = 3x$$

$$\frac{26}{3} = x$$

The solution is $\dfrac{26}{3}$.

Example 4

$$\frac{5}{2x - 3} = \frac{-2}{x + 1}$$

$$\frac{5}{2x - 3}(x + 1)(2x - 3) = \frac{-2}{x + 1}(x + 1)(2x - 3)$$

$$5(x + 1) = -2(2x - 3)$$

$$5x + 5 = -4x + 6$$

$$9x + 5 = 6$$

$$9x = 1$$

$$x = \frac{1}{9}$$

The solution is $\dfrac{1}{9}$.

Example 6

Strategy

To find the cost, write and solve a proportion using x to represent the cost.

Solution

$$\frac{2}{3.10} = \frac{15}{x}$$

$$\frac{2}{3.10} \cdot x(3.10) = \frac{15}{x} \cdot x(3.10)$$

$$2x = 15(3.10)$$

$$2x = 46.50$$

$$x = 23.25$$

The cost of 15 lb of cashews is $23.25.

SECTION 4.6   *pages 197–202*

Example 2

$$\frac{x}{x-2} + x = \frac{6}{x-2}$$

$$(x-2)\left(\frac{x}{x-2} + x\right) = (x-2)\left(\frac{6}{x-2}\right)$$

$$(x-2)\left(\frac{x}{x-2}\right) + (x-2)x = (x-2)\left(\frac{6}{x-2}\right)$$

$$x + x^2 - 2x = 6$$

$$x^2 - x = 6$$

$$x^2 - x - 6 = 0$$

$$(x-3)(x+2) = 0$$

$$x = 3 \qquad x = -2$$

−2 and 3 check as solutions.
The solutions are −2 and 3.

Example 6

Strategy

- Time required for the small pipe to fill the tank: x

	Rate	Time	Part
Large pipe	$\frac{1}{9}$	6	$\frac{6}{9}$
Small pipe	$\frac{1}{x}$	6	$\frac{6}{x}$

- The sum of the part of the task completed by the large pipe and the part of the task completed by the small pipe is 1.

$$\frac{6}{9} + \frac{6}{x} = 1$$

Solution

$$\frac{6}{9} + \frac{6}{x} = 1$$

$$\frac{2}{3} + \frac{6}{x} = 1$$

$$3x\left(\frac{2}{3} + \frac{6}{x}\right) = 3x \cdot 1$$

$$2x + 18 = 3x$$

$$18 = x$$

The small pipe working alone will fill the tank in 18 h.

Example 4

$$\frac{1}{R_1} + \frac{1}{R_2} = \frac{1}{R}$$

$$RR_1R_2\left(\frac{1}{R_1} + \frac{1}{R_2}\right) = RR_1R_2\left(\frac{1}{R}\right)$$

$$RR_1R_2\left(\frac{1}{R_1}\right) + RR_1R_2\left(\frac{1}{R_2}\right) = R_1R_2$$

$$RR_2 + RR_1 = R_1R_2$$

$$R(R_2 + R_1) = R_1R_2$$

$$R = \frac{R_1R_2}{R_2 + R_1}$$

Example 8

Strategy

- Rate of the wind: r

	Distance	Rate	Time
With wind	700	$150 + r$	$\frac{700}{150+r}$
Against wind	500	$150 - r$	$\frac{500}{150-r}$

- The time flying with the wind equals the time flying against the wind.

$$\frac{700}{150+r} = \frac{500}{150-r}$$

Solution

$$\frac{700}{150+r} = \frac{500}{150-r}$$

$$(150+r)(150-r)\left(\frac{700}{150+r}\right) = (150+r)(150-r)\left(\frac{500}{150-r}\right)$$

$$(150-r)700 = (150+r)500$$

$$105{,}000 - 700r = 75{,}000 + 500r$$

$$30{,}000 = 1200r$$

$$25 = r$$

The rate of the wind is 25 mph.

SOLUTIONS to Chapter 5 Examples

SECTION 5.1   *pages 219–224*

Example 2
$$16^{-3/4} = (2^4)^{-3/4}$$
$$= 2^{-3}$$
$$= \frac{1}{2^3} = \frac{1}{8}$$

Example 4 $(-81)^{3/4}$
The base of the exponential expression is negative, while the denominator of the exponent is a positive even number.

Therefore, $(-81)^{3/4}$ is not a real number.

Example 6
$$(x^{3/4}y^{1/2}z^{-2/3})^{-4/3} = x^{-1}y^{-2/3}z^{8/9}$$
$$= \frac{z^{8/9}}{xy^{2/3}}$$

Example 8
$$\left(\frac{16a^{-2}b^{4/3}}{9a^4b^{-2/3}}\right)^{-1/2} = \left(\frac{2^4a^{-6}b^2}{3^2}\right)^{-1/2}$$
$$= \frac{2^{-2}a^3b^{-1}}{3^{-1}}$$
$$= \frac{3a^3}{2^2b} = \frac{3a^3}{4b}$$

Example 10
$$(2x^3)^{3/4} = \sqrt[4]{(2x^3)^3}$$
$$= \sqrt[4]{8x^9}$$

Example 12
$$-5a^{5/6} = -5(a^5)^{1/6}$$
$$= -5\sqrt[6]{a^5}$$

Example 14 $\sqrt[3]{3ab} = (3ab)^{1/3}$

Example 16 $\sqrt[4]{x^4 + y^4} = (x^4 + y^4)^{1/4}$

Example 18
$$\sqrt{121x^{10}y^4} = \sqrt{11^2x^{10}y^4}$$
$$= 11x^5y^2$$

Example 20
$$\sqrt[3]{-125a^6b^9} = \sqrt[3]{(-5)^3a^6b^9}$$
$$= -5a^2b^3$$

SECTION 5.2   *pages 229–236*

Example 2
$$\sqrt[5]{x^7} = \sqrt[5]{x^5 \cdot x^2}$$
$$= \sqrt[5]{x^5}\sqrt[5]{x^2}$$
$$= x\sqrt[5]{x^2}$$

Example 4
$$\sqrt[3]{-64x^8y^{18}} = \sqrt[3]{(-4)^3x^8y^{18}}$$
$$= \sqrt[3]{(-4)^3x^6y^{18}(x^2)}$$
$$= \sqrt[3]{(-4)^3x^6y^{18}}\sqrt[3]{x^2}$$
$$= -4x^2y^6\sqrt[3]{x^2}$$

Example 6
$$\sqrt{216} = \sqrt{2^3 \cdot 3^3}$$
$$= \sqrt{2^2 \cdot 3^2(2 \cdot 3)}$$
$$= \sqrt{2^2 \cdot 3^2}\sqrt{2 \cdot 3}$$
$$= 2 \cdot 3\sqrt{6}$$
$$= 6\sqrt{6} \approx 6(2.449) \approx 14.697$$

Example 8
$$3xy\sqrt[3]{81x^5y} - \sqrt[3]{192x^8y^4}$$
$$= 3xy\sqrt[3]{3^4x^5y} - \sqrt[3]{2^6 \cdot 3x^8y^4}$$
$$= 3xy\sqrt[3]{3^3x^3}\sqrt[3]{3x^2y} - \sqrt[3]{2^6x^6y^3}\sqrt[3]{3x^2y}$$
$$= 3xy \cdot 3x\sqrt[3]{3x^2y} - 2^2x^2y\sqrt[3]{3x^2y}$$
$$= 9x^2y\sqrt[3]{3x^2y} - 4x^2y\sqrt[3]{3x^2y} = 5x^2y\sqrt[3]{3x^2y}$$

Example 10
$$\sqrt{5b}(\sqrt{3b} - \sqrt{10}) = \sqrt{15b^2} - \sqrt{50b}$$
$$= \sqrt{3 \cdot 5b^2} - \sqrt{2 \cdot 5^2b}$$
$$= \sqrt{b^2}\sqrt{3 \cdot 5} - \sqrt{5^2}\sqrt{2b}$$
$$= b\sqrt{15} - 5\sqrt{2b}$$

Example 12
$$(2\sqrt[3]{2x} - 3)(\sqrt[3]{2x} - 5)$$
$$= 2\sqrt[3]{4x^2} - 10\sqrt[3]{2x} - 3\sqrt[3]{2x} + 15$$
$$= 2\sqrt[3]{4x^2} - 13\sqrt[3]{2x} + 15$$

Example 14
$$(\sqrt{a} - 3\sqrt{y})(\sqrt{a} + 3\sqrt{y}) = (\sqrt{a})^2 - (3\sqrt{y})^2$$
$$= a - 9y$$

Example 16
$$\frac{y}{\sqrt{3y}} = \frac{y}{\sqrt{3y}} \cdot \frac{\sqrt{3y}}{\sqrt{3y}} = \frac{y\sqrt{3y}}{\sqrt{3^2y^2}} = \frac{y\sqrt{3y}}{3y} = \frac{\sqrt{3y}}{3}$$

Example 18
$$\frac{3}{\sqrt[3]{3x^2}} = \frac{3}{\sqrt[3]{3x^2}} \cdot \frac{\sqrt[3]{3^2x}}{\sqrt[3]{3^2x}} = \frac{3\sqrt[3]{9x}}{\sqrt[3]{3^3x^3}}$$
$$= \frac{3\sqrt[3]{9x}}{3x} = \frac{\sqrt[3]{9x}}{x}$$

Example 20
$$\frac{3 + \sqrt{6}}{2 - \sqrt{6}} = \frac{3 + \sqrt{6}}{2 - \sqrt{6}} \cdot \frac{2 + \sqrt{6}}{2 + \sqrt{6}} = \frac{6 + 3\sqrt{6} + 2\sqrt{6} + (\sqrt{6})^2}{2^2 - (\sqrt{6})^2}$$
$$= \frac{6 + 5\sqrt{6} + 6}{4 - 6} = \frac{12 + 5\sqrt{6}}{-2} = -\frac{12 + 5\sqrt{6}}{2}$$

Example 22

$$\frac{\sqrt{2} + \sqrt{x}}{\sqrt{2} - \sqrt{x}} = \frac{\sqrt{2} + \sqrt{x}}{\sqrt{2} - \sqrt{x}} \cdot \frac{\sqrt{2} + \sqrt{x}}{\sqrt{2} + \sqrt{x}}$$

$$= \frac{(\sqrt{2})^2 + \sqrt{2x} + \sqrt{2x} + (\sqrt{x})^2}{(\sqrt{2})^2 - (\sqrt{x})^2}$$

$$= \frac{2 + 2\sqrt{2x} + x}{2 - x}$$

SECTION 5.3   *pages 241–246*

Example 2

$$\sqrt{-45} = i\sqrt{45} = i\sqrt{3^2 \cdot 5} = 3i\sqrt{5}$$

Example 4

$$\sqrt{98} - \sqrt{-60} = \sqrt{98} - i\sqrt{60}$$
$$= \sqrt{2 \cdot 7^2} - i\sqrt{2^2 \cdot 3 \cdot 5}$$
$$= 7\sqrt{2} - 2i\sqrt{15}$$

Example 6

$$(-4 + 2i) - (6 - 8i) = -10 + 10i$$

Example 8

$$(16 - \sqrt{-45}) - (3 + \sqrt{-20})$$
$$= (16 - i\sqrt{45}) - (3 + i\sqrt{20})$$
$$= (16 - i\sqrt{3^2 \cdot 5}) - (3 + i\sqrt{2^2 \cdot 5})$$
$$= (16 - 3i\sqrt{5}) - (3 + 2i\sqrt{5})$$
$$= 13 - 5i\sqrt{5}$$

Example 10

$$(3 - 2i) + (-3 + 2i) = 0 + 0i = 0$$

Example 12

$$(-3i)(-10i) = 30i^2 = 30(-1) = -30$$

Example 14

$$-\sqrt{-8} \cdot \sqrt{-5} = -i\sqrt{8} \cdot i\sqrt{5} = -i^2\sqrt{40}$$
$$= (-1)\sqrt{40} = \sqrt{2^3 \cdot 5} = 2\sqrt{10}$$

Example 16

$$-6i(3 + 4i) = -18i - 24i^2$$
$$= -18i - 24(-1) = 24 - 18i$$

Example 18

$$\sqrt{-3}(\sqrt{27} - \sqrt{-6}) = i\sqrt{3}(\sqrt{27} - i\sqrt{6})$$
$$= i\sqrt{81} - i^2\sqrt{18}$$
$$= i\sqrt{3^4} - (-1)\sqrt{2 \cdot 3^2}$$
$$= 9i + 3\sqrt{2}$$
$$= 3\sqrt{2} + 9i$$

Example 20

$$(4 - 3i)(2 - i) = 8 - 4i - 6i + 3i^2$$
$$= 8 - 10i + 3i^2$$
$$= 8 - 10i + 3(-1)$$
$$= 5 - 10i$$

Example 22

$$(3 + 6i)(3 - 6i) = 3^2 + 6^2$$
$$= 9 + 36$$
$$= 45$$

Example 24

$$(3 - i)\left(\frac{3}{10} + \frac{1}{10}i\right) = \frac{9}{10} + \frac{3}{10}i - \frac{3}{10}i - \frac{1}{10}i^2$$
$$= \frac{9}{10} - \frac{1}{10}i^2 = \frac{9}{10} - \frac{1}{10}(-1)$$
$$= \frac{9}{10} + \frac{1}{10} = 1$$

Example 26

$$\frac{2 - 3i}{4i} = \frac{2 - 3i}{4i} \cdot \frac{i}{i}$$

$$= \frac{2i - 3i^2}{4i^2}$$

$$= \frac{2i - 3(-1)}{4(-1)}$$

$$= \frac{3 + 2i}{-4} = -\frac{3}{4} - \frac{1}{2}i$$

Example 28

$$\frac{2 + 5i}{3 - 2i} = \frac{2 + 5i}{3 - 2i} \cdot \frac{3 + 2i}{3 + 2i} = \frac{6 + 4i + 15i + 10i^2}{3^2 + 2^2}$$

$$= \frac{6 + 19i + 10(-1)}{13} = \frac{-4 + 19i}{13}$$

$$= -\frac{4}{13} + \frac{19}{13}i$$

SECTION 5.4   *pages 249–252*

Example 2

$$\sqrt{x} - \sqrt{x+5} = 1$$
$$\sqrt{x} = 1 + \sqrt{x+5}$$
$$(\sqrt{x})^2 = (1 + \sqrt{x+5})^2$$
$$x = 1 + 2\sqrt{x+5} + x + 5$$
$$-3 = \sqrt{x+5}$$
$$(-3)^2 = (\sqrt{x+5})^2$$
$$9 = x + 5$$
$$4 = x$$

4 does not check as a solution. The equation has no solution.

Example 4

$$\sqrt[4]{x-8} = 3$$
$$(\sqrt[4]{x-8})^4 = 3^4$$
$$x - 8 = 81$$
$$x = 89$$

Check:
$$\sqrt[4]{x-8} = 3$$
$$\begin{array}{c|c} \sqrt[4]{89-8} & 3 \\ \sqrt[4]{81} & 3 \\ & 3 = 3 \end{array}$$

The solution is 89.

Example 6

Strategy

To find the diagonal, use the Pythagorean Theorem. One leg is the length of the rectangle. The second leg is the width of the rectangle. The hypotenuse is the diagonal of the rectangle.

Solution

$$c^2 = a^2 + b^2$$
$$c^2 = (6)^2 + (3)^2$$
$$c^2 = 36 + 9$$
$$c^2 = 45$$
$$(c^2)^{1/2} = (45)^{1/2}$$
$$c = \sqrt{45}$$
$$c \approx 6.7$$

The diagonal is 6.7 cm.

Example 8

Strategy

To find the height above water, replace d in the equation with the given value and solve for h.

Solution

$$d = 1.4\sqrt{h}$$
$$5.5 = 1.4\sqrt{h}$$
$$\frac{5.5}{1.4} = \sqrt{h}$$
$$\left(\frac{5.5}{1.4}\right)^2 = (\sqrt{h})^2$$
$$\frac{30.25}{1.96} = h$$
$$15.434 \approx h$$

The periscope must be 15.434 ft above the water.

Example 10

Strategy

To find the distance, replace the variables v and a in the equation by their given values and solve for s.

Solution

$$v = \sqrt{2as}$$
$$88 = \sqrt{2 \cdot 22s}$$
$$88 = \sqrt{44s}$$
$$(88)^2 = (\sqrt{44s})^2$$
$$7744 = 44s$$
$$176 = s$$

The distance required is 176 ft.

SOLUTIONS to Chapter 6 Examples

SECTION 6.1   *pages 265–270*

Example 2

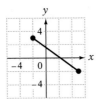

Example 4 $A(-3, 1)$
$B(3, -4)$
$C(-2, -2)$
$D(4, 1)$

Example 6

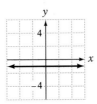

Example 8 $y = -2x + 5$

$\quad = -2\left(\frac{1}{3}\right) + 5$

$\quad = -\frac{2}{3} + 5$

$\quad = \frac{13}{3}$

The ordered pair solution
is $\left(\frac{1}{3}, \frac{13}{3}\right)$.

Example 10

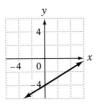

Example 12

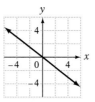

Example 14 $-3x + 2y = 4$
$\qquad 2y = 3x + 4$
$\qquad y = \frac{3}{2}x + 2$

Example 16

SECTION 6.2   *pages 275–280*

Example 2   Let $P_1 = (4, -3)$ and $P_2 = (2, 7)$.

$m = \frac{y_2 - y_1}{x_2 - x_1} = \frac{7 - (-3)}{2 - 4} = \frac{10}{-2} = -5$

The slope is -5.

Example 4 x-intercept: y-intercept:
$$3x - y = 2 \qquad 3x - y = 2$$
$$3x - 0 = 2 \qquad 3(0) - y = 2$$
$$3x = 2 \qquad\qquad -y = 2$$
$$x = \frac{2}{3} \qquad\qquad y = -2$$

$$\left(\frac{2}{3}, 0\right) \qquad\qquad (0, -2)$$

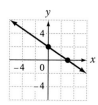

Example 6 x-intercept: y-intercept:
$$y = \frac{1}{4}x + 1 \qquad (0, b)$$
$$\qquad\qquad\qquad b = 1$$
$$0 = \frac{1}{4}x + 1 \qquad (0, 1)$$

$$-\frac{1}{4}x = 1$$
$$x = -4$$
$$(-4, 0)$$

Example 8 $2x + 3y = 6$
$$3y = -2x + 6$$
$$y = -\frac{2}{3}x + 2$$
$$m = -\frac{2}{3} = \frac{-2}{3}$$
$$y\text{-intercept} = (0, 2)$$

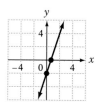

Example 10 $(x_1, y_1) = (-3, -2)$
$$m = 3$$

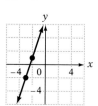

SECTION 6.3   *pages 287–290*

Example 2 $m = -\frac{5}{4}$ $b = 3$
$$y = mx + b$$
$$y = -\frac{5}{4}x + 3$$
The equation of the line is
$$y = -\frac{5}{4}x + 3.$$

Example 4 $m = -\frac{1}{3}$ $(x_1, y_1) = (-3, -2)$
$$y - y_1 = m(x - x_1)$$
$$y - (-2) = -\frac{1}{3}[x - (-3)]$$
$$y + 2 = -\frac{1}{3}(x + 3)$$
$$y + 2 = -\frac{1}{3}x - 1$$
$$y = -\frac{1}{3}x - 1 - 2$$
$$y = -\frac{1}{3}x - 3$$

The equation of the line is
$$y = -\frac{1}{3}x - 3.$$

Example 6 $m = -3$ $(x_1, y_1) = (4, -3)$

$$y - y_1 = m(x - x_1)$$
$$y - (-3) = -3(x - 4)$$
$$y + 3 = -3(x - 4)$$
$$y + 3 = -3x + 12$$
$$y = -3x + 12 - 3$$
$$y = -3x + 9$$

The equation of the line is
$y = -3x + 9$.

Example 8   Let $(x_1, y_1) = (2, 0)$ and $(x_2, y_2) = (5, 3)$.

$$m = \frac{y_2 - y_1}{x_2 - x_1} = \frac{3 - 0}{5 - 2} = \frac{3}{3} = 1$$
$$y - y_1 = m(x - x_1)$$
$$y - 0 = 1(x - 2)$$
$$y = 1(x - 2)$$
$$y = x - 2$$

The equation of the line is
$y = x - 2$.

Example 10   Let $(x_1, y_1) = (4, -2)$ and $(x_2, y_2) = (-1, -7)$.

$$m = \frac{y_2 - y_1}{x_2 - x_1} = \frac{-7 - (-2)}{-1 - 4} = \frac{-5}{-5} = 1$$
$$y - y_1 = m(x - x_1)$$
$$y - (-2) = 1(x - 4)$$
$$y + 2 = 1(x - 4)$$
$$y + 2 = x - 4$$
$$y = x - 6$$

The equation of the line is
$y = x - 6$.

Example 12   Let $(x_1, y_1) = (2, 3)$ and $(x_2, y_2) = (-5, 3)$.

$$m = \frac{y_2 - y_1}{x_2 - x_1} = \frac{3 - 3}{-5 - 2} = \frac{0}{-7} = 0$$

The line has zero slope.
The line is a horizontal line.
All points on the line have an ordinate of 3.
The equation of the line is $y = 3$.

SECTION 6.4   *pages 295–298*

Example 2 $m_1 = \frac{1 - (-3)}{7 - (-2)} = \frac{4}{9}$

$m_2 = \frac{-5 - 1}{6 - 4} = \frac{-6}{2} = -3$

$m_1 \cdot m_2 = \frac{4}{9} \cdot -3 = -\frac{4}{3}$

No, the lines are not perpendicular.

Example 4
$$5x + 2y = 2$$
$$2y = -5x + 2$$
$$y = -\frac{5}{2}x + 1$$

$m_1 = -\frac{5}{2}$

$$5x + 2y = -6$$
$$2y = -5x - 6$$
$$y = -\frac{5}{2}x - 3$$

$m_2 = -\frac{5}{2}$

$m_1 = m_2 = -\frac{5}{2}$

The lines are parallel.

Example 6
$$x - 4y = 3$$
$$-4y = -x + 3$$
$$y = \frac{1}{4}x - \frac{3}{4}$$

$m_1 = \frac{1}{4}$

$m_1 \cdot m_2 = -1$

$\frac{1}{4} \cdot m_2 = -1$

$m_2 = -4$

$$y - y_1 = m(x - x_1)$$
$$y - 2 = -4[x - (-2)]$$
$$y - 2 = -4(x + 2)$$
$$y - 2 = -4x - 8$$
$$y = -4x - 6$$

The equation of the line is
$y = -4x - 6$.

SECTION 6.5   *pages 301–302*

Example 2

Strategy

To write the equation:
- Use two points on the graph to find the slope of the line.
- Locate the *y*-intercept of the line on the graph.
- Use the slope-intercept form of an equation to write the equation of the line.

To find the Fahrenheit temperature, substitute 40° for *x* in the equation and solve for *y*.

Solution

$(x_1, y_1) = (0, 32)$ $(x_2, y_2) = (100, 212)$

$m = \dfrac{y_2 - y_1}{x_2 - x_1} = \dfrac{212 - 32}{100 - 0} = \dfrac{180}{100} = \dfrac{9}{5}$

The *y*-intercept is (0, 32).

$y = mx + b$

$y = \dfrac{9}{5}x + 32$

The equation of the line is $y = \dfrac{9}{5}x + 32$.

$y = \dfrac{9}{5}x + 32$

$y = \dfrac{9}{5}(40) + 32 = 72 + 32 = 104$

The Fahrenheit temperature is 104°.

SECTION 6.6   *pages 305–306*

Example 2

$x + 3y > 6$

$3y > -x + 6$

$y > -\dfrac{1}{3}x + 2$

Example 4 $y < 2$

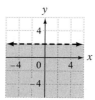

SOLUTIONS to Chapter 7 Examples

SECTION 7.1   *pages 319–322*

Example 2

$$2x^2 = 7x - 3$$
$$2x^2 - 7x + 3 = 0$$
$$(2x - 1)(x - 3) = 0$$

$$2x - 1 = 0 \qquad x - 3 = 0$$
$$2x = 1 \qquad\qquad x = 3$$
$$x = \tfrac{1}{2}$$

The solutions are $\tfrac{1}{2}$ and 3.

Example 4

$$x^2 - 3ax - 4a^2 = 0$$
$$(x + a)(x - 4a) = 0$$

$$x + a = 0 \qquad x - 4a = 0$$
$$x = -a \qquad\qquad x = 4a$$

The solutions are $-a$ and $4a$.

Example 6

$$(x - r_1)(x - r_2) = 0$$
$$(x - 3)\left[x - \left(-\tfrac{1}{2}\right)\right] = 0$$
$$(x - 3)\left(x + \tfrac{1}{2}\right) = 0$$
$$x^2 - \tfrac{5}{2}x - \tfrac{3}{2} = 0$$
$$2\left(x^2 - \tfrac{5}{2}x - \tfrac{3}{2}\right) = 2 \cdot 0$$
$$2x^2 - 5x - 3 = 0$$

Example 8

$$2(x + 1)^2 - 24 = 0$$
$$2(x + 1)^2 = 24$$
$$(x + 1)^2 = 12$$
$$\sqrt{(x + 1)^2} = \sqrt{12}$$
$$x + 1 = \pm\sqrt{12} = \pm 2\sqrt{3}$$

$$x + 1 = 2\sqrt{3} \qquad x + 1 = -2\sqrt{3}$$
$$x = -1 + 2\sqrt{3} \qquad x = -1 - 2\sqrt{3}$$

The solutions are $-1 + 2\sqrt{3}$ and $-1 - 2\sqrt{3}$.

SECTION 7.2   *pages 327–330*

Example 2

$$4x^2 - 4x - 1 = 0$$
$$4x^2 - 4x = 1$$
$$\tfrac{1}{4}(4x^2 - 4x) = \tfrac{1}{4} \cdot 1$$
$$x^2 - x = \tfrac{1}{4}$$

Complete the square.

$$x^2 - x + \tfrac{1}{4} = \tfrac{1}{4} + \tfrac{1}{4}$$
$$\left(x - \tfrac{1}{2}\right)^2 = \tfrac{2}{4}$$
$$\sqrt{\left(x - \tfrac{1}{2}\right)^2} = \sqrt{\tfrac{2}{4}}$$
$$x - \tfrac{1}{2} = \pm\tfrac{\sqrt{2}}{2}$$

$$x - \tfrac{1}{2} = \tfrac{\sqrt{2}}{2} \qquad x - \tfrac{1}{2} = -\tfrac{\sqrt{2}}{2}$$
$$x = \tfrac{1}{2} + \tfrac{\sqrt{2}}{2} \qquad x = \tfrac{1}{2} - \tfrac{\sqrt{2}}{2}$$

The solutions are $\dfrac{1 + \sqrt{2}}{2}$ and $\dfrac{1 - \sqrt{2}}{2}$.

Example 4

$$2x^2 + x - 5 = 0$$
$$2x^2 + x = 5$$
$$\tfrac{1}{2}(2x^2 + x) = \tfrac{1}{2} \cdot 5$$
$$x^2 + \tfrac{1}{2}x = \tfrac{5}{2}$$

Complete the square.

$$x^2 + \tfrac{1}{2}x + \tfrac{1}{16} = \tfrac{5}{2} + \tfrac{1}{16}$$
$$\left(x + \tfrac{1}{4}\right)^2 = \tfrac{41}{16}$$
$$\sqrt{\left(x + \tfrac{1}{4}\right)^2} = \sqrt{\tfrac{41}{16}}$$
$$x + \tfrac{1}{4} = \pm\tfrac{\sqrt{41}}{4}$$

$$x + \tfrac{1}{4} = \tfrac{\sqrt{41}}{4} \qquad x + \tfrac{1}{4} = -\tfrac{\sqrt{41}}{4}$$
$$x = -\tfrac{1}{4} + \tfrac{\sqrt{41}}{4} \qquad x = -\tfrac{1}{4} - \tfrac{\sqrt{41}}{4}$$

The solutions are $\dfrac{-1 + \sqrt{41}}{4}$ and $\dfrac{-1 - \sqrt{41}}{4}$.

SECTION 7.3   *pages 333–336*

Example 2 $x^2 + 6x - 9 = 0$

$a = 1, b = 6, c = -9$

$x = \dfrac{-b \pm \sqrt{b^2 - 4ac}}{2a}$

$= \dfrac{-6 \pm \sqrt{6^2 - 4(1)(-9)}}{2 \cdot 1}$

$= \dfrac{-6 \pm \sqrt{36 + 36}}{2}$

$= \dfrac{-6 \pm \sqrt{72}}{2} = \dfrac{-6 \pm 6\sqrt{2}}{2}$

$= -3 \pm 3\sqrt{2}$

The solutions are $-3 + 3\sqrt{2}$ and $-3 - 3\sqrt{2}$.

Example 6 $4x^2 = 4x - 1$

$4x^2 - 4x + 1 = 0$

$a = 4, b = -4, c = 1$

$x = \dfrac{-b \pm \sqrt{b^2 - 4ac}}{2a}$

$= \dfrac{-(-4) \pm \sqrt{(-4)^2 - 4(4)(1)}}{2 \cdot 4}$

$= \dfrac{4 \pm \sqrt{16 - 16}}{8} = \dfrac{4 \pm \sqrt{0}}{8}$

$= \dfrac{4}{8} = \dfrac{1}{2}$

The solution is $\dfrac{1}{2}$.

Example 4 $x^2 - 2x + 10 = 0$

$a = 1, b = -2, c = 10$

$x = \dfrac{-b \pm \sqrt{b^2 - 4ac}}{2a}$

$= \dfrac{-(-2) \pm \sqrt{(-2)^2 - 4(1)(10)}}{2 \cdot 1}$

$= \dfrac{2 \pm \sqrt{4 - 40}}{2} = \dfrac{2 \pm \sqrt{-36}}{2}$

$= \dfrac{2 \pm 6i}{2} = 1 \pm 3i$

The solutions are $1 + 3i$ and $1 - 3i$.

Example 8 $3x^2 - x - 1 = 0$

$a = 3, b = -1, c = -1$

$b^2 - 4ac =$
$(-1)^2 - 4(3)(-1) = 1 + 12 = 13$

$13 > 0$

Since the discriminant is greater than zero, the equation has two real number solutions.

SECTION 7.4   *pages 339–342*

Example 2

$x - 5x^{1/2} + 6 = 0$

$(x^{1/2})^2 - 5(x^{1/2}) + 6 = 0$

$u^2 - 5u + 6 = 0$

$(u - 2)(u - 3) = 0$

$u - 2 = 0 \qquad u - 3 = 0$

$u = 2 \qquad u = 3$

Replace u by $x^{1/2}$.

$x^{1/2} = 2 \qquad x^{1/2} = 3$

$\sqrt{x} = 2 \qquad \sqrt{x} = 3$

$(\sqrt{x})^2 = 2^2 \qquad (\sqrt{x})^2 = 3^2$

$x = 4 \qquad x = 9$

The solutions are 4 and 9.

Example 4

$\sqrt{2x + 1} + x = 7$

$\sqrt{2x + 1} = 7 - x$

$(\sqrt{2x + 1})^2 = (7 - x)^2$

$2x + 1 = 49 - 14x + x^2$

$0 = x^2 - 16x + 48$

$0 = (x - 4)(x - 12)$

$x - 4 = 0 \qquad x - 12 = 0$

$x = 4 \qquad x = 12$

4 checks as a solution.
12 does not check as a solution.

The solution is 4.

Example 6

$\sqrt{2x - 1} + \sqrt{x} = 2$

Solve for one of the radical expressions.

$\sqrt{2x - 1} = 2 - \sqrt{x}$
$(\sqrt{2x - 1})^2 = (2 - \sqrt{x})^2$
$2x - 1 = 4 - 4\sqrt{x} + x$
$x - 5 = -4\sqrt{x}$

Square each side of the equation.

$(x - 5)^2 = (-4\sqrt{x})^2$
$x^2 - 10x + 25 = 16x$
$x^2 - 26x + 25 = 0$
$(x - 1)(x - 25) = 0$

$x - 1 = 0 \qquad x - 25 = 0$
$\qquad x = 1 \qquad\qquad x = 25$

1 checks as a solution.
25 does not check as a solution.

The solution is 1.

Example 8

$$3y + \frac{25}{3y - 2} = -8$$

$$(3y - 2)\left(3y + \frac{25}{3y - 2}\right) = (3y - 2)(-8)$$

$$(3y - 2)(3y) + (3y - 2)\left(\frac{25}{3y - 2}\right) = (3y - 2)(-8)$$

$$9y^2 - 6y + 25 = -24y + 16$$
$$9y^2 + 18y + 9 = 0$$
$$9(y^2 + 2y + 1) = 0$$
$$9(y + 1)(y + 1) = 0$$

$$y + 1 = 0 \qquad y + 1 = 0$$
$$y = -1 \qquad\quad y = -1$$

The solution is -1.

SECTION 7.5   *pages 345–346*

Example 2

$2x^2 - x - 10 \leq 0$
$(2x - 5)(x + 2) \leq 0$

$\left\{x\,\middle|\,-2 \leq x \leq \frac{5}{2}\right\}$

SECTION 7.6   *pages 349–350*

Example 2

Strategy

- This is a geometry problem.
- Width of the rectangle: w
  Length of the rectangle: $w + 3$
- Use the equation $A = L \cdot w$.

Solution

$A = L \cdot w$
$54 = (w + 3)(w)$
$54 = w^2 + 3w$
$0 = w^2 + 3w - 54$
$0 = (w + 9)(w - 6)$

$w + 9 = 0 \qquad w - 6 = 0$
$\qquad w = -9 \qquad\quad w = 6$

The solution -9 is not possible.

$w + 3 = 6 + 3 = 9$

The length is 9 m.

SOLUTIONS to Chapter 8 Examples

SECTION 8.1   *pages 363–368*

Example 2

$s(t) = \dfrac{t}{t^2 + 1}$

$s(-1) = \dfrac{-1}{(-1)^2 + 1}$

$s(-1) = -\dfrac{1}{2}$

Example 6

$f(a + h) - f(a) = [2(a + h)^2 + 3] - (2a^2 + 3)$
$\qquad\qquad\quad = (2a^2 + 4ah + 2h^2 + 3) - (2a^2 + 3)$
$f(a + h) - f(a) = 2h^2 + 4ah$

Example 10

$g(x) = x^2 - 1$ with domain $\{1, 2, 3, 4\}$

$g(x) = x^2 - 1$
$g(1) = 1^2 - 1 = 1 - 1 = 0$
$g(2) = 2^2 - 1 = 4 - 1 = 3$
$g(3) = 3^2 - 1 = 9 - 1 = 8$
$g(4) = 4^2 - 1 = 16 - 1 = 15$

The range is $\{0, 3, 8, 15\}$.

Example 4

$f(x) - g(x) = 2x^2 - (4x - 1)$
$f(-1) - g(-1) = 2(-1)^2 - [4(-1) - 1] = 2 - (-5)$
$f(-1) - g(-1) = 7$

Example 8

For $\{(0, 2), (1, 0), (2, 0), (3, 2)\}$
The domain is $\{0, 1, 2, 3\}$.
The range is $\{0, 2\}$.

Example 12

Because $x^2 + 4 \geq 0$ for all values of x, $\sqrt{x^2 + 4}$ is a real number for all values of x. The domain of the function is all real numbers.

SECTION 8.2   *pages 373–380*

Example 2

This is a linear function.
The slope is -2, and the y-intercept is $(0, 1)$.
The domain is the set of real numbers.
The range is the set of real numbers.

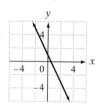

Example 4

x	y
-3	4
-2	3
-1	2
0	1
1	0
2	1
3	2

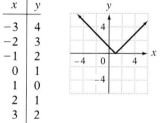

The domain is the set of real numbers.
The range is $\{y | y \geq 0\}$.

Example 6

x	y
-2	10
-1	3
0	2
1	1
2	-6

The domain is the set of real numbers.
The range is the set of real numbers.

Example 8

No member of the set has the same first component and a different second component. The set of ordered pairs is a function.

Example 10

A vertical line intersects the graph at most once. The graph is the graph of a function.

Example 12

Since any horizontal line will intersect the graph no more than once, the graph is the graph of a 1–1 function.

SECTION 8.3   *pages 387–392*

Example 2

$g(-1) = (-1)^2 = 1$

$f(g(-1)) = f(1)$
$\qquad = 1 - 2(1)$
$\qquad = 1 - 2$
$\qquad = -1$

$f(-1) = 1 - 2(-1)$
$\qquad = 1 + 2$
$\qquad = 3$

$g(f(-1)) = g(3)$
$\qquad = 3^2$
$\qquad = 9$

Example 4

$g(f(x)) = \dfrac{1}{(2x)^2 - 2}$

$\qquad\quad = \dfrac{1}{4x^2 - 2}$

Example 6

$f(x) = \frac{1}{2}x + 1$

$y = \frac{1}{2}x + 1$

$x = \frac{1}{2}y + 1$ $\qquad$ Interchange x and y.

$x - 1 = \frac{1}{2}y$ $\qquad$ Solve for y.

$2x - 2 = y$

The inverse function is $f^{-1}(x) = 2x - 2$.

Example 8

$f(g(x)) = f\left(\frac{1}{3}x - 4\right)$

$\qquad\quad = 3\left(\frac{1}{3}x - 4\right) + 12$

$\qquad\quad = x - 12 + 12$
$\qquad\quad = x$

$g(f(x)) = g(3x + 12)$

$\qquad\quad = \frac{1}{3}(3x + 12) - 4$

$\qquad\quad = x + 4 - 4$
$\qquad\quad = x$

$f(g(x)) = g(f(x)) = x$. Therefore the functions are inverses of each other.

SECTION 8.4   *pages 395–398*

Example 2

Strategy

To find the distance:
- Write the basic direct variation equation, replace the variables by the given values, and solve for k.
- Write the direct variation equation, replacing k by its value. Substitute 5 for t and solve for s.

Solution

$s = kt^2$
$64 = k(2)^2$
$64 = k \cdot 4$
$16 = k$

$s = 16t^2 = 16(5)^2 = 400$

The object will fall 400 ft in 5 s.

Example 4

Strategy

To find the resistance:
- Write the basic inverse variation equation, replace the variables by the given values and solve for k.
- Write the inverse variation equation, replacing k by its value. Substitute 0.02 for d and solve for R.

Solution

$$R = \frac{k}{d^2}$$

$$0.5 = \frac{k}{(0.01)^2}$$

$$0.5 = \frac{k}{0.0001}$$

$$0.00005 = k$$

$$R = \frac{0.00005}{d^2} = \frac{0.00005}{(0.02)^2} = 0.125$$

The resistance is 0.125 ohms.

Example 6

Strategy

To find the strength:
- Write the basic combined variation equation, replace the variables by the given values and solve for k.
- Write the combined variation equation, replacing k by its value. Substitute 4 for w and 8 for d, and solve for s.

Solution

$$s = \frac{kw}{d^2}$$

$$1200 = \frac{k \cdot 2}{(12)^2}$$

$$1200 = \frac{k \cdot 2}{144}$$

$$1200 = \frac{k}{72}$$

$$86,400 = k$$

$$s = \frac{86,400w}{d^2} = \frac{86,400 \cdot 4}{8^2} = 5400$$

The strength is 5400 lb.

SOLUTIONS to Chapter 9 Examples

SECTION 9.1   *pages 413–420*

Example 2

$y = x^2 + 2x + 1$

$-\dfrac{b}{2a} = -\dfrac{2}{2(1)} = -1$

$y = (-1)^2 + 2(-1) + 1$

$\quad = 0$

Vertex: $(-1, 0)$

Axis of symmetry:

$\quad x = -1$

Example 4

$x = -y^2 - 2y + 2$

$-\dfrac{b}{2a} = -\dfrac{-2}{2(-1)} = -1$

$x = -(-1)^2 - 2(-1) + 2$

$\quad = 3$

Vertex: $(3, -1)$

Axis of symmetry:

$\quad y = -1$

Example 6

$y = x^2 - 2x - 1$

$-\dfrac{b}{2a} = -\dfrac{-2}{2(1)} = 1$

$y = 1^2 - 2(1) - 1$

$\quad = -2$

Vertex: $(1, -2)$

Axis of symmetry:

$\quad x = 1$

Example 10

$y = x^2 + 4x + 5$

$0 = x^2 + 4x + 5$

$a = 1, b = 4, c = 5$

$x = \dfrac{-b \pm \sqrt{b^2 - 4ac}}{2a}$

$\quad = \dfrac{-4 \pm \sqrt{4^2 - 4(1)(5)}}{2 \cdot 1}$

$\quad = \dfrac{-4 \pm \sqrt{16 - 20}}{2}$

$\quad = \dfrac{-4 \pm \sqrt{-4}}{2} = \dfrac{-4 \pm 2i}{2}$

$\quad = -2 \pm i$

The equation has no real number solutions. There are no x-intercepts.

Example 14

Strategy

- To find the time it takes the ball to reach its maximum height, find the t-coordinate of the vertex.
- To find the maximum height, evaluate the function at the t-coordinate of the vertex.

Solution

$t = -\dfrac{b}{2a} = -\dfrac{64}{2(-16)} = 2$

The ball reaches its maximum height in 2 s.

$s(t) = -16t^2 + 64t$

$s(2) = -16(2)^2 + 64(2) = -64 + 128 = 64$

The maximum height is 64 ft.

Example 8

$f(x) = -3x^2 + 4x - 1$

$x = -\dfrac{b}{2a} = -\dfrac{4}{2(-3)} = \dfrac{2}{3}$

$f(x) = -3x^2 + 4x - 1$

$f\left(\dfrac{2}{3}\right) = -3\left(\dfrac{2}{3}\right)^2 + 4\left(\dfrac{2}{3}\right) - 1 = -\dfrac{4}{3} + \dfrac{8}{3} - 1 = \dfrac{1}{3}$

Since a is negative, the function has a maximum value.

The maximum value of the function is $\dfrac{1}{3}$.

Example 12

$y = x^2 - x - 6$

$a = 1, b = -1 \ c = -6$

$b^2 - 4ac$

$(-1)^2 - 4(1)(-6) = 1 + 24 = 25$

Since the discriminant is greater than zero the parabola has two x-intercepts.

Example 16

Strategy

The perimeter is 44 ft.

$$44 = 2L + 2w$$
$$22 = L + w$$
$$22 - L = w$$

The area is $L \cdot w = L(22 - L) = 22L - L^2$.

- To find the length, find the L-coordinate of the vertex of the function $f(L) = -L^2 + 22L$.
- To find the width, replace L in $22 - L$ by the L-coordinate of the vertex and evaluate.

Solution

$L = -\dfrac{b}{2a} = -\dfrac{22}{2(-1)} = 11$

The length is 11 ft.

$22 - L = 22 - 11 = 11$

The width is 11 ft.

SECTION 9.2   *pages 427–430*

Example 2

$(x_1, y_1) = (3, -2)$ $(x_2, y_2) = (-1, -5)$

$d = \sqrt{(x_1 - x_2)^2 + (y_1 - y_2)^2}$

$\quad = \sqrt{[3 - (-1)]^2 + [(-2) - (-5)]^2}$

$\quad = \sqrt{4^2 + 3^2} = \sqrt{16 + 9} = \sqrt{25} = 5$

Example 4

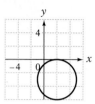

Example 6

$(x - h)^2 + (y - k)^2 = r^2$
$(x - 2)^2 + [y - (-3)]^2 = 4^2$
$(x - 2)^2 + (y + 3)^2 = 16$

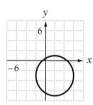

Example 8

$x^2 + y^2 - 4x + 8y + 15 = 0$
$(x^2 - 4x) + (y^2 + 8y) = -15$
$(x^2 - 4x + 4) + (y^2 + 8y + 16) = -15 + 4 + 16$
$(x - 2)^2 + (y + 4)^2 = 5$

Center: $(2, -4)$
Radius: $\sqrt{5}$

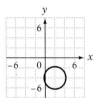

SECTION 9.3   *pages 433–436*

Example 2

x-intercepts:
$(2, 0)$ and $(-2, 0)$

y-intercepts:
$(0, 5)$ and $(0, -5)$

Example 4

x-intercepts:
$(3\sqrt{2}, 0)$ and $(-3\sqrt{2}, 0)$

y-intercepts:
$(0, 3)$ and $(0, -3)$

$\left(3\sqrt{2} \approx 4\frac{1}{4}\right)$

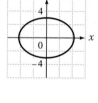

Example 6

Axis of symmetry:
x-axis

Vertices:
$(3, 0)$ and $(-3, 0)$

Asymptotes:
$y = \frac{5}{3}x$ and $y = -\frac{5}{3}x$

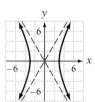

Example 8

Axis of symmetry:
y-axis

Vertices:
$(0, 3)$ and $(0, -3)$

Asymptotes:
$y = x$ and $y = -x$

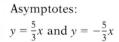

SECTION 9.4   *pages 439–440*

Example 2

$\dfrac{x^2}{9} - \dfrac{y^2}{16} < 1$

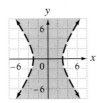

Example 4

$\dfrac{x^2}{16} + \dfrac{y^2}{4} > 1$

SOLUTIONS to Chapter 10 Examples

SECTION 10.1   *pages 453–456*

Example 2

The solution is $(-1, 2)$.

Example 4

The two equations represent the same line. The two equations are dependent. The solutions are the ordered pairs $\left(x, \dfrac{3}{4}x - 3\right)$.

Example 6

$3x - y = 3$
$6x + 3y = -4$

Solve equation (1) for y.
$3x - y = 3$
$\qquad -y = -3x + 3$
$\qquad y = 3x - 3$

Substitute into equation (2).
$\qquad 6x + 3y = -4$
$6x + 3(3x - 3) = -4$
$6x + 9x - 9 = -4$
$\qquad 15x - 9 = -4$
$\qquad 15x = 5$
$\qquad x = \dfrac{5}{15} = \dfrac{1}{3}$

Substitute into equation (1).
$\qquad 3x - y = 3$
$3\left(\dfrac{1}{3}\right) - y = 3$
$\qquad 1 - y = 3$
$\qquad -y = 2$
$\qquad y = -2$

The solution is $\left(\dfrac{1}{3}, -2\right)$.

Example 8

$y = 2x - 3$
$2x - 4y = 5$

$\qquad 2x - 4y = 5$
$2x - 4(2x - 3) = 5$
$2x - 8x + 12 = 5$
$\qquad -6x + 12 = 5$
$\qquad -6x = -7$
$\qquad x = \dfrac{-7}{-6} = \dfrac{7}{6}$

Substitute the value of x into equation (1).
$y = 2x - 3$
$y = 2\left(\dfrac{7}{6}\right) - 3$
$y = \dfrac{7}{3} - 3$
$y = -\dfrac{2}{3}$

The solution is $\left(\dfrac{7}{6}, -\dfrac{2}{3}\right)$.

Example 10 $6x - 3y = 6$
$2x - y = 2$

Solve equation (2) for y.
$2x - y = 2$
$-y = -2x + 2$
$y = 2x - 2$

Substitute into equation (1).
$6x - 3y = 6$
$6x - 3(2x - 2) = 6$
$6x - 6x + 6 = 6$
$6 = 6$

The equations are dependent.
The solutions are the ordered
pairs $(x, 2x - 2)$.

SECTION 10.2   *pages 461–466*

Example 2   (1) $2x + 5y = 6$
(2) $3x - 2y = 6x + 2$

Write equation (2) in the form
$Ax + By = C$.
$3x - 2y = 6x + 2$
$-3x - 2y = 2$

Solve the system $2x + 5y = 6$
$-3x - 2y = 2$.

Eliminate y.
$2(2x + 5y) = 2(6)$
$5(-3x - 2y) = 5(2)$

$4x + 10y = 12$
$-15x - 10y = 10$
Add the equations.
$-11x = 22$
$x = -2$

Replace x in equation (1).
$2x + 5y = 6$
$2(-2) + 5y = 6$
$-4 + 5y = 6$
$5y = 10$
$y = 2$

The solution is $(-2, 2)$.

Example 4 $2x + y = 5$
$4x + 2y = 6$

Eliminate y.
$-2(2x + y) = -2(5)$
$4x + 2y = 6$

$-4x - 2y = -10$
$4x + 2y = 6$

Add the equations.
$0x + 0y = -4$
$0 = -4$

This is not a true equation. The
system is inconsistent and therefore
has no solution.

Example 6   (1) $x - y + z = 6$
(2) $2x + 3y - z = 1$
(3) $x + 2y + 2z = 5$

Eliminate z. Add equations (1) and (2).

$$x - y + z = 6$$
$$2x + 3y - z = 1$$
(4) $3x + 2y = 7$

Multiply equation (2) by 2 and add to equation (3).

$$4x + 6y - 2z = 2$$
$$x + 2y + 2z = 5$$
(5) $5x + 8y = 7$

Solve the system of two equations.
(4) $3x + 2y = 7$
(5) $5x + 8y = 7$

Multiply equation (4) by -4 and add to equation (5).

$$-12x - 8y = -28$$
$$5x + 8y = 7$$
$$-7x = -21$$
$$x = 3$$

Replace x by 3 in equation (4).

$$3x + 2y = 7$$
$$3(3) + 2y = 7$$
$$9 + 2y = 7$$
$$2y = -2$$
$$y = -1$$

Replace x by 3 and y by -1 in equation (1).

$$x - y + z = 6$$
$$3 - (-1) + z = 6$$
$$4 + z = 6$$
$$z = 2$$

The solution is $(3, -1, 2)$.

SECTION 10.3   *pages 471–476*

Example 2

$$\begin{vmatrix} -1 & -4 \\ 3 & -5 \end{vmatrix} = -1(-5) - (-4)(3) = 5 + 12 = 17$$

The value of the determinant is 17.

Example 4

Expand by cofactors of the first row.

$$\begin{vmatrix} 1 & 4 & -2 \\ 3 & 1 & 1 \\ 0 & -2 & 2 \end{vmatrix}$$

$$= 1 \begin{vmatrix} 1 & 1 \\ -2 & 2 \end{vmatrix} - 4 \begin{vmatrix} 3 & 1 \\ 0 & 2 \end{vmatrix} + (-2) \begin{vmatrix} 3 & 1 \\ 0 & -2 \end{vmatrix}$$

$$= 1(2 + 2) - 4(6 - 0) - 2(-6 - 0)$$
$$= 4 - 24 + 12$$
$$= -8$$

The value of the determinant is -8.

Example 6

$$\begin{vmatrix} 3 & -2 & 0 \\ 1 & 4 & 2 \\ -2 & 1 & 3 \end{vmatrix}$$

$$= 3\begin{vmatrix} 4 & 2 \\ 1 & 3 \end{vmatrix} - (-2)\begin{vmatrix} 1 & 2 \\ -2 & 3 \end{vmatrix} + 0\begin{vmatrix} 1 & 4 \\ -2 & 1 \end{vmatrix}$$

$$= 3(12 - 2) + 2(3 + 4) + 0$$
$$= 3(10) + 2(7)$$
$$= 30 + 14$$
$$= 44$$

The value of the determinant is 44.

Example 10

$$3x - y = 4$$
$$6x - 2y = 5$$

$$D = \begin{vmatrix} 3 & -1 \\ 6 & -2 \end{vmatrix} = 0$$

Since $D = 0$, $\frac{D_x}{D}$ is undefined. Therefore, the equations are dependent or inconsistent.

Example 8

$$6x - 6y = 5$$
$$2x - 10y = -1$$

$$D = \begin{vmatrix} 6 & -6 \\ 2 & -10 \end{vmatrix} = -48$$

$$D_x = \begin{vmatrix} 5 & -6 \\ -1 & -10 \end{vmatrix} = -56$$

$$D_y = \begin{vmatrix} 6 & 5 \\ 2 & -1 \end{vmatrix} = -16$$

$$x = \frac{D_x}{D} = \frac{-56}{-48} = \frac{7}{6}$$

$$y = \frac{D_y}{D} = \frac{-16}{-48} = \frac{1}{3}$$

The solution is $\left(\frac{7}{6}, \frac{1}{3}\right)$.

Example 12

$$2x - y + z = -1$$
$$3x + 2y - z = 3$$
$$x + 3y + z = -2$$

$$D = \begin{vmatrix} 2 & -1 & 1 \\ 3 & 2 & -1 \\ 1 & 3 & 1 \end{vmatrix} = 21$$

$$D_x = \begin{vmatrix} -1 & -1 & 1 \\ 3 & 2 & -1 \\ -2 & 3 & 1 \end{vmatrix} = 9$$

$$D_y = \begin{vmatrix} 2 & -1 & 1 \\ 3 & 3 & -1 \\ 1 & -2 & 1 \end{vmatrix} = -3$$

$$D_z = \begin{vmatrix} 2 & -1 & -1 \\ 3 & 2 & 3 \\ 1 & 3 & -2 \end{vmatrix} = -42$$

$$x = \frac{D_x}{D} = \frac{9}{21} = \frac{3}{7}$$

$$y = \frac{D_y}{D} = \frac{-3}{21} = -\frac{1}{7}$$

$$z = \frac{D_z}{D} = \frac{-42}{21} = -2$$

The solution is $\left(\frac{3}{7}, -\frac{1}{7}, -2\right)$.

SECTION 10.4   *pages 479–482*

Example 2

Strategy

- Rate of the rowing team in calm water: t
  Rate of the current: c

	Rate	Time	Distance
With current	$t + c$	2	$2(t + c)$
Against current	$t - c$	2	$2(t - c)$

- The distance traveled with the current is 18 mi.
  The distance traveled against the current is 10 mi.

  $2(t + c) = 18$
  $2(t - c) = 10$

Solution

$2(t + c) = 18$ $\frac{1}{2} \cdot 2(t + c) = \frac{1}{2} \cdot 18$

$2(t - c) = 10$ $\frac{1}{2} \cdot 2(t - c) = \frac{1}{2} \cdot 10$

$\qquad\qquad\qquad\qquad t + c = 9$
$\qquad\qquad\qquad\qquad t - c = 5$

$\qquad\qquad\qquad\qquad\qquad 2t = 14$
$\qquad\qquad\qquad\qquad\qquad\ t = 7$

$t + c = 9$
$7 + c = 9$
$\quad\ \ c = 2$

The rate of the rowing team in calm water is 7 mph.
The rate of the current is 2 mph.

Example 4

Strategy

- Cost of an orange tree: x
  Cost of a grapefruit tree: y

First purchase:

	Amount	Unit Cost	Value
Orange trees	25	x	$25x$
Grapefruit trees	20	y	$20y$

Second purchase:

	Amount	Unit Cost	Value
Orange trees	20	x	$20x$
Grapefruit trees	30	y	$30y$

- The total of the first purchase was $290.
  The total of the second purchase was $330.

  $25x + 20y = 290$
  $20x + 30y = 330$

Solution

$25x + 20y = 290$ $4(25x + 20y) = 4 \cdot 290$
$20x + 30y = 330$ $-5(20x + 30y) = -5 \cdot 330$

$\qquad\qquad\qquad\qquad\qquad 100x + 80y = 1160$
$\qquad\qquad\qquad\qquad\ -100x - 150y = -1650$

$\qquad\qquad\qquad\qquad\qquad\qquad -70y = -490$
$\qquad\qquad\qquad\qquad\qquad\qquad\qquad y = 7$

$25x + 20y = 290$
$25x + 20(7) = 290$
$25x + 140 = 290$
$\qquad 25x = 150$
$\qquad\quad x = 6$

The cost of an orange tree is $6.
The cost of a grapefruit tree is $7.

SECTION 10.5   *pages 487–492*

Example 2   (1) $y = 2x^2 + x - 3$
(2) $y = 2x^2 - 2x + 9$

Use the substitution method.

$$y = 2x^2 + x - 3$$
$$2x^2 - 2x + 9 = 2x^2 + x - 3$$
$$-3x + 12 = 0$$
$$-3x = -12$$
$$x = 4$$

Substitute into equation (1).
$$y = 2x^2 + x - 3$$
$$y = 2(4)^2 + 4 - 3$$
$$y = 32 + 4 - 3$$
$$y = 33$$

The solution is (4, 33).

Example 6   (1) $2x^2 - y^2 = 14$
(2) $y = x - 1$

Use the substitution method.
$$2x^2 - y^2 = 14$$
$$2x^2 - (x - 1)^2 = 14$$
$$2x^2 - (x^2 - 2x + 1) = 14$$
$$2x^2 - x^2 + 2x - 1 = 14$$
$$x^2 + 2x - 1 = 14$$
$$x^2 + 2x - 15 = 0$$
$$(x + 5)(x - 3) = 0$$

$$x + 5 = 0 \qquad x - 3 = 0$$
$$x = -5 \qquad x = 3$$

Substitute into equation (2).
$$y = x - 1 \qquad y = x - 1$$
$$y = (-5) - 1 \qquad y = 3 - 1$$
$$y = -6 \qquad y = 2$$

The solutions are (−5, −6) and (3, 2).

Example 10 $y \geq x - 1$
$y < -2x$

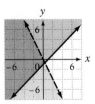

Example 4   (1) $x^2 - y^2 = 10$
(2) $x^2 + y^2 = 8$

Use the addition method.
$$2x^2 = 18$$
$$x^2 = 9$$
$$x = \pm\sqrt{9} = \pm 3$$

Substitute into equation (2).
$$x^2 + y^2 = 8 \qquad\qquad x^2 + y^2 = 8$$
$$3^2 + y^2 = 8 \qquad\qquad (-3)^2 + y^2 = 8$$
$$9 + y^2 = 8 \qquad\qquad 9 + y^2 = 8$$
$$y^2 = -1 \qquad\qquad y^2 = -1$$
$$y = \pm\sqrt{-1} \qquad\qquad y = \pm\sqrt{-1}$$

y is not a real number. Therefore, the system of equations has no real solution. The graphs do not intersect.

Example 8 $x^2 + y^2 < 16$
$y^2 > x$

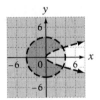

SOLUTIONS to Chapter 11 Examples

SECTION 11.1   *pages 509–514*

Example 2 $f(x) = \left(\frac{2}{3}\right)^x$

$f(3) = \left(\frac{2}{3}\right)^3 = \frac{8}{27}$

$f(-2) = \left(\frac{2}{3}\right)^{-2} = \left(\frac{3}{2}\right)^2 = \frac{9}{4}$

Example 4 $f(x) = 2^{2x+1}$

$f(0) = 2^{2(0)+1} = 2^1 = 2$

$f(-2) = 2^{2(-2)+1} = 2^{-3} = \frac{1}{2^3} = \frac{1}{8}$

Example 6

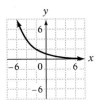

Example 8

Example 10 $10^3 = 1000$ is equivalent to $\log_{10} 1000 = 3.$

Example 12 $\log_4 64 = x$

$64 = 4^x$

$4^3 = 4^x$

$3 = x$

$\log_4 64 = 3$

Example 14 $\log_2 x = -4$

$2^{-4} = x$

$\frac{1}{2^4} = x$

$\frac{1}{16} = x$

The solution is $\frac{1}{16}$.

Example 16 $f(x) = \log_2(x - 1)$
$y = \log_2(x - 1)$

$y = \log_2(x - 1)$ is equivalent to $2^y = x - 1.$

$2^y + 1 = x$

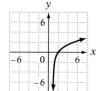

Example 18 $f(x) = \log_3 2x$
$y = \log_3 2x$

$y = \log_3 2x$ is equivalent to $3^y = 2x.$

$\frac{3^y}{2} = x$

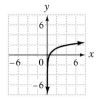

SECTION 11.2   *pages 519–522*

Example 2

$\log_6 y^{1/3} z^3 = \log_6 y^{1/3} + \log_6 z^3$

$= \frac{1}{3}\log_6 y + 3\log_6 z$

Example 4

$\log_8 \sqrt[3]{xy^2} = \log_8 (xy^2)^{1/3} = \frac{1}{3}\log_8 xy^2$

$= \frac{1}{3}(\log_8 x + \log_8 y^2) = \frac{1}{3}(\log_8 x + 2\log_8 y)$

$= \frac{1}{3}\log_8 x + \frac{2}{3}\log_8 y$

Example 6

$3\log_b x + 2\log_b y - \log_b z$
$= \log_b x^3 + \log_b y^2 - \log_b z$

$= \log_b x^3 y^2 - \log_b z = \log_b \frac{x^3 y^2}{z}$

Example 8

$3\log_4 x + 2\log_4 y = \log_4 x^3 + \log_4 y^2$
$= \log_4 x^3 y^2$

Example 10

$\frac{1}{3}(\log_4 x - 2\log_4 y + \log_4 z)$

$\quad = \frac{1}{3}(\log_4 x - \log_4 y^2 + \log_4 z)$

$\quad = \frac{1}{3}\left(\log_4 \frac{x}{y^2} + \log_4 z\right) = \frac{1}{3}\left(\log_4 \frac{xz}{y^2}\right)$

$\quad = \log_4\left(\frac{xz}{y^2}\right)^{1/3} = \log_4 \sqrt[3]{\frac{xz}{y^2}}$

Example 12

Since $\log_b 1 = 0$, $\log_9 1 = 0$.

SECTION 11.3   *pages 527–534*

Example 2

$\log 93{,}000 = \log (9.3 \times 10^4)$
$\qquad\qquad = \log 9.3 + \log 10^4 = 0.9685 + 4$
$\qquad\qquad = 4.9685$

Example 4

$\log 0.0006 = \log (6 \times 10^{-4})$
$\qquad\qquad = \log 6 + \log 10^{-4} = 0.7782 + (-4)$
$\qquad\qquad = 6.7782 - 10$

Example 6

antilog $2.3365 = 2.17 \times 10^2 = 217$

Example 8

antilog $(9.7846 - 10) = 6.09 \times 10^{-1} = 0.609$

Example 10

$\log 31{,}560 = \log (3.156 \times 10^4)$
$\qquad\qquad = \log 3.156 + \log 10^4$

		Number	Mantissa	
0.01	0.006	3.150	0.4983	
		3.156	$0.4983 + d$	d 0.0014
		3.160	0.4997	

$\frac{0.006}{0.01} = \frac{d}{0.0014}$

$0.0008 = d$

$\log 3.156 + \log 10^4 = (0.4983 + d) + 4$
$\qquad\qquad\qquad = (0.4983 + 0.0008) + 4$
$\qquad\qquad\qquad = 0.4991 + 4 = 4.4991$

Example 12

antilog 1.4907

		Mantissa	Antilog	
0.0014	0.0007	0.4900	3.09	
		0.4907	$3.09 + d$	d 0.01
		0.4914	3.10	

$\frac{0.0007}{0.0014} = \frac{d}{0.01}$

$0.005 = d$

antilog $1.4907 = (3.09 + d) \times 10^1$
$\qquad\qquad\quad = (3.09 + 0.005) \times 10$
$\qquad\qquad\quad = 3.095 \times 10 = 30.95$

Example 14

$N = \sqrt[3]{\frac{14.1}{3.7}} = \left(\frac{14.1}{3.7}\right)^{1/3}$

$\log N = \log \left(\frac{14.1}{3.7}\right)^{1/3}$

$\qquad = \frac{1}{3}\log \left(\frac{14.1}{3.7}\right)$

$\qquad = \frac{1}{3}(\log 14.1 - \log 3.7)$

$\qquad = \frac{1}{3}(1.1492 - 0.5682)$

$\qquad = \frac{1}{3}(0.5810) = 0.1937$

Interpolate.

$N = $ antilog $0.1937 = 1.562 \times 10^0 = 1.562$

SECTION 11.4   *pages 537–540*

Example 2

$$4^{3x} = 25$$
$$\log 4^{3x} = \log 25$$
$$3x \log 4 = \log 25$$
$$3x = \frac{\log 25}{\log 4} = \frac{1.3979}{0.6021}$$
$$3x = 2.3217$$
$$x = 0.7739$$

The solution is 0.7739.

Example 6

$$\log_4(x^2 - 3x) = 1$$

Rewrite in exponential form.

$$4^1 = x^2 - 3x$$
$$4 = x^2 - 3x$$
$$0 = x^2 - 3x - 4$$
$$0 = (x + 1)(x - 4)$$

$$x + 1 = 0 \qquad x - 4 = 0$$
$$x = -1 \qquad x = 4$$

The solutions are −1 and 4.

Example 10 $\log_9 23 = \frac{\log_{10} 23}{\log_{10} 9} = \frac{1.3617}{0.9542} = 1.4271$

Example 4

$$(1.06)^n = 1.5$$
$$\log(1.06)^n = \log 1.5$$
$$n \log 1.06 = \log 1.5$$
$$n = \frac{\log 1.5}{\log 1.06} = \frac{0.1761}{0.0253}$$
$$n = 6.9605$$

The solution is 6.9605.

Example 8

$$\log_3 x + \log_3(x + 3) = \log_3 4$$
$$\log_3[x(x + 3)] = \log_3 4$$

Use the fact that if
$\log_b u = \log_b v$, then $u = v$.

$$x(x + 3) = 4$$
$$x^2 + 3x = 4$$
$$x^2 + 3x - 4 = 0$$
$$(x + 4)(x - 1) = 0$$

$$x + 4 = 0 \qquad x - 1 = 0$$
$$x = -4 \qquad x = 1$$

−4 does not check as a solution.
The solution is 1.

SECTION 11.5   *pages 543–546*

Example 2

Strategy

To find the pH, replace H^+ by 2.51×10^{-12} in the equation $pH = -\log (H^+)$ and solve for pH.

Solution

$$pH = -\log (H^+) = -\log (2.51 \times 10^{-12})$$
$$= -(\log 2.51 + \log 10^{-12})$$
$$= -[0.3997 + (-12)] = 11.6003$$

The pH of sodium carbonate is 11.6.

Example 4

Strategy

To find how many times stronger the San Francisco earthquake was, use the Richter equation to write a system of equations. Solve the system of equations for the ratio $\frac{I_1}{I_2}$.

Solution

$$7.8 = \log \frac{I_1}{I_0}$$
$$5 = \log \frac{I_2}{I_0}$$
$$7.8 = \log I_1 - \log I_0$$
$$5 = \log I_2 - \log I_0$$
$$2.8 = \log I_1 - \log I_2$$
$$2.8 = \log \frac{I_1}{I_2}$$
$$\frac{I_1}{I_2} = 10^{2.8} = 630.957$$

The San Francisco earthquake was 631 times stronger than one which can cause serious damage.

SOLUTIONS to Chapter 12 Examples

SECTION 12.1   *pages 561–564*

Example 2

$a_n = n(n + 1)$

$a_1 = 1(1 + 1) = 2$ The first term is 2.

$a_2 = 2(2 + 1) = 6$ The second term is 6.

$a_3 = 3(3 + 1) = 12$ The third term is 12.

$a_4 = 4(4 + 1) = 20$ The fourth term is 20.

Example 4

$a_n = \dfrac{1}{n(n + 2)}$

$a_6 = \dfrac{1}{6(6 + 2)} = \dfrac{1}{48}$ The sixth term is $\dfrac{1}{48}$.

$a_9 = \dfrac{1}{9(9 + 2)} = \dfrac{1}{99}$ The ninth term is $\dfrac{1}{99}$.

Example 6

$\displaystyle\sum_{i=3}^{6}(i^2 - 2) = (3^2 - 2) + (4^2 - 2) + (5^2 - 2) + (6^2 - 2)$

$\qquad = 7 + 14 + 23 + 34 = 78$

Example 8

$\displaystyle\sum_{n=1}^{4}\dfrac{x}{n} = \dfrac{x}{1} + \dfrac{x}{2} + \dfrac{x}{3} + \dfrac{x}{4}$

$\qquad = x + \dfrac{x}{2} + \dfrac{x}{3} + \dfrac{x}{4}$

SECTION 12.2   *pages 567–570*

Example 2

$9, 3, -3, -9, \ldots$

$d = a_2 - a_1 = 3 - 9 = -6$

$a_n = a_1 + (n - 1)d$
$a_{15} = 9 + (15 - 1)(-6) = 9 + (14)(-6) = 9 - 84$
$a_{15} = -75$

Example 4

$7, 9, 11, \ldots, 59$

$d = a_2 - a_1 = 9 - 7 = 2$

$a_n = a_1 + (n - 1)d$
$59 = 7 + (n - 1)2$
$59 = 7 + 2n - 2$
$59 = 5 + 2n$
$54 = 2n$
$27 = n$

There are 27 terms in the sequence.

Example 6

$-4, -2, 0, 2, 4, \ldots$

$d = a_2 - a_1 = -2 - (-4) = 2$

$a_n = a_1 + (n - 1)d$
$a_{25} = -4 + (25 - 1)2 = -4 + (24)2 = -4 + 48$
$a_{25} = 44$

$S_n = \dfrac{n}{2}(a_1 + a_n)$

$S_{25} = \dfrac{25}{2}(-4 + 44) = \dfrac{25}{2}(40) = 25(20)$

$S_{25} = 500$

Example 8

$\displaystyle\sum_{n=1}^{18}(3n - 2)$

$a_n = 3n - 2$
$a_1 = 3(1) - 2 = 1$
$a_{18} = 3(18) - 2 = 52$

$S_n = \dfrac{n}{2}(a_1 + a_n)$

$S_{18} = \dfrac{18}{2}(1 + 52) = 9(53) = 477$

Example 10

Strategy

To find the value of the 20th-place prize:
- Write the equation for the nth-place prize.
- Find the 20th term of the sequence.

To find the total amount of prize money being awarded, use the Formula for Sum of n Terms of an Arithmetic Sequence.

Solution

$10{,}000, 9700, \ldots$

$d = a_2 - a_1 = 9700 - 10{,}000 = -300$

$$\begin{aligned} a_n &= a_1 + (n-1)d \\ &= 10{,}000 + (n-1)(-300) \\ &= 10{,}000 - 300n + 300 \\ &= -300n + 10{,}300 \end{aligned}$$

$a_{20} = -300(20) + 10{,}300 = -6000 + 10{,}300 = 4300$

$$S_n = \frac{n}{2}(a_1 + a_n)$$

$$S_{20} = \frac{20}{2}(10{,}000 + 4300) = 10(14{,}300) = 143{,}000$$

The value of the 20th-place price is $4300.

The total amount of prize money being awarded is $143,000.

SECTION 12.3   *pages 573–580*

Example 2

$5, 2, \dfrac{4}{5}, \ldots$

$r = \dfrac{a_2}{a_1} = \dfrac{2}{5}$

$a_n = a_1 r^{n-1}$

$a_5 = 5\left(\dfrac{2}{5}\right)^{5-1} = 5\left(\dfrac{2}{5}\right)^4 = 5\left(\dfrac{16}{625}\right)$

$a_5 = \dfrac{16}{125}$

Example 6

$1, -\dfrac{1}{3}, \dfrac{1}{9}, -\dfrac{1}{27}$

$r = \dfrac{a_2}{a_1} = \dfrac{-\frac{1}{3}}{1} = -\dfrac{1}{3}$

$S_n = \dfrac{a_1(1 - r^n)}{1 - r}$

$$S_4 = \frac{1\left[1 - \left(-\frac{1}{3}\right)^4\right]}{1 - \left(-\frac{1}{3}\right)} = \frac{1 - \frac{1}{81}}{\frac{4}{3}} = \frac{\frac{80}{81}}{\frac{4}{3}}$$

$$= \frac{80}{81} \cdot \frac{3}{4} = \frac{20}{27}$$

Example 4

$3, a_2, a_3, -192, \ldots$

$$\begin{aligned} a_n &= a_1 r^{n-1} \\ a_4 &= 3r^{4-1} \\ -192 &= 3r^{4-1} \\ -192 &= 3r^3 \\ -64 &= r^3 \\ -4 &= r \end{aligned}$$

$a_n = a_1 r^{n-1}$

$a_3 = 3(-4)^{3-1} = 3(-4)^2 = 3(16) = 48$

Example 8

$\displaystyle\sum_{n=1}^{5} \left(\dfrac{1}{2}\right)^n$

$a_n = \left(\dfrac{1}{2}\right)^n$

$a_1 = \left(\dfrac{1}{2}\right)^1 = \dfrac{1}{2}$

$a_2 = \left(\dfrac{1}{2}\right)^2 = \dfrac{1}{4}$

$r = \dfrac{a_2}{a_1} = \dfrac{\frac{1}{4}}{\frac{1}{2}} = \dfrac{1}{4} \cdot \dfrac{2}{1} = \dfrac{1}{2}$

$S_n = \dfrac{a_1(1 - r^n)}{1 - r}$

$$S_5 = \frac{\frac{1}{2}\left[1 - \left(\frac{1}{2}\right)^5\right]}{1 - \frac{1}{2}} = \frac{\frac{1}{2}\left(1 - \frac{1}{32}\right)}{\frac{1}{2}} = \frac{\frac{1}{2}\left(\frac{31}{32}\right)}{\frac{1}{2}}$$

$$= \frac{\frac{31}{64}}{\frac{1}{2}} = \frac{31}{64} \cdot \frac{2}{1} = \frac{31}{32}$$

Example 10

$3, -2, \frac{4}{3}, -\frac{8}{3}, \ldots$

$r = \frac{a_2}{a_1} = \frac{-2}{3}$

$S = \frac{a_1}{1-r} = \frac{3}{1-\left(-\frac{2}{3}\right)} = \frac{3}{1+\frac{2}{3}}$

$= \frac{3}{\frac{5}{3}} = \frac{9}{5}$

Example 12

$0.\overline{66} = 0.6 + 0.06 + 0.006 + \ldots$

$\quad = \frac{6}{10} + \frac{6}{100} + \frac{6}{1000} + \ldots$

$S = \frac{a_1}{1-r} = \frac{\frac{6}{10}}{1-\frac{1}{10}} = \frac{\frac{6}{10}}{\frac{9}{10}} = \frac{6}{9} = \frac{2}{3}$

An equivalent fraction is $\frac{2}{3}$.

Example 14

Strategy

To find the total number of letters mailed, use the Formula for the Sum of n Terms of a Finite Geometric Series.

Solution

$n = 6, a_1 = 3, r = 3$

$S_n = \frac{a_1(1-r^n)}{1-r}$

$S_6 = \frac{3(1-3^6)}{1-3} = \frac{3(1-729)}{1-3}$

$= \frac{3(-728)}{-2} = \frac{-2184}{-2} = 1092$

From the first through the sixth mailings, 1092 letters will have been mailed.

SECTION 12.4 *pages 583–586*

Example 2

$(3m - n)^4$

$= \binom{4}{0}(3m)^4 + \binom{4}{1}(3m)^3(-n) + \binom{4}{2}(3m)^2(-n)^2 + \binom{4}{3}(3m)(-n)^3 + \binom{4}{4}(-n)^4$

$= 1(81m^4) + 4(27m^3)(-n) + 6(9m^2)(n^2) + 4(3m)(-n^3) + 1(n^4)$

$= 81m^4 - 108m^3n + 54m^2n^2 - 12mn^3 + n^4$

Example 4

$(y - 2)^{10} = \binom{10}{0}y^{10} + \binom{10}{1}y^9(-2) + \binom{10}{2}y^8(-2)^2 + \ldots$

$\quad = 1(y^{10}) + 10y^9(-2) + 45y^8(4)$

$\quad = y^{10} - 20y^9 + 180y^8$

Example 6

$(r - 2s)^7$

$n = 7, a = r, b = -2s, r = 3$

$\binom{7}{3-1}(r)^{7-3+1}(-2s)^{3-1} = \binom{7}{2}(r)^5(-2s)^2$

$\quad = 21r^5(4s^2) = 84r^5s^2$

ANSWERS to Chapter 1 Exercises

SECTION 1.1   *pages 7–8*

1. 2   **3.** −4   **5.** 75   **7.** −51   **9.** 83   **11.** 3   **13.** 4   **15.** −32   **17.** −22   **19.** 436
21. −30   **23.** −17   **25.** −96   **27.** −6   **29.** 5   **31.** 11,200   **33.** 98,915   **35.** 96   **37.** 23
39. 7   **41.** 17   **43.** −4   **45.** −62   **47.** −25   **49.** 7   **51.** 368   **53.** 26   **55.** −83
57. 81   **59.** −64   **61.** 64   **63.** 432   **65.** −1125   **67.** 512   **69.** −576   **71.** 60
73. −50,000   **75.** −2,239,488

SECTION 1.2   *pages 15–16*

1. $\frac{43}{48}$ **3.** $-\frac{67}{45}$ **5.** $-\frac{13}{36}$ **7.** $\frac{11}{24}$ **9.** $\frac{13}{24}$ **11.** $-\frac{3}{56}$ **13.** $-\frac{2}{3}$ **15.** $-\frac{11}{14}$ **17.** $-\frac{1}{24}$
19. −12.974   **21.** −6.008   **23.** 1.9215   **25.** −6.02   **27.** −6.7   **29.** −44.585   **31.** 18.90
33. −1802.87   **35.** −11   **37.** $\frac{1}{4}$ **39.** $\frac{1}{4}$ **41.** 12   **43.** 25   **45.** 44   **47.** $\frac{109}{150}$ **49.** $\frac{91}{36}$
51. $\frac{28}{5}$ **53.** 4.4   **55.** −0.91   **57.** −19.637832

SECTION 1.3   *pages 23–26*

1. 10   **3.** 4   **5.** 0   **7.** $-\frac{1}{7}$ **9.** $\frac{9}{2}$ **11.** $\frac{1}{2}$ **13.** 2   **15.** 3   **17.** −12   **19.** 2   **21.** 6
23. −2   **25.** 6.51   **27.** −19.71   **29.** $7 + 15 = 15 + 7$ **31.** $(3 \cdot 4) \cdot 5 = 3 \cdot (4 \cdot 5)$ **33.** $5 \cdot 0 = 0$
35. $5(y + 4) = 5y + 20$ **37.** $(x + y) + [-(x + y)] = 0$ **39.** $x \cdot 1 = x$ **41.** $ab + bc = bc + ab$ **43.** The
Inverse Property of Addition   **45.** The Commutative Property of Multiplication   **47.** The Associative
Property of Addition   **49.** The Distributive Property   **51.** The Multiplication Property of Zero   **53.** The
Commutative Property of Addition   **55.** $13x$ **57.** $-4x$ **59.** $7a + 7b$ **61.** x **63.** $-3x + 6$
65. $5x + 10$ **67.** $x + y$ **69.** $3x - 6y - 5$ **71.** $-11a + 21$ **73.** $-x + 6y$ **75.** $140 - 30a$
77. $-10y + 30x$ **79.** $-10a + 2b$ **81.** $-12a + b$ **83.** $-2x - 144y - 96$ **85.** $5x - 32 + 3y$
87. $x + 6$ **89.** $n - (5 - n); 2n - 5$ **91.** $\frac{3}{8}n - \frac{1}{6}n; \frac{5}{24}n$ **93.** $n + \frac{2}{3}n; \frac{5}{3}n$ **95.** $\frac{1}{2}(6n + 22); 3n + 11$
97. $(n + 1) + (n + 2); 2n + 3$ **99.** $4(n + 1) + 12; 4n + 16$ **101.** $3\left(\frac{2n}{6}\right); n$ **103.** $2(n + 11) - 4; 2n + 18$
105. $20 - (n + 4)(12); -12n - 28$ **107.** $n + (n - 12)(3); 4n - 36$

SECTION 1.4   *pages 31–32*

1. $A = \{-2, -1, 0, 1, 2, 3, 4\}$ **3.** $A = \{2, 4, 6, 8, 10, 12\}$ **5.** $A = \{3\}$ **7.** $A = \{3, 6, 9, 12, 15, 18\}$
9. $\{x|x > 4, x \text{ is an integer}\}$ **11.** $\{x|x \geq 1, x \in \text{real numbers}\}$ **13.** $\{x|-2 < x < 5, x \text{ is an integer}\}$
15. $\{x|0 < x < 1, x \in \text{real numbers}\}$ **17.** $A \cup B = \{1, 2, 4, 6, 9\}$ **19.** $A \cup B = \{2, 3, 5, 8, 9, 10\}$
21. $A \cup B = \{-4, -2, 0, 2, 4, 8\}$ **23.** $A \cup B = \{1, 2, 3, 4, 5\}$ **25.** $A \cap B = \{6\}$ **27.** $A \cap B = \{5, 10, 20\}$

29. $A \cap B = \varnothing$ **31.** $A \cap B = \{4, 6\}$ **33.**   **35.**

37.   **39.**   **41.**

43.

CHAPTER REVIEW   *pages 37–38*

1. −23   **2.** −5   **3.** −15   **4.** 12   **5.** 14   **6.** −288   **7.** $\frac{7}{120}$ **8.** $\frac{2}{15}$ **9.** $-\frac{5}{8}$ **10.** −2.84
11. 3.1   **12.** 52   **13.** 20   **14.** $\frac{7}{6}$ **15.** 3   **16.** y **17.** ab **18.** 4   **19.** Inverse Property of
Addition   **20.** The Associative Property of Multiplication   **21.** $-6x + 14$ **22.** $16y - 15x + 18$

23. $x + (x + 2) + (x + 4)$; $3x + 6$ **24.** $12 - \frac{x+3}{x}$; $\frac{11x-3}{x}$ **25.** $A \cup B = \{1, 2, 3, 4, 5, 6, 7, 8\}$

26. $A \cap B = \{2, 3\}$ **27.** **28.**

29.   **30.**

CHAPTER TEST   *pages 39–40*

1. 12 (1.1A)   **2.** -2 (1.1A)   **3.** 12 (1.1B)   **4.** -30 (1.1B)   **5.** -15 (1.1B)   **6.** 2 (1.1B)

7. -100 (1.1C)   **8.** -72 (1.1C)   **9.** $\frac{25}{36}$ (1.2A)   **10.** $-\frac{4}{27}$ (1.2A)   **11.** -1.41 (1.2A)

12. -4.9 (1.2A)   **13.** 10 (1.2B)   **14.** 6 (1.2B)   **15.** -5 (1.3A)   **16.** 2 (1.3A)   **17.** 4 (1.3B)
18. The Distributive Property (1.3B)   **19.** $13x - y$ (1.3C)   **20.** $14x + 48y$ (1.3C)   **21.** $13 - (n - 3)(9)$;
$40 - 9n$ (1.3D)   **22.** $5[n + (n + 1)]$; $10n + 5$ (1.3D)   **23.** $A \cup B = \{1, 2, 3, 4, 5, 7\}$ (1.4A)
24. $A \cup B = \{-2, -1, 0, 1, 2, 3\}$ (1.4A)   **25.** $A \cap B = \{5, 7\}$ (1.4A)   **26.** $A \cap B = \{-1, 0, 1\}$ (1.4A)

27. (1.4B)   **28.** (1.4B)   **29.** (1.4B)

30. (1.4B)

ANSWERS to Chapter 2 Exercises

SECTION 2.1   *pages 49–52*

1. The solution is 9.   **3.** The solution is -10.   **5.** The solution is -2.   **7.** The solution is -15.

9. The solution is 4.   **11.** The solution is $-\frac{2}{3}$.   **13.** The solution is $\frac{17}{6}$.   **15.** The solution is $\frac{1}{6}$.

17. The solution is $\frac{15}{2}$.   **19.** The solution is $-\frac{32}{25}$.   **21.** The solution is 1.   **23.** The solution is $\frac{15}{16}$.

25. The solution is -64.   **27.** The solution is $\frac{5}{6}$.   **29.** The solution is -3.73.   **31.** The solution is

-0.1814.   **33.** The solution is 1.34.   **35.** The solution is $\frac{3}{2}$.   **37.** The solution is 8.   **39.** The solution

is 0.   **41.** The solution is 6.   **43.** The solution is 6.   **45.** The equation has no solution.   **47.** The

solution is -3.   **49.** The solution is 1.   **51.** The solution is $\frac{4}{3}$.   **53.** The solution is 1.   **55.** The

solution is $-\frac{4}{3}$.   **57.** The solution is $\frac{2}{5}$.   **59.** The solution is 12.   **61.** The solution is -1.25.   **63.** The

solution is 1.   **65.** The solution is $\frac{23}{15}$.   **67.** The equation has no solution.   **69.** The solution is 2.

71. The solution is -15.   **73.** The solution is -1.07.   **75.** The solution is 2.   **77.** The solution is $\frac{7}{3}$.

79. The solution is $\frac{8}{3}$.   **81.** The solution is $\frac{7}{3}$.   **83.** The solution is -1.   **85.** The solution is $\frac{4}{5}$.

87. The solution is $\frac{4}{3}$.   **89.** The solution is $\frac{33}{5}$.   **91.** The solution is -3.   **93.** The solution is 10.

95. The solution is 6.   **97.** The solution is 3.   **99.** The solution is $-\frac{50}{9}$.   **101.** The solution is 1.

103. The solution is $-\frac{40}{17}$.   **105.** The solution is $\frac{40}{43}$.   **107.** The solution is 1.444267.   **109.** The solution

is 25.   **111.** The solution is 15.   **113.** The solution is 35.   **115.** The solution is $\frac{3}{2}$.

SECTION 2.2 *pages 59–64*

1. $\{x|x < 5\}$ **3.** $\{x|x \le 2\}$ **5.** $\{x|x < -4\}$ **7.** $\{x|x > 3\}$ **9.** $\{x|x > 4\}$ **11.** $\{x|x \le 2\}$ **13.** $\{x|x > -2\}$
15. $\{x|x \ge 2\}$ **17.** $\{x|x > -2\}$ **19.** $\{x|x \le 3\}$ **21.** $\{x|x < 2\}$ **23.** $\{x|x < -3\}$ **25.** $\{x|x \le 5\}$
27. $\left\{x\middle|x \ge -\frac{1}{2}\right\}$ **29.** $\left\{x\middle|x < \frac{23}{16}\right\}$ **31.** $\left\{x\middle|x < \frac{8}{3}\right\}$ **33.** $\{x|x > 1\}$ **35.** $\left\{x\middle|x > \frac{14}{11}\right\}$ **37.** $\{x|x \le 1\}$
39. $\left\{x\middle|x \le \frac{3}{8}\right\}$ **41.** $\left\{x\middle|x \le \frac{7}{4}\right\}$ **43.** $\left\{x\middle|x \ge -\frac{5}{4}\right\}$ **45.** $\{x|x \le 2\}$ **47.** $\{x|-2 \le x \le 4\}$ **49.** $\{x|x < 3 \text{ or } x > 5\}$
51. $\{x|-4 < x < 2\}$ **53.** $\{x|x > 6 \text{ or } x < -4\}$ **55.** $\{x|x < -3\}$ **57.** $\{x|-3 < x < 2\}$ **59.** $\{x|-2 < x < 1\}$
61. $\{x|x < -2 \text{ or } x > 2\}$ **63.** $\{x|2 < x < 6\}$ **65.** $\{x|-3 < x < -2\}$ **67.** $\left\{x\middle|x > 5 \text{ or } x < -\frac{5}{3}\right\}$ **69.** $\{x|x < 3\}$

71. The solution set is the empty set. **73.** The solution set is the set of real numbers. **75.** $\left\{x\middle|\frac{17}{7} \le x \le \frac{45}{7}\right\}$

77. $\left\{x\middle|-5 < x < \frac{17}{3}\right\}$ **79.** The solution set is the set of real numbers. **81.** The solution set is the empty
set. **83.** The smallest number is -12. **85.** The maximum width as an integer is 11 cm. **87.** You can
use the cellular phone for less than 45 minutes. **89.** You can drive less than 350 mi. **91.** The temperature
range is $32° < F < 86°$. **93.** The amount of sales is $44,000 or more. **95.** The customer can write less than
50 checks. **97.** The range of scores is $58 \le x \le 100$. **99.** Three consecutive integers can be 10, 12, 14, or
12, 14, 16, or 14, 16, 18.

SECTION 2.3 *pages 71–74*

1. The solutions arc 7 and -7. **3.** The solutions are 4 and -4. **5.** The solutions are 6 and -6.
7. The solutions are 7 and -7. **9.** There is no solution. **11.** There is no solution. **13.** The solutions
are -5 and 1. **15.** The solutions are 2 and 8. **17.** The solution is 2. **19.** There is no solution.
21. The solutions are $\frac{1}{2}$ and $\frac{9}{2}$. **23.** The solutions are $-\frac{5}{3}$ and 3. **25.** The solutions are $\frac{3}{2}$ and $-\frac{13}{2}$.
27. The solutions are 5 and -1. **29.** The solutions are 2 and $-\frac{2}{3}$. **31.** The solutions are $\frac{8}{3}$ and 0.
33. The solutions are 3 and $-\frac{3}{2}$. **35.** The solution is $\frac{3}{2}$. **37.** There is no solution. **39.** The solutions
are -3 and 7. **41.** The solutions are $-\frac{10}{3}$ and 2. **43.** The solutions are 3 and 1. **45.** The solution is $\frac{3}{2}$.
47. There is no solution. **49.** The solutions are $-\frac{1}{6}$ and $\frac{11}{6}$. **51.** The solutions are -1 and $-\frac{1}{3}$.
53. There is no solution. **55.** The solutions are 0 and 3. **57.** There is no solution. **59.** The solutions
are $\frac{13}{3}$ and 1. **61.** There is no solution. **63.** The solutions are $\frac{1}{3}$ and $\frac{7}{3}$. **65.** The solution is $-\frac{1}{2}$.
67. The solutions are $-\frac{7}{2}$ and $-\frac{1}{2}$. **69.** The solutions are $\frac{10}{3}$ and $-\frac{8}{3}$. **71.** There is no solution.
73. $\{x|x > 3 \text{ or } x < -3\}$ **75.** $\{x|x > 1 \text{ or } x < -3\}$ **77.** $\{x|4 \le x \le 6\}$ **79.** $\{x|x \ge 5 \text{ or } x \le -1\}$
81. $\{x|-3 < x < 2\}$ **83.** $\left\{x\middle|x > 2 \text{ or } x < -\frac{14}{5}\right\}$ **85.** The solution set is the empty set. **87.** The solution set
is the set of real numbers. **89.** $\left\{x\middle|x \le -\frac{1}{3} \text{ or } x \ge 3\right\}$ **91.** $\left\{x\middle|-2 \le x \le \frac{9}{2}\right\}$ **93.** $\{x|x = 2\}$
95. $\left\{x\middle|x < -2 \text{ or } x > \frac{22}{9}\right\}$ **97.** $\left\{x\middle|-\frac{3}{2} < x < \frac{9}{2}\right\}$ **99.** $\left\{x\middle|x < 0 \text{ or } x > \frac{4}{5}\right\}$ **101.** $\{x|x > 5 \text{ or } x < 0\}$
103. lower limit: 3.95 cc; upper limit: 4.05 cc **105.** lower limit: 2.648 in.; upper limit: 2.652 in.
107. lower limit: $9\frac{19}{32}$ in.; upper limit: $9\frac{21}{32}$ in. **109.** lower limit: 28,420 ohms; upper limit: 29,580 ohms
111. lower limit: 23,750 ohms; upper limit: 26,250 ohms

SECTION 2.4 *pages 79–80*

1. The number is 3. **3.** The integers are 3 and 7. **5.** The integers are 21 and 29. **7.** The three
numbers are 21, 44, and 58. **9.** The integers are -20, -19, and -18. **11.** There is no solution.
13. The integers are 5, 7, and 9. **15.** There are 21 dimes in the collection. **17.** There are 12 dimes in the
bank. **19.** The cashier has 26 twenty-dollar bills. **21.** There are eight 20¢ stamps and sixteen 15¢ stamps
in the collection. **23.** There are five 3¢ stamps in the collection. **25.** There are seven 18¢ stamps in the
collection.

SECTION 2.5   *pages 85–86*

1. The selling price of the mixture is $2.95 per lb.   **3.** There were 330 adult tickets sold.   **5.** 225 L of imitation maple syrup is needed.   **7.** The goldsmith used 20 oz of pure gold, and 30 oz of the gold alloy.   **9.** The selling price of the tea mixture is $4.11 per lb.   **11.** 23.8 gal of cranberry juice was used.   **13.** The car will overtake the cyclist 28 mi from the starting point.   **15.** The speed of the first plane is 480 mph. The speed of the second plane is 600 mph.   **17.** The island is 37.5 mi from the harbor.   **19.** The rate of the first plane is 255 mph. The rate of the second plane is 305 mph.   **21.** The student's home is 2.4 mi from the bicycle shop.   **23.** The second plane traveled 1050 mi.

SECTION 2.6   *pages 91–92*

1. The amount which should be invested at 7.2% is $6000. The amount which should be invested at 9.8% is $6500.   **3.** $2000 more must be invested at 10.5%.   **5.** The amount which should be invested at 7% is $24,500. The amount which should be invested at 9.8% is $17,500.   **7.** The total amount to be invested is $25,000.   **9.** The amount invested at 12.6% was $4000.   **11.** 3 quarts of water must be added.   **13.** The percent concentration of the resulting alloy is 22%.   **15.** 45 oz of pure water must be added.   **17.** 6 L of the 75% disinfectant solution and 14 L of the 25% disinfectant solution were used.   **19.** $2\frac{2}{3}$ gallons of water must be evaporated to obtain a 12% solution.   **21.** 4 qt will have to be replaced with pure antifreeze.

CHAPTER REVIEW   *pages 95–96*

1. The solution is -9.   **2.** The solution is $-\frac{2}{3}$.   **3.** The solution is 6.   **4.** The solution is $-\frac{40}{7}$.   **5.** The solution is $\frac{26}{17}$.   **6.** The solution is $\frac{5}{2}$.   **7.** The solution is $-\frac{17}{2}$.   **8.** The solution is $-\frac{9}{19}$.   **9.** $\left\{x \middle| x > \frac{5}{3}\right\}$ **10.** $\left\{x \middle| -3 < x < \frac{4}{3}\right\}$ **11.** $\{x | x > 2 \text{ or } x < -2\}$ **12.** The solution set is the set of real numbers.   **13.** The amount of sales must be $55,000 or more.   **14.** The range of scores is $82 \le x \le 100$.   **15.** The solutions are -1 and 9.   **16.** There is no solution.   **17.** $\{x | 1 < x < 4\}$ **18.** $\left\{x \middle| x \ge 2 \text{ or } x \le \frac{1}{2}\right\}$ **19.** lower limit: 2.747 in.; upper limit: 2.753 in.   **20.** lower limit: 1.75 cc; upper limit: 2.25 cc   **21.** The integers are 6 and 14.   **22.** There are 10 quarters in the collection.   **23.** The grocer must use 52 gal of apple juice.   **24.** The speed of the first plane is 520 mph. The speed of the second plane is 440 mph.   **25.** The amount which should be invested at 10.5% is $3000. The amount which should be invested at 6.4% is $5000.   **26.** 375 lb of 30% tin and 125 lb of 70% tin were used.

CHAPTER TEST   *pages 97–98*

1. The solution is -2. (2.1A)   **2.** The solution is $-\frac{1}{8}$. (2.1A)   **3.** The solution is $\frac{5}{6}$. (2.1A)   **4.** The solution is 4. (2.1A)   **5.** The solution is $\frac{32}{3}$. (2.1B)   **6.** The solution is $-\frac{1}{5}$. (2.1B)   **7.** The solution is 1. (2.1C)   **8.** The solution is -24. (2.1B)   **9.** The solution is $\frac{12}{7}$. (2.1C)   **10.** $\{x | x \le -3\}$ (2.2A)   **11.** $\{x | x > -1\}$ (2.2A)   **12.** $\{x | x > -2\}$ (2.2B)   **13.** The solution set is the empty set. (2.2B)   **14.** The solutions are 3 and $-\frac{9}{5}$. (2.3A)   **15.** The solutions are 7 and -2. (2.3A)   **16.** $\left\{x \middle| \frac{1}{3} \le x \le 3\right\}$ (2.3B)   **17.** $\left\{x \middle| x > 2 \text{ or } x < -\frac{1}{2}\right\}$ (2.3B)   **18.** You can drive less than 120 mi. (2.2C)   **19.** lower limit: 2.90 cc; upper limit: 3.10 cc (2.3C)   **20.** The integers are 4 and 11. (2.4A)   **21.** There are six 24¢ stamps in the collection. (2.4B)   **22.** The selling price of the hamburger mixture is $2.20. (2.5A)   **23.** The total distance the jogger ran was 12 mi. (2.5B)   **24.** The amount invested at 7.8% was $5000. The amount invested at 9% was $7000. (2.6A)   **25.** 100 oz of pure water must be added. (2.6B)

CUMULATIVE REVIEW   *pages 99–100*

1. −11 (1.1B)   **2.** −108 (1.1C)   **3.** 3 (1.1C)   **4.** −64 (1.2A)   **5.** −8 (1.3A)   **6.** 1 (1.3A)
7. Commutative Property of Addition (1.3B)   **8.** $(3n + 6) + 3n$; $6n + 6$ (1.3D)   **9.** $-17x + 2$ (1.3C)

10. $25y$ (1.3C)   **11.** $\{-4, 0\}$ (1.4A)   **12.** (1.4B)   **13.** $x = 2$ (2.1A)

14. $b = \frac{1}{2}$ (2.1A)   **15.** $x = 1$ (2.1B)   **16.** $x = 24$ (2.1B)   **17.** $x = 2$ (2.1C)   **18.** $x = 2$ (2.1C)

19. $y = 1$ (2.1C)   **20.** $x = -\frac{13}{5}$ (2.1C)   **21.** $\{x | x \leq 1\}$ (2.2A)   **22.** $\{x | -4 \leq x \leq 1\}$ (2.2B)   **23.** The

solutions are −1 and 4. (2.3A)   **24.** The solutions are 7 and −4. (2.3A)   **25.** $\left\{x \middle| x < -\frac{4}{3} \text{ or } x > 2\right\}$ (2.3B)
26. $\{x | -2 < x < 6\}$ (2.3B)   **27.** The customer can write less than 50 checks. (2.2C)   **28.** The first integer
is −3. (2.4A)   **29.** There are 15 dimes in the coin purse. (2.4B)   **30.** 80 oz of pure silver were
used. (2.5A)   **31.** The speed of the slower plane is 220 mph. (2.5B)   **32.** The amount invested in the
13.5% account was $6000. (2.6A)   **33.** 3 L of 12% acid solution must be used. (2.6B)

ANSWERS to Chapter 3 Exercises

SECTION 3.1   *pages 115–120*

1. $6x^2 - 6x + 5$ **3.** $3x^2 - 3xy - 2y^2$ **5.** $-x^2 + 1$ **7.** $9x^n + 5$ **9.** $3y^3 + 2y^2 - 15y + 2$
11. $7a^2 - a + 2$ **13.** $3x^4 - 8x^2 + 2x$ **15.** $-5a^2 - 6a - 4$ **17.** $-b^{2n} + 2b^n - 7$
19. $-3x^3 + 10x^2 - 18x - 11$ **21.** a^4b^4 **23.** $-18x^3y^4$ **25.** x^8y^{16} **27.** $81x^8y^{12}$ **29.** $729a^{10}b^6$
31. x^5y^{11} **33.** $729x^6$ **35.** $a^{18}b^{18}$ **37.** $4096x^{12}y^{12}$ **39.** $64a^{24}b^{18}$ **41.** x^{2n+1} **43.** y^{6n-2}
45. a^{2n^2-6n} **47.** x^{15n+10} **49.** $-6x^5y^5z^4$ **51.** $-12a^2b^9c^2$ **53.** $-6x^4y^4z^5$ **55.** $-432a^7b^{11}$
57. $54a^{13}b^{17}$ **59.** 243   **61.** y^3 **63.** $\frac{a^3b^2}{4}$ **65.** $\frac{x}{y^4}$ **67.** $\frac{1}{2}$ **69.** $-\frac{1}{9}$ **71.** $\frac{1}{125x^6}$ **73.** $\frac{1}{x^8}$
75. a^2 **77.** x^9 **79.** $\frac{1}{y^8}$ **81.** $\frac{1}{x^6y^{10}}$ **83.** x^5y^5 **85.** $\frac{a^8}{b^9}$ **87.** $\frac{1}{2187a}$ **89.** $\frac{y^6}{x^3}$ **91.** $\frac{y^4}{x^3}$
93. $-\frac{1}{2x^2}$ **95.** $\frac{5b^5}{2a^3}$ **97.** $\frac{b^2c^2}{a^4}$ **99.** $\frac{x^2}{2y}$ **101.** x^2y **103.** $-\frac{1}{243a^5b^{10}}$ **105.** $\frac{16x^8}{81y^4z^{12}}$ **107.** $\frac{3a^4}{4b^3}$
109. $-\frac{9a}{8b^6}$ **111.** $\frac{16x^2}{y^6}$ **113.** $\frac{1}{b^{4n}}$ **115.** $-\frac{1}{y^{6n}}$ **117.** y^{n-2} **119.** $\frac{1}{y^{2n}}$ **121.** $\frac{x^{n-5}}{y^6}$ **123.** $\frac{8b^{15}}{3a^{18}}$
125. $\frac{8}{5}$ **127.** $\frac{y+x}{y-x}$ **129.** $\frac{1}{a+b}$ **131.** $\frac{a^2+b}{a}$ **133.** $\frac{2y}{x}$ **135.** $\frac{2}{xy}$ **137.** 5×10^{-8} **139.** 4.3×10^6
141. 9.8×10^9 **143.** 0.0000000000062   **145.** 634,000   **147.** 4,350,000,000   **149.** 0.0000000002848
151. 0.004608   **153.** 9,000,000,000   **155.** 20,000,000,000   **157.** 0.692   **159.** 1.38
161. 0.000000000000006   **163.** Light travels 8.64×10^{12} m in 8 h.   **165.** The computer can perform
4×10^9 operations in 1 h.   **167.** The centrifuge makes one revolution in $6.\overline{6} \times 10^{-7}$ s.   **169.** The sun is
3.38983×10^5 times heavier than the earth.   **171.** The weight of one orchid seed is 3.225806×10^{-8} oz.
173. Alpha Centauri is 2.4876979×10^{13} mi from earth.

SECTION 3.2   *pages 127–130*

1. $2x^2 - 6x$ **3.** $6x^4 - 3x^3$ **5.** $6x^2y - 9xy^2$ **7.** $x^{n+1} + x^n$ **9.** $x^{2n} + x^ny^n$ **11.** $-4b^2 + 10b$
13. $-6a^4 + 4a^3 - 6a^2$ **15.** $9b^5 - 9b^3 + 24b$ **17.** $-3y^4 - 4y^3 + 2y^2$ **19.** $-20x^2 + 15x^3 - 15x^4 - 20x^5$
21. $-2x^4y + 6x^3y^2 - 4x^2y^3$ **23.** $x^{3n} + x^{2n} + x^{n+1}$ **25.** $a^{2n+1} - 3a^{n+2} + 2a^{n+1}$ **27.** $5y^2 - 11y$
29. $6y^2 - 31y$ **31.** $x^2 + 5x - 14$ **33.** $8y^2 + 2y - 21$ **35.** $8x^2 + 8xy - 30y^2$ **37.** $-30a^2 + 21ab + 36b^2$
39. $15a^2 + 41ab + 14b^2$ **41.** $21x^2 - 43xy - 14y^2$ **43.** $x^2y^2 + xy - 12$ **45.** $2x^4 - 15x^2 + 25$
47. $10x^4 - 15x^2y + 5y^2$ **49.** $x^{2n} - x^n - 6$ **51.** $6a^{2n} + a^n - 15$ **53.** $6a^{2n} + a^nb^n - 2b^{2n}$
55. $x^3 - 5x^2 + 13x - 14$ **57.** $x^4 + 5x^3 - 3x^2 - 11x + 20$ **59.** $10a^3 - 27a^2b + 26ab^2 - 12b^3$
61. $2y^5 - 10y^4 - y^3 - y^2 + 3$ **63.** $4x^5 - 16x^4 + 15x^3 - 4x^2 + 28x - 45$ **65.** $x^4 - 3x^3 - 6x^2 + 29x - 21$
67. $2a^3 + 7a^2 - 43a + 42$ **69.** $x^{3n} + 2x^{2n} + 2x^n + 1$ **71.** $x^{2n} - 2x^{2n}y^n + 4x^ny^n - 2x^ny^{2n} + 3y^{2n}$
73. $9x^2 - 4$ **75.** $36 - x^2$ **77.** $4a^2 - 9b^2$ **79.** $x^4 - 1$ **81.** $x^{2n} - 9$ **83.** $x^2 - 10x + 25$
85. $9a^2 + 30ab + 25b^2$ **87.** $x^4 - 6x^2 + 9$ **89.** $4x^4 - 12x^2y^2 + 9y^4$ **91.** $a^{2n} - 2a^nb^n + b^{2n}$
93. $-x^2 + 2xy$ **95.** $-4xy$ **97.** $(3x^2 + 10x - 8)$ ft^2 **99.** $(x^2 + 3x)$ m^2 **101.** $(x^3 + 9x^2 + 27x + 27)$ cm^3
103. $(2x^3)$ in.^{3} **105.** $(78.5x^2 + 125.6x + 50.24)$ in.^{2}

SECTION 3.3   *pages 139–142*

1. $3a(2a - 5)$ **3.** $x^2(4x - 3)$ **5.** nonfactorable over the integers   **7.** $x(x^4 - x^2 - 1)$ **9.** $4(4x^2 - 3x + 6)$
11. $5b^2(1 - 2b + 5b^2)$ **13.** $x^n(x^n - 1)$ **15.** $x^{2n}(x^n - 1)$ **17.** $a^2(a^{2n} + 1)$ **19.** $6x^2y(2y - 3x + 4)$
21. $4a^2b^2(6a - 1 - 4b^2)$ **23.** $y^2(y^{2n} + y^n - 1)$ **25.** $(a + 2)(x - 2)$ **27.** $(x - 2)(a + b)$ **29.** $(a - 2b)(x - y)$
31. $(x + 4)(y - 2)$ **33.** $(a + b)(x - y)$ **35.** $(y - 3)(x^2 - 2)$ **37.** $(3 + y)(2 + x^2)$ **39.** $(2a + b)(x^2 - 2y)$
41. $(2b + a)(3x - 2y)$ **43.** $(y - 5)(x^n + 1)$ **45.** $(x - 3)(x - 5)$ **47.** $(a + 1)(a + 11)$ **49.** $(b + 7)(b - 5)$
51. $(y - 3)(y - 13)$ **53.** $(b + 8)(b - 4)$ **55.** $(a - 7)(a - 8)$ **57.** $(y + 1)(y + 12)$ **59.** $(x + 5)(x - 1)$
61. $(a + 5b)(a + 6b)$ **63.** $(x - 2y)(x - 12y)$ **65.** $(y + 9x)(y - 7x)$ **67.** $(7 + x)(3 - x)$ **69.** $(5 + a)(10 - a)$
71. nonfactorable over the integers   **73.** $(2x + 5)(x - 8)$ **75.** $(4y - 3)(y - 3)$ **77.** $(2a + 1)(a + 6)$
79. nonfactorable over the integers   **81.** $(5x + 1)(x + 5)$ **83.** $(11x - 1)(x - 11)$ **85.** nonfactorable over
the integers   **87.** $(2x - 3y)(3x + 7y)$ **89.** $(4a + 7b)(a + 9b)$ **91.** $(2x - 3y)(5x - 4y)$ **93.** $(3 + 2x)(8 - x)$
95. nonfactorable over the integers   **97.** $(3 - 4a)(5 + 2a)$ **99.** $(3x - 7)(2x - 5)$ **101.** $3a(3a - 2)(a + 4)$
103. $y^2(5y - 4)(y - 5)$ **105.** $2x(2x - 3y)(x + 4y)$ **107.** $5(5 + x)(4 - x)$ **109.** $4x(10 + x)(8 - x)$
111. $2x^2(2 - 5x)(5 + 3x)$ **113.** $a^2b^2(ab + 2)(ab - 5)$ **115.** $5(18a^2b^2 + 9ab + 2)$ **117.** nonfactorable over
the integers   **119.** $2a(a^2 + 5)(a^2 + 2)$ **121.** $3y^2(x^2 - 5)(x^2 - 8)$ **123.** $3xy(xy - 2)(xy + 4)$
125. $y(y^2 - 3)(y^2 - 5)$ **127.** $y(3 - 4y)(1 - 4y)$ **129.** $3(y^n - 3)(4y^n - 5)$ **131.** $x^n(x^n + 2)(x^n + 8)$

SECTION 3.4   *pages 149–152*

1. $(x + 4)(x - 4)$ **3.** $(2x + 1)(2x - 1)$ **5.** $(4x + 11)(4x - 11)$ **7.** $(1 + 3a)(1 - 3a)$ **9.** $(xy + 10)(xy - 10)$
11. nonfactorable over the integers   **13.** $(5 + ab)(5 - ab)$ **15.** $(a^n + 1)(a^n - 1)$ **17.** $(x - 6)^2$
19. $(b - 1)^2$ **21.** $(4x - 5)^2$ **23.** nonfactorable over the integers   **25.** nonfactorable over the integers
27. $(x + 3y)^2$ **29.** $(5a - 4b)^2$ **31.** $(x^n + 3)^2$ **33.** $(x - 7)(x - 1)$ **35.** $(x - y + a + b)(x - y - a - b)$
37. $(x - 3)(x^2 + 3x + 9)$ **39.** $(2x - 1)(4x^2 + 2x + 1)$ **41.** $(x - y)(x^2 + xy + y^2)$ **43.** $(m + n)(m^2 - mn + n^2)$
45. $(4x + 1)(16x^2 - 4x + 1)$ **47.** $(3x - 2y)(9x^2 + 6xy + 4y^2)$ **49.** $(xy + 4)(x^2y^2 - 4xy + 16)$
51. nonfactorable over the integers   **53.** nonfactorable over the integers   **55.** $(a - 2b)(a^2 - ab + b^2)$
57. $(x^{2n} + y^n)(x^{4n} - x^{2n}y^n + y^{2n})$ **59.** $(x^n + 2)(x^{2n} - 2x^n + 4)$ **61.** $(xy - 3)(xy - 5)$ **63.** $(xy - 5)(xy - 12)$
65. $(x^2 - 3)(x^2 - 6)$ **67.** $(b^2 + 5)(b^2 - 18)$ **69.** $(x^2y^2 - 2)(x^2y^2 - 6)$ **71.** $(x^n + 1)(x^n + 2)$
73. $(3xy - 5)(xy - 3)$ **75.** $(2ab - 3)(3ab - 7)$ **77.** $(2x^2 - 15)(x^2 + 1)$ **79.** $(2x^n - 1)(x^n - 3)$
81. $(2a^n + 5)(3a^n + 2)$ **83.** $5(x + 1)^2$ **85.** $3x(x - 3)(x^2 + 3x + 9)$ **87.** $7(x + 2)(x - 2)$
89. $y^2(y - 3)(y - 7)$ **91.** $(x^2 + 4)(x + 2)(x - 2)$ **93.** $2x^3(2x + 7)(2x - 7)$ **95.** $x^3(y - 1)(y^2 + y + 1)$
97. $x^3y^3(xy - 1)(x^2y^2 + xy + 1)$ **99.** $x^2(x^2 + 15x - 56)$ **101.** $2x^2(2x - 5)^2$ **103.** $(x^2 + y^2)(x + y)(x - y)$
105. $(x^2 + y^2)(x^4 - x^2y^2 + y^4)$ **107.** nonfactorable over the integers   **109.** $2a(2a - 1)(4a^2 + 2a + 1)$
111. $a^2b^2(a + 4b)(a - 12b)$ **113.** $2b^2(3a + 5b)(4a - 9b)$ **115.** $(x - 2)^2(x + 2)$ **117.** $4(y + 2)(x + 1)$
119. $(x + y)(x - y)(2x + 1)(2x - 1)$ **121.** $(x - 1)(x^2 + x + 1)(xy + 1)(x^2y^2 - xy + 1)$ **123.** $x(x^n + 1)^2$
125. $b^n(3b - 2)(b + 2)$

SECTION 3.5   *pages 155–156*

1. 5 and -3 **3.** -7 and 8   **5.** $-\frac{5}{3}$ and 4   **7.** $-4, -1,$ and 3   **9.** $-\frac{4}{3}, \frac{1}{2},$ and 3   **11.** $-4, 0,$ and $\frac{2}{3}$

13. -5 and 3   **15.** 1 and 3   **17.** -1 and -2 **19.** 3   **21.** 0 and $-\frac{2}{3}$ **23.** -2 and 5   **25.** $\frac{2}{3}$ and 1

27. $\frac{1}{3}$ and -3 **29.** $\frac{2}{3}$ **31.** $-\frac{1}{2}$ and $\frac{3}{2}$ **33.** $\frac{1}{2}$ **35.** -3 and 3   **37.** $-\frac{1}{2}$ and $\frac{1}{2}$ **39.** -3 and 5

41. $-\frac{3}{2}$ and $-\frac{4}{3}$ **43.** -3 and 6   **45.** -1 and $\frac{13}{2}$ **47.** -2 and $\frac{4}{3}$ **49.** $-5, 0,$ and 2   **51.** $-2, 0,$ and 6

53. $-\frac{1}{2}, 0,$ and 2   **55.** $-3, -1, 1,$ and 3   **57.** -2 and 2   **59.** $-3, -1,$ and 3   **61.** $-2, -\frac{2}{3},$ and 2

63. $-2, -\frac{2}{5},$ and 2

CHAPTER REVIEW   *pages 159–160*

1. $4x^2 - 8xy + 5y^2$ **2.** $-24a^7b^{14}$ **3.** $144x^2y^{10}z^{14}$ **4.** $\frac{x^2 + y^2}{xy}$ **5.** $\frac{c^{10}}{2b^{17}}$ **6.** 0.00254

7. 8.103×10^{19} tons   **8.** 1.27×10^{19} miles   **9.** $12x^5y^3 + 8xy^2 - 28x^2y^4$ **10.** $9x^2 + 8x$
11. $6x^3 - 29x^2 + 14x + 24$ **12.** $25a^2 - 4b^2$ **13.** $16x^2 - 24xy + 9y^2$ **14.** $(10x^2 - 29x - 21)$ cm^2

15. $6a^2b(3a^3b - 2ab^2 + 5)$ **16.** $(a + 2b)(2x - 3y)$ **17.** $(4 - x)(3 + x)$ **18.** $(x - 8)(x + 5)$
19. $(3x - 2)(2x - 9)$ **20.** $(6x + 5)(4x + 3)$ **21.** $(xy + 3)(xy - 3)$ **22.** $(2x + 3y)^2$ **23.** $(x^n - 6)^2$
24. $(4a - 3b)(16a^2 + 12ab + 9b^2)$ **25.** $(a + b + 1)(a^2 + 2ab + b^2 - a - b + 1)$ **26.** $(3x^2 + 2)(5x^2 - 3)$
27. $(7x^2y^2 + 3)(3x^2y^2 + 2)$ **28.** $3a^2(a^2 + 1)(a^2 - 6)$ **29.** $\frac{5}{2}$ and 4   **30.** 1, 4, and -4

CHAPTER TEST   *pages 161–162*

1. $2x^3 - 4x^2 + 6x - 14$ (3.1A)   **2.** $64a^7b^7$ (3.1B)   **3.** $\frac{2b^7}{a^{10}}$ (3.1C)   **4.** 5.01×10^{-7} (3.1D)
5. 6.048×10^5 s   (3.1E)   **6.** $-18r^2s^2 + 16rs^3 + 18rs^2$ (3.2A)   **7.** $35x^2 - 55x$ (3.2A)
8. $6a^2 - 13ab - 28b^2$ (3.2B)   **9.** $6t^5 - 8t^4 - 15t^3 + 22t^2 - 5$ (3.2B)   **10.** $49 - 25x^2$ (3.2C)
11. $9z^2 - 30z + 25$ (3.2C)   **12.** $(10x^2 - 3x - 1)$ ft^2 (3.2D)   **13.** $2ab^2(3a^2 - 2a + 2b^2)$ (3.3A)
14. $(y + 12)(y - 6)$ (3.3C)   **15.** $(3x - 2)(4x - 1)$ (3.3D)   **16.** $(3 - 2x)(4 - 3x)$ (3.3D)
17. $(2a^2 - 5)(3a^2 + 1)$ (3.4C)   **18.** $3x(2x - 3)(2x + 5)$ (3.4D)   **19.** $(4x - 5)(4x + 5)$ (3.4A)
20. $(4t + 3)^2$ (3.4A)   **21.** $(3x - 2)(9x^2 + 6x + 4)$ (3.4B)   **22.** $(3x - 2)(2x - a)$ (3.3B)
23. $(3x^2 + 4)(x - 3)(x + 3)$ (3.4C)   **24.** $-\frac{1}{3}$ and $\frac{1}{2}$ (3.5A)   **25.** $-1, -\frac{1}{6}$, and 1   (3.5A)

CUMULATIVE REVIEW   *pages 163–164*

1. 6   (1.2B)   **2.** $-\frac{5}{4}$ (1.3A)   **3.** Inverse Property of Addition   (1.3B)   **4.** $-18x + 8$ (1.3C)
5. $y = -\frac{1}{6}$ (2.1A)   **6.** $x = -\frac{11}{4}$ (2.1B)   **7.** $x = \frac{35}{3}$ (2.1C)   **8.** -1 and $\frac{7}{3}$ (2.3A)
9. $8x^2 - 5xy - y^2$ (3.1A)   **10.** $324a^{10}b^8$ (3.1B)   **11.** $-\frac{27x^2}{16y^7}$ (3.1C)   **12.** $\frac{b^5}{a^8}$ (3.1C)   **13.** $\frac{y^2}{25x^6}$ (3.1C)
14. $\frac{21}{8}$ (3.1C)   **15.** $-3x^2 + 60x + 8$ (3.2A)   **16.** $4x^3 - 7x + 3$ (3.2B)   **17.** $x^{2n} + 2x^n + 1$ (3.2C)
18. $(4x - 3)(2x - 5)$ (3.3D)   **19.** $-2x(2x - 3)(x - 2)$ (3.3D)   **20.** $(2x - 5y)^2$ (3.4A)
21. $(x - y)(a + b)$ (3.3B)   **22.** $(x - 2)(x + 2)(x^2 + 4)$ (3.4D)   **23.** $2(x - 2)(x^2 + 2x + 4)$ (3.4D)
24. $(a + 4)(a^2 - 4a + 16)$ (3.4B)   **25.** The integers are 9 and 15.   (2.4A)   **26.** 40 oz of pure gold must be
used.   (2.5A)   **27.** slower cyclist: 5 mph; faster cyclist: 7.5 mph (2.5B)   **28.** \$4500 more must be invested
at 10%. (2.6A)

ANSWERS to Chapter 4 Exercises

SECTION 4.1   *pages 171–174*

1. $1 - 2x$ **3.** $3x - 1$ **5.** $2x$ **7.** $-\frac{2}{x}$ **9.** $\frac{x}{2}$ **11.** The expression is in simplest form.

13. $4x^2 - 2x + 3$ **15.** $a^2 + 2a - 3$ **17.** $\frac{x^n - 3}{4}$ **19.** $\frac{x^n}{x^n - y^n}$ **21.** $\frac{x - 3}{x - 5}$ **23.** $\frac{x + y}{x - y}$ **25.** $-\frac{x + 3}{3x - 4}$

27. $-\frac{x + 7}{x - 7}$ **29.** The expression is in simplest form.   **31.** $\frac{a - b}{a^2 - ab + b^2}$ **33.** $\frac{4x^2 + 2xy + y^2}{2x + y}$ **35.** $\frac{x + y}{3x}$

37. $\frac{x - 2}{2x(x + 1)}$ **39.** $\frac{x + 2y}{3x + 4y}$ **41.** $-\frac{2x - 3}{2(2x + 3)}$ **43.** $\frac{x - 2}{a - b}$ **45.** $\frac{x^2 + 2}{(x - 1)(x + 1)}$ **47.** $\frac{xy + 7}{xy - 7}$ **49.** $\frac{a^n - 2}{a^n + 2}$

51. $\frac{a^n + 1}{a^n - 1}$ **53.** $-\frac{x - 3}{x + 3}$ **55.** $\frac{15b^4xy}{4}$ **57.** 1   **59.** $\frac{2x - 3}{x^2(x + 1)}$ **61.** $-\frac{x - 1}{x - 5}$ **63.** $\frac{2(x + 2)}{x - 1}$ **65.** $-\frac{x - 3}{2x + 3}$

67. $\frac{x^n - 6}{x^n - 1}$ **69.** $x + y$ **71.** $\frac{x - 7}{2(x + 7)}$ **73.** $\frac{a^2y}{10x}$ **75.** $\frac{3}{2}$ **77.** $\frac{5(x - y)}{xy(x + y)}$ **79.** $-\frac{x + 2}{x + 5}$ **81.** $\frac{x^{2n}}{8}$

83. -1 **85.** 1   **87.** $-\frac{2(x^2 + x + 1)}{(x + 1)^2}$ **89.** $\frac{x + 2}{x - 5}$

SECTION 4.2   *pages 179–180*

1. $x + 8$ **3.** $x^2 + \frac{2}{x - 3}$ **5.** $3x + 5 + \frac{3}{2x + 1}$ **7.** $5x + 7 + \frac{2}{2x - 1}$ **9.** $4x^2 + 6x + 9 + \frac{18}{2x - 3}$

11. $3x^2 + 1 + \frac{1}{2x^2 - 5}$ **13.** $x^2 - 3x - 10$ **15.** $x^2 - 2x + 1 - \frac{1}{x - 3}$ **17.** $2x^3 - 3x^2 + x - 4$ **19.** $2x - 8$

21. $3x - 8$ **23.** $3x + 3 - \dfrac{1}{x-1}$ **25.** $2x^2 - 3x + 9$ **27.** $4x^2 + 8x + 15 + \dfrac{12}{x-2}$ **29.** $2x^2 - 3x + 7 - \dfrac{8}{x+4}$

31. $3x^3 + 2x^2 + 12x + 19 + \dfrac{33}{x-2}$ **33.** $3x^3 - x + 4 - \dfrac{2}{x+1}$ **35.** $2x^3 + 6x^2 + 17x + 51 + \dfrac{155}{x-3}$

SECTION 4.3   pages 185–188

1. $\dfrac{9y^3}{12x^2y^4}, \dfrac{17x}{12x^2y^4}$ **3.** $\dfrac{2x^2-4x}{6x^2(x-2)}, \dfrac{3x-6}{6x^2(x-2)}$ **5.** $\dfrac{3x-1}{2x(x-5)}, -\dfrac{6x^3-30x^2}{2x(x-5)}$ **7.** $\dfrac{6x^2+9x}{(2x-3)(2x+3)}, \dfrac{10x^2-15x}{(2x-3)(2x+3)}$

9. $\dfrac{2x}{(x+3)(x-3)}, \dfrac{x^2+4x+3}{(x+3)(x-3)}$ **11.** $\dfrac{6}{6(x+2y)(x-2y)}, \dfrac{5x+10y}{6(x+2y)(x-2y)}$ **13.** $\dfrac{3x^2-3x}{(x+1)(x-1)^2}, \dfrac{5x^2+5x}{(x+1)(x-1)^2}$

15. $-\dfrac{x-3}{(x-2)(x^2+2x+4)}, \dfrac{2x-4}{(x-2)(x^2+2x+4)}$ **17.** $\dfrac{2x^2+6x}{(x-1)(x+3)^2}, -\dfrac{x^2-x}{(x-1)(x+3)^2}$ **19.** $-\dfrac{12x^2-8x}{(2x-3)(2x-5)(3x-2)},$

$\dfrac{6x^2-9x}{(2x-3)(2x-5)(3x-2)}$ **21.** $\dfrac{5}{(3x-4)(2x-3)}, -\dfrac{4x^2-6x}{(3x-4)(2x-3)}, \dfrac{3x^2-x-4}{(3x-4)(2x-3)}$ **23.** $\dfrac{2x^2+10x}{(x-3)(x+5)}, -\dfrac{2x-6}{(x-3)(x+5)}, -\dfrac{x-1}{(x-3)(x+5)}$

25. $\dfrac{x-5}{(x^n+1)(x^n+2)}, \dfrac{2x^{n+1}+2x}{(x^n+1)(x^n+2)}$ **27.** $\dfrac{1}{2x^2}$ **29.** $\dfrac{1}{x+2}$ **31.** $\dfrac{12ab-9b+8a}{30a^2b^2}$ **33.** $\dfrac{5-16b+12a}{40ab}$ **35.** $\dfrac{7}{12x}$

37. $\dfrac{2xy-8x+3y}{10x^2y^2}$ **39.** $-\dfrac{2a^2-13a}{(a-2)(a+1)}$ **41.** $\dfrac{5x^2-6x+10}{(2x-5)(5x-2)}$ **43.** $\dfrac{a}{b(a-b)}$ **45.** $\dfrac{a^2+18a-9}{a(a-3)}$ **47.** $\dfrac{17x^2+20x-25}{x(6x-5)}$

49. $\dfrac{6}{(x+3)(x-3)^2}$ **51.** $-\dfrac{2x-2}{(x+2)^2}$ **53.** $-\dfrac{5x^2-17x+8}{(x+4)(x-2)}$ **55.** $\dfrac{3x^n+2}{(x^n+1)(x^n-1)}$ **57.** 1 **59.** $\dfrac{x^2-52x+160}{4(x+3)(x-3)}$

61. $\dfrac{3x-1}{4x+1}$ **63.** $\dfrac{10x-6}{(x+3)(x+4)(x-3)}$ **65.** $-\dfrac{x^2+4x-5}{(2x+3)(2x-1)(3x+2)}$ **67.** $\dfrac{x^2-x+10}{(x+2)(x-1)}$ **69.** $\dfrac{x-2}{x+3}$

71. $\dfrac{x+4}{4x-1}$ **73.** 1 **75.** $\dfrac{x+1}{2x-1}$ **77.** $\dfrac{1}{2x-1}$ **79.** $\dfrac{1}{x^2+4}$ **81.** $\dfrac{3-a}{3a}$ **83.** $-\dfrac{2x^2+5x-2}{(x+2)(x+1)}$ **85.** $\dfrac{b-a}{b+2a}$

87. $\dfrac{2}{x+2}$

SECTION 4.4   pages 191–192

1. $\dfrac{5}{23}$ **3.** $\dfrac{2}{5}$ **5.** $\dfrac{x}{x-1}$ **7.** $-\dfrac{a}{a+2}$ **9.** $-\dfrac{a-1}{a+1}$ **11.** $\dfrac{2}{5}$ **13.** $-\dfrac{1}{2}$ **15.** $\dfrac{x^2-x-1}{x^2+x+1}$ **17.** $\dfrac{x+2}{x-1}$

19. $-\dfrac{x+4}{2x+3}$ **21.** $\dfrac{2(x+2)}{x-4}$ **23.** $\dfrac{x+1}{x-4}$ **25.** $\dfrac{x-2}{x-3}$ **27.** $\dfrac{x-3}{x+4}$ **29.** $\dfrac{3x+1}{x-5}$ **31.** $-\dfrac{2a+2}{7a-4}$ **33.** $\dfrac{x+y}{x-y}$

35. $-\dfrac{2x}{x^2+1}$ **37.** $\dfrac{a^3+a^2+a}{a^2+1}$ **39.** $-\dfrac{a^2}{1-2a}$ **41.** $-\dfrac{3x+2}{x-2}$

SECTION 4.5   pages 195–196

1. The solution is 9.   **3.** The solution is $\dfrac{15}{2}$.   **5.** The solution is 3.   **7.** The solution is $\dfrac{10}{3}$.   **9.** The solution is -2.   **11.** The solution is $\dfrac{2}{5}$.   **13.** The solution is -4.   **15.** The solution is $\dfrac{3}{4}$.   **17.** The solution is $-\dfrac{6}{5}$.   **19.** The solution is -2.   **21.** The value of the house is $115,000.   **23.** There are 3000 ducks in the preserve.   **25.** In a shipment of 24,000, 64 transistors would be defective.   **27.** The dimensions of the room are 18 ft by 24 ft.   **29.** An additional 0.75 oz of medication is required.   **31.** To earn a dividend of $600, 300 additional shares are needed.   **33.** To harvest 11,000 bushels of corn, 50 additional acres are needed.   **35.** To make a 150 ft^2 floor, an additional 10 ft^3 of cement are needed.

SECTION 4.6   pages 203–208

1. The solution is -5.   **3.** The solution is $\dfrac{5}{2}$.   **5.** The solution is -1.   **7.** The solution is 8.   **9.** The solution is -3.   **11.** The solution is -4.   **13.** The equation has no solution.   **15.** The solutions are -1 and $-\dfrac{5}{3}$.   **17.** The solutions are $-\dfrac{1}{3}$ and 5.   **19.** The solution is $-\dfrac{11}{3}$.   **21.** The equation has no solution.

23. $C = \dfrac{5F - 160}{9}$ **25.** $P = \dfrac{A}{1+rt}$ **27.** $h = \dfrac{2A}{b}$ **29.** $P_2 = \dfrac{P_1V_1T_2}{T_1V_2}$ **31.** $V_0 = \dfrac{S + 16t^2}{t}$ **33.** $h = \dfrac{3V}{\pi r^2}$

35. $b = \dfrac{af}{a-f}$ **37.** $R = Pn + C$ **39.** $h = \dfrac{S - 2\pi r^2}{2\pi r}$ **41.** With both panels working, it would take 24 min to raise the temperature 1°.   **43.** It would take 48 min to print the checks with both printers operating.

45. Working alone, it would take the new machine 8 h to package the transitors. **47.** With both members working together, it would take 3.6 h to wire the telephone lines. **49.** It would take the apprentice 16 h to shingle the roof. **51.** Working alone, it would take the smaller printer 75 min to print the payroll. **53.** It would take 40 min for the tub to overflow. **55.** When all three computers are working, it would take 10 min to print out the task. **57.** When all three pipes are open, it would take 30 h to fill the tank. **59.** It would take the third painter 2 h to paint the room alone. **61.** The rate of the passenger train is 60 mph. The rate of the freight train is 42 mph. **63.** The rate of the bus is 62 mph. **65.** The cyclist was riding at 12 mph. **67.** The rate of the cabin cruiser is 8 mph. **69.** The rate of the train is 70 mph. **71.** The rate of the current is 3 mph. **73.** The rate of the wind is 30 mph. **75.** The rate of the single-engine plane is 120 mph. The rate of the jet is 480 mph. **77.** The rate of the wind is 51.43 mph. **79.** The rate of the wind is 25 mph.

CHAPTER REVIEW *pages 211–212*

1. $\dfrac{3x^2 - 1}{3x^2 + 1}$ **2.** $\dfrac{x^2 + 3x + 9}{x + 3}$ **3.** $\dfrac{1}{a - 1}$ **4.** x **5.** $\dfrac{3x + 2}{x}$ **6.** $\dfrac{1}{x}$ **7.** $5x + 4 + \dfrac{6}{3x - 2}$ **8.** $2x - 3 - \dfrac{4}{6x + 1}$

9. $4x^2 + 3x - 8 + \dfrac{50}{x + 6}$ **10.** $x^3 + 4x^2 + 16x + 64 + \dfrac{252}{x - 4}$ **11.** $\dfrac{16x^2 + 4x}{(4x - 1)(4x + 1)} ; \dfrac{12x^2 - 7x + 1}{(4x - 1)(4x + 1)}$

12. $\dfrac{x^2 - 7x + 12}{(x - 5)(x - 4)} ; \dfrac{x}{(x - 5)(x - 4)} ; -\dfrac{x - 5}{(x - 5)(x - 4)}$ **13.** $\dfrac{21a + 40b}{24a^2b^4}$ **14.** $\dfrac{4x - 1}{x + 2}$ **15.** $\dfrac{9x^2 + x + 4}{(3x - 1)(x - 2)}$ **16.** $-\dfrac{5x^2 - 17x - 9}{(x - 3)(x + 2)}$

17. $\dfrac{x - 4}{x + 5}$ **18.** $\dfrac{x}{x - 3}$ **19.** $\dfrac{7x + 4}{2x + 1}$ **20.** $-\dfrac{9x^2 + 16}{24x}$ **21.** The solution is $\dfrac{2}{5}$. **22.** The solution is 4.

23. 48 miles would be represented by 12 inches. **24.** It will take the student 375 min to read 150 pages.

25. The equation has no solution. **26.** The solution is 10. **27.** $N = \dfrac{S}{1 - Q}$ **28.** $r = \dfrac{S - a}{S}$ **29.** It would take the apprentice 104 min working alone. **30.** The rate of the helicopter is 180 mph.

CHAPTER TEST *pages 213–214*

1. $\dfrac{v(v + 2)}{2v - 1}$ (4.1A) **2.** $-\dfrac{2(a - 2)}{3a + 2}$ (4.1A) **3.** $\dfrac{6(x - 2)}{5}$ (4.1B) **4.** $\dfrac{x + 2}{x - 1}$ (4.1C) **5.** $\dfrac{x + 1}{3x - 4}$ (4.1C)

6. $3t^2 + t - 1 - \dfrac{5}{4t + 4}$ (4.2A) **7.** $3x^3 + 3x^2 + x + 2 + \dfrac{4}{x - 1}$ (4.2B) **8.** $\dfrac{x^2 - 2x - 3}{(x + 3)(x - 3)(x + 2)} ; \dfrac{2x^2 - 4x}{(x + 3)(x - 3)(x + 2)}$ (4.3A)

9. $\dfrac{x^2 - 9x + 3}{(x + 2)(x - 3)}$ (4.3B) **10.** $\dfrac{-x^2 - 5x + 2}{(x + 1)(x - 4)(x - 1)}$ (4.3B) **11.** $\dfrac{x - 4}{x + 3}$ (4.4A) **12.** $\dfrac{x + 4}{x + 2}$ (4.4A)

13. The solution is 2 (4.5A) **14.** The solutions are 1 and 2. (4.6A) **15.** There is no solution. (4.6A)

16. $x = \dfrac{c}{a - b}$ (4.6B) **17.** $t = \dfrac{4r}{r - 2}$ (4.6B) **18.** The designer needs 14 rolls of wallpaper. (4.5B) **19.** It would take both landscapers 10 min to till the soil. (4.6C) **20.** The rate of the cyclist is 10 mph. (4.6D)

CUMULATIVE REVIEW *pages 215–216*

1. $\dfrac{36}{5}$ (1.2B) **2.** -52 (1.3A) **3.** $y = \dfrac{2}{3}$ (2.1A) **4.** $x = 7\dfrac{1}{2}$ (2.1C) **5.** 7 and 1 (2.3A)

6. -2 and 3 (3.5A) **7.** $\dfrac{a^5}{b^6}$ (3.1C) **8.** $\{x \,|\, x \leq 4\}$ (2.2A) **9.** $-4a^4 + 6a^3 - 2a^2$ (3.2A)

10. $(2x^n - 1)(x^n + 2)$ (3.4C) **11.** $(xy - 3)(x^2y^2 + 3xy + 9)$ (3.4B) **12.** $x + 3y$ (4.1A)

13. $4x^2 + 10x + 10 + \dfrac{21}{x - 2}$ (4.2B) **14.** $\dfrac{3xy}{x - y}$ (4.1C) **15.** $\dfrac{x^3y - 2x^2y}{2x^2(x + 1)(x - 2)} ; \dfrac{2}{2x^2(x + 1)(x - 2)}$ (4.3A)

16. $-\dfrac{x}{(3x + 2)(x + 1)}$ (4.3B) **17.** $\dfrac{x - 3}{x + 3}$ (4.4A) **18.** $x = 9$ (4.5A) **19.** $x = 13$ (4.6A)

20. $r = \dfrac{E - IR}{I}$ (4.6B) **21.** $\dfrac{x}{x - 1}$ (3.1C) **22.** The integers are 5 and 10. (2.4A) **23.** The mixture must contain 50 lb of almonds. (2.5A) **24.** The amount of people expected to vote is 75,000. (4.5B) **25.** The new computer would complete the job in 14 minutes. (4.6C) **26.** The rate of the wind is 60 mph. (4.6D)

ANSWERS to Chapter 5 Exercises

SECTION 5.1 *pages 225–228*

1. 2   **3.** 27   **5.** $\frac{1}{9}$ **7.** 4   **9.** $(-25)^{5/2}$ is not a real number.   **11.** $\frac{343}{125}$ **13.** x **15.** $y^{1/2}$

17. $x^{1/12}$ **19.** $a^{7/12}$ **21.** $\frac{1}{a}$ **23.** $\frac{1}{y}$ **25.** $y^{3/2}$ **27.** $\frac{1}{x}$ **29.** $\frac{1}{x^4}$ **31.** a **33.** $x^{3/10}$ **35.** a^3

37. $\frac{1}{x^{1/2}}$ **39.** $y^{2/3}$ **41.** $x^4 y$ **43.** $x^6 y^3 z^9$ **45.** $\frac{x}{y^2}$ **47.** $\frac{x^{3/2}}{y^{1/4}}$ **49.** $x^2 y^8$ **51.** $\frac{1}{x^{11/12}}$ **53.** $\frac{1}{y^{5/2}}$

55. $\frac{1}{b^{7/8}}$ **57.** $a^5 b^{13}$ **59.** $\frac{m^2}{4n^{3/2}}$ **61.** $\frac{y^{17/2}}{x^3}$ **63.** $\frac{16b^2}{a^{1/3}}$ **65.** $y^2 - y$ **67.** $a - a^2$ **69.** x^{4n}

71. $x^{3n/2}$ **73.** $y^{3n/2}$ **75.** x^{2n^2} **77.** $x^{2n}y^n$ **79.** $\sqrt[4]{3}$ **81.** $\sqrt{a^3}$ **83.** $\sqrt{32t^5}$ **85.** $-2\sqrt[3]{x^2}$

87. $\sqrt[3]{a^4 b^2}$ **89.** $\sqrt[5]{a^6 b^{12}}$ **91.** $\sqrt[4]{(4x-3)^3}$ **93.** $\frac{1}{\sqrt[3]{x^2}}$ **95.** $14^{1/2}$ **97.** $x^{1/3}$ **99.** $x^{4/3}$ **101.** $b^{3/5}$

103. $(2x^2)^{1/3}$ **105.** $-(3x^5)^{1/2}$ **107.** $3xy^{2/3}$ **109.** $(a^2-2)^{1/2}$ **111.** x^8 **113.** $-x^4$ **115.** xy^3

117. $-x^5 y$ **119.** $4a^2 b^6$ **121.** $\sqrt{-16x^4 y^2}$ is not a real number.   **123.** $3x^3$ **125.** $-4x^3 y^4$

127. $-x^2 y^3$ **129.** $x^4 y^2$ **131.** $3xy^5$ **133.** $2ab^2$

SECTION 5.2 *pages 237–240*

1. $x^2 yz^2 \sqrt{yz}$ **3.** $2ab^4 \sqrt{2a}$ **5.** $3xyz^2 \sqrt{5yz}$ **7.** $\sqrt{-9x^3}$ is not a real number.   **9.** $a^5 b^2 \sqrt[3]{ab^2}$

11. $-5y\sqrt[3]{x^2 y}$ **13.** $abc^2 \sqrt[3]{ab^2}$ **15.** $2x^2 y\sqrt[4]{xy}$ **17.** 24.49   **19.** 23.219   **21.** 29.068   **23.** 774.6

25. $-6\sqrt{x}$ **27.** $-2\sqrt{2}$ **29.** $\sqrt{2x}$ **31.** $3\sqrt{3a} - 2\sqrt{2a}$ **33.** $10x\sqrt{2x}$ **35.** $2x\sqrt{3xy}$ **37.** $xy\sqrt{2y}$

39. $5a^2 b^2 \sqrt{ab}$ **41.** $9\sqrt[3]{2}$ **43.** $-7a\sqrt[3]{3a}$ **45.** $-5xy^2 \sqrt[3]{x^2 y}$ **47.** $16ab\sqrt[4]{ab}$ **49.** $6\sqrt{3} - 6\sqrt{2}$

51. $6x^3 y^2 \sqrt{xy}$ **53.** $-9a^2 \sqrt{3ab}$ **55.** $10xy\sqrt[3]{3y}$ **57.** $2xy\sqrt[4]{3x}$ **59.** $7\sqrt{10}$ **61.** 6   **63.** $a^2 b^2 \sqrt{b}$

65. $5x^3 y^2 \sqrt{2y}$ **67.** $2ab^2 \sqrt[3]{4b^2}$ **69.** $2ab\sqrt[4]{27a^3 b^3}$ **71.** $10 - 5\sqrt{2}$ **73.** $y - \sqrt{5y}$ **75.** $9a\sqrt{a} - a\sqrt{3}$

77. $2x + 8\sqrt{2x} + 16$ **79.** $36x^3 y^2 \sqrt{6}$ **81.** $2ab^2 \sqrt[3]{36ab^2}$ **83.** $-10 + \sqrt{2}$ **85.** $y - 4$ **87.** $2x - 9y$

89. $4\sqrt{x}$ **91.** $b^2 \sqrt{3a}$ **93.** $\frac{\sqrt{5}}{5}$ **95.** $\frac{\sqrt{2x}}{2x}$ **97.** $\frac{\sqrt{5x}}{x}$ **99.** $\frac{\sqrt{5x}}{5}$ **101.** $\frac{3\sqrt[3]{4}}{2}$ **103.** $\frac{3\sqrt[3]{2x}}{2x}$

105. $\frac{\sqrt{2xy}}{2y}$ **107.** $\frac{2\sqrt{3ab}}{3b^2}$ **109.** $2\sqrt{5} - 4$ **111.** $\frac{3\sqrt{y} + 6}{y - 4}$ **113.** $-5 + 2\sqrt{6}$ **115.** $-\frac{52 - 19\sqrt{7}}{3}$

117. $\frac{20 - 7\sqrt{3}}{23}$ **119.** $\frac{2x - 8\sqrt{x} + 8}{x - 4}$ **121.** $\frac{9x - 6\sqrt{xy} - 8y}{9x - 4y}$

SECTION 5.3 *pages 247–248*

1. $2i$ **3.** $7i\sqrt{2}$ **5.** $3i\sqrt{3}$ **7.** $4 + 2i$ **9.** $2\sqrt{3} - 3i\sqrt{2}$ **11.** $4\sqrt{10} - 7i\sqrt{3}$ **13.** $8 - i$

15. $-8 + 4i$ **17.** $6 - 6i$ **19.** $19 - 7i\sqrt{2}$ **21.** $6\sqrt{2} - 3i\sqrt{2}$ **23.** $5 - i$ **25.** 0   **27.** 63

29. -4 **31.** $-3\sqrt{2}$ **33.** $-4 + 12i$ **35.** $-2 + 4i$ **37.** $17 - i$ **39.** $8 + 27i$ **41.** 1   **43.** 1

45. $-3i$ **47.** $\frac{3}{4} + \frac{1}{2}i$ **49.** $\frac{10}{13} - \frac{2}{13}i$ **51.** $\frac{4}{5} + \frac{2}{5}i$ **53.** $-i$ **55.** $-\frac{\sqrt{5}}{5} + \frac{2\sqrt{5}}{5}i$ **57.** $\frac{3}{10} - \frac{11}{10}i$

59. $\frac{6}{5} + \frac{7}{5}i$

SECTION 5.4 *pages 253–254*

1. The solution is -2.   **3.** The solution is 9.   **5.** There is no solution.   **7.** The solution is 45.

9. The solution is 11.   **11.** The solution is 29.   **13.** The solution is 21.   **15.** The solution is $\frac{15}{4}$.

17. The solution is 6.   **19.** The solution is -12.   **21.** The solution is 9.   **23.** The solution is 2.
25. The solution is 2.   **27.** The solution is 1.   **29.** The object has fallen 156.25 ft.   **31.** The width is 6 ft.   **33.** The periscope must be 9 ft above the water.   **35.** The distance required is 80 m.   **37.** The length of the pendulum is 7.30 ft.

CHAPTER REVIEW *pages 257–258*

1. $\frac{1}{x^5}$ **2.** $20x^2 y^2$ **3.** $3\sqrt[4]{x^3}$ **4.** $7yx^{2/3}$ **5.** $3a^2 b^3$ **6.** $-2a^2 b^4$ **7.** $3ab^3 \sqrt{2a}$ **8.** $-2ab^2 \sqrt[5]{2a^3 b^2}$

9. $2a^2 b\sqrt{2b}$ **10.** $5x^3 y^3 \sqrt[3]{2x^2 y}$ **11.** $6x^2 \sqrt{3y}$ **12.** $4xy^2 \sqrt[3]{x^2}$ **13.** $31 - 10\sqrt{6}$ **14.** $6\sqrt{3} - 13$

15. $\frac{8\sqrt{3y}}{3y}$ **16.** $\frac{x\sqrt{x} - x\sqrt{2} + 2\sqrt{x} - 2\sqrt{2}}{x - 2}$ **17.** $5i\sqrt{2}$ **18.** $-12 + 10i$ **19.** $-4\sqrt{2} + 8i\sqrt{2}$ **20.** $7 + 3i$

21. $-6\sqrt{2}$ **22.** $39 - 2i$ **23.** $\frac{2}{3} - \frac{5}{3}i$ **24.** $-2 + 7i$ **25.** $x = 7$ **26.** $x = -2$ **27.** $x = 30$

28. The amount of power generated is 120 watts.   **29.** The distance required to reach a velocity of 88 ft/s is 242 ft.   **30.** The bottom of the ladder is 6.63 ft from the building.

CHAPTER TEST   *pages 259–260*

1. $r^{1/6}$ (5.1A)   **2.** $\frac{64x^3}{y^6}$ (5.1A)   **3.** $\frac{b^3}{8a^6}$ (5.1A)   **4.** $3\sqrt[5]{y^2}$ (5.1B)   **5.** $\frac{1}{2}x^{3/4}$ (5.1B)   **6.** $2xy^2$ (5.1C)

7. $4x^2y^3\sqrt{2y}$ (5.2A)   **8.** $3abc^2\sqrt[3]{ac}$ (5.2A)   **9.** $8a\sqrt{2a}$ (5.2B)   **10.** $-2x^2y\sqrt[3]{2x}$ (5.2B)

11. $-4x\sqrt{3}$ (5.2C)   **12.** $14 + 10\sqrt{3}$ (5.2C)   **13.** $2a - \sqrt{ab} - 15b$ (5.2C)   **14.** $4x + 4\sqrt{xy} + y$ (5.2C)

15. $\frac{4x^2}{y}$ (5.2D)   **16.** 2 (5.2D)   **17.** $\frac{x + \sqrt{xy}}{x - y}$ (5.2D)   **18.** -4 (5.3C)   **19.** $-3 + 2i$ (5.3B)

20. $18 + 16i$ (5.3C)   **21.** $-\frac{4}{5} + \frac{7}{5}i$ (5.3D)   **22.** $10 + 2i$ (5.3C)   **23.** The solution is 4. (5.4A)

24. The solution is -3. (5.4A)   **25.** The object has fallen 576 ft. (5.4B)

CUMULATIVE REVIEW   *pages 261–262*

1. Distributive Property (1.3B)   **2.** $-x - 24$ (1.3C)   **3.** $x = \frac{3}{2}$ (2.1B)

4. $x = \frac{2}{3}$ (2.1C)   **5.** The solutions are $\frac{1}{3}$ and $\frac{7}{3}$. (2.3A)   **6.** $\{x | x > 1\}$ (2.2A)   **7.** $\{x | -6 \le x \le 3\}$ (2.3B)

8. $(9x + y)(9x - y)$ (3.4A)   **9.** $x(x^2 + 3)(x + 1)(x - 1)$ (3.4D)   **10.** $\frac{4}{a - 1}$ (4.1A)   **11.** $a - 1$ (4.4A)

12. $C = R - nP$ (4.6B)   **13.** $2x^2y^2$ (5.1A)   **14.** $\frac{y^5}{x^4}$ (5.1A)   **15.** $-x\sqrt{10x}$ (5.2A)

16. $13 - 7\sqrt{3}$ (5.2C)   **17.** $\sqrt{10} + \sqrt{3}$ (5.2D)   **18.** $7 + i$ (5.3A)   **19.** $-\frac{1}{5} + \frac{3}{5}i$ (5.3D)

20. $x = -20$ (5.4A)   **21.** $\sqrt{6} + \sqrt{2}$ (5.2D)   **22.** There are nineteen 18¢ stamps in the collection. (2.4B)
23. The amount which must be invested at 8.4% is $4000. (2.6A)   **24.** The rate of the plane was 250 mph. (4.6D)   **25.** Light travels to the earth from the moon in 1.25 s. (3.1E)   **26.** The periscope would have to be 25 feet above the surface of the water. (5.4B)

ANSWERS to Chapter 6 Exercises

SECTION 6.1   *pages 271–274*

1. 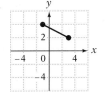   **3.** 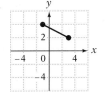   **5.** A is $(0, 3)$, B is $(1, 1)$, C is $(3, -4)$, D is $(-4, 4)$.

7. A is $(4, 0)$, B is $(-1, 1)$, C is $(-4, -2)$, D is $(1, 2)$.   **9.** 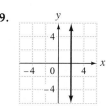   **11.**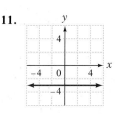

13. The ordered pair solution is $(-3, -6)$.   **15.** The ordered pair solution is $(-6, 8)$.   **17.** The ordered pair solution is $(-4, -3)$.   **19.** The ordered pair solution is $(-2, 1)$.   **21.** The ordered pair solution is $\left(-2, -\frac{1}{3}\right)$.

23. 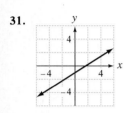   **25.** 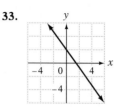   **27.** 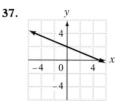   **29.**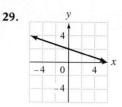

31. 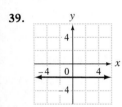   **33.** 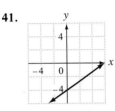   **35.** 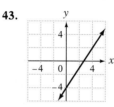   **37.**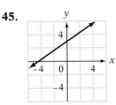

39. 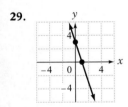   **41.** 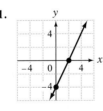   **43.** 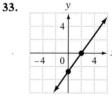   **45.**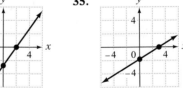

SECTION 6.2   *pages 281–286*

1. The slope is -1.   **3.** The slope is $\frac{1}{3}$.   **5.** The slope is $-\frac{2}{3}$.   **7.** The slope is $-\frac{3}{4}$.

9. The slope is undefined.   **11.** The slope is $\frac{7}{5}$.   **13.** The line has zero slope.   **15.** The slope is $-\frac{1}{2}$.

17. The slope is -3.   **19.** The slope is undefined.   **21.** The slope is -2.   **23.** The line has zero slope.

25. The slope is $\frac{1}{4}$.   **27.** The slope is $\frac{5}{2}$.

29. 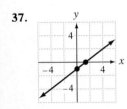   **31.** 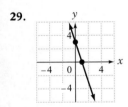   **33.** 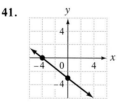   **35.**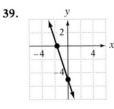

37. 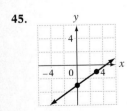   **39.** 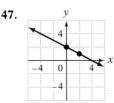   **41.** 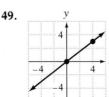   **43.**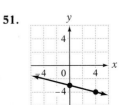

45. 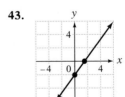   **47.** 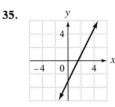   **49.** 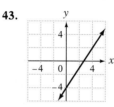   **51.**

53.   **55.** 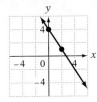   **57.** 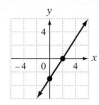   **59.**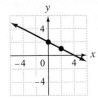

61. 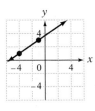   **63.** 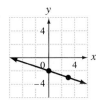   **65.** 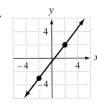   **67.**

SECTION 6.3   *pages 291–294*

1. The equation of the line is $y = 2x + 5$.   **3.** The equation of the line is $y = \frac{1}{2}x + 2$.   **5.** The equation of the line is $y = \frac{5}{4}x + \frac{21}{4}$.   **7.** The equation of the line is $y = -\frac{5}{3}x + 5$.   **9.** The equation of the line is $y = -3x + 9$.

11. The equation of the line is $y = -3x + 4$.   **13.** The equation of the line is $y = \frac{2}{3}x - \frac{7}{3}$.   **15.** The equation of the line is $y = \frac{1}{2}x$.   **17.** The equation of the line is $y = 3x - 9$.   **19.** The equation of the line is $y = -\frac{2}{3}x + 7$.

21. The equation of the line is $y = -x - 3$.   **23.** The equation of the line is $y = \frac{7}{5}x - \frac{27}{5}$.   **25.** The equation of the line is $y = 0$.   **27.** The equation of the line is $y = \frac{2}{3}x - \frac{5}{3}$.   **29.** The equation of the line is $x = 0$.

31. The equation of the line is $y = 3x - 3$.   **33.** The equation of the line is $y = -\frac{5}{2}x - \frac{25}{2}$.   **35.** The equation of the line is $y = \frac{4}{3}x + 5$.   **37.** The equation of the line is $y = -\frac{2}{5}x + \frac{3}{5}$.   **39.** The equation of the line is $x = 3$.

41. The equation of the line is $y = -\frac{5}{4}x - \frac{15}{2}$.   **43.** The equation of the line is $y = -3$.   **45.** The equation of the line is $y = -2x + 3$.   **47.** The equation of the line is $x = -5$.   **49.** The equation of the line is $y = x + 2$.   **51.** The equation of the line is $y = -2x - 3$.   **53.** The equation of the line is $y = \frac{2}{3}x + \frac{5}{3}$.

55. The equation of the line is $y = \frac{1}{3}x + \frac{10}{3}$.   **57.** The equation of the line is $y = \frac{3}{2}x - \frac{1}{2}$.   **59.** The equation of the line is $y = -\frac{3}{2}x + 3$.   **61.** The equation of the line is $y = -1$.   **63.** The equation of the line is $y = x - 1$.

65. The equation of the line is $y = -2x - 1$.   **67.** The equation of the line is $x = -2$.   **69.** The equation of the line is $y = -\frac{7}{2}x - 2$.   **71.** The equation of the line is $y = -\frac{4}{3}x + \frac{20}{3}$.   **73.** The equation of the line is $y = -\frac{3}{4}x + \frac{17}{4}$.   **75.** The equation of the line is $y = 3x - 10$.   **77.** The equation of the line is $y = -x + 1$.

79. The equation of the line is $y = \frac{2}{3}x + \frac{5}{3}$.   **81.** The equation of the line is $y = \frac{1}{2}x - 1$.   **83.** The equation of the line is $y = -4$.   **85.** The equation of the line is $y = \frac{3}{4}x$.   **87.** The equation of the line is $y = -\frac{4}{3}x + \frac{5}{3}$.   **89.** The equation of the line is $x = -2$.   **91.** The equation of the line is $y = x - 1$.   **93.** The equation of the line is $y = \frac{4}{3}x + \frac{7}{3}$.   **95.** The equation of the line is $y = -x + 3$.

SECTION 6.4 *pages 299–300*

1. The lines are perpendicular. **3.** No, the lines are not parallel. **5.** No, the lines are not parallel. **7.** The lines are perpendicular. **9.** The lines are parallel. **11.** The lines are perpendicular. **13.** No, the lines are not parallel. **15.** The lines are perpendicular. **17.** The lines are parallel. **19.** The equation of the line is $y = \frac{2}{3}x - \frac{8}{3}$. **21.** The equation of the line is $y = \frac{1}{3}x - \frac{1}{3}$. **23.** The equation of the line is $y = -\frac{5}{3}x - \frac{14}{3}$. **25.** The equation of the line is $y = \frac{2}{3}x + 3$. **27.** The equation of the line is $y = -\frac{5}{3}x - \frac{13}{3}$. **29.** The equation of the line is $y = \frac{5}{2}x + 8$. **31.** The equation of the line is $y = -\frac{1}{5}x - \frac{11}{5}$. **33.** The equation of the line is $y = -\frac{2}{3}x - 5$.

SECTION 6.5 *pages 303–304*

1. The equation of the line is $y = 0.08x$. The annual income is $320. **3.** The equation of the line is $y = 10x$. The amount of milk in the tank is 600 gal. **5.** The equation of the line is $y = -\frac{25,000}{3}x + 250,000$. The value of the building is $150,000. **7.** The equation of the line is $y = 120x + 3000$. The cost of manufacturing 35 cameras is $7200. **9.** The equation of the line is $y = -\frac{1}{2}x + 20$. The demand for the product is 24 units.

11. The equation of the line is $y = -\frac{1}{4}x + 5$. The demand for the product is 8 units.

SECTION 6.6 *pages 307–308*

1. 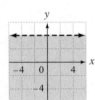   **3.**   **5.**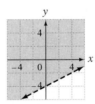

7.   **9.** 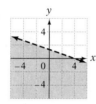   **11.**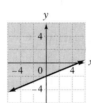

13.   **15.**   **17.**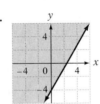

CHAPTER REVIEW *pages 311–312*

1.   **2.** 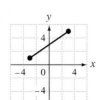   **3.** The ordered pair solution is (2, 1).

4. The ordered pair solution is $(-6, 0)$.   **5.** 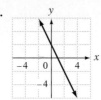   **6.**

7. The slope is -1.   **8.** The slope is undefined.

9. 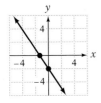   **10.** 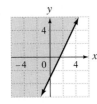   **11.** The equation of the line is $y = \frac{5}{2}x + \frac{23}{2}$.

12. The equation of the line is $y = -\frac{4}{3}x + 9$.   **13.** The equation of the line is $y = -\frac{7}{6}x + \frac{5}{3}$.   **14.** The equation of the line is $y = 4$.   **15.** The equation of the line is $y = -3x + 7$.   **16.** The equation of the line is $y = \frac{2}{3}x - \frac{8}{3}$.

17. The equation of the line is $y = \frac{3}{2}x + 2$.   **18.** The equation of the line is $y = -\frac{1}{2}x - \frac{5}{2}$.

19. The equation of the line is $y = 20x + 2000$. The cost of manufacturing 125 calculators is \$4500.
20. The equation of the line is $y = 400x$. The distance traveled in two and one-half hours is 1000 mi.

21. 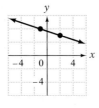   **22.**

CHAPTER TEST   *pages 313–314*

1. 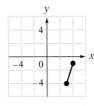   (6.1A)   **2.** The ordered pair solution is $(-3, 0)$.   (6.1B)   **3.** 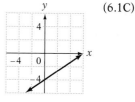   (6.1C)

4.   (6.1D)   **5.** The equation of the line is $x = -2$.   (6.3A)

6. The equation of the line is $y = -3$.   (6.3A)   **7.** The slope is $-\frac{1}{6}$.   (6.2A)

8. The line has zero slope. (6.2A) **9.** (6.2B) 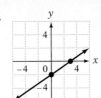 **10.** (6.2C)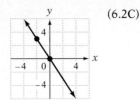

11. The equation of the line is $y = \frac{2}{5}x + 4$. (6.3A) **12.** The equation of the line is $y = -\frac{3}{4}x - 2$. (6.3A)

13. The equation of the line is $y = -\frac{7}{5}x + \frac{1}{5}$. (6.3B) **14.** The equation of the line is $x = 2$. (6.3B)

15. The equation of the line is $y = 3$. (6.4A) **16.** The equation of the line is $y = -\frac{3}{2}x + \frac{7}{2}$. (6.4A)

17. The equation of the line is $y = 2x + 1$. (6.4A) **18.** The equation of the line is $y = -\frac{2}{3}x + 2$. (6.4A)

19. The equation for the depreciation is $y = -\frac{10,000}{3}x + 50,000$. (6.5A) **20.** (6.6A)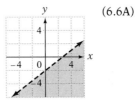

CUMULATIVE REVIEW *pages 315–316*

1. Commutative Property of Multiplication (1.3B) **2.** $x = \frac{9}{2}$ (2.1B) **3.** $y = \frac{8}{9}$ (2.1C) **4.** $x = -\frac{1}{14}$ (2.1C)

5. $\left\{x \mid x < -1 \text{ or } x > \frac{1}{2}\right\}$ (2.2B) **6.** The solutions are $\frac{5}{2}$ and $-\frac{3}{2}$. (2.3A) **7.** $\left\{x \mid 0 < x < \frac{10}{3}\right\}$ (2.3B)

8. $(x - 4y)(x^2 + 4xy + 16y^2)$ (3.4B) **9.** $(a - b)(b - a)(b + a)$ (3.4D) **10.** $\frac{x+1}{x(x+y)}$ (4.1C)

11. $C = R - Pn$ (4.6B) **12.** $\frac{x\sqrt{x} - 2x - 2\sqrt{x} + 4}{x - 4}$ (5.2D) **13.** $-4x^2y^3\sqrt{2x}$ (5.2B) **14.** $-\frac{1}{5} + \frac{2}{5}i$ (5.3D)

15. The ordered pair solution is $(-8, 13)$. (6.1B) **16.** The slope is $-\frac{7}{4}$. (6.2A)

17. The equation of the line is $y = \frac{3}{2}x + \frac{13}{2}$. (6.3A) **18.** The equation of the line is $y = -\frac{5}{4}x + 3$. (6.3B)

19. The equation of the line is $y = -\frac{3}{2}x + 7$. (6.4A) **20.** The equation of the line is $y = -\frac{2}{3}x + \frac{8}{3}$. (6.4A)

21. There are 7 dimes in the purse. (2.4B) **22.** first plane: 200 mph; second plane: 400 mph (2.5B)

23. 32 lb of $8 coffee and 48 lb of $3 coffee should be used. (2.5A)

24. (6.2B) 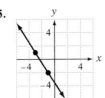 **25.** (6.2C) 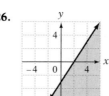 **26.** (6.6A)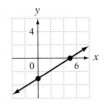

27. The equation of the line is $y = -2500x + 15,000$. The value of the truck is $6250. (6.5A)

ANSWERS to Chapter 7 Exercises

SECTION 7.1   *pages 323–326*

1. The solutions are 0 and 4.   **3.** The solutions are 5 and −5.   **5.** The solutions are 3 and −2.   **7.** The solution is 3.   **9.** The solutions are 0 and 2.   **11.** The solutions are 5 and −2.   **13.** The solutions are 2 and 5.   **15.** The solutions are 6 and $-\frac{3}{2}$.   **17.** The solutions are 2 and $\frac{1}{4}$.   **19.** The solutions are $\frac{1}{3}$ and −4.   **21.** The solutions are $\frac{9}{2}$ and $-\frac{2}{3}$.   **23.** The solutions are $\frac{1}{4}$ and −4.   **25.** The solutions are 9 and −2.   **27.** The solutions are −2 and $-\frac{3}{4}$.   **29.** The solutions are 2 and −5.   **31.** The solutions are −4 and $-\frac{3}{2}$.   **33.** The solutions are $2b$ and $7b$.   **35.** The solutions are $7c$ and $-c$.   **37.** The solutions are $-\frac{b}{2}$ and $-b$.   **39.** The solutions are $4a$ and $\frac{2a}{3}$.   **41.** The solutions are $-\frac{a}{3}$ and $3a$.   **43.** The solutions are $-\frac{3y}{2}$ and $-\frac{y}{2}$.   **45.** The solutions are $-\frac{a}{2}$ and $-\frac{4a}{3}$.   **47.** $x^2 - 7x + 10 = 0$ **49.** $x^2 + 6x + 8 = 0$ **51.** $x^2 - 5x - 6 = 0$ **53.** $x^2 - 9 = 0$ **55.** $x^2 - 8x + 16 = 0$ **57.** $x^2 - 5x = 0$ **59.** $x^2 - 3x = 0$ **61.** $2x^2 - 7x + 3 = 0$ **63.** $4x^2 - 5x - 6 = 0$ **65.** $3x^2 + 11x + 10 = 0$ **67.** $9x^2 - 4 = 0$ **69.** $6x^2 - 5x + 1 = 0$ **71.** $10x^2 - 7x - 6 = 0$ **73.** $8x^2 + 6x + 1 = 0$ **75.** $50x^2 - 25x - 3 = 0$ **77.** The solutions are 7 and −7.   **79.** The solutions are $2i$ and $-2i$.   **81.** The solutions are 2 and −2.   **83.** The solutions are $\frac{9}{2}$ and $-\frac{9}{2}$.   **85.** The solutions are $7i$ and $-7i$.   **87.** The solutions are $4\sqrt{3}$ and $-4\sqrt{3}$.   **89.** The solutions are $5\sqrt{3}$ and $-5\sqrt{3}$.   **91.** The solutions are $3i\sqrt{2}$ and $-3i\sqrt{2}$.   **93.** The solutions are 7 and −5.   **95.** The solutions are 0 and −6.   **97.** The solutions are −7 and 3.   **99.** The solutions are 1 and 0.   **101.** The solutions are $-5 + \sqrt{6}$ and $-5 - \sqrt{6}$.   **103.** The solutions are $2 + 2\sqrt{6}$ and $2 - 2\sqrt{6}$.   **105.** The solutions are $-1 + 2i\sqrt{3}$ and $-1 - 2i\sqrt{3}$.   **107.** The solutions are $3 + 3i\sqrt{5}$ and $3 - 3i\sqrt{5}$.   **109.** The solutions are $\frac{-2 + 9\sqrt{2}}{3}$ and $\frac{-2 - 9\sqrt{2}}{3}$.   **111.** The solutions are $\frac{3 + 12\sqrt{3}}{4}$ and $\frac{3 - 12\sqrt{3}}{4}$.   **113.** The solutions are $-\frac{1}{2} + 2i\sqrt{10}$ and $-\frac{1}{2} - 2i\sqrt{10}$.

SECTION 7.2   *pages 331–332*

1. The solutions are 5 and −1.   **3.** The solutions are −9 and 1.   **5.** The solution is 3.   **7.** The solutions are $-2 + \sqrt{11}$ and $-2 - \sqrt{11}$.   **9.** The solutions are $3 + \sqrt{2}$ and $3 - \sqrt{2}$.   **11.** The solutions are $1 + i$ and $1 - i$.   **13.** The solutions are 8 and −3.   **15.** The solutions are 4 and −9.   **17.** The solutions are $\frac{3 + \sqrt{5}}{2}$ and $\frac{3 - \sqrt{5}}{2}$.   **19.** The solutions are $\frac{1 + \sqrt{5}}{2}$ and $\frac{1 - \sqrt{5}}{2}$.   **21.** The solutions are $3 + \sqrt{13}$ and $3 - \sqrt{13}$.   **23.** The solutions are 5 and 3.   **25.** The solutions are $2 + 3i$ and $2 - 3i$.   **27.** The solutions are $-3 + 2i$ and $-3 - 2i$.   **29.** The solutions are $1 + 3\sqrt{2}$ and $1 - 3\sqrt{2}$.   **31.** The solutions are $\frac{1 + \sqrt{17}}{2}$ and $\frac{1 - \sqrt{17}}{2}$.   **33.** The solutions are $1 + 2i\sqrt{3}$ and $1 - 2i\sqrt{3}$.   **35.** The solutions are −1 and $-\frac{1}{2}$.   **37.** The solutions are $\frac{1}{2}$ and $\frac{3}{2}$.   **39.** The solutions are $\frac{4}{3}$ and $-\frac{1}{2}$.   **41.** The solutions are $\frac{1}{2} + i$ and $\frac{1}{2} - i$.   **43.** The solutions are $\frac{1}{3} + \frac{1}{3}i$ and $\frac{1}{3} - \frac{1}{3}i$.   **45.** The solutions are $\frac{2 + \sqrt{14}}{2}$ and $\frac{2 - \sqrt{14}}{2}$.   **47.** The solutions are 1 and $-\frac{3}{2}$.   **49.** The solutions are $1 + \sqrt{5}$ and $1 - \sqrt{5}$.   **51.** The solutions are $\frac{1}{2}$ and 5.   **53.** The solutions are $2 + \sqrt{5}$ and $2 - \sqrt{5}$.   **55.** The solutions are $3 + i$ and $3 - i$.   **57.** The solutions are approximately 1.236 and −3.236.   **59.** The solutions are approximately 1.707 and 0.293.   **61.** The solutions are approximately 0.309 and −0.809.

SECTION 7.3   *pages 337–338*

1. The solutions are 5 and -2.   **3.** The solutions are 4 and -9.   **5.** The solutions are $4 + 2\sqrt{22}$ and $4 - 2\sqrt{22}$.   **7.** The solutions are 3 and -8.   **9.** The solutions are -3 and $\frac{1}{2}$.   **11.** The solutions are $\frac{3}{2}$ and $-\frac{1}{4}$.   **13.** The solutions are $-\frac{3}{2}$ and $\frac{2}{3}$.   **15.** The solutions are $2 + \sqrt{2}$ and $2 - \sqrt{2}$.   **17.** The solutions are $1 + 2\sqrt{3}$ and $1 - 2\sqrt{3}$.   **19.** The solutions are 2 and 12.   **21.** The solutions are $\frac{1 + \sqrt{3}}{2}$ and $\frac{1 - \sqrt{3}}{2}$.   **23.** The solutions are $\frac{4 + \sqrt{6}}{2}$ and $\frac{4 - \sqrt{6}}{2}$.   **25.** The solutions are $\frac{1 + \sqrt{2}}{2}$ and $\frac{1 - \sqrt{2}}{2}$.   **27.** The solutions are $\frac{3 + \sqrt{3}}{3}$ and $\frac{3 - \sqrt{3}}{3}$.   **29.** The solutions are $-1 + i$ and $-1 - i$.   **31.** The solutions are $1 + 2i$ and $1 - 2i$.   **33.** The solutions are $2 + 3i$ and $2 - 3i$.   **35.** The solutions are $\frac{1}{2} + \frac{3}{2}i$ and $\frac{1}{2} - \frac{3}{2}i$.   **37.** The solutions are $-\frac{3}{2} + \frac{1}{2}i$ and $-\frac{3}{2} - \frac{1}{2}i$.   **39.** The solutions are $\frac{3}{4} + \frac{3\sqrt{3}}{4}i$ and $\frac{3}{4} - \frac{3\sqrt{3}}{4}i$.   **41.** The solutions are $1 + \frac{3\sqrt{2}}{2}i$ and $1 - \frac{3\sqrt{2}}{2}i$.   **43.** The solutions are 4 and $-\frac{3}{2}$.   **45.** The solutions are $-\frac{4}{3}$ and 1.   **47.** The quadratic equation has two complex number solutions.   **49.** The quadratic equation has one real number solution.   **51.** The quadratic equation has two real number solutions.   **53.** The quadratic equation has two real number solutions.   **55.** The quadratic equation has two complex number solutions.   **57.** The solutions are approximately 7.606 and 0.394.   **59.** The solutions are approximately 0.236 and -4.236.   **61.** The solutions are approximately 1.851 and -1.351.

SECTION 7.4   *pages 343–344*

1. The solutions are 3, -3, 2, and -2.   **3.** The solutions are $\sqrt{2}$, $-\sqrt{2}$, 2, and -2.   **5.** The solutions are 1 and 4.   **7.** The solution is 16.   **9.** The solutions are $2i$, $-2i$, 1, and -1.   **11.** The solutions are $4i$, $-4i$, 2, and -2.   **13.** The solution is 16.   **15.** The solutions are 1 and 512.   **17.** The solutions are $\frac{2}{3}$, $-\frac{2}{3}$, 1, and -1.   **19.** The solution is 3.   **21.** The solution is 9.   **23.** The solutions are 2 and -1.   **25.** The solutions are 0 and 2.   **27.** The solutions are 2 and $-\frac{1}{2}$.   **29.** The solution is -2.   **31.** The solution is 1.   **33.** The solution is 1.   **35.** The solution is -3.   **37.** The solutions are 10 and -1.   **39.** The solutions are $-\frac{1}{2} + \frac{\sqrt{7}}{2}i$ and $-\frac{1}{2} - \frac{\sqrt{7}}{2}i$.   **41.** The solutions are 1 and -3.   **43.** The solutions are 0 and -1.   **45.** The solutions are $\frac{1}{2}$ and $-\frac{1}{3}$.   **47.** The solutions are $-\frac{2}{3}$ and 6.   **49.** The solutions are $\frac{4}{3}$ and 3.   **51.** The solutions are $-\frac{1}{4}$ and 3.   **53.** The solutions are $\frac{-5 - \sqrt{33}}{2}$ and $\frac{-5 + \sqrt{33}}{2}$.

SECTION 7.5   *pages 347–348*

1. $\{x \mid x < -2 \text{ or } x > 4\}$ **3.** $\{x \mid x \le 1 \text{ or } x \ge 2\}$

5. $\{x \mid -3 < x < 4\}$ **7.** $\{x \mid x < -2 \text{ or } 1 < x < 3\}$

9. $\{x \mid -4 \le x \le 1 \text{ or } x \ge 2\}$ **11.** $\{x \mid x < -2 \text{ or } x > 4\}$

13. $\{x \mid -1 < x \le 3\}$ **15.** $\{x \mid x \le -2 \text{ or } 1 \le x < 3\}$

17. $\{x \mid x > 4 \text{ or } x < -4\}$ **19.** $\{x \mid -3 \le x \le 12\}$ **21.** $\{x \mid x \ge 2 \text{ or } x \le \frac{1}{2}\}$ **23.** $\{x \mid \frac{1}{2} < x < \frac{3}{2}\}$

25. $\{x \mid x \le -3 \text{ or } 2 \le x \le 6\}$ **27.** $\{x \mid -\frac{3}{2} < x < \frac{1}{2} \text{ or } x > 4\}$ **29.** $\{x \mid x < -1 \text{ or } \frac{7}{2} < x < 5\}$

31. $\{x \mid x < -1 \text{ or } x > 2\}$ **33.** $\{x \mid -1 < x \le 0\}$ **35.** $\{x \mid x > 1 \text{ or } -2 < x \le 0\}$ **37.** $\{x \mid x > \frac{1}{2} \text{ or } x < 0\}$

39. $\{x \mid x < \frac{3}{2} \text{ or } x \ge 3\}$ **41.** $\{x \mid x > 2 \text{ or } -10 < x < -2\}$

SECTION 7.6   *pages 351–352*

1. The height is 3 cm. The base is 14 cm.   **3.** The length is 13 ft. The width is 5 ft.   **5.** The integers are 3 and 5 or -5 and -3.   **7.** The integer is 3 or 8.   **9.** The time for a projectile to return is 12.5 s.   **11.** The time for the larger pump to evacuate the chamber is 5 s.   **13.** The larger air conditioner would take 8 min. The smaller air conditioner would take 24 min.   **15.** The new sorter would take 14 min. The old sorter would take 35 min.   **17.** The rate of the cruise ship was 10 mph.   **19.** The rate of the wind is 20 mph.   **21.** The rowing rate in calm water is 6 mph.

CHAPTER REVIEW   *pages 355–356*

1. The solutions are 0 and $\frac{3}{2}$.   **2.** The solutions are $\frac{c}{2}$ and $-2c$.   **3.** $x^2 + 3x = 0$ **4.** $12x^2 - x - 6 = 0$

5. The solutions are $4\sqrt{3}$ and $-4\sqrt{3}$.   **6.** The solutions are $-\frac{1}{2} + 2i$ and $-\frac{1}{2} - 2i$.   **7.** The solutions are -3 and -1.   **8.** The solutions are $\frac{7 + 2\sqrt{7}}{7}$ and $\frac{7 - 2\sqrt{7}}{7}$.   **9.** The solutions are $1 + i\sqrt{7}$ and $1 - i\sqrt{7}$.   **10.** The solutions are $2i$ and $-2i$.   **11.** The solutions are $\frac{3}{4}$ and $\frac{4}{3}$.   **12.** The solutions are $\frac{1}{2} + \frac{\sqrt{31}}{2}i$ and $\frac{1}{2} - \frac{\sqrt{31}}{2}i$.

13. The solutions are $\frac{11 + \sqrt{73}}{6}$ and $\frac{11 - \sqrt{73}}{6}$.   **14.** The equation has two real number solutions.   **15.** The solutions are 27 and -64.   **16.** The solution is $\frac{5}{4}$.   **17.** The solution is 4.   **18.** The solution is 5.

19. The solutions are 3 and -1.   **20.** The solution is -1.   **21.** The solutions are $\frac{-3 + \sqrt{249}}{10}$ and $\frac{-3 - \sqrt{249}}{10}$.

22. The solutions are $\frac{-11 + \sqrt{129}}{2}$ and $\frac{-11 - \sqrt{129}}{2}$.   **23.** $\left\{x \mid -3 < x < \frac{5}{2}\right\}$ **24.** $\left\{x \mid x \le -4 \text{ or } -\frac{3}{2} \le x \le 2\right\}$

25. $\left\{x \mid x < \frac{3}{2} \text{ or } x \ge 2\right\}$ **26.** $\left\{x \mid x \le -3 \text{ or } \frac{1}{2} \le x < 4\right\}$

27. The length of the rectangle is 12 cm. The width is 5 cm.

28. The integers are 2, 4, and 6, or -6, -4, and -2.   **29.** The time for the new computer to complete the payroll is 12 min.   **30.** The speed of the first car is 40 mph. The speed of the second car is 50 mph.

CHAPTER TEST   *pages 357–358*

1. The solutions are $\frac{2}{3}$ and -4. (7.1A)   **2.** The solutions are $\frac{3}{2}$ and $-\frac{2}{3}$. (7.1A)   **3.** $x^2 - 9 = 0$ (7.1B)

4. $2x^2 + 7x - 4 = 0$ (7.1B)   **5.** The solutions are $2 + 2\sqrt{2}$ and $2 - 2\sqrt{2}$. (7.1C)   **6.** The solutions are $3 + \sqrt{11}$ and $3 - \sqrt{11}$. (7.2A)   **7.** The solutions are $\frac{3 + \sqrt{15}}{3}$ and $\frac{3 - \sqrt{15}}{3}$. (7.2A)   **8.** The solutions are $\frac{1 + \sqrt{3}}{2}$ and $\frac{1 - \sqrt{3}}{2}$. (7.3A)   **9.** The solutions are $-2 + 2i\sqrt{2}$ and $-2 - 2i\sqrt{2}$. (7.3A)   **10.** The quadratic equation has two real number solutions. (7.3A)   **11.** The quadratic equation has two complex number solutions. (7.3A)   **12.** The solution is $\frac{1}{4}$. (7.4A)   **13.** The solutions are $1, -1, \sqrt{3}$, and $-\sqrt{3}$. (7.4A)   **14.** The solution is 4. (7.4B)   **15.** The equation has no solution. (7.4B)

16. The solutions are 2 and -9. (7.4C)   **17.** $\{x \mid x < -4 \text{ or } 2 < x < 4\}$ (7.5A)

18. $\left\{x \mid -4 < x \le \frac{3}{2}\right\}$ (7.5A)   **19.** The base is 15 ft. The height is 4 ft. (7.6A)

20. The rowing rate of the canoe in calm water is 4 mph. (7.6A)

CUMULATIVE REVIEW   *pages 359–360*

1. 14 (1.3A)   **2.** The solution is −28. (2.1C)   **3.** The slope is $-\frac{3}{2}$. (6.2A)   **4.** The equation of the line is $y = x + 1$. (6.4A)   **5.** $-3xy(x^2 - 2xy + 3y^2)$ (3.3A)   **6.** $(2x - 5)(3x + 4)$ (3.3D)   **7.** $(x + y)(a^n - 2)$ (3.3B)   **8.** $x^2 - 3x - 4 - \frac{6}{3x - 4}$ (4.2A)   **9.** $\frac{x}{2}$ (4.1B)   **10.** The solution is $-\frac{1}{2}$. (4.6A)   **11.** $b = \frac{2S - an}{n}$ (4.6B)   **12.** $-8 - 14i$ (5.3C)   **13.** $1 - a$ (5.1A)   **14.** $\frac{x\sqrt[3]{4y^2}}{2y}$ (5.2D)   **15.** The solutions are $\frac{2}{3}$ and −3. (7.1A)   **16.** The solutions are $-3 + i$ and $-3 - i$. (7.3A)   **17.** The solutions are 2, −2, $\sqrt{2}$, and $-\sqrt{2}$. (7.4A)   **18.** The solutions are 0 and 1. (7.4B)   **19.** The solutions are $-\frac{3}{2}$ and −1. (7.4C)   **20.** $\left(\frac{5}{2}, 0\right)$ and $(0, -3)$ (6.2B)   **21.** lower limit: $9\frac{23}{64}$ in.; upper limit: $9\frac{25}{64}$ in. (2.3C)   **22.** The area of the triangle is $(x^2 + 6x - 16)$ ft^2. (3.2D)   **23.** $\{x \mid x < -3 \text{ or } 0 < x < 2\}$ (7.5A)

24. $\{x \mid -3 < x \le 1 \text{ or } x \ge 5\}$ (7.5A)   **25.** The quadratic equation has two complex number solutions. (7.3A)   **26.** The rate of the wind is 50 mph. (7.6A)

ANSWERS to Chapter 8 Exercises

SECTION 8.1   *pages 369–372*

1. $f(2) = 12$ **3.** $f(-1) = 3$ **5.** $f(a) = 3a^2$ **7.** $g(0) = 1$ **9.** $g(-2) = 7$ **11.** $g(t) = t^2 - t + 1$
13. $f(-2) = -11$ **15.** $f(2 + h) = 4h + 5$ **17.** $f(1 + h) - f(1) = 4h$ **19.** $g(2) = 3$ **21.** $g(1 + h) = h^2 + 2h$
23. $g(3 + h) - g(3) = h^2 + 6h$ **25.** $f(2) - g(2) = 6$ **27.** $f(0) + g(0) = 0$ **29.** $f(1 + h) - f(1) = 2h^2 + 4h$
31. $\frac{g(1 + h) - g(1)}{h} = -2$ **33.** $\frac{f(3 + h) - f(3)}{h} = 2h + 12$ **35.** $\frac{g(a + h) - g(a)}{h} = -2$ **37.** domain: {1, 2, 3, 4, 5}; range: {1, 4, 7, 10, 13}   **39.** domain: {0, 2, 4, 6}; range: {1, 2, 3, 4}   **41.** domain: {1, 3, 5, 7, 9}; range: {0}
43. domain: {−2, −1, 0, 1, 2}; range: {0, 1, 2}   **45.** {−3, 1, 5, 9, 13}   **47.** {1, 2, 3, 4, 5}   **49.** $\left\{\frac{2}{5}, \frac{1}{2}, \frac{2}{3}, 1, 2\right\}$
51. $\left\{-1, -\frac{1}{2}, 1\right\}$ **53.** −8; (−8, −3)   **55.** −1; (−1, −1)   **57.** 1; (1, 1)   **59.** ± 2; (2, 7) or (−2, 7)
61. 34; (34, 7)   **63.** None   **65.** 0 and $\frac{1}{3}$ **67.** −3 and 2   **69.** $\left\{x \mid x < \frac{1}{3}\right\}$ **71.** None

SECTION 8.2   *pages 381–386*

1.   **3.**   **5.**

7. 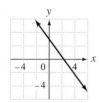   **9.**   **11.**

13.   **15.**   **17.**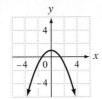

19. 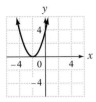   **21.**   **23.**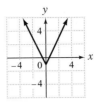

25.   **27.** 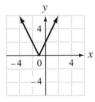   **29.**

31. 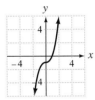   **33.**   **35.**

37. The set of ordered pairs is a function. The function is not 1–1.   **39.** The set of ordered pairs is not a function.   **41.** The set of ordered pairs is a 1–1 function.   **43.** The set of ordered pairs is not a function.
45. It is the graph of a function.   **47.** It is the graph of a function.   **49.** It is the graph of a function.
51. It is not the graph of a function.   **53.** It is the graph of a 1–1 function.   **55.** It is the graph of a 1–1 function.   **57.** It is not the graph of a 1–1 function.   **59.** It is not the graph of a 1–1 function.

SECTION 8.3   *pages 393–394*

1. -5 **3.** 11 **5.** $8x - 5$ **7.** 8 **9.** 10 **11.** $x + 8$ **13.** 7 **15.** 7 **17.** $x^2 - 4x + 7$
19. 7 **21.** 1 **23.** $9x^2 + 15x + 7$ **25.** $\{(0, 1), (3, 2), (8, 3), (15, 4)\}$ **27.** no inverse
29. $f^{-1}(x) = \frac{1}{4}x + 2$ **31.** no inverse   **33.** $f^{-1}(x) = x + 5$ **35.** $f^{-1}(x) = 3x - 6$ **37.** $f^{-1}(x) = -\frac{1}{3}x - 3$
39. $f^{-1}(x) = \frac{3}{2}x - 6$ **41.** $f^{-1}(x) = -3x + 3$ **43.** $f^{-1}(x) = \frac{1}{2}x + \frac{5}{2}$ **45.** no inverse   **47.** $f^{-1}(x) = \frac{1}{4}x + \frac{1}{2}$
49. $f^{-1}(x) = -\frac{1}{8}x + \frac{1}{2}$ **51.** $f^{-1}(x) = \frac{1}{8}x - \frac{3}{4}$ **53.** yes   **55.** no   **57.** yes   **59.** yes   **61.** range

---

SECTION 8.4   *pages 399–400*

1. The profit is $80,000.   **3.** The pressure is 6.75 lb/in.²   **5.** The distance to the horizon is 24.04 mi.
7. The ball will roll 54 ft in 3 s.   **9.** The length is 10 ft when the width is 4 ft.   **11.** The pressure will be
75 lb/in.²   **13.** The pressure will be 112.5 lb/in.²   **15.** The repulsive force will be 80 lb.
17. The resistance will be 56.25 ohms.   **19.** The force will be 180 lb.

CHAPTER REVIEW   *pages 403–406*

1. $f(-2) = -16$ **2.** $f(4 + h) - f(4) = h^2 + 9h$ **3.** $\dfrac{f(2 + h) - f(2)}{h} = 3h + 12$ **4.** domain: $\{3, 5, 7\}$; range: $\{8\}$

5. range: $\{0, 6, 12, 18\}$ **6.** -2 or 2 **7.** -1 and 3 **8.** $\{x \mid x < 3\}$ **9.** 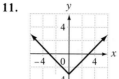 domain and range are all real numbers

10. 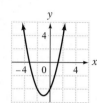 domain: all real numbers; range: $\{y \mid y \ge -5\}$ **11.** 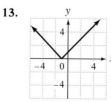 domain: all real numbers; range: $\{y \mid y \ge -3\}$

12. The graph is the graph of a function.   **13.** The set of ordered pairs is not a function.

14. The set of ordered pairs is a 1–1 function.   **15.** It is the graph of a 1–1 function.
16. It is not the graph of a 1–1 function.   **17.** It is not the graph of a 1–1 function.
18. The graph is not the graph of a function.   **19.** $12x^2 + 12x - 1$ **20.** 5 **21.** 10

22. $6x^2 + 3x - 16$ **23.** $f^{-1}(x) = 2x - 16$ **24.** $f^{-1}(x) = -\dfrac{1}{6}x + \dfrac{2}{3}$ **25.** $f^{-1}(x) = \dfrac{3}{2}x + 18$ **26.** yes

27. The pressure will be 40 lb.   **28.** The illumination 2 ft from the light will be 300 footcandles.
29. The resistance will be 0.4 ohm.   **30.** The pressure will be 50 lb/in.^{2}.

CHAPTER TEST   *pages 407–408*

1. $f(4) = 15$ (8.1A)   **2.** $g(-2) = -12$ (8.1A)   **3.** The set of ordered pairs is a function.   (8.1B)   **4.** 0 and 8   (8.1B)   **5.** $\{-11, -7, -3, 1, 5\}$ (8.1B)   **6.** $\{3, 5, 7, 9\}$ (8.1B)   **7.** $g(1 + h) = 1 - 2h$ (8.1A)

8. $\{x \mid x < 7\}$ (8.1B)   **9.** $-\dfrac{4}{3}$ (8.1B)   **10.** $g(0) + h(0) = 5$ (8.1A)   **11.**   (8.2A)

12. 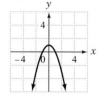   (8.2A)   **13.** 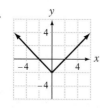 domain: all real numbers; range: $\{y \mid y \ge 0\}$ (8.2A)

14. 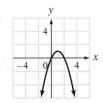 domain: all real numbers; range: $\{y \mid y \le 1\}$ (8.2A)   **15.** domain: all real numbers; range: $\{y \mid y \ge -2\}$ (8.2A)

16. 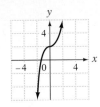   domain: all real numbers;   **17.** It is not the graph of a function.   (8.2B)
range: all real numbers   (8.2A)

18. It is the graph of a 1–1 function.   (8.2C)   **19.** $f(g(0)) = -1$ (8.3A)

20. $f^{-1}(x) = 2x + 2$ (8.3B)   **21.** $g(h(x)) = -3x + 14$ (8.3A)   **22.** no   (8.3B)   **23.** $f^{-1}(x) = \frac{1}{3}x - 2$ (8.3B)

24. The stopping distance is 61.2 ft.   (8.4A)   **25.** The current in the circuit is 2 amps.   (8.4A)

CUMULATIVE REVIEW   *pages 409–410*

1. $-\frac{23}{4}$ (1.3A)   **2.** 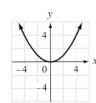   (1.4A)   **3.** 3   (2.1C)   **4.** $\{x \mid x < -2 \text{ or } x > 3\}$ (2.2B)

5. The set of all real numbers.   (2.3B)   **6.** $-\frac{a^{10}}{12b^4}$ (3.1C)   **7.** $2x^3 - 4x^2 - 17x + 4$ (3.2B)

8. $(a + 2)(a - 2)(a^2 + 2)$ (3.4D)   **9.** $xy(x - 2y)(x + 3y)$ (3.3D)   **10.** The solutions are -3 and 8.   (3.5A)

11. $\{x \mid x < -3 \text{ or } x > 5\}$ (7.5A)   **12.** $\frac{5}{2x - 1}$ (4.3B)   **13.** The solution is -2.   (4.6A)   **14.** $-3 - 2i$ (5.3D)

15.   (8.2A)   **16.**   (6.6A)   **17.** $y = -2x - 2$ (6.3B)

18. $y = -\frac{3}{2}x - \frac{7}{2}$ (6.4A)   **19.** $\frac{1}{2} + \frac{\sqrt{3}}{6}i$ and $\frac{1}{2} - \frac{\sqrt{3}}{6}i$ (7.3A)   **20.** 3   (7.4B)   **21.** $f(-2) = 5$ (8.1A)

22. $\{1, 2, 4, 5\}$ (8.1B)   **23.** The set of ordered pairs is a function.   (8.2B)   **24.** The solution is 2.   (5.4A)

25. $g(h(2)) = 10$ (8.3A)   **26.** $f^{-1}(x) = -\frac{1}{3}x + 3$ (8.3B)   **27.** The cost per pound is $3.96.   (2.5A)

28. There must be 25 lb of 80% copper alloy.   (2.6B)   **29.** There must be an additional 4.5 oz.   (4.5B)

30. It would take the larger pipe 4 min.   (7.6A)   **31.** The spring will stretch 24 in.   (8.4A)

32. The frequency in the pipe is 80 vibrations/min.   (8.4A)

ANSWERS to Chapter 9 Exercises

SECTION 9.1   *pages 421–426*

1.

Vertex: $(1, -5)$
Axis of symmetry: $x = 1$

3.

Vertex: $(1, -2)$
Axis of symmetry: $x = 1$

5.

Vertex: $(-4, -3)$
Axis of symmetry: $y = -3$

7.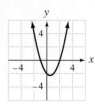

Vertex: $\left(\frac{1}{2}, -\frac{9}{4}\right)$
Axis of symmetry: $x = \frac{1}{2}$

9.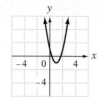

Vertex: $(1, -1)$
Axis of symmetry: $x = 1$

11.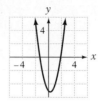

Vertex: $\left(\frac{1}{4}, -\frac{41}{8}\right)$
Axis of symmetry: $x = \frac{1}{4}$

13.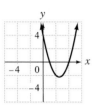

Vertex: $\left(\frac{5}{2}, -\frac{9}{4}\right)$
Axis of symmetry: $x = \frac{5}{2}$

15.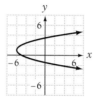

Vertex: $(-6, 1)$
Axis of symmetry: $y = 1$

17.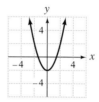

Vertex: $(0, -2)$
Axis of symmetry: $x = 0$

19.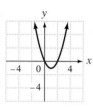

Vertex: $(1, -1)$
Axis of symmetry: $x = 1$

21.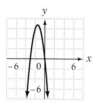

Vertex: $\left(-\frac{3}{2}, \frac{27}{4}\right)$
Axis of symmetry: $x = -\frac{3}{2}$

23.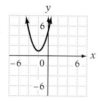

Vertex: $(-2, 1)$
Axis of symmetry: $x = -2$

25.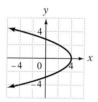

Vertex: $(4, 0)$
Axis of symmetry: $y = 0$

27.

Vertex: $\left(\frac{1}{2}, 1\right)$
Axis of symmetry: $y = 1$

29.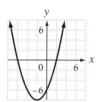

Vertex: $(-2, -8)$
Axis of symmetry: $x = -2$

31. The minimum value of the function is 2.   **33.** The maximum value of the function is -1.

35. The minimum value of the function is $-\frac{11}{4}$.   **37.** The maximum value of the function is $-\frac{2}{3}$.

39. The minimum value of the function is $-\frac{1}{12}$.   **41.** The x-intercepts are $(3, 0)$ and $(-3, 0)$.

43. The x-intercepts are $(0, 0)$ and $(-2, 0)$.   **45.** The x-intercepts are $(4, 0)$ and $(-2, 0)$.

47. The x-intercepts are $\left(-\frac{1}{2}, 0\right)$ and $(3, 0)$.   **49.** The x-intercepts are $(-2 + \sqrt{7}, 0)$ and $(-2 - \sqrt{7}, 0)$.

51. The parabola has no x-intercepts.   **53.** The x-intercepts are $(-1 + \sqrt{2}, 0)$ and $(-1 - \sqrt{2}, 0)$.

55. The parabola has no x-intercepts.   **57.** The parabola has two x-intercepts.

59. The parabola has one x-intercept.   **61.** The parabola has no x-intercepts.   **63.** The parabola has two x-intercepts.   **65.** The parabola has no x-intercepts.   **67.** The maximum height is 114 ft.   **69.** The pool will have the least amount of algae 5 days after treatment.   **71.** A price of \$250 will give the maximum revenue.   **73.** The numbers are 10 and 10.   **75.** The numbers are 12 and -12.   **77.** The dimensions are 15 ft by 15 ft. The maximum area is 225 ft².

SECTION 9.2 *pages 431–432*

1. $d = \sqrt{10}$ **3.** $d = \sqrt{34}$ **5.** $d = \sqrt{53}$ **7.** $d = 5$ **9.**

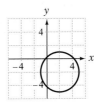

11.

13.

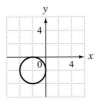

15. $(x - 2)^2 + (y + 1)^2 = 4$

17. $(x + 1)^2 + (y - 1)^2 = 5$ **19.** $(x - 1)^2 + (y + 2)^2 = 25$ **21.** $(x + 3)^2 + (y + 4)^2 = 16$

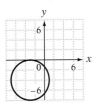

SECTION 9.3 *pages 437–438*

1.

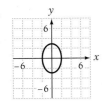

3.

5.

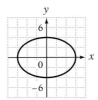

7.

9.

11.

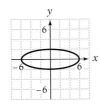

13.

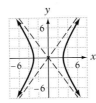

15.

17.

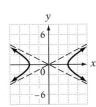

19.

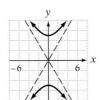

21.

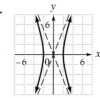

23.

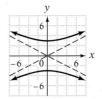

SECTION 9.4   *pages 441–442*

1.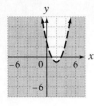

3.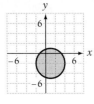

5.

7.

9.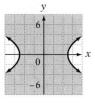

11.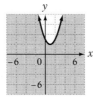

13.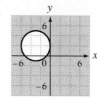

15.

17.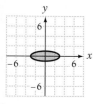

19.

21.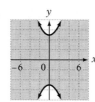

23.

CHAPTER REVIEW   *pages 445–446*

1. vertex: $(2, 4)$; axis of symmetry: $x = 2$ **2.** vertex: $\left(\frac{7}{2}, \frac{17}{4}\right)$; axis of symmetry: $x = \frac{7}{2}$

3.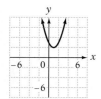

4.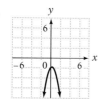

5.

6.

7. minimum value: $-\frac{17}{4}$ **8.** maximum value: 9   **9.** x-intercepts: $(0, 0)$ and $(-3, 0)$ **10.** There are no x-intercepts.   **11.** There are two x-intercepts.   **12.** There are no x-intercepts.   **13.** The maximum product is 100.   **14.** 7 ft by 7 ft   **15.** The distance is $5\sqrt{2}$.   **16.** The distance is $3\sqrt{5}$.

17.

18.

19. $(x + 1)^2 + (y - 5)^2 = 36$

20. $(x + 3)^2 + (y + 3)^2 = 16$ **21.** $x^2 + (y + 3)^2 = 97$ **22.** $(x + 2)^2 + (y - 1)^2 = 9$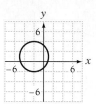

23.   **24.** 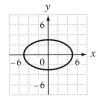   **25.**   **26.**

27. 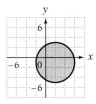   **28.** 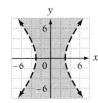   **29.** 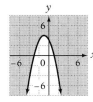   **30.**

CHAPTER TEST   *pages 447–448*

1. axis of symmetry: $x = 3$ (9.1A)   **2.** vertex: $\left(\frac{3}{2}, \frac{1}{4}\right)$ (9.1A)

3. 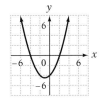   (9.1A)   **4.**   (9.1A)   **5.** The maximum value is 4.   (9.1B)

6. The minimum value is $-\frac{7}{2}$.   (9.1B)   **7.** The parabola has no x-intercepts.   (9.1C)

8. There are two x-intercepts.   (9.1C)   **9.** The numbers are 15 and -15.   (9.1D)

10. 12 cm by 12 cm   (9.1D)   **11.** $\sqrt{58}$ (9.2A)   **12.**   (9.2B)

13. $(x + 2)^2 + (y - 4)^2 = 9$ (9.2B)   **14.** $(x + 2)^2 + (y - 1)^2 = 32$ (9.2B)   **15.** $(x - 2)^2 + (y + 1)^2 = 4$ (9.2C)

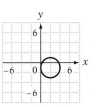

16. 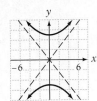 (9.3B) **17.** 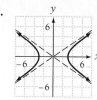 (9.3B) **18.** 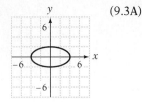 (9.3A)

19. 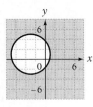 (9.4A) **20.** 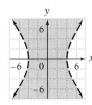 (9.4A)

CUMULATIVE REVIEW *pages 449–450*

1. (1.4B) **2.** $\frac{38}{53}$ (2.1C) **3.** $\left\{x\middle|-\frac{4}{3}<x<0\right\}$ (2.3A) **4.** $y=-\frac{3}{2}x$ (6.3A)

5. $y=x-6$ (6.4A) **6.** $x^{4n}+2x^{3n}-3x^{2n+1}$ (3.2A) **7.** $(x-y-1)(x^2-2x+1+xy-y+y^2)$ (3.4B)

8. $\{x|-4<x\le 3\}$ (7.5A) **9.** $\frac{x}{x+y}$ (4.1A) **10.** $\frac{x-5}{3x-2}$ (4.3B) **11.** The solution is $\frac{5}{2}$. (4.6A)

12. (6.6A) **13.** $\frac{4a}{3b^8}$ (3.1C) **14.** $2x^{3/4}$ (5.1B) **15.** $3\sqrt{2}-5i$ (5.3A)

16. The solutions are $\frac{-1+\sqrt{7}}{2}$ and $\frac{-1-\sqrt{7}}{2}$. (7.3A) **17.** The solutions are 27 and -1. (7.4A)

18. The solution is 6. (7.4B) **19.** $f(-3)=-20$ (8.1A) **20.** $f^{-1}(x)=\frac{1}{4}x-2$ (8.3B) **21.** maximum value: 0 (9.1B) **22.** The maximum product is 400. (9.1D) **23.** The distance is 5. (9.2A)

24. $(x+1)^2+(y-2)^2=17$ (9.2B) **25.** 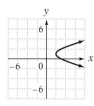 (9.1A) **26.** 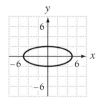 (9.3A)

27. 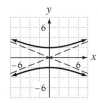 (9.3B) **28.** 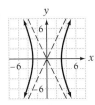 (9.3B)

29. The number of adult tickets sold was 82. (2.5A) **30.** The rate of the motorcycle is 60 mph. (4.6D)
31. The rowing rate in calm water is 4.5 mph. (4.6D) **32.** The gear will make 18 revolutions/min. (8.4A)

ANSWERS to Chapter 10 Exercises

SECTION 10.1   *pages 457–460*

1.
The solution
is (3, −1).

3.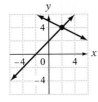
The solution
is (2, 4).

5.
The solution
is (4, 3).

7.
The solution
is (4, −1).

9.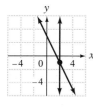
The solution
is (2, −1).

11.
The solution
is (3, −1).

13.
The solution
is (3, −2).

15.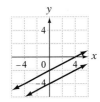
The lines are parallel
and therefore do not
intersect. The system
of equations has no solution.

17.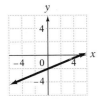
The equations are
dependent. The
solutions are the
ordered pairs
$\left(x, \frac{2}{5}x - 2\right)$.

19.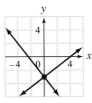
The solution
is (0, −3).

21.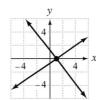
The solution
is (1, 0).

23.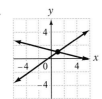
The solution
is (1, 1).

25. The solution is (2, 1).   **27.** The solution is (1, 1).   **29.** The solution is (16, 5).   **31.** The solution is (2, 1).   **33.** The solution is (−1, −7).   **35.** The solution is (−2, −3).   **37.** The solution is (3, −4).

39. The solution is (3, 3).   **41.** The solution is (0, −1).   **43.** The solution is $\left(\frac{1}{2}, 3\right)$.   **45.** The solution is (1, 2).   **47.** The solution is (−1, 2).   **49.** The system is inconsistent and therefore has no solution.

51. The solution is (−2, 5).   **53.** The solution is (0, 0).   **55.** The solutions are $\left(x, \frac{1}{2}x - 4\right)$.

57. The solution is (1, 5).   **59.** The solution is $\left(\frac{2}{3}, 3\right)$.   **61.** The solution is (5, 2).

63. The solution is (1, 4).   **65.** The solution is (−1, 6).

SECTION 10.2   *pages 467–470*

1. The solution is (6, 1).   **3.** The solution is (1, 1).   **5.** The solution is (2, 1).   **7.** The solution is (−2, 1).

9. The equations are dependent. The solutions are (x, 3x − 4).   **11.** The solution is $\left(-\frac{1}{2}, 2\right)$.   **13.** The system is inconsistent and therefore has no solution.   **15.** The solution is (−1, −2).   **17.** The solution is (−5, 4).   **19.** The solution is (2, 5).   **21.** The solution is $\left(\frac{1}{2}, \frac{3}{4}\right)$.   **23.** The solution is (0, 0).   **25.** The solution is (−1, 3).   **27.** The solution is $\left(\frac{2}{3}, -\frac{2}{3}\right)$.   **29.** The solution is (1, −1).   **31.** The equations are dependent. The solutions are $\left(x, \frac{4}{3}x - 6\right)$.   **33.** The solution is (5, 3).   **35.** The solution is $\left(\frac{1}{3}, -1\right)$.

37. The solution is $\left(\frac{5}{3}, \frac{1}{3}\right)$.   **39.** The system is inconsistent and therefore has no solution.   **41.** The solution is (1, −1).   **43.** The solution is (2, 1, 3).   **45.** The solution is (1, −1, 2).   **47.** The solution is (1, 2, 4). **49.** The solution is (2, −1, −2).   **51.** The solution is (−2, −1, 3).   **53.** The system is inconsistent and therefore has no solution.   **55.** The solution is (1, 4, 1).   **57.** The solution is (1, 3, 2).   **59.** The solution is (1, −1, 3).   **61.** The solution is (0, 2, 0).   **63.** The solution is (1, 5, 2).   **65.** The solution is (−2, 1, 1).

SECTION 10.3   *pages 477–478*

1. The value of the determinant is 11.   **3.** The value of the determinant is 18.   **5.** The value of the determinant is 0.   **7.** The value of the determinant is 15.   **9.** The value of the determinant is −30. **11.** The value of the determinant is 0.   **13.** The solution is (3, −4).   **15.** The solution is (4, −1).

17. The solution is $\left(\frac{11}{14}, \frac{17}{21}\right)$.   **19.** The solution is $\left(\frac{1}{2}, 1\right)$.   **21.** The system of equations is not independent.

23. The solution is (−1, 0).   **25.** The solution is (1, −1, 2).   **27.** The solution is (2, −2, 3).

29. The system of equations is not independent.   **31.** The solution is $\left(\frac{68}{25}, \frac{56}{25}, -\frac{8}{25}\right)$.

SECTION 10.4   *pages 483–486*

1. The rate of the boat is 15 mph. The rate of the current is 3 mph.   **3.** The rate of the plane is 502.5 mph. The rate of the wind is 47.5 mph.   **5.** The rate of the team is 8 km/h. The rate of the current is 2 km/h. **7.** The rate of the plane is 180 mph. The rate of the wind is 20 mph.   **9.** The cost of the cinnamon tea is $2.50/lb. The cost of the spice tea is $3/lb.   **11.** There are 30 nickels in the bank.   **13.** The rate of the plane is 210 mph. The rate of the wind is 30 mph.   **15.** The cost of the wool carpet is $26/yd.   **17.** The company plans to manufacture 25 ten-speed bicycles.   **19.** There are 480 g of the first alloy and 40 g of the second alloy.

SECTION 10.5   *pages 493–496*

1. The solutions are (−2, 5) and (5, 19).   **3.** The solutions are (−1, −2) and (2, 1).   **5.** The solution is (2, −2).   **7.** The solutions are (2, 2) and $\left(-\frac{2}{9}, -\frac{22}{9}\right)$.   **9.** The solutions are (3, 2) and (2, 3).   **11.** The solutions are $\left(\frac{\sqrt{3}}{2}, 3\right)$ and $\left(-\frac{\sqrt{3}}{2}, 3\right)$.   **13.** The system of equations has no real solution.   **15.** The solutions are (1, 2), (1, −2), (−1, 2), and (−1, −2).   **17.** The solutions are (3, 2), (3, −2), (−3, 2), and (−3, −2). **19.** The solutions are (2, 7) and (−2, 11).   **21.** The solutions are (3, $\sqrt{2}$), (3, $-\sqrt{2}$), (−3, $\sqrt{2}$), and (−3, $-\sqrt{2}$). **23.** The system of equations has no real solution.   **25.** The solutions are ($\sqrt{2}$, 3), ($\sqrt{2}$, −3), ($-\sqrt{2}$, 3), and ($-\sqrt{2}$, −3).   **27.** The solutions are (2, −1) and (8, 11).   **29.** The solution is (1, 0).   **31.** The solutions are (2, −1) and (1, −2).   **33.** The solutions are $\left(-\frac{1}{2}, \frac{9}{4}\right)$ and (−4, 4).   **35.** The solutions are (9, −2) and (4, 3).

37.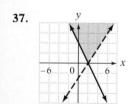

39.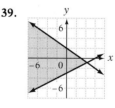

41.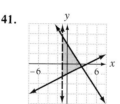

43.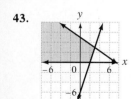

45.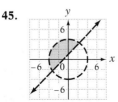

47.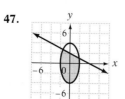

49.   **51.**   **53.**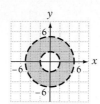

CHAPTER REVIEW   *pages 499–502*

1. 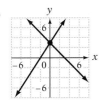   The solution is (0, 3).   **2.** 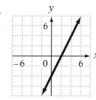   The solutions are $(x, 2x - 4)$.

3. The system is inconsistent and therefore has no solution.   **4.** The solutions are $\left(x, -\frac{1}{4}x + \frac{3}{2}\right)$.   **5.** The

solution is (−4, 7).   **6.** The solutions are $\left(x, \frac{1}{3}x - 2\right)$.   **7.** The solution is $\left(\frac{1}{4}, -3\right)$.   **8.** The solution is

(−1, 2, 3).   **9.** The solution is (5, −2, 3).   **10.** The solution is (3, −1, −2).   **11.** The value of the

determinant is 28.   **12.** The value of the determinant is 0.   **13.** The solution is (3, −1).   **14.** The

solution is $\left(\frac{110}{23}, \frac{25}{23}\right)$.   **15.** The solution is (−1, −3, 4).   **16.** The solution is (2, 3, −5).   **17.** The rate of the

cabin cruiser is 16 mph. The rate of the current is 4 mph.   **18.** The rate of the plane is 175 mph. The rate of

the wind is 25 mph.   **19.** There were 100 children who attended the movie.   **20.** The milk chocolate costs

\$5.50/lb. The semi-sweet chocolate costs \$4.50/lb.   **21.** The price per pencil is \$.20.   **22.** There are 14

one-dollar bills.   **23.** The solution is (−2, −12).   **24.** The solutions are (4, 2), (−4, 2), (4, −2), and (−4, −2).

25. The solutions are (5, 1), (−5, 1), (5, −1), and (−5, −1).   **26.** The solutions are $(\sqrt{5}, 3)$, $(-\sqrt{5}, 3)$, $(\sqrt{5}, -3)$,

and $(-\sqrt{5}, -3)$.

27.   **28.**   **29.**   **30.**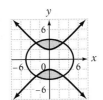

CHAPTER TEST   *pages 503–504*

1. 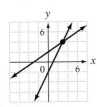   The solution is (3, 4).   (10.1A)   **2.**   The solution is (−5, 0).   (10.1A)

3. 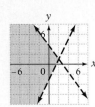 (10.5B) **4.** 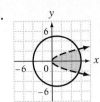 (10.5B) **5.** The solution is $\left(\frac{3}{4}, \frac{7}{8}\right)$. (10.1B)

6. The solution is $(-3, -4)$. (10.1B) **7.** The solution is $(2, -1)$. (10.1B) **8.** The solution is $(-2, 1)$. (10.2A) **9.** The system is inconsistent and therefore has no solution. (10.2A) **10.** The solution is $(1, 1)$. (10.2A) **11.** The system is inconsistent and therefore has no solution. (10.2B) **12.** The solution is $(2, -1, -2)$. (10.2B) **13.** The value of the determinant is 10. (10.3A) **14.** The value of the determinant is -32. (10.3A) **15.** The solution is $\left(\frac{59}{19}, -\frac{62}{19}\right)$. (10.3B) **16.** The solution is $(0, -2, 3)$. (10.3B)

17. The solutions are $(4, -1)$ and $\left(-\frac{7}{2}, -\frac{19}{4}\right)$. (10.5A) **18.** The solutions are $(2, 2), (2, -2), (-2, 2)$, and $(-2, -2)$. (10.5A) **19.** The rate of the plane is 150 mph. The rate of the wind is 25 mph. (10.4A)
20. The cost of the cotton is \$4.50/yd. The cost of the wool is \$7/yd. (10.4B)

CUMULATIVE REVIEW *pages 505–506*

1. The solution is $-\frac{11}{28}$. (2.1C) **2.** The equation of the line is $y = 5x - 11$. (6.3B)

3. $(2x - 5)(3x - 2)$ (3.3D) **4.** $\frac{2x^2 - x + 6}{(x-3)(x-2)(x+1)}$ (4.3B) **5.** The solution is -3. (4.6A) **6.** $\frac{1+b}{a}$ (3.1C)
7. $-2ab^2\sqrt[3]{ab^2}$ (5.2C) **8.** The solution is -3. (5.4A) **9.** The solutions are $\frac{1}{2}$ and -5. (7.1A) **10.** The solutions are $\frac{1 + \sqrt{7}}{3}$ and $\frac{1 - \sqrt{7}}{3}$. (7.3A) **11.** 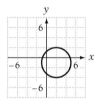 (8.2A) **12.** $(x - 2)^2 + (y + 1)^2 = 9$ (9.2C)

13. $d = 5\sqrt{2}$ (9.2A) **14.** The parabola has no x-intercepts. (9.1C)

15. (10.1A) **16.** The solution is $(-5, -11)$. (10.1B)

The solution is $(2, 0)$.

17. The solution is $(1, 0, -1)$. (10.2A) **18.** The value of the determinant is 3. (10.3A) **19.** The solution is $(2, -3)$. (10.3B) **20.** The solutions are $(1, 2)$ and $(1, -2)$. (10.5A) **21.** There are 16 nickels in the coin purse. (2.4B) **22.** 60 ml of pure water must be used. (2.6B) **23.** The base is 6 m. (7.6A)
24. A force of 40 lb will stretch the spring 24 in. (8.4A) **25.** The rate of the wind is 12.5 mph. (8.4A)
26. One pound of steak costs \$3.00. (10.4B)

ANSWERS To Chapter 11 Exercises

SECTION 11.1   *pages 515–518*

1. $f(2) = 9$ **3.** $f(-2) = \frac{1}{9}$ **5.** $f(0) = 1$ **7.** $f(3) = 16$ **9.** $f(-3) = \frac{1}{4}$ **11.** $f(-1) = 1$ **13.** $f(0) = 1$

15. $f(-2) = 16$ **17.** $f\left(\frac{3}{2}\right) = \frac{1}{8}$ **19.** $f(2) = \frac{1}{3}$ **21.** $f(-1) = 9$ **23.** $f(0) = 3$ **25.** $f(1) = 2$

27. $f(2) = 16$ **29.** $f(-1) = 2$ **31.**   **33.**

35. 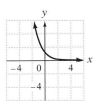   **37.** 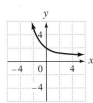   **39.** 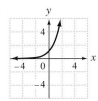   **41.**

43. $\log_2 32 = 5$ **45.** $\log_5 25 = 2$ **47.** $\log_4 \frac{1}{16} = -2$ **49.** $\log_{1/2}\frac{1}{4} = 2$ **51.** $\log_3 1 = 0$ **53.** $\log_a w = x$

55. $3^2 = 9$ **57.** $4^1 = 4$ **59.** $10^0 = 1$ **61.** $10^{-2} = 0.01$ **63.** $\left(\frac{1}{3}\right)^2 = \frac{1}{9}$ **65.** $b^v = u$ **67.** $\log_4 16 = 2$

69. $\log_2 32 = 5$ **71.** $\log_{10} 100 = 2$ **73.** $\log_6 216 = 3$ **75.** $\log_8 1 = 0$ **77.** $\log_5 625 = 4$

79. The solution is 9.   **81.** The solution is 64.   **83.** The solution is $\frac{1}{7}$.   **85.** The solution is 216.

87. The solution is 1.   **89.** The solution is $\frac{1}{27}$.   **91.** 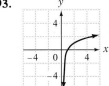   **93.**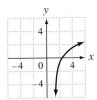

95. 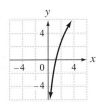   **97.** 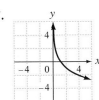   **99.** 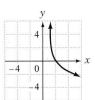   **101.**

SECTION 11.2   *pages 523–526*

1. $\log_8 x + \log_8 z$ **3.** $5\log_3 x$ **5.** $\log_b r - \log_b s$ **7.** $2\log_3 x + 6\log_3 y$ **9.** $3\log_7 u - 4\log_7 v$

11. $2\log_2 r + 2\log_2 s$ **13.** $2\log_9 x + \log_9 y + \log_9 z$ **15.** $\log_5 x + 2\log_5 y - 4\log_5 z$ **17.** $2\log_8 x - \log_8 y - 2\log_8 z$

19. $\frac{1}{2}\log_7 x + \frac{1}{2}\log_7 y$ **21.** $\frac{1}{2}\log_2 x - \frac{1}{2}\log_2 y$ **23.** $\frac{3}{2}\log_4 x + \frac{1}{2}\log_4 y$ **25.** $\frac{3}{2}\log_7 x - \frac{1}{2}\log_7 y$

27. $\log_b x + \frac{1}{2}\log_b y - \frac{1}{2}\log_b z$ **29.** $\log_3 t - \frac{1}{2}\log_3 x$ **31.** $\log_3 \frac{x^3}{y}$ **33.** $\log_8 x^4 y^2$ **35.** $\log_7 x^3$

37. $\log_5 x^3 y^4$ **39.** $\log_4 \frac{1}{x^2}$ **41.** $\log_3 \frac{x^2 z^2}{y}$ **43.** $\log_b \frac{x}{y^2 z}$ **45.** $\log_4 x^2 y^2$ **47.** $\log_6 \sqrt{\frac{x}{y}}$

49. $\log_4 \frac{s^2 r^2}{t^4}$ **51.** $\log_5 \frac{x}{y^2 z^2}$ **53.** $\log_2 \frac{t^3 v^2}{r^2}$ **55.** $\log_4 \sqrt{\frac{x^3 z}{y^2}}$

SECTION 11.3   *pages 535–536*

1. 0.7686   **3.** 1.5378   **5.** 2.5899   **7.** 3.0128   **9.** 4.6749   **11.** 9.0569 − 10   **13.** 7.3010 − 10
15. 7.2430 − 10   **17.** 9.9930 − 10   **19.** 5.49   **21.** 34.3   **23.** 7970   **25.** 632   **27.** 0.455
29. 0.0245   **31.** 0.00742   **33.** 79.9   **35.** 0.141   **37.** 0.5849   **39.** 1.9270   **41.** 2.6984
43. 8.8495 − 10   **45.** 4.8804   **47.** 9.9247 − 10   **49.** 7.954   **51.** 544.5   **53.** 0.5505   **55.** 398.4
57. 0.02036   **59.** 44.93   **61.** 316.8   **63.** 0.2749   **65.** 2.948   **67.** 16.73   **69.** 658.4
71. 32.65   **73.** 12.32   **75.** 0.01089   **77.** 0.8958

SECTION 11.4   *pages 541–542*

1. The solution is $-\frac{1}{3}$.   **3.** The solution is −3.   **5.** The solution is 1.1133.   **7.** The solution is 0.7211.

9. The solution is −1.5850.   **11.** The solution is 1.7093.   **13.** The solution is 1.3222.

15. The solution is −2.8076.   **17.** The solution is 3.5854.   **19.** The solution is 1.1310.   **21.** The solution is 1.

23. The solution is 6.   **25.** The solution is $\frac{4}{7}$.   **27.** The solution is $\frac{9}{8}$.   **29.** The solution is 8.

31. The solution is $\frac{11}{2}$.   **33.** The solutions are 2 and −4.   **35.** The solution is $\frac{5}{3}$.   **37.** The solution is $\frac{1}{2}$.

39. The solution is 5.   **41.** The solution is 2.   **43.** The solution is 2.   **45.** The equation has no solution.

47. The solution is 2.   **49.** $\log_2 9 = 3.1701$ **51.** $\log_5 10 = 1.4306$ **53.** $\log_7 6 = 0.9208$
55. $\log_{12} 9 = 0.8842$ **57.** $\log_4 8 = 1.4999$ **59.** $\log_6 28 = 1.8597$ **61.** $\log_5 9.2 - 1.3788$

63. $\log_8 5 = 0.7740$ **65.** $\log_5(0.6) = -0.3173$ **67.** $\log_6(2.3) = 0.4648$

SECTION 11.5   *pages 547–550*

1. The value of the investment is $5982 after two years.   **3.** The investment should double in 7 years.
5. The company must deposit $16,786.   **7.** It will take 936 years to decay to 10 mg.   **9.** The half-life of
the material is 16 h.   **11.** It will take 6 weeks of practice to make 80% of the welds correctly.   **13.** The pH
of the solution is 8.1.   **15.** 79% of the light will pass through the glass.   **17.** The earthquake in China was
20 times stronger.   **19.** There are 65 decibels in normal conversation.

CHAPTER REVIEW   *pages 553–554*

1. $f(2) = 1$ **2.** $f(-3) = \frac{3}{2}$ **3.** 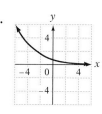   **4.** 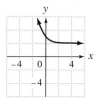   **5.** $\log_2 32 = 5$ **6.** $5^2 = 25$

7. The solution is 32.   **8.** The solution is 1000.   **9.** 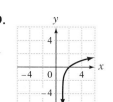   **10.**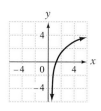

11. $\frac{2}{5}\log_3 x + \frac{4}{5}\log_3 y$ **12.** $\frac{1}{2}(5\log_8 x - 3\log_8 y)$ **13.** $\log_3 \frac{x^2}{y^5}$ **14.** $\log_7 \sqrt[3]{xy^4}$ **15.** 5.7536
16. 0.00514   **17.** 1.35×10^{-6} **18.** 5.923   **19.** The solution is −12.   **20.** The solution is −0.5350.
21. The solution is 79.   **22.** The solution is $\frac{1}{4}$.   **23.** The solution is $\frac{2}{3}$.   **24.** The solution is 2.
25. 1.7251   **26.** 0.4278   **27.** The value of the investment after two years is $4692.   **28.** The earthquake
is 1000 times stronger.   **29.** The half-life of the material is 27 days.   **30.** There are 107 decibels.

CHAPTER TEST   *pages 555–556*

1. $f(0) = 1$ (11.1A)   **2.** $f(-2) = \frac{1}{3}$ (11.1A)   **3.** (11.1B)   **4.** (11.1B)

5. $\log_4 16 = 2$ (11.1C)   **6.** The solution is $\frac{1}{9}$.   (11.1C)

7. (11.1D)   **8.** 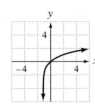 (11.1D)   **9.** $\frac{1}{2}(\log_6 x + 3\log_6 y)$ (11.2A)

10. $\log_3 \sqrt{\frac{x}{y}}$ (11.2A)   **11.** $7.5647 - 10$ (11.3A)   **12.** 0.0404 (11.3B)   **13.** 711.7 (11.3C)

14. 0.06433 (11.3D)   **15.** 0.7988 (11.3D)   **16.** The solution is -2.   (11.4A)   **17.** The solution is -3.   (11.4A)   **18.** The solution is 6.   (11.4B)   **19.** $\log_3 19 = 2.6804$ (11.4C)   **20.** The half-life is 33 h.   (11.5A)

CUMULATIVE REVIEW   *pages 557–558*

1. The solution is $\frac{8}{7}$.   (2.1C)   **2.** The equation of the line is $y = 2x - 6$.   (6.4A)

3. $(4x^n + 3)(x^n + 1)$ (3.4C)   **4.** $\frac{x-3}{x+3}$ (4.4A)   **5.** $\frac{x\sqrt{y} + y\sqrt{x}}{x - y}$ (5.2D)

6. The solutions are $2 + \sqrt{10}$ and $2 - \sqrt{10}$.   (7.2A)   **7.** $3x^2 + 8x - 3 = 0$ (7.1B)

8. 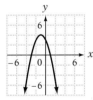 (9.1A)   **9.** The solution is $(0, -1, 2)$.   (10.2B)

10. The solutions are $(2, 3)$ and $(-1, -3)$.   (10.5A)   **11.** $\{x | -5 \le x \le 1\}$ (7.5A)   **12.** $\{x | 1 \le x \le 4\}$ (2.3B)

13. 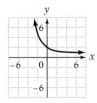 (11.1C)   **14.** (11.1D)

15. $f(-3) = 4$ (11.1A)   **16.** The solution is 125.   (11.1C)

17. $\log_b \frac{x^3}{y^5}$ (11.2A)   **18.** 0.719 (11.3D)   **19.** The solution is 2.   (11.4A)   **20.** The solution is $\frac{1}{2}$.   (11.4B)

21. The customer can write 49 checks.   (2.2C)   **22.** The mixture costs \$3.10/lb.   (2.5A)   **23.** The rate of the wind is 25 mph.   (4.6D)   **24.** The force will stretch the spring 10.2 in.   (8.4A)   **25.** The redwood costs 40¢ per foot. The fir costs 25¢ per foot.   (10.4B)   **26.** The investment should double in 8 years.   (11.5A)

ANSWERS to Chapter 12 Exercises

SECTION 12.1 *pages 565–566*

1. The first four terms are 2, 3, 4, and 5. **3.** The first four terms are 3, 5, 7, and 9. **5.** The first four terms are 0, −2, −4, and −6. **7.** The first four terms are 2, 4, 8, and 16. **9.** The first four terms are 2, 5, 10, and 17. **11.** The first four terms are $\frac{1}{2}, \frac{2}{5}, \frac{3}{10}$, and $\frac{4}{17}$. **13.** The first four terms are $0, \frac{3}{2}, \frac{8}{3}$, and $\frac{15}{4}$. **15.** The first four terms are 1, −2, 3, and −4. **17.** The tenth term is 15. **19.** The twelfth term is $\frac{12}{13}$. **21.** The twenty-fifth term is 24. **23.** The fifth term is $\frac{32}{243}$. **25.** The seventh term is 88. **27.** The sixteenth term is $\frac{1}{20}$. **29.** 42 **31.** 28 **33.** 60 **35.** $\frac{25}{24}$ **37.** 28 **39.** $\frac{99}{20}$ **41.** $\frac{137}{120}$ **43.** −2 **45.** $-\frac{7}{12}$ **47.** $\frac{2}{x} + \frac{4}{x} + \frac{6}{x} + \frac{8}{x}$ **49.** $\frac{x}{2} + \frac{x^2}{3} + \frac{x^3}{4} + \frac{x^4}{5}$ **51.** $\frac{x^2}{3} + \frac{x^3}{5} + \frac{x^4}{7}$ **53.** $x + x^3 + x^5 + x^7$ **55.** $2x + 4x^2 + 6x^3 + 8x^4$ **57.** $\frac{1}{x} + \frac{1}{x^2} + \frac{1}{x^3} + \frac{1}{x^4} + \frac{1}{x^5}$

SECTION 12.2 *pages 571–572*

1. $a_{15} = 141$ **3.** $a_{15} = 50$ **5.** $a_{31} = 17$ **7.** $a_{12} = 12\frac{1}{2}$ **9.** $a_{40} = -109$ **11.** $a_{10} = 3.75$ **13.** $a_n = n$ **15.** $a_n = -4n + 10$ **17.** $a_n = \frac{3n+1}{2}$ **19.** $a_n = -5n - 3$ **21.** $a_n = -10n + 36$ **23.** There are 20 terms in the sequence. **25.** There are 20 terms in the sequence. **27.** There are 13 terms in the sequence. **29.** There are 20 terms in the sequence. **31.** 2500 **33.** −760 **35.** 189 **37.** 345 **39.** $\frac{187}{2}$ **41.** −180 **43.** The object will fall 576 ft. **45.** The total number of cans in the display is 108 cans. **47.** There are 4077 seats in the loge seating section.

SECTION 12.3 *pages 581–582*

1. $a_9 = 131{,}072$ **3.** $a_7 = \frac{128}{243}$ **5.** $a_{10} = 32$ **7.** $a_9 = 16$ **9.** $a_{11} = -8\sqrt{2}$ **11.** $a_2 = 6, a_3 = \frac{9}{2}$ **13.** $a_2 = -12, a_3 = 24$ **15.** $a_2 = 25, a_3 = 125$ **17.** −2188 **19.** $189 + 189\sqrt{2}$ **21.** $\frac{1995}{64}$ **23.** $\frac{1330}{729}$ **25.** $\frac{16}{9}$ **27.** $\frac{1}{9}$ **29.** $\frac{5}{99}$ **31.** An equivalent fraction is $\frac{5}{9}$. **33.** An equivalent fraction is 1. **35.** An equivalent fraction is $\frac{2}{11}$. **37.** An equivalent fraction is $\frac{5}{6}$. **39.** The total distance the pendulum travels is 59 in. **41.** The temperature of the spa is 102.1°F. **43.** The total amount earned is $10,737,418.23.

SECTION 12.4 *pages 587–588*

1. 6 **3.** 40,320 **5.** 1 **7.** 10 **9.** 1 **11.** 84 **13.** 21 **15.** 45 **17.** 1 **19.** 20 **21.** 11 **23.** 6 **25.** $x^4 + 4x^3y + 6x^2y^2 + 4xy^3 + y^4$ **27.** $x^5 - 5x^4y + 10x^3y^2 - 10x^2y^3 + 5xy^4 - y^5$ **29.** $16m^4 + 32m^3 + 24m^2 + 8m + 1$ **31.** $32r^5 - 240r^4 + 720r^3 - 1080r^2 + 810r - 243$ **33.** $a^{10} + 10a^9b + 45a^8b^2$ **35.** $a^{11} - 11a^{10}b + 55a^9b^2$ **37.** $256x^8 + 1024x^7y + 1792x^6y^2$ **39.** $65{,}536x^8 - 393{,}216x^7y + 1{,}032{,}192x^6y^2$ **41.** $x^7 + 7x^5 + 21x^3$ **43.** $x^{10} + 15x^8 + 90x^6$ **45.** $-560x^4$ **47.** $-6x^{10}y^2$ **49.** $126y^5$ **51.** $5n^3$ **53.** $\frac{x^5}{32}$

CHAPTER REVIEW *pages 591–592*

1. $a_{10} = 28$ **2.** $a_5 = -\frac{5}{27}$ **3.** 35   **4.** $-\frac{13}{60}$ **5.** $2 + 2x + 2x^2 + 2x^3$ **6.** $2x + 2x^2 + \frac{8x^3}{3} + 4x^4 + \frac{32x^5}{5}$

7. The 20th term is 77.   **8.** $a_n = 5n - 12$ **9.** There are 22 terms.   **10.** The sum is 2000.   **11.** The sum is 1830.   **12.** The total salary is $12,240.   **13.** $a_8 = \frac{1}{729}$ **14.** $a_{12} = 234\sqrt{3}$ **15.** The sum is 1127.

16. The sum is 726.   **17.** The sum is 0.996.   **18.** The sum is 6.   **19.** The sum is $\frac{1}{3}$.

20. The sum is $\frac{16}{5}$.   **21.** $\frac{23}{99}$ **22.** $\frac{19}{30}$ **23.** The temperature of the spa is 67.7°F.

24. The amount of radioactive material is 3.125 mg.   **25.** 99   **26.** 220   **27.** 84

28. $x^5 - 15x^4y^2 + 90x^3y^4 - 270x^2y^6 + 405xy^8 - 243y^{10}$ **29.** $5670x^4y^4$ **30.** $-42{,}240x^4y^7$

CHAPTER TEST *pages 593–594*

1. $a_{14} = \frac{1}{2}$ (12.1A)   **2.** $a_6 = \frac{7}{6}, a_7 = \frac{8}{7}$ (12.1A)   **3.** 34 (12.1B)   **4.** $3x + 3x^2 + 3x^3 + 3x^4$ (12.1B)

5. $a_{35} = -115$ (12.2A)   **6.** $a_n = -3n + 15$ (12.2A)   **7.** There are 16 terms in the sequence. (12.2A)

8. 468 (12.2B)   **9.** 1575 (12.2B)   **10.** There were 2550 yd of material in stock after the shipment on October 1. (12.2C)   **11.** $a_7 = 32$ (12.3A)   **12.** $a_5 = \frac{2}{27}$ (12.3A)   **13.** $\frac{665}{32}$ (12.3B)   **14.** -66 (12.3B)

15. 16 (12.3B)   **16.** An equivalent fraction is $\frac{7}{30}$. (12.3C)   **17.** There will be 20 mg of radioactive material in the sample at the beginning of the fifth day. (12.3D)   **18.** 70 (12.4A)   **19.** 84 (12.4A)

20. $-280x^4y^3$ (12.4A)

CUMULATIVE REVIEW *pages 595–596*

1.   (6.1D)   **2.** $2(x^2 + 2)(x^4 - 2x^2 + 4)$ (3.4D)   **3.** $\frac{x^2 + 5x - 2}{(x + 2)(x - 1)}$ (4.3B)   **4.** $\frac{1}{x^{26}}$ (5.1A)

5. $4y\sqrt{x} - y\sqrt{2}$ (5.2C)   **6.** The solutions are $\frac{1}{4} + \frac{\sqrt{55}}{4}i$ and $\frac{1}{4} - \frac{\sqrt{55}}{4}i$. (7.3A)   **7.** The solution is 4. (5.4A)

8. The equation of the circle is $(x + 1)^2 + (y + 1)^2 = 34$ (9.2B)   **9.** The solution is $\left(\frac{7}{6}, \frac{1}{2}\right)$. (10.2A)   **10.** The value of the determinant is -10. (10.3A)   **11.** $\{x \mid x < -2 \text{ or } x > 2\}$ (2.2B)   **12.** 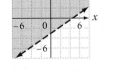 (6.6A)

13. $\frac{1}{2}\log_5 x - \frac{1}{2}\log_5 y$ (11.2A)   **14.** The solution is 3. (11.4A)   **15.** The fifth term is 20. The sixth term is 30. (12.1A)   **16.** 6 (12.1B)   **17.** The 33rd term is -103. (12.2A)   **18.** $\frac{9}{5}$ (12.3C)   **19.** An equivalent fraction is $\frac{7}{15}$. (12.3C)   **20.** The sixth term is $12xy^5$. (12.4A)   **21.** There must be 120 oz of pure water added. (2.6B)   **22.** The new computer takes 24 min. The old computer takes 40 min. (4.6C)

23. The rate of the boat is 6.25 mph. The rate of the current is 1.25 mph. (4.6D)   **24.** The half-life of the sample is 55 days. (11.5A)   **25.** There are 1536 seats. (12.2C)   **26.** The ball bounces 2.6 ft on the fifth bounce. (12.3D)

FINAL EXAM *pages 597–602*

1. -31 (1.2B)   **2.** -1 (1.3A)   **3.** $33 - 10x$ (1.3C)   **4.** The solution is 8. (2.1B)   **5.** The solution is $\frac{2}{3}$. (2.1C)   **6.** The solutions are 4 and $-\frac{2}{3}$. (2.3A)   **7.**

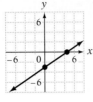

(6.2B)

x-intercept: $\left(\frac{9}{2}, 0\right)$

y-intercept: $(0, -3)$

8. The equation of the line is $y = -3x + 7$. (6.3B)   **9.** The equation of the line is $y = -\frac{2}{3}x - \frac{1}{3}$. (6.4A)

10. $6a^3 - 5a^2 + 10a$ (3.2A)   **11.** $(2 - xy)(4 + 2xy + x^2y^2)$ (3.4B)   **12.** $(x - y)(1 + x)(1 - x)$ (3.4D)

13. $x^2 - 2x - 3 - \frac{5}{2x - 3}$ (4.2A)   **14.** $\frac{x(x - 1)}{2x - 5}$ (4.1C)   **15.** $\frac{-10x}{(x + 2)(x - 3)}$ (4.3B)   **16.** $\frac{x + 3}{x + 1}$ (4.4A)

17. The solution is $-\frac{7}{4}$. (4.6A)   **18.** $d = \frac{a_n - a_1}{n - 1}$ (4.6B)   **19.** $\frac{y^4}{162x^3}$ (3.1C)   **20.** $\frac{1}{64x^8y^5}$ (5.1A)

21. $-2x^2y\sqrt{2y}$ (5.2B)   **22.** $\frac{x^2\sqrt{2y}}{2y^2}$ (5.2D)   **23.** $\frac{6}{5} - \frac{3}{5}i$ (5.3D)   **24.** $2x^2 - 3x - 2 = 0$ (7.1B)

25. The solutions are $\frac{3 + \sqrt{17}}{4}$ and $\frac{3 - \sqrt{17}}{4}$. (7.3A)   **26.** The solutions are 27 and -8. (7.4A)

27. The solutions are $\frac{3}{2}$ and -2. (7.4C)

28.

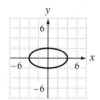

(9.1A)   **29.**

(9.3A)

30. The inverse function is $f^{-1}(x) = \frac{3}{2}x + 6$. (8.3B)   **31.** The solution is $(3, 4)$. (10.2A)

32. The value of the determinant is 10. (10.3A)   **33.** The solution is $\left(\frac{5}{2}, -\frac{3}{2}\right)$. (10.5A)

34. $\left\{x \mid x > \frac{3}{2}\right\}$ (2.2B)   **35.** $\{x \mid -4 < x < -1\}$ (2.3B)

36.

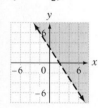

(6.6A)   **37.**

(11.1D)   **38.** $\log_2 \frac{a^2}{b^2}$ (11.2A)

39. The solution is 6. (11.4B)   **40.** $2y + 2y^2 + 2y^3 + 2y^4 + 2y^5$ (12.1B)

41. An equivalent fraction is $\frac{23}{45}$. (12.3C)   **42.** The third term is $144x^7y^2$. (12.4A)

43. The range of scores is $69 \le x \le 100$. (2.2C)   **44.** The cyclist rode 40 mi. (2.5B)
45. There is $8000 invested at 8.5% and $4000 invested at 6.4%. (2.6A)   **46.** The length is 20 ft. The width is 7 ft. (7.6A)   **47.** An additional 200 shares are needed. (8.4A)   **48.** The rate of the plane was 420 mph. (10.4B)   **49.** The object has fallen 88 ft when the speed reaches 75 ft/s. (5.4B)
50. The rate of the plane for the first 360 mi was 120 mph. (7.6A)   **51.** The intensity is 200 lumens. (8.4A)
52. The rate of the boat is 12.5 mph. The rate of the current is 2.5 mph. (10.4A)   **53.** The value of the investment after two years would be $4785.65. (11.5A)   **54.** The house would be worth $256,571 in twenty years. (12.3D)

Index

Content and Format © 1991 HMCo.